Telecommunications Engineering

Telecommunications Engineering

Third Edition

J. Dunlop

and

D. G. Smith

Department of Electronic and Electrical Engineering, University of Strathclyde, Glasgow

CHAPMAN & HALL

University and Professional Division

London · Glasgow · Weinheim · New York · Tokyo · Melbourne · Madras

Publishel by Chapman & Hall, 2–6 Boundary Row, London SE1 8HN, UK

Chapman & Hall, 2–6 Boundary Row, London SE1 8HN, UK

Blackie Academic & Professional, Wester Cleddens Road, Bishopbriggs, Glasgow G64 2NZ, UK

Chapman & Hall GmbH, Pappelallee 3, 69469 Weinheim, Germany

Chapman & Hall USA, One Penn Plaza, 41st Floor, New York NY 10119, USA

Chapman & Hall Japan, Thomson Publishing Japan, Kyowa Building, 3F, 2-2-1 Hirakawa-cho, Chiyoda-ku, Tokyo 102, Japan

Chapman & Hall Australia, Thomas Nelson Australia, 102 Dodds Street, South Melbourne, Victoria 3205, Australia

Chapman & Hall India, R. Seshadri, 32 Second Main Road, CIT East, Madras 600 035, India

First language edition 1984

Reprinted 1984, 1986, 1987 (twice)

Second edition 1989

Reprinted 1990, 1991

Third edition 1994

© 1984, 1989, 1994 J. Dunlop and D. G. Smith

Typeset in $10\frac{1}{2}/12$ pts Times by Thomson Press (India) Ltd., New Delhi

Printed in England by Clays Ltd, St Ives plc

ISBN 0 412 56270 7

∞ Printed on permanent acid-free text paper, manufactured in accordance with ANSI/NISO Z39.48-1992 and ANSI/NISO Z39.48-1984 (Permanence of Paper).

Contents

Preface to the third edition

During the lifetime of the second edition telecommunications systems have continued to develop rapidly, with major advances occurring in mobile communications and broadband digital networks and services. Digital systems have become more even more prevalent and sophisticated signal processing techniques are now common at increasinlgy higher bit rates. These advances have prompted the inclusion of new chapters on mobile communications and broadband ISDN, and a considerable revision of many others.

Several readers have commented that a discussion of the transient performance of the phase locked loop would be useful and this has now been included in Chapter 2.

Chapter 3 has been extended to give a deeper insight into non-linear coding of voice waveforms for PCM and a section has been included on NICAM, which has been adopted for the digital transmission of television sound.

Chapter 5 now includes an introduction to coding techniques for burst errors which dominate in modern mobile communications systems.

Chapter 11 has been completely revised to address the major developments which have occurred in television since the second edition and a new section on satellite television has been included.

Chapter 15 is a completely new chapter on Mobile Communication Systems covering first generation analogue and second generation digital systems.

Geoffrey Smith is Professor of Communication Engineering at the University of Strathclyde. His main research interests include performance evaluation of networks, mobile communications, integrated local communications and broadband optical systems.

John Dunlop is Professor of Electronic Systems Engineering at the University of Strathclyde and has been involved in international research projects in digital mobile communications for a number of years. He also has research interests in speech coding, local and wide area networks and underwater communications.

Preface to
the second edition

Since the first edition telecommunications have changed considerably. In particular, digital systems have become more common. To reflect such changes two new chapters have been added to the text and several have been modified, some considerably.

Chapter 3 has been extended to give a detailed description of the *A* law compression characteristic used in PCM systems and substantial extra material is included on PCM transmission techniques. A section is also included on other digital transmission formats such as differential PCM and delta modulation.

Chapter 10 introduces several elements of switching, both analogue and digital. In the latter case the fundamentals of both time and space switching are discussed. The methods used to analyse the switching performance are covered although the discussion is limited to simple cases in order to highlight the underlying concepts. Some of the material on older signalling systems has been omitted from this edition.

Chapter 13 deals with the topic of packet transmission which is finding increasing use in both wide area and local networks. This chapter provides the theoretical background for scheduled and random access transmission and draws attention to the limitations of these theoretical descriptions and to the need for using reliable computer models for estimating the performance of practical systems. This chapter also introduces the concept of the 7-layer Open Systems Interconnection reference model and illustrates how some of the OSI layers are incorporated in packet switched systems.

An introduction to satellite communications is given in Chapter 14. So as not to over-complicate the concepts involved, discussion is limited to geo-stationary systems. At the time of writing, the development of direct broadcasting of television programmes by satellite is at a very early stage, but there is little doubt that it will become an increasingly important application.

The authors are most grateful to the many readers who have made constructive suggestions for improvement to the text and who have identified several errors that existed in the first edition. It is hoped that the errors have been corrected and that a number of areas of difficulty have been removed by additional explanation or discussion.

Preface to the first edition

The influence of telecommunications has increased steadily since the introduction of telegraphy, radio and telephony. Now, most people are directly dependent on one or more of its many facets for the efficient execution of their work, at home and in leisure.

Consequently, as a subject for study it has become more and more important, finding its way into a large range of higher education courses, given at a variety of levels. For many students, telecommunications will be presented as an area of which they should be aware. The course they follow will include the essential features and principles of communicating by electromagnetic energy, without developing them to any great depth. For others, however, the subject is of more specialized interest; they will start with an overview course and proceed to specialize in some aspects at a later time. This book has been written with both types of student in mind. It brings together a broader range of material than is usually found in one text, and combines an analytical approach to important concepts with a descriptive account of system design. In several places the approximate nature of analysis has been stressed, and also the need to exercise engineering judgement in its application. The intention has been to avoid too much detail, so that the text will stand on its own as a general undergraduate-level introduction, and it will also provide a strong foundation for those who will eventually develop more specialized interests.

It has been assumed that the reader is familiar with basic concepts in electronic engineering, electromagnetic theory, probability theory and differential calculus.

Chapter 1 begins with the theoretical description of signals and the channels through which they are transmitted. Emphasis is placed on numerical methods of analysis such as the discrete Fourier transform, and the relationship between the time and frequency domain representations is covered in detail. This chapter also deals with the description and transmission of information-bearing signals.

Chapter 2 is concerned with analogue modulation theory. In this chapter there is a strong link between the theoretical concepts of modulation theory and the practical significance of this theory. The chapter assumes that the reader has a realistic knowledge of electronic circuit techniques.

Chapter 3 is devoted to discrete signals and in particular the coding and transmission of analogue signals in digital format. This chapter also emphasizes the relationship between the theoretical concepts and their practical significance.

Chapters 4 and 5 are concerned with the performance of telecommunications systems in noise. Chapter 4 covers the performance of analogue systems and concentrates on the spectral properties of noise. Chapter 5 covers the performance of digital systems and is based on the statistical properties of noise. This chapter also deals in detail with the practical implication of error correcting codes, a topic which is often ignored by more specialized texts in digital communications.

In Chapter 6 the elements of high-frequency transmission-line theory are discussed, with particular emphasis on lossless lines. The purpose is to introduce the concepts of impedance, reflection and standing waves, and to show how the designer can influence the behaviour of the line.

Basic antenna analysis, and examples of some commonly used arrays and microwave antennas, are introduced in Chapter 7, while Chapters 8 and 9 describe the essential features of waveguide-based microwave components. A fairly full treatment of the propagation of signals along waveguide is considered from both the descriptive and field-theory analysis points of view.

Telephone system equipment represents the largest part of a country's investment in telecommunications, yet teletraffic theory and basic system design do not always form part of a telecommunications class syllabus. Chapter 10 is a comprehensive chapter on traditional switching systems and the techniques used in their analysis. Care has been taken to limit the theoretical discussion to simple cases, to enable the underlying concepts to be emphasized.

Chapter 11 is devoted to television systems. In a text of this nature such a coverage must be selective. We have endeavoured to cover the main topics in modern colour television systems from the measurement of light to the transmission of teletext information. The three main television systems, NTSC, PAL and SECAM, are covered but the major part of this chapter is devoted to the PAL system.

One of the outstanding major developments in recent years has been the production of optical fibres of extremely low loss, making optical communication systems very attractive, both technically and commercially. Chapter 12 discusses the main features of these systems, without introducing any of the analytical techniques used by specialists. The chapter is intended to give an impression of the exciting future for this new technology.

It cannot be claimed that this is a universal text; some omissions will not turn out to be justified, and topics which appear to be of only specialized interest now may suddenly assume a much more general importance. However, it is hoped that a coverage has been provided in one volume which will find acceptance by many students who are taking an interest in this stimulating and expanding field of engineering.

List of symbols and abbreviations

a	normalized propagation delay
A	telephone traffic, in erlangs
A_e	effective aperture of an antenna
AAL	ATM adaption layer
ADC	analogue to digital conversion
ADPCM	adaptive differential pulse code modulation
AGCH	GSM Access Grant Channel
AM	amplitude modulation
AMI	alternate mark inversion
AMPS	advanced mobile phone system
ARQ	automatic repeat request
ASK	amplitude shift keying
ATDM	asynchronous time division multiplexing
ATM	asynchronous transfer mode
α	attenuation coefficient of a transmission line
α	traffic offered per free source
B	bandwidth of a signal or channel
	speech (64 kb/sec) channel in ISDN
B	call congestion
B_c	coherence bandwidth
BCCH	GSM Broadcast Control Channel
B-ISDN	broadband ISDN
BSC	base station controller
BST	base station transceiver
β	phase constant of a transmission line
β	modulation index
c	velocity of light in free space
C	capacitance per unit length of transmission line
CCCH	GSM Common Control Channel
CDMA	code division multiple access
CFP	cordless fixed part
CIR	carrier to interference ratio
$C(n)$	discrete spectrum
C_n	nth harmonic in a Fourier series
CPP	cordless portable part

CRC	cyclic redundancy check
CSMA/CD	carrier sense multiple access with collision detection
CVSDM	continuously variable slope delta modulation
D	re-use distance
	diameter, or largest dimension, of an antenna
	signalling channel in ISDN
DECT	digital European cordless telecommunication standard
DFT	discrete Fourier transform
δ_s	skin depth
$\partial(t - t_0)$	impulse function at t_0
Δ	delay spread
Δf	elemental bandwidth
Δf_c	carrier deviation
Δf_n	noise bandwidth
DPSK	differential phase shift keying
DQDB	distributed queue dual bus
DSB-AM	double sideband amplitude modulation
DSB-SC-AM	double sideband supressed carrier modulation
DUP	data user part
E	time congestion
$E(f)$	energy density spectrum
ERP	effective radiated power
EIRP	effective isotropic radiated power
$E_n(A)$	Erlang's loss function
$E(N, s, \alpha)$	Engset's loss function
ε	permittivity
ε_0	permittivity of free space
ε_r	relative permittivity of dielectric
$\varepsilon(t)$	phase error
$\varepsilon(s)$	Laplace transform of $\varepsilon(t)$
η_0	free space characteristic
η	single-sided power spectral density of white noise
F	noise figure of a network
f_0	fundamental frequency of a periodic wave
f_c	cut-off frequency, carrier frequency
FACCH	GSM Fast Associated Control Channel
FCCH	GSM Frequency Correction Channel
FDM	frequency division multiplex
FDMA	frequency division multiple access
FET	field effect transistor
FISU	fill in signalling unit
FFT	fast Fourier transform
FM	frequency modulation
FOCC	TACS Forward Control Channel
FSK	frequency shift keying
FVC	TACS Forward Voice Channel
$F_v(v)$	cumulative distribution function

G	conductance per unit length of transmission line
	normalized offered traffic
	generator polynomial
$G(f)$	power spectral density
$G(i)$	probability of any i devices being busy
GMSK	Gaussian minimum shift keying
GSM	global system for mobile communications
γ	propagation coefficient of a transmission line
n	thickness of dielectric in microstrip line
	height of an antenna
H	magnetic field
H_{av}	entropy of a message (bits/symbol)
$H(f)$	Fourier transform of $h(t)$
$h(k)$	discrete signal
$H(s)$	Laplace transform of $h(t)$
$h(t)$	general function of time
HLR	home location register
$[i]$	probability that a network is in state i
I_k	interference power
IDFT	inverse discrete Fourier transform
IF	intermediate frequency
ISDN	integrated services digital network
ISUP	ISDN user part
$I_0(x)$	modified Bessel function
$J_n(\beta)$	Bessel functions of the first kind
K	cluster size
k_c	$2\pi/\lambda_c$
L	inductance per unit length of transmission line
LAN	local area network
LED	light emitting diode
LAP	link access procedure
LAP-D	local access protocol for D Channel
λ	likelihood ratio
	mean packet arrival rate
	wavelength
λ_c	cut-off wavelength
λ_{cmn}	cut-off wavelength of the TE_{mn} or TM_{mn} mode
λ_g	guide wavelength
λ_i	call arrival rate in state i
λ_0	free space wavelength
$L(f)$	frequency domain output of a network
$l(t)$	time domain output of a network
m	depth of modulation
MAC	mixed analogue components TV standard
	medium access control
	TACS mobile attenuation code
MAN	metropolitan area network

MAP	manufacturing automation protocol
MSC	mobile switching centre
MSU	message signalling unit
MTP	message transfer part
μ	permeability
μ_i	call departure rate in state i
μ_0	permeability of free space
n	refractive index of glass fibre
N	number of devices
	electron density in ionosphere (electrons/m^3)
n_0	refractive index of freespace
N_i	normalized noise power at the input of a network
N_0	normalized noise power at the output of a network
$N(A_0)$	level crossing rate
NICAM	nearly instantaneously companded audio multiplex
NMT	nordic mobile telephone system
NNI	network–network interface
NTP	network service part
NTSC	National Television Systems Committee
$n(t)$	elemental noise voltage
OSI	open systems interconnection reference model
ω_n	natural (radian) frequency
P	power in a signal
P_a	power/unit area
P_c	error probability
P_r	received power
P_t	transmitted power
p_{nm}	root of $J'_n(k_c, a)$
PAD	packet assembler/disassembler
PAL	phase alternation line by line
PCM	pulse code modulation
PCH	GSM paging channel
PDH	plesiochronous digital hierarchy
PDU	ATM protocol data unit
$P(f)$	transfer function of a network
PM	phase modulation
POTS	plain old telephone system
PSK	phase shift keying
$p(t)$	impulse response of a network
$pv^{(v)}$	probability density function
ψ	angle of reflection coefficient
q	co-channel interference reduction factor
Q	resonator quality factor
QPSK	quaternary phase shift keying
QPSX	queued packet switched exchange
R	resistance per unit length of transmission line
RACH	GSM access channel
RECC	TACS Reverse Control Channel

$R_h(\tau)$	autocorrelation function of $h(t)$
RVC	TACS Reverse Voice Channel
ρ	reflection coefficient at transmission line load
s	mean call holding time
S	normalized throughput
	number of traffic sources
S	voltage standing wave ratio (VSWR)
S_c	normalized carrier power
S_i	normalized signal power at the input of a network
S_0	normalized signal power at the output of a network
SACCH	GSM Slow Associated Control Channel
SAPI	service access point identifier
SAT	TACS supervisory audio tone
SCART	Syndicat des Constructeurs d'Appareils Radio récepteurs et Teléviseurs
SCCP	signalling connection control part
SCH	GSM Synchronization Channel
SDCCH	GSM Slow Dedicated Control Channel
SDH	synchronous digital hierarchy
SECAM	séquential couleur à mémoire
SQNR	single to quantization noise ratio (power)
SNR	signal to noise ratio (power)
SONET	synchronous optical network
SP	signalling point
SS7	CCITT signalling system number 7
SSB-AM	signal sideband amplitude modulation
ST	TACS signalling tone
STM-n	synchronous transport mode-level n
STP	signalling transfer point
σ	rms voltage of a random signal
T	period of a periodic wave
θ_i	angle of incidence of radio wave to ionosphere
t_a	mean access delay
t_d	mean packet delivery time
t_p	end to end propagation delay
t_r	token rotation time
t_t	token transmission time
t_s	scan time
t_{sl}	walk time
T_c	effective noise temperature of a network or antenna
T_s	standard noise temperature (290K)
TACS	total access communication system
TCH	GSM Traffic Channel
TDD	time division duplex
TDM	time division multiplexing
TDMA	time division multiple access
TE_{mn}	transverse electric waveguide mode
TEI	terminal end point identifier

TM_{mn}	transverse magnetic waveguide mode
TU	tributary unit
TUP	telephone user part
τ	dummy time variable
$\tau(A_0)$	average fade duration below the level A_0
$u(t)$	unit step function
UNI	user-network interface
v	transmission line wave velocity
V	peak voltage of a waveform
V_1	incident (forward) voltage on a transmission line
V_2	reflected (backward) voltage on a transmission line
v_g	group velocity
v_{ph}	phase velocity
VAD	voice activity detector
$v(t)$	general function of time
$v_m(t)$	modulating waveform
$v_c(t)$	carrier waveform
$V_n(t)$	bandlimited noise voltage
VC	virtual container
VCI	virtual circuit identifier
VPI	virtual path identifier
VSB-AM	vestigial sideband amplitude modulation
W	highest frequency component in a signal
	width of strip in microstrip line
WAN	wide area network
X^n	power of 2 in a generator polynomial
$x(t)$	amplitude of in-phase noise component
y	mean call arrival rate
$y(t)$	amplitude of quadrature noise component
Z_L	transmission line load impedance
Z_O	characteristic impedance of transmission line
Z_O	characteristic impedance of microstrip line

Signals and channels $\boxed{1}$

1.1 INTRODUCTION

Telecommunication engineering is concerned with the transmission of information between two distant points. Intuitively we may say that a signal contains information if it tells us something we did not already know. This definition is too imprecise for telecommunications studies, and we shall devote a section of this chapter to a formal description of information. For the present it is sufficient to say that a signal that contains information varies in an unpredictable or random manner. We have thus specified a primary characteristic of the signals in telecommunications systems; they are random in nature.

These random signals can be broadly subdivided into discrete signals that have a fixed number of possible values, and continuous signals that have any value between given limits. Whichever type of signal we deal with, the telecommunication system that it uses can be represented by the generalized model of Fig. 1.1. The central feature of this model is the transmission medium or channel. Some examples of channels are coaxial cables, radio links, optical fibres and ultrasonic transmission through solids and liquids. It is clear from these examples that the characteristics of channels can vary widely. The common feature of all channels, however, is that they modify or distort the waveform of the transmitted signal. In some cases the distortion can be so severe that the signal becomes totally unrecognizable.

In many instances it is possible to minimize distortion by careful choice of the transmitted signal waveform. To do this the telecommunications engineer must be able to define and analyse the properties of both the signals and the channels over which they are transmitted. In this chapter we shall concentrate on the techniques used in signal and linear systems analysis, although we should point out that many telecommunications systems do have non-linear characteristics.

1.2 THE FREQUENCY AND TIME DOMAINS

The analysis of linear systems is relatively straightforward if the applied signals are sinusoidal. We have already indicated that the signals encountered in telecommunications systems are random in nature and, as such, are non-deterministic. It is often possible to approximate such signals by periodic

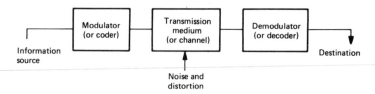

Fig. 1.1 Basic elements of a telecommunications system.

functions that themselves can be decomposed into a sum of sinusoidal components. The signal waveforms are functions of time and the variation of signal amplitude with time is known as the 'time domain representation' of the signal. Alternatively, if a signal is decomposed into a sum of sinusoidal components, the amplitude and phase of these components can be expressed as a function of frequency. This leads us to the 'frequency domain representation' of the signal.

The relationship between frequnecy domain and time domain is an extremely important one and is specified by Fourier's theorem. The response of a linear system to a signal can be determined in the time domain by using the principle of convolution, and in the frequency domain by applying the principle of superposition to the responses produced by the individual sinusoidal components. We will consider the frequency domain first, as this makes use of the theorems of linear network analysis which will be familiar to readers with an electronics background. Time domain analysis is considered in detail in Section 1.11. Frequency domain analysis will be introduced using traditional Fourier methods and we will then develop the discrete Fourier transform (DFT) which is now an essential tool in computer aided analysis of modern telecommunications systems.

1.3   CONTINUOUS FOURIER ANALYSIS

Fourier's theorem states that any single-valued periodic function, which has a repetition interval T, can be represented by an infinite series of sine and cosine terms which are harmonics of $f_0 = 1/T$. The theorem is given by Eqn (1.1).

$$h(t) = \frac{a_0}{T} + \frac{2}{T} \sum_{n=1}^{\infty} (a_n \cos 2\pi n f_0 t + b_n \sin 2\pi n f_0 t) \qquad (1.1)$$

where $f_0 = 1/T$ is the fundamental frequency. The response of a linear system to a waveform $h(t)$ that is not a simple harmonic function is found by summing the responses produced by the individual sinusoidal components of which $h(t)$ is composed. The term a_0/T is known as the dc component and is the mean value of $h(t)$.

$$\frac{a_0}{T} = \frac{1}{T} \int_{-T/2}^{T/2} h(t)\,dt$$

i.e.

$$a_0 = \int_{-T/2}^{T/2} h(t)\,dt \qquad (1.2)$$

The amplitudes of the sine and cosine terms are given by

$$a_n = \int_{-T/2}^{T/2} h(t) \cos(2\pi n f_0 t)\, dt$$

$$b_n = \int_{-T/2}^{T/2} h(t) \sin(2\pi n f_0 t)\, dt \qquad (1.3)$$

The Fourier series thus contains an infinite number of sine and cosine terms. This can be reduced to a more compact form as follows; let

$$x(t) = a_n \cos(2\pi n f_0 t) + b_n \sin(2\pi n f_0 t)$$

$$\cos \phi_n = \frac{a_n}{\sqrt{(a_n^2 + b_n^2)}}$$

$$\sin \phi_n = \frac{-b_n}{\sqrt{(a_n^2 + b_n^2)}}$$

Hence $\phi_n = \tan^{-1}[-b_n/a_n]$ and

$$x(t) = (a_n^2 + b_n^2)^{1/2} [\cos(2\pi n f_0 t) \cos \phi_n - \sin(2\pi n f_0 t) \sin \phi_n]$$

i.e.

$$x(t) = (a_n^2 + b_n^2)^{1/2} \cos(2\pi n f_0 t + \phi_n)$$

Hence the Fourier series can be modified to

$$h(t) = \frac{a_0}{T} + \frac{2}{T} \sum_{n=1}^{\infty} C_n \cos(2\pi n f_0 t + \phi_n)$$

where

$$C_n = (a_n^2 + b_n^2)^{1/2} \quad \text{and} \quad \phi_n = \tan^{-1}\left[\frac{-b_n}{a_n}\right] \qquad (1.4)$$

A graph of C_n against frequency is known as the **amplitude spectrum** of $h(t)$ and a graph of ϕ_n against frequency is known as the **phase spectrum** of $h(t)$. Note that if the voltage developed across a $1\,\Omega$ resistance is

$$v(t) = \frac{2C_n}{T} \cos(2\pi n f_0 t + \phi_n)$$

the average power dissipated in the resistance is

$$P = \frac{2}{T^2} C_n^2 = \frac{2}{T^2}(a_n^2 + b_n^2) \qquad (1.5)$$

A graph of C_n^2 against frequency is known as the **power spectrum** of $h(t)$. The total power developed in a $1\,\Omega$ resistance by $h(t)$ is thus given by

$$P_T = \frac{a_2^0}{T^2} + \frac{2}{T^2} \sum_{n=1}^{\infty} C_n^2 \qquad (1.6)$$

The mean square value of $h(t)$ is given by

$$\sigma^2 = \frac{1}{T} \int_{-T/2}^{T/2} |h(t)|^2 \, dt$$

This is effectively the power dissipated when a voltage equal to $h(t)$ is developed across a resistance of $1\,\Omega$. The frequency and time domain representations of $h(t)$ are thus related by Eqn (1.7). This equation is formally known as Parseval's theorem.

$$\frac{1}{T} \int_{-T/2}^{T/2} |h(t)|^2 \, dt = \frac{a_0^2}{T^2} + \frac{2}{T^2} \cdot \sum_{n=1}^{\infty} C_n^2 \qquad (1.7)$$

EXAMPLE: Find the amplitude and power spectrum of the periodic rectangular pulse train of Fig. 1.2.

The zero frequency (mean) value is a_0/T where

$$a_0 \int_{-T/2}^{T/2} h(t) \, dt = \int_{-t_{1/2}}^{t_{1/2}} A \, dt = A t_1$$

$$a_n \int_{-T/2}^{T/2} h(t) \cos(2\pi n f_0 t) \, dt = \int_{-t_{1/2}}^{t_{1/2}} A \cos(2\pi n f_0 t) \, dt$$

i.e.

$$a_n = \frac{A}{2\pi n f_0} [\sin(\pi n f_0 t_1) - \sin(-\pi n f_0 t_1)]$$

hence

$$a_n = \frac{A}{\pi n f} \sin(\pi n f_0 t_1)$$

similarly

$$b_n = \frac{A}{2\pi n f_0} [\cos(\pi n f_0 t_1) - \cos(-\pi n f_0 t_1)] = 0$$

Hence in this example $C_n = a_n$.

The amplitude spectrum is $C_n = A t_1 (\sin \pi n f_0 t_1)/\pi n f_0 t_1$, which is often written $C_n = A t_1 \, \mathrm{sinc}\,(\pi n f_0 t_1)$.

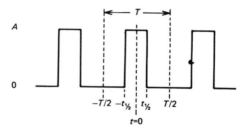

Fig. 1.2 Rectangular periodic pulse train.

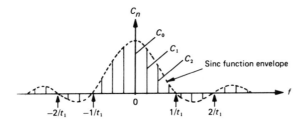

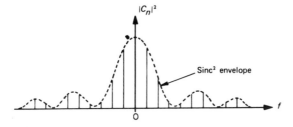

Fig. 1.3 Amplitude and power spectrum of a periodic pulse train.

The amplitude spectrum is plotted in Fig. 1.3 and it should be noted that the envelope of this spectrum is a sinc function that has unity value when $\pi f t_1 = 0$ and zero value when $\pi f t_1 = m\pi$, i.e. when $f = m/t_1$. In this particular example $\phi_n = 0$ indicating all harmonics are in phase. The power spectrum of $h(t)$ is simply the square of the amplitude spectrum.

A convenient alternative form of Eqn (1.1) can be developed by writing the sine and cosine terms in exponential notation, i.e.

$$\cos(2\pi n f_0 t) = [\exp(j2\pi n f_0 t) - \exp(-j2\pi n f_0 t)]/2$$

$$\sin(2\pi n f_0 t) = [\exp(j2\pi n f_0 t) - \exp(-j2\pi n f_0 t)]/2j$$

Substitution in Eqn (1.1) gives

$$h(t) = \frac{a_0}{T} + \frac{1}{T}\sum_{n=1}^{\infty}(a_n - jb_n)\exp(j2\pi n f_0 t) + (a_n + jb_n)\exp(-j2\pi n f_0 t)$$

$$(a_n - jb_n) = \int_{-T/2}^{T/2} h(t)[\cos(2\pi n f_0 t) - j\sin(2\pi n f_0 t)]\,dt$$

i.e. if $C_n = (a_n - jb_n)$, then

$$C_n = \int_{-T/2}^{T/2} h(t)\exp(-j2\pi n f_0 t)\,dt \qquad (1.8)$$

The complex conjugate of C_n is $C_n^* = (a_n + jb_n)$, and

$$C_n^* = \int_{-T/2}^{T/2} h(t)\exp(j2\pi n f_0 t)\,dt$$

i.e. $C_n^* = C_{-n}$.

Hence

$$h(t) = \frac{a_0}{T} + \frac{1}{T} \sum_{n=1}^{\infty} [C_n \exp(j2\pi n f_0 t) + C_{-n} \exp(-j2\pi n f_0 t)]$$

and

$$C_0 = \int_{-T/2}^{T/2} h(t) \exp(j^0) \, dt = a_0$$

so that $h(t)$ can be written

$$h(t) = \frac{1}{T} \sum_{n=-\infty}^{\infty} C_n \exp(j2\pi n f_0 t) \tag{1.9}$$

This is the exponential form of the Fourier series and the limits of the summation are now $n = \pm\infty$. The spectrum that contains both positive and negative components is known as a double-sided spectrum.

The negative frequencies are a direct result of expressing sine and cosine in complex exponential form. In Eqn (1.8) C_n is a complex quantity and can be separated into a magnitude and phase characteristic, i.e. $C_n = |C_n| \exp(j\phi_n)$; i.e.

$$h(t) = \frac{1}{T} \sum_{n=-\infty}^{\infty} |C_n| \exp[j(2\pi n f_0 t + \phi_n)]$$

But since $|C_n| = |C_{-n}|$ then

$$h(t) = \frac{C_0}{T} + \frac{1}{T} \sum_{n=1}^{\infty} |C_n| \exp[j(2\pi n f_0 t + \phi_n)] + |C_n| \exp[-j(2\pi n f_0 t + \phi_n)]$$

i.e.

$$h(t) = \frac{C_0}{T} + \frac{2}{T} \sum_{n=1}^{\infty} |C_n| \cos(2\pi n f_0 t + \phi_n)$$

Equations (1.9) and (1.4) are therefore equivalent, but some care is required in interpreting Eqn (1.9). The harmonic amplitude C_n/T is exactly half the value given by Eqn (1.4), but it is defined for both negative and positive values of n. The correct amplitude is obtained by summing the equal coefficients which are obtained for negative and positive values of n. This is quite reasonable because only one frequency component actually exists.

The power of any frequency is derived from Eqn (1.9) in a similar way. Since C_n is a complex quantity the power at any value of n is $(C_n/T) \cdot (C_n^*/T)$, i.e.

$$(C_n/T) \cdot (C_n^*/T) = (a_n - jb_n)(a_n + jb_n)/T^2 = (a_n^2 + b_n^2)/T$$

Both negative and positive values of n will contribute an equal amount of power; the total power at any frequency is thus $2(a_n^2 + b_n^2)/T^2$. This of course agrees with Eqn (1.5), since physically only a single component exists at any one frequency. We can write Parseval's theorem for the exponential series as

$$\frac{1}{T} \int_{-T/2}^{T/2} |h(t)|^2 \, dt = \frac{1}{T^2} \sum_{n=-\infty}^{\infty} |C_n|^2 \tag{1.10}$$

Note that only a single value of n appears at $n = 0$.

EXAMPLE: Evaluate the amplitude spectrum of the waveform in Fig. 1.2 using the exponential series

$$h(t) = \frac{1}{T} \sum_{n=-\infty}^{\infty} C_n \exp j(2\pi n f_0 t) \quad \text{where} \quad C_n = \int_{-t_{1/2}}^{t_{1/2}} A \exp(-j2\pi n f_0 t) \, dt$$

i.e.

$$C_n = \left[\frac{-A}{j2\pi n f_0} \exp(-j2\pi n f_0 t) \right]_{-t_{1/2}}^{t_{1/2}}$$

$$= \frac{A}{\pi n f_0} \left[\frac{\exp(-j\pi n f_0 t_1) - \exp(-j\pi n f_0 t_1)}{2j} \right]$$

i.e.

$$C_n = \frac{A}{\pi n f_0} \sin(\pi n f_0 t_1)$$

or

$$C_n = At_1 \, \text{sinc}(\pi n f_0 t_1)$$

This is identical to the equation obtained from the cosine series.

1.4 ODD AND EVEN FUNCTIONS

The waveform $h(t)$ is defined as an even function if $h(t) = h(-t)$; it has the property of being symmetrical about the $t = 0$ axis. If $h(t) = -h(-t)$ the waveform is an odd function and has skew symmetry about the $t = 0$ axis. A function that has no symmetry about the $t = 0$ axis is neither odd nor even; the sawtooth waveform of Fig. 1.4(c) is an example of such a waveform.

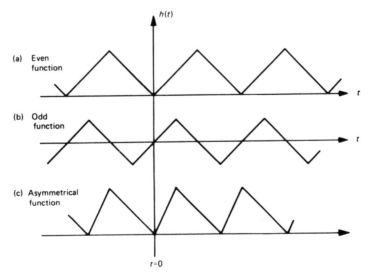

(a) Even function

(b) Odd function

(c) Asymmetrical function

Fig. 1.4 Examples of periodic functions.

Odd and even functions have properties that may be used to simplify Fourier analysis, e.g.

if

$$h(t) = \frac{a_0}{T} + \frac{2}{T} \sum_{n=1}^{\infty} [a_n \cos(2\pi n f_0 t) + b_n (\sin 2\pi n f_0 t)]$$

then

$$h(-t) = \frac{a_0}{T} + \frac{2}{T} \sum_{n=1}^{\infty} [a_n \cos(2\pi n f_0 t) - b_n \sin(2\pi n f_0 t)]$$

If $h(t) = h(-t)$ this can only be true if $b_n = 0$, i.e. the Fourier series of an even function has cosine terms only. Alternatively, all phase angles ϕ_n in Eqn (1.4) are 0 or $\pm \pi$ and all values of C_n in Eqn (1.8) are real. If $h(t) = -h(-t)$ then $a_n = 0$, i.e. the Fourier series of an odd function contains only sine terms. Alternatively, all phase angles ϕ_n in Eqn (1.4) are $\pm \pi/2$ and all values of C_n in Eqn (1.8) are imaginary. If $h(t)$ has no symmetry about $t = 0$ the Fourier series contains both sines and cosines, the phase angles ϕ_n of Eqn (1.8) are given by $\tan^{-1}(-b_n/a_n)$, and all values of C_n in Eqn (1.8) are complex.

Many waveforms that are not symmetrical about $t = 0$ can be made either odd or even by shifting the waveform relative to the $t = 0$ axis. The shifting process is illustrated in Fig. 1.5. It is of interest to examine the effect of such a shift on the Fourier series. We shall consider the Fourier series of the shifted waveform $h(t - t_s)$. Let $(t - t_s) = t_x$; the amplitude spectrum is thus given by

$$C_n = \int_{-t_{1/2}}^{t_{1/2}} A \exp(-j2\pi n f_0 t_x) \, dt_x$$

This evaluates to $C_n = A t_1 \operatorname{sinc}(\pi n f_0 t_1)$; hence shifting the time axis does not affect the amplitude spectrum. It does, however, affect the phase spectrum:

$$h(t - t_s) = \frac{1}{T} \sum_{n=-\infty}^{\infty} C_n \exp(j2\pi n f_0 t) \exp(-j2\pi n f_0 t_s)$$

This effectively adds a phase shift of $\phi_s = -2\pi n f_0 t_s$ to each component in the series. Interpreted in another way, a time delay of t_s is equivalent to a phase shift of $2\pi f_0 t_s$ in the fundamental, $4\pi f_0 t_s$ in the second harmonic, etc.

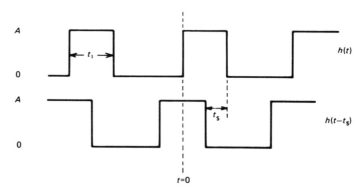

Fig. 1.5 Time shifting.

1.5 WAVEFORM SYNTHESIS

This can be regarded as the inverse of Fourier analysis. In effect the Fourier series indicates that any periodic waveform can be synthesized by adding an infinite number of cosine waves with specific amplitudes and phases. In most practical cases a very good approximation to a given periodic waveform can be obtained by truncating the series to only a few terms. As the number of terms in the series is increased the mean square error between the synthesized waveform and the desired waveform decreases. A difficulty does arise in the vicinity of a discontinuity, however. As the number of terms in the series tends to infinity the mean value of the synthesized waveform approaches the mean value of the desired waveform at the discontinuity. The amplitude of the synthesized waveform on either side of the discontinuity is subject to error, which is not reduced when the number of terms in the series is increased. This error is known as Gibb's phenomenon, and is illustrated for a synthesized rectangular wave in Fig. 1.6.

This is in fact a convergence property of the Fourier series. The Fourier series converges to the mean value of $h(t)$ at discontinuities in the waveform of $h(t)$. The conditions required for the convergence of the series are

(i) $h(t)$ must have a finite number of maxima and minima in the interval T;
(ii) $h(t)$ must have a finite number of discontinuities in the interval T;
(iii) $h(t)$ must satisfy the inequality $\int_0^T |h(t)|\, dt < \infty$.

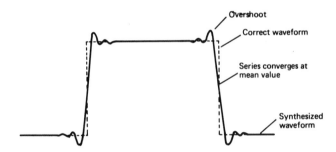

Fig. 1.6 Gibb's phenomenon in waveform synthesis.

1.6 THE FOURIER INTEGRAL

The Fourier series representation of $h(t)$ is only valid when $h(t)$ is periodic. We have already indicated that information-bearing signals change in a random fashion and do not therefore belong to this category. The amplitude spectra of non-periodic signals are obtained from the Fourier integral. The Fourier integral may be developed from the Fourier series by allowing the period T to approach infinity. In Fig. 1.2 allowing $T \to \infty$ means that $h(t)$ becomes a single pulse of width t_1 seconds.

$$h(t) = \frac{1}{T} \sum_{n=-\infty}^{\infty} C_n \exp(j2\pi n f_0 t) \quad \text{where } f_0 = 1/T$$

and

$$C_n = \int_{-T/2}^{T/2} h(t) \exp(-j2\pi n f_0 t) \, dt$$

If we let Δf be the spacing between harmonics in the Fourier series then $\Delta f = (n+1)f_0 - nf_0 = 1/T$.

The Fourier series may thus be written

$$h(t) = \sum_{n=-\infty}^{\infty} C_n \exp(j2\pi n f_0 t) \, \Delta f$$

As $T \to \infty$ then $\Delta f \to 0$ and the discrete harmonics in the series merge, and an amplitude spectrum that is a continuous function of frequency results, i.e.

$$\lim_{T \to \infty} C_n = H(f)$$

The harmonic number n now has all possible values and the summation of the series can thus be replaced by an integral, i.e. nf_0 is replaced by a continuous function f and

$$h(t) = \int_{-\infty}^{\infty} H(f) \exp(j2\pi ft) \, df$$

$$H(f) = \int_{-\infty}^{\infty} h(t) \exp(-j2\pi ft) \, dt$$

These two integrals are known as the **Fourier transform pair**. To illustrate the use of the Fourier transform, assume $h(t)$ is a single pulse of amplitude A and duration t_1 seconds.

$$H(f) = \int_{-\infty}^{\infty} h(t) \exp(-j2\pi ft) \, dt$$

$$= \int_{-t_{1/2}}^{t_{1/2}} A \exp(j2\pi ft) \, dt \tag{1.11}$$

$$= \frac{A}{\pi f} \sin \pi f t_1$$

i.e.

$$H(f) = A t_1 \, \text{sinc}\,(\pi f t_1)$$

A rectangular pulse in the time domain thus has a Fourier transform that is a sinc function in the frequency domain. The converse is also true, i.e. a sinc pulse in the time domain has a Fourier transform that is a rectangular function in the frequency domain. Consider the sinc pulse of Fig. 1.7(b).

$$h(t) = V \frac{\sin(2\pi f_1 t)}{2\pi f_1 t} \quad \text{where } f_1 = 1/t_1$$

The Fourier transform is

$$H(f) = V \int_{-\infty}^{\infty} \frac{\sin(2\pi f_1 t)}{2\pi f_1 t} \exp(-j2\pi ft) \, dt$$

(a)

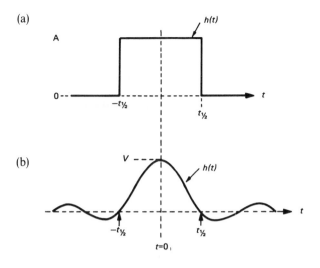

(b)

(c)

(d)

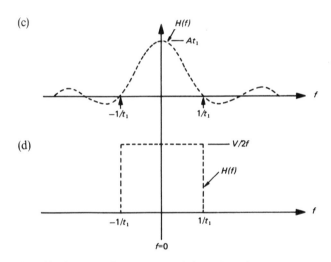

Fig. 1.7 Relationship between frequency and time domain.

Recalling that $\exp(-j2\pi ft) = \cos(2\pi ft) - j\sin(2\pi ft)$, we may write

$$H(f) = \frac{V}{2\pi f_1} \int_{-\infty}^{\infty} \frac{\sin(2\pi f_1 t)\cos(2\pi ft)}{t}\,dt$$

$$-j\frac{V}{2\pi f_1}\int_{-\infty}^{\infty}\frac{\sin(2\pi f_1 t)\sin(2\pi ft)}{t}\,dt \qquad (1.12)$$

The integral of an odd function betwen $\pm\infty$ is zero; thus the second integral of Eqn (1.12) vanishes. Using the trigonometric relationship $\cos\phi\sin\theta = \frac{1}{2}[\sin(\phi+\theta) - \sin(\phi-\theta)]$ we can say

$$H(f) = \frac{V}{2\pi f_1}\int_0^{\infty}\frac{\sin[2\pi(f+f_1)t]}{t}\,dt - \frac{V}{2\pi f_1}\int_0^{\infty}\frac{\sin[2\pi(f-f_1)t]}{t}\,dt \qquad (1.13)$$

At this point we make use of the standard integral

$$\int_0^\infty \frac{\sin ax}{x}\,dx = \pi/2 \quad \text{for } a > 0$$

$$0 \quad \text{for } a = 0$$

$$-\pi/2 \quad \text{for } a < 0$$

Re-writing Eqn (1.13) as

$$H(f)\frac{V}{2\pi f_1}(I_1 - I_2)$$

there are three frequency ranges of interest:

$$-\infty < f < f_1 \text{ gives } I_1 = -\pi/2, \quad I_2 = -\pi/2$$
$$-f_1 < f < f_1 \text{ gives } I_1 = \pi/2, \quad I_2 = -\pi/2$$
$$f_1 < f < \infty \text{ gives } I_1 = \pi/2, \quad I_2 = \pi/2$$

Hence $H(f) = V/2f_1$ for $-f_1 < f < f_1 = 0$ else. The resulting $H(f)$ is shown in Fig. 1.7(d).

Comparing Fig. 1.7(a) with Fig. 1.3 shows another important relationship. The envelope of the amplitude spectrum of a single pulse is identical to the envelope of the amplitude spectrum of a periodic pulse train of the same pulse width. This relationship is not restricted to rectangular pulses and is useful in determining the spectral envelope of signals composed of randomly occurring pulses, such as are encountered in digital communications systems.

The fact that all frequencies are present in the amplitude spectrum of a non-periodic signal requires careful interpretation when considering the power dissipated by such signals.

1.7 POWER AND ENERGY DENSITY SPECTRUM

The power density spectrum of a non-periodic signal is developed in a similar way to the amplitude spectrum. If we assume that $h(t)$ is a periodic function we may write Eqn (1.10) as

$$\frac{1}{T}\int_{-T/2}^{T/2}|h(t)|^2\,dt = \frac{1}{T}\sum_{n=-\infty}^{\infty}|C_n|^2\,\Delta f$$

As $T \to \infty$ for non-periodic signals, this equation can be written in the limit as

$$\frac{1}{T}\int_{-\infty}^{\infty}|h(t)|^2\,dt = \frac{1}{T}\int_{-\infty}^{\infty}|H(f)|^2\,df \tag{1.14}$$

The power spectrum of a non-periodic signal is then defined as

$$G(f) = \frac{|H(f)|^2}{T} \tag{1.15}$$

The power spectral density is a measure of the distribution of power as a function of frequency. It is a useful concept for random signals, such as noise, that have a finite power and are eternal; T is then the period of measurement. When T is large the power is independent of the value of T. If signals exist for a finite time only, the power spectrum approaches zero as $T \to \infty$. When dealing with such signals the concept of energy density spectrum is more meaningful. Equation (1.14) can also be written

$$\frac{1}{T}\int_{-\infty}^{\infty}|h(t)|^2\,\mathrm{d}t = \int_{-\infty}^{\infty}|H(f)|^2\,\mathrm{d}f \qquad (1.16)$$

This is Parseval's theorem for non-periodic signals. The LHS of the equation represents the total energy dissipated in a $1\,\Omega$ resistance by a voltage equal in amplitude to $h(t)$. It is clear, therefore, that $|H(f)|^2$ is an energy density, that is, a measure of the distribution of the energy of $h(t)$ with frequency. The double-sided (defined for $\pm f$) energy density spectrum of $h(t)$ is

$$E(f) = |H(f)|^2 \qquad (1.17)$$

The total energy within the frequency range f_1 to f_2 is

$$E\int_{-f_2}^{-f_1}E(f)\,\mathrm{d}f + \int_{f_1}^{f_2}E(f)\,\mathrm{d}f \quad \text{joules} \qquad (1.18)$$

i.e. half the energy is contributed by the negative components. In particular, as $f_2 \to f_1$, the total energy $\to 0$. Thus although the energy density spectrum of a non-periodic signal is continuous, the energy at a specific frequency is zero.

1.8   SIGNAL TRANSMISSION THROUGH LINEAR SYSTEMS

We noted in Section 1.1 that all communications channels have the common feature of modifying or distorting the waveforms of signals transmitted through them. The amount of distortion produced by a channel with a given transfer function (attenuation and phase shift as a function of frequency) is readily calculated using Fourier transform techniques.

If we assume that $P(f)$ is the channel transfer function (often a voltage ratio in electrical networks) we can obtain the amplitude spectrum of the signal at the channel output by multiplying the amplitude spectrum of the input signal by the network transfer function, i.e.

$$L(f) = H(f) \cdot P(f) \qquad (1.19)$$

We can then obtain the output signal $l(t)$ by taking the Fourier transform of Eqn (1.19), i.e.

$$l(t) = \int_{-\infty}^{\infty}H(f) \cdot P(f)\exp(j2\pi ft)\,\mathrm{d}t \qquad (1.20)$$

Note that $P(f) = |P(f)|\exp(-j2\pi ft)$ where $|P(f)|$ represents attenuation as a function of frequency (i.e. the frequency response of the channel) and $\phi(f)$ represents the phase shift produced. Both $|P(f)|$ and $\phi(f)$ produce signal distortion. Phase distortion is normally neglected when speech and music signals are transmitted over a channel, but it assumes special significance for digital transmission. This topic is covered further in Chapter 3.

When considering signal transmission through networks, we are often concerned with the loss of signal power or energy that occurs during transmission. If $G(f)$ is the power spectral density of a signal and $P(f)$ is the transfer function of a channel the power spectral density at the output is

$$G_0(f) = G_i(f) \cdot |P(f)|^2 \qquad (1.21)$$

The total power in a given frequency range f_1 to f_2 at the channel output is

$$W = \int_{-f_2}^{-f_1} G_0(f)\,\mathrm{d}f + \int_{f_1}^{f_2} G_0(f)\,\mathrm{d}f \quad \text{watts}$$

i.e.

$$W = \int_{-f_2}^{-f_1} G_i(f)|P(f)|^2\,\mathrm{d}f + \int_{f_1}^{f_2} G_i(f)|P(f)|^2\,\mathrm{d}f \qquad (1.22)$$

In most practical cases Eqn (1.22) can only be solved by numerical integration. In such cases the use of the discrete Fourier transform, which is discussed in Section 1.10, is particularly useful. To illustrate the application of Eqn (1.22) we will consider a specific example in which the integrals can be evaluated in closed form.

EXAMPLE: A sinc pulse of amplitude V and zero crossings at intervals of $\pm nt_1/2$ is passed through a low-pass RC network of the type shown in Fig. 1.8. If the value of t_1 for the pulse is 2 ms, find the cut-off frequency of the filter in order that 60% of the pulse energy is transmitted.

We calculate the incident energy of the pulse using Parseval's theorem, i.e.

$$E = \int_{-\infty}^{\infty} |h(t)|^2\,\mathrm{d}t = \int_{-\infty}^{\infty} |H(f)|^2\,\mathrm{d}f$$

We have shown that for the sinc pulse $H(f) = V/2f_1$ for $-f_1 < f < f_1$ and zero otherwise.

In this example $f_1 = 1/t_1$ and the energy is

$$E_i = \int_{-f_1}^{f_1} \left(\frac{V}{2f_1}\right)^2\,\mathrm{d}f = \frac{V^2}{2f_1} \quad \text{joules}$$

The network transfer function is $P(f) = 1/[1 + j(f/f_c)]$ where $f_c = 1/2\pi RC$ is the network cut-off frequency. We note from Fig. 1.8 that the network response extends to negative frequencies because $H(f)$ is a double-sided spectrum. For this network $|P(f)|^2 = f_c^2/(f_c^2 + f^2)$.

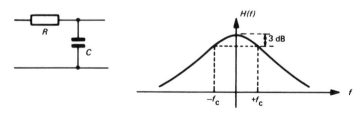

Fig. 1.8 Low-pass RC network.

The energy in the transmitted signal is then

$$E_0 = \int_{-f_1}^{f_1} \left(\frac{V}{2f_1}\right)^2 \frac{f_c^2}{f_c^2 + f^2} \, df$$

$$= 2f_c^2 \left(\frac{V}{2f_1}\right)^2 \frac{1}{f_c} \tan^{-1}\left(\frac{f_1}{f_c}\right)$$

But $E_0 = 0.6 \, E_i$. Hence

$$\frac{V^2}{2f_1^2} f_c \tan^{-1}\left(\frac{f_1}{f_c}\right) = \frac{0.6 \, V^2}{2f_1}$$

or

$$\tan^{-1}\left(\frac{f_1}{f_c}\right) = 0.6 \frac{f_1}{f_c}$$

The solution to this equation is $f_1/f_c \simeq 1.755$, i.e.

$$f_c = \frac{1}{1.755 t_1} = 285 \, \text{Hz}$$

1.9   THE IMPULSE FUNCTION

If we consider the rectangular pulse of Fig. 1.7(a) and let $A = 1/t_1$, the pulse area becomes unity, i.e. area $= (1/t_1) \cdot t_1 = 1$. If t_1 is allowed to approach zero, then in order to preserve unit area the pulse amplitude A is allowed to approach infinity. Such a pulse cannot be produced practically, but it is extremely useful for analytical purposes and is known as the unit impulse. The unit impulse function is formally defined by Eqn (1.23):

$$\int_a^b \delta(t - t_0) \, dt = 1 \quad \text{for } a < t_0 < b, \quad = 0 \text{ else} \tag{1.23}$$

The impulse exists only at time t_0 and has zero value for all other values of t. The amplitude of the impulse at $t = t_0$ is undefined; instead the impulse is defined in terms of its area (or weight) at time $t = t_0$. If any continuous function $h(t)$ is multiplied by an impulse with unit weight at time $t = t_0$ the resulting function is given by

$$\int_a^b h(t) \, \delta(t - t_0) \, dt = h(t_0) \quad \text{for } a < t_0 < b, \quad = 0 \text{ else} \tag{1.24}$$

Hence multiplying $h(t)$ by an impulse function at $t = t_0$ and performing the integration of Eqn (1.24) is equivalent to taking an instantaneous sample of $h(t)$ at $t = t_0$. The impulse function is defined only in integral form and expressions such as $h(t_0) = h(t) \cdot \delta(t - t_0)$ are strictly meaningless. However, it is common practice to express the integral equation (1.24) in this form, the process of integration being implicit. The Fourier transform of an impulse function is particularly important. The Fourier transform of the unit impulse defined by Eqn (1.23) is

$$\Delta(f) = \int_{-\infty}^{\infty} \delta(t - t_0)\exp(-j2\pi ft)\,dt = \exp(-j2\pi ft_0) \qquad (1.25)$$

This means that $|\Delta(f)|$ has unity value for all values of f. The function $\exp(-j2\pi ft)$ represents the phase of each component in $\Delta(f)$, i.e. $\phi = 2\pi ft_0$.

If, instead of a single impulse, we consider a periodic train of impulses separated by a period T, the amplitude spectrum is obtained from the Fourier series. The amplitude of the nth harmonic is then

$$C_n = \int_{-\infty}^{\infty} h(t)\exp(-j2\pi f_0 t)\,dt \quad \text{where } h(t) = \delta(t - nT)$$

i.e.

$$C_n = \int_{-\infty}^{\infty} \delta(t - nT)\exp(-j2\pi f_0 t)\,dt = \exp(-j2\pi) \qquad (1.26)$$

Hence each component in the Fourier series has unity value and a phase of 2π radians. This periodic train of impulses is used to obtain regularly spaced samples of a continuous waveform $h(t)$ and is of fundamental importance in the digital transmission of analogue signals.

Now that we have defined the impulse (or delta) function we can show that the Fourier integral can also be used to define the amplitude spectrum of a periodic signal and is therefore a general transform. If

$$h(t) = \frac{1}{T}\sum_{n=-\infty}^{\infty} C_n\exp(j2\pi nf_0 t)$$

the Fourier transform of $h(t)$ is

$$H(f) = \frac{1}{T}\int_{-\infty}^{\infty}\sum_{n=-\infty}^{\infty} C_n\exp(j2\pi(f - nf_0)t)\,dt$$

$$= \frac{1}{T}\sum_{n=-\infty}^{\infty} C_n\int_{-\infty}^{\infty} \exp[j2\pi(f - nf_0)t]\,dt$$

$$= \frac{1}{T}\sum_{n=-\infty}^{\infty} C_n\delta(f - nf_0) \qquad (1.27)$$

The Fourier transform of a periodic signal is thus a set of impulses located at harmonics of the fundamental frequency $f_0 = 1/T$.

1.10 THE DISCRETE FOURIER TRANSFORM (DFT)

We pointed out in Section 1.8 the extensive use made of computer-aided analysis in the study of modern telecommunication systems. Computers cannot handle continuous signals but can process signals that are defined at discrete intervals of time. The DFT is an extension of the continuous Fourier transform designed specifically to operate on signals that have been sampled at regular intervals of time. The sampling process may be regarded as multiplying the continuous signal by a periodic series of impulses. We have shown [Eqn (1.27)] that the spectrum of such a periodic signal is a series of harmonics all of equal amplitude. When such a spectrum is multiplied by the spectrum of a continuous signal, each component in the continuous signal will form sum and difference frequencies with the harmonics of the periodic impulse train. If we assume that the impulses are separated by an interval T_s and that the maximum frequency component of the continuous signal is W Hz, the amplitude spectrum of the sampled signal will take the form of Fig. 1.9. It will be noted from Fig. 1.9 that, provided the sampling frequency $f_s(= 1/T_s)$ is at least $2W$, there will be no overlap (aliasing) between the signal spectrum and the first lower sideband of the sampled signal spectrum. The original signal is defined by its amplitude spectrum which is preserved in the sampled version provided that $f_s \geqslant 2W$. This is in fact a statement of the 'sampling theorem' that we consider in more detail in Chaper 3.

The DFT is developed for a periodic signal $h(t)$ with no components at or above a frequency $f_x = x/T$, x being an integer and T being the period of the signal waveform. An example of such a signal and its spectrum is given in Fig. 1.10, which also contains the sampled version of $h(t)$, denoted $h(k\Delta t)$, and its amplitude spectrum $C(n\Delta f)$. The sampling frequency $(f_s = 1/\Delta T)$ is chosen to equal $2f_x$ which avoids aliasing.

The Fourier series for $h(k\Delta t)$ is

$$h(k\Delta t) = \frac{1}{T} \sum_{n=-(x-1)}^{x-1} C_n \exp\left(j2\pi nk\Delta t/T\right) \tag{1.28}$$

but over the range $-(x-1) \leqslant n \leqslant (x-1)$ the coefficients C_n are identical to $C(n\Delta f)$; hence

$$h(k\Delta t) = \frac{1}{T} \sum_{n=-(x-1)}^{(x-1)} C(n\Delta f) \exp\left(j2\pi nk\Delta t/T\right) \tag{1.29}$$

If there is a total of N samples in the interval T, then $T = N\Delta t$ and the range of k is $0, \pm 1, \pm 2 \ldots \pm [(N/2) - 1]$. Since we can also write $t = 1/2f_x = T/2x$ then $N = 2x$ and Eqn (1.29) becomes

$$h(k\Delta t) = \frac{1}{N\Delta t} \sum_{n=-N/2+1}^{N/2+1} C(n\Delta f) \exp\left(j2\pi nk/N\right) \tag{1.30}$$

We observe from Fig. 1.10 that $C(n\Delta f)$ is periodic and thus we can change the

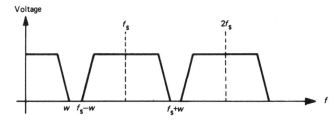

Fig. 1.9 Spectrum of a sampled signal.

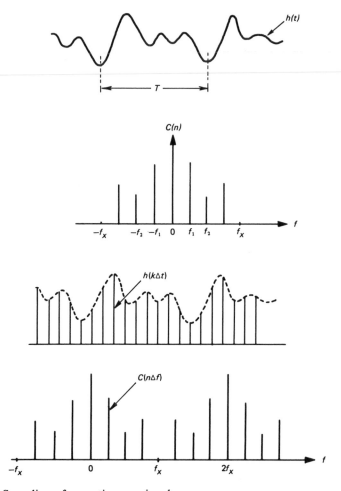

Fig. 1.10 Sampling of a continuous signal.

range of n in Eqn (1.30) to make $C(n\Delta f)$ symmetrical about the frequency f_x; thus

$$h(k\Delta t) = \frac{1}{N\Delta t} \sum_{n=0}^{N-1} C(n\Delta f) \exp(j2\pi nk/N)$$

This equation is usually written using the notation of Eqn (1.31), i.e.

$$h(k) = \frac{1}{N} \sum_{n=0}^{N-1} C(n) \exp(j2\pi nk/N) \qquad (1.31)$$

The multiplying factor $1/\Delta t$ is often omitted, as in Eqn (1.31). This does not affect the relative values of $h(k)$, but it should be included for an accurate representation of $h(k)$. The amplitude spectrum $C(n\Delta f)$ is obtained using Eqn (1.8) and noting that, as $h(t)$ exists only for discrete values of t, the integral can be replaced by a summation:

$$C(n\Delta f) = \sum_{k=-N/2-1}^{N/2-1} h(k\Delta t)\exp(-j2\pi nk\Delta t/T)\Delta t \tag{1.32}$$

Here we note that $\Delta t = T/N$ and $h(k\Delta t)$ is a periodic function so that the limits of the summation may be changed to give

$$C(n\Delta f) = \Delta t \sum_{k=0}^{N-1} h(k\Delta t)\exp(-j2\pi nk/N) \tag{1.33}$$

This equation is usually written in the notation of Eqn (1.34) and it should be noted that once again it is customary to omit the multiplying factor, which in this case is Δt:

$$C(n) = \sum_{k=0}^{N-1} h(k)\exp(-j2\pi nk/N) \tag{1.34}$$

Equation (1.34) is known as the discrete Fourier transform (DFT) of $h(t)$, and Eqn (1.31) is known as the inverse discrete Fourier transform (IDFT) of $C(n\Delta f)$. It should be noted that in both Eqns (1.31) and (1.34) there is no explicit frequency or time scale as the coefficients k, n and N simply have numerical values.

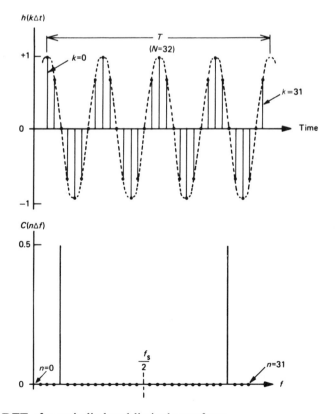

Fig. 1.11 DFT of a periodic band-limited waveform.

Some care is required in the use of the DFT because, as we have shown, it is valid only for the special case of a band-limited periodic signal. A waveform of this type and its DFT is shown in Fig. 1.11. In this figure $h(t)$ is a single tone with four complete cycles in the interval T. The DFT has a component at a value of $n = 4$ (i.e. the fourth harmonic of the fundamental frequency $f_0 = 1/T$) and a second component at a value of $n = 28$. This second component is the equivalent of $n = -4$ resulting from the change of range in Eqn (1.30). We note, therefore, that in the special case of a band-limited periodic function the DFT produces the correct spectrum of $h(t)$. In all other cases the DFT will produce only an approximation to the amplitude spectrum of $h(t)$.

Consider next the DFT of the waveform of Fig. 1.12. In this case the interval T contains 3.5 cycles of $h(t)$. The DFT requires the signal to be periodic with period T, and this means that discontinuities must now exist at the extremities of the interval T.

In other words, the periodic signal is no longer band-limited, and a form of distortion known as leakage is introduced into the spectrum. This form of distortion is considered in more detail in Section 1.11 after the concept of convolution has been introduced. It suffices here to note that sampling a non-band-limited signal produces a discrete spectrum of the form shown in Fig. 1.12. This is clearly an approximation to the original spectrum; the approximation can be made more accurate by increasing the

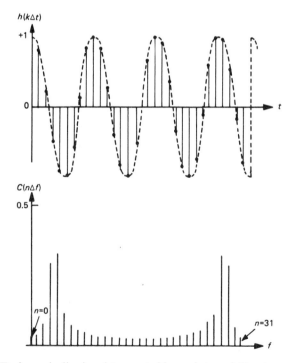

Fig. 1.12 DFT of a periodic signal truncated by an interval T not equal to a multiple of the signal period.

interval of observation T (for non-periodic signals) or by making T equal to a multiple of the period (for periodic signals). In addition, increasing the sampling frequency always reduces aliasing which is produced by sampling non-band-limited signals.

The DFT can be calculated directly from Eqn (1.34) but it will be noticed that each coefficient $C(n)$ requires N complex multiplications and additions. There are N spectral coefficients so that a total of N^2 complex multiplications and additions will be required in the complete DFT calculation. Multiplication is a relatively slow process in a general purpose digital processor, and for this reason the DFT is usually calculated by using the 'fast Fourier transform' algorithm. This is an algorithm designed to reduce the number of multiplications required to evaluate the DFT. The algorithm achieves a reduction from N^2 to $N \log_2 N$ multiplications by dividing Eqn (1.34) into the sum of several smaller sequences.[1] This reduction can be very significant when N is a very large number. (It should be noted that special purpose processors are now available which can perform multiplication in one machine cycle.)

1.11 TIME DOMAIN ANALYSIS

The time domain and frequency domain are uniquely linked by the Fourier transform and consequently the frequency domain analysis of the previous sections can also be undertaken in the time domain. To illustrate this point, consider Eqn (1.19) which relates the spectrum at a network output to the product of the input spectrum and the network transfer function. If the input to the network $h(t)$ is a unit impulse we have shown in Eqn (1.25) that the spectrum $\Delta(f)$ has unity value for all f. Hence the spectrum at the network output is simply $L(f) = P(f)$ where $P(f)$ is the network transfer function. The response in the time domain is the Fourier transform of $P(f)$, and is known as the impulse response, i.e.

$$p(t) = \int_{-\infty}^{\infty} P(f) \exp(j\,2\pi f t)\,\mathrm{d}f \qquad (1.35)$$

Having defined impulse response we now make use of Eqn (1.24), which states that the value of a signal $h(t)$ at any time t_0 is obtained by multiplying $h(t)$ by a unit impulse centred at t_0. The signal $h(t)$ can thus be regarded as an infinite number of impulses, the weight of each impulse being equal to the instantaneous value of $h(t)$. Each of these impulses will produce an impulse response and the network output is then obtained by the superposition of the individual impulse responses. The response of a linear network $l(t)$ to an input signal $h(t)$ is given in terms of the network impulse response by Eqn (1.36):

$$l(t) = \int_{-\infty}^{\infty} h(\tau)p(t - \tau)\,\mathrm{d}\tau \qquad (1.36)$$

In this equation τ is a dummy time variable and both $h(\tau)$ and $p(\tau)$ are continuous functions. Equation (1.36) therefore states that the output of a linear network at time t is given by the sum of all values of the input $h(\tau)$

weighted by the appropriate value of $p(\tau)$ at time t. The integral in Eqn (1.36) is known as the convolution integral and the equation is often written as

$$l(\tau) = h(t) * p(t) \tag{1.37}$$

where the symbol $*$ denotes convolution. Comparing Eqn (1.37) with Eqn (1.19) we note the important relationship that multiplication in the frequency domain is equivalent to convolution in the time domain. The converse is also true; that is, multiplication in the time domain is equivalent to convolution in the frequency domain. We will now consider some examples of convolution.

The first example concerns frequency domain convolution. If we consider the waveform of Fig. 1.12 we note that in selecting a time window of T s we are in effect multiplying the continuous signal $h(t)$ by a rectangular pulse of unity amplitude and duration T. This is equivalent to convolving the amplitude spectrum of $h(t)$ with the spectrum of the rectangular window function which, as we have already seen, is a sinc function. The spectrum of $h(t)$ is actually a delta function at $\pm f_0$ since only a single frequency is present. The convolution integral is thus

$$\int_{-\infty}^{\infty} H(f)\delta(f-f_0)\,\mathrm{d}f + \int_{-\infty}^{\infty} H(f)\delta(f+f_0)\,\mathrm{d}f = H(-f_0) + H(f_0) \tag{1.38}$$

The original spectrum centred at $f = 0$ is thus transferred to frequencies $\pm f_0$. The procedure is illustrated in Fig. 1.13 and it is interesting to compare this spectrum with the DFT of Fig. 1.12. The effect of truncating the signal $h(t)$ in the time domain causes a spreading of the spectrum (leakage) in the frequency domain.

We next consider the transmission of a rectangular pulse through a low-pass RC network of the form shown in Fig. 1.8. The impulse response of this network is $p(t) = (1/RC)\exp(-t/RC)$. This may be proved as follows:

$$P(f) = \int_{-\infty}^{\infty} p(t)\exp(-j2\pi ft)\,\mathrm{d}t$$

Since $p(t)$ is the impulse response of a real network it must have a value of zero for $t < 0$; hence

$$P(f) = \frac{1}{RC}\int_{0}^{\infty} \exp(-t/RC)\exp(-j2\pi ft)\,\mathrm{d}t$$

$$= \frac{-1}{RC(j2\pi f + 1/RC)}[\exp\{-(j2\pi f + 1/RC)t\}]_0^\infty$$

i.e.

$$P(f) = \frac{1}{1 + j2\pi fRC} = \frac{1}{1 + j(f/f_c)}$$

which agrees with the expression obtained by network analysis.

Before proceeding further we shall consider the physical interpretation of Eqn (1.36). In this equation t represents the present instant in time and we

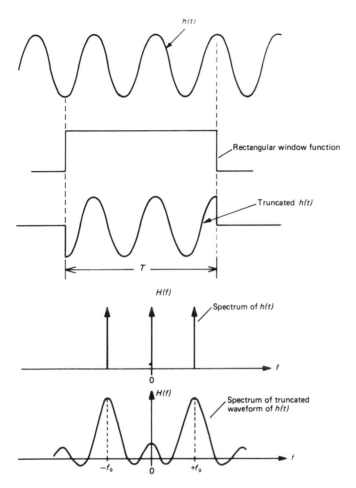

Fig. 1.13 Convolution in the frequency domain.

note that for practical filters $p(t - \tau)$ must be zero for $t < \tau$; in other words, the impulse response must be zero for all time before an impulse occurs. The impulse response of an ideal filter exists for all values of τ and is therefore unrealizable (see Section 3.3). If we confine our interest to practical networks, then it is clear that the impulse response $p(t - \tau)$ scans the signal $h(\tau)$ and produces a weighted sum of past inputs. The values of $h(\tau)$ closest to the present (i.e. $\tau \simeq t$) will have a greater effect on the output than values occurring a long time in the past ($\tau \ll t$). For practical networks Eqn (1.36) becomes

$$l(t) = \int_{-\infty}^{t} h(t)\, p(t - \tau)\, d\tau \tag{1.39}$$

We can now consider the response of the RC network to a rectangular pulse of width t_1. This is split into a positive step of unit amplitude at $\tau = 0$ followed by a negative step of unit amplitude at $\tau = t_1$, as shown in Fig. 1.14. Since the

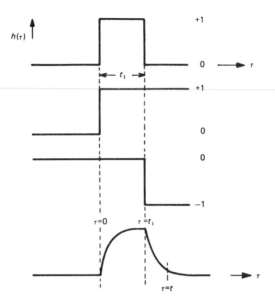

Fig. 1.14 Response of an RC network to a rectangular pulse.

system is linear the output is obtained by superposition. Considering the positive step first, since $h(\tau) = 0$ for $\tau < 0$, then $l(t) = \int_0^t p(t - \tau)\,d\tau$, i.e.

$$l(t) = \frac{1}{RC} \int_0^t \exp[-(t - \tau)/RC]\,dt$$

$$= \exp\frac{(-t/RC)}{RC}\,d\tau \int_0^t \exp(\tau/RC)\,d\tau$$

Therefore

$$l(t) = -\exp(-t/RC)[\exp(t/RC) - 1] = 1 - \exp(-t/RC)$$

This is the step response of the filter, i.e. the output for $0 < t < t_1$. For the negative step

$$l(t) = -\int_{-t_1}^t \exp[-(t - \tau)/RC]\,d\tau$$

i.e.

$$l(t) = \exp[(t_1 - t)/RC] - 1$$

The output of the filter at times in excess of $t = t_1$ is therefore

$$l(t) = \exp[(t_1 - t)/RC] - \exp(-t/RC)$$

As we have already pointed out, time domain analysis is equivalent to frequency domain analysis and it is not possible to give a general rule as to which technique is more appropriate to particular situations. Time domain analysis is frequently used in digital systems, especially in specifying network characteristics to minimize signal distortion.

1.12 CORRELATION FUNCTIONS

These functions have particular application in the time domain specification of signals that vary in an unpredictable manner (i.e. information-bearing signals, noise, etc.). The autocorrelation function of a waveform $h(t)$ is defined as

$$R_h(\tau) = \lim_{T \to \infty} \frac{1}{T} \int_{-T/2}^{T/2} h(t) \, h(t + \tau) \, dt \qquad (1.40)$$

The autocorrelation function is the average of the product $h(t) \, h(t + \tau)$ and will clearly depend on the value of τ. The function $h(t + \tau)$ is a replica of $h(t)$ delayed by an interval τ. The numerical value of $R_h(\tau)$ is a measure of the similarity (or correlation) of $h(t)$ and $h(t + \tau)$. If there is no similarity between $h(t)$ and $h(t + \tau)$ and each has zero mean value then $R_h(\tau)$ is zero. Completely random signals, such as white noise, have an autocorrelation function equal to zero. The maximum value of $R_h(\tau)$ for any signal will occur when $\tau = 0$. In these circumstances

$$R_h(0) = \lim_{T \to \infty} \frac{1}{T} \int_{-T/2}^{T/2} h^2(t) \, dt$$

which is the mean square value of $h(t)$.

The autocorrelation function is widely used in signal analysis for recognizing signals in the presence of noise and also for estimating the power spectral density of random signals. We will derive, as an example, the autocorrelation function of a periodic pulse train and a random binary data signal. Consider the periodic waveform of Fig. 1.15: it is only necessary to average the signal over one period and thus

$$R_h(\tau) = \frac{1}{T} \int_{-T/2}^{T/2} h(t) \, h(t + \tau) \, dt$$

The waveform of $h(t) \, h(t + \tau)$ is shown in Fig. 1.15. It is a periodic pulse train of amplitude A^2, pulse duration $t_1 - |\tau|$ and period T. The autocorrelation function is thus

$$R_h(\tau) = \frac{A^2}{T}(t_1 - |\tau|) \quad \text{for} - t_1 < \tau < t_1$$

but because $h(t + T) = h(t)$ the autocorrelation function is periodic. The value of $R_h(\tau)$ is plotted as a function of τ in Fig. 1.15.

We note that if $h(t)$ is a single pulse the autocorrelation function is modified to

$$R_h(\tau) = \frac{1}{t_1} \int_{-t_{1/2}}^{t_{1/2}} h(t) \, h(t + \tau) \, dt$$

which evaluates to

$$R_h(\tau) = A^2 \left(1 - \frac{|\tau|}{t_1}\right) \quad \text{for} - t_1 < \tau < t_1$$

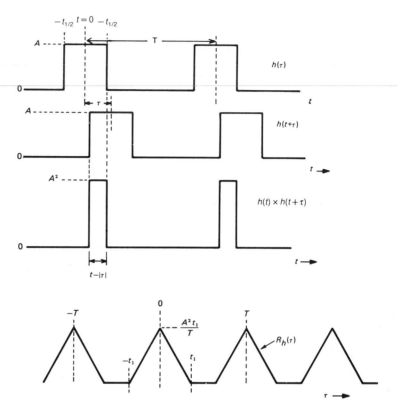

Fig. 1.15 Autocorrelation function of a periodic pulse train.

In this case $R_h(\tau)$ is a single triangular pulse of peak amplitude A^2. The autocorrelation function of a random binary pulse train is obtained in a similar way. Assume such a signal is composed of pulses of amplitude A volts and 0 volts and each of the duration t_1, both of equal probability. During any time interval T there will be an equal number of pulses of amplitude A volts and 0 volts. The average value of $h(t)\,h(t + \tau)$ must therefore be

$$R_h(\tau) = \frac{A^2}{2}\left(1 - \frac{|\tau|}{t_1}\right) \quad \text{for} - t_1 < \tau < t_1$$

and this is the autocorrelation function of a random binary pulse train.

Equation (1.40) is very similar to the convolution integral of Eqn (1.36). Remembering that convolution in the time domain is equivalent to multiplication in the frequency domain, the Fourier transform of $R_h(\tau)$ is equal to the Fourier transform of $h(t)$ multiplied by the Fourier transform of $h(t + \tau)$. The amplitude spectrum of $h(t)$ is identical to the amplitude spectrum of $h(t + \tau)$, i.e.

$$\int_{-\infty}^{\infty} R_h(\tau)\exp{(j2\pi f\tau)}\,\mathrm{d}\tau = |H(f)|^2 \qquad (1.41)$$

Thus the power spectral density of any signal is the Fourier transform of its autocorrelation function. The power spectral density of the random data signal is thus

$$G(f) = \int_{-\infty}^{\infty} \frac{A^2}{2}\left(1 - \frac{|\tau|}{t_1}\right) \exp(j2\pi f\tau)\, d\tau$$

i.e.

$$G(f) = \int_{-\infty}^{\infty} \frac{A^2}{t_1}\left(1 - \frac{|\tau|}{t_1}\right) [\cos(2\pi f\tau) + j\sin(2\pi f\tau)]\, d\tau \qquad (1.42)$$

Since $R_h(\tau)$ is an even function of τ the imaginary terms in Eqn (1.42) vanish and

$$G(f) = \int_{-\infty}^{\infty} \frac{A^2}{2}\left(1 - \frac{|\tau|}{t_1}\right) \cos(2\pi f\tau)\, d\tau$$

which evaluates to $G(f) = (At_1)^2 \operatorname{sinc}^2(\pi ft_1)$.

The autocorrelation function is a measure of the degree of similarity between $h(t)$ and a delayed version of the same waveform. The cross-correlation function is a measure of the degree of similarity between two different waveforms $h(t)$ and $g(t)$. The cross-correlation function is defined as

$$R_{hg}(\tau) = \lim_{T \to \infty} \frac{1}{T} \int_{-T/2}^{T/2} h(t)\, g(t + \tau)\, dt \qquad (1.43)$$

This function finds specific application in the detection of signals at low signal-to-noise ratios. Correlation detection is considered in detail in Chapter 5.

It should be clear from the previous two sections that both time domain and frequency domain techniques are important tools in the analysis of telecommunications systems. They should be regarded as complementary, as it is not possible to give a general rule as to which technique is more appropriate to a particular situation.

1.13 INFORMATION CONTENT OF SIGNALS

In previous sections we have considered signals in terms of waveforms and spectra. In this section we consider the information content of signals and show how it is related to the information capacity of communication channels.

Information is conveyed by a signal that changes in an unpredictable fashion. It is important to have some method of evaluating the information content of a signal because this will determine whether or not the signal can be transmitted over a particular channel. The information content of a signal is measured in bits which, as we shall show later, is not necessarily related to the number of binary digits required to transmit it. The information capacity of a communication channel is limited by bandwidth, which determines the maximum signalling speed, and by noise, which determines the number of distinguishable signal levels.

We shall consider the specific example of a teleprinter which is restricted to transmitting the four signals *ABCD*. If one of these symbols is transmitted there are four (4^1) possible messages which are *A* or *B* or *C* or *D*. If two symbols are sent there are 16 (4^2) possible messages, viz.

$$AA \quad \text{or} \quad AB \quad \text{or} \quad AC \quad \text{or} \quad AD$$
$$\text{or} \quad BA \quad \text{or} \quad BB \quad \text{or} \quad BC \quad \text{or} \quad BD$$
$$\text{or} \quad CA \quad \text{or} \quad CB \quad \text{or} \quad CC \quad \text{or} \quad CD$$
$$\text{or} \quad DA \quad \text{or} \quad DB \quad \text{or} \quad DC \quad \text{or} \quad DD$$

If *P* symbols are sent the number of possible messages is 4^P. If the teleprinter can transmit *n* different symbols, the number of different messages that could be transmitted when *P* symbols are sent is n^P. Obviously the greater the number of possible messages the less predictable is any particular message. Intuitively we would argue that the more unpredictable a particular message the more information it contains. It is reasonable to assume that the information content is a function of the **unpredictability** of a message. In algebraic form the information content *H* is

$$H \propto f(n^P) \tag{1.44}$$

Equation (1.44) can be made a function of time by assuming that one symbol is transmitted every t_1 seconds. The total number of symbols transmitted in *T* seconds is thus T/t_1 and the information content of such a message of *T* seconds duration would be

$$H \propto f(n^{T/t_1}) \tag{1.45}$$

It is reasonable to assume that a similar message of duration 2*T* seconds would contain twice as much information as a message of duration *T* seconds; in other words $f(n^{T/t_1})$ should be linearly related to *T*. This defines the function *f* as a logarithm, i.e.

$$H \propto \log_x(n^{T/t_1})$$

or

$$H = K\frac{T}{t_1}\log_x n \tag{1.46}$$

We are still required to define the numerical values of *K* and *x*. The constant of proportionality is taken as unity and the base of the logarithm is specified by defining the unit of information. To illustrate this idea consider the simplest possible system, i.e. a source that can send only two possible symbols, *A* or *B*. The simplest possible message will occur when only one of the two possible symbols is sent. The information content of such a message is defined as 1 bit.

In such a system, T/t_1 symbols are sent in *T* seconds and the information transmitted is T/t_1 bits, i.e.

$$H = \frac{T}{t_1}\log_x n \quad \text{where } n = 2$$

or

$$\frac{T}{t_1}\log_x 2 = \frac{T}{t_1}$$

Hence

$$x = 2$$

The information transmitted by a source that can send n different symbols is

$$H = \frac{T}{t_1}\log_2 n \quad \text{bits} \tag{1.47}$$

The information rate is $(1/t_1)\log_2 n$ bits/s (also b/s) and the information per symbol is $\log_2 n$ bits. In arriving at this result we have made the implicit assumption that each of the n different symbols has equal probability of being sent. When the probability is not equal our definition of information informs us that symbols that occur least frequently contain a greater amount of information than symbols that occur very frequently. Before considering probability in detail it is important to note that the symbol example chosen is not restricted to alphabetic characters.

Consider the example of a voltage pulse, and assume that each pulse can have any one of eight different voltages. A typical signal is shown in Fig. 1.16. If we assume that each of the eight levels is equi-probable the information per pulse is

$$H = \log_2 8 = 3 \text{ bits/pulse}$$

It is not always convenient to use base 2 logarithms so, making use of the relationship

$$\log_2 n = \frac{\log_{10} n}{\log_{10} 2}$$

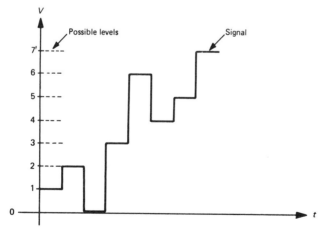

Fig. 1.16 Information-bearing signal.

then

$$H = 3.32 \log_{10} n \quad \text{bits/pulse} \tag{1.48}$$

The information capacity of a channel must be greater than the information rate of the transmitted signal in order for reliable communication to occur. We show in Section 3.7 that an ideal low-pass channel of bandwidth B can transmit pulses at a maximum rate of $2B$ per second. If we use the pulse analogy and assume that m different pulse levels can be distinguished at the channel output, the maximum rate at which information can be transmitted over the channel becomes

$$C = 2B \log_2 m \quad \text{bits/s} \tag{1.49}$$

This relationship is known as Hartley's law. If the capacity of a channel is known it is possible to determine the rate at which information can be transmitted and, consequently, the time required to transmit a given amount of information. Hartley's law does not give any indication of how the value of m is determined. This depends on the signal-to-noise ratio at the channel output, and to pursue this further it is necessary to introduce the significance of probability in information theory.

This may be illustrated by assuming that an information source can send n equi-likely symbols, each of which belongs to one of two groups. It is further assumed that the receiver is not interested in the value of a particular received symbol, rather it is concerned with knowing only to which group the received symbol belongs. If group 1 contains n_1 symbols and group 2 contains n_2 symbols there are two messages which are of interest to the receiver and these have a probability of occurrence of $P_1 = n_1/n$ and $P_2 = n_2/n$, respectively. The information per symbol for n equi-likely symbols is $\log_2 n$ bits, hence the total information in n symbols is $n \log_2 n$. It follows that the total information in group 1 (which is not of interest to the receiver) is $n_1 \log_2 n_1$ and that the total information in group 2 (which is also not of interest to the receiver) is $n_2 \log_2 n_2$. Thus the useful information H may be defined as the total information less the information which is not of interest, i.e.

$$H = n \log_2 n - n_1 \log_2 n_1 - n_2 \log_2 n_2 \tag{1.50}$$

The average information is thus $H_{\text{av}} = H/n$ or

$$H_{\text{av}} = \left(\frac{n_1 + n_2}{n} \right) \log_2 n - \frac{n_1}{n} \log_2 n_1 - \frac{n_2}{n} \log_2 n_2$$

or

$$H_{\text{av}} = \frac{n_1}{n} (\log_2 n - \log_2 n_1) + \frac{n_2}{n} (\log_2 n - \text{long}_2 n_2)$$

i.e.

$$H_{\text{av}} = \frac{-n_1}{n} \log_2 \frac{n_1}{n} - \frac{n_2}{n} \log_2 \frac{n_2}{n}$$

or

$$H_{av} = -P_1 \log_2 P_1 - P_2 \log_2 P_2 \qquad (1.51)$$

The more unpredictable an event the more information it contains; for instance let $P_1 = 0.8$, which means that $P_2 = 0.2$ since $P_1 + P_2 = 1$.

The information associated with the first event is

$$H_1 = -\log_2 0.8 = -3.32 \log_{10} 0.8 = 0.32 \text{ bits}$$

The information associated with the second event is

$$H_2 = -\log_2 0.2 = -3.32 \log_{10} 0.2 = 2.32 \text{ bits}$$

This agrees with our concept of information. The average information in this case would be

$$H_{av} = (0.8 \times 0.32) + (0.2 \times 2.32) = 0.72 \text{ bits/symbol}$$

which is considerably less than the information transmitted by the symbol with lower probability. It is of interest to determine the maximum value of H_{av}, and to do this we eliminate P_2 from Eqn. (1.51), i.e.

$$H_{av} = -P_1 \log_2 P_1 + (P_1 - 1) \log_2 (1 - P_1) \qquad (1.52)$$

To find the maximum value of H_{av} we differentiate with respect to P_1 and set the result equal to zero:

$$\frac{dH_{av}}{dP_1} = -P_1 \frac{1}{P_1} - \log_2 P_1 + (P_1 - 1) \frac{-1}{1 - P_1} + \log_2 (1 - P_1)$$

i.e.

$$\frac{dH_{av}}{dP_1} = \log_2 (1 - P_1) - \log_2 (P_1)$$

which is zero when $(1 - P_1) = P_2$ or $P_1 = P_2 = 0.5$. The average information is a maximum when the symbols are equi-probable.

In developing Eqn (1.50), the original n symbols were divided into two separate groups. This idea can be extended for any number of groups up to a maximum of n. When the number of groups equals the number of symbols, we are in effect saying that each individual symbol has its own probability of occurrence and the average information is

$$H_{av} = -\sum_{i=1}^{n} P_i \log_2 P_i \qquad (1.53)$$

Equation (1.53) is similar to an equation in statistical mechanics that defines a quantity known as 'entropy'. For this reason H_{av} is usually known as the entropy of a message. In particular, if all symbols are equi-probable, $P_i = 1/n$ and Eqn (1.53) becomes

$$H_{av} = -\sum_{i=1}^{n} P_i \log_2 n = \log_2 n \quad \text{since} \sum_{i=1}^{n} P_i = 1$$

Extension of the analysis for maximum entropy produces the same result as for the two-symbol case; that is, the entropy of a message is a maximum when all symbols are equi-probable. In any other situation the entropy will be less than the maximum and the message is said to contain 'redundancy'.

When all symbols are equi-probable, the average information is a maximum and it is not possible to make other than a pure guess at what the next symbol will be after a number have been received. In certain circumstances this can be a serious problem because if an error occurs during transmission the receiver will not be aware of it. When all symbols are not equi-probable, it becomes feasible to predict what the next symbol in a received sequence should be. The redundancy in such a message is defined as

$$R = \frac{H_{av(max)} - H_{av}}{H_{av(max)}} \times 100\% \qquad (1.54)$$

The significance of redundancy in a message will be illustrated by reference to the English language. If we assumed that all letters in the English alphabet were equi-probable, the average information per letter would be $\log_2 26 = 4.7$ bits. If the relative frequencies of occurrence of individual letters are taken into account (E has a probability of 0.1073, Z has a probability of 0.006) the figure works out as $H_{av} = 4.15$ bits/letter. This gives a redundancy of 11.7%. The redundancy is actually much higher than this because of the interdependence between letters, words and groups of words within English text. For example, if the letter Q occurs in a message it is almost certain that the next letter will be U. The U contains no information because it can be guessed with almost 100% certainty. If the letters IN have been received, the probability that the next letter will be G is much higher than the probability that it will be Z. There are many examples of this interdependence, which can be extended to words and sentences. When all these issues are considered the redundancy of English is estimated at 47%. The overall effect of redundancy is twofold; it reduces the rate of transmission of information but at the same time it allows the receiver to detect, and sometimes correct, errors.

Consider the received message

Thi ship wilp arrive on September 28

It is clear that we can detect and correct errors in the alphabetic section of the message. There is no way that we can detect an error in the date (unless it is a number greater than 30), however. The numerical part of message thus contains no redundancy. This is very important because data transmission occurs as a sequence of binary numbers that has no inherent redundancy. It is important to detect occasional errors when they occur, and in data systems a form of redundancy known as 'parity checking' is often employed. The binary digits are divided into groups of 7, e.g., 1000001, and an extra digit is added to make the total number of 1s in the group of 8 either even or odd depending on the system. The receiver then checks each group of 8 digits to determine whether an odd or even parity has been preserved. If the parity check is not valid an error is detected. (This topic is covered in more detail in Section 5.6.)

1.14 INFORMATION TRANSMISSION

The information capacity of a communications channel is specified by Hartley's law [Eqn (1.49)], but this equation does not tell us how to evaluate the number of detectable levels, m. All signals in telecommunications are subject to corruption by noise, and we can make a qualitative statement to the effect that the difference between detectable levels must be greater than the noise present during transmission. If this were not the case, signal plus noise could produce a false level indication. Noise is a random disturbance that may be analysed either in terms of its statistical properties or in terms of its spectral properties. From either description we are able to define a mean square value for the noise that is equivalent to the power developed by the noise voltage in a resistance of $1\,\Omega$.

The relationship between the number of messages that can be transmitted and noise power was obtained by Shannon[2] in 1948 using the mathematics of n-dimensional space. The mathematics of n-dimensional space is a theoretical extension of the familiar mathematics of two- and three-dimensional space. An n-dimensional space is termed a 'hyperspace' and is defined by a set of n mutually perpendicular axes. If q is a point in this hyperspace, its distance from the origin (the point of intersection of the n mutually perpendicular axes) is d where

$$d^2 = x_1^2 + x_2^2 + x_3^2 + \cdots\cdots\cdots x_n^2 \tag{1.55}$$

x_n being the perpendicular distance from the point to the nth axis. When dealing wih hyperspace, the 'volume' of an n-dimensional figure is defined as the product of the lengths of its sides. If we are considering an n-dimensional cube (hypercube) in which all the sides have length L units, the volume is given by L^n. (Note that when $n = 2$ the hypercube is actually a square, and the 'volume' is interpreted as an area.)

A circle is a two-dimensional figure whose volume (i.e. area) is πr^2. If two concentric circles are drawn, one with a radius of 1 and the other with a radius of $\frac{1}{2}$, then one-quarter of the total area is enclosed within the inner circle which has half the total radius. The volume of a three-dimensional sphere is $\frac{4}{3}\pi r^3$. This means that a sphere of radius $\frac{1}{2}$ would contain only one-eighth of the volume of a sphere of radius 1. An n-dimensional sphere (hypersphere) has a volume proportional to r^n where r is its radius. The volume of such a hypersphere of radius $\frac{1}{2}$ will thus be equal to $(\frac{1}{2})^n$ of the volume of a hypersphere of radius 1. In other words, the larger the value of n the smaller is the percentage of total volume contained within the hypersphere of radius $\frac{1}{2}$. When $n = 7$ only $1/128$ of the total volume lies within the hypersphere of radius $\frac{1}{2}$. For a hypersphere for which $n = 100$ it turns out that only 0.004% of the total volume of a hypersphere of radius 1 lies within a hypersphere of radius 0.99. This leads to the important conclusion that for a sphere of n dimensions (where $n \gg 1$) practically all of the volume lies very close to the surface of the sphere. This property is fundamental to Shannon's derivation of the information capacity of a channel.

Shannon postulates a signal source that can send out messages of duration

T seconds; the source is limited to a bandwidth of B Hz and has a statistically stationary output (the properties of the output averaged over a long period are constant). We have already indicated in Section 1.10 that any signal band limited to B Hz can be represented by $2B$ samples/s. The energy in each of these samples is proportional to the square of the sample amplitude and the total signal energy will therefore be proportional to the sum of these mean square values. If the signal amplitudes are denoted by x_1, x_2, etc., then the total normalized signal energy is

$$E = x_1^2 + x_2^2 + \cdots + x_n^2 \qquad (1.56)$$

This is identical to the expression for the $(distance)^2$ of a point from the origin of the hyperspace. If the average energy/sample is S the total energy in the message is $2BTS$ joules. Hence all messages of energy $2BTS$ joules may be represented as a point in a hyperspace of n dimensions. As $n \to \infty$ virtually all messages will be points very close to the surface of a hypersphere of radius $(2BTS)^{1/2}$.

Received signals are always accompanied by noise, which in an ideal case occupies the same bandwidth as the signal. This means that the noise waveform can also be represented by $2BT$ samples. If the average energy per noise sample is N joules the total noise energy is $2BTN$ joules. We have stated that each message is represented by a point in the hypersphere of radius $(2BTS)^{1/2}$ and therefore the sum of message and noise will be a point whose distance is $(2BTN)^{1/2}$ from each point representing the message alone. This means that all possible message + noise combinations are represented by points that are very close to the surface of a hypersphere of radius $(2BTN)^{1/2}$ centred on the point representing the noise alone. The receiver actually receives message + noise with a total energy of $2BT(S + N)$ and the points representing each combination must lie within a hypersphere of radius $[2BT(S + N)]^{1/2}$.

Because each message must be a distance $(2BTN)^{1/2}$ from points representing message + noise, we can represent the system as a hypersphere of radius $[2BT(S + N)]^{1/2}$ filled with non-overlapping hyperspheres of radius $(2BTN)^{1/2}$. The centres of the small hyperspheres represent a distinguishable message that could have been sent. The total number of distinguishable messages is given by the number of non-overlapping hyperspheres of radius $(2BTN)^{1/2}$ which can exist within a hypersphere of radius $[2BT(S + N)]^{1/2}$. This will be equal to the ratio of the volumes of the two hyperspheres. The physical picture is given in Fig. 1.17. If the number of dimensions is $2BT$, the ratio of the volumes of the two hyperspheres is

$$\left[\frac{(2BT(S + N))^{1/2}}{(2BTN)^{1/2}} \right]^{2BT} = \left[\frac{S + N}{N} \right]^{BT} \qquad (1.57)$$

This gives the number of distinguishable messages that can be sent. Assuming each message to be equi-probable, the information transmitted is

$$\log_2 \left[\frac{S + N}{N} \right]^{BT} = BT \log_2 \left(1 + \frac{S}{N} \right) \quad \text{bits} \qquad (1.58)$$

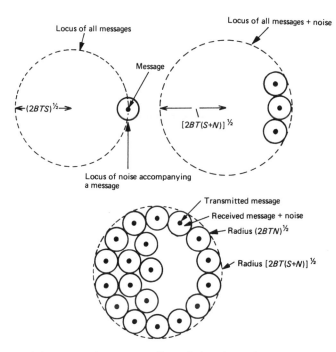

Fig. 1.17 Multidimensional representation of signal + noise.

The channel capacity is thus

$$C = B \log_2\left(1 + \frac{S}{N}\right) \quad \text{bits/s}$$

The ratio of signal energy to noise energy is identical to the corresponding power ratio so that Shannon's law for channel capacity is

$$C = B \log_2\left(1 + \frac{S_p}{N_p}\right) \quad \text{bits/s} \tag{1.59}$$

This law is a fundamental law of telecommunications, and states that if a channel has a bandwidth B and the mean SNR is S_p/N_p the maximum rate at which information may be transmitted is C bits/s. In other words there is a theoretical limit on the amount of information that can be transmitted over any telecommunications channel and, as we shall show in later chapters, all practical systems fall short of this theoretical limit by varying degrees. As an illustration of the theoretical application of Shannon's law we will assume that a link of effective bandwidth 3 kHz is to be used for pulse transmission. This link is to be used for six identical channels, the data rate being 1 kb/s per channel. The SNR on the link is $2/\log_{10} D$ where D is the length of the link in kilometres. We are required to determine the maximum distance over which reliable communication is possible.

The total data rate is $6 \times 1000 = 6\,\text{kb/s}$. Hence from Shannon's law

$$6 \times 10^3 = 3 \times 10^3 \log_2\left(1 + \frac{2}{\log_{10} D}\right)$$

which reduces to

$$1 + \frac{2}{\log_{10} D} = 4.0$$

or

$$D = 4.64 \text{ km}$$

This example does not consider the way in which the information is actually transmitted. In order to achieve the theoretical channel capacity Shannon postulated that each message should be represented by a large number of samples (data points). The receiver, in effect, compares the received message, corrupted by noise, with each possible uncorrupted message that could have been transmitted. These uncorrupted messages are the centres of the hyperspheres of radius $(2BTN)^{1/2}$. The receiver then decides that the message which was actually transmitted is the one (in the n-dimensional space) closest to the point representing the received message + noise. As the number of data points describing each message increases, the closer is any individual message to the surface of the hypersphere of radius $(2BTS)^{1/2}$. Alternatively, the larger the number of data points, the more accurate is Shannon's law.

The practical drawback is that the larger the number of data points the longer is the time taken to transmit them over a link of fixed bandwidth. The receiver cannot decide which message was transmitted until all data points have been received. Thus in attempting to realize a communications system that obeys Shannon's law, long delays would be introduced between transmission and reception (the information rate is not affected by a delay).

Shannon's law would not seem to be a practical proposition, but it is extremely useful as a standard for comparing the performance of various telecommunications systems. Of particular importance is the possibility of exchanging bandwidth and SNR in order to achieve a given information transmission rate. We will now examine this possibility in some detail. It will be noted from Eqn (1.59) that the value C can remain fixed when the bandwidth B is changed, provided that the SNR is modified accordingly. This means that SNR and bandwidth can be exchanged without affecting the channel capacity. Figure 1.18 shows bandwidth plotted against SNR for a constant channel capacity of 1 b/s.

Point A on this graph shows that it is possible to transmit 1 b/s in a bandwidth of 0.25 Hz provided that the signal power is 15 times the noise power. If the bandwidth is now doubled (point B on the graph) the same rate of transmission is possible with a SNR of 3. If we assume that the noise power is constant then evidently much less signal power is required at point B. As a simple physical example assume that 1 b/s is being transmitted at point B on the curve using binary pulses. If the bandwidth is decreased then clearly fewer binary pulses can be transmitted per second. To maintain the same information rate, the binary pulses must be replaced by pulses with more than two levels, which consequently increases the mean signal power. For large SNRs, Shannon's law approximates to

$$C = B \log_2(S_p/N_p) \tag{1.60}$$

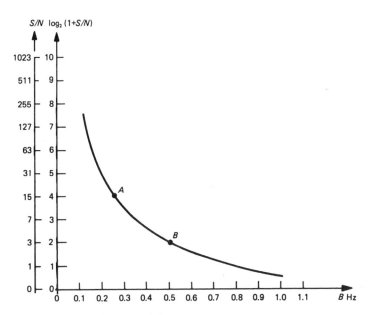

Fig. 1.18 Shannon's law for $C = 1$ b/s.

This equation states that in order to maintain a given rate of transmission of information, a linear change in bandwidth must be accompanied by an exponential change in SNR. Thus very large increases in signal power are required to compensate for relatively small reductions in the channel bandwidth B. The assumption that noise power remains constant is not in fact justified. In telecommunications systems the noise power is linearly related to bandwidth, i.e.

$$N_p = \eta B \qquad (1.61)$$

η is known as the noise power spectral density. This means that when the bandwidth is doubled the noise power increases by 3 dB; hence the reduction of required signal power that accompanies an increase in bandwidth is partially offset by the extra noise power. At low SNRs, the saving in received signal power will be cancelled by the extra noise. We can write Shannon's law as

$$C = 1.45\, B \ln(1 + S_p/N_p)$$

If S_p/N_p is very small, $\ln(1 + S_p/N_p) \simeq S_p/N_p$. Hence $C = 1.45\, BS_p/N_p$.
But $N_p = \eta B$; thus

$$C = 1.45\, S_p/\eta \quad \text{bits/s} \qquad (1.62)$$

The channel capacity now becomes independent of the bandwidth and a further increase in bandwidth has no effect. Hence there is a limit to the amount of information that can be transmitted with a fixed signal power regardless of the bandwidth. Once again this is a theoretical limit and all practical systems fall short of this limit. In the following chapters we consider both analogue and digital communications systems and compare their relative performance in terms of the theoretical limits imposed by Shannon's law.

1.15 CONCLUSION

In this chapter we have considered, in a general way, methods of describing both signal and channel characteristics met in telecommunications systems. We have introduced the idea that the channel characteristic intimately influences the signal waveform of the transmitted information. In defining information, which is the basic entity to be transmitted, we were able to show once again an intimate relationship between channel characteristics and the rate at which this information can be transmitted.

In the following chapters we will be considering specific telecommunications systems in detail and showing that the relationships outlined in this chapter are common to all systems. Of particular importance to the telecommunications engineer is the effect of the omnipresent noise which corrupts the information-bearing signal, in addition to any distortion produced by channel characteristics. The effect of noise is quite different in analogue and digital systems, and we will show that this is one reason why digital systems have a performance much closer to the theoretical performance of Shannon's law than their analogue counterparts.

REFERENCES

1. Brigham, O., *The Fast Fourier Transform*, Prentice-Hall, London and New Jersey, 1974.
2. Shannon, C. E., 'A mathematical theory of communication', *Bell Systems Technical Journal*, **27**, 379–423 (1948).

PROBLEMS

1.1 Each of the pulse trains shown in the figure represents a voltage across a $1\,\Omega$

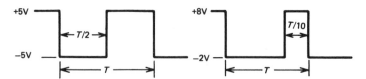

resistance. Find the total average power dissipated by each voltage and also the percentage of the total average power contributed by the first harmonic in each case.

Answer: 25 W; 10 W; 19.2%.

1.2 Assuming the integrator shown in the figure is ideal (the output voltage increases at a rate of $-1/RC$ volts/second when the input is 1 volt), sketch the output voltage when the input is a square wave of amplitude $+A$ volts or $-A$ volts and period T.

If $1/RC = 2/T$, plot the amplitude spectrum of the input and output signals.

Answer: $(AT/2) \operatorname{sinc}(\pi n f T/2); \, AT \operatorname{sinc}^2 (\pi n f T/2)$.

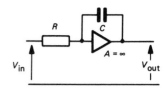

1.3 Using the principle of superposition, or otherwise, obtain an expression for the amplitude of the nth harmonic of the exponential Fourier series of the waveform shown in the figure.

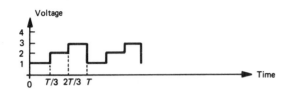

Answer: $[\cos(2n\pi/3) - 1]\pi n f$.

1.4 Plot the single-sided and double-sided power spectral densities for a pulsed waveform with period 0.1 ms and pulse width of 0.01 ms. The amplitude of the narrow pulse is 10 V and the amplitude of the wide pulse is 0 V.

What fraction of the total average power is contained within a bandwidth extending from 0 to 100 kHz?

Answer: 91%.

1.5 A rectangular pulse train with amplitude 0 V or 10 V is applied to the input of the RC network shown in the figure, the pulse repitition rate being 20 000 per second. If the half-power frequency of the network is 500 kHz and the pulse train has a duty cycle of 0.01, sketch the signal power spectrum at the network output. How much power is contained in the second harmonic?

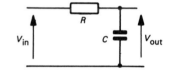

Answer: 127 μW.

1.6 Find the Fourier transform of the function $V(t)$ which is zero for negative values of its argument and is equal to $\exp(-t)$ for positive values. Find the transform when $\exp(-t)$ is replaced by $t * \exp(-t)$.

Answer: $1/(1 + j2\pi f); \, 1/(1 + j2\pi f)^2$

1.7 Show that the response of a linear system with a transfer function $H(jf)$ to a unit impulse function is

$$v(t) = 2 \int_0^\infty m \cos(2\pi f t + \phi) \mathrm{d}f$$

where $m = |H|$ and $\phi = \arg(H)$. (*Hint*: observe that m is an even function and ϕ is an odd function.)

1.8 Show that the autocorrelation function of a periodic waveform has the same period as the waveform. A voltage waveform is odd with period $2p$ and has the value A volts over half its period and 0 volt over the remainder of its period. Show that for $0 < t < P$ its autocorrelation function is given by

$$R(\tau) = (1 - 2\tau/p)A^2$$

1.9 A voltage waveform $e(t)$ has an arithmetric mean given by

$$\bar{e} = \lim_{T \to \infty} 1/2T \int_{-T}^{T} e(t)\,\mathrm{d}t$$

and $R(\tau)$ is its autocorrelation function. Deduce that the autocorrelation function of $e(t) + C$ (where C is a constant) is

$$R(\tau) + 2C\bar{e} + C^2$$

1.10 A system can send out a group of four pulses, each of 1 ms width and with equal probability of having an amplitude of 0, 1, 2 or 3 V. The four pulses are always followed by a pulse of amplitude -1 V to separate the groups. What is the average rate of information transmitted by this system?

Answer: 1600 b/s.

1.11 The probabilities of the previous question are altered such that the 0 V level occurs one-half of the time on average, the 1 V level occurs one-quarter of the time on average, the remaining levels occurring one-eighth of the time each. Find the average rate of transmission of information and determine the redundancy.

Answer: 1400 b/s; 12.5%.

1.12 An alphabet consists of the symbol A, B, C, D. For transmission, each symbol is coded into a sequence of binary pulses. The A is represented by 00, the B by 01, the C by 10, and the D by 11. Each individual pulse interval is 5 ms.
  Calculate the average rate of transmission of information if the different symbols have equal probability of occurrence.
  Find the average rate of transmission of information when the probability of occurrence of each symbol is $P(A) = 1/5$, $P(B) = 1/4$, $P(C) = 1/4$, $P(D) = 3/10$.

Answer: 200 b/s; 198 b/s.

1.13 A telemetering system can transmit eight different characters which are coded for this purpose into pulses of varying duration. The width of each coded pulse is inversely proportional to the probability of the character it represents. The transmitted pulses have durations of 1, 2, 3, 4, 5, 6, 7 and 8 ms, respectively. Find the average rate of information transmitted by this system.
  If the eight characters are coded into 3-digit binary words find the necessary digit rate to maintain the same transmitted information rate.

Answer: 888.8 b/s; 1019.4 digits/s.

1.14 A space vehicle at a distance of 381 000 km from the Earth's surface is equipped with a transmitter with a power of 6 W and a bandwidth of 9 kHz. The attenuation of the signal between the transmitter and a receiver on the Earth is given by $10 + 7 \log_{10}(X)$ dB, where X is the distance measured in kilometres. The noise power at the receiver input is $0 \cdot 1$ μW. If the receiver requires an input signal-to-noise ratio 12 dB above the value given by Shannon's law, find the maximum rate of transmission of information.

Answer: 50.41 kb/s.

Analogue modulation theory 2

By definition, an information-bearing signal is non-deterministic, i.e. it changes in an unpredictable manner. Such a signal cannot be defined in terms of a specific amplitude and phase spectrum, but it is usually possible to specify its power spectrum.

The characteristics of the channel over which the signal is to be transmitted may be specified in terms of a frequency and phase response. For efficient transmission to occur, the parameters of the signal must match the characteristics of the channel. When this match does not occur, the signal must be modified or processed. The processing is termed modulation, and the need for it may be made clearer by considering two specific examples, viz. frequency multiplexing and electromagnetic radiation from an antenna (aerial).

Frequency multiplexing is commonly used in long-distance telephone transmission, in which many narrowband voice channels are accommodated in a wideband coaxial cable. The bandwidth of such a cable is typically 4 MHz, and the bandwidth of each voice channel is about 3 kHz. The 4 MHz bandwidth is divided up into intervals of 4 kHz, and one voice channel is transmitted in each interval. Hence each voice channel must be processed (modulated) in order to shift its amplitude spectrum into the appropriate frequency slot. This form of processing is termed frequency division multiplexing (FDM), and is discussed in more detail in Section 2.19.

In Chapter 7 we show that, for efficient radiation of electromagnetic energy to occur from an antenna, the wavelength of the radiated signal must be comparable with the physical dimensions of the antenna. For audio-frequency signals, antennas of several hundred kilometres length would be required – clearly a practical impossibility. For convenient antenna dimensions the radiated signal must be of a very high frequency. In this particular example the high-frequency signal would be varied (modulated) in some way to obtain efficient transmission of the low-frequency information.

This chapter is concerned only with continuous wave (i.e. sinusoidal) modulation and assumes a noise-free environment. It should be stressed at this point that real communication is always accompanied by noise, but it is desirable to consider the effects of noise after the basic ideas of modulation have been presented. The general expression for a sinusoidal carrier is

$$v_c(t) = A \cos(2\pi f_c t + \phi) \tag{2.1}$$

The three parameters A, f_c and ϕ may be varied for the purpose of transmitting information giving respectively amplitude, frequency and phase modulation.

2.1 DOUBLE SIDEBAND AMPLITUDE MODULATION (DSB-AM)

With this type of modulation the carrier amplitude is made proportional to the instantaneous amplitude of the modulating signal. Figure 2.1 shows that the original modulating signal is reproduced as an envelope variation of the carrier. Let $A = K + v_m(t)$ where K is the unmodulated carrier amplitude and $v_m(t) = a\cos(2\pi f_m t)$ is the modulating signal. The modulated carrier, assuming $\phi = 0$, is

$$v_c(t) = [K + a\cos(2\pi f_m t)]\cos(2\pi f_c t)$$

i.e.

$$v_c(t) = K[1 + m\cos(2\pi f_m t)]\cos(2\pi f_c t) \tag{2.2}$$

m is the depth of modulation and is defined as

$$m = \frac{\text{modulating signal amplitude}}{\text{unmodulated carrier amplitude}} = \frac{a}{K}$$

For an undistorted envelope, $m \leqslant 1$. If this condition is not observed the envelope becomes distorted, and a carrier phase reversal occurs as shown in Fig. 2.2.

Equation (2.2) may be expanded to yield

$$v_c(t) = K\left[\cos(2\pi f_c t) + \frac{m}{2}\cos\{2\pi(f_c - f_m)t\} + \frac{m}{2}\cos\{2\pi(f_c + f_m)t\}\right] \tag{2.3}$$

The amplitude spectrum of the modulated carrier clearly consists of three components: the carrier frequency f_c and the lower and upper side frequencies $(f_c - f_m)$ and $(f_c + f_m)$. If $v_m(t)$ is a multitone signal, e.g.

$$v_m(t) = a_1\cos(2\pi f_1 t) + a_2\cos(2\pi f_2 t) + a_3\cos(2\pi f_3 t) \tag{2.4}$$

the modulated carrier becomes

$$v_c(t) = K[1 + m_1\cos(2\pi f_1 t) + m_2\cos(2\pi f_2 t)$$
$$+ m_3\cos(2\pi f_3 t)]\cos 2\pi f_c t \tag{2.5}$$

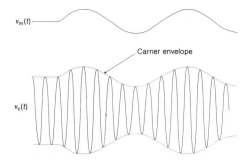

$v_m(t)$

Carrier envelope

$v_c(t)$

Fig. 2.1 DSB-AM.

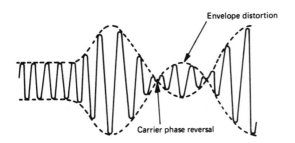

Envelope distortion

Carrier phase reversal

Fig. 2.2 Effect of overmodulation.

The depth of modulation is $m = m_1 + m_2 + m_3$ and, once again for an undistorted envelope, $m \leqslant 1$. The amplitude spectrum of $v_c(t)$ now contains a band of frequencies, termed sidebands, above and below the carrier. If the modulating signal is expressed in terms of a two-sided amplitude spectrum the process of full amplitude modulation (AM) (which is an alternative name for this type of modulation) reproduces the spectrum of $v_m(t)$ centred at frequencies $\pm f_c$. This process is illustrated in Fig. 2.3.

The phasor representation of Eqn (2.3) is a single component of length K rotating with angular velocity $\omega_c (= 2\pi f_c)$ rad/s representing the carrier added to two phasors of length $Km/2$ rotating in opposite directions with angular velocity ω_m rad/s relative to the carrier. This is illustrated in Fig. 2.4. The resultant is a single phasor rotating with angular velocity ω_c rad/s with an amplitude varying between the limits $K(1 - m)$ and $K(1 + m)$.

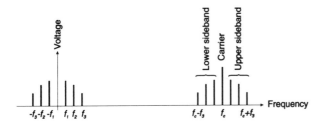

Fig. 2.3 Amplitude spectrum of DSB-AM.

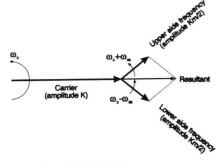

Fig. 2.4 Phasor representation of DSB-AM.

2.2 DOUBLE SIDEBAND SUPPRESSED CARRIER AMPLITUDE MODULATION (DSB-SC-AM)

The total power transmitted in a DSB-AM signal is the sum of the carrier power and the power in the sidebands. The power in a sinusoidal signal is proportional to the square of its amplitude. If the modulating signal is a single tone the total transmitted power is proportional to

$$K^2 + \left(\frac{Km}{2}\right)^2 + \left(\frac{Km}{2}\right)^2 \quad \text{watts}$$

The useful power can be regarded as the power in the sidebands as the carrier component carries no information. The ratio of useful power to total power is therefore

$$\frac{m^2}{2} : \left(1 + \frac{m^2}{2}\right)$$

For peak modulation ($m = 1$) thus ratio has a maximum value of $\frac{1}{3}$. If the carrier can be suppressed, or at least reduced in amplitude, practically all the transmitted power is then useful power. This can be important when the received signal is distorted by noise, as it produces a higher effective SNR than when the full carrier power is transmitted.

The price paid for removing the carrier is an increase in the complexity of the detector. This is discussed in more detail in Section 2.10. The amplitude spectrum of DSB-SC-AM may be derived by assuming a carrier $A \cos(2\pi f_c t)$ and a modulating signal given by $a \cos(2\pi f_m t)$. The modulated signal is simply the product of these two component, i.e.

$$v_c(t) = A \cos(2\pi f_c t)\, a \cos(2\pi f_m t)$$

i.e.

$$v_c(t) = \frac{aA}{2} \cos[2\pi(f_c - f_m)t] + \frac{aA}{2}\cos[2\pi(f_c + f_m)t] \tag{2.6}$$

When the carrier is suppressed the envelope no longer represents the mod-

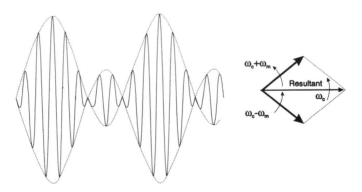

Fig. 2.5 Waveform and phasor representation.

ulating signal. The modulated carrier and the phasor representation of Eqn. (2.6) are illustrated in Fig. 2.5. For this specific case the resultant phasor is a single component with angular velocity ω_c rad/s varying in amplitude between the limits $\pm (aA)$.

2.3 SINGLE SIDEBAND AMPLITUDE MODULATION (SSB-AM)

The reason for.suppressing the carrier component was that it contained no information. It can also be observed that the signal information transmitted in the lower sideband is the same as is transmitted in the upper sideband. If either one is suppressed, therefore, it is possible to transmit the same information, but the bandwidth requirement is halved. The SSB signal that is transmitted can be either sideband, e.g.

$$v_c(t) = \frac{aA}{2} \cos[2\pi(f_c + f_m)t] \qquad (2.7)$$

It should be noted that Eqn. (2.7) seems to infer that a single sinusoidal component is transmitted. This is a special case since it is applicable to a single modulating tone. When the modulating signal is multitone the SSB signal becomes a band of frequencies. The resultant phasor is then the resultant of several phasors each rotating with different angular velocities and lengths. The price paid for reducing the signal bandwidth is an increase in the complexity of the receiver, which is dealt with in Section 2.11.

It becomes clear that, since SSB-AM has half the bandwidth of DSB-AM, twice as many independent information-bearing signals can be transmitted over a channel of fixed bandwidth when SSB-AM is used to produce frequency multiplexing.

2.4 VESTIGIAL SIDEBAND AMPLITUDE MODULATION (VSB-AM)

This is used for wideband modulating signals, such as television, where the bandwidth of the modulating signal can extend up to 5.5 MHz (for a 625 line system). The required bandwidth for a DSB-AM transmission would therefore be 11 MHz. This is regarded as excessive both from the point of view of transmission bandwidth occupation and of cost. It is generally accepted that the wider the bandwidth of a receiver, the greater the cost.

Since the amplitude spectrum of a video waveform has a dc component it would be extremely difficult to produce SSB-AM for a television signal. As a compromise, part (i.e. the vestige) of one sideband and the whole of the other is transmitted. A typical example is shown in Fig. 2.6.

The VSB-AM signal thus has both lower power and less bandwidth than DSB-AM and higher power and greater bandwidth than SSB-AM. The VSB-AM signal, however, does permit a much simpler receiver than a SSB-AM signal.

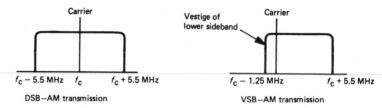

Fig. 2.6 Amplitude spectrum of a television signal.

2.5 DSB-AM MODULATORS

DSB-AM is produced by multiplying together the carrier and the modulating signal. The multiplication is achieved by using a network with a non-linear characteristic. There are basically two types of non-linear networks, one in which the characteristic is continuous and the other in which the characteristic is non-continuous, e.g. a switch.

Non-linear networks are not true multipliers because other components, which have to be filtered off, are produced. The diode modulator is an example of a modulator with a continuous non-linear characteristic of the form shown in Fig. 2.7.

The input/output characteristic of the circuit of Fig. 2.7 can be written in terms of a power series, i.e.

$$V_{\text{out}} = aV_{\text{in}} + bV_{\text{in}}^2 + cV_{\text{in}}^3 + \cdots \qquad (2.8)$$

If

$$V_{\text{in}} = [A\cos(2\pi f_c t) + B\cos(2\pi f_m t)]$$

the output is given by

$$V_{\text{out}} = a[A\cos(2\pi f_c t) + B\cos(2\pi f_m t)] + b[A^2\cos^2(2\pi f_c t)$$
$$+ B^2\cos^2(2\pi f_m t) + 2AB\cos(2\pi f_c t)\cos(2\pi f_m t] + \cdots$$

i.e.

$$V_{\text{out}} = aA\cos(2\pi f_c t)[1 + K_1\cos(2\pi f_m t)] + \text{other terms} \qquad (2.9)$$

where $K_1 = 2B/a$. If the 'other terms' are filtered off, Eqn. (2.9) has the same form as Eqn. (2.2). Hence the diode produces the required modulation.

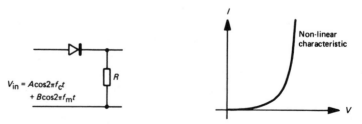

Fig. 2.7 Diode modulator.

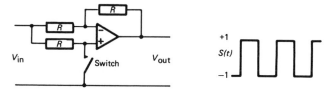

Fig. 2.8 Switching modulator.

Considering the circuit of Fig. 2.8, it can be shown that when the switch is open the circuit has a voltage gain of $+1$ and when the switch is closed the circuit has a voltage gain of -1. If the switch is replaced by a semiconductor device that is pulsed at the carrier frequency, the output of the amplifier is V_{in} multiplied by a square wave of amplitude ± 1. The Fourier series of the square wave is

$$S(t) = C_1 \cos(2\pi f_c t) + C_3 \cos(6\pi f_c t) + C_5 \cos(10\pi f_c t) + \cdots$$

If the input is $V_{in} = V_A + V_B \cos(2\pi f_m t)$ then the output will contain a term

$$C_1 V_A \cos(2\pi f_c t)[1 + K_2 \cos(2\pi f_m t)] \qquad (2.10)$$

where $K_2 = V_B/(C_1 V_A)$, which again has the same form as Eqn. (2.2). Hence the switching modulator also produces DSB-AM. Filtering is required in this case also, to remove the unwanted frequency components.

2.6 DSB-SC-AM MODULATORS

The switching modulator may also be used to produce DSB-SC modulation. If the input of the circuit of Fig. 2.8 is $V_B \cos(2\pi f_m t)$ the output will contain a term

$$C_1 V_B \cos(2\pi f_c t) \cos(2\pi f_m t)$$

i.e.

$$V_{out} = \frac{C_1 V_B}{2} \cos[2\pi(f_c - f_m)t] + \frac{C_1 V_B}{2} \cos[2\pi(f_c + f_m)t]$$

$$+ \text{ higher-frequency terms} \qquad (2.11)$$

This equation, when compared with Eqn. (2.3), will be seen to contain no carrier frequency f_c.

2.7 SSB-AM MODULATORS

SSB-AM can be produced from DSB-SC-AM by filtering off one of the sidebands. The filtering process is relatively difficult to accomplish at high frequency where the sideband separation would be a very small fraction of the filter centre frequency. The problem is eased considerably if the initial modulation takes place at a low carrier frequency. The selected sideband can then be shifted to the required frequency range by a second modulation. The sidebands of the second modulation are widely spaced, and less exact filtering is necessary.

The two-stage modulation process is shown in Fig. 2.9; a typical first modulating frequency would be in the region of 100 kHz. Most modern high-power SSB-AM transmitters working the HF band (3 − 30 MHz) generate signals in this way. An interesting alternative method is based on the Hilbert transformation. The Hilbert transform of a signal $v(t)$ is the sum of the individual frequency components that have all been shifted in phase by $90°$. The SSB-AM signal for a single modulating tone is

$$v_c(t) = A\cos[2\pi(f_c + f_m)t]$$

i.e.

$$v_c(t) = A[\cos(2\pi f_c t)\cos(2\pi f_m t) - \sin(2\pi f_c t)\sin(2\pi f_m t)] \qquad (2.12)$$

Equation (2.12) reveals that $v_c(t)$ is obtained by subtracting the outputs of two DSB-SC-AM modulators, often called balanced modulators. There is a quadrature phase relationship between the inputs of the first modulator relative to the inputs of the second modulator. A block diagram of this type of SSB-AM modulator is given in Fig. 2.10, and it should be noted that sideband filters are not required. The modulator does, however, require the provision of wideband phase-shifting networks.

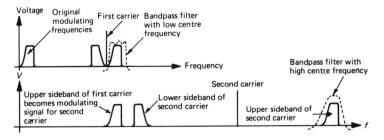

Fig. 2.9 Two-stage production of SSB-AM.

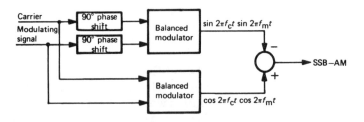

Fig. 2.10 Hilbert transform production of SSB-AM.

2.8 VSB-AM MODULATORS

VSB-AM modulation is produced by filtering the corresponding DSB-AM signal. In a television transmission system the required filtering is usually achieved by a series of cascaded high-frequency tuned amplifiers. The centre frequency of each amplifier in the chain is chosen so that the overall cascaded frequency response is asymmetrically positioned relative to the carrier frequency.

2.9 DSB-AM DETECTION

The detection (i.e. demodulation) of DSB-AM can be considered broadly under two headings; these are non-coherent detection and coherent (synchronous) detection. Traditionally, broadcast receivers have been of the superheterodyne type (see later) and have made use of envelope detection. With the advent of the integrated phase locked loop (PLL), coherent detectors are now attractive. The envelope (non-coherent) detector, as its name suggests, physically reproduces the envelope of the modulated carrier. This detector is basically a half-wave rectifier, and commonly makes use of a silicon diode whose typical current voltage relationship is shown in Fig. 2.11.

The diode acts as a rectifier and its effect on the envelope of a DSB-AM signal is shown in Fig. 2.11. For large carrier amplitudes the envelope is reproduced by the linear part of the characteristic (provided the depth of modulation is less than 100%) and no distortion results. This is termed 'large signal operation'. For small carrier amplitudes the envelope is reproduced by the non-linear part of the characteristic and distortion of the envelope occurs. Assuming the input to the detector is

$$V_{in} = K[1 + m\cos(2\pi f_m t)]\cos(2\pi f_c t)$$

when operating in the large signal mode, the output is given by

$$i = P[1 + m\cos(2\pi f_m t)]V(t) \tag{2.13}$$

where $V(t)$ is a half-wave rectified sinusoid of carrier frequency. Representing $V(t)$ by its Fourier series the output is

$$i = P[1 + m\cos(2\pi f_m t)][C_0 + C_1\cos(2\pi f_c t)$$
$$+ C_2\cos(4\pi f_c t) + C_3\cos(6\pi f_c t) + \cdots] \tag{2.14}$$

i.e.

$$i = PC_0 + PC_0\cos(2\pi f_m t) + \text{unwanted terms} \tag{2.15}$$

The unwanted terms can be filtered off leaving the original modulating signal. To ensure that the envelope remains in the linear region the depth of modulation must be less than 100%. When operating in the small signal mode the diode current is given by

$$i = aV_{in} + bV_{in}^2 + cV_{in}^3 + \cdots \tag{2.16}$$

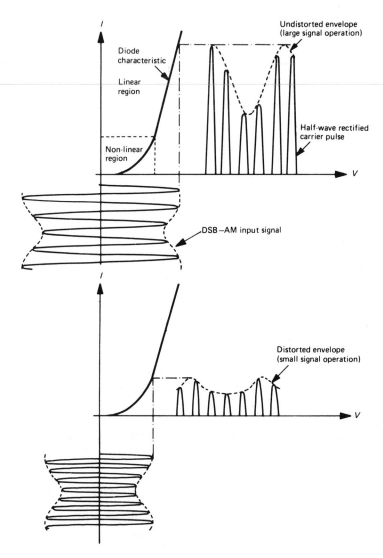

Fig. 2.11 Large and small signal operation of a diode detector.

Letting

$$V_{\text{in}} = K[\cos(2\pi f_c t) + \tfrac{1}{2}m\cos\{2\pi(f_c - f_m)t\} + \tfrac{1}{2}m\cos\{2\pi(f_c + f_m)t\}]$$

The square term gives a contribution to the output current of

$$bK^2\{\cos(2\pi f_c t) + \tfrac{1}{2}m\cos[2\pi(f_c - f_m)t] + \tfrac{1}{2}m\cos[2\pi(f_c + f_m)t]\}^2$$

i.e.

$$b\{K^2 m\cos(2\pi f_m t) + \tfrac{1}{4}K^2 m^2 \cos(4\pi f_m t)\} + \text{unwanted terms} \quad (2.17)$$

The unwanted terms are outside the modulating signal bandwidth and can be filtered off. The output current thus contains the required modulating frequency together with a second harmonic distortion term. For the lower modulating frequencies the second harmonic will be within the modulating signal bandwidth and cannot be filtered off. (When the modulating signal is multitone, cross-modulation between the individual components will also occur.) The ratio of fundamental to second harmonic is

$$K^2 m : K^2 \frac{m^2}{4} \quad \text{i.e.} \quad 1 : \frac{m}{4}$$

Thus, the relative distortion is proportional to the depth of modulation, m. (For a depth of modulation of 30%, the second harmonic distortion is 7%.)

To minimize distortion, the receiver should be designed so that the signal voltage is as large as possible. For modulating signals derived from speech or music, the peak modulation is restricted to a depth of 80%; this restricts the average depth of modulation to about 30%, which limits distortion under both large and small signal operation.

The 'unwanted terms' which appear in Eqns (2.15) and (2.17) have frequencies very much higher than the modulating frequencies and can be removed by a simple RC filter as shown in Fig. 2.12. The physical action of this detector is identical to the action of a half-wave rectifier with capacitive smoothing. The RC time constant is chosen such that the capacitor can follow the highest modulating frequencies in the envelope of the rectified carrier without losing excessive charge between carrier pulses.

In the foregoing analysis we assumed that a single modulated carrier appeared at the input of the envelope detector. In a broadcast environment there will be many different modulated carriers present in the antenna circuits of all receivers. A primary function of the receiver, therefore, will be to select one of these carriers and reject all others. The obvious way to achieve this would be to precede the detector with a bandpass filter designed to pass the required carrier and its sidebands and to reject the rest. In modern receivers the necessary frequency response is obtained by use of ceramic resonators. These circuits are available as hybrid units, and a typical example with its frequency response is shown in Fig. 2.13(a).

The circuits can be designed to have a flat response over the required bandwidth and a very large attenuation on either side. One problem, however, is that the centre frequency has a fixed value. It is not possible to vary this centre

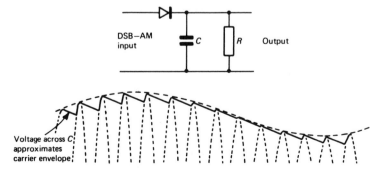

Fig. 2.12 Physical action of envelope detector.

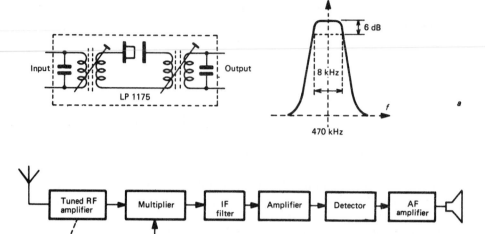

Fig. 2.13 The superheterodyne principle. (a) Ceramic resonator circuit and frequency response; (b) the superheterodyne receiver.

frequency, in order to receive other modulated carriers, and at the same time to maintain the required attenuation on either side of the passband. The superheterodyne receiver was designed specifically to overcome such a problem by transferring the sidebands of the selected carrier frequency to a constant intermediate frequency (IF). A standard IF of 470 kHz has been adopted for receivers working in the medium waveband (540 kHz to 1640 kHz).

A block diagram of the superheterodyne receiver is shown in Fig. 2.13(b) and by examining the functions of each of the blocks the derivation of the term superheterodyne will become apparent. The sidebands of the selected carrier are transferred to a frequency of 470 kHz by multiplying (heterodyning) the carrier with the output of a local oscillator. The output of the multiplier will contain sum and difference frequencies and the difference frequency between the local oscillator and selected carrier is made equal to 470 kHz. To produce this figure, the local oscillator must have a frequency of 470 kHz above or below the required carrier. If the local oscillator is above the carrier frequency the required frequency range of the oscillator is 1010 kHz to 2110 kHz for a receiver operating in the medium waveband. This is a frequency ratio of approximately 2:1 (i.e. one octave). If the local oscillator is below the carrier frequency the corresponding range would be 70 kHz to 1170 kHz, which is a ratio of 17:1. The local oscillator is designed to work above the required carrier frequency as it is then required to have a range of about one octave. Hence the derivation of the term superheterodyne.

It is not the purpose of this text to discuss in detail the design of superheterodyne receivers,[1] but there is one important point worth noting that is a disadvantage of the superheterodyne principle. When the local oscillator is 470 kHz above the required carrier the sidebands will be transferred to 470 kHz. However, if there is a second carrier at a frequency of 940 kHz above the required carrier the difference between this second frequency and the local oscillator frequency will also be 470 kHz. This is known as the 'image frequency' and is twice the intermediate frequency above the required carrier frequency.

To prevent the sidebands of the image frequency from reaching the detector it must be attenuated; hence the radio frequency amplifier in the superheterodyne receiver also has a bandpass frequency response. A single tuned LC circuit is usually adequate for this purpose.

It is worth noting that although Fig. 2.13(b) shows the superheterodyne receiver as a series of separate blocks, these blocks, with the exception of the ceramic resonator, are available on a single chip.

Coherent detection operates on the principle that is the inverse of modulation, i.e. the frequency of the sidebands is translated back to baseband by multiplying the DSB-AM signal by a sinusoid of the same frequency as the carrier. In the case of DSB-AM the carrier is actually transmitted, so the problem is one of extracting this component from the DSB-AM signal and then applying the extracted component to a multiplier along with the original modulated signal. A block diagram of the coherent detector is shown in Fig. 2.14. The output of the coherent detector may be written

$$V_{\text{out}} = K[1 + m\cos(2\pi f_m t)]\cos(2\pi f_c t)\cos(2\pi f_c t)$$

i.e.

$$V_{\text{out}} = K[1 + m\cos(2\pi f_m t)]\cos^2(2\pi f_c t)$$

but

$$\cos^2 2\pi f_c t = \tfrac{1}{2}[1 + \cos(4\pi f_c t)]$$

Therefore

$$V_{\text{out}} = \tfrac{1}{2}Km\cos(2\pi f_m t) + \text{unwanted terms} \qquad (2.18)$$

The unwanted terms will be sidebands of $2f_c$ and are easily filtered off. A

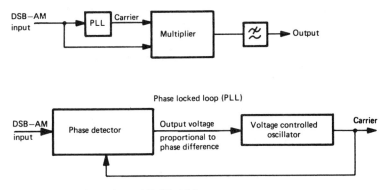

Fig. 2.14 Coherent detection of DSB-AM.

convenient method of extracting the carrier is to use a phase locked loop as shown in Fig. 2.14.

The basic operation of the loop may be described as follows. The input to the phase detector is the DSB-AM signal and a sinusoid from a voltage controlled oscillator (VCO). The output of the multiplier is a signal proportional to the phase difference between the carrier of the DSB-AM signal and the VCO output. This error signal is fed via a low-pass filter to the VCO input. Its effect is to cause the output frequency of the VCO to change in order to minimize the phase error. Hence the PLL is essentially a negative feedback system. When the loop is locked, the frequency of the VCO output is equal in frequency to the incoming carrier.

The PLL is a very important circuit as it also finds use as a frequency modulation (FM) detector. This is described more fully in Section 2.18.

2.10 DSB-SC-AM DETECTION

This type of modulation requires coherent detection, but since there is no carrier transmitted the required component must be generated by a local oscillator. Special circuits are required to ensure that the local oscillator is locked in phase to the incoming signal. If the incoming signal is represented by

$$V_{in} = K \cos(2\pi f_c t) \cos(2\pi f_m t)$$

and the output of the local oscillator is $\cos(2\pi f_c t + \phi)$, the output of the coherent detector will be

$$V_{out} = \tfrac{1}{2} K \cos(2\pi f_m t) [\cos(4\pi f_c t + \phi) + \cos\phi] \qquad (2.19)$$

After filtering off the components centred at $2f_c$ this gives a term $\tfrac{1}{2} K \cos(2\pi f_m t) \cos\phi$. When $\phi = 0$ the output is a maximum and is proportional to the original modulating signal; when $\phi = 90°$ the output is zero. A circuit for demodulating a DSB-SC-AM signal, which makes use of this property, is shown in Fig. 2.15.

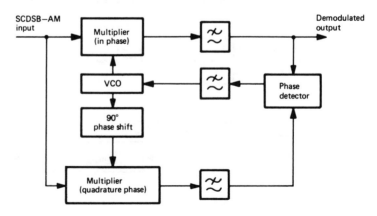

Fig. 2.15 The Costas loop.

If the local oscillator has the same phase as the missing carrier the 'in phase' channel will have the correct output and the 'quadrature channel' will have zero output. The output of the multiplier is a voltage required to maintain the desired value of the VCO phase. If there is a phase error between the VCO output and the missing carrier, the output of the 'in phase' channel will drop and the output of the 'quadrature channel' will become non-zero.

The multiplier now produces an output signal that, when applied to the VCO, will cause the phase error between the missing carrier and the VCO output to tend to zero. Hence the circuit automatically produces a locally generated sinusoid of the correct phase. This circuit was initially developed for data communications and is known as the Costas loop.

2.11 SSB-AM DETECTION

SSB-AM signals require coherent detection and, once again, a local oscillator is required, as no carrier is transmitted.

Assuming the SSB-AM signal is $K \cos 2\pi(f_c + f_m)t$ and the local oscillator signal is $\cos(2\pi f_c t + \phi)$ the output of the coherent detector will be

$$V_{out} = \tfrac{1}{2} K \cos[2\pi(2f_c + f_m)t + \phi] + \tfrac{1}{2} K \cos(2\pi f_m t - \phi) \qquad (2.20)$$

When the term at frequency $(2f_c + f_m)$ is filtered off the remaining components can be written

$$\tfrac{1}{2} K \cos(2\pi f_m t) \cos\phi + \sin(2\pi f_m t)\sin\phi \qquad (2.21)$$

If $\phi = 0$ the output is the $\tfrac{1}{2} K \cos(2\pi f_m t)$ and when $\phi = 90°$ the output is $\tfrac{1}{2} K \sin(2\pi f_m t)$. Hence in contrast to DSB-SC-AM there is also an output (shifted in phase by 90°) when $\phi = 90°$. For human listeners, whose hearing is relatively insensitive to phase distortion, the requirement of phase coherence can be relaxed. It is, however, necessary to have frequency coherence between the missing carrier and the local oscillator output. Modern point-to-point and mobile SSB-AM communications system use crystal-controlled oscillators and frequency synthesizers to achieve the required frequency stability. This is typically 1 part in 10^6 (i.e. a stability of 1 Hz at a frequency of 1 MHz.)

2.12 VSB-AM DETECTION

VSB-AM transmission is virtually exclusive to television systems, and a typical television detection system will therefore be described in this section. Television receivers use a conventional diode detector to demodulate the video carrier but some pre-proccessing of the VSB-AM signal occurs in the amplifiers that precede the detector. To illustrate the detection procedure it will be assumed that a VSB-AM signal has been produced from a two-tone modulating signal, i.e.

$$v_m(t) = q_1 \cos(2\pi f_1 t) + q_2 \cos(2\pi f_2 t)$$

The DSB-AM signal from which the vestigial sideband signal is derived will

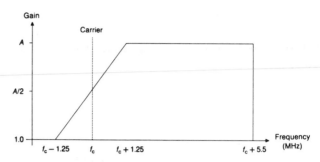

Fig. 2.16 VSB-AM receiver response.

be

$$v_c(t) = K\left[\cos(2\pi f_c t) + \tfrac{1}{2}m_1 \cos\{2\pi(f_c - f_1)t\} + \tfrac{1}{2}m_2 \cos\{2\pi(f_c - f_1)t\}\right.$$
$$\left. + \tfrac{1}{2}m_1 \cos\{2\pi(f_c + f_1)t\} + \tfrac{1}{2}m_2 \cos\{2\pi(f_c + f_2)t\}\right] \qquad (2.22)$$

To produce a VSB-AM signal the component at a frequency of $(f_c - f_2)$ in the lower sideband is assumed suppressed.

It has been shown in Section 2.9 that, if the diode detector is working in the linear region, the input signal is effectively multiplied by the carrier and its harmonics [Eqn (2.14)]. The detector output will therefore contain components

$$K\left[\cos^2(2\pi f_c t) + \tfrac{1}{4}m_1 \cos\{2\pi(2f_c - f_1)t\} + \tfrac{1}{4}m_1 \cos(-2\pi f_1 t)\right.$$
$$+ \tfrac{1}{4}m_1 \cos\{2\pi(2f_c + f_1)t\} + \tfrac{1}{4}m_1 \cos(2\pi f_1 t)$$
$$\left. + \tfrac{1}{4}m_2 \cos 2\pi\{(2f_c + f_2)t\} + \tfrac{1}{4}m_2 \cos(2\pi f_2 t)\right]$$
$$+ \text{unwanted terms} \qquad (2.23)$$

The unwanted terms can be filtered off, and it then becomes clear that the detector output contains components

$$\tfrac{1}{4}K\left[2m_1\cos(2\pi f_1 t) + m_2 \cos(2\pi f_2 t)\right]$$

i.e.

$$\tfrac{1}{4}\left[2a_1\cos(2\pi f_1 t) + a_2 \cos(2\pi f_2 t)\right]$$

The amplitude of the component transmitted in both sidebands is doubled relative to the amplitude of the component transmitted in the single sideband. To compensate for this, the gains of the amplifiers preceding the detector are designed to be asymmetric about the carrier frequency. The characteristic is shown in Fig. 2.16. The response is adjusted in such a way that the components present in both sidebands add to give the same output that would occur if only one sideband was transmitted.

2.13 ECONOMIC FACTORS AFFECTING THE CHOICE OF AM SYSTEMS

It is apparent that the special amplifier responses would not be required if the pre-processing was done at the transmitter, which would have the added advantage of reducing the transmitted power. The present standards were chosen on an economic basis. If the processing was done at the transmitter, the television receiver amplifier (of which there are many) would require a flat response over a bandwidth of 6.75 MHz. When the processing is done in the receiver, the required amplifier bandwidths are reduced to 4.25 MHz. This represents a considerable saving in the production costs of a typical television receiver.

The economic argument also decides to a large extent the standards adopted in other forms of telecommunication. A local broadcast transmission system, for example, will have a single high-power transmitter with many receivers. The unit cost of each receiver is lowest when DSB-AM is used, which dictates the use of DSB-AM in this particular situation. The conditions in a mobile communications system will be somewhat different; there will usually be a limited number of receivers, and transmitted power will be at a premium. In this environment SSB-AM with its lower transmitted power becomes attractive. SSB-AM is also used in frequency multiplexed systems over coaxial cables. The obvious advantage here is that twice as many channels may be transmitted as with DSB-AM for a given cable bandwidth.

2.14 ANGLE MODULATION

This is an alternative to amplitude modulation and, as its name suggests, information is transmitted by varying the phase angle of a sinusoidal signal. The sinusoidal signal can be conveniently written as

$$v_c(t) = A \cos \theta(t)$$

where θ is a phase angle that is made proportional to a function of the modulating signal. The time derivative of $\theta(t)$ is defined as the instantaneous frequency of the sinusoid. Strictly speaking, 'frequency' is only defined when $\theta(t)$ is a linear function of time. It is mathematically convenient to write the derivative of $\theta(t)$ as an 'instantaneous frequency', i.e.

$$\dot{\theta}(t) = 2\pi f_i$$

Angle modulation is itself divided into two categories; for example, let

$$\theta(t) = 2\pi f_c t + \phi + K_1 v_m(t) \tag{2.24}$$

This is termed phase modulation because $\theta(t)$ varies linearly with the amplitude of the modulating signal. Frequency modulation is produced if $\dot{\theta}(t)$ varies linearly with the amplitude of the modulating signal, e.g. let

$$\dot{\theta}(t) = 2\pi f_c + 2\pi K_2 v_m(t) = 2\pi f_i \tag{2.25}$$

Therefore

$$\theta(t) = 2\pi f_c t + \phi + 2\pi K_2 \int_0^t v_m(t) \, dt \tag{2.26}$$

In this case $\theta(t)$ is linearly proportional to the amplitude of the integral of the modulating signal.

2.15 PHASE MODULATION (PM)

It is shown in Section 2.16 that angle modulation may be considered as a non-linear process in terms of the relationship between time and frequency domains. This means that it becomes very difficult to derive expressions for angle-modulated waves unless very simple modulating waveforms are considered. In this context a single modulating tone may be regarded as such a signal. The modulating signal is given by

$$v_m(t) = a\cos(2\pi f_m t)$$

The PM carrier produced by this signal has the form

$$v_c(t) = A\cos[2\pi f_c t + \Delta\theta\cos(2\pi f_m t)] \qquad (2.27)$$

where $\Delta\theta = K_1 a$ is the phase shift produced when the modulating signal has its maximum positive value, i.e.

$$v_c(t) = A\cos(2\pi f_c t)\cos[\Delta\theta\cos(2\pi f_m t)]$$
$$- A\sin(2\pi f_c t)\sin[\Delta\theta\cos(2\pi f_m t)]$$

If $\Delta\theta$ is restricted to a low value ($\Delta\theta \ll 1$) then $\cos[\Delta\theta\cos(2\pi f_m t)] \simeq 1$ and $\sin[\Delta\theta\cos(2\pi f_m t) \simeq \Delta\theta\cos(2\pi f_m t)$. The modulated carrier may then be approximated as

$$v_c(t) = A\left[\cos(2\pi f_c t) - \Delta\theta\sin(2\pi f_c t)\cos(2\pi f_m t)\right]$$

This can be expanded to yield

$$v_c(t) = A\left[\cos(2\pi f_c t) + \tfrac{1}{2}\Delta\theta\sin\{2\pi(f_c - f_m)t\}\right.$$
$$\left. + \tfrac{1}{2}\Delta\theta\sin\{2\pi(f_c + f_m)t\}\right] \qquad (2.28)$$

When this equation is compared with the expression for DSB-AM [Eqn (2.3)] it will be seen that, provided the maximum phase shift it restricted to low values, the PM signal is equivalent to DSB-AM in which the carrier has been shifted in phase by 90° relative to the sidebands. PM may thus be produced from DSB-SC-AM by reintroducing the suppressed carrier in phase quadrature. This process is illustrated in block schematic form in Fig. 2.17. The phasor representation of narrowband PM ($\Delta\theta \ll 1$) is given in Fig. 2.18. For small values of $\Delta\theta$ the resultant phasor has an almost constant amplitude but the phase angle α is a function of time.

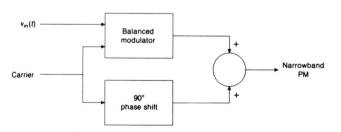

Fig. 2.17 Phase modulation.

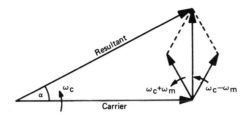

Fig. 2.18 Phasor representation of narrowband PM.

2.16 FREQUENCY MODULATION (FM)

The expression for a frequency modulated carrier is developed in a similar fashion to the expression for a PM carrier. The modulating signal is again assumed to be a single tone, and the phase of the carrier is now related to the time integral of $v_m(t)$, i.e.

$$v_c(t) = A \cos\left[2\pi f_c t + 2\pi K_2 \int_0^t a \cos(2\pi f_m t)\,\mathrm{d}t \right] \qquad (2.29)$$

Therefore

$$v_c(t) = A \cos\left[2\pi f_c t + \frac{K_2 a}{f_m} \sin(2\pi f_m t) \right]$$

i.e.

$$v_c(t) = A \cos\left[2\pi f_c t + \beta \sin(2\pi f_m t) \right] \qquad (2.30)$$

The constant β is termed the modulation index. By comparison with Eqn (2.7), it is clear the $K_2 a$ represents the maximum change in carrier frequency that is produced by the modulating signal. Letting $K_2 a = \Delta f_c$ the modulation index is defined as $\beta = \Delta f_c / f_m$. It is important to note that β depends both on the carrier deviation, which is linearly proportional to the

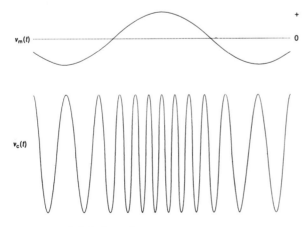

Fig. 2.19 Frequency modulated carrier.

signal amplitude, and also on the frequency of the modulating signal. The phase shift $\Delta\theta$ that occurs in PM is independent of the modulating signal frequency. The FM waveform that results from a single modulating tone is shown in Fig. 2.19.

The amplitude spectrum of the FM carrier is obtained by expanding the trigonometric function of Eqn. (2.30).

$$v_c(t) = A\cos(2\pi f_c t)\cos[\beta\sin(2\pi f_m t)]$$
$$- A\sin(2\pi f_c t)\sin[\beta\sin(2\pi f_m t)] \tag{2.31}$$

It is convenient to consider Eqn (2.31) for two specific types of FM – narrowband FM and wideband FM. The reason for this distinction is made clearer by considering the series expansion of the trigonometric functions $\cos[\beta\sin(2\pi f_m t)]$ and $\sin[\beta\sin(2\pi f_m t)]$.

$$\cos[\beta\sin(2\pi f_m t)] = 1 - \frac{\beta^2\sin^2(2\pi f_m t)}{2!} + \frac{\beta^4\sin^4(2\pi f_m t)}{4!}$$

$$+ \frac{\beta^6\sin^6(2\pi f_m t)}{6!} + \cdots$$

$$\sin[\beta\sin(2\pi f_m t)] = \beta\sin(2\pi f_m t) - \frac{\beta^3\sin^3(2\pi f_m t)}{3!}$$

$$+ \frac{\beta^5\sin^5(2\pi f_m t)}{5!} + \cdots$$

Narrowband FM is produced when $\beta \ll 1$. The series expansions for this case can be approximated by

$$\cos[\beta\sin(2\pi f_m t)] \simeq 1 \quad \text{and} \quad \sin[\beta\sin(2\pi f_m t)] \simeq \beta\sin(2\pi f_m t)$$

Using these approximations, Eqn (2.31) can be rewritten as

$$v_c(t) = A\left[\cos(2\pi f_c t) - \frac{\beta}{2}\cos\{2\pi(f_c - f_m)t\} + \frac{\beta}{2}\cos 2\pi\{(f_c + f_m)t\}\right] \tag{2.32}$$

This expression shows that narrowband FM is equivalent to DSB-AM with a phase shift of 180° in the lower sideband. The bandwidth in this case is identical to the bandwidth of a DSB-AM signal and is exactly twice the bandwidth of the modulating signal. The resultant phasor, which is illustrated in Fig. 2.20, is similar to the resultant for PM, i.e. it has almost constant amplitude and a phase angle that is a function of time.

It is evident that as the value of β increases more terms in the series expansions of $\cos[\beta\sin(2\pi f_m t)]$ and $\sin[\beta\sin(2\pi f_m t)]$ become significant and cannot be ignored. This illustrates the non-linear relationship between the time domain and frequency domain, which is a feature of angle-modulated signals. As an example, consider the series expansions of $\cos\beta\sin(2\pi f_m t)$ and $\sin\beta\sin(2\pi f_m t)$ when $\beta = 0.5$. The new approximation are

$$\cos[\beta\sin(2\pi f_m t)] \simeq 1 - \tfrac{1}{2}\beta^2\sin^2(2\pi f_m t) = 1 - \tfrac{1}{4}\beta^2[1 - \cos(4\pi f_m t)]$$

$$\sin[\beta\sin(2\pi f_m t)] \simeq \beta\sin(2\pi f_m t)$$

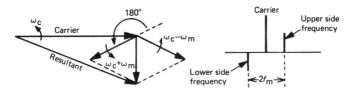

Fig. 2.20 Phasor and spectral representation of narrowband FM.

Equation (2.31) may be rewritten for this example as

$$v_c(t) = A\{(1 - \tfrac{1}{4}\beta^2)\cos 2\pi f_c t$$
$$- \tfrac{1}{2}\beta^2[\cos\{2\pi(f_c - f_m)t\} - \cos\{2\pi(f_c + f_m)t\}]$$
$$+ \tfrac{1}{8}\beta^2[\cos\{2\pi(f_c - 2f_m)t\} + \cos\{2\pi(f_c + 2f_m)t\}]\} \quad (2.33)$$

A simple increase in the value of β (produced by an increase in the amplitude of the modulating signal or a decrease in its frequency) produces a decrease in the amplitude of the component at frequency f_c and results in two additional frequency components spaced at $\pm 2f_m$ relative to f_c. The bandwidth thus effectively doubles, when compared with the previous case, simply as a result of increasing the amplitude of the modulating signal. This non-linear relationship between time and frequency domains becomes apparent when Fig. 2.21 is compared with Fig. 2.20.

Further increases in the value of β will clearly increase the number of significant terms in the series expansion of the two trigonometric functions. Both these functions are, in fact, periodic and can therefore be expressed in terms of a Fourier series, i.e.

$$\cos[\beta\sin(2\pi f_m t)] = C_0 + C_2\cos(4\pi f_m t) + C_4\cos(8\pi f_m t) + \cdots$$
$$\sin[\beta\sin(2\pi f_m t)] = C_1\sin(2\pi f_m t) + C_3\sin(6\pi f_m t) + \cdots$$

The Fourier coefficients C_n are themselves infinite series and may only be evaluated numerically for specific values of n and β. These coefficients are termed Bessel functions of the first kind, i.e. $C_n = J_n(\beta)$. Graphs of $J_n(\beta)$ for several values of n are given in Fig. 2.22. Using the Fourier series representation, Eqn. (2.31) may be rewritten as

$$v_c(t) = J_0(\beta)\cos(2\pi f_c t) - J_1(\beta)[\cos\{2\pi(f_c - f_m)t\} - \cos\{2\pi(f_c + f_m)t\}]$$
$$+ J_2(\beta)[\cos\{2\pi(f_c - 2f_m)t\} + \cos\{2\pi(f_c + 2f_m)t\}]$$
$$- J_3(\beta)[\cos\{2\pi(f_c - 3f_m)t\} - \cos\{2\pi(f_c + 3f_m)t\}]$$
$$+ \cdots \quad (2.34)$$

The amplitudes of the carrier and sidebands, of which there is an infinite number, depend on the value of β which fixes the appropriate Bessel function value. Tables of Bessel functions for a range of n and β are given in Appendix A.

The bandwidth of the wideband FM signal is apparently infinite, but in practice some simplifying approximations are possible. With reference to

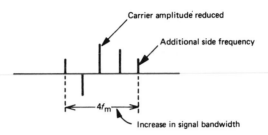

Fig. 2.21 Non-linearity of the FM process.

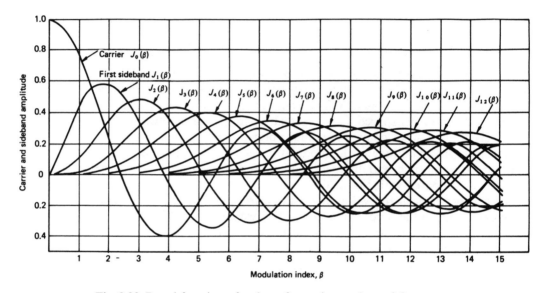

Fig. 2.22 Bessel function of order n for various values of β.

Fig. 2.22, it can be observed that as the value of n is increased then β must be made progressively larger before the appropriate value of $J_n(\beta)$ becomes non-zero. Extrapolating this trend it is found that for large values of n the value of $J_n(\beta)$ is approximately zero when $\beta < n$, e.g. when $\beta = 100$ then $J_{101}(\beta)$, $J_{102}(\beta)$, etc., are approximately zero since $\beta < n$. In this example, therefore, there are 100 values of $J_n(\beta)$ (excluding the carrier $J_0(\beta)$) that have a non-zero value. Hence this leads to the approximation that when β is large there are β values of $J_n(\beta)$ that are non-zero. Under these circumstances there will be β pairs of sidebands, each spaced by f_m, in the amplitude spectrum of the modulated signal.

The signal bandwidth thus becomes

$$B = 2\beta f_m = 2\Delta f_c$$

i.e. for large values of β the signal bandwidth is approximately twice the carrier deviation. An accurate figure for the bandwidth of an FM signal for either very large or very small values of β is given by Carson's rule, which states

$$B = 2f_m(1 + \beta)$$

i.e.

when $\quad \beta \gg 1 \quad B \simeq 2f_m\beta = 2\Delta f_c$

when $\quad \beta \ll 1 \quad B \simeq 2f_m$

For other values of β the signal bandwidth must be computed from Bessel function tables. The spectrum of the FM signal is then defined as containing all sideband components with an amplitude $\geqslant 1\%$ of the unmodulated carrier amplitude.

In the UK, commercial FM broadcasting stations restrict the maximum carrier deviation to $\pm 75\,\text{kHz}$. A single modulating tone of frequency 15 kHz producing peak frequency deviation of the carrier would yield a modulation index $\beta = 5$. The number of sideband terms with an amplitude of 0.01 or greater is obtained from the Bessel function tables as 8. The bandwidth is then

$$B = 2 \times 8 \times f_m = 240\,\text{kHz}$$

If the peak frequency deviation is produced by a 5 kHz modulating tone the modulation index will have a value of $\beta = 15$. The number of significant sidebands increases to 19 and the bandwidth becomes

$$B = 2 \times 19 \times f_m = 190\,\text{kHz}$$

Although the larger value of β produces more significant sidebands, the sidebands are actually closer together and the bandwidth of the FM signal is less.

It will be apparent from the foregoing analysis that FM and PM are very closely related, so much so that an 'instantaneous frequency' and 'modulation index' can also be defined for PM. It is instructive to compare PM and FM from the point of view of these parameters, as the comparison gives some insight into why FM is used almost exclusively in preference to PM. For a single modulating tone the instantaneous frequency in a frequency modulated wave has been defined as

$$2\pi f_i = 2\pi f_c + 2\pi K_2 a \cos(2\pi f_m t)$$

i.e.

$$f_i = f_c + \Delta f_c \cos(2\pi f_m t)$$

which gives

$$v_c(t) = A \cos\left[2\pi f_c t + \frac{\Delta f_c}{f_m} \sin(2\pi f_m t) \right]$$

For PM,

$$\theta(t) = 2\pi f_c t + \Delta\theta \cos(2\pi f_m t)$$

Therefore

$$\dot{\theta}(t) = 2\pi f_i = 2\pi f_c - \Delta\theta 2\pi f_m \sin(2\pi f_m t)$$

i.e.

$$\dot{\theta}(t) = 2\pi f_c - 2\pi \Delta f_p \sin(2\pi f_m t)$$

The PM carrier may therefore be written in terms of a frequency deviation $\Delta f_p = -\Delta\theta f_m$, i.e.

$$v_c(t) = A \cos\left[2\pi f_c t + \frac{\Delta f_p}{f_m}\cos(2\pi f_m t)\right] \tag{2.35}$$

If a modulation index β_p is defined for PM, Eqn. (2.35) can be written

$$v_c(t) = A \cos[2\pi f_c t + \beta_p \cos(2\pi f_m t)]$$

which is similar to Eqn (2.30) except that β_p does not depend on the value of f_m.

Having shown that a PM carrier can be considered in this way, it is then possible to obtain the bandwidth of a wideband PM signal by reference to Bessel function tables. If a frequency deviation $\Delta f_p = \pm 75\,\text{kHz}$ is produced in a PM carrier by a modulating tone of 15 kHz there will be eight significant sidebands and, as with the FM carrier, the bandwidth of the signal will be approximately 240 kHz. However, since β_p does not depend on the value of f_m there will be eight sidebands for all possible values of f_m. A modulating tone of 50 Hz would result in a bandwidth of $2 \times 8 \times 50 = 800\,\text{Hz}$. The value for FM would be approximately 150 kHz. Thus it becomes apparent that PM makes less efficient use of bandwidth than frequency modulation, since the bandwidth of any receiver must be equal to the maximum bandwidth of the received signal. It is shown in Chapter 4 that SNRs in modulated systems are directly related to the bandwidth of the modulate wave. Hence it may be concluded that the signal-to-noise performance of FM will be superior to PM, and for this reason FM is usually chosen in preference to PM. (There is one notable exception to this rule, namely the transmission of wideband data signals, which is dealt with in Chapter 3.)

Because of the inherent non-linearity of the FM process, it is not possible to derive the bandwidth for a multi-tone modulating signal. (The bandwidth of a frequency modulated wave has been derived by Black[2] for a two-tone modulating signal.) One observation that can be made is that if the modulating signal is multitone then no single component can produce peak frequency deviation, or overmodulation would result. Under these circumstances, practical measurements have shown that a reasonably accurate measure of the signal bandwidth is given by a modified form of Carson's rule, i.e. $B = 2(\Delta f_c + D)$, where D is the bandwidth of the modulating signal.

2.17 FREQUENCY MODULATORS

Frequency modulation may be produced directly by varying the frequency of a voltage-controlled oscillator or indirectly from phase modulation. Figure 2.23 illustrates two methods of producing frequency modulation directly. The first method is based on a RC oscillator and the second is based on a parallel resonant LC circuit.

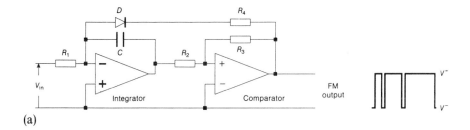

(a)

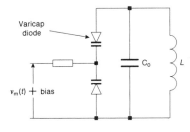

(b)

Fig. 2.23 Direct FM. (a) RC voltage controlled oscillator, (b) Varicap diode tuning of LC tank circuit.

The RC circuit has the advantage of requiring no inductance and operates as described. When the output of the comparator is V^+ the diode is reversed biased and the integrator charges the capacitor C from a current proportional to V_{in}. When the capacitor charges to the switching voltage of the comparator output goes to V^- which forward biases the diode and produces a rapid discharge of C. This causes the comparator output to return to V^+ reverse biasing the diode and causing C to charge once again from V_{in}. The charging time of the capacitor is inversely proportional to charging current and hence the frequency of oscillation is linearly related to V_{in}. The comparator output will be a variable-frequency pulse train; a sinusoidal frequency modulated waveform is obtained by appropriate filtering.

The LC circuit forms the tank circuit of a LC oscillator. A common method of altering the resonant frequency of such a tuned circuit is to use the dependence of capacitance of a reverse biased p–n junction on the reverse bias voltage. Varactor (varicap) diodes are specifically designed for this purpose and Fig. 2.23 shows a typical arrangement of these diodes in a parallel resonant circuit. The total capacitance in this circuit will be $C = C_0 + \Delta C$ where $\Delta C = K v_m(t)$. The resonant frequency of the circuit is given by

$$2\pi f_r = \frac{1}{(LC)^{1/2}} = \frac{1}{(LC_0)^{1/2}} \times \frac{1}{[1 + (\Delta C/C_0)]^{1/2}}$$

If the change in capacitance is small then $\Delta C \ll C_0$ and the binomial expansion of $[1 + (\Delta C/C_0)]^{-1/2}$ is approximated by $(1 - \Delta C/2C_0)$, i.e.

$$2\pi f_r = \frac{1}{(LC_0)^{1/2}}\left(1 - \frac{\Delta C}{2C_0}\right)$$

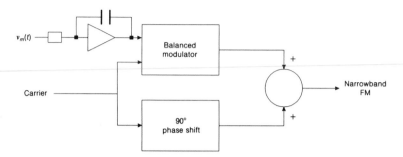

Fig. 2.24 Armstrong frequency modulator.

Therefore

$$2\pi f_r = 2\pi f_c \{1 - [K v_m(t)/2C_0]\}$$

or

$$f_r = f_c[1 - K_c v_m(t)] \qquad (2.36)$$

The resonant frequency of the tuned circuit is thus directly related to the amplitude of the modulating signal. The tuned circuit is used as the frequency determining network in a feedback oscillator thus producing FM directly. Such a modulator is restricted to narrowband FM since $\Delta C \ll C_0$ and frequency multiplication is required to produce wideband FM.

The analysis of the previous section has shown that FM is equivalent to PM by the time integral of the modulating signal. Narrowband PM is itself equivalent to DSB-SC-AM with the carrier reinserted in phase quadrature. The Armstrong indirect frequency modulator combines both these properties, and a block diagram is given in Fig. 2.24.

The maximum value of β that can be produced by this modulator is about 0.2. This means that several stages of frequency multiplication are required to produce wideband FM. The advantage of the Armstrong modulator is that the carrier is produced by a stable crystal oscillator. It should be noted that when the frequency of the carrier is multiplied by n (by means of a non-linear device) the frequency deviation, and hence the value of β, is also multiplied by the same factor.

2.18 DEMODULATION OF A FREQUENCY MODULATED WAVE

The primary function of a frequency modulation detector is to produce a voltage proportional to the instantaneous frequency f_i of the modulated wave. The FM waveform is given by

$$v_c(t) = A \cos\left(2\pi f_c t + K \int_0^t v_m(t)\, dt\right)$$

If this waveform is differentiated, the resultant waveform is

$$\dot{v}_c(t) = - A\left[2\pi f_c + K v_m(t)\right] \sin\left(2\pi f_c t + K \int_0^t v_m(t)\,dt\right)$$

This is a frequency modulated wave that now has an envelope of magnitude proportional to the amplitude of the modulating signal $v_m(t)$. The modulating signal may then be recovered by envelope detection. The envelope detector will ignore the frequency variations of the carrier. Traditionally FM detectors have relied upon the properties of tuned circuits to perform the required differentiation. This is demonstrated in Fig. 2.25, where the resonant frequency of the tuned circuit is chosen such that the carrier frequency f_c is on the slope of the circuit response.

The linearity of a single tuned circuit is limited to a relatively small frequency range. This range may be extended by introducing a second tuned circuit with a slightly different resonant frequency. A typical circuit and its response is shown in Fig. 2.26. Each circuit is fed in antiphase by the tuned secondary of the high frequency transformer. Several FM discriminators are based upon this type of circuit; the circuit of Fig. 2.26 is known as a balanced discriminator. Circuits of this type are also sensitive to fluctuations of the amplitude of the FM wave. To avoid this the FM waveform is 'hard limited', as shown in Fig. 2.27, before being applied to the discriminator.

Modern detection techniques are based upon integrated circuit technology, where the emphasis is on inductorless circuits. The hard limited FM signal is in fact a variable frequency pulse waveform. The instantaneous frequency is preserved in the zero crossings of this pulsed signal. A completely digital circuit based upon a zero crossing detector is shown in Fig. 2.28. The number of zero crossings in a fixed interval are gated to the input of a binary counter. The counter outputs are then used as inputs to a digital-to-analogue converter (DAC). The analogue output voltage is thus proportional to the number of zero crossings that occur during the gating interval. Thus the output of the DAC is proportional to the original modulating signal $v_m(t)$.

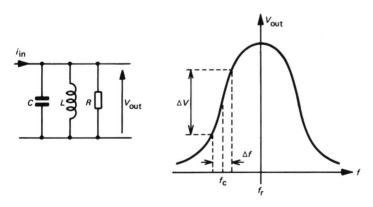

Fig. 2.25 Detection of FM with a single tuned circuit.

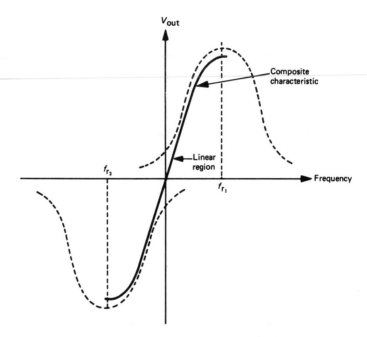

Extended range frequency-to-voltage conversion

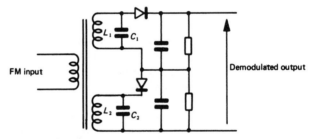

Fig. 2.26 Balanced discriminator.

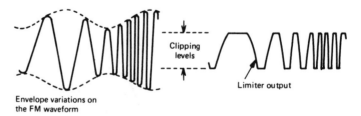

Fig. 2.27 Action of the limiter.

2.18.1 The phase locked loop

This is an important class of inductorless frequency modulation detector which is widely available in integrated form. The circuit is a feedback network with a voltage-controlled oscillator (VCO) in the feedback path. Feedback is arranged so that the output frequency of the VCO is equal to the frequency of

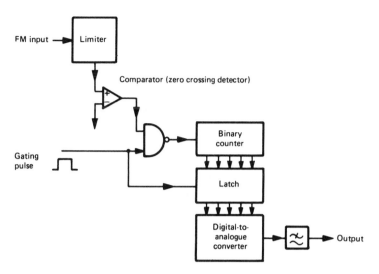

Fig. 2.28 Digital (zero crossing) FM detector.

the input waveform. If the input frequency is modulated by a voltage and the output frequency of the VCO tracks the variation in the input frequency, then the voltage at the VCO input must be equal to the voltage which produced the frequency modulation. The phase locked loop therefore demodulates the frequency modulated input.

The basic phase locked loop is shown in Fig. 2.29, and in the initial analysis it will be assumed that the transfer function of the loop filter $H(f) = 1$. It is convenient to represent the frequency modulated input as

$$v_c(t) = A \cos(2\pi f_c t + \phi_1(t))$$

where

$$\phi_1(t) = k_m \int_0^t v_m(t) \, dt$$

i.e. $\dot{\phi}_1(t) = k_m v_m(t)$, $v_m(t)$ being the modulating voltage.

The output of the VCO may be written as

$$v_r(t) = B \cos(2\pi f_c t + \phi_2(t))$$

The phase comparator is essentially a multiplier followed by a low-pass filter

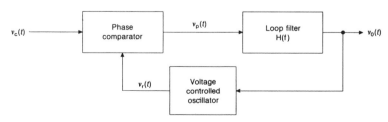

Fig. 2.29 The basic phase locked loop.

and produces an output proportional to the difference in phase between the input waveform and the output of the VCO. The output of the multiplier component of the phase comparator will be the product of $v_c(t)$ and $v_r(t)$ and is given by

$$\tfrac{1}{2}k_d AB[\sin\{\phi_2(t) - \phi_1(t)\} + \sin\{4\pi f_c t + \phi_2(t) + \phi_1(t)\}]$$

The high-frequency components are removed by the low-pass filter component and hence the output of the phase comparator may be written

$$v_p(t) = \tfrac{1}{2}k_d AB \sin\{\phi_1(t) - \phi_2(t)\}$$

When the loop is locked the phase error will be very small, i.e. $\{\phi_1(t) - \phi_2(t)\} \to 0$.

Thus

$$v_p(t) \approx \tfrac{1}{2}k_d\{\phi_1(t) - \phi_2(t)\}$$

If $H(f) = 1$ then

$$v_o(t) \approx \tfrac{1}{2}k_d\{\phi_1(t) - \phi_2(t)\} \tag{2.37}$$

where k_d is the gain of the phase comparator and has units of volts per radian. It should be noted that when the loop is locked there is actually a phase difference of $90°$ between $v_c(t)$ and $v_r(t)$. The output frequency of the VCO may be written as $2\pi f_c + \dot{\phi}_2(t)$ where $\dot{\phi}_2(t) = k_0 v_o(t)$, k_0 being the gain of the VCO with units of radians/second/volt, i.e.

$$\phi_2(t) = k_0 \int_0^t v_o(t)\,dt$$

Equation (2.37) can thus be written

$$v_o(t) = \tfrac{1}{2}k_d AB\left[k_0 \int_0^t v_o(t)\,dt - \phi_1(t)\right] \tag{2.38}$$

Differentiation of Eqn (2.38) yields

$$\dot{\phi}_1(t) = k_0\left[\frac{2\dot{v}_o(t)}{k_0 k_d AB} - v_o(t)\right] \tag{2.39}$$

It should be noted that the product $k_0 k_d$ has dimensions of seconds^{-1}, which is a frequency. Thus if the **frequency** of $v_o(t) \ll k_0 k_d$ then

$$\frac{2\dot{v}_o(t)}{k_0 k_d AB} \to 0 \quad \text{and} \quad \dot{\phi}_1(t) = -k_0 v_o(t)$$

But $\dot{\phi}_1(t) = k_m v_m(t)$, therefore

$$v_o(t) = -\frac{k_m}{k_0}v_m(t)$$

This shows that the PLL demodulates the frequency modulated input waveform.

An insight into the physical action of the PLL can be gained by considering the loop when the input is an unmodulated carrier. The phase of $v_c(t)$ will be a linear function of time, i.e. $2\pi f_c t$, which t can then be regarded as a ramp function. To maintain a small error at the phase detector output the phase of the VCO output must also be a ramp of the same slope. If the frequency of $v_c(t)$ increases the slope of the 'phase ramp' also increases. The error between the phase ramp of $v_c(t)$ and the phase ramp of the VCO output then begins to increase producing an increase in the phase comparator output. This increase modifies the frequency of the VCO output and a new state of equilibrium is reached when the two slopes are again equal, indicating that the two frequencies are also equal. There will, however, be a different output from the phase comparator indicating a different, but constant, phase error.

The analysis presented above deals only with the steady-state response of the PLL the assumption being that the loop is locked. If this is not the case the PLL will behave as a negative feedback system with the usual transient properties. The transient response of the phase locked loop is an important consideration as it determines the range of frequencies over which the loop will acquire lock and also the range of frequencies over which a frequency locked loop will remain in lock. The transient response of the phase locked loop depends on the frequency response of the loop filter and when $H(f)=1$ the circuit is known as a first-order phase locked loop.

In order to analyse the transient response of the phase locked loop it is convenient to represent the device as a linear system as shown in Fig. 2.30. The frequency domain equivalent is derived in terms of the Laplace transform and is also given in Fig. 2.30.

Considering the frequency domain model

$$V_o(s) = k_d H(s)\left[\phi_1(s) - V_o(s)\frac{k_0}{s}\right]$$

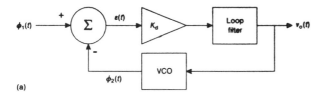

(a)

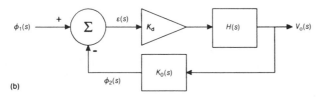

(b)

Fig. 2.30 Alternative time (a) and frequency (b) models of the phase locked loop.

Hence

$$\frac{V_o(s)}{\phi_1(s)} = \frac{sk_dH(s)}{k_ds + k_0k_dH(s)} \tag{2.40}$$

but

$$V_o(s) = \varepsilon(s)k_dH(s)$$

thus

$$\frac{\varepsilon(s)}{\phi_1(s)} = \frac{s}{s + k_0k_dH(s)} \tag{2.41}$$

If $H(s) = 1$ the loop is known as a first-order loop and Eqn (2.40) becomes

$$\frac{V_o(s)}{\phi_1(s)} = \frac{k_ds}{s + k_0k_d} \tag{2.42}$$

The transient response of the loop is examined by applying a unit impulse in frequency at the input, which is equivalent to $\phi_1(s) = 1/s$. Substituting into Eqn (2.42) gives the frequency response of the loop as

$$V_o(s) = \frac{k_d}{s + k_0k_d} \tag{2.43}$$

The 3 dB cut-off frequency is then

$$f_{3\,\text{dB}} = \frac{k_0k_d}{2\pi} \tag{2.44}$$

Hence in order to avoid attenuation of high frequency components in the output signal it is necessary that the bandwidth of this signal $< f_{3\,\text{dB}}$, which agrees with constraints applied to Eqn 2.39. For a step change in input frequency $\phi_1(s) = 2\pi f/s^2$ which may be substituted in Eqn (2.41) to give the phase error as

$$\varepsilon(s) = \frac{2\pi f}{s(s + k_0k_d)}$$

or

$$\varepsilon(t) = \frac{2\pi f}{k_0k_d}(1 - e^{-k_0k_dt})u(t) \tag{2.45}$$

The steady-state error, after the transient interval, is

$$\varepsilon(\infty) = \frac{2\pi f}{k_0k_d} \tag{2.46}$$

In order to minimize the phase error k_0k_d should be large, which will result in a large value of $f_{3\,\text{dB}}$. In practice the bandwidth of the loop should be sufficient only to avoid attenuation of components in the modulating signal. If the bandwidth is significantly greater than this the noise performance of the PLL, as a frequency demodulator deteriorates (see Chapter 4 for noise in frequency modulation transmission). Hence the first-order PLL is seldom used in practice.

The second-order PLL is derived by making the transfer function of the loop filter

$$H(s) = \frac{1 + \tau_2 s}{1 + \tau_1 s} \qquad (2.46)$$

A typical loop filter is shown in Fig. 2.31 and it should be noted that when $R_1 \gg R_2$ the transfer function of the loop filter over the range of frequencies in $V_o(s)$ is approximated by

$$H'(s) = \frac{1 + s\tau_2}{s\tau_1}$$

Substituting for $H'(s)$ in Eqn (2.40) gives

$$\frac{V_o(s)}{\phi_1(s)} = \frac{k_d s (s\tau_2 + 1)}{\tau_1 (s^2 + 2\zeta\omega_n s + \omega_n^2)} \qquad (2.47)$$

where

$$\omega_n = 2\pi f_n = \sqrt{\frac{k_o k_d}{\tau_1}}$$

is known as the natural (radian) frequency and

$$\zeta = \frac{\omega_n \tau_2}{2}$$

is known as the damping factor. The bandwidth of the second-order loop depends on the value of ζ and it is usual to operate the loop with $\zeta < 1$. Under

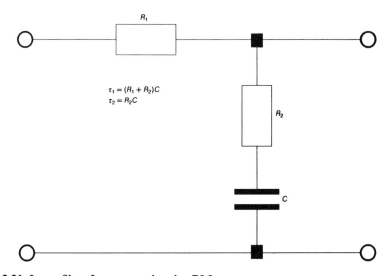

$$\tau_1 = (R_1 + R_2)C$$
$$\tau_2 = R_2 C$$

Fig. 2.31 Loop filter for a second-order PLL.

these circumstances the bandwidth of the loop is approximately

$$f_{3\,dB} = \frac{2\omega_n\zeta}{2\pi} = \frac{k_0 k_d \tau_2}{2\pi\tau_1}$$

The transfer function of the second-order PLL is given by

$$\frac{\varepsilon(s)}{\phi_1(s)} = \frac{s^2}{s^2 + 2\zeta\omega_n s + \omega_n^2} \tag{2.48}$$

For a step change in input frequency $\phi_1(s) = 2\pi f/s^2$ which may be substituted in Eqn (2.48) to give the phase error as

$$\varepsilon(s) = \frac{2\pi f}{s^2 + 2\zeta\omega_n s + \omega_n^2} \tag{2.49}$$

or

$$\varepsilon(t) = \frac{f e^{-\zeta\omega_n t}}{f_n \sqrt{1-\zeta^2}} \sin\left(\omega_n t \sqrt{1-\zeta^2}\right) u(t) \tag{2.50}$$

The steady-state error, after the transient interval, is

$$\varepsilon(\infty) = 0 \tag{2.51}$$

Hence the second-order PLL has zero steady-sate error and the transient response and bandwidth can be determined by appropriate choice of natural frequency and damping factor. The second-order PLL is therefore the preferred choice in most practical applications.

2.19 FREQUENCY DIVISION MULTIPLEX (FDM) TRANSMISSION

This is the form of transmission used extensively in telephone systems for the simultaneous transmission of several separate telephone circuits over a wideband link. Each of the telephone signals has a bandwidth limited to

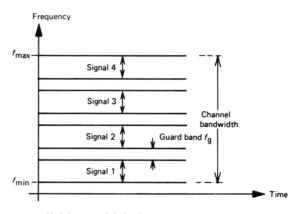

Fig. 2.32 Frequency division multiplexing.

3.4 kHz, and therefore many such signals can be transmitted over, say, a coaxial cable with a bandwidth of several megahertz.

Each signal modulates a sinusoidal carrier of different frequency, single sideband modulation being used. The signals are therefore effectively stacked one above the other throughout the transmission bandwidth. To facilitate separation of the signals at the far end of the link, adjacent signals are separated by a guard band f_g (Fig. 2.32). Each signal in a frequency division multiplex transmission system thus occupies part of transmission bandwidth for the whole of the transmission time.

2.20 CONCLUSION

The economic argument for use of the various amplitude modulated systems was given in Section 2.13. Frequency division multiplexing of telephone circuits is based on SSB-AM because here the emphasis is on packing as many channels as possible into a finite bandwidth.

The arguments for the use of frequency modulation are less well defined. It is shown in Chapter 4 that for a fixed transmitter power frequency modulation has a superior signal-to-noise performance over all types of amplitude modulation. Consequently, frequency modulation has been adopted as the standard for high-quality sound broadcast transmission in many countries. Frequency modulation is also an extremely important technique used in both analogue and digital cellular radio in which the capture effect (covered in Section 4.11) is utilized to increase system capacity.

REFERENCES

1. Scroggie, M. G., *Foundations of Wireless*, 8th edn, Illife, London, 1971.
2. Black, H. S., *Modulation Theory*, Van Nostrand Reinhold, Wokingham, UK, 1953.
3. Stremler, F. G., *Introduction to Communication Systems*, Addison-Wesley, London, 1976.

PROBLEMS

2.1 A transmitter radiates a DSB-AM signals with a total power of 5 kW at a depth of modulation of 60%. Calculate the power transmitted in the carrier and also in each sideband.

Answer: 4.24 kW; 0.38 kW.

2.2 A carrier wave represented by $10\cos(2\pi 10^6 t)$ V is amplitude modulated by a second wave represented by $3\cos(2\pi 10^3 t)$ V. Calculate
   (a) the depth of modulation;
   (b) the upper and lower side frequencies;
   (c) the amplitude of the side frequencies;
   (d) the fraction of the power transmitted in the sidebands.

Answer: (a) 30%; (b) 1.001 MHz, 0.999 MHz; (c) 1.5 V; (d) 4.3%.

2.3 A DSB-AM transmitter produces a total output of 24 kW when modulated to a depth of 100%. Determine the power output when
(a) the carrier is unmodulated;
(b) the carrier is modulated to a depth of 60%, one sideband is suppressed and the carrier component is attenuated by 26 dB.

Answer: (a) 16 kW; (b) 1.48 kW.

2.4 A DSB-AM receiver uses a square law detector. What is the maximum depth of modulation that may be used if the second harmonic distortion of the modulating signal, produced by the detector, is restricted to 10% of the fundamental?

Answer: 40%.

2.5 A carrier of 5 V rms and frequency 1 MHz is added to a modulating signal of 2 V rms and frequency 1 kHz. The composite signal is applied to a biased diode rectifier in which the relationship between current and voltage over the range ± 10 V is $i = (5 + v + 0.05 v^2) \, \mu A$ where v is the instantaneous voltage. Find the depth of modulation of the resulting DSB-AM signal and the frequency of each component in the diode current.

Answer: 28.3%; 0 Hz, 1 kHz, 2 kHz, 1 MHz, 2 MHz, 0.999 MHz, 1·001 MHz.

2.6 Narrowband FM is produced indirectly by varying the phase of a carrier of frequency 13 MHz, the maximum phase shift being 0.5 rad. Show that wideband FM may be produced by multiplying the frequency of the modulated carrier. If the modulating signal has a frequency of 1.5 kHz find the frequency deviation of the carrier after a frequency multiplication of 15.

Answer: 11.25 kHz.

2.7 A frequency modulated wave has a total bandwidth of 165 kHz when the modulating signal is a single tone of frequency 10 kHz. Using Bessel function tables, or otherwise, find the maximum carrier frequency deviation produced by the modulating signal.

Answer: 50 kHz.

2.8 In a direct FM transmitter an inductance of 10 μH is tuned by a capacitor whose capacitance is a function of the amplitude of the modulating signal. When the modulating signal is zero the effective capacitance is 1000 pF. An input signal of $4.5 \cos(3\pi 10^3 t)$ V produces a maximum change in capacitance of 6 pF. Assuming the resultant FM signal is multiplied in frequency by 5, calculate the bandwidth of the eventual output.

Answer: 60 kHz.

2.9 A single tone of frequency 7.5 kHz forms the modulating signal for both a DSB-AM and a FM transmission. When modulated the peak frequency deviation of the FM signals is 60 kHz. Assuming that the same total power is transmitted for each of the modulated signals, find the depth of modulation for the DSB-AM signal when the amplitude of the first pair of sidebands of the FM wave equals the amplitude of the sidebands of the DSB-AM wave.

Answer: 50%.

Discrete signals 3

The signals considered in Chapter 2 were continuous functions of time. There are many advantages that result from the conversion of analogue signals into a binary coded format. Two such advantages that are easily identified are that the transmission and processing of binary signals are generally much easier to achieve than the transmission and processing of analogue signals.

It is not possible to code a continuous analogue signal into binary format because there is an infinite number of values of the continuous signal. Instead the continuous signal is coded at fixed instants of time. These instants are known as sampling instants, and it is important to determine the effect of the sampling action on the properties of the original continuous signal. The rules governing the sampling of continuous signals are specified by the sampling theorem.

3.1 SAMPLING OF CONTINUOUS SIGNALS

The sampling theorem states that if a signal has a maximum frequency of W Hz it is completely defined by samples which occur at intervals of $1/2W$ s. The sampling theorem can be proved by assuming that $h(t)$ is a non-periodic signal band limited to W Hz. The amplitude spectrum is given by

$$H(f) = \int_{-\infty}^{\infty} h(t) \exp(-j2\pi ft) \, dt \qquad (3.1)$$

Since $H(f)$ is band limited to $\pm W$ Hz it is convenient to make $H(f)$ a periodic function of frequency (Fig. 3.1). The value of $H(f)$ in the region $-W$ to $+W$ can be expressed in terms of a Fourier series in the frequency domain.

The time domain Fourier series is

$$h(t) = 1/T \sum_{n=-\infty}^{\infty} C_n(\exp j \, 2\pi nt/T)$$

and the corresponding series in the frequency domain is

$$H(f) = 1/2W \sum_{n=-\infty}^{\infty} X_n \exp(j \, 2\pi nf/2W) \qquad (3.2)$$

The values of the Fourier coefficients are give by

$$X_n = \int_{-W}^{W} H(f) \exp(-j\pi nf/W) \, df$$

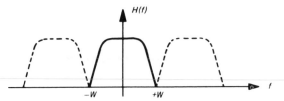

Fig. 3.1 Representation of $H(f)$ as periodic function.

But $H(f)$ is the Fourier transform of $h(t)$, i.e.

$$h(t) = \int_{-W}^{W} H(f) \exp(j2\pi ft) \, df \tag{3.3}$$

Hence if $t = -n/2W$

$$h(-n/2W) = \int_{-W}^{W} H(f) \exp(-j\pi nf/W) \, df$$

i.e.

$$h(-n/2W) = X_n \tag{3.4}$$

The values $h(-n/2W)$ are samples of $h(t)$ taken at equally spaced instants of time, the time between samples being $1/2W$ s. These samples define X_n, which in turn completely defines $H(f)$. Since $H(f)$ is the Fourier transform of $h(t)$, then $H(f)$ defines $h(t)$ for all values of t. Hence $h(-n/2W)$ completely defines $h(t)$ for all values of t.

The physical interpretation of the sampling process is shown in Fig. 3.2 The continuous signal $h(t)$ is multiplied by a periodic pulse train $S(t)$ in which the pulse width is much less than the pulse period.

Since the sampling pulse train is periodic it can be expanded in a Fourier series as

$$S(t) = a_0 + a_1 \cos \omega_s t + a_2 \cos 2\omega_s t + \cdots$$

where $\omega_s = 2\pi/T_s$. If the continuous signal $h(t)$ is assumed to be a single tone $\cos \omega_m t$ then the sampled waveform $h_s(t)$ is given by

$$h_s(t) = a_0 \cos \omega_m t + (a_1/2) \cos(\omega_s - \omega_m)t + (a_1/2) \cos(\omega_s + \omega_m)t$$
$$+ (a_2/2) \cos(2\omega_s - \omega_m)t + (a_2/2) \cos(2\omega_s + \omega_m)t + \cdots \tag{3.5}$$

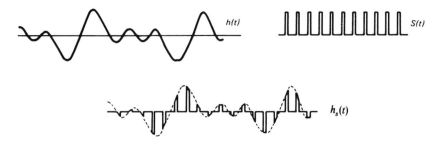

Fig. 3.2 The sampling process.

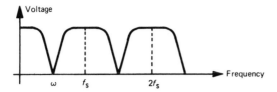

Fig. 3.3 Amplitude spectrum of a sampled signal.

The spectrum of $h_s(t)$ thus contains the original spectrum of $h(t)$ and upper and lower sidebands centred at harmonics of the sampling frequency. The amplitude spectrum for $h_s(t)$ when $h(t)$ is a multitone signal band limited to W Hz is shown in Fig. 3.3. It can be seen from this figure that if the sampling frequency $f_s = 2W$, the sidebands just fail to overlap. If $f_s < 2W$, overlap (aliasing) occurs and distortion of the spectrum of $h(t)$ results.

3.2 RECONSTRUCTION OF THE CONTINUOUS SIGNAL

In order to reproduce a continuous signal from the samples some form of interpolation is required. The output from a sample-and-hold circuit is shown in Fig. 3.4. This circuit holds the level of the last sample until a new sample arrives and then assumes the value of the new sample. It can be seen from the figure that there is a considerable error between the output of the sample-and-hold circuit and the original value of $h(t)$. The exact interpolation function is obtained by considering Eqn (3.3).

If the Fourier series representation of $H(f)$ [Eqn (3.2)] is substituted in Eqn (3.3), then

$$h(t) = \int_{-W}^{W} \left[1/2W \sum_{n=-\infty}^{\infty} X_n \exp(j\pi nf/W) \right] \exp(j2\pi ft)\,df$$

Changing the order of summation and integration gives

$$h(t) = \sum_{n=-\infty}^{\infty} X_n/2W \int_{-W}^{W} \exp[j2\pi nf(t+n/2W)]\,df$$

i.e.

$$h(t) = \sum_{n=-\infty}^{\infty} X_n \operatorname{sinc}[2\pi W(t+n/2W)]$$

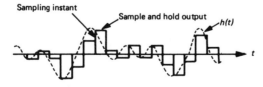

Fig. 3.4 Sample and hold interpolation.

but

$$X_n = h(-n/2W)$$

Therefore

$$h(t) = \sum_{n=-\infty}^{\infty} h(-n/2W)\,\text{sinc}\,[2\pi W(t + n/2W)]$$

or alternatively

$$h(t) = \sum_{n=-\infty}^{\infty} h(n/2W)\,\text{sinc}\,[2\pi W(t - n/2W)] \tag{3.6}$$

The values $h(n/2W)$ are the samples of $h(t)$ and sinc $[2\pi W(t - n/2W)]$ is the required interpolation function. This function is centred (has unity value) at intervals of time spaced at $1/2W$. The function has zeros at instants of time equal to $(n + p)/2W$ where p has integer values, except zero, between $\pm \infty$. The value of $h(t)$ at a sampling instant, therefore, is equal to the amplitude of the sample only, since the weights of the other samples are zero. The value of $h(t)$ between the sampling instants is given by the summation of the corresponding sinc functions and this process is shown in Fig. 3.5.

3.3 LOW-PASS FILTERING OF A SAMPLED SIGNAL

The previous section has shown that the required interpolation function for perfect signal reconstruction is a sinc function. This function is the impulse response of an ideal low-pass filter. If $P(f)$ is the transfer function of a network and $G(f)$ is the spectrum of an input signal, the spectrum of the signal at the network output $L(f)$ is given by

$$L(f) = P(f)G(f)$$

If the input to the network is a very narrow pulse (which approximates to a unit impulse), $G(f) \simeq 1$ and the spectrum of the output is then $L(f) = P(f)$.

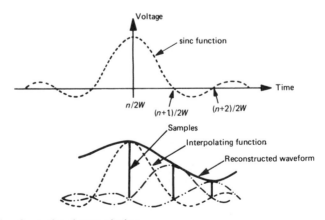

Fig. 3.5 Sinc fuunction interpolation.

The response in the time domain is obtained by taking the Fourier transform of $P(f)$. If the network is an ideal low-pass filter with cut-off frequency f_c the transfer function is given by

$$P(f) = |P(f)| \exp(-j2\pi n f t_0)$$

where

$$|P(f)| = 1 \quad \text{for } |f| \leqslant f_c, \quad = 0 \text{ else}$$

The factor $\exp(-j2\pi f t_0)$ is the linear phase characteristic of the ideal filter. The impulse response $p(t)$ is

$$p(t) = \int_{-f_c}^{f_c} \exp[-j2\pi f(t_0 - t)]\,\mathrm{d}f$$

i.e.

$$p(t) = 2f_c \frac{\sin[2\pi f_c(t - t_0)]}{2\pi f_c(t - t_0)} \tag{3.7}$$

This function has the required form when $f_c = W$.

In the frequency domain the effect of the ideal low-pass filter is to remove completely all spectral components above W Hz (Fig. 3.6).

In practice the ideal low-pass filter is not physically realizable. This may be seen from the fact that the impulse response exists for $t < 0$ which means that an output exists before the impulse is applied, which is physically impossible. The ideal filter may be approximated by a physically realizable network, the approximation becoming more exact as the complexity of the network is increased. One such approximation with a sharp cut-off characteristic is the Chebyshev approximation. The impulse response of a fifth-order Chebyshev low-pass filter is

$$p(t) = Ae^{-at} + Be^{-bt}(\cos\theta t + \sin\theta t) + Ce^{-ct}(\cos\phi t + \sin\phi t) \tag{3.8}$$

This impulse response will clearly produce some distortion, which is most easily evaluated in the frequency domain as shown in Fig. 3.7.

Since the filter has a finite transmission above W Hz some distortion components outside the signal bandwidth will appear in the filter output. To minimize this distortion the sampling frequency is increased above $2W$ Hz. A practical telephone channel has an upper cut-off frequency of 3.4 kHz and the sampling rate employed is 8 kHz, that is 2.35 times the maximum signal frequency.

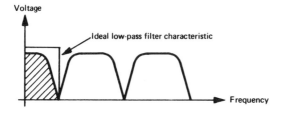

Fig. 3.6 Ideal filtering of a sampled signal.

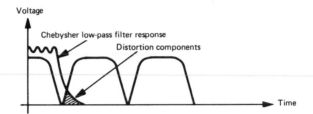

Fig. 3.7 Practical filtering of a sampled signal.

3.4 TIME DIVISION MULTIPLEX (TDM) TRANSMISSION

When a signal is sampled by narrow pulses there are large intervals between the samples in which no signal exists. It is possible during these intervals to transmit the samples of other signals. This process is shown in Fig. 3.8 and is called 'time division multiplex' (TDM) transmission. Since each sampled signal gives rise to a continuous signal after filtering, TDM transmission allows simultaneous transmission of several signals over a single wideband link. It is therefore an alternative to FDM transmission, described in Section 2.19.

The switches at transmitter and receiver (which would be solid-state devices) are synchronized and perform the sampling and interlacing. The samples themselves are very narrow and consequently have a large bandwidth. When transmitted over a link with a fixed bandwidth the samples

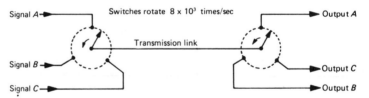

Fig. 3.8 TDM transmission.

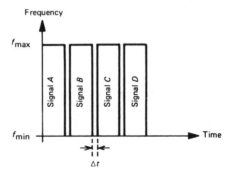

Fig. 3.9 Spectrum/time diagram for TDM.

spread and can overlap adjacent samples. To minimise this, guard intervals Δt are allowed between adjacent signals. In the case of TDM transmission each signal therefore occupies the whole of the transmission bandwidth for part of the transmission time (Fig. 3.9).

3.5 PULSE CODE MODULATION (PCM)

Although it is possible to transmit the samples directly in a TDM system there is a considerable advantage in coding each sample into a binary word before transmission. This is because binary signals have a much greater immunity to noise than analogue signals. A widely used form of digital transmission is pulse code modulation, the essentials of which are shown in Fig 3.10.

In this system, 30 speech channnels are each limited in frequency to 3.4 kHz and sampled at 8 kHz. The sampled signals are converted into binary form for transmission. In addition to the coded signals other signals are sent over the link for synchronization and identification. Once each sample has been converted into a binary code this effectively means that it has been quantized into one of a fixed number of levels. The greater the number of quantization levels the greater is the accuracy of the quantized representation, but also the greater is the number of binary digits (bits) that are required to represent the sample. Since more bits require a higher transmission bandwidth a balance must be struck between accuracy and bandwidth.

The quantization process is shown in Fig 3.11 and it is clear that once quantized the precise amplitude of the original sample can never be restored. This gives rise to an error in the recovered analogue signal, known as quantization error. The quantization error for an eight-level PCM system can be calculated by reference to Fig 3.11. This shows that the peak-to-peak input is I volts. If this value is quantized into M equally spaced levels the spacing between each level is $\delta V = I/M$. The amplitudes reproduced after decoding are usually the mid-point of each quantizing interval, which gives a maximum decoded peak to-peak value of $A = (M - 1)\delta V$. If a level V_j is reproduced by the decoder the true amplitude will be anywhere in the range $V_j \pm \partial V/2$; hence

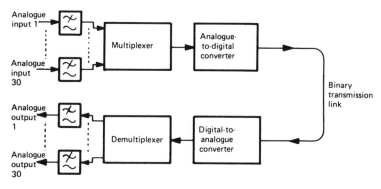

Fig. 3.10 Pulse code modulation system.

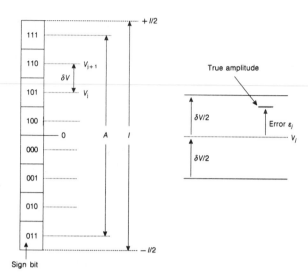

Fig. 3.11 Linear quantization (8 level).

a maximum error of $\partial V/2$ will be present on any decoded output. The error is random in nature and is called quantization noise. In order to calculate the magnitude of this quantization noise it is necessary to know the amplitude probability density function of the signal to be coded. From this knowledge it is possible to calculate the probability that the signal will be in any quantizing interval and the mean square error for each of the quantizing steps.

The simplest case to evaluate is a signal with a uniformly distributed amplitude density function. Such a signal is a triangular wave of amplitude $\pm I/2$. The density function for both a triangular wave and a sine wave are shown in Fig 3.12. A signal with a uniform probability density has equal probability of being in any of the quantizing intervals and also has equal probability of having any particular amplitude within a given quantizing interval. The error produced is shown in Fig 3.11 and has a mean square value given by

$$\bar{\varepsilon}^2 = \int_{-\partial V_j/2}^{\partial V_j/2} \varepsilon_j^2 \, d\varepsilon = \frac{(\partial V_j)^2}{12} \tag{3.9}$$

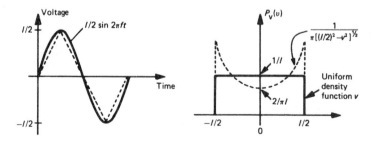

Fig. 3.12 Voltage waveforms and amplitude density functions.

The total mean square error throughout the range of coder is

$$\bar{\sigma}^2 = \sum_j \bar{\varepsilon}_j^2 \, p(V_j)$$

where $p(V_j)$ is the probability that the signal will be in the jth quantization interval, i.e.

$$\bar{\sigma}^2 = \frac{1}{12} \sum_j (\partial V_j)^2 \, p(V_j) \tag{3.10}$$

But $\sum_j (\partial V_j)^2 \, p(V_j)$ is the mean square value of ∂V_j, i.e.

$$\bar{\sigma}^2 = \frac{\overline{(\partial V_j)^2}}{12} \tag{3.11}$$

If the signal has a uniform amplitude distribution and the quantization is linear, then all values of ∂V_j are equal, i.e.

$$\bar{\sigma}^2 = \frac{(\partial V)^2}{12} \tag{3.12}$$

$\bar{\sigma}^2$ is effectively the quantization noise power. If the uniformly distributed signal has a maximum amplitude of $\pm I/2$, it has a mean square value of

$$\bar{S}^2 = \int_{-I/2}^{I/2} V^2 \, p_v(V) \, dV = \int_{-I/2}^{I/2} \frac{V^2}{I} \, dV = \frac{I^2}{12} \tag{3.13}$$

The mean signal to quantization noise power ratio (SQNR) is thus

$$\text{SQNR} = \frac{\bar{S}^2}{\bar{\sigma}^2} = \frac{I^2}{(\partial V)^2} = M^2$$

In a binary system the number of bits m required to code M levels must satisfy the relationship $M = 2^m$; hence

$$\text{SQNR} = 10 \log_{10} 2^{2m} \quad \text{dB}$$

i.e.

$$\text{SQNR} = 20 \, m \log_{10} 2 \quad \text{dB}$$

or

$$\text{mean SQNR} = 6m \quad \text{dB} \tag{3.14}$$

The maximum signal power is $(I/2)^2$, which gives a maximum SQNR of $3M^2$, i.e.

$$\text{maximum SQNR} = (4.8 + 6m) \quad \text{dB} \tag{3.15}$$

If the input signal is a sine wave which occupies the full coder range then $V_{in}(t) = (I/2) \sin 2\pi ft$ and the mean square value is $\bar{S}^2 = I^2/8$, which results in

$$\text{mean SQNR} = (1.8 + 6m) \quad \text{dB} \tag{3.16}$$

Equations (3.14) to (3.16) show that the SQNR of the decoded signal, measured in dB, increases linearly with m which itself is linearly related to transmission bandwidth. This is a similar relationship between SNR and bandwidth to that expressed by Shannon's law. However, in the case of PCM, the noise is quantization noise rather than fluctuation noise. The effect of fluctuation noise on PCM transmissions is covered in Section 5.3.

It is appropriate at this point to investigate the effect of sampling frequency on SQNR by considering the sampling and quantization of the waveform shown in Fig. 3.13. The error between the original waveform and the quantized version is also shown in this figure and is seen to be approximately similar to a sawtooth waveform of period $1/f_s$ where f_s is the sampling frequency. It is clear that as the sampling frequency is increased the frequency of the error waveform also increases. However, the quantization noise power is fixed at $(\partial V)^2/12$ and, as the sampling rate is increased, this power will be distributed over a wider frequency range. Consequently the proportion of quantization noise power occupying the same bandwidth as the original signal, and passing through the reconstruction filter, will decrease as the sampling rate is increased. Thus oversampling a bandlimited waveform will increase the SQNR at the output of the reconstruction filter. This principle is often used to advantage in compact disc players, for example.

In deriving Eqn (3.16) it was assumed that the input waveform fully occupied the range of the coder, which would correspond to the loudest talker in a telephone system. It follows that the SQNR for the quietest talker would be considerably lower than this value. PCM encoders are designed to make the SQNR constant over as wide a range of input amplitudes as possible and this is achieved by making the quantization step size a function of the input

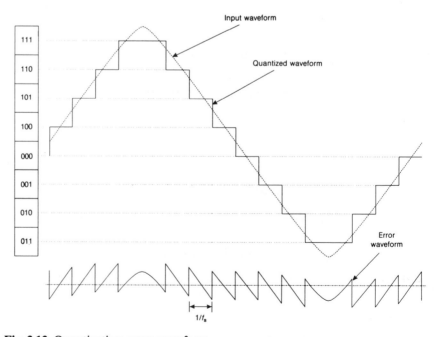

Fig. 3.13 Quantization error waveform.

waveform amplitude. This is known as non-linear (or non-uniform) quantization. Essentially the step size of a non-linear coder is reduced near zero and increases towards the maximum input level. Non-linear coding is often achieved in practice by first COMpressing the signal then coding the compressed signal in a linear coder and finally exPANDING the decoded signal with the inverse of the compression characteristic. The combined process is known as COMPANDING.

The compression characteristic is required to give a constant signal to quantization noise ratio for all levels of input. The quantization noise is calculated by considering the linear quantization of the output of the compression circuit. This is equivalent to the non-linear quantization of the input signal. The linear coder and equivalent non-linear coder are shown in Fig. 3.14. From this figure it is clear that the step size for the linear coder is $\partial V_{out} = 2/M$ where M is the number of quantization levels. Considering the jth quantizing interval of the non-linear coder this is assumed to have a range of $\partial V_{in(j)}$ centred at $V_{in(j)}$. The range of voltages for the jth interval is $V_{in(j)} \pm V_{in(j)}/2$.

Assuming the M is large, then from Fig. 3.14,

$$\frac{\mathrm{d}V_{out}}{\mathrm{d}V_{in}} = \frac{2/M}{\partial V_{in(j)}} \text{ or } \partial V_{in(j)} = \frac{2/M}{(\mathrm{d}V_{out}/\mathrm{d}V_{in})|_j}$$

The mean square error produced by quantizing a voltage in the jth interval is then

$$\bar{\varepsilon}^2 = \int_{V^-}^{V^+} (V_{in} - V_{in(j)})^2 p_v(V_{in}) \, \mathrm{d}V_{in} \qquad (3.17)$$

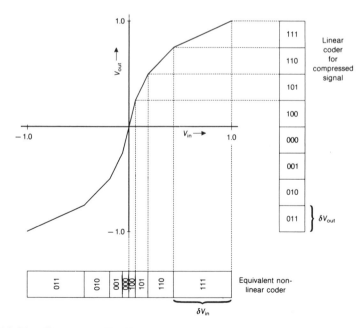

Fig. 3.14 Non-linear quantization.

where

$$V^+ = V_{\text{in}(j)} + \frac{\partial V_{\text{in}(j)}}{2} \quad \text{and} \quad V^- = V_{\text{in}(j)} - \frac{\partial V_{\text{in}(j)}}{2}$$

$p_v(V_{\text{in}})$ is the amplitude distribution function of the input waveform and when M is large then $p_v(V_{\text{in}})$ may be considered constant over the jth interval: i.e.

$$p_v(V_{\text{in}}) = p_{vj}(V_{\text{in}}) = \text{ constant}$$

The mean square error is then

$$\bar{\varepsilon}^2 = p_{vj}(V_{\text{in}}) \frac{\partial V_{\text{in}(j)}^3}{12}$$

which may be written

$$\bar{\varepsilon}^2 = p_{vj}(V_{\text{in}}) \frac{\partial V_{\text{in}(j)}^2}{12} \partial V_{\text{in}(j)}$$

but

$$\partial V_{\text{in}(j)} = \frac{2/M}{(\mathrm{d}V_{\text{out}}/\mathrm{d}V_{\text{in}})|_j}$$

hence the mean square error is

$$\bar{\varepsilon}^2 = p_{vj}(V_{\text{in}}) \cdot \frac{(2/M)^2}{(\mathrm{d}V_{\text{out}}/\mathrm{d}V_{\text{in}})^2|_j} \cdot \frac{\partial V_{\text{in}(j)}}{12} \tag{3.18}$$

The total mean square error is found by summing the contributions from each interval: i.e.

$$\bar{\sigma}_n^2 = 2 \sum_{j=1}^{M/2} p_{vj}(V_{\text{in}}) \cdot \frac{(2/M)^2}{(\mathrm{d}V_{\text{out}}/\mathrm{d}V_{\text{in}})^2|_j} \cdot \frac{\partial V_{\text{in}(j)}}{12}$$

For large values of M the summation may be replaced by an integral over the range ± 1, the total MSE is then

$$\bar{\sigma}_n^2 = 2 \int_0^1 \frac{p_v(V_{\text{in}})}{3M^2} \cdot \frac{1}{(\mathrm{d}V_{\text{out}}/\mathrm{d}V_{\text{in}})^2} \cdot \mathrm{d}V_{\text{in}} \tag{3.19}$$

The mean square signal value is

$$\bar{\sigma}_s^2 = 2 \int_0^1 V_{\text{in}}^2 p_v(V_{\text{in}}) \, \mathrm{d}V_{\text{in}}$$

Hence

$$\text{SQNR} = \frac{2 \displaystyle\int_0^1 V_{\text{in}}^2 p_v(V_{\text{in}}) \, \mathrm{d}V_{\text{in}}}{2 \displaystyle\int_0^1 \frac{p_v(V_{\text{in}})}{3M^2} \cdot \frac{1}{(\mathrm{d}V_{\text{out}}/\mathrm{d}V_{\text{in}})^2} \cdot \mathrm{d}V_{\text{in}}} \tag{3.20}$$

If $V_{\text{out}} = (1 \, k\ln V_{\text{in}})$ then $\mathrm{d}V_{\text{out}}/\mathrm{d}V_{\text{in}} = k/V_{\text{in}}$ and the equation for SQNR has a constant value which is independent of the amplitude of V_{in}, which is the required result: i.e.

$$SQNR = 3M^2 \qquad (3.21)$$

Unfortunately such a characteristic cannot be used in practice because when $V_{in} = 0$ then $V_{out} = (1 + k\ln 0) = -\infty$. Under these circumstance a practical approximation to the theoretical compression characteristic is required.

3.6 PRACTICAL COMPRESSION CHARACTERISTICS

Two practical characteristics which overcome this disadvantage are the A law characteristic (Europe) and the μ law characteristic (North America). The A law characteristic is divided into two regions and is given by

$$V_{out} = \frac{AV_{in}}{1 + \ln A} \quad \text{for} \quad 0 \leqslant |V_{in}| \leqslant \frac{1}{A} \quad \text{(linear)}$$

$$V_{out} = \frac{1 + \ln(AV_{in})}{1 + \ln A} \quad \text{for} \quad \frac{1}{A} \leqslant |V_{in}| \leqslant 1 \quad \text{(logarithmic)} \qquad (3.22)$$

The linear region ensures that $V_{out} = 0$ when $V_{in} = 0$ and the logarithmic region is specified so that $|V_{out}| = 1$ when $|V_{in}| = 1$, the characteristic is continuous at $|V_{in}| = 1/A$. A is known as the compression coefficient and for large values of A the characteristic is predominantly logarithmic. The characteristic is shown in Fig. 3.15.

The relationship between input coder (non-linear) and output coder (linear) is derived for both regions. For the linear region

$$dV_{out} = \frac{AdV_{in}}{1 + \ln A} \quad \text{or} \quad dV_{in} = \frac{dV_{out}(1 + \ln A)}{A}$$

For the logarithmic region

$$dV_{out} = \frac{1}{1 + \ln A} \times \frac{dV_{in}}{V_{in}} \quad \left(\text{since} \frac{d}{dx}\{\ln(ax + b)\} = \frac{a}{ax + b} \right)$$

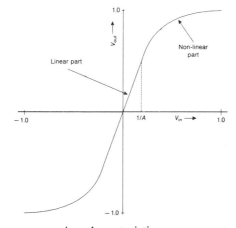

Fig. 3.15 The A law compression characteristic.

or

$$dV_{in} = dV_{out}(1 + \ln A) V_{in}$$

Since the characteristic is normalized with a range of ± 1 then $dV_{out} = M/2$. When the input waveform is in the linear region of the characteristic the quantization noise produced is

$$\sigma^2_{lin} = \frac{(dV_{in})^2}{12} = \frac{\{1 + \ln A\}^2}{3M^2 A^2} = \frac{k^2}{A^2} \tag{3.23}$$

When the input waveform is in the logarithmic region the quantization noise produced is

$$\bar{\sigma}^2_{log} = \frac{(\overline{dV}_{in})^2}{12} = \frac{(1 + \ln A)^2 \bar{V}^2_{log}}{3M^2} = k^2 \bar{V}^2_{log} \tag{3.24}$$

In this case $\bar{V}^2_{log}$ is the mean square value of the input waveform when in the logarithmic region. In practice the input waveform will occupy both linear and logarithmic regions of the characteristic. In order to determine the mean quantization noise it is necessary to know the probability density function of the input waveform. The amplitude probability density function of a typical speech waveform may be approximated by

$$p_v(V) = \frac{1}{\sqrt{2}\sigma_s} \exp\left(\frac{-\sqrt{2}|V|}{\sigma_s}\right) \tag{3.25}$$

where σ_s is the normalized rms voltage of the speech waveform. The probability that the speech waveform is within the linear region is

$$P_{lin} = \int_{-1/A}^{1/A} p_v(V) dV$$

or

$$P_{lin} = 2\int_0^{1/A} \frac{1}{\sqrt{2}\sigma_s} \exp\left(\frac{-\sqrt{2}|V|}{\sigma_s}\right) dV = 1 - \exp\left(\frac{-\sqrt{2}}{A\sigma_s}\right)$$

The mean square value when the waveform is in the logarithmic section is

$$\bar{V}^2_{log} = 2\int_{1/A}^1 V^2 p_v(V) dV$$

Integrating by parts and assuming that $A \gg 1$ gives

$$\bar{V}^2_{log} = \left(\frac{1}{A^2} + \frac{\sqrt{2}\sigma_s}{A} + \sigma_s^2\right) \exp\left(\frac{-\sqrt{2}}{A\sigma_s}\right)$$

The total quantization noise is

$$\sigma_n^2 = \bar{\sigma}^2_{lin} + \bar{\sigma}^2_{log} = k^2 \left[\frac{1}{A^2} P_{lin} + \bar{V}^2_{log}\right]$$

and the signal-to-quantization-noise ratio becomes

$$SQNR = \frac{\sigma_s^2}{\sigma_n^2}$$

Note that if $A \gg 1$ then $P_{\text{lin}} \ll 1$ and $\overline{V}^2_{\text{log}} \approx \sigma^2_s$, i.e.

$$\text{SQNR} \approx \frac{\sigma^2_s}{k\sigma^2_s} = \frac{1}{k^2}$$

Thus for large values of A the SQNR is constant and independent of the value of V_{in}. The full equation for SQNR is

$$\text{SQNR} = \frac{\sigma^2_s}{k^2\left[\frac{1}{A^2}\left\{1 - \exp\left(\frac{-\sqrt{2}}{A\sigma_s}\right)\right\} + \left\{\frac{1}{A^2} + \frac{-\sqrt{2}\sigma_s}{A} + \sigma^2_s\right\}\exp\left(\frac{-\sqrt{2}}{A\sigma_s}\right)\right]}$$

$$(3.26)$$

When $\sigma_s = 1/A$ then

$$\text{SQNR} = \frac{1/A^2}{k^2\left[1/A^2(0.76) + 1/A^2(1 + \sqrt{2} + 1)(0.24)\right]} = \frac{1}{1.58k^2}$$

But $10\log_{10}(1/1.58) = -2\,\text{dB}$, hence the effect of the linear part of the characteristic is to cause a drop in SQNR by 2 dB for input levels below $1/A$. Thus if $A \gg 1$ the input waveform will be in the logarithmic section for most of the time and the SQNR will be approximately constant. The SQNR characteristic for the A law compression characteristic is shown in Figure 3.16.

The system is thus designed so that the rms voltage produced by the quietest talker is equal to $1/A$, which will produce an effective constant SQNR for all users. The SQNR is a function of both A and M which are in turn chosen to give an acceptable performance for a specified dynamic range. The required dynamic range is determined by measurements on a sample of the population. Such measurements reveal that 98% of the population have an rms voice amplitude within $\pm 13\,\text{dB}$ of the median talker, and 99.8% of the population have an rms amplitude within $\pm 17\,\text{dB}$ of the median talker. The ratio of the rms output of the loudest talker to the quietest talker is known as the useful volume range (UVR) and will thus be between 26 dB and 34 dB, depending on which statistic is used.

Further measurements show that talkers have an output amplitude within $\pm 13\,\text{dB}$ of their individual rms values for 99% of the time. The required

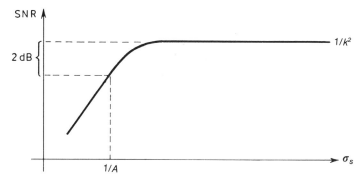

Fig. 3.16 SQNR characteristic for A law compression.

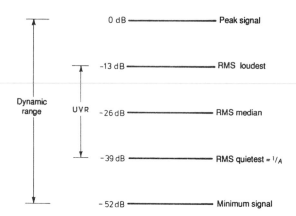

Fig. 3.17 Dynamic range of A law compression.

dynamic range of the coder will therefore exceed the UVR by approximately 26 dB as shown in Fig. 3.17. In this figure the peak signal has been normalized to 0 dB, hence it may be seen that the required value of the compression coefficient is

$$A = \text{UVR} + 13\,\text{dB}$$

Thus if UVR $= 30\,\text{dB}$, then $20\log_{10}(1/A) = -43$ i.e. $A = 141$.

The CCITT recommended value of A is 87.6 which gives a UVR of 26 dB. With this value of A and an 8 bit code ($M = 256$) the signal-to-quantization-noise ratio is 38 dB. The UVR may be increased at the expense of SQNR, e.g. for a UVR of 30 dB ($A = 141$) and an 8 bit code then SQNR $= 36\,\text{dB}$. In order to give perspective to non-linear compression of voice waveforms it is appropriate to compare the A law compression with linear coding throughout (no compression). To do this it is necessary to define the maximum input voltage which may be handled by the linear coder, in terms of probability. The probability that the input voltage will be within the range of the coder is

$$P_{\text{coder}} = \int_{-I/2}^{I/2} p_v(V)\,\mathrm{d}V$$

i.e.

$$P_{\text{lin}} = 2\int_0^{I/2} \frac{1}{\sqrt{2}\,\sigma_s} \exp\left(\frac{-\sqrt{2}|V|}{\sigma_s}\right)\mathrm{d}V = 1 - \exp\left(\frac{-I}{\sqrt{2}\sigma_s}\right)$$

The maximum input is defined in terms of the probability of being within range of the coder 98% of the time; i.e.

$$0.98 = 1 - \exp\left(\frac{-I}{\sqrt{2}\sigma_s}\right)$$

from which $\sigma_s = 0.18I$. The SQNR for the loudest talker is thus

$$\text{SQNR} = \sigma_s^2 \cdot \frac{12}{\partial v^2} = (0.18P)^2 \frac{12M^2}{I^2}$$

i.e.

$$SQNR = 0.388M^2$$

To match the performance of the companded system the SQNR for the quietest talker $= 38\,\text{dB}$. It follows that the SQNR for the loudest talker will be $38 + 26 = 64\,\text{dB}$. Hence

$$10\log_{10}0.388 + 20\log_{10}M = 64$$

but $M = 2m$, hence $SQNR = -4.1 + 20\,m\log_{10}2$, from which $m = 11.3$.

Thus 12 bits are required for a linear coder to have the same performance as the 8 bit coder with A law compression. Hence companding produces a saving in transmission bandwidth of the order of 33%.

The actual A law compression characteristic used in practice is shown in Fig. 3.18. In effect the coder is divided into 14 segments, 7 for positive amplitudes and 7 for negative amplitudes. Within these segments linear quantization is employed, the step size varying with segment so that the step size in segment 7, for example, is 56 times the step size in segment 1. This produces a similar quantization noise characteristic to that given in Eqn (3.26). The total number of input levels in this characteristic is 8192 (2^{13}) and it is clear from segment 1 that 64 input levels are transformed into 32 output levels. Hence the low-level input values are quantized at the equivalent of 12 bits linear quantization.

In the USA companding is carried out using the μ law characteristic which is given by

$$V_{\text{out}} = \ln(1 + \mu V_{\text{in}}) \quad \text{for} \quad 0 < |V_{in}| < 1 \tag{3.27}$$

A typical figure for μ is 255. It should be noted that when V_{in} is low $\ln(1 + \mu V_{\text{in}}) \approx \mu V_{\text{in}}$, hence this characteristic is also approximately linear for low input voltages.

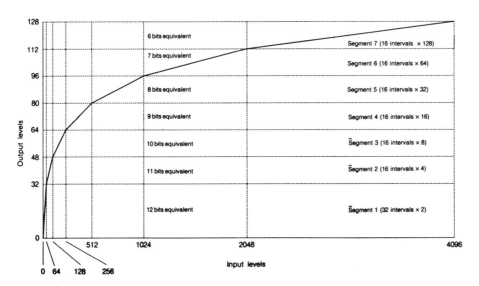

Fig. 3.18 Practical A law quantization characteristic (positive values).

3.7 NICAM

Near instantaneously companded audio multiplex (NICAM) is an alternative compression technique which is used for digital transmission of high-fidelity television stereophonic sound signals. It is essentially a bit reduction technique which is designed to maintain a constant signal-to-quantization-noise ratio over a wide dynamic range. In fact A law compression may also be considered as a bit reduction technique. This is illustrated in Fig. 3.19 which depicts an example of non-linear quantization with 5 to 4 bit compression. This is, in essence, a scaled down version of Fig. 3.18, in which the compression is from 12 to 8 bits.

In Fig. 3.19 the sampled waveform is first quantized linearly into 5 bits and the encoded signal is then processed into a non-linear form to reduce the number of bits to 4. At the receiver the 4 bit codes are converted back into the 5 bit equivalents. This means that the quantization noise for low-level amplitudes is determined by the full resolution of the 5 bit code. It is clearly not possible to reproduce the individual 5 bit code words when several such code words are compressed into a single 4 bit equivalent. In such cases the received 4 bit word is converted into a 5 bit equivalent near the centre of the range. Hence the resolution deteriorates (and quantization noise increases) for the higher signal levels. The advantage of this form of compression is that it is implemented with a simple conversion from 5 to 4 bits, and vice versa.

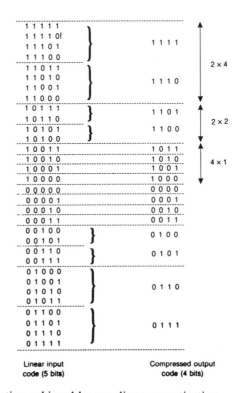

Fig. 3.19 Bit reduction achieved by non-linear quantization.

Table 3.1 Transmitted bits in NICAM

bits	1 MSB	2	3	4	5	6	7	8	9	10	11	12	13	14 LSB
Range 0	*					*	*	*	*	*	*	*	*	*
Range 1	*				*	*	*	*	*	*	*	*	*	
Range 2	*			*	*	*	*	*	*	*	*	*		
Range 3	*		*	*	*	*	*	*	*	*	*			
Range 4	*	*	*	*	*	*	*	*	*	*				

An alternative form of compression can be based on **range coding** in which the input waveform amplitude is divided into a fixed number of ranges. Different groups of bits are then transmitted for different ranges. This is the form of compression used in NICAM transmission and has a superior quantization noise performance, compared to A law, for the higher signal amplitudes. For low input amplitudes only the least significant bits are transmitted, while for high input signal amplitudes only the most significant bits are transmitted. In the case of NICAM, 14 bit resolution is used and this is compressed into 10 bits for transmission. The NICAM signal is divided into, five ranges as shown in Table 3.1. In addition to the bits shown in this table it is also necessary to transmit a 3 bit range code to define the range of the transmitted bits.

If a bit reduction is to be achieved the sum of amplitude bits and range code bits must be less than the uncoded bits. This is achieved by sending a 3 bit range code for a block of amplitude samples, rather than for each individual sample. This has the effect of adding a fraction of a bit per sample. The NICAM system uses a sampling frequency of 32 kHz and transmits one range code for a block of 32 samples. This represents a time interval of 1 ms and may be regarded as **nearly instantaneous** as far as the audio signal is concerned. In order to avoid clipping of the signal the range code corresponds to the largest amplitude in the 1 ms block. A schematic diagram of the NICAM coder is shown in Fig. 3.20.

If one 3 bit range code is transmitted for each 32 samples then this represents a degree of inefficiency as 3 bits can define 8 ranges. This inefficiency is reduced in NICAM by collecting 3×32 sample blocks, which will have $5^3 = 125$ range code combinations and transmitting these combinations as a 7 bit word. This produces an overall delay of 3 ms in the transmitted

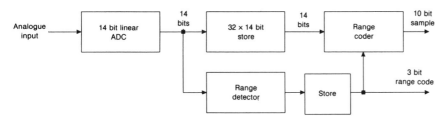

Fig. 3.20 NICAM range coder.

audio signal. Clearly it is important to avoid errors in the range code as this would produce a gross distortion of the reproduced voltage level. To minimize errors 4 parity check digits are added to the 7 bit range code to provide an error correction code with a Hamming distance of 4 (see Section 5.6).

The NICAM system effectively transmits the signal with 10 bits and the relative signal to noise ratio is maintained constant as the quantization noise varies with signal amplitude. It has been found that this system has a SQNR improvement of approximately 12 dB over the equivalent A law system with 10 bits, but requires a more complex coder and decoder. A 10 bit linear quantization would produce a SQNR of approximately 60 dB, but with range coding this is increased to a subjective equivalent of 80 dB.

3.8 PCM LINE SIGNALS

Although it is possible to encode signals at their source this is not in widespread use. TDM transmission with PCM encoding is used extensively between telephone exchanges. This transmission makes use of standard telephone lines that were designed specifically for voice communications. The standards of transmission adopted have been dictated largely by the characteristics of the cables 'already in the ground'. To minimize signal distortion over the audio frequency band these cables are artificially loaded by inductance at intervals of approximately 2 km. This produces very high attenuation of frequencies above about 4 kHz, which makes the line totally unsuitable for digital signals. In addition to the loss of response at high frequency the 2 km lengths of line are transformer coupled and therefore have no dc path. The high-frequency response of the section can be considerably improved by simply removing the loading coil. A typical frequency response for a section of loaded and unloaded cable is shown in Fig. 3.21.

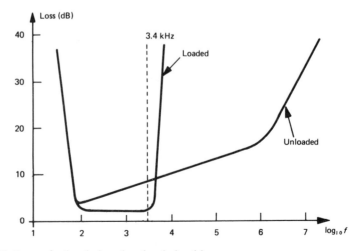

Fig. 3.21 Loss of a loaded and unloaded cable.

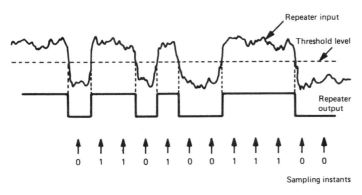

Fig. 3.22 Regenerative repeating of a PCM signal.

In practical PCM systems the loading coils are replaced by regenerative repeaters which effectively isolate each 2 km section of line. The regenerative repeater, in fact, produces a binary signal free from noise at the start of each section. This is shown in Fig. 3.22.

If the repeater input is above the threshold at a timing instant a binary 1 is transmitted to the next section. If the input is below the threshold a binary 0 is transmitted. If the SNR at each repeater input is adequate few decision errors occur and the binary signal is repeated free from noise. This is a distinct advantage of digital transmission over analogue transmission. In the latter case, amplifiers are required at intervals to compensate for signal attenuation. These amplifiers boost the noise as well as the signal. The absence of a dc path presents a serious problem for PCM signals. Its effect on a long sequence of binary 1s is to cause a gradual droop of signal level below the decision threshold, and this will of course produce decision errors. This is illustrated in Fig. 3.23a.

The solution is to remove the dc component in the PCM signal. The first stage in the process is to convert the full-width non-return-to-zero (NRZ) pulses into half-width return-to-zero pulses (RZ). This effectively doubles the signal bandwidth, but is necessary for synchronization purposes (Section 3.8). The dc component in the RZ waveform is removed by inverting alternate binary 1s, the process being called 'alternate mark inversion' (AMI). The original two-level PCM signal is actually transmitted as a three-level signal with zero dc component. The bandwidth of a PCM signal is calculated on the basis of AMI. The processing to produce AMI is illustrated in Fig. 3.23b.

3.9 BANDWIDTH REQUIREMENTS FOR PCM TRANSMISSION

The PCM signal consists of a random sequence of binary 1s and 0s. The transmitted waveform will have maximum bandwidth when the number of transitions per unit time is also a maximum. This corresponds, in the case of NRZ pulses, to an alternating sequence of binary 1s and 0s. The maximum

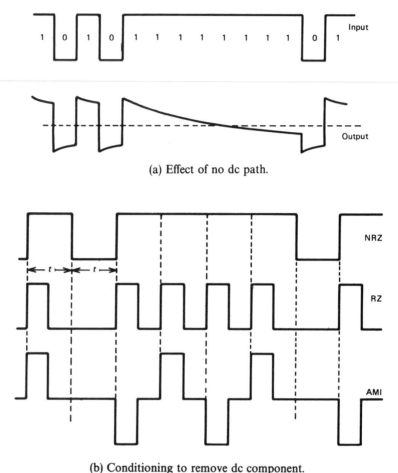

(a) Effect of no dc path.

(b) Conditioning to remove dc component.

Fig. 3.23 Transmission waveforms.

bandwidth in the case of RZ pulses and AMI occurs for a sequence of binary 1s. The transmitted waveforms and their respective amplitude spectra are given in Fig. 3.24.

The CCITT[2] recommendation specifies PCM transmission in terms of time division multiplexing of 32 channels. Each channel is sampled 8×10^3 times per second with each sample represented as an 8 bit code. This gives a total frame bit rate of 2.048 Mb/s. It may be seen from Fig. 3.24 that the minimum bandwidth requirement for AMI is approximately half the bit rate. This is an 'idealized' figure based on the premise that the three levels of the AMI signal could be extracted from a sine wave with a period of twice the bit interval. This bandwidth is insufficient in practice owing to distortion of the pulse waveform, which produces inter-symbol interference. To illustrate this point assume that the line (or channel) has an ideal low-pass response with cut off frequency f_c.

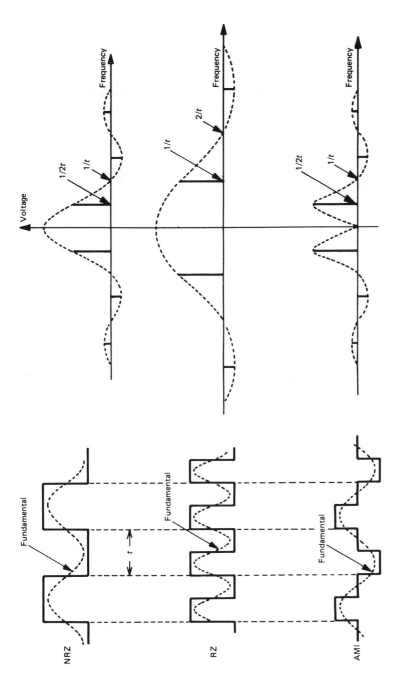

Fig. 3.24 PCM transmission waveforms and their spectra.

The impulse response of such a channel is a sinc function with zeros at intervals $t = n/2f_c$. Hence in theory pulses transmitted at a rate of $2f_c$ per second could be received free of interference from adjacent pulses. It was stated in Chapter 1 that an ideal response cannot be realized in practice; however, the raised cosine response is an approximation to the ideal filter which can be synthesized in practice. The raised cosine response is shown in Fig. 3.25 and it should be observed that the impulse response of such a characteristic also has zeros at intervals of $n/2f_c$.

The raised cosine response is given by

$$H(f) = 0.5 [1 + \cos(\pi f/2f_c)] \quad \text{for} \quad |f| < 2f_c \tag{3.28}$$

The impulse response of the network is given by the Fourier transform of Eqn (3.24) and is

$$h(t) = 2f_c \text{ sinc } (2\pi f_c t) \cdot \cos (2\pi f_c t)/[1 - (4f_c t)^2] \tag{3.29}$$

Thus in order to transmit pulses at a rate of $2f_c$ per second a raised cosine response with cut-off frequency $2f_c$ is required. (The raised cosine response is a linear phase response which may be approximated by Bessel polynomials[3].) Hence for AMI with a data rate of 2.048 Mb/s the overall response between

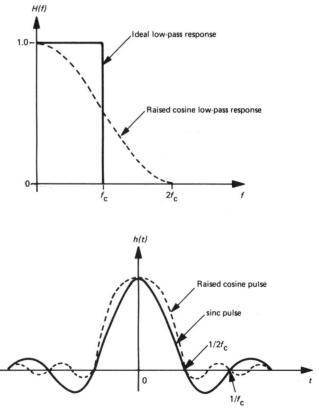

Fig. 3.25 Ideal and raised cosine frequency and phase response.

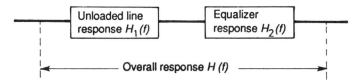

Fig. 3.26 Equalization of line response.

two repeaters should be a raised cosine response with a cut-off frequency of 2.048 MHz. This is achieved by use of equalization as shown in Fig. 3.26.

The overall response $H(f) = H_1(f) H_2(f)$ is designed to be a linear phase approximation of the ideal low-pass filter, i.e. to have a raised cosine amplitude response. The equalizer response $H_2(f)$, which may be realized with digital filters, is thus designed to produce this response.

3.10 SYNCHRONIZATION OF PCM LINKS

A PCM link is essentially made up of many sections, each one being terminated by a regenerative repeater. In such a system it is essential that each repeater operates at the same clock rate as there is no provision for data storage. It is possible to synchronize each repeater by a clock signal inserted at one end of the link, but this would reduce the available bandwidth for data transmission. A more efficient technique is to extract the clock signal for each repeater from the data signal itself. The clock signal is derived from the AMI waveform by full-wave rectification to produce a RZ spectrum. This is necessary because, as illustrated in Fig. 3.24, the AMI waveform has no component at the data rate. The component at t^{-1} is then extracted, possibly by use of a phase locked loop, for repeater timing.

A practical problem arises with timing if a series of binary 0s occurs, as would be the case during pauses in normal speech. To maintain repeater timing during such situations a code known as HDB3 is employed. This code limits the maximum number of successive 0s transmitted in AMI format to three. When four successive 0s occur in the binary (NRZ) signal the AMI waveform, which would be zero, is replaced by a three-level code $(-0+)$. The actual code substituted depends upon the AMI polarity of the previous 1. The receiver must be able to recognize the HDB3 code, and to make this possible the transmitter produces what is termed a bipolar violation. When there is no HDB3 code adjacent 1s in the AMI waveform will have opposite polarity. If adjacent 1s have the same polarity the inclusion of the HDB3 code is detected and hence decoded.

When four successive zeros occur in the binary signal, one of four possible HDB3 line signals is transmitted. The possible three-level codes are $000+, 000-, -00-, +00+$. The actual code transmitted depends on the polarity of the preceding binary 1 and also on whether the number of binary 1s which have occurred since the last HDB3 code (bipolar violation) is odd or even. The substitutions used are given in Table 3.2.

Table 3.2

Polarity of preceding pulse	Number of pulses since last bipolar violation	
	ODD	EVEN
+	000 +	− 00 −
−	000 −	+ 00 +

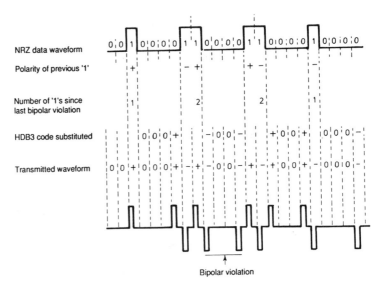

Fig. 3.27 HDB3 coding of line signals.

An example of HDB3 coding is shown in Fig. 3.27 In addition to repeater synchronization it is also necessary to synchronize the multiplexers.

In the CCITT specification two of the 32 PCM channels are reserved for signalling and synchronization (see Section 10.33). The channels are numbered 0 to 31, the 32 channels being called a 'frame'. Frame synchronization is achieved by transmitting a fixed code word in channel 0 on alternate frames. Circuits at the receiver search for this code word and its absence in alternate frames and derive from it a synchronizing signal for the demultiplexer. In this way each 8 bit word in the received frame is routed to its correct destination.

3.11 DELTA MODULATION

This is an alternative binary transmission system using a single digit binary code. Delta modulation does not have the widespread application of standard PCM, it is however used in some rural telephone networks[4] and in digital recording of analogue signals. The fundamental delta modulator, or tracking coder is shown in Fig. 3.28.

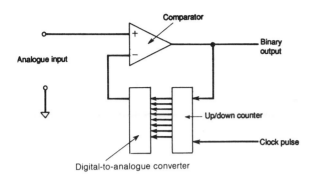

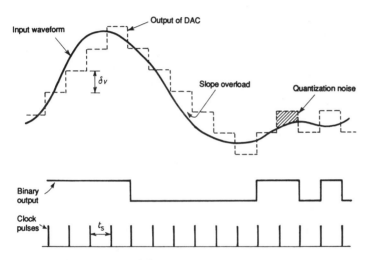

Fig. 3.28 The idealized delta modulator.

The analogue input is compared with the output of a DAC the input of which is derived from an up/down counter. If the amplitude of the analogue input exceeds the output of the DAC the comparator output will be high. This sets the up/down counter to increment on the next clock pulse. If the output of the DAC exceeds the amplitude of the input the comparator output will be low. This sets the up/down counter to decrement on the next clock pulse. The output of the DAC is thus a staircase approximation of the analogue input. The demodulator will consist of the elements in the feedback loop of the modulator. It will be noted from Fig. 3.28 that there are two kinds of distortion produced by this system. Slope overload distortion occurs when the transition from one step to the next fails to cross the input waveform. Quantization distortion occurs due to the finite step size δv.

If t_s is the clocking interval the condition required to prevent slope overload is

$$\dot{h}(t)t_s < \delta v \qquad (3.30)$$

For sinusoidal waveforms $h(t) = A \cos 2\pi f_m t$ and thus $\dot{h}(t)_{max} = A2\pi f_m$, the condition to prevent slope overload is then $A2\pi f_m < \delta v f_s$. Alternatively

$$A_{max} = \frac{\delta v f_s}{2\pi f_m} \qquad (3.31)$$

where f_s is the clocking frequency. It may be noted that either δv or f_s can be increased to avoid overload, but with some penalty. Increasing f_s increases the transmitted bit rate (and hence bandwidth requirement), increasing δv increases the quantization error.

In practice a much simpler circuit than that of Fig. 3.28 is used, which has the additional advantage that the effect of any digit errors decreases to zero after a given interval. (This would not be the case for the circuit of Fig. 3.28.) The practical circuit is based on a simple RC integrator and is shown in Fig. 3.29. The circuit will function if the output of the flip-flop is $\pm V$ volts (which may be easily achieved with a CMOS device).

The voltage across the capacitor will be a series of positive and negative exponential decays since the capacitor charges from either $+V$ or $-V$. The capacitor voltage for a typical input waveform is illustrated in Fig. 3.29. The error voltage in this figure is the difference between the input voltage and the capacitor voltage. This is approximately triangular when the input signal is zero and this is known as the idling voltage.

Considering a single RC network in which V is the charging voltage and v is the instaneous voltage across the capacitor, then

$$\frac{V-v}{R} = C\frac{dv}{dt} \text{ and integrating gives } e^{-t/RC} = \frac{V-v}{V-v_i}$$

where v_i is the initial capacitor voltage. Letting $v_1 = -V$ (i.e. logic 0) the

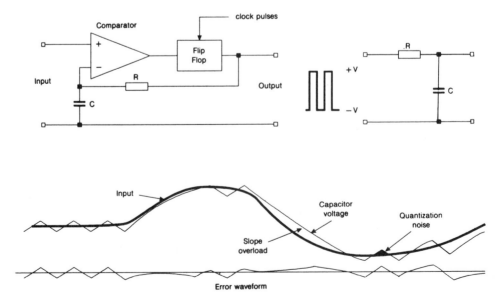

Fig. 3.29 Practical RC delta modulator and waveforms.

capacitor voltage is

$$v = V(1 - 2e^{-t/RC})$$

To avoid overload the slope of the capacitor voltage must be equal to or greater than the slope of the input waveform. If it is assumed that the input $h(t) = A\cos(2\pi f_m t)$ the value of $|\dot{h}(t)|$, when $h(t) = v$, must be less than the slope of the capacitor voltage.

$$|\dot{h}(t)| = A2\pi f_m \sin(2\pi f_m t)$$

or

$$\dot{h}(t) = A2\pi f_m [1 - \cos^2(2\pi f_m t)]^{1/2}$$

but when $h(t) = v$ then $\cos(2\pi f_m t) = v/A$ which means

$$|\dot{h}(t)| = 2\pi f_m (A^2 - v^2)^{1/2}$$

Thus to avoid slope overload

$$\frac{V - v}{RC} > 2\pi f_m (A^2 - v^2)^{1/2} \tag{3.32}$$

The difference between the slope of the capacitor voltage and the input waveform is

$$D = \frac{(V - v)}{RC} - 2\pi f_m (A^2 - v^2)^{1/2}$$

If overload is to be avoided then D must be positive and in the limit $D \to 0$. D changes from a positive to negative value when $dD/dv = 0$: i.e.

$$\frac{dD}{dv} = -\frac{1}{RC} + \frac{2\pi f_m}{2}(A^2 - v^2)^{-1/2} 2v = 0$$

Thus

$$\frac{1}{RC} = \frac{2\pi f_m v}{(A^2 - v^2)^{1/2}}$$

From which

$$v = \frac{A}{[1 + (f_m/f_0)^2]^{1/2}} \quad \text{where } f_0 = \frac{1}{2\pi RC}$$

Substituting this into Eqn (3.32) gives

$$A_{max} = \frac{V}{[1 + (f_m/f_0)^2]^{1/2}} \tag{3.33}$$

The overload characteristic has a frequency response equivalent to that of a single lag characteristic. The optimum value of f_0 is chosen with reference to the amplitude spectrum of normal speech. In practice it is desirable to work as close to overload as possible, and a suitable choice for f_0 is 150 Hz as illustrated in Fig. 3.30. This allows maximum SNR to be achieved at the expense of some overload in the mid-frequency range.

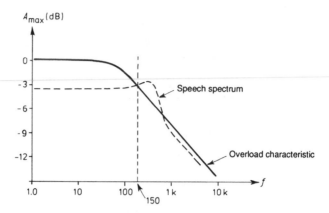

Fig. 3.30 Overload characteristic of RC delta modulator.

3.11.1 Dynamic range and quantization noise

The demodulator in the practical realization is simply the RC network. The quantization noise produced will then be the difference between the input waveform and the voltage across the capacitor of the RC network. When the input signal is zero the error signal is an approximate triangular waveform of period $1/2f_s$ and peak-to-peak amplitude δv. The slope of this waveform at $v = 0$ is $dv/dt = V/RC$ hence $\delta v = V/RC f_s$. This error waveform is shown in Fig. 3.29 and clearly the coder will not deviate from this idling pattern unless the signal exceeds the value $\delta v/2$. The minimum signal level corresponding to the threshold of coding is thus

$$A_{\min} = \frac{V}{2RC f_s} \tag{3.34}$$

The coding range of the delta modulator is thus

$$\frac{A_{\max}}{A_{\min}} = \frac{V}{[1 + (f_m/f_0)^2]^{1/2}} \frac{2RC f_s}{V}$$

i.e.

$$\text{coding range} = \frac{2RC f_s}{[1 + (f_m/f_0)^2]^{1/2}} \tag{3.35}$$

The peak-to-peak error is

$$\bar{\varepsilon}^2 = \frac{2}{t_s} \int_0^t \left(\delta v \frac{t}{t_s} \right)^2 dt = \frac{(\delta v)^2}{12}$$

This is the mean square quantization noise output when there is zero input. The error when the input is a sine wave is shown in Fig. 3.29 and it has been shown[5] that the mean square error in this case is numerically equal to

$$\bar{\varepsilon}^2 = \frac{(\delta v)^2}{6} \tag{3.36}$$

This is constant from the threshold of coding to overload. The error wave-form is actually random but approximately triangular in shape. The power spectrum may thus be approximated by the spectrum of a single triangular pulse (via the autocorrelation function), which is a sinc^2 function with spectral zeros occurring at frequency intervals of $n/2f_s$. The total area under the sinc^2 function is equal to the area under a rectangular figure of the same zero frequency amplitude extending to a frequency of $1/3t_s$, as shown in Fig. 3.31. The total power in the error waveform can thus be represented as $G(f)/3t_s$ where $G(f)$ is a uniform power spectral density, hence

$$\bar{\varepsilon}^2 = G(f)/3t_s \tag{3.37}$$

If the detector output is low-pass filtered, the cut-off frequency of the filter being f_1, then provided $f_1 < 1/3t_s$ the output quantization noise power will be

$$N_q = G(f)f_1 = \frac{3\bar{\varepsilon}^2 f_1}{f_s} = \frac{(\delta v)^2 f_1}{2f_s}$$

But $\delta v = 2\pi V f_0/f_s$, thus

$$N_q = 2\pi V^2 f_1 (f_0)^2 \tag{3.38}$$

The maximum signal power which the system can handle before overload is

$$S = \frac{(A_{max})^2}{2} = \frac{V^2}{2[1 + (f_m/f_0)^2]}$$

hence

$$\text{SQNR}_{max} = \frac{V^2(f_0)^2}{2[(f_0)^2 + (f_m)^2]} \frac{(f_s)^3}{2\pi^2 V^2 f_1 (f_0)^2}$$

Assuming, on average, that $f_m \gg f_0$ then

$$\text{SQNR}_{max} = \frac{(f_s)^3}{4\pi^2 f_1 (f_m)^2} \tag{3.39}$$

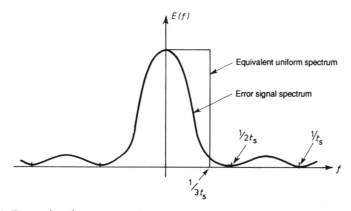

Fig. 3.31 Error signal power spectrum.

This is an approximate result for single tone inputs. With speech waveforms it is found that overload is minimized if the instantaneous speech amplitude is limited to the amplitude of a single tone of frequency 800 Hz, which could be transmitted without overload. Thus if $f_m = 800$ Hz, $f_1 = 3.4 \times 10^3$ and $f_s = 64 \times 10^3$ (which is equivalent to 8 bit linearly quantized PCM) the SQNR is 35 dB. The corresponding figure for 8 bit linearly quantized PCM, obtained from Eqn (3.16) is 49.8 dB. It is usually assumed that the minimum acceptable signal to quantization noise for telephone transmission is about 26 dB. This would give delta modulation a dynamic range of only 9 dB as compared with a figure of 23.8 dB for linearly quantized PCM. (Both of these figures are far short of companded PCM with a UVR of 26 dB at a SQNR of 38 dB.)

The dynamic range of delta modulation may be increased by a form of companding. In the case of the delta modulator the slope of the voltage across the integrating capacitor is varied with input signal amplitude. This gives rise to continuously variable slope delta modulation (CVSDM). Essentially the output of the delta modulator is monitored and when several successive pulses have the value '1' or '0' (indicating that slope overload is occurring) the amplitude of the charging voltage V is increased thereby increasing the slope of the integrator output. When the output returns to 010101 overload is removed and the step size may be progressively reduced. The analysis of several forms of CVSDM is covered in Steele.[6]

3.12 DIFFERENTIAL PCM

A number of analogue waveforms such as speech and video exhibit the property of predictability. This means that the change in value between one sample and the next is small because the rate of change of the analogue waveform is usually low compared with the sampling frequency. Alternatively waveforms such as speech and video have 'instantaneous frequencies' considerably lower than the maximum frequency component on which the sampling frequency is based. The next sample in a waveform can thus be predicted from a knowledge of previous samples. There will be some prediction error, but the peak-to-peak value of the error will be considerably less than the peak-to-peak value of the original waveform. Differential pulse code modulation (DPCM) capitalizes on this fact by coding and transmitting the prediction error. The prediction error requires fewer quantization levels for a given SNR and hence the required transmission bandwidth is less. A simple predictive coder and decoder are shown in Fig. 3.32.

In Fig. 3.32(a), $h(t)$ represents an input sample and $e(t)$ represents the difference between $h(t)$ and the previous sample weighted by the coefficient $a_0(\leqslant 1)$. T is a delay equal to the sampling period. From this figure

$$e(t) = h(t) - a_0 h(t - T)$$

Considering the circuit of Fig. 3.32(b) it is evident that

$$h(t) = e(t) + a_0 h(t - T) \qquad (3.38)$$

This circuit is known as a predictor because the current sample is predicted

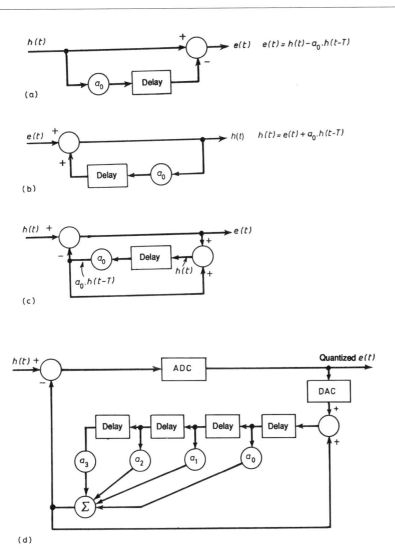

Fig. 3.32 Differential PCM.

from the previous sample and the error signal $e(t)$. The predictor is incorporated into the transmitter as shown in Fig. 3.32(c) and output $e(t)$ is known as the prediction error. There is an advantage in incorporating the predictor in the transmitter as the feedback loop minimizes quantization error when an analogue-to-digital converter is employed. The receiver is simply the feedback loop of the transmitter which, in this case, is the circuit of Fig. 3.32(b).

In practice the estimate of the predictor circuit is based on estimates of the previous four quantized samples as shown in Fig. 3.32(d). This figure also contains an analogue-to-digital converter in the forward path and a digital-to-analogue converter in the feedback path. As stated previously, the feedback action minimizes quantization error. The SQNR for differential PCM is between 5 and 10 dB higher than for PCM without differential coding. This SQNR may be increased further by use of adaptive DPCM (ADPCM). The quantizer step size (i.e. the ADC and DAC) is adapted according to the amplitude of the prediction error. Alternatively, with this type of coder, using a 4 bit quantizer, it is possible to transmit speech with the same quality as 64 kb/s A law compression at a bit rate of 32 kb/s.

3.13 DATA COMMUNICATIONS

PCM and delta modulation have been optimized for the transmission of coded voice signals over trunk routes. Data communications deals primarily with the transmission of digital signals between machines. The bulk of data communications now uses some form of packet switched network, several of which are described in Chapter 13. Traditionally data communications was via the public telephone network and techniques were developed for use on this particular medium and are considered in this section. The most common example of this form of communication is the connection of a terminal to a distant computer via a modem.

When considering data communications, one of the basic parameters that must be defined is the signalling speed. The unit of signalling speed is known as the **baud** after the telegraph engineer Baudot. The signalling speed in baud is in effect the rate at which pulses are transmitted over the communications link. These pulses need not be binary, which means that the data rate, which is usually expressed in bits/s (or b/s), does not necessarily equal the signalling speed.

Unlike PCM, which uses unloaded lines for transmission, data signals which are transmitted over normal telephone circuits must cope with the severely restricted frequency response of such lines. Special problems linked to this response therefore arise in data transmission, and these problems will be considered in some detail. The characteristics of the data signal produced by a VDU keyboard, for example, have two well-defined properties:

(1) low data rate, limited by human typing speed, and
(2) spasmodic output with long periods of no output at all.

To cope with the second of these characteristics, asynchronous communications is used, but it should be noted that dedicated high-speed data links use synchronous communication.

For transmission purposes, each symbol on a keyboard is represented by a unique binary code. The international standard code represents each character by a 7 bit word, i.e.

$$b_6 b_5 b_4 b_3 b_2 b_1 b_0$$

where b_n is a binary digit. Some examples of the 7 bit code are listed in Table 3.3. The 7 bit word has an additional digit, b_7, called a parity check bit, added for error-detection purposes. The parity bit is chosen so that the number of 1s in each 8 bit word is even. If an odd number of 1s occurs at the receiver, the receiver is aware that an error has occurred.

Table 3.3 International 7 bit code

Character	Binary	Octal	Hexadecimal
A	1000001	101	41
B	1000010	102	42
C	1000011	103	43
1	0110001	061	31
2	0110010	062	32
3	0110011	063	33

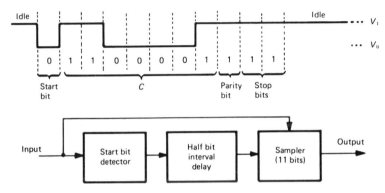

Fig. 3.33 Asynchronous transmission.

The asynchronous transmission system requires extra bits to allow the receiver at either end of the link to determine the beginning and end of each symbol. The transmission of each symbol is preceded by a level change from 1 to 0. The 0 level has a duration of 1 digit interval. The symbol is then transmitted serially, least significant digit (b_0) first, followed by the parity check bit (b_7). The end of each symbol is signalled by a binary 1 which lasts for two digit periods. The idle state (no signal) is thus equivalent to binary 1. A typical digit sequence (10 bits) is illustrated in Fig. 3.33. All timing is initiated by the falling edge of the start bit and the following digits are sampled at their mid-points. This means that timing clocks do not have to be closely matched as synchronization occurs on each start bit. A human operated VDU will produce a maximum of about 10 characters/s, which is, of course, very slow compared with the speed of operation of the computer to which it may be connected. If the transmitted pulses are binary, each pulse has a duration of 9.1 ms, which is equivalent to a signalling speed of 110 baud.

3.14 SPECTRAL PROPERTIES OF DATA SIGNALS

We must know something of the spectral properties of data signals before we can specify the most appropriate form of data transmission over telephone circuits. A typical data signal will consist of a random sequence of pulses of

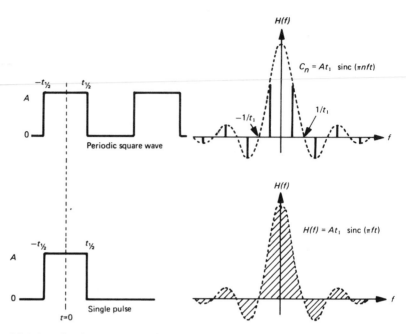

Fig. 3.34 Amplitude spectrum of data signals.

binary 1s and 0s. The power spectral density of such a signal was derived in Section 1.12 from its autocorrelation function. A random binary signal with pulse amplitudes of 0 or A volts and pulse duration t_1 seconds has an amplitude spectrum given by

$$H(f) = At_1 \operatorname{sinc}(\pi f t_1)$$

It can be seen from Fig. 3.34 that most of the energy in the spectral envelope is confined to frequencies below $f = 1/t_1$ hertz. The bandwidth of the data signal is therefore usually approximated by the reciprocal of the pulse width.

The data spectrum, which has a component at zero frequency must be modified for transmission over a telephone circuit which usually has a bandwidth from 300 Hz to 3.4 kHz. Further, since two-way signalling is required over a single circuit, it is necessary to differentiate between transmitted and received data signals. Both these requirements are met by modulating the data signal on to an audio frequency tone. The three possible forms of modulation are AM, FM and PM.

3.15 AMPLITUDE SHIFT KEYING (ASK)

This is the name given to AM when used to transmit data signals. It is not normally used on telephone lines because the large variations in circuit attenuation which can occur make it difficult to fix a threshold for deciding between binary 1 and 0. We shall, however, consider ASK in some detail because it is convenient to represent FM as the sum of two ASK signals.

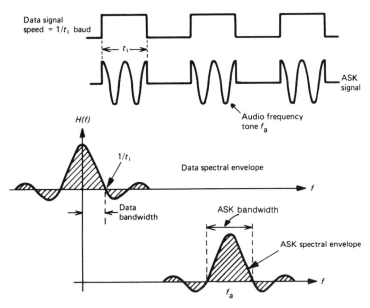

Fig. 3.35 ASK amplitude spectrum.

The ASK signal is generated by multiplying the data signal by an audio tone. This effectively shifts the data spectrum to a centre frequency equal to that of the audio tone. The process is shown in Fig. 3.35. The bandwidth of the modulated signal is twice the bandwidth of the original data signal. This means that the original 110 baud signalling rate requires a transmission bandwidth of 220 Hz using ASK:

$$\text{ASK} \equiv \text{DSBAM} \equiv (\text{carrier} + \text{upper and lower sidebands})$$

3.16 FREQUENCY SHIFT KEYING (FSK)

This is the binary equivalent of FM. In this case a binary 0 is transmitted as an audio frequency tone f_0 and a binary 1 is transmitted as a tone f_1. Hence the binary signal effectively modulates the frequency of a 'carrier'.

Although, strictly speaking, FSK is FM, it is more convenient to consider FSK as the sum of two ASK waveforms with different carrier frequencies. The spectrum of the FSK wave is thus the sum of the spectra of the two ASK waves. This spectrum is shown in Fig. 3.36. Using the FM analogy, it is possible to define a 'carrier frequency' $f_c = f_0 + (f_1 - f_1)/2$ and a 'carrier deviation' $\Delta f = (f_1 - f_0)/2$. The modulation index β is defined as $\beta = \Delta f/B$, where $B = 1/t_1$ is the bandwidth of the data signal. Using these definitions the bandwidth of the FSK signal is

$$B_{\text{FSK}} = 2B(1 + \beta) \tag{3.39}$$

This is similar to Carson's rule for continuous FM. Unlike analogue FM

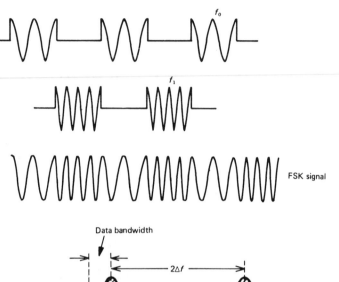

Fig. 3.36 FSK amplitude spectrum.

there is no advantage in increasing Δf beyond the value $\Delta f = B$ since the receiver only needs to differentiate between the two tones f_0 and f_1.

3.17 PHASE SHIFT KEYING (PSK)

This is the binary equivalent of PM, the binary information being transmitted either as zero phase shift or a phase shift of π radians. This is equivalent to multiplying the audio tone by either $+1$ or -1. The bandwidth is thus the same as for ASK. Since there is no dc component in the modulating signal, the carrier in the PSK spectrum will be suppressed. The equivalent modulating signal and the PSK spectral envelope are shown in Fig. 3.37. This form of PSK is sometimes referred to as binary PSK (BPSK) because the phase shift is restricted to two possible values, and it is equivalent to binary DSB-SC-AM.

3.18 PRACTICAL DATA SYSTEMS

We have already indicated the reason for not using ASK for data communications on the public telephone network. The choice between FSK and PSK is determined by the data rate. At low data rates FSK is employed for two-way (duplex) communication. A typical system operating at a signalling rate of

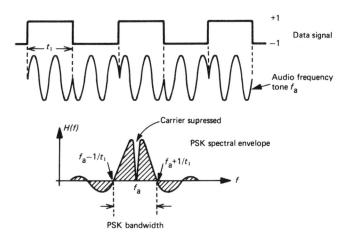

Fig. 3.37 PSK amplitude spectrum.

200 baud uses two tones of 980 Hz and 1180 Hz for binary 1 and 0 in one direction and 1650 Hz and 1850 Hz for binary 1 and 0 in the reverse direction. The incoming FSK is separated into two tones using bandpass filters. Envelope detection is then used to reproduce the binary signal. The combined modulator/demodulator (modem) is illustrated in schematic form in Fig. 3.38.

As the data rate is increased higher carrier frequencies are required; otherwise, each data interval would contain very few cycles of carrier, which would make detection extremely difficult. There is a limit on carrier frequency imposed by the upper cut-off frequency of the telephone line. For this reason

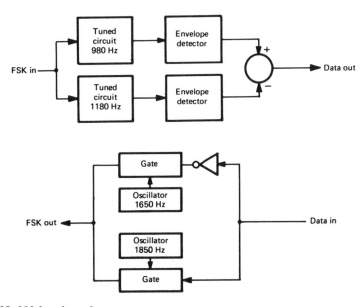

Fig. 3.38 200 baud modem.

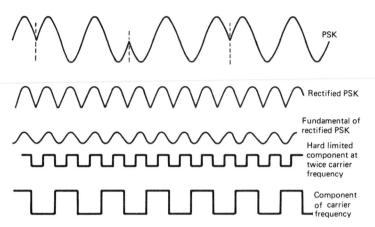

Fig. 3.39 Coherent detection of PSK.

FSK is limited to signalling speeds up to 600 baud. At speeds above this PSK is employed. This type of modulation makes more efficient use of bandwidth but requires more sophisticated coherent detectors. The reference signal for coherent detection is derived from the PSK signal itself. Since the carrier is suppressed in the PSK spectrum the received waveform is first rectified to produce a component at twice carrier frequency. This component is then limited and divided by two to produce the required reference signal. The required signal processing is illustrated in Fig. 3.39.

In the public telephone network any connection between transmitter and receiver will be made via several different paths which will contain several stages of frequency multiplexing and demultiplexing. Imperfections in the various stages of modulation result in random, slowly varying, phase shifts which are introduced into the PSK waveform. This results in phase ambiguity at the receiver and can produce data inversion.[6] The problem is greatly reduced if differential encoding is employed.

3.19 DIFFERENTIAL PHASE SHIFT KEYING (DPSK)

DPSK has the advantage of using the phase of the previous bit interval as the reference for the present bit interval. In order to make this possible, a binary 0

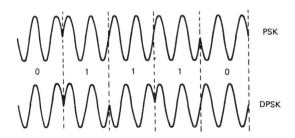

Fig. 3.40 Relationship between PSK and DPSK.

is transmitted as the same phase as the previous digit and a binary 1 is transmitted as a change of phase. The relationship between PSK and DPSK is shown in Fig. 3.40. The receiver compares the phase of the current digit with the phase of the previous digit. If they are the same the current digit is interpreted as a 0; otherwise it is interpreted as a 1. DPSK can be produced by pre-coding the data signal which then modulates the carriers as in standard PSK. If A_n is the present input to the encoder (A_n is binary) and C_{n-1} is the previous output the truth table for the encoder is

A_n	C_{n-1}	C_n
0	0	0
0	1	1
1	0	1
1	1	0

which will be recognized as the exclusive-OR operation

$$C_n = A_n \oplus C_{n-1}$$

We have already noted that pulses can be transmitted at a rate of $1/t_1$ without mutual interference over a channel of cut-off frequency t_1 hertz provided that the channel has a raised cosine frequency response. This applies to the unmodulated signal. A signalling rate of 1200 baud thus requires a raised cosine channel of bandwidth 1200 Hz. The PSK signal will require a band-pass channel with a raised cosine characteristic with a bandwidth of 2400 Hz. A typical PSK signal system would operate at a carrier frequency of 1.8 kHz and a signalling rate of 1200 baud. The bandwidth occupied by this waveform extends from 600 Hz to 3 kHz. Hence a data rate of 1.2 kb/s is an upper limit for BPSK.

3.20 ADVANCED MODULATION METHODS

The data rate of BPSK is sometimes expressed as 1 bit/baud. Since the baud rate is fixed by the channel characteristics the data rate can only be increased by increasing the number of levels per pulse beyond two. If each pulse has four levels the data rate becomes 2 bits/baud and each level can produce a unique phase shift. Thus it is possible to transmit data at a rate of 2.4 kb/s without any increase in bandwidth. It is convenient when considering multiphase PSK to represent the transmitted signal in terms of the sum of two quadrature audi frequency tones, i.e.

$$v_c(t) = a \cos 2\pi f t + b \sin 2\pi f t \tag{3.40}$$

Each of the levels in a four-level pulse can be represented by two binary digits called dibits. Thus it is not actually necessary in practice to produce a four-level pulse; instead the binary signal can be grouped into dibits and each dibit can be used to produce a unique phase shift in multiples of $\pi/2$. When interpreted in this way each dibit represents one pulse, i.e. the signalling rate

Table 3.4

Dibit	Phase shift	In-phase component	Quadrature component
		a	b
00	$\pi/4$	$+1$	$+1$
01	$3\pi/4$	-1	$+1$
11	$-3\pi/4$	-1	-1
10	$-\pi/4$	$+1$	-1

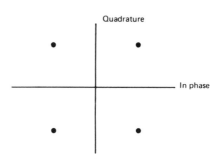

Fig. 3.41 Signal space diagram for QPSK.

in bauds equals half the bit rate. Table 3.4 lists the possible dibits and the values of a and b in Eqn. (3.40) necessary to produce the required phase shifts. The resulting quaternary PSK can be represented on a signal space diagram of the type shown in Fig. 3.41.

The data signal is recovered from the QPSK waveform by using two coherent detectors supplied with locally generated carriers in phase quadrature. The data rate can be increased further by increasing the number of levels of each pulse beyond four. For example, if the number of levels is increased to 16, it is possible to transmit data at a rate of 4 bits/baud. The QPSK signal was characterized by the fact that the coefficients a and b of Eqn. (3.40) always have the same magnitude, thereby producing a resultant of constant amplitude and varying phase. It is also possible for a and b to have different values and the resulting signal is in fact quadrature amplitude modulation. The detection of QAM is covered in Section 11.11 in connection with the transmission of chrominance signals in the PAL colour television system.

The signal space diagram for QAM with 16-1evel pulses is shown in Fig. 3.42. Each individual level is represented by a unique combination of a and b in Eqn (3.40). It is possible with this system to transmit data at a rate of 4.8 kb/s over a raised cosine bandpass channel, with a bandwidth of 2.4 kHz.

At these high data rates, intersymbol interference is a severe problem and elaborate equalization networks (transversal digital filters) are always employed. The lines used are not part of the public telephone network and are maintained to close tolerances in respect of loss and bandwidth. Such lines are often known as leased lines. The sophisticated equalization used on these lines means that signalling rates can approach the Nyquist rate, i.e. pulses can be

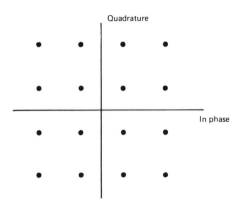

Fig. 3.42 Signal space diagram for 16-level QAM.

transmitted at rates approaching 2400 baud. This means that it is possible to transmit data rates up to 9.6 kb/s using 16-level QAM.

The leased lines referred to in the previous paragraph are basically high grade voice channels. Much higher data rates are possible on specially designed wideband links. These wideband links operate in a synchronous fashion and modern networks are adopting packet switching techniques to maximize the efficiency of usage of these links.

3.21 CONCLUSIONS

This chapter has introduced the basic concepts of digital communications and stressed the advantages of transmission of information in digital format. This is a huge growth area in telecommunications systems engineering and is likely to remain so for the foreseable future. All-digital telephone networks, including cellular mobile telephones, are progressively being installed. These networks will gradually replace analogue systems and will provide enhanced services and increased reliability. A detailed treatment of packetized transmission is given in Chapter 13.

REFERENCES

1. Schwartz, M., *Information Transmission Modulation and Noise*, 3rd edn, McGraw-Hill, 1980.
2. CCITT, *Orange Book*, Vo. III-2, Recommendations G711, G712, G732.
3. Henderson, K. W. and Kautz, W. H., 'Transient Responses of Conventional Filters', *IRE Transactions on Circuit Theory*, Vol. 5, 1958, pp. 333–47.
4. Johnson, F. B., 'Calculating Delta Modulator Performance', *Trans. IEEE*, Vol. AU-16, No. 1 1968, pp. 121–9.
5. Betts, J. A., *Signal Processing Modulation and Noise*, The English Universities Press, 1972, Ch. 7.

6. Steele, R., *Delta Modulation Systems*, Pentech Press, 1975.

7. Oberst, J. P. and Schilling, D.L., 'Performance of Self Synchronized PSK Systems', *Trans. IEEE*, Vol. COM-17, 1969, pp. 666–9.

PROBLEMS

3.1 The sampling theorem is normally applied to signals with a low-pass spectrum. Show that this theorem can also be applied to signals with a bandpass spectrum. What is the statement of the theorem in this case?

A signal with a bandwidth extending from 30 to 34 kHz is to be transmitted through an ideal channel with a low-pass characteristic. Determine the minimum theoretical cut-off frequency of the channel. Give a block diagram of the system and describe how the original signal may be reproduced at the receiver.

Answer: 4 kHz.

3.2 A signal is used to amplitude-modulate a periodic pulse train in which the pulse width is much less than the pulse duration. The sampled signal is to be reconstructed by a sample-and-hold circuit. Sketch the output of this circuit and, by considering its frequency response, find the ratio of the signal frequency amplitude to the amplitude of the lowest frequency distortion component. The signal is a single tone of frequency 2.8 kHz sampled at 8 kHz.

Answer: 1.86:1.

3.3 Explain the difference between the actual bandwidth of a pulse and the bandwidth required for pulse transmission.

Twenty-four speech signals each with a bandwidth of 0 to 4.5 kHz are to be transmitted over a line with a raised cosine frequency response by TDM. Calculate the minimum theoretical bandwidth of the line. Would this bandwidth be adequate in a practical system?

Answer: 216 kHz.

3.4 An equalized line with a raised cosine response has an effective bandwidth of 200 kHz. Six speech channels are to be transmitted over this line using time multiplexed PCM. Assuming linear quantization and a sampling frequency of 8 kHz for each signal, what will be the average signal power-to-noise ratio at the decoder output. Assume quantization noise only is to be considered.

Answer: 26 dB.

3.5 A single information channel carries voice frequencies in the range 50 Hz to 4.3 kHz. The channel is sampled at a 9 kHz rate and the resulting pulses may be transmitted either directly by pulse amplitude modulation (PAM) or by PCM.

Calculate the minimum bandwidth required for the PAM transmission assuming a line with a raised cosine response.

If the pulses are linearly quantized into eight levels and are transmitted as binary digits, find the bandwidth required to transmit the digital signal and compare it with the analogue figure. If the number of levels of quantization is increased to 128 what is the new bandwidth required? Calculate the increase in SNR at the decoder output, assuming the peak-to-peak voltage swing at the quantizer is 2 V.

Answer: 9 kHz; 27 kHz; 63 kHz; 24 dB.

3.6 A PCM system employing uniform quantization and generating a 7 digit code is capable of handling analogue signals of 5 V peak-to-peak. Calculate the mean signal-to-quantizing noise ratio when the analogue waveform has a probability density function given by

$$P(v) = K \exp(-|v|) \quad -2.5 < v < 2.5, \quad = 0 \text{ else}$$

Assume uniform signal distribution within a given quantization interval.

Answer: 38.9 dB.

3.7 Derive an expression for the amplitude spectrum of a single triangular pulse of base width t seconds and amplitude A volts. Hence estimate its bandwidth.

Answer: $2/t$ Hz.

3.8 A data signal consists of a series of binary pulses occurring at a rate of 100 digits/s. This signal is to be transmitted over a telephone line, binary 1 being sent as a 1.5 kHz tone and binary 0 as a 2.8 kHz tone. What is the bandwidth of the transmitted signal?

If the digit rate is increased to 1000 b/s what are the required upper and lower cut-off frequencies of the line in order that it may transmit this signal?

Answer: 1.5 kHz; 500 Hz; 3.8 kHz.

3.9 If the transmission of question 3.8 is by DPSK, what is the maximum data rate that can be transmitted over the telephone line? What is the optimum carrier frequency in this case?

Answer: 1650 b/s; 2.35 kHz.

3.10 Derive an expression for the amplitude spectrum of a FSK transmission when the digit stream is a series of alternate 1s and 0s.

3.11 A sinusoidal signal is switched periodically from 10 MHz to 11 MHz at a rate of 5000 times/s. Sketch the resulting waveform and identify the modulating signal if the switched sine wave is regarded as FSK.

Find the approximate transmission bandwidth of the FSK signal and compare this with the bandwidth required if the modulating signal is approximated by a sine wave producing the same carrier deviation.

Answer: 1.02 MHz; 1.01 MHz.

4 Noise in analogue communications systems

4.1 INTRODUCTION

We have considered, in the previous chapters, various ways of transmitting information from one location to another. We concern ourselves now with the performance of these systems in a noisy environment. This comparative analysis will provide the insight required to determine the suitability, or otherwise, of using a particular form of transmission in a specific environment. The relative performance of various systems in noise is, of course, only one of the factors taken into account when choosing a particular method of information transmission. The economic considerations, e.g. cost, complexity, maintenance, are sometimes of paramount importance, but are outside the scope of this text.

Noise is defined as any spurious signal that tends to mask or obscure the information in the transmitted signal. The ratio of signal power to noise power at any point in a telecommunications system is known as the SNR, and the fundamental exchange possible between signal bandwidth and SNR is given by Shannon's law (see Section 1.14). In this chapter we will consider the performance of systems using analogue transmission and then discuss the significance of Shannon's law for the examples chosen.

4.2 PHYSICAL SOURCES OF NOISE

Noise is usually divided into naturally occurring noise and artificial noise. Artificial noise comes from various sources, the most important types being ignition interference, produced whenever sparks occur at electrical contacts, and crosstalk which is produced by inductive or capacitive coupling between one or more communications channels. Artificial noise can, in theory, be eliminated, although the cost of such elimination is often quite uneconomic. Natural noise is produced by many different phenomena, some examples being lightning discharges, thermal radiation and cosmic radiation. Natural noise cannot be eliminated and communication systems must perform efficiently in the presence of this type of noise, and often in the presence of artificial noise also.

The two most important types of natural noise are thermal noise and shot noise. Thermal noise is produced by random motion of charged particles in resistive materials and by thermal radiation from objects surrounding a telecommunications system, particularly one with an antenna. Shot noise is produced in semiconductor devices and results from the fact that currents flow across p–n junctions in finite quanta rather than continuously.

Thermal noise, caused by random motion of charged particles in resistive materials, produces a mean square noise voltage with a magnitude directly related to the temperature of the resistive material, which is given by

$$\bar{v}^2 = 4kTR\Delta f_n \quad (\text{volt})^2 \tag{4.1}$$

In this expression k is Boltzmann's constant $(1.38 \times 10^{-23}\,\text{J/K})$, T is the absolute temperature in kelvins, R is the resistance of the material in ohms, and Δf_n is the bandwidth of the measurement. The derivation of Eqn (4.1) is given by King.[1] Any noisy resistance can be represented by an equivalent circuit consisting of a noise-free resistance in series with a voltage source whose mean square amplitude is given by Eqn (4.1). This allows the determination of the noise properties of networks using conventional network theorems. We will consider one such calculation, of particular importance in telecommunications, to illustrate the procedures. We wish to calculate the noise power delivered by a resistance to a matched load, i.e. a load with the same resistive value. The circuit is shown in Fig. 4.1.

In this circuit R_1 is the noise source and R_2 is the load. Both source and load will produce a mean square noise voltage which will give rise to a mean square current $\overline{i^2}$. The power delivered by R_1 to R_2 is

$$P_n = \frac{\bar{v}_1^2 R_2}{(R_1 + R_2)^2} \tag{4.2}$$

If $R_1 = R_2$ (i.e. the source and the load are matched) the power delivered by R_1 to R_2 is

$$P_n = \frac{\bar{v}_1^2}{4R_2} = \frac{4kT\Delta f_n R_2}{4R_2} = kT\Delta f_n \quad \text{watts}$$

This is often written

$$P_n = \eta \Delta f_n \tag{4.3}$$

The constant η is the single-sided noise power spectral density and is independent of the range of frequencies Δf_n. If Δf_n is measured over both negative and positive frequencies the double-sided noise spectral density is

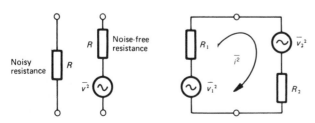

Fig. 4.1 Noise networks. (a) Equivalent circuit of a noisy resistance; (b) noise delivered to a matched load.

$\eta/2$ watts/hertz. This agrees with the Fourier analysis developed in Chapter 1, in which we indicated that half the total power was contributed each by the negative and positive components. When the noise spectral density is independent of frequency the noise is said to be white, by analogy with white light. When white noise is filtered by a frequency-selective network the resulting noise is known as coloured noise.

When thermal noise is derived from an antenna, the temperature of the noise source is not necessarily equal to the antenna temperature. The expression for the noise power delivered by an antenna to a matched load is $P_n = kT_a\Delta f_n$ watts. The effective noise temperature of the antenna T_a is related to the temperature of the radiating bodies surrounding the antenna and is determined by measurement. If the radiating bodies have the same temperature as the antenna, this is the value used for T_a. Antennas pointing into space generally receive far less radiation than antennas directed towards bodies on the Earth's surface and consequently have a much lower effective noise temperature.

4.3 NOISE PROPERTIES OF NETWORKS

All telecommunications systems are characterized by the fact that received signals are always accompanied by noise. The effectiveness of such a system is measured in terms of the ratio of signal power to noise at the system output. The SNR at any point in a telecommunications link is usually expressed in decibels:

$$\text{SNR} = 10\log_{10}[S_p/N_p] \qquad (4.4)$$

The minimum acceptable SNR for reliable communication is normally considered to be about 10 dB. Many systems operate at much higher ratios than this: the minimum SNR for telephone circuits is around 26 dB and for high-quality audio transmissions a figure in excess of 60 dB is typical. Some space systems operate with an SNR much less than 10 dB, but such systems require sophisticated techniques, such as correlation detection at the receiver.

All electrical networks generate noise and it will be clear, therefore, that when a signal passes through such a network the SNR at the network output will always be less than at the network input. The amount of extra noise generated by a network is specified by its noise figure. This is given the symbol F and is defined as

$$F = \text{SNR}_{\text{in}}/\text{SNR}_{\text{out}} \qquad (4.5)$$

A noiseless network has a noise figure of unity and it therefore follows that real networks always have a noise figure with a numerical value greater than unity. The noise figure of any network is derived in terms of the schematic diagram of Fig. 4.2. The input signal power is S_i and the network power gain is A_p (not necessarily restricted to values greater than unity); the signal power at the network output is thus $S_0 = A_pS_i$. If the input noise power is N_i and the noise power generated within the network is N_a the total noise power at the network output is $N_0 = A_pN_i + N_a$. The output SNR is

$$\text{SNR}_0 = A_pS_i/(A_pN_i + N_a)$$

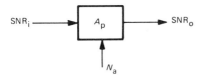

Fig. 4.2 Noise figure of a network.

i.e.

$$F = \frac{S_i}{N_i} \frac{(A_p N_i + N_a)}{A_p S_i} = \frac{A_p N_i + N_a}{A_p N_i} \tag{4.6}$$

The factor $A_p N_i$ is the output noise power of a noise-free network; thus

$$F = \frac{\text{total output noise power}}{\text{output noise power if network was noise free}} \tag{4.7}$$

We see from the definition given as Eqn (4.7) that F is not constant but is related to the noise power at the network input. The value of F is standardized by fixing the input noise power as that produced by a matched source at a standard temperature of 290 K. Use of noise figures in network calculations is thus only valid if the input noise power is $kT_s \Delta f_n$, where $T_s = 290$ K.

The usefulness of noise figures is demonstrated by considering the cascaded networks of Fig. 4.3. Using the definition of Eqn (4.7) we can write the overall noise figure for the cascaded network as

$$F = \frac{(A_{p1} N_i + N_{a1}) A_{p2} + N_{a2}}{A_{p1} A_{p2} N_i}$$

We assume that the networks are matched, i.e. the output resistance of the first network is equal to the input resistance of the second network, and that the input noise N_i is the noise produced by a matched source at 290 K. The noise figure of the first network is therefore

$$F_1 = (A_{p1} N_i + N_{a1})/A_{p1} N_i$$

In defining the noise figure of the second network the input power is also N_i; hence

$$F_2 = (A_{p2} N_i + N_{a2})/A_{p2} N_i$$

Thus the overall noise figure may be written

$$F = \frac{A_{p1} N_i + N_{a1}}{A_{p1} N_i} \frac{N_{a2}}{A_{p1} A_{p2} N_i}$$

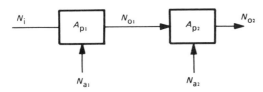

Fig. 4.3 Cascaded noisy networks.

But

$$F_2 - 1 = \left(1 + \frac{N_{a2}}{A_{p2}N_i}\right) - 1$$

Therefore

$$F = F_1 + \frac{F_2 - 1}{A_{p1}} \tag{4.8}$$

We see from Eqn (4.8) that if $A_{p1} \gg 1$, which is usually the case, the major contribution to the overall noise figure is produced by the first network. Evidently it becomes extremely important to ensure that the first network in any cascaded system has as low a noise figure as possible. Equation (4.8) can be expanded to include any number of cascaded networks. The equation for three networks in cascade is

$$F = F_1 + \frac{F_2 - 1}{A_{p1}} + \frac{F_3 - 1}{A_{p1}A_{p2}} \tag{4.9}$$

The effective noise temperature of a network is an alternative method of describing the noise performance of a network. This alternative is especially useful when considering low-noise networks or networks in which the input noise is not produced by a matched source at 290 K. In the latter case the use of noise figure is not valid. The effective noise temperature of a network is determined by replacing the noisy network by a noise-free network with an equivalent noise source at its input. The temperature of the equivalent noise source is chosen to make the noise at the output of the noise-free network equal to the noise at the output of the noisy network. Referring to Fig. 4.2, the noise produced by the network is replaced by an equivalent noise source of

$$N_a = kT_e \Delta f_n A_p$$

The factor $(kT_e \Delta f_n)$ is the noise delivered by an equivalent matched source at a temperature T_e. The temperature T_e is known as the 'effective noise temperature' of the network. If the value of $T_e \ll 290$ K the network itself contributes very little extra noise. The relationship between noise figure and effective noise temperature is

$$F = \frac{A_p N_i + N_a}{A_p N_i} = \frac{A_p k \Delta f_n (T_s + T_e)}{A_p k \Delta f_n T_s}$$

T_s is the standard temperature equal to 290 K. Hence

$$F = 1 + T_e / T_s \quad \text{or} \quad T_e = (F - 1) T_s \tag{4.10}$$

The maser microwave amplifier is an example of a very low noise network; the effective noise temperature of such a device would be between 10 and 30 K, which is equivalent to a noise figure of between 1.03 and 1.11.

It is often convenient to represent cascaded networks in terms of effective noise temperature; substituting Eqn (4.10) into Eqn (4.9) gives

$$T_e = T_1 + \frac{T_2}{A_{p1}} + \frac{T_3}{A_{p1}A_{p2}} \tag{4.11}$$

This equation is particularly useful when considering the noise performance of a cascaded system in which the first element is an antenna with an effective noise temperature not equal to 290 K. The cascaded system, excluding the antenna, is replaced by a noise-free system with an equivalent matched noise source at the input. The antenna noise is then included by adding the effective noise temperature of the antenna to the equivalent noise temperature of the noise source, and using the sum as the noise temperature of the matched source.

We indicated when referring to Fig. 4.2 that the power gain A_p was not necessarily greater than unity; when A_p is less than unity the network is passive and is usually characterized in terms of insertion loss rather than power gain. The insertion loss of a passive network is the reciprocal of power gain:

$$\text{insertion loss } L = \frac{\text{input power}}{\text{output power}} \qquad (4.12)$$

When a passive network is matched at both the input and output the insertion loss has the same numerical value as the network noise figure. The noise power delivered to the network by a matched source is $kT_s\Delta f$ watts. The noise delivered by the network to its load will be the sum of the input noise power multiplied by the network power gain (which is less than unity for passive networks) and the noise power generated within the network. This latter component is represented, as in the determination of effective noise temperature, by an equivalent noise source at the input of the network which is then assumed noise free. The total noise power delivered to this noise-free network is $k\Delta f_n(T_s + T_e)$. The noise power delivered to the load is $N_0 = A_p k\Delta f_n(T_s + T_e)$. However, the network is a matched source for its load, and such a source delivers a noise power $N_0 = k\Delta f_n T_s$. Thus these two values must be equal or

$$A_p k\Delta f_n(T_s + T_e) = k\Delta f_n T_s$$

i.e.

$$T_e = T_s(1 - A_p)/A_p = T_s(L - 1) \qquad (4.13)$$

But $T_e = (F - 1)T_s$. Thus for a matched passive network the insertion loss L has the same numerical value as the network noise figure, i.e. $F = L$. Both F and L are usually measured in decibels.

EXAMPLE: To illustrate the significance of the analysis presented in this section we will consider a typical system involving a domestic television receiver. The receiver, which has a video bandwidth of 5.5 MHz, is coupled via a 70 Ω coaxial cable with an insertion loss of 6 dB to an antenna with an effective noise temperature of $T_a = 290$ K. The noise figure of the receiver, referred to a matched source of 70 Ω at 290 K, is 6 dB. We are requested to find the SNR at the receiver output when the open-circuit signal voltage at the antenna terminals is 1 mV rms.

Each component in this system will cause a degradation of the SNR. The maximum value of SNR will occur at the input to the coaxial feeder. We begin by determining the SNR at this point and to do this we consider the antenna

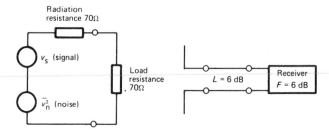

Fig. 4.4 Matched antenna.

as both a matched signal and noise source as shown in Fig. 4.4. We assume that the antenna has a source (see Chapter 7) resistance of $70\,\Omega$, and is therefore matched to the $70\,\Omega$ feeder. We also assume that the receiver has an input resistance of $70\,\Omega$.

Since the antenna open-circuit signal voltage is $1\,\mathrm{mV}$ rms the voltage across the $70\,\Omega$ load will be half this value, i.e. $0.5\,\mathrm{mV}$ rms. The signal power delivered to the coaxial feeder will be $S_p = (0.5 \times 10^{-3})^2/70$. The noise power delivered by the antenna, which acts as a matched noise source, is $N_p = k\Delta f_n T_a$. The SNR at the feeder input is thus

$$\mathrm{SNR} = 10\log_{10}S_p/N_p = 52.5 \quad \mathrm{dB}$$

We can now calculate the overall noise figure for receiver and feeder. This is valid in this example because the antenna effective temperature is $290\,\mathrm{K}$. The overall noise figure is

$$F = F_1 + (F_2 - 1)/A_{p1}$$

F_1 is the noise figure of the feeder $= L = 6\,\mathrm{dB}$, i.e. $F_1 = 3.98$. A_{p1} is the reciprocal of the feeder insertion loss and has a value of 0.251. F_2 is the receiver noise figure $= 6\,\mathrm{dB} = 3.98$.

Hence the overall noise figure is $F = 15.85\,(12\,\mathrm{dB})$. The SNR at the receiver output will then be

$$\mathrm{SNR_{out}} = 52.5\,\mathrm{dB} - 12\,\mathrm{dB} = 40\,\mathrm{dB}$$

This figure is actually lower than the minimum acceptable SNR for reasonable picture quality, which is usually considered to be $47\,\mathrm{dB}$. The solution to this problem would be the use of a low-noise pre-amplifier between antenna and receiver. The pre-amplifier would have a typical power gain of $20\,\mathrm{dB}$ and a typical noise figure of $3\,\mathrm{dB}$. It is possible to connect such an amplifier either directly to the antenna terminals (i.e. before the feeder) or directly to the receiver input (i.e. after the feeder). We suggest that the reader examines both cases and determines the output SNR when the pre-amplifier is included. It will be found that the output SNR will be $6\,\mathrm{dB}$ higher when the pre-amplifier is connected directly to the antenna terminals. This verifies the earlier conclusion that the first network in a cascaded system has the major effect on the overall noise figure.

In this example we represented the domestic television receiver as a

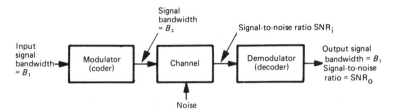

Fig. 4.5 General telecommunications system.

single network with an overall noise figure. In fact the receiver will be divided into several distinct functional blocks including RF and IF amplifiers, vision detector, video frequency amplifier, etc. From our consideration of the super-heterodyne receiver in Chapter 2 it is clear that the signal bandwidth at various points in a receiver (e.g. before and after the detector) is not necessarily constant. This means that possibilities exist for exchanging bandwidth for SNR at various points in a receiver, the theoretical relationship being given by Shannon's law. Thus although it is common practice to specify a single noise figure for a receiver it is more instructive to consider the individual sections of such a system.

A general telecommunications system can be represented in the form shown in Fig. 4.5. In this figure the bandwidth of the signal source is B_1 hertz. The signal enters the modulator and the bandwidth is changed to B_2 hertz. During the transmission noise is added to the signal. It is convenient to show this noise as a single input as in Fig. 4.5, but as we have shown, noise is actually added at every stage of a communications system. The SNR at the detector input is given the symbol SNR_i. At this point the detector transforms the bandwidth of the received signal back to its original value of B_1 hertz and in so doing produces an output signal-to-noise ratio SNR_0. Shannon's law states that whenever there is a change in signal bandwidth there should be an accompanying change in SNR. Shannon's law is, however, a theoretical law in which no allowance is made for any physical constraints that may exist in practical systems. We must therefore examine each transmission system on an individual basis and in order to do this it is necessary to develop an algebraic technique for specifying the effect of noise.

4.4 ALGEBRAIC REPRESENTATION OF BAND-LIMITED NOISE

Equation (4.3) indicates that the noise power delivered by a matched source is $\eta \Delta f_n$ watts. This means that the total noise delivered to the detector in a receiver of bandwidth B hertz is ηB hertz. The algebraic representation of this noise is derived by dividing the bandwidth B into small elements Δf and approximating the noise power within each element by a cosine wave. As the element of bandwith $\Delta f_n \to 0$ this gives a very accurate representation of the noise signal. The technique is illustrated by Fig. 4.6. The noise voltage

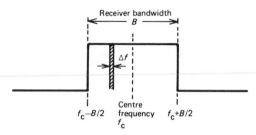

Receiver bandwidth
B
Δf
Centre
frequency
f_c
$f_c - B/2$ $f_c + B/2$

Fig. 4.6 Band-limited white noise.

produced in the elemental bandwidth Δf is represented as

$$n(t) = A_n \cos[2\pi f_k t + \theta_k(t)] \qquad (4.14)$$

where $\frac{1}{2} A_n^2 = \eta \Delta f$ and f_k is the centre frequency of the interval Δf. The phase angle $\theta_k(t)$ is an arbitrary random number. The total noise voltage produced over the entire bandwidth B is calculated by summing the contributions of each element Δf, i.e.

$$V_n(t) = \sum_k A_n \cos[2\pi f_k t + \theta_k(t)] \qquad (4.15)$$

It is convenient at this point to introduce the substitution $f_k = (f_k - f_c) + f_c$ where f_c is the centre frequency of the bandwidth B. Equation (4.14) then becomes

$$n(t) = A_n \cos[2\pi(f_k - f_c)t + \theta_k(t)] \cos(2\pi f_c t)$$
$$- A_n \sin[2\pi(f_k - f_c)t + \theta_k(t)] \sin(2\pi f_c t)$$

which means that Eqn (4.15) can be re-written

$$V_n(t) = x(t) \cos(2\pi f_c t) + y(t) \sin(2\pi f_c t) \qquad (4.16)$$

where

$$x(t) = \sum_k (2\eta \Delta f)^{1/2} \cos[2\pi(f_k - f_c)t + \theta_k(t)] \qquad (4.16a)$$

and

$$y(t) = - \sum_k (2\eta \Delta f)^{1/2} \sin[2\pi(f_k - f_c)t + \theta_k(t)] \qquad (4.16b)$$

We can thus represent the noise voltage in terms of the sum of two amplitude modulated carriers in phase quadrature. The carrier amplitudes are the random variables $x(t)$ and $y(t)$ that have mean square values $\overline{x^2(t)}$ and $\overline{y^2(t)}$, respectively. The total noise power in the bandwidth B is

$$P = \frac{\overline{x^2(t)}}{2} + \frac{\overline{y^2(t)}}{2} \quad \text{watts} \qquad (4.17)$$

but

$$\overline{x^2(t)} = \overline{y^2(t)} = \sum_k \frac{2\eta\Delta f}{2}$$

hence

$$P = \overline{x^2(t)} \quad \text{or} \quad \overline{y^2(t)} \quad \text{i.e.} \quad P = \eta B \quad \text{watts} \qquad (4.18)$$

Having derived this representation of band-limited noise we can now examine the effectiveness of various signal transmission systems in the presence of noise. This will be accomplished by comparing the SNR after detection (or decoding) with the value that exists before detection.

4.5 SNR CHARACTERISTICS OF ENVELOPE-DETECTED DSB-AM

The assumptions made in this section are

(i) the modulating signal is a single tone;
(ii) the envelope detector has an ideal characteristic which infers that its output is directly proportional to instantaneous carrier amplitude.

Some care is required in defining SNR at the detector input; the DSB-AM signal consists of a carrier and two sidebands and we can specify the signal power in terms of combination of these components. The DSB-AM signal plus noise is written as

$$V_{in}(t) = A_c [1 + m\cos(2\pi f_m t)]\cos 2\pi f_c t$$
$$+ x(t)\cos(2\pi f_c t) + y(t)\sin(2\pi f_c t) \qquad (4.19)$$

If we normalize this voltage to be the voltage developed across a resistance of $1\,\Omega$ then the carrier power is $\frac{1}{2}A_c^2$ watts, the sideband power is $\frac{1}{4}(mA_c)^2$ watts and the total power is $\frac{1}{2}A_c^2(1 + \frac{1}{2}m^2)$ watts. The noise power is $\frac{1}{2}\overline{x^2(t)} + \frac{1}{2}\overline{y^2(t)} = \overline{x^2(t)}$ watts. There are three commonly used methods of specifying SNR at the detector input: these are

(1) carrier power-to-noise ratio,
(2) sideband power-to-noise ratio, and
(3) total power-to-noise ratio.

The carrier-to-noise ratio is $S_c/N = A_c^2/2\overline{x^2(t)}$, the sideband-to-noise ratio is $S_{sb}/N = (mA_c)^2/4\overline{x^2(t)}$, and the total SNR is

$$S_t/N = A_c^2\left(1 + \frac{m^2}{2}\right)\bigg/ 2\overline{x^2(t)}$$

The resultant input to the detector is given by the vector sum of the amplitude modulated waveform and the noise components. The graphical addition is shown in Fig. 4.7, from which we can see that if the SNR is very large then the phase angle $\phi \to 0$. The input to the detector for large SNR is given approximately by

$$V_{in}(t) = A_c(1 + m\cos 2\pi f_m t) + x(t) \qquad (4.20)$$

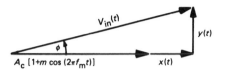

Fig. 4.7 Signal plus noise at the decteor input.

The detector, being ideal, will have an output

$$V_{out}(t) = aA_c(1 + m\cos 2\pi f_m t) + ax(t) \tag{4.21}$$

The output signal power is $\frac{1}{2}(amA_c)^2$ and the output noise power is $a^2\overline{x^2(t)}$. The SNR at the detector output is thus

$$\text{SNR}_{out} = (mA_c)^2/2\overline{x^2(t)} \tag{4.22}$$

If we compare this figure with the carrier-to-noise ratio at the detector input then

$$\text{SNR}_{out} = m^2 S_c/N$$

This has its maximum when $m = 1$, i.e.

$$\text{SNR}_{out(max)} = S_c/N \tag{4.23}$$

This equation states that the SNR at the detector output has the same numerical value as the carrier-to-noise ratio at the detector input. It does not state that the SNR at the detector output is equal to the SNR at the detector input, which is the interpretation sometimes wrongly used in the literature.

If SNR_{out} is compared with the sideband-to-noise ratio at the detector input, then

$$\text{SNR}_{out} = 2S_{sb}/N \tag{4.24}$$

The output SNR is 3 dB greater than the sideband power-to-noise ratio. It is not correct, however, to say that DSB-AM produces a SNR improvement of 3 dB.

To produce a realistic assessment of DSB-AM we must compare the output SNR with the total SNR at the detector input. When $m = 1$ then

$$\text{SNR}_{out} = \frac{2}{3}\frac{S_t}{N} \tag{4.25}$$

The SNR at the output of an ideal envelope detector is actually lower than the SNR at the detector input. This is explained by the fact that even when $m = 1$, 66.6% of the total transmitted power is contained within the carrier component, which does not contribute at all to the signal power at the detector output. Thus DSB-AM does not obey Shannon's law in its strict theoretical statement. The situation is improved somewhat if the carrier is suppressed, as this shows an improvement of 3 dB in SNR as given by Eqn (4.24). If the carrier is suppressed, however, it is not possible to employ envelope detection.

4.6 SNR CHARACTERISTICS OF COHERENTLY DETECTED DSB-AM

Equation (4.20) was derived on the assumption that the SNR at the detector input was very large. If this condition is not met the resulting values of SNR_{out} given by Eqns (4.23), (4.24) and (4.25) are not valid. In fact at low values of SNR the performance of the envelope detector deteriorates rapidly. The envelope detector can therefore be employed only in good SNR conditions. This is usually taken to mean that the envelope detector has a signal-to-noise performance that is acceptable only if the SNR at the detector input is greater than 10 dB.

Coherent detection is an alternative demodulation technique for DSB-AM and an essential technique for suppressed carrier and SSB-AM. The performance of the coherent detector is maintained for all values of input SNR. The coherent detector multiplies the received signal by a locally produced reference signal $E \cos (2\pi f_c t)$. When the received signal is accompanied by noise, the detector output is

$$V_0(t) = EA_c [1 + m \cos (2\pi f_m t)] \cos^2 (2\pi f_c t)$$
$$+ Ex(t) \cos^2 (2\pi f_c t) + Ey(t) \cos (2\pi f_c t) \sin (2\pi f_c t) \quad (4.26)$$

The frequency terms produced by the multiplication, which are outside the bandwidth occupied by the modulating signal, are removed by the filter that follows the coherent detector, the resulting output being

$$V_{out}(t) = \tfrac{1}{2} EA_c + \tfrac{1}{2} EA_c m \cos 2\pi f_m t + \tfrac{1}{2} Ex(t) \quad (4.27)$$

The signal and noise powers in this expression are respectively $S_p = \tfrac{1}{8}(EA_c m)^2$ watts and $N_p = \tfrac{1}{4} E^2 \overline{x^2(t)}$ watts. The SNR at the detector filter output is thus

$$SNR_{out} = (mA_c)^2 / 2\overline{x^2(t)} \quad (4.28)$$

This is the same result as for the envelope detector but there is no precondition that the SNR should be large. In other words, the coherent detector maintains its SNR performance for all values of input SNR and is therefore superior to the envelope detector in poor SNR conditions.

4.7 SNR CHARACTERISTICS OF DSB-SC-AM

The detection of DSB-SC-AM is discussed fully in Section 2.10. If we assume that such a detector has an ideal characteristic the output SNR will be 3 dB greater than input sideband power-to-noise ratio. In other words, DSB-SC-AM produces a 3 dB SNR improvement.

4.8 SNR CHARACTERISTICS OF SSB-AM

This form of modulation requires coherent detection. In this case the input to the detector will be one sideband only, plus noise. If we assume a single modulating tone of frequency f_m, the signal plus noise at the detector input will be

$$V_{in}(t) = A_c \cos\left[2\pi(f_c + f_m)t\right] + x(t)\cos\left(2\pi f_c t\right) + y(t)\sin\left(2\pi f_c t\right) \quad (4.30)$$

The input SNR is thus

$$\text{SNR}_{in} = A_c^2/2\overline{x^2(t)} \quad (4.31)$$

We should point out here that because the bandwidth of a SSB signal is approximately half that of a DSB signal, the noise power at the detector input will be half the equivalent noise power in a DSB system. The output of the SSB coherent detector after filtering is

$$V_{out}(t) = \tfrac{1}{2}EA_c \cos 2\pi f_m t + \tfrac{1}{2}Ex(t) \quad (4.32)$$

The SNR at the filter output is thus $\text{SNR}_{out} = A_c^2/2\overline{x^2(t)}$. Thus for SSB-AM

$$\text{SNR}_{out} = \text{SNR}_{in} \quad (4.33)$$

It is interesting to compare a SSB transmission with a DSB-SC transmission when the transmitted power is the same in each case. We have shown that there is a 3 dB improvement in the DSB case, but because the noise power in the SSB system is only half that of the DSB system the two are actually equivalent in terms of output SNR when the transmitted power in each case is the same.

It would seem on a purely SNR basis that there is little to choose between DSB-SC-AM and SSB-AM. This is not entirely true: in certain situations (when ionospheric reflections are used, for example) severe distortion can be produced in DSB systems because components in the two sidebands can have differing phase velocities resulting in partial cancellation after detection. This problem is not encountered in SSB systems, when used for audio signal transmission, because the ear is insensitive to phase distortion. In these circumstances there is a considerable advantage in using SSB transmission.

4.9 SNR CHARACTERISTICS OF FM

We will establish the SNR properties of FM by again assuming an ideal detector, i.e. one that has an output voltage directly proportional to the 'instantaneous frequency' of the input signal. The algebraic representation of band-limited noise is identical to the AM case, but it should be borne in mind that the bandwidth of a frequency modulated signal is usually considerably greater than the bandwidth of an AM signal, and that noise power is directly proportional to bandwidth.

The voltage at the FM detector input will be the sum of signal and noise, i.e.

$$V_{in}(t) = A_c \cos\left[2\pi f_c t + \beta \sin\left(2\pi f_m t\right)\right]$$
$$+ x(t)\cos\left(2\pi f_c t\right) + y(t)\sin\left(2\pi f_c t\right) \quad (4.34)$$

The FM signal has constant amplitude and the signal power is also constant and independent of the amplitude of the modulating signal. This is a fundamental difference between FM and AM. The signal power at the detector input has a value of $A_c^2/2$ watts. It is convenient to calculate the signal power at the detector output in the absence of noise, and the noise power in the presence of an unmodulated carrier (i.e. in the absence of signal).

If we assume that the modulating signal is a single tone of frequency f_m the instantaneous frequency of the FM waveform is $f_i = f_c + \Delta f_c \cos(2\pi f_m t)$, and this will produce a voltage at the detector output given by

$$V_s = b2\pi\Delta f_c \cos(2\pi f_m t) \qquad (4.35)$$

The signal power at the detector output will thus be

$$S_0 = (b2\pi\Delta f_c)^2/2 \quad \text{watts} \qquad (4.36)$$

We calculate the noise voltage at the detector output using phasor methods. The phasor diagram for the unmodulated carrier and quadrature noise components is shown in Fig. 4.8. The FM detector produces an output proportional to frequency, which is the time differential of the phase angle ϕ. If we assume that the SNR at the detector input is large, then the phase angle is given by

$$\phi(t) \simeq \tan^{-1}\left[\frac{y(t)}{A_c}\right] \simeq \frac{y(t)}{A_c} \qquad (4.37)$$

The 'instantaneous frequency' produced by the noise is

$$\dot{\phi}(t) = \frac{\dot{y}(t)}{A_c} \qquad (4.38)$$

which, if we assume $\dot{y}(t) \ll A_c$, is narrowband FM. The noise voltage at the detector output is

$$V_n = \frac{b}{A_c}\dot{y}(t) \qquad (4.39)$$

where b is the detector constant of proportionality. The noise waveform $y(t)$ is itself the sum of many elemental noise components, i.e.

$$y(t) = \sum_k A_n \sin\left[2\pi(f_k - f_c)t + \theta_k(t)\right]$$

The noise component at a particular frequency f_k is

$$y_k(t) = A_n \sin\left[2\pi f t + \theta_k(t)\right] \qquad (4.40)$$

where $f = (f_k - f_c)$ is the frequency difference between the noise component and the centre frequency, and can have both negative and positive values. The

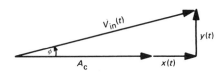

Fig. 4.8 Unmodulated FM carrier plus noise.

noise voltage at the detector output produced by an elemental noise component at the detector input is

$$\delta V_n = \frac{b}{A_c} 2\pi f A_n \cos\left[2\pi ft + \theta_k(t)\right] \tag{4.41}$$

where we have assumed that $\theta_k(t)$ varies very slowly and does not contribute to the output amplitude. The power at the detector output produced by this elemental component is

$$\Delta N_0 = \frac{(b 2\pi A_n f)^2}{2 A_c^2} \quad \text{watts} \tag{4.42}$$

but

$$A_n = (2\eta \Delta f)^{1/2}$$

i.e.

$$\Delta N_0 = \left(\frac{2\pi b}{A_c}\right)^2 \eta f^2 \Delta f$$

Hence

$$\frac{dN_0}{df} = Kf^2 \quad \text{where} \quad K = \eta\left(\frac{2\pi b}{A_c}\right)^2 \tag{4.43}$$

In Eqn (4.43) dN_0/df represents the power spectral density of the noise at the detector output and is no longer white but proportional to f^2. This means that the noise power at the output of an ideal FM detector increases with the square of the frequency difference between the centre frequency and the elemental noise frequency. The noise voltage spectral density $(dN_0/df)^{1/2}$ is plotted as a function of f in Fig. 4.9.

The noise at the output of a frequency modulation detector will be produced by the sum of all components within the passband of the filter that follows the detector. If we assume that this filter has an ideal low-pass characteristic with a cut-off frequency of $\pm f_0$ (the negative figure is required for negative values of $f = (f_k - f_c)$), then since the noise produces narrowband FM we can determine the total noise using superposition. The total noise

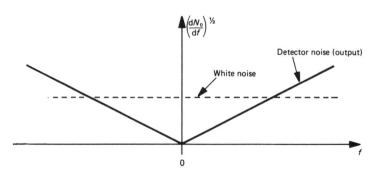

Fig. 4.9 FM detector noise spectral density.

power is

$$N_0 = \int_{-f_0}^{f_0} dN_0$$

i.e.

$$N_0 = 2K \int_0^{f_0} f^2 df$$

which evaluates to

$$N_0 = \eta \left(\frac{2\pi b}{A_c}\right)^2 \tfrac{2}{3} f_0^3 \tag{4.44}$$

The signal power at the detector output is given by Eqn (4.36). Thus the SNR is

$$\text{SNR}_{\text{out}} = \frac{3S_c}{2\eta f_0} \left(\frac{\Delta f_c}{f_0}\right)^2 \tag{4.45}$$

It should be noted that $\Delta f_c / f_0 \neq \beta$ unless $f_0 = f_m$, which is not necessarily the case.

Equation (4.45) is interpreted as stating that the SNR at the output of a frequency modulation detector increases with the square of the carrier deviation, which is independent of the carrier power. It might be argued that the SNR could be increased indefinitely simply by increasing the carrier deviation. This overlooks the fact that increasing Δf_c increases the signal bandwidth with a consequent increase in the noise power, which is itself directly proportional to bandwidth. As Δf_c increases a threshold is reached at which the assumption $y(t) \ll A_c$ is no longer valid. Beyond this threshold a very rapid fall in SNR at the detector output is witnessed. This threshold effect is a characteristic of all wideband systems and is considered in more detail in Section 4.11.

The relative performance of AM and FM systems in the presence of noise can be compared by reference to Eqn (4.45). In this equation, f_0 is the bandwidth of the filter following the detector and will be equal to the bandwidth of the modulating signal. The factor $2\eta f_0$ is thus the noise power that would occur at the output of an AM system with the same modulating signal bandwidth. If we compare FM and AM transmissions in which the AM carrier power equals the FM carrier power, the factor $S_c/2\eta f_0$ of Eqn (4.45) is equal to the output SNR for an AM system when the depth of modulation $m = 100\%$, i.e.

$$\text{SNR}_{\text{out(FM)}} = \text{SNR}_{\text{out(AM)}} 3(\Delta f_c / f_0)^2 \tag{4.46}$$

If we consider a typical FM commercial broadcast system in which $\Delta f_c = 75\,\text{kHz}$ and $f_0 = 15\,\text{kHz}$, then

$$\text{SNR}_{\text{out(FM)}} = 75\,\text{SNR}_{\text{out(AM)}}$$

In fact, in commercial broadcast AM systems the depth of modulation is usually restricted to 30% and the total transmitter power is $(1 + \tfrac{1}{2}m^2)$ times the carrier power. Hence comparing FM and AM on the basis of total transmitted power,

$$\text{SNR}_{\text{out(FM)}} = \frac{3(1 + \frac{1}{2}m^2)}{m^2}\left(\frac{\Delta f}{f_0}\right)^2 \text{SNR}_{\text{out(AM)}} \qquad (4.47)$$

If

$$m = 0.3, \Delta f = 75\,\text{kHz and } f_0 = 15\,\text{kHz},$$

then

$$\text{SNR}_{\text{out(FM)}} = 810.83\,\text{SNR}_{\text{out(AM)}}$$

Alternatively, to produce the same SNR the transmitted power in the FM case is 29 dB less than the required power in the AM case.

One conclusion we may draw from these figures is that for a given radiated power a FM transmitter will have a greater range than a DSB-AM transmitter, provided that the SNR at the FM detector input is sufficiently high for the noise to produce narrowband frequency modulation of the carrier. In other words, the FM system must be operating above the threshold level. This threshold level is related to both the frequency deviation and the SNR at the detector input. For large values of Δf_c the threshold occurs at a SNR of about 13 dB. The conditions required for FM to exhibit its SNR improvement properties are that $\beta > 1/\sqrt{3}$ [assuming $f_0 = f_m$ in Eqn (4.46)] and the SNR at the detector input must exceed 13 dB. If the SNR at the detector input is below the threshold value, the output SNR decreases rapidly and ultimately becomes poorer than the equivalent AM value.

4.10 PRE-EMPHASIS AND DE-EMPHASIS

The analysis of the previous section was based on a single tone modulating signal and we assumed that irrespective of the frequency of this tone the carrier deviation had its maximum value Δf_c. In other words, we assumed that the FM signal occupied the maximum possible bandwidth for a given modulating signal. In reality the situation is somewhat different; the modulating signal is not a single tone but a complex signal with a particular power spectral density.

The power spectral density of natural speech is closely approximated by the graph shown in Fig. 4.10. We can see from this figure that above a certain frequency f_1 the power spectrum decreases at a rate approaching 6 dB/ octave. If such a signal frequency modulates a carrier, the higher-frequency components will produce a lower carrier deviation than the low-frequency components. In other words, the bandwidth occupied by a carrier frequency modulated by a signal of this type will be considerably less than if the power spectrum of the modulating signal were uniformly distributed. In practice, therefore, the FM system would not produce the maximum SNR improvement suggested by Eqn (4.46).

The theoretical SNR can be approached, however, if the power spectrum of

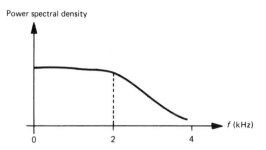

Fig. 4.10 Power spectrum of natural speech.

the modulating signal is made uniform by emphasizing the higher-frequency components before modulation takes place. The spectrum of the modulating signal is restored after detection by applying a corresponding amount of the de-emphasis. The de-emphasis will, of course, operate on the noise produced by the detector as well as the signal, the overall effect being a reduction in the noise power at the detector output. The process of pre-emphasis and de-emphasis is illustrated graphically in Fig. 4.11.

The actual component values used in pre- and de-emphasis networks will vary from circuit to circuit. The amount of pre- and de-emphasis will depend upon the time constant of the filter used, and it is normal to express the pre- and de-emphasis in terms of this time constant. In the UK the value of the time constant used is $50\,\mu s$; in contrast the value used in the USA is $75\,\mu s$.

It is worth stating that the technique of pre- and de-emphasis does not produce SNR increases above the theoretical values specified by Eqn (4.46). The technique is one of conditioning the modulating signal in order that the

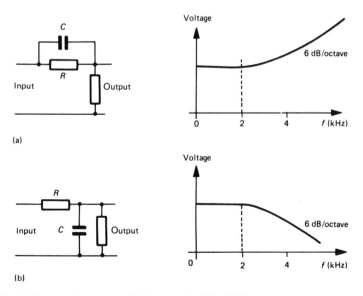

Fig. 4.11 (a) Pre-emphasis and (b) de-emphasis in FM.

maximum theoretical performance can be realized. It is shown by Schwartz[2] that the technique of pre- and de-emphasis is not restricted to FM but can be used in any system in which the power spectral density of the signal decreases more rapidly, as a function of frequency, than the noise power spectral density. In this context, pre- and de-emphasis could be used with AM systems to increase the overall SNR. It is not used in practice with AM system because it is actually advantageous, from the point of view of adjacent channel separation, to have a signal power spectrum that decreases with frequency.

4.11 THE FM CAPTURE (THRESHOLD) EFFECT

Consider the phasor diagram of Fig. 4.12(a) in which X represents the required FM carrier and Y represents an interfering FM carrier. (Y could be due to a second FM station at the receiver image frequency, or Y could represent the frequency modulated carrier produced by noise entering the detector.) It is assumed that the amplitudes X and Y are fixed and that the angle between the phasor ϕ_y is uniformly distributed in the range $-\pi \leqslant \phi_y \leqslant +\pi$. It is required to find the resultant phase angle ϕ_r which will be responsible for any interference generated at the detector output.

From Fig. 4.12(a)

$$\phi_r = \tan^{-1}\left[\frac{Y\sin(\phi_y)}{X + Y\cos(\phi_y)}\right] \tag{4.48}$$

ϕ_r will be a random quantity with mean square value

$$\overline{(\phi_r)^2} = \frac{1}{2\pi}\int_{-\pi}^{+\pi}\left\{\tan^{-1}\left[\frac{Y\sin(\phi_y)}{X + Y\cos(\phi_y)}\right]\right\}^2 d\phi_y \tag{4.49}$$

Equation 4.49 can be solved numerically and the rms value of ϕ_r is plotted against the ratio X/Y in Fig. 4.12(b). It will be noted from this figure that

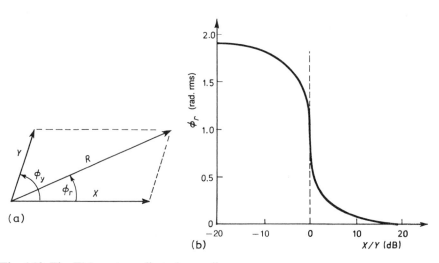

(a)

(b)

Fig. 4.12 The FM capture effect phasor diagram.

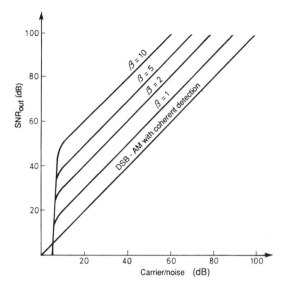

Fig. 4.13 Noise improvement obtained with FM.

a very rapid change in $\sqrt{(\phi_r)^2}$ occurs in the region $X = Y$. When $X/Y = 3$ the phase angle ϕ_r becomes very small. This means that the interference component produced by Y (which is proportional to $d\phi_r/dt$) also becomes very small. Hence when the ratio $X/Y > 3$ the carrier X takes over, or captures, the system. The FM receiver will thus discriminate in favour of the stronger signal. This is also true when $X/Y < 1$. In the latter case component Y captures the system. Hence if Y is due to noise, a very rapid deterioration in performance is observed. The FM threshold effect is clearly demonstrated in Fig. 4.13 which plots the SNR at the detector output as a function of the SNR at the detector input.

The FM capture effect is actually very useful and is employed in mobile cellular radio (see Fig. 13.21) to suppress interference from base stations in co-channel cells operating on the same frequency.

4.12 CONCLUSION

In this chapter we have compared several analogue transmission systems from a SNR point of view. It may be concluded that wideband systems, such as FM, produce SNR improvements, but not to the degree specified by Shannon's law. (This conclusion is also valid if narrowband pulse amplitude modulation (PAM) is compared with wideband pulse frequency modulation. We have omitted a detailed study of analogue pulse modulation because this has been largely overtaken by digital techniques.) If SNR was the sole figure of merit of a telecommunications system, it would be reasonable to conclude that wide-band systems can offer superior performance to narrow-band systems. There are, however, many other factors that influence the choice of modulation system. For example, the wide bandwidth of FM precludes its use in the medium-wave broadcast band because here the main requirement is to accommodate as many individual stations as possible in a relatively small bandwidth. In any case SNR problems are not usually significant in high-power broadcast transmissions.

There are many instances where the lower bandwidth of SSB is used in preference to DSB-AM. One such instance is the frequency multiplexing of telephone circuits for trunk transmission. In such a system it is possible to transmit a synchronizing signal from which the individual carriers required for coherent demultiplexing can be derived.

Traditionally medium-wave (i.e. local) broadcasts have used DSB-AM although bandwidth is at a premium. Historically DSB receivers were cheaper to produce and more reliable than SSB receivers. Modern technology has completely changed the situation where complexity and reliability are no longer closely related to cost. It is now perfectly possible, both technically and economically, to mass-produce reliable SSB receivers. It is unlikely that any moves will be made in this direction for some time because of the problems of compatibility, i.e. existing DSB receivers, of which there are many millions, would be unable to receive acceptable quality SSB transmission.

Thus, as we have stated, SNR is but one of many factors that influence the choice of a particular transmission system.

REFERENCES

1. King, R., *Electrical Noise*, Chapman & Hall, London, 1966, Chapter 3.
2. Schwartz, M., *Information Transmission Modulation and Noise*, 3rd edn. McGraw-Hill, New York, 1980, p. 412.

PROBLEMS

4.1 Calculate the mean square output noise voltage when a signal generator with an output resistance of $600\,\Omega$ is connected to the input of the two-port network shown.

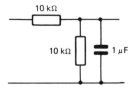

Answer: $4.002 \times 10^{-15}\,\mathrm{V}^2$.

4.2 An amplifier is made up of three identical stages in cascade, each stage having equal input and output resistances. The power gain per stage is 8 dB and the noise figure per stage is 6 dB when the amplifiers are correctly matched. Calculate the overall power gain and noise figure for the cascaded amplifier.

Answer: 24 dB; 6.6 dB.

4.3 The noise figure of a receiver, relative to a matched source at a temperature of 290 K, is 0.9 dB. Calculate the effective noise temperature at the input of the receiver when an antenna of effective noise temperature 200 K is connected.

Answer: 266.6 K.

4.4 The following diagram represents a satellite receiving system coupled by a waveguide to an antenna of effective noise temperature 70 K.

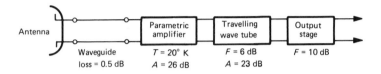

(a) Calculate the equivalent noise temperature of the waveguide and travelling wave tube (a device for amplifying frequencies in the gigahertz range).
(b) Calculate the SNR, at the output of the receiver assuming the antenna radiation resistance to be 50 Ω and the available received power is 10 pW. The bandwidth of the system is 10 MHz. (Hint: find the equivalent SNR referred to the waveguide input.)

Answer: (a) 35 K, 870 K; (b) 27.5 dB.

4.5 A superheterodyne receiver is connected to an antenna with a noise temperature of 100 K by a coaxial feeder having a loss of 2 dB. The receiver characteristics are

RF bandwidth = 5 MHz
IF = 20 MHz
IF bandwidth = 1 MHz
noise figure = 4 dB

Calculate the total system noise temperature and the required signal power delivered by the antenna to give a SNR of 20 dB at the output of the IF stage. Assume the antenna and receiver are both matched to the co-axial line.

Answer: 964 K; 1.33 pW.

4.6 A single tone of amplitude 2 V rms and frequency 5.8 kHz is used to amplitude modulate a carrier of amplitude 5 V rms, the carrier and both sidebands being transmitted. Given that the noise spectral density at the detector input is 0.1 μW/Hz, find the SNR at the detector output. The audio bandwidth of the receiver is 10 kHz and the carrier amplitude at the detector input is 1 V rms.

Answer: 16 dB.

4.7 A vhf transmitter radiates a DSB-AM signal at a depth of modulation of 45% with an audio bandwidth of 15 kHz. This produces a SNR of 40 dB at the output of a receiver at a distance of 3 km from the transmitter. If the transmitter, is switched to FM radiating the same total power, at a carrier deviation of 60 kHz, find the theoretical distance from the transmitter for the same SNR at the output of a FM receiver. Assume the noise spectral density at the receiver input is the same in each case and that the received power decreases as the square of the distance from the transmitter.

Answer: 48.5 km.

4.8 A radio station transmits a DSB-SC-AM signal with a mean power of 1 kW. If SSB-AM is used instead calculate the mean power for (a) the same signal strength and (b) the same SNR at the detector output.

A single tone of frequency 7.5 kHz forms the modulating signal for both a DSB-AM and a FM system, the power transmitted in each case being the same. When modulated, the peak deviation of the FM carrier is 60 kHz and the amplitude of the first pair of FM sidebands is equal to the sideband amplitude of the AM transmission. Assuming an audio bandwidth of 7.5 kHz for both AM and FM receivers, determine the SNR advantage of the FM receiver. It may be assumed that the noise spectral density has a constant value and is the same for each case.

Answer: (a) 2 kW, (b) 1 kW; 29.4 dB.

4.9 A frequency modulation receiver consists of a tuned amplifier of bandwidth 225 kHz that feeds a limiter and an ideal discriminator followed by a low-pass filter with a bandwidth of 10 kHz. The carrier-to-noise ratio at the discriminator input is 40 dB when the modulating signal is a 10 kHz tone, producing a carrier deviation of 5 kHz. Calculate the SNR at the output of the filter.

If the amplitude of the modulating signal is maintained at the same value when the frequency is changed to 1 kHz, find the new SNR at the filter output. What would be the SNR if the amplitude of the modulating signal is halved?

Answer: 69 dB; 69 dB; 63 dB.

4.10 The SNR at the output of a coherent detector is 25 dB when the input is a SSB-AM wave. If the input to the detector is transferred to DSB-AM with $m = 1$, find the increase in total power at the detector input to maintain an output SNR of 25 dB.

Answer: 4.8 dB.

Noise in digital communications systems

5

The primary interest in the study of analogue communications systems is the obscuring or masking of the transmitted signal by additive noise. This effect is most conveniently analysed by considering the spectral properties of the noise waveform. The situation in digital systems is quite different since only fixed signal levels are allowed. When these levels are obscured by additive noise the receiver is required to decide which of the allowed levels the noisy signal represents. If the receiver decides correctly the noise has no effect on the received signal whatever. If the receiver makes an incorrect decision the results can be catastrophic. Decision theory is based upon the statistical rather than the spectral properties of noise although, as one might expect, these properties are related. The most important statistical property of white noise, with respect to decision theory, is its amplitude distribution function.

5.1 THE AMPLITUDE DISTRIBUTION FUNCTION OF WHITE NOISE

White noise is a naturally occurring phenomenon produced by the superposition of many randomly occurring events. Consequently it is not possible to specify the instantaneous amplitude of the noise waveform at any given instant. The alternative is to determine the probability that the noise waveform amplitude will exceed a given value. This is possible and requires a knowledge of the amplitude distribution function of the noise waveform. Suppose that n independent samples are taken of a noise waveform. (In order for the samples to be statistically independent, these samples must be spaced in time by an interval not less than $1/B$ seconds, where B is the noise bandwidth.) Suppose also that n_v of these samples have amplitudes between v_1 and v_2, i.e. the total number of samples in the range $\Delta v = (v_1 - v_2)$ is n_v. The probability that any noise sample will be in this range is simply n_v/n. If the interval $\Delta v \to 0$, a continuous distribution results. However, as this limit is approached, the probability that any sample will be in the range Δv also approaches zero. The problem is avoided if we represent probability as an area. The probability that a sample lies within the amplitude range Δv is represented as the area of a rectangle of base width Δv. The area of the rectangle thus has the value n_v/n and its height is $n_v/n\Delta v$ which approaches a finite value as $\Delta v \to 0$ and $n \to \infty$. If the height of the rectangle is plotted for all values of Δv the resulting curve is termed the probability density function.

The probability that a sample will be within the range v_1 to v_2 is the area under the probability density function between these two limits. The probability density function $p_v(v)$ is formally defined by

$$p_v(v) = \lim_{\substack{\Delta v \to 0 \\ n \to \infty}} \left[\frac{n_0}{n\Delta v} \right]$$

Referring to Fig. 5.1, the probability that v has a value between v_1 and v_2 is

$$P = \int_{v_1}^{v_2} p_v(v)\,dv \tag{5.1}$$

The central limit theorem is an important theorem in statistics which states that the probability density function (PDF) of the sum of many random variables approaches a Gaussian distribution regardless of the distributions of the individual variables. The PDF (also known as the amplitude distribution function) of white noise is therefore Gaussian and is given by

$$p_v(v) = \frac{\exp[-(v-a)^2/2\sigma^2]}{\sqrt{(2\pi\sigma^2)}} \tag{5.2}$$

In Eqn (5.2) a is known as the mean or expected value of v and σ is the standard deviation. The Gaussian PDF has a characteristic bell shape with a width proportional to σ. The mean value is given by

$$a = \int_{-\infty}^{\infty} v p_v(v)\,dv \tag{5.3}$$

and the variance (σ^2) is given by

$$\sigma^2 = \int_{-\infty}^{\infty} (v-a)^2 p_v(v)\,dv \tag{5.4}$$

The total area under the $p_v(v)$ curve must be unity, i.e. the probability that a noise sample has an amplitude between $-\infty$ and $+\infty$ is 1, hence

$$\int_{-\infty}^{\infty} p_v(v)\,dv = 1 \tag{5.5}$$

This gives the Gaussian curve a peak value of $1/\sqrt{(2\pi\sigma^2)}$ at a value of $v = a$ (Fig. 5.2).

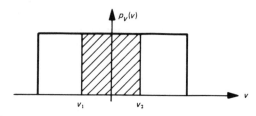

Fig. 5.1 Uniform PDF.

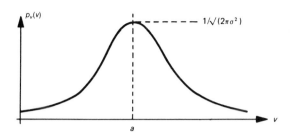

Fig. 5.2 Gaussian PDF.

The mean value of white noise is zero; hence the variance may be interpreted as the mean square value. Alternatively σ = rms noise voltage.

A second distribution function, known as the cumulative distribution function, is defined by Eqn (5.6):

$$F_v(v) = \int_{-\infty}^{v} p_v(v)\,dv \tag{5.6}$$

i.e.

$$F_v(v) = \int_{-\infty}^{v} \frac{\exp[-(v-a)^2/2\sigma^2]}{\sqrt{(2\pi\sigma^2)}}\,dv \tag{5.7}$$

$F_v(v)$ is the probability that the noise will be less than some value v. Since $p_v(v)$ is symmetrical about $v = a$ it follows that $F_v(v) = 0.5$ when $v = a$. As shown in Fig. 5.3 the cumulative density function for white noise is symmetrical about $v = 0$.

The probability that the noise amplitude is less than some value $K\sigma$ is

$$\text{prob}(-K\sigma \leqslant v \leqslant K\sigma) = \int_{-K\sigma}^{K\sigma} \frac{\exp(-v^2/2\sigma^2)}{\sqrt{(2\pi\sigma^2)}}\,dv \tag{5.8}$$

Substituting $y = v/\sqrt{(2\sigma^2)}$ and noting that the Gaussian function is symmetrical about $v = 0$ gives

$$\text{prob}(-K\sigma \leqslant v \leqslant K\sigma) = \frac{2}{\sqrt{\pi}} \int_{0}^{K/\sqrt{2}} \exp(-y^2)\,dy \tag{5.9}$$

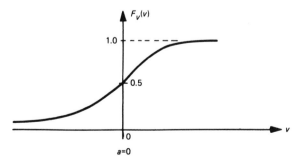

Fig. 5.3 Cumulative density function for white noise with zero mean.

This integral can be evaluated numerically for values of K and is known as the error function, i.e.

$$\text{erf}(P) = \frac{2}{\sqrt{\pi}} \int_0^P \exp(-y^2)\,dy \tag{5.10}$$

The error function is tabulated in Appendix C. The probability that v will exceed $K\sigma$ is

$$1 - \frac{2}{\sqrt{\pi}} \int_0^{K/\sqrt{2}} \exp(-y^2)\,dy$$

The complementary error function is defined as

$$\text{erfc}(P) = 1 - \frac{2}{\sqrt{\pi}} \int_0^P \exp(-y^2)\,dy \tag{5.11}$$

Hence the probability that the noise voltage will be less than its rms voltage is, from Eqn (5.9),

$$\text{prob}(-\sigma \leqslant v \leqslant \sigma) = \text{erf}(1/\sqrt{2}) = 0.683$$

Alternatively, the probability that the noise voltage will exceed its rms value is $\text{erfc}(1/\sqrt{2}) = 0.317$. Equations (5.9) and (5.11) are of fundamental importance in determining the probability of decision errors in digital communications systems.

5.2 STATISTICAL DECISION THEORY

This topic deals with the problem of developing statistical tests to determine which of M possible signals was transmitted over a communications link. In the case of binary communications $M = 2$, i.e. it is possible to transmit either a binary 0 or a binary 1. The receiver takes a single sample during each bit interval and then decides that 0 was transmitted or that 1 was transmitted. There are clearly two types of error that can occur, i.e. deciding 1 when 0 was transmitted or deciding 0 when 1 was transmitted. If the total probability of error is to be minimized, the minimization must be based on both types of error.

The decision rule is based on splitting all values of the received voltage v into two regions V_0 and V_1. The boundary between these regions is then chosen to minimize the total error probability. To illustrate this procedure we will assume that P_0 represents the probability of transmitting 0 and P_1 represents the probability of transmitting 1, clearly $P_0 + P_1 = 1$.

The probability that v will fall into region V_0 when a 1 is transmitted is

$$\int_{V_0} p_1(v)\,dv$$

where $p_1(v)$ is the probability density function of the received voltage when a 1 is transmitted. The probability that v will fall into the region V_1 when a 0 is transmitted is

$$\int_{V_1} p_0(v)dv$$

where $p_0(v)$ is the probability density function of the received voltage when a 0 is transmitted. The overall probability of error is then

$$P_e = P_0 \int_{V_1} p_0(v)dv + P_1 \int_{V_0} p_1(v)dv \qquad (5.12)$$

The region $V_0 + V_1$ covers all values of v; hence

$$\int_{V_0+V_1} p_1(v)dv = \int_{V_0} p_1(v)dv + \int_{V_1} p_1(v)dv = 1$$

Hence we can eliminate the integral over V_0 from Eqn (5.12) to give

$$P_e = P_1 + \int_{V_1} [P_0 p_0(v) - P_1 p_1(v)]dv \qquad (5.13)$$

If P_1 is known then P_e can be minimized by making the integral of Eqn (5.13) negative and as large as possible, i.e.

$$P_1 p_1(v) > P_0 p_0(v)$$

The required decision rule is

$$\frac{p_1(v)}{p_0(v)} > \frac{P_0}{P_1} \qquad (5.14)$$

$\lambda = p_1(v)/p_0(v)$ is known as the likelihood ratio. In the binary case, if 0 is represented by zero volts and 1 is represented by A volts and the noise is Gaussian, the two density functions will be

$$p_0(v) = \frac{1}{(2\pi\sigma^2)^{1/2}} \exp[-(v)^2/2\sigma^2] \qquad (5.15)$$

$$p_1(v) = \frac{1}{(2\pi\sigma^2)^{1/2}} \exp[-(v-A)^2/2\sigma^2]$$

The region V_1 is defined by all values of v for which $\lambda > P_0/P_1$, i.e.

$$\frac{\exp[-(v-A)^2/2\sigma^2]}{\exp[-v^2/2\sigma^2]} > \frac{P_0}{P_1} \qquad (5.16)$$

Taking logarithms of Eqn (5.16)

$$v^2 - (v-A)^2 > 2\sigma^2 \ln P_0/P_1$$

The desired region V_1 is thus defined by all values of v corresponding to

$$v > \frac{A}{2} + \frac{\sigma^2}{A} \ln \frac{P_0}{P_1} \qquad (5.17)$$

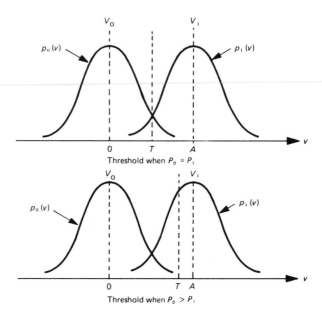

Fig. 5.4 Determination of decision thresholds.

or the boundary between the regions V_0 and V_1 is

$$T = \frac{A}{2} + \frac{\sigma^2}{A} \ln \frac{P_0}{P_1} \tag{5.18}$$

If binary 1 and 0 are equi-probable the decision threshold becomes $T = A/2$ which is exactly half-way between the voltage levels representing 1 and 0. If 0 is transmitted more often than 1 the threshold moves towards V_1. This is illustrated in Fig. 5.4.

Once the decision threshold has been established the total probability of error can then be calculated. If binary 0 (0 volts) is sent, the probability that it will be received as a 1 is the probability that the noise will exceed $+ A/2$ volts (assuming $P_0 = P_1$) i.e.

$$P_{e0} = \int_{A/2}^{\infty} \frac{\exp(-v^2/2\sigma^2)}{\sqrt{(2\pi\sigma^2)}} \, dv \tag{5.19}$$

If a 1 is sent (A volts) the probability that it will be received as a 0 is the probability that the noise voltage will be between $- A/2$ and $- \infty$, i.e.

$$P_{e1} = \int_{-\infty}^{-A/2} \frac{\exp(-v^2/2\sigma^2)}{\sqrt{(2\pi\sigma^2)}} \, dv \tag{5.20}$$

These types of error are mutually exclusive, since sending a 0 precludes sending a 1. Because of the symmetry of the Gaussian function $P_{e0} = P_{e1}$, and the total probability of error, which is $P_0 P_{e0} + P_1 P_{e1}$, can then be written

$$P_e = P_{e1}(P_0 + P_1) = P_{e1}$$

i.e.

$$P_e = \int_{-\infty}^{-A/2} \frac{\exp(-v^2/2\sigma^2)}{\sqrt{(2\pi\sigma^2)}} \, dv \qquad (5.21)$$

This equation can be written in terms of two integrals

$$P_e = \int_{-\infty}^{0} \frac{\exp(-v^2/2\sigma^2)}{\sqrt{(2\pi\sigma^2)}} \, dv - \int_{-A/2}^{0} \frac{\exp(-v^2/2\sigma^2)}{\sqrt{(2\pi\sigma^2)}} \, dv \qquad (5.22)$$

Using the fact that the Gaussian distribution is symmetrical about its mean value, Eqn (5.22) reduces to

$$P_e = \frac{1}{2} + \int_{0}^{-A/2} \frac{\exp(-v^2/2\sigma^2)}{\sqrt{(2\pi\sigma^2)}} \, dv$$

If we substitute $y = v/\sqrt{(2\sigma^2)}$ this becomes

$$P_e = \frac{1}{2} + \frac{1}{\sqrt{\pi}} \int_{0}^{A/(2\sqrt{2}\sigma)} \exp(-y^2) \, dy$$

i.e.

$$P_e = \tfrac{1}{2}\{1 - \operatorname{erf}[A/(2\sqrt{2}\sigma)]\} \qquad (5.23)$$

The error probability therefore depends solely on the ratio of the peak pulse voltage A to the rms noise voltage σ. A graph of error probability against A/σ is plotted in Fig. 5.5.

This figures shows that when $A/\sigma = 17.4\,\mathrm{dB}$ (a voltage ratio of 7.4:1), $P_e = 10^{-4}$, i.e. on average 1 digit in 10^4 will be in error. If A/σ is increased to 21 dB the error probability drops to $P_e = 10^{-8}$. Hence a very large decrease in error probability occurs for an increase in A/σ of only 3.6 dB. This decrease is much smaller for values of A/σ less than 14 dB. The characteristic thus exhibits a threshold effect for values of A/σ around 18 dB. An error probability of

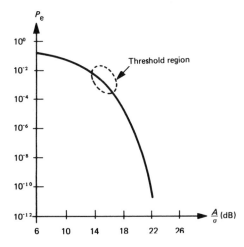

Fig. 5.5 Error probability in binary transmission.

about 10^{-5} is usually considered acceptable for practical data communications systems. This corresponds to a pulse amplitude of about ten times the rms noise voltage.

5.3 DECISION ERRORS IN PCM

In Section 3.5 we made a somewhat simplistic assumption that the only source of signal distortion in PCM systems was due to quantization noise. This is only true if the probability of error during transmission is zero. In any PCM system decision errors will occur with a low but finite probability at each regenerative repeater. We will examine the effect of errors in a linearly quantized PCM system with a word length of 8 digits, i.e. 256 levels.

If an error occurs in the least significant digit, this will result in an error equal to the quantization step size δV after the 8 digit code word is converted to an analogue signal (see Fig. 3.11). If the next most significant digit is in error, the voltage error of the analogue signal will be $2\delta V$. An error in the rth digit will produce a voltage error of $\delta V 2^{r-1}$ volts. If we assume that only 1 digit in the 8 digit code is in error (a reasonable assumption if $P_e = 10^{-5}$) the mean square voltage error will be

$$\sum_{r=1}^{8} \frac{(\delta V 2^{r-1})^2}{8} = W \qquad (5.24)$$

This is the mean 'noise power' associated with each error. Since the error probability is P_e, the mean noise power at the output of a PCM link due to decision errors is

$$W \times P_e = \frac{1}{8} \sum_{r=1}^{8} P_e/(\delta V 2^{r-1})^2 \text{ watts.}$$

This noise will of course be added to the quantization noise already present on the PCM signal. In the case of PCM an error in the most significant digit is much more serious than an error in the least significant digit. In the transmission of high fidelity sound signals by PCM, the most significant digit is encoded in such a way that errors can be detected, and thus corrected before the digital signal is decoded.

5.4 DECISION ERRORS IN CARRIER-MODULATED DATA SIGNALS

In this section we will examine the relative performance of ASK, PSK and FSK in noisy conditions. Unlike the comparison of the analogue modulation systems of Chapter 4, which was done on an SNR basis, we compare modulation methods for digital systems in terms of probability of error. All three modulation systems can, in fact, be detected using coherent detection, although it is more usual to use envelope detection for ASK and FSK. It is convenient, initially, to compare the performance of all three systems assuming coherent detection, because the coherent detector does not alter the PDF of the noise signal.

We consider first of all coherent detection of ASK. The input to the detector will be

$$V_{in}(t) = h(t)\cos(2\pi f_c t) + x(t)\cos(2\pi f_c t) + y(t)\sin(2\pi f_c t) \qquad (5.25)$$

Where $h(t)$ is a binary function with value A or 0 and $[x(t)\cos(2\pi f_c t) + y(t)\sin(2\pi f_c t)]$ represents the band-limited white noise. It was noted in Section 4.4 that both $x(t)$ and $y(t)$ are random variables. We can apply the central limit theorem to Eqn (4.16a) which defines $x(t)$ as the sum of many sinusoidal components with random frequency and phase. The result is that the PDF of both $x(t)$ and $y(t)$ is Gaussian. If the ASK signal plus noise is coherently detected (i.e. multiplied by $\cos[2\pi f_c t]$ and filtered), the baseband output will be

$$V_{out}(t) = \tfrac{1}{2}h(t) + \tfrac{1}{2}x(t) \qquad (5.26)$$

This is similar to Eqn (4.27) which describes the coherent detection of DSB-AM. Ignoring the factor of $\tfrac{1}{2}$ in Eqn (5.26) the output of the coherent detector is

$$V_{out} = (A \text{ or } 0) + x(t) \qquad (5.27)$$

If binary 1 and 0 are equi-probable and σ^2 represents the variance of $x(t)$ the probability of error, assuming a decision threshold of $A/2$, is

$$P_{e(ASK)} = \tfrac{1}{2}\{1 - \text{erf}\,[A/(2\sqrt{2}\sigma)]\} \qquad (5.28)$$

which is identical to Eqn (5.23) for baseband signalling.

For PSK the value of $h(t)$ in Eqn (5.26) will be $\pm A$ and the decision threshold will be 0 volts. The error probability is then

$$P_{e(PSK)} = \tfrac{1}{2}[1 - \text{erf}(A/\sqrt{2}\sigma)] \qquad (5.29)$$

The peak signal-to-noise power ratio for PSK is 6 dB less than for ASK for a given error rate. If we bear in mind that ASK transmits no signal at all for binary 0(i.e. there is zero signal for half the time) it is evident that to produce the same error rate PSK requires a mean signal-to-noise ratio 3 dB below the value required for ASK. In other words, PSK has a 3 dB advantage, in terms of signal-to-noise power ratio, over ASK and, as is seen from Fig. 5.6, this can represent a significant improvement in terms of error probability.

FSK modems usually employ envelope detection and the block schematic of such a modem is given as Fig. 3.27. A similar schematic for a FSK modem employing coherent detection is presented in Fig. 5.7. The pre-detection filters, which are assumed to have an ideal bandpass characteristic, determine the bandwidth and hence the input noise power for each coherent detector. The input voltage at each detector will be

$$V_1(t) = h_1(t)\cos(2\pi f_1 t) + x_1(t)\cos(2\pi f_1 t) + y_1(t)\sin(2\pi f_1 t) \quad (5.30a)$$

and

$$V_0(t) = h_0(t)\cos(2\pi f_0 t) + x_0(t)\cos(2\pi f_0 t) + y_0(t)\sin(2\pi f_0 t) \quad (5.30b)$$

where $h_1(t)$ is the logical complement of $h_0(t)$ and $x_1(t)$ and $x_0(t)$ are independent random variables, provided that the passband of each predetection filter does not overlap that of the other.

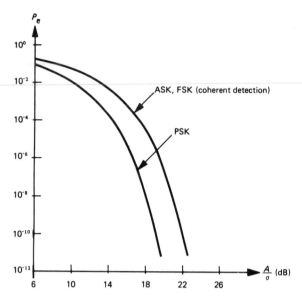

Fig. 5.6 Error probabilities for constant mean signal power.

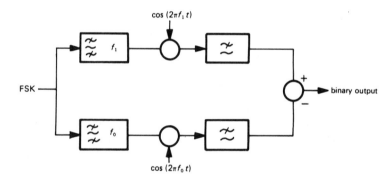

Fig. 5.7 Coherent FSK modem.

The output of the composite detector will be

$$V_{\text{out}}(t) = (+A \text{ or } -A) + [x_1(t) - x_0(t)] \tag{5.31}$$

Since $x_1(t)$ and $x_0(t)$ are independent Gaussian variables, the variable $[x_1(t) - x_0(t)]$ will also be Gaussian with a variance equal to the sum of the original variances. (This might seem to be a surprising result, but remember that if we substract two sinusoids of **different** frequencies the power of the resultant waveform is the sum of the powers in the individual sinusoids.) The noise variance at the detector output will thus be $2\sigma^2$, which gives an effective peak signal-to-rms noise ratio of $2A/\sqrt{2}\sigma$. The probability of error in this case is then

$$P_{\text{e(FSK)}} = \tfrac{1}{2}[1 - \text{erf}(A/2\sigma)] \tag{5.32}$$

This means that if FSK and PSK transmissions are to have the same error probability the FSK transmission requires a SNR which is 3 dB above the value required for PSK.

Thus, compared on a peak power basis, for the same error probability PSK requires an SNR that is 3 dB below that of FSK which in turn requires a SNR which is 3 dB below that of ASK. Compared on a mean power basis the performance of ASK and FSK are identical, but the same error rate can be obtained with PSK with a mean SNR which is 3 dB lower. A comparison of the performance of these three systems is given in Fig. 5.6, where it has been assumed that coherent detection has been used throughout. In practice it is more likely that envelope detection would be used for ASK and FSK. The analysis for envelope detection is slightly different because the envelope detector modifies the statistical properties of the noise. The envelope detector output voltage is proportional to the instantaneous carrier envelope, which is equal to the phasor sum of the inphase and quadrature components of Eqn (5.25). The instantaneous detector output is thus

$$V_{out} = k\left[\{h(t) + x(t)\}^2 + \{y(t)^2\}\right]^{1/2} \tag{5.33}$$

To calculate the error probability we need to know the PDF of Eqn (5.33) when $h(t) = 0$ (i.e. binary 0) and when $h(t) = A$ (binary 1). In the former case the density function has a Rayleigh distribution given by

$$p_0(v) = (v/\sigma^2)\exp(-v^2/2\sigma^2) \quad \text{for} \quad v \geqslant 0$$

and in the latter case the density function has a Rician distribution given by

$$p_1(v) = (v/\sigma^2)\exp[-(v^2 + A^2)/2\sigma^2]I_0(vA/\sigma^2) \quad \text{for} \quad v \leqslant 0$$

$I_0(x)$ is the modified Bessel function of zero order. The Rician distribution approaches the Gaussian distribution for $A/\sigma \gg 1$. The envelope probability density functions are illustrated in Fig. 5.8.

It has been shown[1] that the decision threshold for the envelope detector is

$$T = A^2/2\sigma^2 + \ln(P_0/P_1) \tag{5.34}$$

This reference also shows that for high SNRs envelope detection is marginally poorer for both ASK and FSK than coherent detection. For lower SNRs there is a significant deterioration in the performance of the envelope detection receiver. This, of course, is in agreement with the results for analogue transmission considered in Chapter 4.

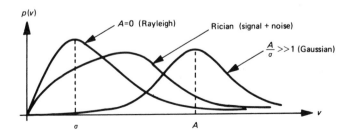

Fig. 5.8 Envelope distribution for signal + noise.

5.5 MATCHED FILTERING AND CORRELATION DETECTION

In all digital communication systems, baseband or carrier modulated, the probability of error ultimately depends upon the SNR. (This, of course, ignores the influence of intersymbol interference, which we considered in Section 3.7.) Thus it is reasonable to consider techniques that may be used to maximize the ratio A/σ. Matched filtering is one such technique, which emphasizes the signal voltage relative to the noise. In so doing, considerable distortion of the signal waveform usually results. Hence where waveform fidelity is important (as in analogue systems) matched-filtering techniques are not applicable. This is not the case with digital transmission, where the receiver is required only to decide between signal and noise.

The matched filter is designed to maximize the SNR at a precise interval t_0 after the signal is applied to its input. We will consider a general voltage waveform $h(t)$ with Fourier transform $H(f)$ which is applied to a filter with frequency response $P(f)$. The amplitude spectrum of the output is given by $G(f) = P(f)H(f)$ and this has a Fourier transform

$$g(t) = \int_{-\infty}^{\infty} P(f)H(f)\exp(j2\pi ft)\,df \qquad (5.35)$$

The magnitude of this signal at time t_0 is

$$|g(t_0)| = \left| \int_{-\infty}^{\infty} P(f)H(f)\exp(j2\pi ft_0)\,df \right|$$

Hence the signal power (normalized) at time t_0 is

$$S_{\mathrm{p}} = |g(t_0)|^2 = \left| \int_{-\infty}^{\infty} P(f)H(f)\exp(j2\pi ft_0)\,df \right|^2 \qquad (5.36)$$

$P(f)$ is defined for both negative and positive frequencies and it is therefore necessary to use the double-sided power spectral density of $\eta/2$ watts/hertz for the white noise.

The noise at the filter output will no longer be white but will have a power spectrum given by

$$N(f) = \frac{\eta}{2}|P(f)|^2$$

The noise power at the filter output is thus

$$N_{\mathrm{p}} = (\eta/2)\int_{-\infty}^{\infty} |P(f)|^2 df \qquad (5.37)$$

The SNR at time t_0 is thus

$$\mathrm{SNR} = \frac{\left| \int_{-\infty}^{\infty} P(f)H(f)\exp(j2\pi ft_0)df \right|^2}{(\eta/2)\int_{-\infty}^{\infty} |P(f)|^2 df} \qquad (5.38)$$

The filter frequency response is then chosen to maximize the RHS of Eqn (5.38). This is accomplished by applying the Schwartz inequality,[2] which effectively states that the sum of two sides of a triangle must always be greater than or equal to the third side; in integral form this is usually written

$$\int_{-\infty}^{\infty} A^*(x)\,A(x)\,dx \int_{-\infty}^{\infty} B^*(x)\,B(x)\,dx \geqslant \left| \int_{-\infty}^{\infty} A^*(x)\,B(x)\,dx \right|^2 \quad (5.39)$$

The inequality becomes an equality when $A(x)$ and $B(x)$ are co-linear, i.e. $A(x) = KB(x)$. If $A^*(x) = H(f)\exp(j2\pi f t_0)$ then its complex conjugate $A(x) = H^*(f)\exp(-j2\pi f t_0)$, and if $B(x) = P(f)$ then $B^*(x) = P^*(f)$.

Equation (5.39) can then be written

$$\int_{-\infty}^{\infty} |H(f)|^2 df \int_{-\infty}^{\infty} |P(f)|^2 df \geqslant \left| \int_{-\infty}^{\infty} P(f)\,H(f)\exp(j2\pi f t_0)\,df \right|^2 \quad (5.40)$$

$$\int_{-\infty}^{\infty} |H(f)|^2 df \geqslant \frac{\left| \int_{-\infty}^{\infty} P(f)\,H(f)\exp(j2\pi f t_0)\,df \right|^2}{\int_{-\infty}^{\infty} |P(f)|^2 df}$$

Substituting in Eqn (5.38) we obtain

$$\frac{2}{\eta} \int_{-\infty}^{\infty} |H(f)|^2 df \geqslant \text{SNR}$$

But $\int_{-\infty}^{\infty} |H(f)|^2 df$ is the energy of the signal, E. The maximum value of SNR is therefore

$$\text{SNR}_{\text{max}} = 2E/\eta \quad (5.41)$$

This value depends only on the ratio of signal energy to noise spectral density and is independent of the shape of the signal waveform. The SNR is a maximum when $A(x) = KB(x)$, i.e.

$$H^*(f)\exp(-j2\pi f t_0) = KP(f) \quad (5.42)$$

The impulse response of the matched filter is thus

$$p(t) = \frac{1}{K} \int_{-\infty}^{\infty} H^*(f)\exp(-j2\pi f t_0)\exp(j2\pi f t)\,df$$

i.e.

$$p(t) = \frac{1}{K} h(t_0 - t) \quad (5.43)$$

The impulse response is therefore the time reverse of the input signal waveform $h(t)$ with respect to t_0. Clearly the value of t_0 must be greater than the duration of the signal to which the filter is matched. This is the condition for physical realizability; in other words, the impulse response must be zero for negative values of t.

We will derive, as an example, the matched filter for a rectangular pulse of amplitude A and duration t_1. The Fourier transform of this pulse is

$$H(f) = \int_{-t_{1/2}}^{t_{1/2}} A \exp(-j2\pi f t)\,dt$$

Hence from Eqn (5.42) $P(f) = KH^*(f)\exp(-j2\pi ft_0)$ i.e.

$$P(f) = [\text{sinc}(\pi ft_1)\exp(-j2\pi ft_0)]/K \tag{5.44}$$

This is a low-pass filter with a linear phase shift. Matched filters can be specified for any signal shape, e.g. when a FSK waveform is the input to a matched filter the SNR at the output is given by Eqn (5.41).

If we write Eqn (5.35) representing the matched filtering operation in terms of the convolution integral it becomes

$$g(t) = \int_{-\infty}^{\infty} h(t) h(t_0 - t)\, dt$$

$h(t_0 - t)$ being the appropriate substitution for $p(t)$. We now compare this equation with the equation for the cross-correlation between two waveforms $h(t)$ and $y(t)$.

$$R_{hy}(\tau) = \int_{-\infty}^{\infty} h(t)\, y(t + \tau)\, dt$$

If we make $y(t + \tau) = h(t_0 - t)$ we see that matched filtering is equivalent to cross-correlating the 'noisy signal' $h(t)$ with a noise-free signal $h(t_0 - t)$, which is the impulse response of the matched filter. The correlation detector is based upon this principle.

The correlation coefficient between the waveforms $h(t)$ and $y(t)$ is defined as

$$R(\tau) = \frac{\displaystyle\int_{-\infty}^{\infty} h(t)y(t + \tau)\,dt}{\left|\displaystyle\int_{-\infty}^{\infty} h^2(t)\,dt \int_{-\infty}^{\infty} y^2(t)\,dt\right|^{1/2}} \tag{5.45}$$

The denominator of this equation is a normalizing factor that makes $R(\tau)$ independent of the actual mean square values of $h(t)$ and $y(t)$ and restricted to the range ± 1. Note when $h(t) = ky(t)$ then $R(\tau) = +1$, and when $h(t) = -ky(t)$ then $R(\tau) = -1$.

$R(\tau)$ is thus a measure of the similarity of the two waveforms $h(t)$ and $y(t)$. Correlation detection is frequently used in situations where the SNR < 1, e.g. space applications. The received waveform is correlated with several noise-free waveforms stored at the receiver. The one that produces the largest correlation coefficient is then assumed to have been the transmitted waveform.

It is found in most practical situations that the improvement obtained using matched filters, rather than conventional low- or band-pass filters, is marginal (usually less than 3 dB). Hence filters are usually designed to minimize intersymbol interference rather than maximize SNR. However, where SNR is of prime importance (e.g. radar signals) matched filtering and correlation techniques are employed.

5.6 ERROR DETECTION AND CORRECTION

The probability of error in a digital transmission system depends ultimately upon the SNR at the receiver input. This ratio is a maximum when matched filtering is used and this produces the lowest probability of error. A significant decrease in this figure can be produced by employing a type of code known as a block code for the digital signal. The purpose of coding the signal in this fashion is to introduce redundancy into the transmitted signal. We observed in Section 1.13 that there are many sources of redundancy in the English language and we showed that the redundancy present reduced the information content but at the same time allowed the receiver to identify and correct errors. Redundancy has the same effect in digital signals. The simplest form of block code is the parity check 8 bit international standard code described in Section 3.11.

The international code represents each character by seven information digits and a single parity check digit. Either an even or an odd parity check can be used; in the former case the parity digit is chosen to make the number of 1s in the 8 bit word even, and in the latter case the parity digit is chosen to make the number of 1s in the 8 bit word odd. The receiver then checks each 8 bit word for odd or even parity depending upon the system in use. If the parity check fails the receiver notes that an error has occurred and requests the transmitter to repeat the 8 bit word. The parity check is carried out simply by exclusive-ORing the eight digits in each code word. If the number of 1s in the code word is odd the result of the exclusive-OR operation is binary 1. If the number of 1s is even, the result of the exclusive-OR operation is 0. A parity check circuit is given in Fig. 5.9.

The error-detecting properties of this code can be demonstrated by comparing the probability of error in the uncoded case with the probability of an undetected error in the coded case. We assume that the probability of error in the uncoded case is P_e. The probability of an undetected error in the coded case will be the probability of an even number of errors. The parity check at

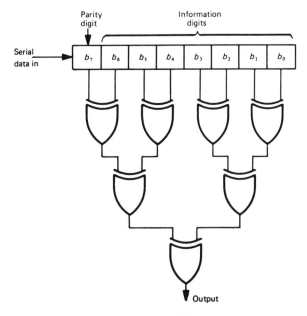

Fig. 5.9 Error detection using a single parity digit.

the receiver will fail, and therefore indicate an error, only if an odd number of errors occurs. We therefore have to compare the probability of error in the uncoded case with the probability of an even number of errors in the coded case.

If there are n digits in each code word the probability of one digit being in error is the joint probability of one incorrect digit and $(n-1)$ correct digits, i.e. the joint probability is

$$P_j = P_e(1 - P_e)^{n-1} \tag{5.46}$$

There are n digits in the group and Eqn (5.46) gives the probability that any one of these digits will be received incorrectly. The total probability in the n digits is thus

$$P = nP_e(1 - P_e)^{n-1}$$

As an example consider the uncoded case where $n = 7$. If $P_e = 10^{-4}$ the probability of one error in seven digits is

$$P_t = 7 \times 10^{-4} \times (0.9999)^6 \simeq 7 \times 10^{-4}$$

The probability of r errors in a group of n digits is the joint probability that r digits will be received incorrectly and $(n-r)$ will be received correctly, i.e.

$$P_j = P_e^r(1 - P_e)^{n-r}$$

There is a total of ^nC_r possible ways of receiving r digits incorrectly in a total of n digits where

$$^nC_r = \frac{n!}{r!\,(n-r)!}$$

The total probability of error in this case is thus

$$P_t = {}^nC_r P_e^r(1 - P_e)^{n-r} \tag{5.47}$$

In the coded case, an extra digit is added to the original seven so that we require the probability of an even number of errors in eight digits.

Probability of 2 errors $= {}^8C_2(10^{-4})^2(1 - 10^{-4})^6 = 2.8 \times 10^{-7}$
Probability of 4 errors $= {}^8C_4(10^{-4})^4(1 - 10^{-4})^4 = 7 \times 10^{-15}$
Probability of 6 errors $= {}^8C_6(10^{-4})^6(1 - 10^{-4})^2 = 2.8 \times 10^{-23}$
Probability of 8 errors $= {}^8C_8(10^{-4})^8 \qquad\qquad = 10^{-32}$

Hence the total probability of an undetected error in the coded case is $2.8 \times 10^{-7} + 7.0 \times 10^{-15} + 2.8 \times 10^{-23} \simeq 2.8 \times 10^{-7}$. This is considerably less than the probability of error in the original uncoded group of seven digits. These numerical values show that a significant reduction in error probability is possible even with a very rudimentary error-checking system.

We have assumed in this example that the digit transmission rate is unchanged. This means that the rate of information transmission must necessarily be reduced to $\frac{7}{8}$ths of its value in the uncoded system. It would be more realistic to maintain a constant information rate which means that the transmission bandwidth should be increased to $\frac{8}{7}$ of its original value to accommodate the parity check digit. Such an increase would increase the noise power by the same ratio, i.e. the rms noise would be increased to 1.069 times its previous value. The effect of this increased noise on the error probability must be calculated. Assuming an uncoded probability of error $P_e = 10^{-4}$ for illustration then

$$\mathrm{erf}\,[(A/(2\sqrt{2}\sigma))] = 1 - 2 \times 10^{-4}$$

i.e.

$$A/(2\sqrt{2}\sigma) = 2.63$$

If the rms noise increases by a factor of 1.069, there is a corresponding decrease in this ratio, i.e. the ratio becomes 2.46, which produces an uncoded error probability of $P_e' = 2.51 \times 10^{-4}$. This value must now be a used in Eqn (5.47) to calculate the probability of an undetected error in the coded case, which is $P_t' = 1.75 \times 10^{-6}$. Thus even when the extra noise is taken into account the reduction in undetected errors is still considerable.

This is not always the case, however, if several parity digits are added to the information digits. The extra noise introduced by the increased bandwidth can cancel any advantage obtained by the error-detecting code. In such cases the error probability can sometimes be decreased by increasing the signal power. There are many cases where error-detecting codes do produce a significant advantage, and bearing in mind the threshold region shown by Fig. 5.5 we can only say that each case must be examined separately.

The error-detecting properties of a code can be enhanced to include error correction by increasing the redundancy. This becomes clear by considering an example that codes 16 information digits with 8 parity check digits, i.e. the code redundancy is 33%. In this example the 16 information digits are grouped into a 4 × 4 matrix. An even or odd parity check digit is transmitted for each row and column of the matrix

```
         Column parity digits                    │ Column check fails
                                                  ↓
  0  0  1  0 ↙                          0  0  1  0
 ───────────                           ───────────
  0  1  1  1 │ 1                        0  0  1  1 │ 1  ← Row check fails
  0  1  1  0 │ 0                        0  1  1  0 │ 0
  1  0  0  0 │ 1                        1  0  0  0 │ 1
  1  0  1  1 │ 1                        1  0  1  1 │ 1
             ↑
      Row parity digits
```

If a single error occurs in the 16 information digits then a row and column parity check will both fail. The point of intersection of row and column then indicates which digit was in error. If a single parity digit is in error then only one column or one row will fail the parity check, and again the incorrect digit can be identified and therefore corrected. This type of code is generally known as a forward error correcting code (FEC).

The above example was chosen to illustrate the possibility of correcting errors, and was not a particularly efficient code. There are many more efficient error-correcting codes in use. The block codes are based upon the work of Hamming.[3] An example of Hamming's single-error-correcting code will now be considered. We assume that the original information digits are split up into blocks of m digits to which are added c parity check digits. For illustration assume that $m = 4$ and $c = 3$. Each coded word therefore contains seven digits and there are $2^7 = 128$ different words. Of these 128 words only $2^4 = 16$ words are required to transmit the original information. The 16 code words are chosen from the possible 128 to give a single-error-correcting capability. A possible choice of code words is shown in Table 5.1.

Each of the code words in Table 5.1 differs from any other in at least three positions. Such a code is said to have a Hamming distance of three. This means that at least three errors are required to convert any code word into one of the others. If a single error occurs in a received code word it will differ from the correct code word in one digit only and from all other code words in at least two digits. Hence the correct code word is chosen as the one with least difference from the received code word. This implies that the receiver requires a list of all allowed code words and then compares each received code word until a match is achieved. In practice this process can be reduced to one of parity checking.

The mathematical relationship between information digits and check digits is based on modulo 2 addition (exclusive-OR) and was devised by Hamming. The rules for a 7 digit block code containing four information digits and three check digits are

$$C_1 = M_1 \oplus M_2 \oplus M_3$$
$$C_2 = M_1 \oplus M_2 \oplus M_4$$
$$C_3 = M_1 \oplus M_3 \oplus M_4 \qquad (5.48)$$

Remembering that $C_1 \oplus C_1 = 0$, etc., these equations can be written

$$C_1 \oplus M_1 \oplus M_2 \oplus M_3 = 0$$
$$C_2 \oplus M_1 \oplus M_2 \oplus M_4 = 0$$
$$C_3 \oplus M_1 \oplus M_3 \oplus M_4 = 0 \qquad (5.49)$$

Table 5.1

Original message				Transmitted code						
M_1	M_2	M_3	M_4	M_1	M_2	M_3	M_4	C_1	C_2	C_3
0	0	0	0	0	0	0	0	0	0	0
0	0	0	1	0	0	0	1	0	1	1
0	0	1	0	0	0	1	0	1	0	1
0	0	1	1	0	0	1	1	1	1	0
0	1	0	0	0	1	0	0	1	1	0
0	1	0	1	0	1	0	1	1	0	1
0	1	1	0	0	1	1	0	0	1	1
0	1	1	1	0	1	1	1	0	0	0
1	0	0	0	1	0	0	0	1	1	1
1	0	0	1	1	0	0	1	1	0	0
1	0	1	0	1	0	1	0	0	1	0
1	0	1	1	1	0	1	1	0	0	1
1	1	0	0	1	1	0	0	0	0	1
1	1	0	1	1	1	0	1	0	1	0
1	1	1	0	1	1	1	0	1	0	0
1	1	1	1	1	1	1	1	1	1	1

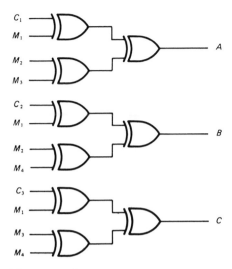

Fig. 5.10 Parity checking network.

If there are no errors at the receiver the modulo 2 additions represented by Eqns (5.49) will all produce a binary 0 result. If there is a single error in any of the information or check digits then one or more of Eqns (5.49) will produce a binary 1. The parity checking circuit for the 7 digit block code is shown in Fig. 5.10.

If we represent a correct digit by Y and an incorrect digit by N the possible values of A, B, C for a single error are listed in Table 5.2.

The outputs A, B, C of the parity check circuits can be regarded as a 3×8 matrix. The number of columns in the matrix equals the number of check digits and each row, except the first, uniquely defines a single error position. This means that c check digits can define $2^c - 1$ error positions assuming only a single error occurs.

If there are m information digits and c check digits then for a single-error correcting code the relationship between m and c is

$$m + c \leqslant 2^c - 1 \qquad (5.50)$$

Thus for an ASCII character with $m = 7$ it is necessary to add $c = 4$ check

Table 5.2

M_1	M_2	M_3	M_4	C_1	C_2	C_3	A	B	C
Y	Y	Y	Y	Y	Y	Y	0	0	0
Y	Y	Y	Y	Y	Y	N	1	0	0
Y	Y	Y	Y	Y	N	Y	0	1	0
Y	Y	Y	Y	N	Y	Y	0	0	1
Y	Y	Y	N	Y	Y	Y	1	1	1
Y	Y	N	Y	Y	Y	Y	1	1	0
Y	N	Y	Y	Y	Y	Y	1	0	1
N	Y	Y	Y	Y	Y	Y	0	1	1

Table 5.3 Single-error-correcting codes

m	c	Code type $(m+c, m)$	Efficiency $m/(m+c)$
1	2	$(3, 1)$	0.33
4	3	$(7, 4)$	0.57
7	4	$(11, 7)$	0.64
11	4	$(15, 11)$	0.73
26	5	$(31, 26)$	0.83
57	6	$(63, 57)$	0.90
120	7	$(127, 120)$	0.94
247	8	$(255, 247)$	0.97

digits for a single-error-correcting capability. Possible single-error-correcting codes are listed in Table 5.3, and it may be seen from this table that the larger the value of c the more efficient the code.

Returning to Table 5.2 it is clear that in a practical system it is necessary only to correct errors in the information digits. This means that for error-correcting purposes only the last four rows of the parity check matrix need to be considered. Noting the following relationship:

$$X \oplus 1 = \bar{X} \quad \text{and} \quad X \oplus 0 = X$$

Then if a received digit is in error, exclusive-ORing it with binary 1 produces the correct value. If a received digit is correct, exclusive-ORing it with binary 0 will preserve the received value. With these conditions it is possible to devise a very simple error-correcting circuit consisting of exclusive-OR gates. This circuit is shown in Fig. 5.11. The binary inputs w, x, y, z are derived from the parity check matrix of Table 5.2 and are given by

$$w = \bar{A}BC$$
$$x = A\bar{B}C$$
$$y = AB\bar{C}$$
$$z = ABC$$

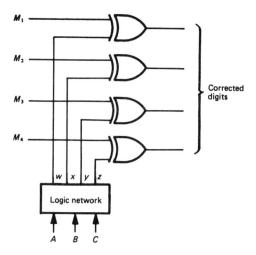

Fig. 5.11 Error-checking network.

The $(7, 4)$ code has a Hamming distance of 3 and can correct single errors. If it is assumed, for illustrative purposes, that an error occurs in both M_1 and M_2 the parity checking network will produce an output of $A = 0$, $B = 0$, $C = 1$, which is equivalent to a single error in c_1. The fact that the parity checking network does not give $A = 0$, $B = 0$, $C = 0$ means that a double error has been detected, but this will be incorrectly interpreted by the circuit of Fig. 5.11.

Thus, the $(7, 4)$ code can detect a double error provided it is not required to correct a single error also. If it is assumed that the $(7, 4)$ code is designed to correct a single error (using the hardware of Fig. 5.11) a double error will be ignored. In order to evaluate the effectiveness of the $(7, 4)$ code it is necessary to compare the probability of two errors in seven digits with the probability of one error in four digits. If it is desired to maintain the same information rate the bandwidth must be increased in the ratio of 7:4, which increases the rms noise by a factor of 1.322.

Two examples will be considered: in the first the uncoded error probability is assumed to be $P_e = 10^{-4}$ and in the second case $P_e = 10^{-5}$. The probability of a single error in four digits is thus either $P_t = 4 \times 10^{-4}$ or $P_t = 4 \times 10^{-5}$. If the rms noise is increased by a factor of 1.322 the corresponding error probability for an uncoded system is $P'_e = 2.4 \times 10^{-3}$ or $P'_e = 6.31 \times 10^{-3}$. The probability of an undetected error in the coded case is the probability of two or more errors, i.e.

$$P'_t = {}^7C_2 (P'_e)^2 (1 - P'_e)^5 + {}^7C_3 (P'_e)^3 (1 - P'_e)^4 + {}^7C_4 (P'_e)^4 (1 - P'_e)^3$$
$$+ {}^7C_5 (P'_e)^5 (1 - P'_e)^2 + {}^7C_6 (P'_e)^6 (1 - P'_e) + {}^7C_7 (P'_e)^7$$

This evaluates to $P'_t = 1.29 \times 10^{-4}$ or $P'_t = 8.36 \times 10^{-6}$. In the first case the error rate is reduced from 4×10^{-4} to 1.29×10^{-4} (a ratio of 3.1:1) and in the second case from 4×10^{-5} to 8.36×10^{-6} (a greater ratio of 4.7:1). These are fairly modest decreases, the improvement being greater the smaller the initial error probability. If more efficient single-error-detecting codes are used (e.g. 127, 120 code) a very significant reduction in undetected error probability is obtained.

If the code is required to correct more than one error, then clearly the Hamming distance must be increased. The Hamming distance for a double-error-correcting code is 5, which means that each code word must differ from all other code words in at least five digits. If two errors do occur, the received code word will differ by two digits from the correct code and at least three digits from all other code words.

If it is assumed that there are up to y errors in each code word the number of ways that these errors can occur in $(m + c)$ digits is

$$\sum_{t=1}^{y} {}^{m+c}C_t$$

The c parity digits define $(2^c - 1)$ rows in the parity matrix so that the relationship between m and c for a code that corrects up to y errors is

$$\sum_{t=1}^{y} {}^{m+c}C_t \leqslant 2^c - 1 \qquad (5.51)$$

which is known as the Hamming bound. Clearly the greater the error-

Table 5.4 Block code properties

Hamming distance	Code properties
1	None
2	Single-error detecting
3	Single-error correcting **or** double-error detecting
4	Single-error correcting **and** double-error detecting
5	Double-error correcting **or** triple-error detecting
6	Double-error correcting **and** triple-error detecting
7	Triple-error detecting

correcting capability of the code the greater is the redundancy. The relationship between the Hamming distance of a code and its error-combating capabilities is listed in Table 5.4 for Hamming distances up to 7.

Table 5.4 indicates that block codes are best suited to correcting a small number of errors in each block. As the number of errors per block rises then block codes become less and less attractive. The interference encountered on most communications links tends to be impulsive rather than Gaussian and this leads to errors occurring in bursts rather than individually. Under such circumstances several errors can occur in a single block and hence simple block codes are not an effective means of error correction. Some care is necessary in defining the duration of an error burst as such a burst often contains some digits which may not be in error. An error burst is bounded by two erroneous digits and must be separated from the next error burst by a number of correct digits greater than or equal to the number of digits in the error burst. This apparently convoluted definition may be clarified by the example shown in Fig. 5.12. In this figure it becomes clear that errors 1 and 3 could not be used to define a single burst of length 13 digits because error 4 would occur within the following 13 digits. Hence it becomes apparent that errors 1 and 2 define an error burst of five digits and this is followed by errors 3 and 4, which define an error burst of six digits.

The unsuitability of the simple block code may be demonstrated by considering the transmission of 7 bit ASCII characters to which have been added four check bits to provide a single error correcting ability. If such characters are transmitted over a channel subject to burst errors of up to 5 bits duration, then clearly the resulting (11,7) block code would not be able to correct the errors even when the bursts occurred infrequently. This problem can be overcome by a process known as interleaving in which the

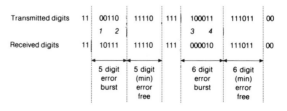

Fig. 5.12 Error bursts in a transmitted sequence.

transmitter assembles five successive 11 bit blocks and transmits the first bit of each block in sequence, followed by the second bit of each block and so on (i.e. the bits of the individual 11 bit blocks are interleaved). The receiver performs the opposite de-interleaving process to reproduce the original 11 bit blocks. In this system if an error burst of up to 5 bits occurs during the transmission of 55 interleaved bits then a maximum of one error per block will occur after the de-interleaving process. Such errors can be corrected by the (11, 7) block code.

5.7 THE CYCLIC REDUNDANCY CHECK MECHANISM

In many instances error coding is restricted to error detection rather than error correction, this is particularly true of packet transmission (see Chapter 11). When an error is detected the transmitter is requested to repeat the data, which clearly requires the provision of a return path between receiver and transmitter. In such circumstances cyclic codes are often employed and these codes are able to detect errors which occur in bursts. Cyclic codes are used to append a cyclic redundancy check (CRC) sequence to a block of information digits and are therefore a form of block code. In general the addition of c parity check digits to a block of m information digits enables any burst of c digits or less to be detected, irrespective of the length of the information block. In addition the fraction of bursts of length $b > c$ which remain undetected by a cyclic code is 2^{-c} if $b > c + 1$. The length of the CRC can be altered to suit the anticipated error statistics but 16 and 32 bits are commonly used. The Ethernet packet shown in Fig. 13.1 appends a CRC of 32 bits to a block of information the length of which varies between 480 and 12 112 bits, which means that if $b > 33$ the fraction of bursts remaining undetected is 2^{-32}, which is a very small number.

The operation of the CRC mechanism can be demonstrated as follows: assume that the information to be transmitted is a binary number M which contains m bits. This is appended by a CRC which may be regarded as a binary number R containing c bits. The CRC is actually the remainder of a division by a another binary number G, known as the generator, which contains $c + 1$ bits. The value of R is obtained from

$$\frac{M \times 2^c}{G} = Q + \frac{R}{G} \tag{5.52}$$

where Q is the quotient. If R/G is added modulo 2 to each side of this equation, then

$$\frac{M \times 2^c + R}{G} = Q \tag{5.53}$$

This indicates that if the number represented by $M \times 2^c + R$ is divided by G, the remainder will be zero. The number $M \times 2^c + R$ is formed by adding the c bit CRC to the original m information bits which are shifted c places to the right. The number so formed has a length $m + c$ bits and is the transmitted code word. If this code word is divided by the same number G at the receiver then, provided that there have been no errors, the remainder will be zero. If errors have occurred the division will result in a non-zero remainder.

It is common practice to represent M, G and R in polynomial form in which case G is known as the generator polynomial. In the case of the Ethernet packet G is 33 bits long and is represented by:

$$G = X^{32} + X^{26} + X^{23} + X^{16} + X^{12} + X^{11} + X^{10} + X^8 + X^7 + X^5$$
$$+ X^4 + X^2 + X + 1 \tag{5.54}$$

In effect the powers of X shown in this polynomial are the powers of 2 in the number G. Thus $G = 100000100100000010001110110110111$.

In theory any polynomial of order $c + 1$ can act as the generator polynomial. However, it is found in practice that not all such polynomials provide good error-detecting characteristics and selection is made by extensive computer simulation.

5.8   CONVOLUTIONAL CODES

Convolutional codes are an alternative to block codes which are more suited to correcting errors in environments in which errors occur in bursts. In a block code the codeword depends only on the m information digits being coded and is therefore memoryless. In convolutional codes a continuous stream of information digits is processed to produce a continuous stream of encoded digits each of which has a value depending on the value of several previous information digits, implying a form of memory. In practice, convolutional codes are generated using a shift register of specified length L, known as the constraint length, and a number of modulo 2 adders (exclusive OR gates). An encoder with a constraint length of 3 and two exclusive OR gates is shown in Fig. 5.13. Clearly, in this example, the output digit rate is double the input digit rate and the code is known as a $\frac{1}{2}$ rate code. (In general the code may be a m/n rate code where m is the number of input digits and n is the number of output digits.)

The constraint length, the number of exclusive OR gates and the way in which they are connected determine the error-correcting properties of the code produced. The goal of the decoder is to determine the most likely input data stream of the encoder given a knowledge of the encoder characteristics

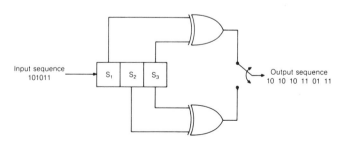

Fig. 5.13 Convolutional encoder.

and the received bit stream. The decoding procedure is equivalent to measuring the distance, in a multi-dimensional space, between the point representing the received sequence and the points representing all possible sequences which could have been produced by the coder. The one with the minimum distance is then chosen as the correct sequence. This minimum distance criterion is similar to that used in the physical interpretation of Shannon's law, presented in Chapter 1. The actual algorithm used in the decoder is known as the Viterbi[4] algorithm and is designed to minimize receiver complexity.

5.9 MULTIPLE DECISIONS

We have assumed in this chapter that the receiver in a system is the only point at which a decision is made. This is not necessarily the case; in a PCM system, for example, a decision is made at each regenerative repeater. Hence when evaluating error probability in a multistage link, the probability of error at each stage must be taken into account. For simplicity we shall assume that the probability of error at each stage in such a link is the same, although this restriction need not apply in a real situation. A typical PCM system is illustrated in Fig. 5.14.

It is evident that errors that may occur at one repeater can be corrected by a second error in the same digit at a subsequent stage. If we consider a particular digit in its passage through a multistage link when it is subject to an even number of incorrect decisions it will be received correctly at the destination.

If there are L decision stages the probability of making U incorrect and $L - U$ correct decision is

$$P = P_e^{U} \times (1 - P_e)^{L - U}$$

but there are ${}^L C_U$ possible ways of making U incorrect decisions in a total of L decisions. Hence for a single digit the probability of making U incorrect decisions in an L stage link is

$$P_t = {}^L C_U P_e \times (1 - P_e)^{L - U} \tag{5.55}$$

The probability of a digit being incorrect at the receiver is found by summing Eqn (5.55) over all odd values of U, i.e.

$$P_m = \sum_{U(\text{odd})} {}^L C_U P_e^{U} (1 - P_e)^{L - U} \tag{5.56}$$

If $P_e \ll 1$ this approximates to $P_m \simeq L P_e$. Thus if P_e is small the probability of a single digit received incorrectly at more than one stage is negligible.

Fig. 5.14 Transmission system with multiple decision points.

However, the use of L decision points makes the individual error L times as likely.

If error coding is employed in a multistage link, equations such as Eqn (5.47) should reflect the higher individual digit error probability, e.g.

$$P_t = {}^nC_r(LP_e)^r(1 - LP_e)^{n-r} \tag{5.57}$$

5.10 CONCLUSION

In this chapter specific attention has been directed towards the effects of Gaussian noise on digital communications and it has been shown that the probability of error is related to the SNR at the receiver. This probability can be reduced either by employing SNR enhancement techniques such as matched filtering or by employing error-coding techniques. The major source of signal impairment on data networks is due to inter-symbol interference, and the noise encountered tends to be impulsive rather than Gaussian. This leads to errors which occur in bursts and the correction of such errors requires the use of specialized codes and techniques of interleaving. This is also the situation in digital cellular radio, described in Chapter 15, where burst errors result from multipath propagation which produces signal fading.

One example where block codes are used is in satellite transmissions, where signal power is at a premium and the received noise is predominantly Gaussian. A second example is covered in detail in Section 11.19, which deals with teletext transmission. In this case page and row address digits are error protected using a code with a Hamming distance of 4 which can correct a single error and detect a double error.

REFERENCES

1. Schwartz, M., Bennett, W. R. and Stein, S., *Communication Systems and Techniques*, McGraw-Hill, New York, 1966.
2. Spiegel, M. R., *Advanced Calculus*, Schaum, 1963.
3. Hamming, R. W., 'Error detecting and correcting codes', *Bell Systems Technical Journal*, **29**, 147 (1950).
4. Virterbi, A. J., 'Convolutional codes and their performance in communications systems', *IEEE Transactions on Communication Technology*, COM **19**(5), (1971).

PROBLEMS

5.1 A binary waveform of amplitude $+2V$ or $-V$ is added to Gaussian noise with zero mean and variance σ^2. The *a priori* signal probabilities are $P(+2V) = \frac{1}{3}$ and $P(-V) = \frac{2}{3}$. If a single sample of signal-plus-noise is taken, determine the decision threshold that minimizes overall error probability.

Answer: $T = V/2 + 0.23\sigma^2/V$.

5.2 A teleprinter system represents each character by a 5 digit binary code that is transmitted as either $A/2$ volts or $-A/2$ volts. The binary signal is received in the presence of Gaussian noise with zero mean at a SNR of 11 dB. Determine the optimum decision level and the probability that the receiver will make an error.

Answer: $T = 0\,V$; $P = 5 \times 0.000\,193$.

5.3 In a binary transmission system, 1 is represented by a raised cosine pulse given by

$$v(t) = V[1 + \cos(2\pi t/T)] \quad \text{for} \quad -T/2 < t < T/2$$

where T is the pulse duration. A binary 0 is represented by the same raised cosine waveform with amplitude $-V$. After transmission over a noisy channel the signal-to-noise power ratio at the detector is 6 dB. If the binary signal is reformed by sampling the received signal at the middle of each pulse interval, find the probability of error. Assume 1 and 0 are equi-probable and that the noise is white.

Answer: 0.000 572.

5.4 Two binary communication links are connected in series, each link having a transmitter and a receiver. If the probability of error in each link is 0.000 01, find the overall probability that:
(a) a 0 is received when a 1 is transmitted;
(b) a 0 is received when a 0 is transmitted;
(c) a 1 is received when a 0 is transmitted;
(d) a 1 is received when a 1 is transmitted.

Answer: (a) 0.000 02; (b) 0.999 98; (c) 0.000 02; (d) 0.999 98.

5.5 If a simple coding scheme is used in the previous question such that each individual digit is repeated three times, what is the probability of deciding incorrectly at the receiver if the following rule is used?
Decide 0 if the received code is 000, 001, 010, 100
Decide 1 if the received code is 111, 110, 101, 011

Answer: 1.2×10^{-9}.

5.6 The noise level on a channel produces an error probability of 0.001 during a binary transmission. In an effort to overcome the effect of noise each digit is repeated five times, and a majority decision is made at the receiver. Find the probability of error in this system if

(a) the digit rate remains constant;
(b) the information rate is half its original value.

Assume white Gaussian noise.

Answer: (a) 9.94×10^{-9}; (b) 1.445×10^{-4}.

5.7 Use Eqns (5.35) and (5.41) to determine the maximum signal and mean square noise voltage at the output of a matched filter.

Answer: E, $E\eta/2$.

5.8 The matched filter is a linear network. This means that when the input to such a filter is Gaussian noise the output noise will also be Gaussian but with a modified variance.
Using this fact, calculate the probability of error in a binary transmission system when the threshold detector is preceded by a matched filter. The transmission

rate is 64 kb/s with 0 and 1 having equal probability. The received waveform has a mean value (normalized) of 0 or 25 mV, and is accompanied by noise with a power spectral density of 1.68×10^{-9} W/Hz.

Answer: 0.000 020 5.

5.9 A signal which is a single pulse of amplitude A volts and duration T seconds is received masked by white Gaussian noise. Calculate the improvement in SNR produced by a matched filter as compared with the SNR at the output of a single-stage RC low-pass filter. Assume the 3 dB cut-off frequency of the RC network is $1/T$ hertz.

Answer: 0.9 dB.

5.10 A FSK receiver consists of two matched filters, one for each of the tone bursts f_0 and f_1. The FSK waveform is fed to both filters, each output being sampled at the instant of maximum signal. The samples are subtracted and then fed to a threshold detector.
Calculate the error probability in terms of the signal energy and noise power spectral density.

Answer: $1 - \text{erf} \sqrt{(E/\eta)}$.

High-frequency transmission lines $\boxed{6}$

The study of transmission lines is the investigation of the properties of the system of conductors used to carry electromagnetic waves from one point to another. Here, however, our attention will be limited to high-frequency applications, i.e. when the length of the transmission line is of at least the same order of magnitude as the wavelength of the signal. In this chapter an idealized model of the line will be used to represent the many different forms found in practice, ranging from twisted pairs to coaxial cables. The theory of transmission lines, which was developed in the early years of the study of electromagnetic propagation, is strictly applicable only to systems of conductors that have a 'go' and 'return' path, or that, in electromagnetic field terms, can support a TEM wave. Hollow-tube waveguides do not fall into this category, although, as we discuss in the chapter on microwaves, many of the concepts of transmission line theory can be applied to them. Transmission line theory is important to communications engineers because it gives them the means for making the most efficient use of the power and equipment at their disposal. By applying their knowledge correctly they can ensure that a transmitting system is designed to transfer as much power as possible from the feeder line into the antenna, or they can take steps to ensure that a receiving antenna is correctly matched to the line that connects it to the receiver itself, so that no power is wasted. The range of systems over which transmission lines are used is as extensive as the subject of communications engineering itself, and the general theory we develop in this chapter may be used to solve a very wide variety of problems.

6.1 VOLTAGE AND CURRENT RELATIONSHIPS ON THE LINE

A transmission line consists of continuous conductors with a cross-sectional configuration that is constant throughout its length. A voltage is placed across the line at the sending end, and it is necessary to be able to determine how it changes with distance so that its value at the load, or at some other point of interest, can be found. Similarly, knowledge of the current on the line may also be required.

To calculate current and voltage values the line must be represented in some way that will allow circuit analysis to be used. However, in such analysis the parameters are discrete and considered to exist at one point, whereas in a transmission line they are evenly distributed throughout its length. This difficulty is overcome by considering a very short length of the line, as we shall see in the next few paragraphs.

The parameters used to describe the line are:

(i) Resistance (R). The conductors making up the line offer some resistance to the flow of current. R is usually made to include the total resistance of both conductors.

(ii) Inductance (L). The signal on the line is time varying and so there will be an inductive reactance associated with the line. The value of L depends on the cross-sectional geometry of the conductors.

(iii) Conductance (G). The conductors that form the line are held in position by a dielectric material that, because it cannot be a perfect insulator, will allow some leakage current to pass between them.

(iv) Capacitance (C). The conductors, and the dielectric between them, form a capacitor, and there will therefore be a capacitive reactance to a time-varying signal.

These four parameters are usually given per length of line, consequently they must be multiplied by the line length to find the total resistance, inductance, conductance and capacitance of the line.

The continuous distribution of these parameters is approximated by representing the line as a cascaded network of elements, each element having a very short length δz, as shown in Fig. 6.1.

By considering only one of these elements, as shown in Fig. 6.2, the voltage and current dependence on z, the distance from the sending end, can be calculated.

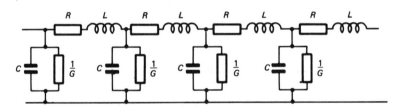

Fig. 6.1 Cascaded network representation of transmission line.

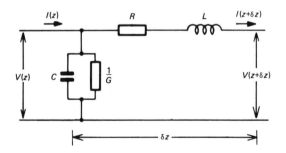

Fig. 6.2 Equivalent circuit for element of length δz.

The current through the shunt circuit is

$$I(z) - I(z + \delta z) = G\delta z V(z) + C\delta z \frac{\partial}{\partial t} V(z) \tag{6.1}$$

and the voltage drop in the series circuit is

$$V(z) - V(z + \delta z) = R\delta z I(z + \delta z) + L\delta z \frac{\partial}{\partial t} I(z + \delta z) \tag{6.2}$$

Note that in these equations the impedance terms have been multiplied by the length of the element, δz. For small δz

$$I(z + \delta z) \simeq I(z) + \frac{\partial I}{\partial z}(z)\delta z$$

Then Eqn (6.1) becomes

$$\frac{\partial I}{\partial z}(z) = -\left(G + C\frac{\partial}{\partial t}\right)V(z) \tag{6.3}$$

Similarly, from Eqn (6.2),

$$\frac{\partial V(z)}{\partial z} = -\left(R + L\frac{\partial}{\partial t}\right)I(z + \delta z) \tag{6.4}$$

i.e.

$$\frac{\partial V(z)}{\partial z} = -\left(R + L\frac{\partial}{\partial t}\right)\left[I(z) + \frac{\partial I(z)}{\partial z}\delta z\right] \tag{6.5}$$

Assuming that the last term, $[\partial I(z)/\partial z]\,\delta z$, is very small compared with the other terms in Eqn (6.5), it can be ignored, and the equation written as

$$\frac{\partial V(z)}{\partial z} = -\left(R + L\frac{\partial}{\partial t}\right)I(z) \tag{6.6}$$

Equations (6.3) and (6.6) may now be combined to give solutions for voltage or current as functions of z. Further simplifications may be made to these equations if it is assumed that the signal on the line is sinusoidal, in which case the time dependence can be expressed by the term $\exp(j\omega t)$, and the time derivative, $\partial/\partial t$, replaced by $j\omega$. If this is done in Eqns (6.3) and (6.6), they become

$$\frac{dI}{dz} = -(G + j\omega C)V \tag{6.7}$$

$$\frac{dV}{dz} = -(R + j\omega L)I \tag{6.8}$$

where the partial differentials have been replaced by total differentials, and V and I are implicitly assumed to be functions of both z and t.
Differentiating Eqn (6.8) gives

$$\frac{d^2V}{dz^2} = -(R + j\omega L)\frac{dI}{dz} \tag{6.9}$$

and substituting $\mathrm{d}I/\mathrm{d}z$ from Eqn (6.7)

$$\frac{\mathrm{d}^2 V}{\mathrm{d}z^2} = (R + j\omega L)(G + j\omega C)V \tag{6.10}$$

More conveniently, Eqn (6.10) is expressed as

$$\mathrm{d}^2 V/\mathrm{d}z^2 = \gamma^2 V \tag{6.11}$$

where

$$\gamma^2 = (R + j\omega L)(G + j\omega C) \tag{6.12}$$

Equation (6.11), which is a standard form of differential equation, has the solution

$$V = V_1 \exp(-\gamma z) + V_2 \exp(\gamma z) \tag{6.13}$$

V_1 and V_2 are determined by the applied voltage and the condition of the line. By a similar analysis, the current is related to z by

$$I = I_1 \exp(-\gamma z) + I_2 \exp(\gamma z) \tag{6.14}$$

Earlier we assumed a time dependence $\exp(j\omega t)$. It is conventional to omit that term from Eqns (6.13) and (6.14), and indeed from the equations in the rest of this chapter, but it is always implicitly assumed because of the time varying signal on the line. We shall return to this later, in Section 6.20.

Equation (6.13) can be expressed in a slightly different form if Eqn (6.12) is examined more closely. γ is a complex quantity and it can therefore be separated into real and imaginary parts, i.e.

$$\gamma = \sqrt{[(R + j\omega L)(G + j\omega C)]} = \alpha + j\beta \tag{6.15}$$

Then, if Eqn (6.15) is used in Eqn (6.13), the total voltage on the line is

$$V = V_1 \exp[-(\alpha + j\beta)z] + V_2 \exp(\alpha + j\beta)z \tag{6.16}$$

$$= V_1 \exp(-\alpha z)\exp(-j\beta z) + V_2 \exp(\alpha z)\exp(j\beta z) \tag{6.17}$$

Thus, the voltage at any point z from the sending end is the sum of two components:

(i) $V_1 \exp(-\alpha z)\exp(-j\beta z)$. The initial voltage V_1 at $z = 0$, is attenuated as it travels down the line, i.e. the amplitude decreases with z as $\exp(-\alpha z)$. The term $\exp(-j\beta z)$ is a phase term and does not influence the amplitude of the voltage. This component is called the forward, or incident, wave.

(ii) $V_2 \exp(\alpha z)\exp(j\beta z)$. Here the amplitude term is of the form $V_2 \exp(\alpha z)$. The component $\exp(\alpha z)$ increases with increasing z, so z must decrease because the voltage must be attenuated as it travels along the line. This component is called the backward, or reflected, wave. It is, as we shall see, produced by a mismatch between the transmission line and the load.

We can therefore conclude that the voltage at a point on the line a distance z from the sending end is the sum of the voltages of the forward and reflected waves at that point.

6.2   LINE PARAMETERS

Referring again to Eqn (6.16), α and β are parameters determined by the characteristics of the line itself, and they have the following interpretation:

α *Attenuation coefficient.* Both the forward and reflected waves are attenuated exponentially at a rate α with distance travelled. α, the real part of Eqn (6.15), is a function of all the line parameters. If R, L, G and C are in their normal units, α is in nepers/m, although it is usually expressed in dB/m.

β *Phase constant.* β, the imaginary part of Eqn (6.15), shows the phase dependence with z of both the forward and backward waves. If z changes from z_1, by a wavelength λ, to $z_1 + \lambda$, the phase of the wave must change by 2π. Therefore

$$\beta(z_1 + \lambda) - \beta z_1 = 2\pi$$

and

$$\beta = 2\pi/\lambda \tag{6.18}$$

γ *Propagation constant.* The complex sum of the attenuation and phase coefficients, as given by Eqn (6.15), is called the propagation constant because it determines, from Eqn (6.13), how the voltage on the line changes with z.

6.3   CHARACTERISTIC IMPEDANCE

In Eqn (6.14) the current on the line is given as the sum of two component current waves, but by using Eqns (6.8), (6.13) and (6.15) it can be expressed in terms of voltage and the line parameters. From Eqn (6.8)

$$I = - \frac{1}{R + j\omega L} \frac{dV}{dz}$$

From Eqn (6.13)

$$\frac{dV}{dz} = \gamma[V_2 \exp(\gamma z) - V_1 \exp(-\gamma z)]$$

then

$$I = \frac{\gamma}{R + j\omega L}[V_1 \exp(-\gamma z) - V_2 \exp(\gamma z)] \tag{6.19}$$

Substituting for γ from Eqn (6.15) gives

$$I = \sqrt{\left(\frac{G + j\omega C}{R + j\omega L}\right)}[V_1 \exp(-\gamma z) - V_2 \exp(\gamma z)] \tag{6.20}$$

This equation is of the form: current $= K \times$ voltage and, by analogy with Ohm's Law, $[(R + j\omega L)/(G + j\omega L)]^{1/2}$ is an impedance. Its value at any particular frequency is determined entirely by the line parameters R, L, G and

C, so it is called the characteristic impedance of the line, Z_0, where

$$Z_0 = \sqrt{[(R + j\omega L)/(G + j\omega C)]} \qquad (6.21)$$

It is sometimes helpful to think of Z_0 in one of the following ways:

(i)   it is the value which the load impedance must have to match the load to the line;

(ii) it is the impedance seen from the sending end of an infinitely long line;

(iii) it is the impedance seen looking towards the load at any point on a matched line – moving along the line produces no change in the impedance towards the load.

The concept of matching is explained further in the following section.

Having produced the fundamental analysis, we need to look at real problems on transmission lines, and the remainder of this chapter will make considerable use of the equations developed. It is therefore worthwhile to note the assumptions that have been made as we have built up our model of a transmission line:

(i)   the line is uniform, homogeneous and straight;

(ii) the line parameters R, L, G and C do not vary with change in ambient conditions, such as temperature and humidity;

(iii) the line parameters are not frequency dependent;

(iv) the analysis applies only between the junctions on the line because the circuit model used in Fig. 6.3 is not valid across a junction.

It is important to be aware that these assumptions have been made and to appreciate that there may be occasions when they should be taken into account.

6.4   REFLECTION FROM THE LOAD

In the last section we noted that a line is matched if it is terminated in a load equal to its characteristic impedance. When we say the line is matched we mean that the forward wave is totally absorbed by the load; consequently there is no reflected wave, and V_2 is zero.

Usually, however, the load will be different from Z_0, say Z_L, and some of the incident wave will be reflected back down the line. The size of the reflected wave will depend on the difference between Z_0 and Z_L as will be shown in the next section.

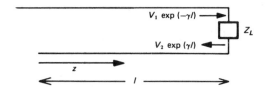

Fig. 6.3 Reflection at the load.

6.5   REFLECTION COEFFICIENT ρ

The amount of reflection caused by the load is expressed in terms of the voltage reflection coefficient ρ. It is defined as the ratio of the reflected voltage to the incident voltage at the load terminals.

The load is at the position $z = l$, as shown in Fig. 6.3, and from Eqn (6.13) the voltage there is

$$V_L = V_1 \exp(-\gamma l) + V_1 \exp(\gamma l) \qquad (6.22)$$

The terms $V_1 \exp(-\gamma l)$ and $V_2 \exp(\gamma l)$ are the incident and reflected voltages, respectively, at the load.

So, from the definition of ρ,

$$\rho = V_1 \exp(\gamma l)/V_1 \exp(-\gamma l)$$
$$= (V_2/V_1)\exp(2\gamma l) \qquad (6.23)$$

ρ will usually be a complex quantity and it is often convenient to express it in a polar form as

$$\rho = |\rho|\exp(j\psi) \qquad (6.24)$$

ψ is referred to as the angle of the reflection coefficient. Now we can find an expression for ρ in terms of the load and the characteristic impedance of the line.

The current at the load, from Eqns (6.20) and (6.21), is

$$I_L = \frac{V_1}{Z_0}\exp(-\gamma l) - \frac{V_2}{Z_0}\exp(\gamma l) \qquad (6.25)$$

and the load impedance is

$$Z_L = \frac{V_L}{I_L}$$

Therefore, from Eqns (6.25) and (6.22),

$$Z_L = \frac{V_1 \exp(-\gamma l) + V_2 \exp(\gamma l)}{(V_1/Z_0)\exp(-\gamma l) - (V_2/Z_0)\exp(\gamma l)}$$

Dividing by $V_1 \exp(-\gamma l)$

$$Z_L = Z_0 \left[\frac{1 + (V_2/V_1)\exp(2\gamma l)}{1 - (V_2/V_1)\exp(2\gamma l)}\right]$$

and from Eqn (6.23)

$$Z_L = Z_0 \left(\frac{1+\rho}{1-\rho}\right) \qquad (6.26)$$

or

$$\rho = (Z_L - Z_0)/(Z_L + Z_0) \qquad (6.27)$$

Consider the following three cases:

(i) *Short-circuit load,* $Z_L = 0$
From Eqn. (6.27), $\rho = -1$. Hence, from Eqn (6.24), $|\rho| = 1$ and $\psi = \pi$.

(ii) *Open-circuit load,* $Z_L = \infty$
From Eqn (6.27), $\rho = 1$, so $|\rho| = 1$ and $\psi = 0$.
(iii) $Z_L = 150 + j\,100\,\Omega$, $Z_0 = 50\,\Omega$
Then

$$\rho = (100 + j\,100)/(200 + j\,100)$$

from which $\rho = 0.6 + j\,0.2$
which gives $|\rho| = 0.63$ and $\psi = 18.43$.

Each of these cases is shown diagrammatically in Fig. 6.4 in terms of incident and reflected voltages.

So far we have been interested in the voltage on the line and the voltage reflection coefficient. However, we would expect the current wave to be

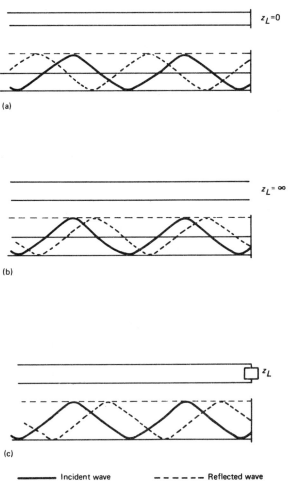

Fig. 6.4 (a) Voltage reflections at a short circuit, $|\rho| = 1$, $\psi = \pi$; (b) voltage reflections at an open circuit, $|\rho| = 1$, $\psi = 0$; (c) voltage reflections at a typical load z_L, $\rho = |\rho| \exp(j\psi)$.

reflected also, and by an analysis similar to that used above the amount of reflection can be seen to depend on the load impedance, as in the voltage case. This time, however, the current reflection coefficient, $\rho_I = |\rho_I| \exp(j\psi_I)$ is given by

$$\rho_I = (Z_0 - Z_L)/(Z_0 + Z_L)$$

For a short-circuit load $\rho_I = 1$ ($|\rho_I| = 1$, $\psi = 0$), which means that there is no change of phase between the incident and reflected waves – as we would expect.

From the discussion about reflection coefficients it is evident that the phase of the reflected wave relative to the incident wave is determined by Z_L and Z_0. However, we have not justified our implicit assumption that the incident current and voltage waves are in phase. Their phase relationship depends on Z_0, as can be seen by considering the case where there are no reflected waves, i.e. $V_2 = I_2 = 0$. Then

$$Z_0 = V_1 \exp(-\gamma z)/I_1 \exp(-\gamma z)$$

and if Z_0 is real, V_1 and I_1 will be in phase. In some applications Z_0 cannot be assumed to be real and then V_1 and I_1 will be out of phase.

In the following part of this chapter, the term reflection coefficient will apply to the voltage reflection coefficient given by Eqns (6.23), (6.24) and (6.27).

6.6 SENDING-END IMPEDANCE

It is often of interest to know the impedance that the combination of transmission line and load presents to the source, so that the degree of mismatch between source and line can be determined. The impedance looking into the line from the source, or generator, is called the sending-end impedance. It can be found in terms of the characteristic impedance Z_0, the load impedance Z_L, and the length of the line, l, as follows.

The impedance at some point A (Fig. 6.5) a distance z from the generator is, from Eqns (6.13), (6.20) and (6.21),

$$Z_A = \frac{V_A}{I_A} = Z_0 \left[\frac{V_1 \exp(-\gamma z) + V_2 \exp(\gamma z)}{V_1 \exp(-\gamma z) - V_2 \exp(\gamma z)} \right]$$

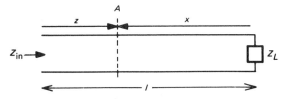

Fig. 6.5 Reference distance to calculate Z_{in}.

Substituting for V_2/V_1 from Eqn (6.23),

$$Z_A = Z_0 \left[\frac{\exp(-\gamma z) + \rho \exp(-2\gamma l)\exp(\gamma z)}{\exp(-\gamma z) - \rho \exp(-2\gamma l)\exp(\gamma z)} \right]$$

or

$$Z_A = Z_0 \left[\frac{\exp\gamma(l-z) + \rho\exp-\gamma(l-z)}{\exp\gamma(l-z) - \rho\exp-\gamma(l-z)} \right]$$

From Fig. 6.5, $x = l - z$; then

$$Z_A = Z_0 \left[\frac{\exp(\gamma x) + \rho\exp(-\gamma x)}{\exp(\gamma x) - \rho\exp(-\gamma x)} \right]$$

Replacing ρ from Eqn (6.27)

$$Z_A = Z_0 \left[\frac{(Z_L + Z_0)\exp(\gamma x) + (Z_L - Z_0)\exp(-\gamma x)}{(Z_L + Z_0)\exp(\gamma x) - (Z_L - Z_0)\exp(-\gamma x)} \right]$$

Rearranging terms

$$Z_A = Z_0 \left[\frac{Z_L[\exp(\gamma x) + \exp(-\gamma x)] + Z_0[\exp(\gamma x) - \exp(-\gamma x)]}{Z_L[\exp(\gamma x) - \exp(-\gamma x)] + Z_0[\exp(\gamma x) + \exp(-\gamma x)]} \right]$$

The exponentials can be replaced by hyperbolic functions to give

$$Z_A = Z_0 \left[\frac{Z_L \cosh \gamma x + Z_0 \sinh \gamma x}{Z_L \sinh \gamma x + Z_0 \cosh \gamma x} \right]$$

or

$$Z_A = Z_0 \left[\frac{Z_L + Z_0 \tanh \gamma x}{Z_L \tanh \gamma x + Z_0} \right] \tag{6.28}$$

The sending-end impedance, Z_{in}, is obtained from Eqn (6.28) by putting $x = l$, i.e.

$$Z_{in} = Z_0 \left[\frac{Z_L + Z_0 \tanh \gamma l}{Z_L \tanh \gamma l + Z_0} \right] \tag{6.29}$$

For general application of this equation it is often more useful if it is normalized to the characteristic impedance of the line. Normalized impedances will be shown as lower-case letters, e.g.

$$z_{in} = Z_{in}/Z_0 \quad \text{and} \quad z_L = Z_L/Z_0$$

From Eqn (6.29),

$$\frac{Z_{in}}{Z_0} = \left[\frac{(Z_L/Z_0) + \tanh \gamma l}{1 + (Z_L/Z_0)\tanh \gamma l} \right]$$

or

$$z_{in} = \frac{z_L + \tanh \gamma l}{1 + z_L \tanh \gamma l} \tag{6.30}$$

6.7 LINES OF LOW LOSS

In Eqn (6.12) we noted that the propagation constant is related to the line parameters by

$$\gamma = (R + j\omega L)^{1/2}(G + j\omega C)^{1/2}$$

$$= j\omega \sqrt{(LC)} \left(1 + \frac{R}{j\omega L}\right)^{1/2} \left(1 + \frac{G}{j\omega C}\right)^{1/2} \qquad (6.31)$$

Using the binomial series, this equation can be expanded to

$$\gamma = j\omega \sqrt{(LC)} \left(1 + \frac{R}{2\,j\omega L} - \frac{1}{4}\frac{R^2}{(j\omega L)^2} + \cdots\right)$$

$$\times \left(1 + \frac{G}{2\,j\omega C} - \frac{G^2}{4(j\omega C)^2} + \cdots\right) \qquad (6.32)$$

Now, in low-loss lines the impedance parameters R and G will be very small, allowing all terms in R^2 and G^2, and higher powers, to be ignored. Then

$$\gamma \simeq j\omega \sqrt{(LC)} \left[1 - \frac{RG}{4\omega^2 LC} - j\frac{R}{2\omega L} - j\frac{G}{2\omega C}\right] \qquad (6.33)$$

and since $\gamma = \alpha + j\beta$ we find that

$$\alpha \simeq \frac{R}{2}\sqrt{\frac{C}{L}} + \frac{G}{2}\sqrt{\frac{L}{C}} \qquad (6.34)$$

and

$$\beta \simeq j\omega \sqrt{(LC)} \left[1 - \frac{RG}{4\omega^2 LC}\right] \qquad (6.35)$$

Since R and G are small, we can reasonably assume that at high frequencies the second term in the bracket is negligible. Then

$$\beta \simeq j\omega \sqrt{(LC)} \qquad (6.36)$$

By a similar argument

$$Z_0 = (R + j\omega L)^{1/2}(G + j\omega C)^{-1/2}$$

$$= \sqrt{\left(\frac{j\omega L}{j\omega C}\right)\left(1 + \frac{R}{j\omega L}\right)^{1/2}\left(1 + \frac{G}{j\omega C}\right)^{-1/2}}$$

$$\simeq \sqrt{\left(\frac{L}{C}\right)\left[\left(1 + \frac{R}{2\,j\omega L}\right)\left(1 - \frac{G}{2\,j\omega C}\right)\right]}$$

if, as before, all terms in R^2, G^2 and above are ignored.

Expanding, and again assuming that the term in RG/ω^2 is negligible,

$$Z_0 \simeq \sqrt{\left(\frac{L}{C}\right)\left[1 + \frac{jR}{2\omega L} - \frac{jG}{2\omega C}\right]}$$

For many applications $R/\omega L$ and $G/\omega C$ will be very small, and a reasonable approximation for Z_0 is

$$Z_0 \simeq \sqrt{(L/C)} \tag{6.37}$$

Note that when this approximation is valid, Z_0 is real. Using Eqn (6.37) in Eqn (6.34) gives

$$\alpha \simeq \frac{R}{2Z_0} + \frac{GZ_0}{2} \tag{6.38}$$

6.8 LOSSLESS LINES

In many communications problems concerning transmission lines it is often a reasonable approximation to assume that the line is lossless, i.e. the attenuation is zero. The advantage of making this assumption is that it greatly simplifies the calculations involved, and the disadvantage of a loss in accuracy is not serious when the line is relatively short and operating at very high frequencies.

The effect of zero attenuation is to make the characteristic impedance real, as in Eqn (6.37), and to make the propagation constant imaginary, i.e.

$$\gamma = j\beta$$

We can see immediately the simplification that can result if we replace γ in Eqn (6.29) by $j\beta$. The sending-end impedance is then given by

$$Z_{in} = Z_0 \left[\frac{Z_L + jZ_0 \tan \beta l}{Z_0 + j Z_L \tan \beta l} \right] \tag{6.39}$$

using

$$\tanh j\beta l = j \tan \beta l.$$

Similarly, Eqn (6.30) simplifies to

$$z_{in} = \frac{z_L + j \tan \beta l}{1 + jz_L \tan \beta l} \tag{6.40}$$

6.9 QUARTER-WAVE TRANSFORMER

The relationship between the input and load impedances of a quarter-wave lossless line can be found easily from Eqn (6.39) expressed as

$$Z_{in} = Z_0 \left[\frac{Z_L/\tan \beta l + jZ_0}{Z_0/\tan \beta l + jZ_L} \right] \tag{6.41}$$

When l is a quarter-wavelength long, $\tan \beta l = \tan \pi/2 = \infty$. Then

$$Z_{in} = Z_0^2/Z_L \tag{6.42}$$

or

$$Z_{in}Z_L = Z_0^2 \tag{6.43}$$

We therefore have a method for matching a known load to a known input (line) impedance by placing between them a quarter-wave section of lossless line of characteristic impedance calculated from Eqn (6.43).

This type of transformer is very frequency sensitive because $\tan\theta$ varies rapidly about $\theta = \pi/2$. Problem 6.10 looks further into this sensitivity.

6.10 STUBS

Froms Eqns (6.30) and (6.40) we can see that the input impedance of a transmission line varies with its length. This property can be used in short lengths of line, known as stubs, to provide an adjustable impedance (or admittance) for use in matching applications, as we shall discuss later. Stubs are terminated in either a short-circuit or open-circuit load. If the load is a short circuit, $z_L = 0$, and from Eqn (6.40), if we assume that the stub is formed from a lossless length of line, the input impedance of the line is

$$z_{in} = j\tan\beta l \qquad (6.44)$$

If the load is an open circuit, $z_L = \infty$, and

$$z_{in} = -j\cot\beta l \qquad (6.45)$$

The variations of the input impedance with the length of the stub is shown in Fig. 6.6. It can be seen that the impedance is always reactive and that its value is repeated at half-wavelength intervals.

The use of stubs as matching devices will be discussed in a later section.

6.11 STANDING WAVES

The total voltage on a lossless line at some point z from the sending end is, from Eqn (6.17),

$$V = V_1\exp(-j\beta z) + V_2\exp(j\beta z)$$

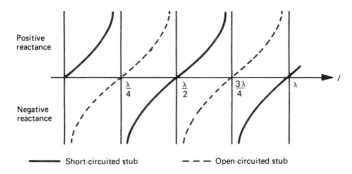

Fig. 6.6 Stub reactance.

and, remembering the time dependence,

$$V = [V_1 \exp(-j\beta z) + V_2 \exp(j\beta z)] \exp(j\omega t)$$

Rerranging the right-hand side, and substituting for V_2/V_1 from Eqn (6.23),

$$V = V_1 \exp(j\omega t) \exp(-j\beta l) [\exp(j\beta x) + \rho \exp(-j\beta x)] \quad (6.46)$$

where $x = l - z$ as shown in Fig. 6.5.

This equation represents a voltage standing wave, i.e. a stationary wave composed of two component travelling waves, one in the forward direction and the other in the backward direction, reflected from the load.

The precise shape of the voltage standing wave defined by Eqn (6.46) will depend on the value of ρ, which may be complex. In practice, a simple analytical solution of that equation may not be easy to obtain. However, as we noted earlier, when the load is a short or open circuit, ρ has the values -1 or $+1$, respectively, and in each of these cases simple trigonometric solutions of Eqn (6.46) can be found. Without the time dependence, Eqn (6.46) is

$$V = V_1 \exp(-j\beta l) [\exp(j\beta x) + \rho \exp(-j\beta x)] \quad (6.47)$$

which, for a short-circuit load ($\rho = -1$), becomes

$$V = V_1 \exp(-j\beta l) [\exp(j\beta x) - \exp(-j\beta x)]$$
$$= j2V_1 \exp(-j\beta l) \sin \beta x \quad (6.48)$$

This standing wave could be measured by using a standing-wave detector which displays the real part of the modulus of Eqn (6.48), i.e.

$$|V| = \text{Re}\,[2V_1 \exp(-j\beta l)|\sin \beta x|] \quad (6.49)$$

as shown in Fig. 6.7(a).

From Eqn (6.49) we can see that minima will occur on the line when $\beta x = (n-1)\pi$, where n is any positive integer. From this expression we can see that adjacent minima will be separated by π/β; that is, by $\lambda/2$, because

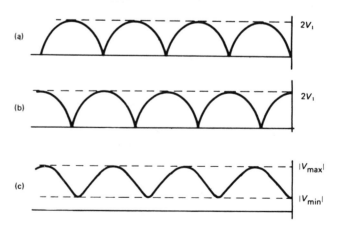

Fig. 6.7 Standing waves: (a) produced by a short circuit; (b) produced by an open circuit; (c) produced by a load z, with reflection coefficient $0.6 + j0.3$.

$\beta = 2\pi/\lambda$. The first voltage minimum will be at the load terminals, $x = 0$, as we have concluded earlier.

In a similar way we can use Eqn (6.47) to show that if there is an open circuit at the load the detected voltage will be

$$|V| = \text{Re}\,[2\,V_1\exp(-j\beta l)|\cos\beta\,x|]$$

which again has minima separated by $\lambda/2$, but with a maximum at the load. The standing wave due to an open circuit is shown in Fig. 6.7(b).

For any other load, ρ may be complex and, as an example, the standing wave for $\rho = 0.6 + j0.3$ is shown in Fig. 6.7(c). It can be seen that the minima do not fall to zero and the maxima do not rise to $2V_1$. However, adjacent minima are still half a wavelength apart, and indeed they will be for all values of ρ.

Current standing waves also exist and they can be discussed in exactly the same way but they will not be considered any further here.

6.12   VOLTAGE STANDING WAVE RATIO (VSWR)

The general shape of the voltage standing wave pattern on a lossless line is shown in Fig. 6.7(c). By using a detector, and moving it along the line, values for V_{max} and V_{min} can be obtained. The ratio of these two is called the voltage standing wave ratio, or VSWR. We shall use the symbol S to denote VSWR and define it as

$$S = \frac{|V_{max}|}{|V_{min}|} \tag{6.50}$$

which allows S to take values in the range $1 \leqslant S \leqslant \infty$. The value of S can often be determined experimentally without too much difficulty by using a slotted line.[1] Its value depends on the degree of mismatch at the load, i.e. on the reflection coefficient, as can be seen if the complex form of ρ given in Eqn (6.24) is put into Eqn (6.47):

$$V = V_1\exp(-j\beta l)\,[\exp(j\beta\,x) + |\rho|\exp(j(\psi - \beta\,x))]$$
$$= V_1\exp(-j\beta(l-x))\,[1 + |\rho|\exp(j(\psi - 2\beta\,x))] \tag{6.51}$$

so that

$$|V_{max}| = V_1[1 + |\rho|] \tag{6.52}$$

when

$$(\psi - 2\beta x) = 2(m-1)\pi \quad m = 1, 2, 3\dots$$

and

$$|V_{min}| = V_1[1 - |\rho|] \tag{6.53}$$

when

$$(\psi - 2\beta\,x) = 2(m-1)\pi \quad m = 1, 2, 3\dots$$

Therefore, from Eqns (6.52) and (6.53), the standing wave ratio is

$$S = \frac{|V_{min}|}{|V_{min}|} = \frac{1+|\rho|}{1-|\rho|} \tag{6.54}$$

We will find it convenient later to use this expression in the form

$$|\rho| = \frac{S-1}{S+1} \tag{6.55}$$

6.13 IMPEDANCE AT A VOLTAGE MINIMUM AND AT A VOLTAGE MAXIMUM

The normalized impedance at a voltage minimum can be derived from Eqn (6.51):

$$V = V_1 \exp[-j\beta(l-x)]\{1+|\rho|\exp[j(\psi-2\beta x)]\}$$

A voltage minimum occurs when

$$\exp[j(\psi-2\beta x)] = \exp(j\pi)$$

and this coincides with an impedance minimum.

At the first voltage minimum from the load, when $x = x_{min}$,

$$\psi = \pi + 2\beta x_{min}$$

and at x_{min}

$$Z = Z_{min} = \left(\frac{V}{I}\right)_{x_{min}}$$

Putting $x = x_{min}$ in Eqn (6.51) gives $V_{x_{min}}$ and $I_{x_{min}}$ is found by putting $z = l - x_{min}$ in Eqn (6.20).

Then

$$Z_{min} = \frac{\{V_1\exp[-j\beta(l-x_{min})]\}\{1+|\rho|\exp[j(\psi-2\beta x_{min})]\}}{\{V_1\exp[-j\beta(l-x_{min})]\}\{1-|\rho|\exp[j(\psi-2\beta x_{min})]\}}$$

$$= Z_0\frac{[1+|\rho|\exp(j\pi)]}{[1-|\rho|\exp(j\pi)]} = Z_0\left[\frac{1-|\rho|}{1+|\rho|}\right] = Z_0/S \tag{6.56}$$

or

$$z_{min} = 1/S \tag{6.57}$$

By a similar argument, we can show that

$$Z_{max} = Z_0 S \tag{6.58}$$

and

$$z_{max} = S \tag{6.59}$$

6.14 LOAD IMPEDANCE ON A LOSSLESS LINE

There are often occasions when the load impedance is not known, but its value is required. It can be found, if the line is assumed to be lossless, by measuring the VSWR, the wavelength and the distance from the load to the nearest voltage minimum.

Re-writing Eqn (6.51)

$$V = V_1 \exp[-j\beta(l-x)]\{1 + |\rho| \exp[j(\psi - 2\beta x)]\}$$

and, as we noted in the last section, this is a minimum when

$$\exp[j(\psi - 2\beta x)] = -1$$

i.e. when

$$\psi - 2\beta x = (2m-1)\pi \quad m = 1, 2, 3 \ldots$$

The solution of interest is that for the smallest value of x, which we have called $x_{\min}$. It occurs when $m = 1$; then

$$\psi - 2\beta x_{\min} = \pi$$

or

$$\psi = \pi + 2\beta x_{\min} \tag{6.60}$$

The load impedance and the reflection coefficient are related by

$$Z_L = Z_0 \left(\frac{1+\rho}{1-\rho} \right)$$

i.e. using Eqn (6.24),

$$Z_L = Z_0 \left(\frac{1 + |\rho| \exp(j\psi)}{1 - |\rho| \exp(j\psi)} \right)$$

and substituting for $|\rho|$ from Eqn (6.55),

$$Z_L = Z_0 \left(\frac{1 + [(S-1)/(S+1)] \exp(j\psi)}{1 - [(S-1)/(S+1)] \exp(j\psi)} \right)$$

ψ can be replaced by Eqn (6.60):

$$Z_L = Z_0 \left[\frac{1 + [(S-1)/(S+1)] \exp[j(\pi + 2\beta x_{\min})]}{1 - [(S-1)/(S+1)] \exp[j(\pi + 2\beta x_{\min})]} \right] \tag{6.61}$$

Rearranging Eqn (6.61), and remembering that $e^{j\pi} = -1$,

$$Z_L = Z_0 \left[\frac{S[1 - \exp(j2\beta x_{\min})] + [1 + \exp(j2\beta x_{\min})]}{S[1 + \exp(j2\beta x_{\min})] + [1 - \exp(j2\beta x_{\min})]} \right]$$

Dividing throughout by $\exp(j\beta x_{\min})$ gives

$$Z_L = Z_0 \left[\frac{S[\exp(-j\beta x_{\min}) - \exp(j\beta x_{\min})] + [\exp(-j\beta x_{\min}) + \exp(j\beta x_{\min})]}{S[\exp(-j\beta x_{\min}) + \exp(j\beta x_{\min})] + [\exp(-j\beta x_{\min}) - \exp(j\beta x_{\min})]} \right]$$

or

$$Z_L = Z_0 \left[\frac{-jS \sin \beta x_{min} + \cos \beta x_{min}}{S \cos \beta x_{min} - j \sin \beta x_{min}} \right]$$

$$= Z_0 \left[\frac{1 - jS \tan \beta x_{min}}{S - j \tan \beta x_{min}} \right] \tag{6.62}$$

The normalized load impedance is therefore

$$z_L = \frac{1 - jS \tan \beta x_{min}}{S - j \tan \beta x_{min}} \tag{6.63}$$

6.15 SMITH TRANSMISSION LINE CHART

Transmission line problems can be solved using the equations we have derived in the earlier part of this chapter, but often there is a considerable amount of tedious algebraic manipulation involved that can make calculations rather lengthy. To overcome this problem, several graphical techniques have been developed. We will study only one – the Smith chart, named after P. H. Smith who introduced it in 1944.[2] It is the best known and most comprehensive of all graphical methods, and compared with analysis it has the considerable additional benefit of providing a picture of the conditions on the line. This allows us to assess what sort of adjustments might be necessary to produce the conditions we require.

The derivation of the chart is not difficult to follow, but it is easier to appreciate after some familiarity with its use has been developed, so a discussion on why it has its particular shape will be deferred until a later section. The most commonly used version of the Smith chart is shown in Fig. 6.8. Some time should be spent in reading the various legends inside the chart and on the circumferential scales. You will notice that there are three sets of circular arcs in the chart, and one straight line. The chart is designed to allow any value of impedance or admittance to be represented, and in the examples that we study later it will be used in both modes. The horizontal line dividing the chart into an upper and a lower half is the locus of impedances and admittances having no reactive or susceptive component. Along that line the following equations are satisfied

$$z = r + j0$$

$$y = g + j0$$

Figure 6.9(a) shows one set of circles. They all have their centres on the horizontal line, and they are contiguous at the right-hand end of the chart. These circles give the normalized resistive or conductive component of impedance or admittance, i.e. the r or g in the above equations. The centre of the chart, where the circle representing unity normalized resistance or conductance crosses the horizontal line, has the coordinates $(1, 0)$. It therefore represents the impedance of the line under consideration. When we discuss

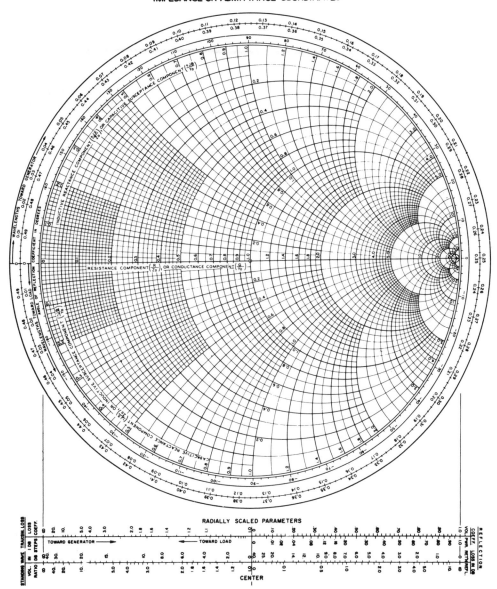

IMPEDANCE OR ADMITTANCE COORDINATES

Fig. 6.8 Smith chart. Reproduced by permission of Phillip H. Smith, Analog Instruments Company, New Providence, N.J.

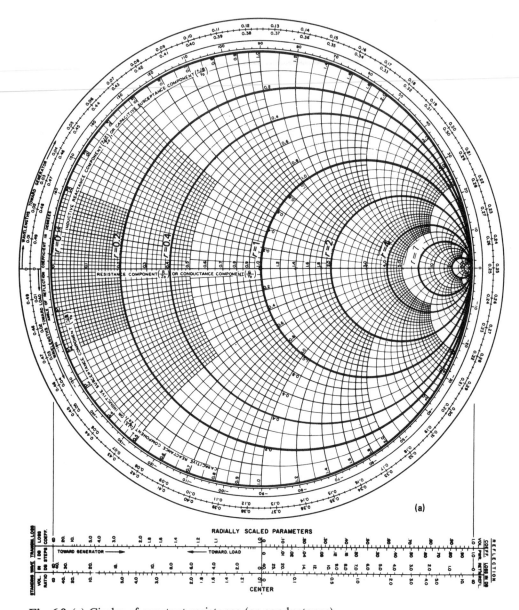

(a)

Fig. 6.9 (a) Circles of constant resistance (or conductance).

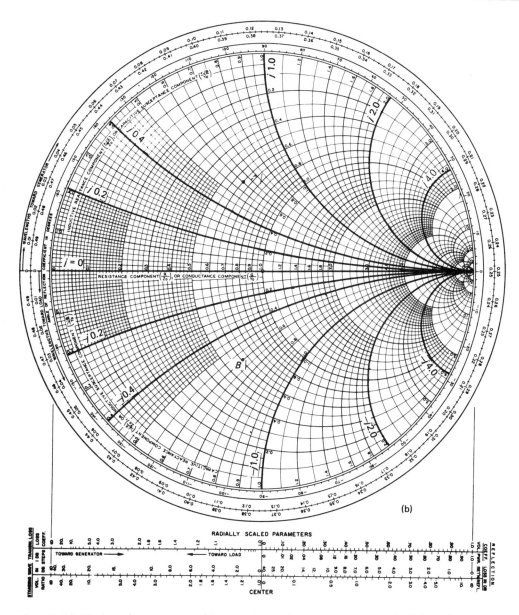

Fig. 6.9 (b) Circles of constant positive reactance (or susceptance), and circles of constant negative reactance (or susceptance).

methods of matching, the object will be to change the impedance at the matching point to the value represented by $(1, 0)$.

The resistance, or conductance, circles range from zero at the outer periphery to infinity at the right-hand side, although the highest value of normalized resistance or conductance shown is 50. Another set of circular arcs is shown in Fig. 6.9(b). These are lines that represent the positive imaginary components of impedance or admittance. For example, the point A has an impedance or admittance of $0.6 + j0.6$.

The third set of arcs, shown in Fig. 6.9(b), is in the lower half of the chart and represents negative imaginary components of impedance or admittance. Point B has an impedance or admittance of $0.6 - j0.6$.

The arcs for the imaginary components are parts of circles that have their centres on a vertical line through the right-hand end of the horizontal line and they all touch at that point (see Section 6.19). Only those parts of the arcs that occur in the range of positive real components are shown.

Another feature that will be used regularly can be seen from Fig. 6.10. Plot $z = 1.6 - j2.0$ on the chart. Then draw a circle, with centre at $(1, 0)$ through that point. From z draw a diameter to cut the circle at the opposite side. The point of intersection has a value $y = 0.25 + j0.31$ and comparing y and z it is found that $z = 1/y$. This relationship applies in general, i.e. the relative impedance and admittance are at opposite ends of a diameter of the circle, with centre $(1,0)$ passing through them. The inner scale round the chart is marked 'angle of reflection coefficient in degrees'. It relates the characteristic impedance of the line, at $(1, 0)$, to any load impedance, e.g. in Fig. 6.10, if point A represents a load of $1.2 + j1.4$, the angle of the reflection coefficient can be read from the intersection of a line from $(1, 0)$ through A, and the inner scale at B, to be $\psi = 49.2°$.

The magnitude of the reflection coefficient, $|\rho|$, is related to the distance between $(1,0)$ and A. Below the chart there are eight radial scales, and the uppermost one on the right is marked 'Voltage Reflection Coefficient'. $|\rho|$ is found by measuring the distance from $(1,0)$ to A along that scale from the centre. In the example given $|\rho| = 0.541$.

The other important radial scale, which is related through Eqn (6.54) to that of $|\rho|$, is the lowest one on the left, marked 'Standing Wave Voltage Ratio'. We know that if $|\rho|$ is constant, S is also represented by a circle. As an example, Fig. 6.10 shows the circle for $S = 4.5$. Its radius is found from the bottom-left radial scale, but we can see from the figure that the circle passes through the point 4.5 on the horizontal line across the chart. The resistance scale to the right of the centre of the chart is identical with the VSWR scale, and either may be used.

It will help if we consider an example of how these various aspects of the Smith chart are used.

EXAMPLE: A lossless transmission line of characteristic impedance $50\,\Omega$ is terminated in a load of $150 + j75\,\Omega$. Find the reflection coefficient, the load admittance, the VSWR, the distance between the load and the nearest voltage minimum to it, and the input impedance, if the line is 92 cm long and the wavelength of the signal on the line is 40 cm.

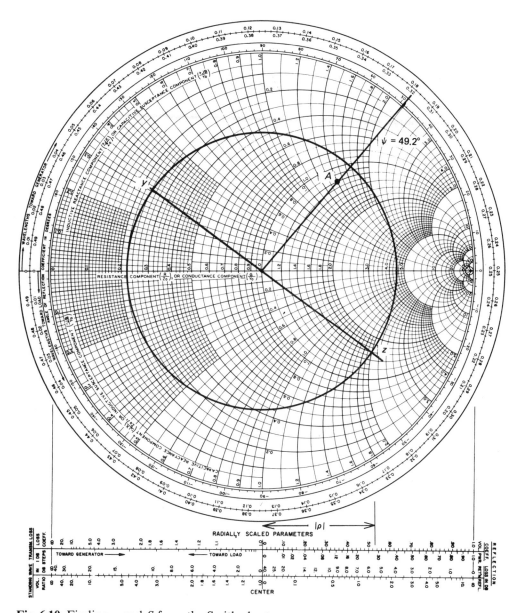

Fig. 6.10 Finding ρ and S from the Smith chart.

The normalized load impedance is z_L where

$$z_L = \frac{150 + j75}{50} = 3 + j1.5$$

and it is plotted on the chart as shown in Fig. 6.11. Then the S circle is drawn through z_L. We can now read the value of VSWR to be $S = 3.8$. By drawing a line from $(1, 0)$ through z_L to the Angle of Reflection Coefficient scale we can read that $\psi = 16°$; $|\rho|$ is measured on the appropriate radial scale to be 0.585. Thus we have found that $\rho = 0.585 \exp(j0.28)$.

The load admittance y_L is on the same S circle as z_L, and at the opposite end of a diameter. Thus, from the chart, $y_L = 0.27 - 0.13$.

The line is 92 cm long which is equivalent to $(92/40)\lambda = 2.3\lambda$. Therefore, to find the input impedance z_{in} we must go from the load a distance 2.3λ towards the generator, on a constant S circle. This we can do by noting where the line from the centre through z cuts the Wavelengths Towards Generator scale. Reading from the chart, this occurs at 0.227λ. The input impedance is therefore at $(2.3 + 0.227)\lambda$ on the Wavelengths Towards Generator scale, i.e. at 2.527λ. One complete revolution on an S circle is 0.5λ; therefore in 2.527λ there are five revolutions $+ 0.027\lambda$. Hence the input impedance is at 0.027λ on the Wavelengths Towards Generator scale and reading from the chart, $z_{in} = 0.27 + j0.16$.

Finally, we want to know the distance of the nearest voltage minimum to the load from the load itself, x_{min}. The normalized impedance at a voltage minimum, which has been shown to be $1/S$ [Eqn (6.57)], must lie on the horizontal line to the left of $(1, 0)$. Then x_{min} is found from the chart by reading the distance from z_L to V_{min} on the Wavelengths Towards Generator scale, and V_{min} will be at point C on the chart which reads 0.0 on the scale. Therefore V_{min} is 0.273λ from the load.

All of these results can be obtained by using the equations which were developed earlier, and to do so would be a very useful exercise.

The decision on whether to use the chart in the admittance or impedance mode depends on the application. If lines are joined in series then, because the impedances at the junction can be added, the impedance chart should be used. Alternatively, if there are parallel junctions on the line, then admittances would be used. The differences in the use of the chart can be summarized as follows (see Fig. 6.12).

1. When the chart is used for impedance
   (a) The intersection of the circle S with the line $r + j0$ gives the maximum impedance on the line at the intersection to the right of $(1,0)$, and the minimum impedance at the intersection to the left.
   (b) Voltage minima are at impedance minima.
2. When the chart is used for admittance
   (a) The intersection of the circle S with the line $g + j0$ gives the maximum admittance at the intersection to the right of $(1, 0)$, and the minimum admittance at the intersection to the left.
   (b) Voltage minima are at admittance maxima.

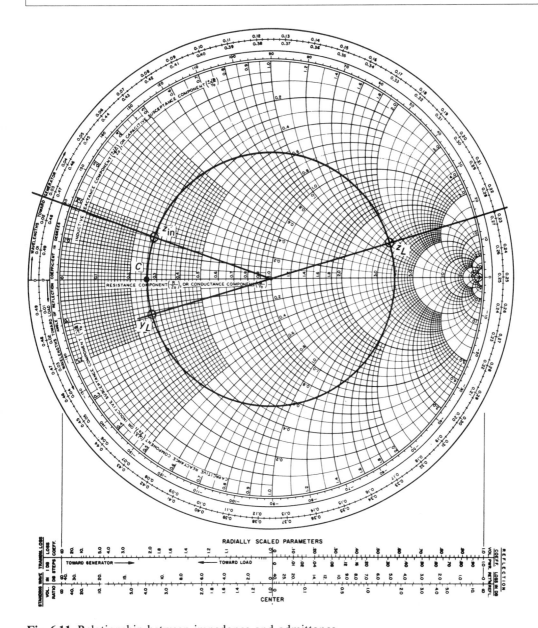

Fig. 6.11 Relationship between impedance and admittance.

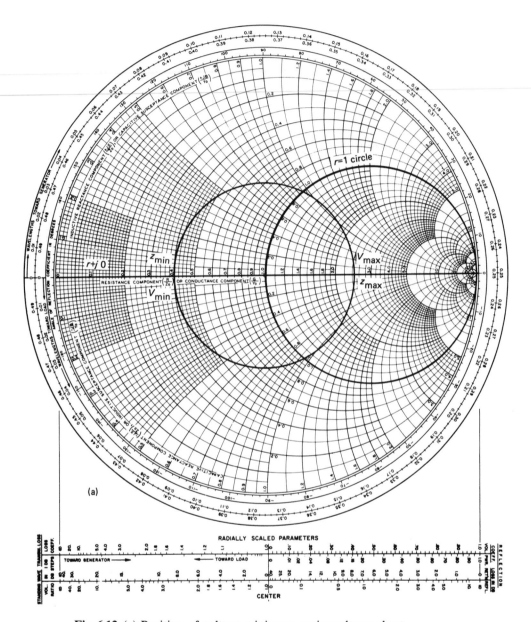

Fig. 6.12 (a) Position of voltage minimum on impedance chart.

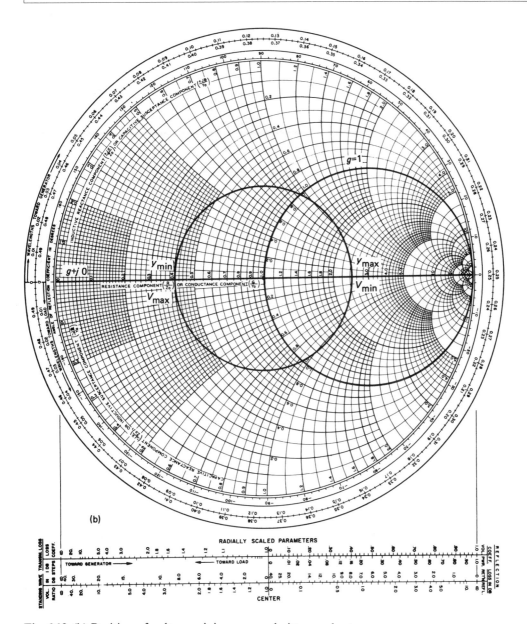

Fig. 6.12 (b) Position of voltage minimum on admittance chart.

6.16   STUB MATCHING

The presence of a reflected wave and a high VSWR, due to a mismatched load, can have undesirable effects on both the generator and the line. The reflected wave may either interfere with the performance of the generator or cause it some damage. The voltage maximum of the standing wave may, in high-voltage applications, produce breakdown stresses in the dielectric. Therefore there is some value in arranging for as much of the line as possible to have a VSWR of unity so that there is no reflected wave at the generator, and this can be done by using stubs to match the line to the load at some point near the load.

6.17   SINGLE-STUB MATCHING

Moving from the load towards the generator the immittance (impedance or admittance) of the line will, at some point, have a normalized real part of unity. If a stub is inserted there and its length adjusted so that it presents an imaginary imittance equal to, and of opposite sign from, that of the line at that point the line will be matched on the generator side of the stub. In this method there are two distances to be determined: first, the position of the stub relative to the load and, second, the length of the stub itself. The stub, as described in Section 6.8, may be either open or short circuited, and may be joined to the line in either series or parallel.

EXAMPLE: Consider a line terminated in an unknown load. Measurements show that the VSWR is 3.5, adjacent minima are 15 cm apart, and $x_{min} = 1.2$ cm. A short-circuit shunt stub is to be used to match the line, as shown in Fig. 6.13(a).

Because a shunt stub is being used, from the information given, calculations should be made in admittances. The VSWR from the load to y_1 is given as 3.5, and this is drawn on the Smith chart as shown in Fig. 6.13(b). y_{in} provides a match to the line, i.e.

$$y_{in} = 1 + j0$$

The stub can only add susceptance, shown as jb, and therefore

$$y_{in} = y_1 + jb$$

or

$$1 + j0 = y_1 + jb$$

and

$$y_1 = 1 - jb$$

Hence y_1 must be on the locus of unit conductance, i.e. the $g = 1$ circle, and as it must also lie on the $S = 3.5$ circle it is situated at the points at which these two intersect, shown in Fig. 6.13(b) as a and b.

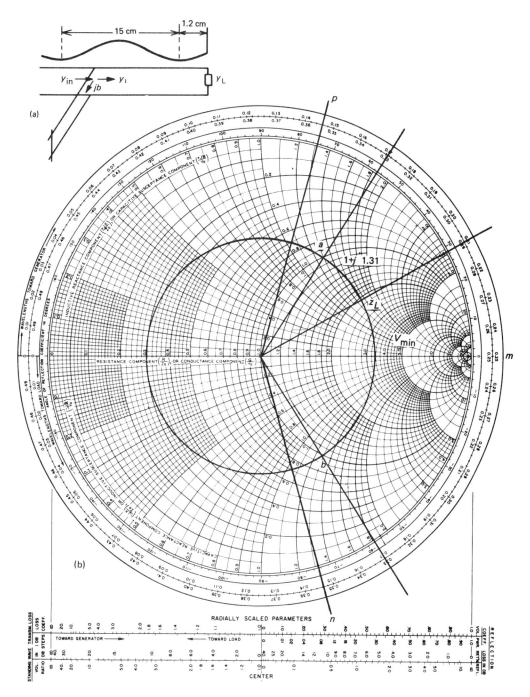

Fig. 6.13 Single-stub matching. (a) Example of shunt stub; (b) Smith chart representation.

Consider point a

The admittance of a is, from the chart, $1 + j1.31$. Therefore the stub must supply a susceptance of $-j1.31$. The position of V_{min} is known – it is at the point shown on the chart, which is $0.421\lambda\,(= 12.63\,\text{cm})$ towards the load from a. The voltage minimum is given as $1.2\,\text{cm}$ from the load, making point a $13.83\,\text{cm}$ from the load.

The length of the stub can be found from the chart with reference to Fig. 6.13(b). The short-circuit load produces a VSWR of ∞ in the stub, which is represented on the chart by the outermost circle of the chart proper. The load admittance of the stub is $y_L = \infty$ which is at the right-hand end of the horizontal line, point m in Fig. 6.13(b). The stub length is therefore determined by moving from m towards the generator on the $S = \infty$ circle to the point at which susceptance is $-j1.31$, shown as point n in the diagram. Thus mn is the stub length required.

From the chart, $mn = 0.104\,\lambda$, and since $\lambda = 30\,\text{cm}$, $mn = 3.12\,\text{cm}$

Hence one solution to the problem is to make the short-circuited stub $3.12\,\text{cm}$ long and place it $13.83\,\text{cm}$ from the load.

Alternatively, consider point b

The susceptance at b is $-j1.31$ so the stub susceptance must be $j1.31$.

Then, following the same procedure as before, the stub length is given by the distance from m to the point on the $S = \infty$ circle where susceptance is $j1.31$, i.e. point p.

From the chart, $mp = 0.396\,\lambda = 11.88\,\text{cm}$

The stub positions, from V_{min} to b, is the same distance as a from V_{min}, but towards the generator. V_{min} to b is $0.079\,\lambda$, making the total distance from the load to the stub $(0.079 + 0.04)\,\lambda = 3.57\,\text{cm}$.

The other solution, therefore, is to make the stub $11.88\,\text{cm}$ long, and place it $3.57\,\text{cm}$ from the load.

6.18 DOUBLE-STUB MATCHING

The use of a single stub may not be convenient for some applications because of the need to be able to adjust the position of the stub relative to the load if the load impedance is changed. By using two stubs the two variables necessary to produce a match are the lengths of the stubs, and the stub positions can be fixed. Their position is not entirely arbitrary, as will be seen later. This method of matching is referred to as double stub, and the basic idea behind it can be understood with the help of Fig. 6.14. Admittances are used again, because short-circuited shunt stubs are shown. If series stubs were being used, it would be necessary to work in impedances.

The input admittance of the stub nearest to the generator, y_{in}, must match the line, so

$$y_{in} = 1 + j0$$

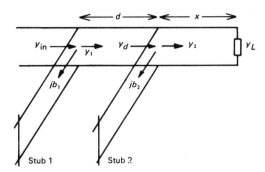

Fig. 6.14 Diagram of two-stub matching.

which is at the centre of the chart. The admittance on the load side of the first stub, y_1, must be given by

$$y_{in} = y_1 + jb_1$$

or

$$y_1 = 1 - jb_1$$

which lies somewhere on the unity conductance $g = 1$ circle. y_d, the admittance on the generator side of stub 2, must lie on the same VSWR circle as y_1, since it belongs to the same section of line, and it can be found by moving from y_1 on a constant S circle a distance d towards the load. However, the precise position of y_1 is not known. All that we know is that it lies on the $g = 1$ circle, so y_d must lie on the $g = 1$ circle translated through distance d towards the load. This circle we will call, after Everitt and Anner,[3] the L circle.

If y_d is on the L circle, y_1 lies on the $g = 1$ circle, as required, because each point on the L circle has a corresponding point, distance d towards the generator, on the $g = 1$ circle. The position of the L circle is determined by the stub separation d, and it can be drawn on the chart, once a value for d has been fixed, by drawing a circle of the same radius as the $g = 1$ circle on a centre equal to the $g = 1$ circle moved a distance d, on a constant S circle, towards the load.

The function of the first stub is to add the susceptance or reactance required to cancel that of y_1, in much the same way as was done with the single stub in the previous method.

EXAMPLE: A lossless transmission line, of characteristic impedance $50\,\Omega$, is terminated in a load of $100 + j25\,\Omega$. Two short-circuited series stubs are to be used to match the line. One stub is 8 cm and the other 32 cm from the load. The signal on the line is at 750 MHz. Find the lengths of the stubs.

Figure 6.15(a) shows the arrangement. Because series stubs are used, the problem will be solved by using impedances.

First, the stub positions must be fixed in terms of wavelength. 750 MHz is equivalent to a wavelength of 40 cm. So the first stub is $0.2\,\lambda$ from the load, and the distance between the stubs is $0.6\,\lambda$.

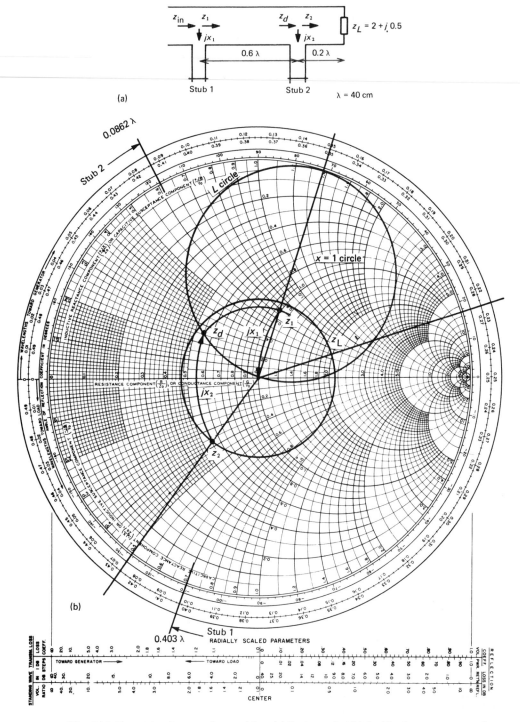

Fig. 6.15 Example of two-stub matching. (a) Series stubs; (b) Smith chart representation.

The L circle is drawn with its centre $0.6\,\lambda(\equiv 0.1\,\lambda)$ towards the load from the $x = 1$ circle, as shown in Fig. 6.15(b). The normalized load, $2 + j0.5$, is fixed on the chart, and the VSWR circle is drawn.

The value of z_2 can be found by moving on the constant S circle a distance of $0.2\,\lambda$ towards the generator from z_L as shown in Fig. 6.15(b). Thus the impedance z_2, at the load side of the terminals of stub 2 is

$$z_2 = 0.55 - j0.36$$

The stub must add a reactance to z_2 which is sufficient to put z_d on the L circle. Moving on a constant resistance circle from z_2, z_d is on the L circle at the point shown. From the chart

$$z_d = 0.55 + j0.25$$

The difference in reactance between z_d and z_2 is that provided by stub 2. From the chart, then, stub 2 adds reactance of $j0.61$ to z_2 to give z_d on the L circle.

z_1 is found by moving towards the generator a distance d on a constant S circle from z_d, i.e. by moving on a constant S circle from z_d, on the L circle, to the corresponding point, d wavelengths towards the generator, on the $x = 1$ circle. Note that once z_d is fixed there is only one point on the $x = 1$ circle that corresponds to z_1. From Fig. 6.15, $z_1 = 1 + j0.7$ and stub 1 must add the reactance required to reduce the reactance of z_1 to zero, i.e. $- j0.7$.

The stubs are short-circuited, so to find their length it is necessary to start at point $z = 0$, and go round the $S = \infty$ circle towards the generator to the required reactance.

Stub 1 is $0.8\,\lambda$ from the load, and 16.12 cm long.
Stub 2 is $0.2\,\lambda$ from the load, and 3.45 cm long.

Earlier, it was stated that the choice of stub separation cannot be entirely arbitrary. In the above example, if the first stub were $0.08\,\lambda$ and the second $0.29\,\lambda$, from the load the L circle would be drawn as shown in Fig. 6.16. In that case it would not be possible to place z_d on the L circle by adding positive or negative reactance to z_2, and a match could not be obtained. In that event, different stub positions would have to be used.

6.19   DERIVATION OF THE SMITH CHART

The construction of the chart is based on a conformal transformation of impedance from the z plane to the ρ plane. Earlier, we derived the relationship between the load impedance and the reflection coefficient.

$$z_L = \frac{1 + \rho}{1 - \rho} \tag{6.26}$$

In general, both z_L and ρ are complex.

$$\text{Let } z_L = r + jx \quad \rho = u + jv$$

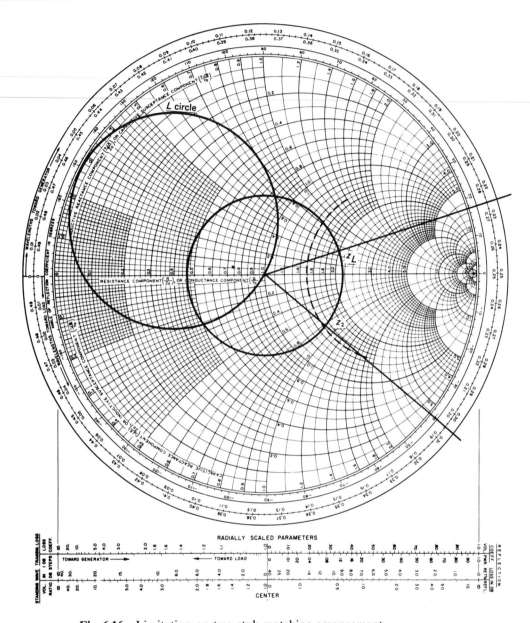

Fig. 6.16 Limitation on two-stub matching arrangement.

Substituting these values into Eqn (6.26) gives

$$r + jx = \frac{1 + u + jv}{1 - (u + jv)} = \frac{1 - u^2 - v^2 + 2jv}{(1 - u)^2 + v^2}$$

and from this expression

$$r = \frac{1 - u^2 - v^2}{(1 - u)^2 + v^2} \qquad x = \frac{2v}{(1 - u)^2 + v^2}$$

These equations can be put in the form

$$\left(u - \frac{r}{1 + r}\right)^2 + v^2 = \frac{1}{(1 + r)^2} \qquad (6.64)$$

$$(u - 1)^2 + \left(v - \frac{1}{x}\right)^2 = \frac{1}{x^2} \qquad (6.65)$$

Equation (6.64) is the equation of a circle in the ρ plane if r is constant, with the centre of the circle at

$$u = \frac{r}{1 + r} \qquad v = 0$$

r cannot be negative, so the centre of the circle must lie on the horizontal axis between $u = 0$ and $+1$. The radius of the circle is $1/(1 + r)$.

Similarly, Eqn (6.65) is also the equation of a circle in the ρ plane, when x is constant. Now the centre of the circle is at

$$u = 1, \quad v = 1/x$$

which lies on a vertical line, one unit to the right of the origin. If x is positive, the centre points will be above $v = 0$, and if x is negative they will lie below that point.

The Smith chart therefore consists of plots of Eqns (6.64) and (6.65) for various values of r and x. The result is a series of orthogonal circles, with resistance circles having their centres on the horizontal line, and all passing through the point $u = 1$, $v = 0$, at the right-hand edge of the chart; and reactance circles having their centres on the vertical line $u = 1$, and their perimeters also tangential at the point $u = 1, v = 0$. As we noted earlier, all $r =$ constant circles are shown, but only those segments of the $x =$ constant circles situated in the region $r \geqslant 0$. On the chart, apart from $v = 0$, which corresponds to the line $x = 0$, there are no u or v values shown. The numbers given on the circles are for r and x, but if it is remembered that the radius of the circle $r = 0$, is one unit of u, then Eqns (6.64) and (6.65) are consistent with the normalized values on the chart.

The assumption that $\rho = u + jv$ means that $\tan \psi = v/u$. Therefore, any line from the centre of the chart to the outer circumference will be of constant ψ. Angular movement round the chart therefore represents a movement with ψ, and so the inner circumferential scale is marked 'Angle of Reflection Coefficient in Degrees'.

Any circle with centre at the origin must satisfy the equation

$$u^2 + v^2 = |\rho|^2$$

and therefore it has radius $|\rho|$. From Eqn (6.54) the VSWR and $|\rho|$ are directly related, so constant $|\rho|$ means constant S.

6.20 TRAVELLING WAVES ON A LOSSLESS LINE

In Section 6.1 we assumed that forward and backward travelling waves would be produced by the time-varying input signal, and in deriving the voltage equations we ignored the effect of time. Now we will consider again our earlier ideas, and show that we do indeed have waves travelling along the line.

Referring to Eqns (6.3) and (6.6), i.e.

$$\frac{\partial I}{\partial z} = -\left(G + C\frac{\partial}{\partial t}\right)V \tag{6.3}$$

and

$$\frac{\partial V}{\partial z} = -\left(R + L\frac{\partial}{\partial t}\right)I \tag{6.6}$$

Differentiating Eqn (6.3) with respect to t, and Eqn (6.6) with respect to z,

$$\frac{\partial^2 I}{\partial t\partial z} = -\left(G + C\frac{\partial}{\partial t}\right)\frac{\partial V}{\partial t} \tag{6.66}$$

and

$$\frac{\partial^2 V}{\partial z^2} = -\left(R + L\frac{\partial}{\partial t}\right)\frac{\partial I}{\partial z} \tag{6.67}$$

6.21 WAVE EQUATION FOR A LOSSLESS TRANSMISSION LINE

In a lossless line it is assumed that R and G can be ignored. Then Eqns (6.66) and (6.67) become

$$\frac{\partial^2 I}{\partial t\partial z} = -C\frac{\partial^2 V}{\partial t^2} \tag{6.68}$$

and

$$\frac{\partial^2 V}{\partial z^2} = -L\frac{\partial^2 I}{\partial t\partial z} \tag{6.69}$$

Substituting in Eqn (6.69) for $\partial^2 I/(\partial t\partial z)$ from Eqn (6.68),

$$\frac{\partial^2 V}{\partial z^2} = LC\frac{\partial^2 V}{\partial t^2} \tag{6.70}$$

This is the usual form of what is known as the wave equation. It is satisfied by any function of the form

$$f\left(t - \frac{z}{v}\right)$$

so that, if

$$V = V_1 f\left(t - \frac{z}{v}\right)$$

then

$$\frac{\partial^2 V}{\partial z^2} = \frac{V_1}{v^2} \quad \text{and} \quad \frac{\partial^2 V}{\partial t^2} = V_1$$

and Eqn (6.70) is satisfied if

$$1/v^2 = LC$$

A similar argument shows that with the same condition on V_2, the function $f[t + (z/v)]$ is also a solution of Eqn (6.70), so that the total voltage on the line is

$$V = V_1 f\left(t - \frac{z}{v}\right) + V_2 f\left(t + \frac{z}{v}\right) \tag{6.71}$$

The function $f[t - (z/v)]$ can be seen to represent a travelling wave if it is considered that in order to retain a position of constant phase on the line it is necessary that $[t - (z/v)]$ is constant, i.e. that a movement in the z direction of δz must be related to the time it takes to make the movement δt, in such a way that

$$t - \frac{z}{v} = t + \delta t - \frac{z + \delta z}{v}$$

or

$$\delta t = \frac{\delta z}{v}$$

Hence the velocity of propagation, defined as the change in z in unit time, is

$$v = \frac{\delta z}{\delta t}$$

Thus v in Eqn (6.71) is the velocity of the wave and is given by

$$v = 1/\sqrt{(LC)} \tag{6.72}$$

v is referred to as the **phase velocity** of the wave.

If a sinusoidal signal is placed on the line, the functions $f[t \pm (z/v)]$ are written

$$\exp[j\omega(t \pm \beta z)] \tag{6.73}$$

for a lossless line, as noted earlier. Then,

$$V = V_1 \exp[j\omega(t - \beta z)] + V_2 \exp[j\omega(t + \beta z)] \tag{6.74}$$

Comparing Eqns (6.71) and (6.74),

$$\frac{\beta}{\omega} = \frac{1}{v} \quad \text{i.e.} \quad v = \frac{\omega}{\beta} \tag{6.75}$$

so, from Eqn (6.72),

$$\beta = \omega \sqrt{(LC)} \tag{6.76}$$

6.22   CONCLUSION

In this chapter we have introduced and examined several concepts such as reflection, standing waves, line impedance, characteristic impedance, matching and travelling waves. We shall see in some of the following chapters that these same ideas occur in different contexts, but with similar meaning, and it is this wide applicability of the main features of transmission line theory that makes it so important to the study and design of communications transmission systems in general.

REFERENCES

1. Chipman, R. A., *Transmission Lines*, Schaum — McGraw-Hill, New York, 1968.
2. Smith, P. H., 'An improved transmission line calculator', *Electronics*, Jan 1944, 130.
3. Everitt, W. E. and Anner, G. E., *Communication Engineering*, McGraw-Hill, New York, 1956.
4. Davidson, C. W., *Transmission Lines for Communications*, Macmillan, London, 1978.

PROBLEMS

Note: In all problems assume velocity of wave $= c$.

6.1   A lossless transmission line of $Z_0 = 100\,\Omega$ is terminated by an unknown impedance. The termination is found to be at a maximum of the voltage standing wave, and the VSWR is 5. What is the value of the terminating impedance?

Answer: $500\,\Omega$.

6.2   A load of $90 - j120\,\Omega$ terminates a $50\,\Omega$ lossless transmission line. Find, both analytically and by using the Smith chart,

(a) the reflection coefficient;
(b) the VSWR;
(c) the distance from the load to the first voltage minimum
if the operating frequency is 750 MHz.

Answer: (a) $0.69e^{-0.54}$; (b) 5.4; (c) 8.28 cm.

6.3   Calculate the impedance at terminals *AA* of the lossless transmission line system shown. All three lines have a characteristic impedance of $50\,\Omega$. The reflection coefficients, at an operating frequency of 750 MHz, are $\rho_1 = 0.6 - j0.5$ and $\rho_2 = 0.8 + j0.6$.

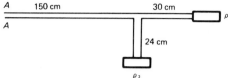

Answer: $0.9 + j17.75 \, \Omega$.

6.4 A section of line $\lambda/4$ long at 100 MHz is to be used to match a $50 \, \Omega$ resistive load to a transmission line having $Z_0 = 200 \, \Omega$.
(a) Find the characteristic impedance of the matching section.
(b) What is the input impedance of the matching section at 12 MHz?
(c) Repeat part (b) for a frequency of 80 MHz.

Answer: (a) $100 \, \Omega$; (b) $155.5 - j68.5 \, \Omega$; (c) $67.5 + j47.3 \, \Omega$.

6.5 A load Z produces a VSWR of 5 on a $50 \, \Omega$ lossless transmission line. The first minimum is 0.2λ from the load and the frequency of operation is 750 MHz. Without using a Smith chart, find the position and length of a short-circuit shunt stub that will match the load to the line.

Answer: Either 3.252 cm long, 5.608 cm from load, or 16.478 cm long, 10.948 cm from load.

6.6 A $50 \, \Omega$ lossless transmission line is terminated by a load $60 + j90 \, \Omega$. Find the position and length of an open-circuit series stub that will provide a match at 800 MHz.

Answer: Either 15.4 cm from load, 4.95 cm long or 3.26 cm from load, 18.4 cm long.

6.7 A lossless transmission line is carrying a signal of 47 cm wavelength, $S = 4.5$ and $x_{\min}$ is 5 cm from the load. Find the lengths of the two open-cricuited series stubs that will match the line if one is placed 0.43λ and the other 0.22λ from the load.

Answer: 11.49 cm at 0.22λ, 19.27 cm at 0.43λ.

6.8 A lossless transmission line with $Z_0 = 200 \, \Omega$ is terminated by a load Z_L. If the VSWR is 6.5, and $x_{\min}$ is 0.168λ, find Z_L and the length of the two short-circuited shunt stubs required to match the line to the load if one stub is placed at the load and the other one-quarter of a wavelength away.

Answer: $Z_L = 120 - j320 \, \Omega$, $l_1 = 0.423 \, \lambda$, $l_2 = 0.226 \, \lambda$.

6.9 A double-stub matching system is to be used on a lossless transmission line of load impedance z_L. If the distance from the load to the stubs is s and $d + s$, show, using a Smith chart if necessary, that there are values of s for which the system cannot be used.

6.10 A transmission line is matched by two stubs, one open cirucuited and placed in parallel with the line 0.15λ from the load, and the other short circuited, 0.3λ long, and placed in series with the line a distance d towards the generator from the first stub. Find the length of the parallel stub, and the distance d, if the line is assumed to be lossless and the wavelength of the signal is 40 cm. The first voltage minimum is 0.07λ from the load, and the VSWR at the load is 3.

Answer: Stub length $= 12.78$ cm, $d = 6.47$ cm.

7 | Antennas

In the last chapter we discussed transmission lines, assuming that each line was terminated in a load. That load was drawn as a lumped circuit, and we did not enquire about its purpose, or its form. In practical systems, a transmission line is often used to connect either a power source to a transmitting antenna, or a receiving antenna to a detecting circuit. Antennas are reciprocal devices, so the ideas we will develop in this chapter apply to both transmitting and receiving systems. However, to avoid unnecessary repetition we will only discuss antennas in the transmitting mode.

When a transmission line feeds an antenna, it 'sees' the impedance of the antenna, which will be frequency dependent. If the antenna is matched to the transmission line, all the energy transmitted along the line is absorbed by the antenna and, apart from losses within the antenna itself, is radiated.

From an antenna designer's point of view, there are several parameters that contribute to a design specification. Those of particular interest are polar diagram, beamwidth, bandwidth and impedance. Other quantities relating to the mechanical specification of the antenna will also be important, such as its size, weight, shape and material.

Mechanisms involved in the propagation of radio waves have an important effect on the thinking of the antenna designer. This chapter concludes with some of the fundamental features of radio propagation and indicates the extent to which they are dependent on the frequency of operation.

7.1 RADIATION PATTERN (POLAR DIAGRAM)

The radiation from an antenna is an electromagnetic wave, and for most applications our interest is in the strength of this wave at a distant point. In terms of metres, what constitutes a distant point depends on the size of the antenna and the wavelength of the radiation, but in principle it refers to the region in space in which the wave can be considered to be plane, and normal to the direction of the antenna. There will be just two field components, an E field vector and an H field vector, as shown in Fig. 7.1.

The radiation pattern is the variation of this distant electric field as a function of angle; an example is shown in Fig. 7.2. A detector is placed at a fixed distance R from the antenna, and the E field is measured [Fig. 7.2(a)]. The value of this E field varies with the angle, θ, relative to some reference line, shown as $\theta = 0$. The variation of E with θ is the radiation pattern of the antenna in the particular plane in which the measurement is taken. The

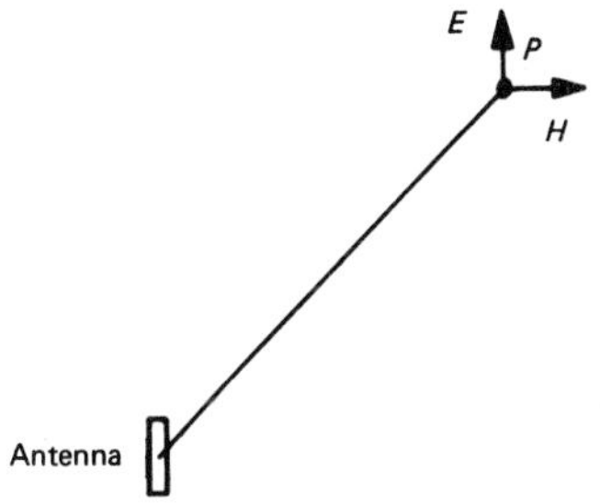

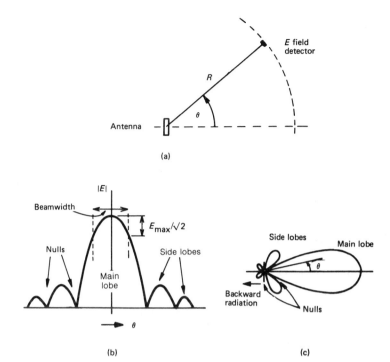

Fig. 7.1 Field components due to an antenna at a distant point.

Fig. 7.2 (a) Reference for radiation patterns; (b) parameters at a radiation pattern; (c) polar diagram.

radiation pattern can be represented in either Cartesian coordinates [Fig. 7.2(b)], or in polar coordinates, when it is usually referred to as a polar diagram [Fig. 7.2(c)]. Clearly, the radiation pattern of any real antenna is three-dimensional and therefore, to represent it fully, measurements must be made in at least two orthogonal planes. The traditional planes used are the vertical and the horizontal (which are sometimes, especially in radar applications, called elevation and azimuth, respectively), and these are related to the orientation in which the antenna is to be used. The reference direction, $\theta = 0$, is generally assumed to be normal to the plane of the antenna, as shown. For symmetrical antennas this is the direction of the main lobe. The various terms associated with the radiation pattern are shown in Fig. 7.2(b) and (c).

7.2 LOBES

We are sometimes interested in the absolute power radiated in any particular direction; for example, the actual power output from a broadcast transmitter will determine the range over which the transmission can be detected. More usually, however, we are more interested in the shape of the radiation pattern. Then we use a normalized pattern, with magnitudes plotted in relation to the main lobe maximum, which is given the value of unity. The finite size of the wavelength, relative to the dimensions of the antenna, produces diffraction effects, causing side lobes. These side lobes, separated from the main lobe and from each other by field minima, generally represent inefficiencies in the antenna. For most applications the designer wishes to keep side lobe levels to a minimum.

7.3 BEAMWIDTH

For most applications we are interested to know the width of the main lobe, i.e. over what angular distance the main lobe is spread. A definition of the beamwidth could be the distance between the nulls on either side of the main lobe. Although appealing, because it should be easy to read from a radiation pattern, in practice there are three disadvantages to its use:

(i) there may not be a null;
(ii) if a null exists, its precise location may be difficult to determine accurately;
(iii) the detectable level for any practical measurement is well above zero field.

However, some convention must be agreed upon, and that normally used is to define the beamwidth as the angular separation, in a given plane, between the points on either side of the main lobe that are 3 dB in power below that of the maximum. On an E field radiation plot, which has a normalized maximum of unity, the beamwidth is between those points on each side of the main lobe which have field values of $1/\sqrt{2}$.

7.4 ANTENNA IMPEDANCE AND BANDWIDTH

Antennas can be considered to belong to one of two categories, resonant or non-resonant. Resonant antennas have a dimension that is near to a half-wavelength, or one of its multiples, whereas non-resonant systems do not. The difference between the two is particularly noticeable in respect of antenna

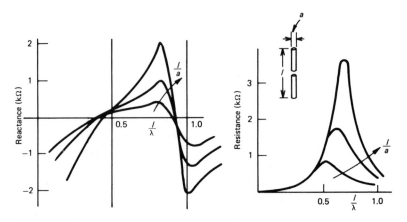

Fig. 7.3 Dipole impedance variation with dipole length.

impedance and bandwidth. The bandwidth is the frequency range over which the input impedance is substantially constant. In resonant systems, the impedance and bandwidth vary significantly with small changes in operating frequency; the impedance changes from negative reactance, through pure resistance, to positive reactance, as the frequency changes from below to above resonance. Figure 7.3 shows the impedance as a function of wavelength for a dipole. From it we can see that at about resonance, $l \simeq \lambda/2$, there is a rapid change of impedance with frequency resulting in a small bandwidth, which means that the characteristics of the antenna are strongly frequency dependent. In non-resonant antennas the radiation pattern and the input impedance are not so variable with frequency, and these antennas are sometimes referred to as broadband.

7.5 GAIN

The final parameter of interest to the systems designer is antenna gain. In any physically realisable antenna there will be one direction in which the power density is highest. To give a quantitative value to this maximum, it is compared with some reference, the most common being the power density from an isotropic antenna. An isotropic antenna has a spherical radiation pattern. There is no practical device that can produce such a pattern, but it is a useful notional tool for gain comparisons. There is a distinction to be made between two types of gain. Understandably, the power at both the isotropic antenna and the antenna under test must be the same. However, that may refer to the input power, or to the total radiated power. For the isotropic antenna, which has no loss mechanisms, the input and radiated powers are identical, but for the test antenna there will be losses that reduce its efficiency by several per cent, resulting in a difference between these two values of gain. In some of the literature they are distinguished by the terms 'directive gain' (D) and 'power gain' (G), where D assumes equal radiated powers from, and G assumes equal input powers to, the test and isotropic antennas. Thus, we can establish the relationship

$$G = kD \qquad (7.1)$$

where k, an efficiency factor, is the ratio of the total radiated power to the total input power for the antenna under test.

For applications in which the power transmitted in a specified direction is of particular interest (e.g. in a microwave link, or a satellite earth station), the gain is an indicator of the antenna's effectiveness.

7.6 FIELD DUE TO A FILAMENT OF CURRENT

By considering the simplest possible antenna, we can obtain some insight into radiated fields, and derive some general ideas that are applicable to all antennas. In Fig. 7.4 we see a short, very thin element of length l, carrying a sinusoidal current $I \sin \omega t$. The current flowing along the filament induces a

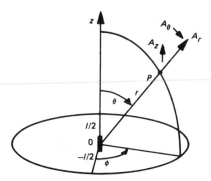

Fig. 7.4 Coordinate system relating fields at distant point P to antenna current.

magnetic field in the surrounding air. This magnetic field has an associated electric field and the result is a movement of electromagnetic energy away from the filament. We will examine the field at some point P, a distance r from the centre of the filament at O. Imagine that P is situated on a sphere centre O. Then P is related to O by r, and ϕ. In spherical coordinates, the field components at P are E_r, E_θ, E_ϕ and H_r, H_θ, H_ϕ. Allowing the current through the filament to be $I \sin \omega t$, the field at P will be produced by the current flowing through the filament at time $t - r/c$, where r/c is the time taken for the field to travel from the filament to P at the velocity of light, c.

To find the field at P we use the concept of a retarded magnetic vector potential, $\bar{A}$, which is related to the current in the filament by[1]

$$\bar{A} = \frac{1}{4\pi} \int_V \frac{i[t - (r/c)]}{r} \mathrm{d}V \tag{7.2}$$

In this example, the current has only one component, in the z direction (see Fig. 7.4), which is constant throughout the length of the filament. We can therefore replace $\bar{A}$ by A_z, and the volume integral by the linear integral over the range $-l/2 \leqslant z \leqslant l/2$, giving

$$A_z = \frac{1}{4\pi} \int_{-l/2}^{l/2} \frac{I \sin \omega[t - (r/c)]}{r} \mathrm{d}z \tag{7.3}$$

$$= \frac{Il}{4\pi r} \sin \omega [t - (r/c)] \tag{7.4}$$

The vector potential is a mathematical device. It allows us, in some cases, to find the electromagnetic field components more easily than would be possible if we were to try to relate the fields directly to the electric and magnetic charge on the antenna. The vector potential we are using here is related to the current in the antenna (the thin filament) through Eqn (7.4), and also to the magnetic field $\bar{H}$. The quantities $\bar{A}$ and $\bar{H}$ are, by definition, related by

$$\nabla \times \bar{A} = \bar{H} \tag{7.5}$$

Hence, knowing the components of $\bar{A}$, those of $\bar{H}$ can be found. From

Eqn (7.4) and Fig. 7.4, the spherical coordinate components of $\bar{A}$ are

$$A_r = A_z \cos \theta = \frac{Il}{4\pi r} \sin \omega \left(t - \frac{r}{c} \right) \cos \theta$$

$$A_\theta = -A_z \sin \theta = \frac{-Il}{4\pi r} \sin \omega \left(t - \frac{r}{c} \right) \sin \theta \qquad (7.6)$$

$$A_\phi = 0$$

By expanding Eqn (7.5), we can see that the only component of $\bar{H}$ is H_ϕ. Both H_θ and H_r are zero. Appendix 7.1 gives a fuller derivation of the electric and magnetic field components at P. We can see from there that

$$H_\phi = \frac{Il \sin \theta}{4\pi} \left[\frac{\omega}{cr} \cos \omega \left(t - \frac{r}{c} \right) + \frac{1}{r^2} \sin \omega \left(t - \frac{r}{c} \right) \right] \qquad (7.7)$$

$$E_\theta = \frac{Il \sin \theta}{4\pi\varepsilon} \left[\frac{\omega}{c^2 r} \cos \omega \left(t - \frac{r}{c} \right) + \frac{1}{cr^2} \sin \omega \left(t - \frac{r}{c} \right) - \frac{1}{\omega r^3} \cos \omega \left(t - \frac{r}{c} \right) \right]$$

$$(7.8)$$

$$E_r = \frac{Il \cos \theta}{2\pi\varepsilon} \left[\frac{1}{cr^2} \sin \omega \left(t - \frac{r}{c} \right) - \frac{1}{\omega r^3} \cos \omega \left(t - \frac{r}{c} \right) \right] \qquad (7.9)$$

We are not really interested in the magnitudes of these expressions, but in their form. Within each of the brackets there are terms related to powers of $1/r$, and these allow us to divide the space surrounding the current-carrying element into two parts. In the region near the antenna, terms in $1/r^3$ and $1/r^2$ will predominate, but at large distances these terms will be negligible, leaving only the terms in $1/r$. The fields in these two regions are referred to as the induction and radiation fields, respectively.

7.7 INDUCTION FIELD

The near-field region, or Fresnel zone, is occupied chiefly by the induction field, which does not radiate. Energy is stored in the field terms having $1/r^2$ coefficients for one half-cycle, and passes back to the antenna for the other half-cycle. The $1/r^3$ terms are confined to a region very close to the antenna before becoming negligible as r increases.

When considered in detail, the induction field can be subdivided into two parts. That nearest the antenna, known as the reactive near field, consists of strong reactive terms and is confined to a radius of less than $0.62\sqrt{D^2/\lambda}$, where D is the largest dimension of the antenna and is much greater than the wavelength λ.

Beyond the reactive near field is the radiating near field, or Fresnel zone, and that extends to the boundary with the far field. In this radiation near field region the distribution of the field as a function of angle from the antenna varies with radius.

7.8 RADIATION FIELD

In the far field, or Fraunhofer region, only the $1/r$ terms are significant, and, as we shall see below, it is those that provide the energy outflow from the antenna.

We noted earlier that in most applications it is the far field that is of interest, and radiation patterns refer to that area. The distance from the antenna at which the radiation field predominates can be found from the above equations by comparing the coefficients of the $1/r$ and $1/r^2$ terms for various values of r at the wavelength of interest.

A figure often used as the near/far field boundary is a radius of $2D^2/\lambda$.

Clearly, since the field is a continuous function of radius, the boundary expressions given above are merely guides, and in a particular application the transition into the far field will be at a radius determined by the dimensions of the antenna, and the wavelength of the signal.

7.9 POWER RADIATED FROM A CURRENT ELEMENT

We have called the fields in the Fraunhofer region the radiation fields, thus suggesting that it is the $1/r$ terms in Eqns (7.7) and (7.9) that are the components of power flow. We can justify this claim by considering the Poynting vector which, by taking the vector cross-product of the electric and magnetic fields, gives the value and direction of the net power flow from the antenna.

In vector notation, the Poynting vector $\bar{P}$ gives the power density in terms of the electric and magnetic field vectors:

$$\bar{P} = \bar{E} \times \bar{H} \tag{7.10}$$

For the short current element, the fields that exist at a point outside the antenna are E_θ, E_r, and H_ϕ; therefore

$$P_\phi = 0$$

and

$$P_\theta = -E_r H_\phi$$

i.e. from Eqns (7.7) and (7.9),

$$P_\theta = -\frac{Il\cos\theta}{2\pi\varepsilon}\left[\frac{\sin\omega[t-(r/c)]}{r^2 c} - \frac{\cos\omega[t-(r/c)]}{\omega r^3}\right]$$

$$\times \frac{Il\sin\theta}{4\pi}\left[\frac{\omega\cos\omega[t-(r/c)]}{rc} + \frac{\sin\omega[t-(r/c)]}{r^2}\right]$$

Since the time average of terms in $\cos 2\omega[t-(r/c)]$ and $\sin 2\omega[t-(r/c)]$ is zero, P_θ is zero. This leaves the radial power density component

$$P_r = E_\theta H_\phi \tag{7.11}$$

i.e. from Eqns (7.7) and (7.8),

$$P_r = \frac{I^2 l^2 \sin^2 \theta}{16\pi^2 \varepsilon} \left\{ \left[\frac{\omega}{rc} \cos \omega [t - (r/c)] + \frac{1}{r^2} \sin \omega [t - (r/c)] \right] \right.$$

$$\left. \times \left[\frac{\omega}{rc^2} \cos [t - (r/c)] - \frac{1}{\omega r^3} \cos \omega [t - (r/c)] + \frac{1}{cr^2} \sin \omega [t - (r/c)] \right] \right\}$$

(7.12)

Again, the terms in $\sin 2\omega [t - (r/c)]$ and $\cos 2\omega[t - (r/c)]$ have zero time average, leaving

$$P_r = \frac{\omega^2 I^2 l^2 \sin^2 \theta}{16\pi^2 \varepsilon r^2 c^3} \frac{1}{2}$$

giving

$$P_r = \frac{\eta I^2 l^2 \sin^2 \theta}{8\lambda^2 r^2} \tag{7.13}$$

where η = the space impedance, $\sqrt{(\mu_0/\varepsilon_0)}$, f = the frequency of the radiation = $\omega/2\pi = c/\lambda$ and c = the velocity of the radiated wave = $1/\sqrt{(\mu_0\varepsilon_0)}$.

This remaining component of the Poynting vector gives the power density flowing from the antenna. Before finding the actual power from the filament, it is worth nothing that P_r in Eqn (7.13) is formed from the $1/r$ terms in E_θ and H_ϕ, which is why they are called the radiation terms. The positive sign of P_r indicates a flow of power outwards from the antenna.

We can find the total radiated power from the current element by integrating the power density at a distant point P over the surface of a sphere through P. Referring to Fig. 7.5 the total power P_T is given by

$$P_T = \int P_r \, da$$

integrated over the area of the sphere.

From the diagram, $da = 2\pi r^3 \sin \theta \, d\theta$. Hence, from Eqn (7.13),

$$P_T = \int_0^\pi \frac{\eta I^2 l^2 \sin^2 \theta}{8\lambda^2 r^2} 2\pi r^2 \sin \theta \, d\theta$$

$$= \frac{\eta I^2 l^2 \pi}{4\lambda^2} \int_0^\pi \sin^3 \theta \, d\theta$$

$$= \frac{\eta I^2 l^2 \pi}{8\lambda^2} \left[-\frac{\cos \theta}{3} (\sin^2 \theta + 2) \right]_0^\pi$$

$$= \frac{\eta I^2 l^2 \pi}{3\lambda^2} \tag{7.14}$$

I is the peak value of current in the filament. If, instead, the rms value is used,

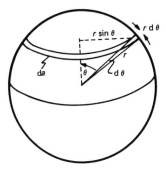

Fig. 7.5 Power radiated from an antenna at the centre of a sphere.

the total power from the filament is

$$P = \frac{2\eta I_{\text{rms}}^2 l^2 \pi}{3\lambda^2} \quad \text{watts} \tag{7.15}$$

In free space, the impedance η has the value $120\pi = 377\,\Omega$, and using this value

$$P = I_{\text{rms}}^2 \, 80\pi^2 \, (l/\lambda)^2 \quad \text{watts} \tag{7.16}$$

We can see immediately that the quantity $80\pi^2(l/\lambda)^2$ has the dimensions of resistance. It is known usually as the radiation resistance of the antenna, and for this particular antenna it is related to the electrical length of the element, l/λ.

The idea of radiation resistance is applicable to any antenna. Its value for a particular antenna can be found if the current distribution over the antenna surface is known.

In the short element considered above, we assumed that the current was constant over the length of the antenna. Usually, however, this assumption is not valid, and in order to determine the amount of power radiated, the current distribution must be known. There are, then, two difficulties associated with finding the power radiated from a practical antenna.

(i) The current distribution cannot be determined very readily. Experimental measurements are not easy to obtain, and theoretical analysis, in which Maxwell's equations are solved for the antenna surface, is very complex. Current distributions have been calculated for simple, idealized shapes, but even then the solutions are complicated and difficult to use. In any practical analysis, unless lengthy numerical methods are employed, the usual approach is to build up an approximate model from simpler shapes, but even then the analysis is not easy.

(ii) Once the current distribution is known, the radiation fields must be calculated. That, too, may be a very difficult analytical process which must sometimes be approached by making realistic simplifications.

We will not concern ourselves with the solution of such problems, but we will accept that any antenna has

(i) a radiation field at some distance from the antenna, appearing as a plane wave moving radially outwards,
(ii) an induction field near to the antenna,
(iii) a radiation resistance that gives a measure of the amount of power radiated by the antenna,

and that the distance from the antenna at which the near field merges into the far field depends on the electrical length of the antenna.

7.10 RADIATION PATTERN OF A SHORT DIPOLE

The short current element we have just considered is like a short dipole. Equation (7.41) gives E_θ, the electric field component at a distant point. If the

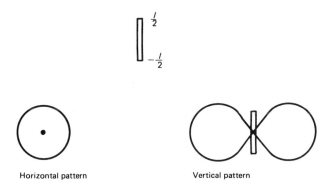

Fig. 7.6 Polar diagram of short dipole.

range from the antenna to the point of interest is in the far field, only the radiation term will be significant, i.e.

$$E_\theta = \frac{\eta Il}{2\lambda\pi}\sin\theta\cos\omega\left[t - (r/c)\right]$$

which has the magnitude

$$\frac{60\pi I}{r}\frac{l}{\lambda}\sin\theta \qquad (7.17)$$

where, as before, η has been replaced by 120π.

Referring again to Fig. 7.4 we can see that for fixed θ the magnitude will be constant for all values of ϕ. As θ is varied, the magnitude will vary from zero at $\theta = 0$ and π, to a maximum at $\theta = \pi/2$ from Eqn (7.17). If the dipole is placed vertically, the horizontal radiation pattern, with θ constant, is a circle, and the vertical radiation pattern, with ϕ constant, is a figure of eight. We can see that the three-dimensional radiation pattern is a solid of revolution around the dipole axis (Fig. 7.6). From a practical point of view this type of antenna could be used where no preference was to be given to any direction, or where nulls were required, as in some direction-finding methods. The basic shape of the radiation pattern in Fig. 7.6 is maintained as the value of l increases into the resonant region of $\lambda/2$, the half-wave dipole. The impedance of the antenna changes quite dramatically as l is increased, as we saw in Fig. 7.3.

7.11 ANTENNA ARRAYS

The circular horizontal radiation pattern of a dipole antenna may not be satisfactory for some applications. By placing several dipoles in line, the radiation pattern can be modified to give a large variety of radiation patterns depending on the parameters of the array. Here we shall start with the simplest array of two elements, and then consider arrays of several elements, equally spaced and in line. Other, more complex, arrays have been studied, and details can be found in, for example, Kraus.[2]

7.12 TWO-ELEMENT ARRAY

In the arrangement of Fig. 7.7, the elements are fed with currents I_0 and I_1, which we will assume to be of equal magnitude, but out of phase, i.e.

$$I_1 = I_0 \underline{/\alpha}$$

If P, the point of observation, is well into the far field, the path-length difference to P from the elements will be approximately equal to $d \cos \phi$, where d is the element spacing.

In the position shown, the phase of the radiation at P from element 1 will lead that from element 0 by ψ, where

$$\psi = \beta d \cos \phi + \alpha$$

$\beta (= 2\pi/\lambda)$ is the phase constant of the transmitted wave.

Given that the field at P due to element 0 is E_0, then the total field E_p is given by

$$E_p = E_0 [1 + \exp(j\psi)]$$
$$= E_0 [\exp(-\tfrac{1}{2} j\psi) + \exp(\tfrac{1}{2} j\psi)] \exp(j\psi/2)$$

which has a magnitude

$$E = 2E_0 \cos \tfrac{1}{2}\psi \tag{7.18}$$

i.e.

$$E = 2E_0 \cos \tfrac{1}{2}(\beta d \cos \phi + \alpha)$$
$$= 2E_0 \cos \left(\frac{\pi d}{\lambda} \cos \phi + \frac{\alpha}{2} \right) \tag{7.19}$$

Thus, for any given separation and phase difference, the radiation pattern, which is the variation of E with ϕ, can be found. Major changes in the polar diagram can be produced by varying d/λ and α. Problem 7.1 clearly demonstrates these changes, and it is a useful exercise to plot the radiation pattern for each value, and then to compare the results. Jordan,[1] Kraus[2] and Jasik[3] show many of these patterns.

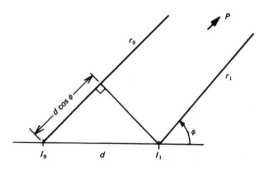

Fig. 7.7 Two-element array.

7.13 LINEAR ARRAYS

Several elements in line create a linear array. In this section we restrict our attention to the simplest form of array, in which n equally spaced elements are fed with currents of equal magnitude, and the phase difference between adjacent elements is constant.

From Fig. 7.8 the field at some distant point P is

$$E_p = E_0[1 + \exp(j\psi) + \exp(j2\psi) + \cdots \exp[j(n-1)\psi]] \qquad (7.20)$$

where, as before, E_0 is the field at P due to radiation from element 0, and ψ is the phase difference at P due to radiation from element 1 compared with that from element 0, i.e.

$$\psi = \beta d \cos \phi + \alpha \qquad (7.21)$$

To find the magnitude of E_p we can use the relationship

$$1 + \exp(j\psi) + \cdots \exp[j(n+1)\psi] = \frac{1 - \exp(jn\psi)}{1 - \exp(j\psi)} \qquad (7.22)$$

Then, the magnitude of Eqn (7.22) is

$$E = E_0 \left| \frac{1 - \exp(jn\psi)}{1 - \exp(j\psi)} \right|$$

$$= E_0 \left| \frac{\sin n\psi/2}{\sin \psi/2} \right| \qquad (7.23)$$

The quantity $|(\sin n\psi/2)/(\sin \psi/2)|$ is known as the array factor because it determines the shape of the radiation pattern. It has a minimum, as we could verify by using L'Hospital's Rule, when $\psi = 0$, i.e. when

$$\beta d \cos \phi = -\alpha$$

By choosing α correctly we can place the maximum where required. Consider two examples.

Broadside array. The maximum is normal to the line of the elements, in the direction $\phi = \pi/2$. The radiation from the elements must be in phase in this

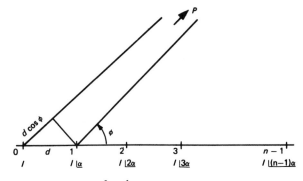

Fig. 7.8 Uniform linear array of n elements.

direction, and from Eqn (7.21) this requires, as we would expect, that the elements are fed in phase, i.e. $\alpha = 0$.

End-fire array. In this case the maximum is placed along the line of the elements, i.e. in the direction $\phi = 0$. Here the radiation from adjacent elements will add in one direction, and subtract in the reverse, backward, direction. Again, we are not surprised to find from Eqn (7.21) that the phase relationship between adjacent elements is

$$\alpha = -\beta d$$

so that the phase difference in the currents between, for example, elements 0 and 1 is cancelled by their separation.

In both these cases, and for any linear array, the width of the main lobe is reduced by increasing the number of elements, which is similar to making the array physically longer.

To find the shape of the radiation pattern of a particular array it is often sufficient to establish the direction of the maximum, from Eqn (7.23), and to find the position of the nulls. These null directions, where the field at P is zero, can be determined much more easily than the directions of the maxima of the secondary lobes. From Eqn (7.23) we can see that a zero field occurs when the array factor is zero, i.e. when

$$\frac{n\psi}{2} = \pm k\pi$$

or

$$\tfrac{1}{2}n(\beta d \cos \phi + \alpha) = \pm k\pi \tag{7.24}$$

where $k = 1, 2, 3 \cdots$. Further insight into the behaviour of these arrays can be found in Problems 7.3 and 7.4 at the end of the chapter.

A detailed evaluation of the array factor is necessary if a complete radiation pattern is required. Often, however, it is sufficient to estimate the minor lobes by calculating one or two values for sample angles between adjacent nulls.

7.14 PATTERN MULTIPLICATION

There are some more complex linear arrangements of elements that can be analysed by decomposing the array into components that have a known, or easily found, radiation pattern, and then recombining these results to find the overall pattern. This method of pattern multiplication can be described most easily by using an example. Consider the array of elements shown in Fig. 7.9(a). Assuming that the elements are isotropic, we can decompose the array into

(i) a three-element linear array with $d = 3\lambda/4$, and $\alpha = 30°$ [Fig. 7.9(b)];
(ii) each of these three elements is in fact a two-element array with

$$d = 0.6\lambda \quad \text{and} \quad \alpha = 0 \quad \text{[Fig. 7.9(c)].}$$

In this example the six elements of the initial array, $A\,B\,C\,D\,E\,F$ have been

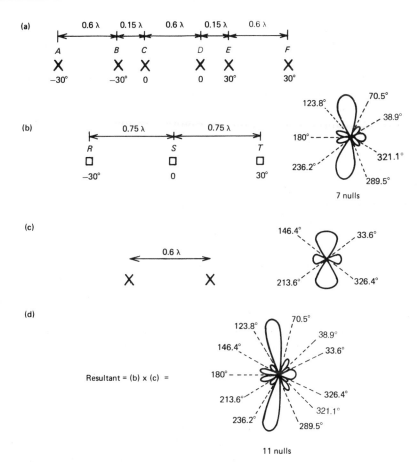

Fig. 7.9 Application of pattern multiplication.

replaced by three elements, $R\,S\,T$, where R has replaced A and B, S has replaced C and D, and T has replaced E and F.

The radiation pattern of $R\,S\,T$, assuming each element to be isotropic, is derived from Section 7.13. The result is shown in Fig. 7.9(b). The radiation pattern of R (or S or T) is found from Eqn (7.19) and it has the shape given in Fig. 7.9(c).

The resultant radiation pattern for the six-element array is therefore the product of the array pattern [Fig. 7.9(b)] and the group pattern [Fig. 7.9(c)]. We can see that the number of nulls in the resultant pattern is the sum of the nulls in the component patterns, i.e. four nulls from the group pattern and seven nulls form the array pattern, giving eleven in all [Fig. 7.9(d)].

7.15 ANTENNA MATCHING

Efficient connection between the feeder transmission line and the antenna depends on matching the impedance of one to the other. For narrowband

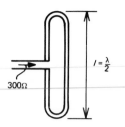

Fig. 7.10 Folded dipole.

applications, the antenna impedance can be calculated or measured, and an appropriate transmission line, with matching unit if necessary, can be used. In wideband applications, however, such matching may not be possible. Non-resonant antennas have a wide bandwidth capability because their impedance variation with frequency is less marked than that of resonant structures. We can see the effect of resonance on antenna terminal impedance by considering a simple half-wave dipole. Figure 7.3 shows the typical form of the variation for several thicknesses of antenna rod.

The resonant length is that which produces a purely resistive impedance, and we can see from the diagram that the reactance is zero at length slightly less than $\lambda/2$, depending on the thickness of the dipole rod. As the rod diameter falls to zero the resonant length approaches $\lambda/2$. This is due in part to the capacitive effect of the rod ends.

The half-wave dipole impedance at resonance is about 73 Ω, and therefore a cable having a nominal characteristic impedance of 75 Ω will provide a good match. For some applications, however, a 300 Ω balanced cable is more appropriate, and this does not match with the 73 Ω dipole. If the dipole is folded, as in Fig. 7.10, the impedance is approximately 300 Ω, which will match the characteristic impedance of the line. There are other advantages of using a folded dipole; it is mechanically more robust, it can be attached more easily to a support, and the feed area is free from supporting brackets.

7.16 PARASITIC ELEMENTS

In discussing the dipole, we established that its horizontal polar diagram is circular, and therefore has no directive properties. By using several dipoles in line a more directive antenna can be produced, but that is not always convenient. The feeding arrangements, requiring specified magnitude and phase of the current at each element, add to the complexity and hence the expense of the antenna.

For domestic broadcasting, in the frequency ranges where a half-wave dipole is a manageable size, parasitic elements are used to distort the dipole field to produce a directive radiation pattern. A parasitic element does not have any feeder cable. It is placed near to the driven element, and picks up current from it by mutual inductance. The parasitic then acts as a separate, weaker, radiator and the resulting radiation pattern is a combination of the radiation from both the driven and parasitic dipoles. The position of the parasitic element, relative to the direction of the main beam, determines its function. In Fig. 7.11 we see the parasitic element acting as a director, being placed at such a distance in front of the driven element that the fields from the two add along the line joining them, in the direction of the required maximum. Alternatively, a parasitic element placed behind the driven element acts as a reflector, with the reradiated energy adding to that of the driven element, again in the forward direction (Fig. 7.12).

By using many director elements, as is common in a VHF antenna (Fig. 7.13), a highly directional beam results. When used as a receiver, this type of antenna is most sensitive in the forward direction, and therefore

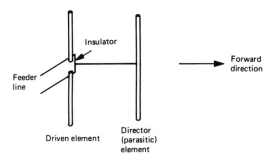

Fig. 7.11 Dipole with passive director element.

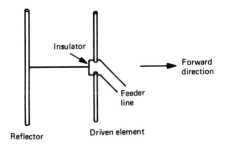

Fig. 7.12 Dipole with passive reflector element.

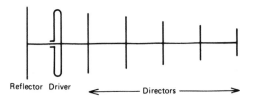

Fig. 7.13 Yagi array.

discriminates against signals from other directions. The beamwidth decreases as the number of elements increases.

The antenna is named after Hidetsugu Yagi, who was active in the general field of high-frequency communications. A description of the antenna he developed was published in the Proceedings of the Institute of Radio Engineers in 1925. Emphasizing the high directivity of the array, he referred to it as a wave projector, or wave canal.

We are not attempting to analyse these antennas. The mutual coupling is complex, depending on the spacing, length, diameter, and material of each element, and it exists not only between the driven and parasitic elements, but between the parasitic elements themselves. The design of multi-element antennas is based on empirical experience, rather than analysis, because of the difficulty in modelling the interaction between the elements.

7.17 MICROWAVE ANTENNAS

At microwave frequencies the antennas are generally significantly larger than a wavelength, and this allows techniques akin to those of optical devices, such as reflectors and lenses, to be used, thus making some microwave antennas different in kind from those possible at lower frequencies. The range of applications of microwaves in radar and communications is very large, and growing, and hence the demand for efficient, robust and inexpensive antennas, particularly in the millimetre-wave range, will increase.

Section 14.5 discusses some of the design problems related to satellite antennas.

7.18 PARABOLIC REFLECTOR

Probably the most commonly used microwave antenna is the parabolic dish. It reflects energy from a primary source placed at its focus. The principle is the same as that used in an optical mirror; when the source is at the focus, on the axis of the parabola, the reflected beam is parallel to the axis. Figure 7.14 shows the main features. Assuming that ray theory applies, in order to produce a parallel beam, all rays leaving the focus F and striking the reflecting surface must be in phase in some plane normal to the axis of the reflector, such as AA in the diagram. Typical rays would be FCD, FGH, FJK and FLM as shown. The requirement that DHK and M are in phase means that

$$FC + CD = FG + GH = FJ + JK = FL + LM$$

and this relationship is satisfied by a parabolic reflecting surface. The usual shape of antenna is the paraboloid, formed by revolving a parabola on its main axis, and designed to give a symmetrical radiation pattern. Two main factors prevent the reflected beam from being parallel with the axis: first, the feed cannot be a point source; and second, the wavelength of the radiation is not so small that diffraction effects can be ignored.

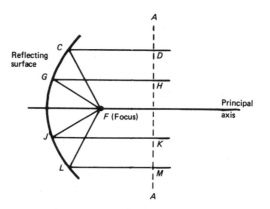

Fig. 7.14 Geometry of a parabolic dish antenna.

The feed introduces some problems. It has a comparatively large size and therefore blocks energy reflected from the paraboloid, and its own radiation pattern may cause destructive interference with energy from the reflector.

The comparatively large wavelength of microwave radiation produces diffraction effects at the edge of the reflector, allowing energy to spill round the back of the dish, and be wasted. One way of limiting such loss is to ensure, by design of the primary feed, that the reflector is not illuminated uniformly. A typical design figure is that the energy from the source, incident at the edge of the reflector, is only 10 % of that along the axis.

The primary feeds normally used are shown in Fig. 7.15. The dipole, with parasitic reflector to prevent radiation in the forward direction, has a less directive radiation pattern than the horn antenna, and therefore illuminates the edge of the reflector too strongly, but it has a much smaller supporting structure than that required for the horn, and so does not block as much of the reflected wave. To reduce this blocking by the support, a Cassegrain antenna is sometimes used, particularly in receiving applications. The Cassegrain antenna (Fig. 7.16) consists of a primary horn feed, set in the surface of the reflector, and a sub-reflector of hyperbolic shape placed just in front of the focus of the parabola. Energy from the horn strikes the sub-reflector and is

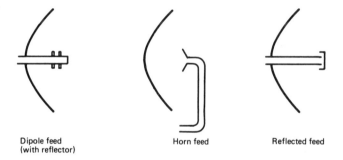

Dipole feed
(with reflector)

Horn feed

Reflected feed

Fig. 7.15 Parabolic dish feeds.

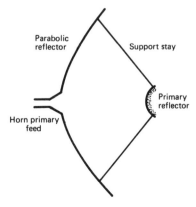

Parabolic
reflector

Support stay

Primary
reflector

Horn primary
feed

Fig. 7.16 Cassegrain antenna.

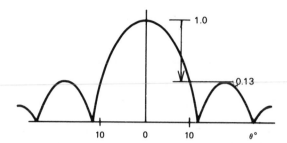

Fig. 7.17 Radiation pattern of parabolic dish with uniform illumination.

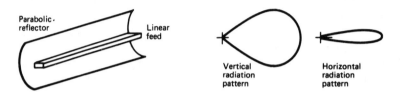

Fig. 7.18 Parabolic cylinder antenna and its polar diagrams.

then redirected to the surface of the parabolic dish. The supports used for the sub-reflector are much less robust than the waveguide feed necessary for a horn-type primary source.

The beamwidth of the parabolic reflector is related to the wavelength of the signal, and the diameter of the dish at its rim, as well as to the illumination variation across the reflector surface. A typical radiation pattern, assuming uniform illumination, is shown in Fig. 7.17. The first side lobes rise to only 13% of the main lobe intensity, and the beamwidth is a few degrees. As the diameter is increased, the beamwidth decreases, according to the relationship

$$\theta = k\lambda/D \tag{7.25}$$

where θ is the beamwidth, D is the diameter of the reflector at the rim and λ is the radiation wavelength. k is a constant depending on the illumination from the primary feed, with a value of about 60. Thus, for $\theta = 10°$, D will need to be about 6λ.

Other shapes of reflector are used, usually to produce beamshapes that are not equal in the horizontal and vertical planes. For example, the parabolic cylinder shown in Fig. 7.18 is fed by a line source, such as an array of dipoles, and produces a polar diagram that has a very narrow horizontal beamwidth and a broad vertical beamwidth. Used as a radar antenna it can therefore identify the horizontal direction of a target with reasonable accuracy, but it will have no discrimination in elevation.

7.19 HORN ANTENNAS

If a waveguide is unterminated it will radiate energy, producing a broad radiation pattern, as in Fig. 7.19. By flaring the end of the guide a more directional pattern is produced.

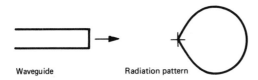

Fig. 7.19 Radiation pattern from waveguide end.

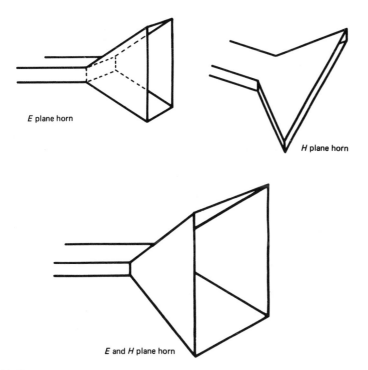

Fig. 7.20 Horn antennas.

The flare can be in the *E* plane [Fig. 7.20(a)], the *H* plane [Fig. 7.20(b)] or both [Fig. 7.20(c)], producing a beam which is narrow in the direction of the flare. It is worth noting that radiation from a horn antenna is polarized, that is the direction of the *E* field is related to the orientation of the waveguide, hence the examples shown in Fig. 7.20 are all vertically polarized.

The shape of the polar diagram from a waveguide horn will depend on two factors – the flare angle θ and the length of the horn L (Fig. 7.21). By using a large flare angle, the horn aperture can be made large in a short distance from the throat. However, a large θ will result in a large phase difference across the wavefront at the horn aperture, shown as δ in Fig. 7.21. A reasonable limit for δ is $\lambda/4$, in which case from the geometry of the horn the relationship between L, D and the wavelength λ is

$$L = D^2/2\lambda \qquad (7.26)$$

At small wavelengths, in the millimetre range for example, Eqn (7.26) does not

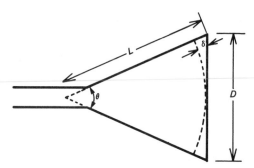

Fig. 7.21 Horn antenna dimensions.

impose difficult conditions on the size of the horn, but at longer wavelengths the value of L is so large that any practical horn would be too long, too bulky and too heavy. The dependence of beamwidth on horn aperture is approximately the same as that given in Eqn (7.25) for the paraboloidal reflector.

7.20 DIELECTRIC LENS

The length required to produce a horn with an aperture large enough to give a narrow main beam can be reduced by placing a perspex lens in the mouth of the horn.

The lens works in exactly the same way as an optical lens. It is made from a dielectric with a permittivity greater than that of air and therefore slows a wave travelling through it. If we refer to Fig. 7.22 we can see a ray theory approximation of the lens. Consider the ray that starts at the point F on the convex side of a convex-planar lens, strikes the lens at A, and passes through the lens to leave at B. We can compare it with the ray that travels along the axis from F to T, leaving the lens at P.

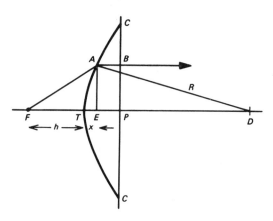

Fig. 7.22 Plano-convex lens antenna.

Given the dimensions in the diagram, for the two rays to have equal travelling times from F to the plane CC,

$$FA/c + AB/v = FT/c + TP/v \qquad (7.27)$$

where c is the velocity of the wave in air and v is the velocity of the wave in the dielectric.

Equation (7.27) can be written

$$FA/c = FT/c + (TP - AB)/v$$

From Fig. 7.22, FA, FT, and $TP - AB$ can be replaced by r, h and x, giving

$$r/c = h/c + x/v \qquad (7.28)$$

The velocity v is related to c by

$$v = c/\sqrt{(\varepsilon_r)}$$

where ε_r is the relative permittivity of the lens material. The quantity $\sqrt{(\varepsilon_r)}$ is called the refractive index, n, so $v = c/n$. Using this expression in Eqn (7.28) allows the velocity of the wave in air, c, to be cancelled, leaving

$$r = h + nx \qquad (7.29)$$

From the triangle TAF, x is related to r, θ and h by

$$x = r \cos \theta - h$$

Hence Eqn (7.29) becomes

$$r = h + n(r \cos \theta - h)$$

Rearranging terms

$$r(n \cos \theta - 1) = h(n - 1)$$

or

$$r = h(n - 1)/(n \cos \theta - 1)$$

This equation is the polar coordinate expression for a hyperbola, and it defines the curve of the convex surface of the lens.

If the focal length h is short, the thickness of the lens at the axis (TP) will be substantial, resulting in a device which is both bulky and heavy. Provided that the application is at a fixed frequency, the weight and structure of the lens

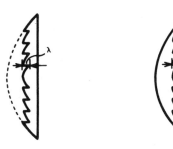

Fig. 7.23 Stepped lens antennas.

can be improved by taking away zones that are a wavelength deep. The zones can be removed from either the convex or plane side of the lens, as shown in Fig. 7.23.

7.21 MICROSTRIP ANTENNAS

Antennas that are physically robust, that can be mounted in the surface of a body such as an aeroplane, that are relatively cheap to produce and that can be built into steerable one- or two-dimensional arrays, are extremely attractive to the antenna designer. Such antennas can be made using the microstrip approach; in this case a conducting patch is mounted on to a dielectric ground plane (Fig. 7.24).

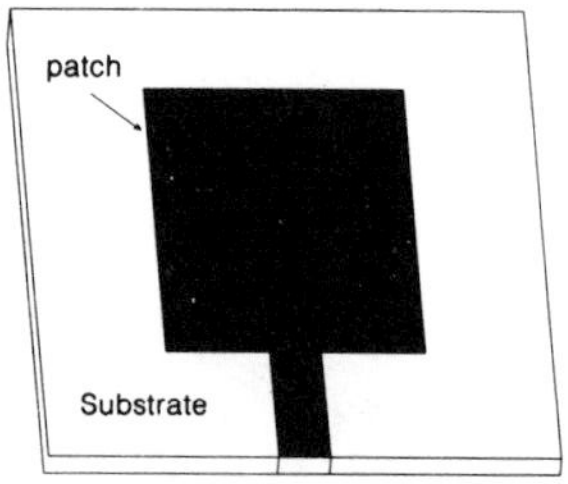

The patch can have any geometry, although rectangular and circular are the most common. The feed is either directly from a stripline, or via a coaxial probe in which the centre conductor is connected to the patch and the outer to the ground on which the dielectric sits.

Although the microstrip feed would appear to be the obvious choice, in fact the function of a microstrip line is rather different from that of a microstrip antenna, in that the first is designed to constrain and guide the transmitted power, whereas the second is designed to radiate it. This results in different design parameters for the dielectric in each case. In the guide the dielectric is thick and made from a material with a low permittivity; in the antenna the dielectric is thin and has a high permittivity.

Fig. 7.24 Microstrip rectangular patch antenna.

7.21.1 Modelling

The designer is keen to be able to correlate the parameters of the antenna with its electrical properties such as bandwidth, impedance, radiation pattern and efficiency, so that the best balance between the parameters can be chosen to meet the specifications of a particular application. To relate properties to physical parameters, some analysis of the behaviour of microstrip antennas is required, and if possible the analysis should provide a guide to appropriate CAD tools.

Two approaches to modelling have been used, in much the same way as happens in many of the other systems considered in this book. An approximate approach is used to provide intuitive insights into the operation of the antennas, with the hope that the results obtained are reasonably accurate, and sufficient for a design in some cases. The other is to develop a rigorous, accurate analysis that gives a detailed assessment of the performance of the system, taking into account as many elements and parameters as possible. The cost of this approach is long and complex computation, following some difficult analysis related to the particular system being studied.

For accurate results, and as a basis for building CAD packages that are not dependent on too many assumptions, the second approach is the best, but it requires some powerful computing resource. For many initial designs, the more approximate method will be adequate, provided its limitations are clearly understood.

There can be an interpretative problem with rigorous analysis. Because of its complexity it is difficult to sustain an awareness of the effects of each of the parameters of the system on the final result, and there are many applications where the designer needs to assess what trade-offs are available. Intuition is more easily guided by an approximate analysis in which the main elements of the system involved in determining a particular characteristic are explicitly available, and others are supressed. A helpful initial review is contained in reference 8.

7.21.2 Types of antenna

The polar diagram of a patch antenna is generally broadside and symmetrical in relation to the plane of the patch; most of the energy is radiated normally to the patch, and the beamwidth itself is broad. The particular arrangement of the feed, and the thickness of the antenna, will influence the radiation pattern and the input impedance at the feed point. In the case of a coaxial feed, this impedance is a function of the position of the probe relative to the patch. Towards the edge the impedance is high and it falls towards the centre. Some degree of matching is therefore possible by choosing the position of the feed point.

The patch is frequency sensitive. It can be modelled as a microwave cavity. To a first approximation, there is zero tangential E-field along the patch surface, and the edge of the patch can be considered as a magnetic wall from which the magnetic field is reflected. This approach leads to expressions similar to those in Section 9.19. Patch antennas have a high Q and hence a narrow bandwidth.

In addition, the resonant frequency is high for small patches. For example, for use as a mobile cellular car antenna, a reasonably sized device, fitted in the car roof, would resonate at about 2.2 GHz, whereas cellular radio frequencies are presently below 1 GHz. An antenna with a natural resonant frequency as low as that would be too large for the application, so one approach is to modify the patch itself by cutting a series of overlapping concentric slots into it. These oblige the current to go from the probe, at the centre, to the patch edge via a longer path since it has to travel round the slots. In this way the resonant frequency is moved into the region required.

There are many applications, particularly in moving vehicles, or in structures that are exposed to extreme conditions, where conformal surface antennas are extremely attractive, and as communication technology is increasingly related to mobile use, microstrip antennas will continue to attract considerable attention from theorists and designers. This is one of several areas in antenna technology that has yet to realize its full potential.

7.22 RADIO WAVE PROPAGATION

The mechanism whereby a radio signal transmits through the air between transmitters depends chiefly on the frequency of the wave. In the electromag-

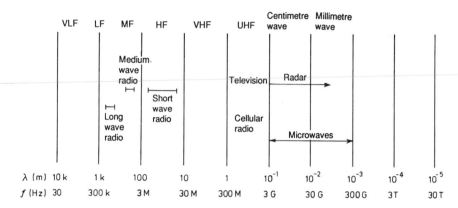

Fig. 7.25 Radio spectrum.

netic spectrum shown in Fig. 7.25 the bands of frequencies used for various applications are indicated. For domestic radio broadcasts the medium and long wave bands are used, and broadcast television frequencies are normally in the VHF and UHF bands. Between these two ranges there is the commonly used shortwave radio or HF band between 1 and 30 MHz. Other bands of common interest are also indicated in the diagram. It is noteworthy that there is comparatively little interest at present in the enormous range between millimetre radar (100 GHz) and optical fibre transmission frequencies. This is due mainly to a lack of suitable sources and detectors, and the problems of transmission.

The propagation mechanisms involved in radio transmission are (a) ground-wave (b) skywave (c) troposphere scatter and (d) line of sight. At low frequencies the groundwave is the most important; at HF frequencies it is the skywave that is predominant and at higher frequencies still, line of sight becomes the most important mechanism.

At very low frequencies the atmosphere between the earth's surface and the ionosphere acts as a waveguide which allows the low-frequency energy to be guided around the surface of the earth. This mechanism supports propagation distances of thousands of kilometres and is therefore greatly used for long distance navigation.

The groundwave becomes increasingly attenuated as the frequency rises. It is still predominant at the radio broadcast frequencies but at the upper end of the MF band the groundwave has become a surface wave which attenuates rapidly if the surface has high conductivity. In practice the conductivity of the surface of the earth is complicated by, first, the large range of materials of which the earth's surface is constituted, and secondly the presence of water contaminated by salt solutions. Both the dielectric constant and the conductivity of the earth's surface are much dependent on the moisture content and the nature of the moisture. This has the effect that propagation over the sea is better than over the land and the depth of penetration of the radiowave over the earth is highly related to the dryness of the ground, its constituent materials, and the presence of water below its surface. A particulary useful discussion of groundwaves is given in reference 1 to which the reader is referred for further information.

7.22.1 Skywaves

In the HF frequency band the main mode of propagation is by skywave. Radiation from the earth is propagated outwards into the sky and reflected or refracted back to earth where it can be detected. The mechanism which causes the refraction or reflection is the ionosphere. The ionosphere consists of a number of layers of ionized gas which are produced at various heights due to the absorption of radiation from the sun. The predominant layers are known as D, E, Fl and F2, with the D layer at the lowest altitude, of round about 50 km, extending to approximately 300 km for the F2 layer. Because these layers are generated by the sun's activities the lower layers are in place only during the day; at night the ionization disperses and there are no free electrons. At the upper reaches of the atmosphere the sun is effective throughout the 24 hours. The ionized gases in the various layers act like a refractive medium which, depending on the frequency and angle of presentation of the radiowave, will cause reflection or refraction to a degree dependent on both the density of the free electrons and the frequency of the wave.

We can see from Fig. 7.26 that there is a distance measured along the ground known as the skip distance; this is the distance between the transmitting antenna and the first position at which the wave can be received. To some extent the area around the transmitter will be covered by the groundwave but there will be a significant region which does not receive any signal.

A simple argument for the refraction mechanism, based on geometric optics and a very simplified model of the ionospheric layers is as follows.

To a first approximation the refractive index of an ionized medium, relative to that of free space, is related to the electron density, the electron mass and charge, and the frequency of the incident wave by

$$\varepsilon_r \simeq 1 - \frac{Ne^2 \cdot}{\varepsilon_0 m \omega^2}$$

$$\simeq 1 - \frac{81N}{f^2}$$

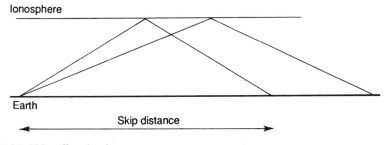

Fig. 7.26 Skip effect in skywave.

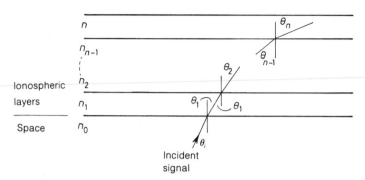

Fig. 7.27 Simple ionospheric model.

using $e = 1.59 \times 10^{-19}\,\text{C}$, $m = 9 \times 10^{-31}\,\text{kg}$, $\varepsilon_0 = 10^{-9}/36\pi$ farads/m and $\omega = 2\pi f$. N is measured in electron/m³.

We can model the ionosphere as a series of layers, each with an increasing electron density, as shown in Fig. 7.27. The thickness of each layer can be made arbitrarily small. The angle of incidence θ_i at the ionosphere is related to the refractive index of the first layer by Snell's Law of Refraction:

$$n_0 \sin \theta_i = n_1 \sin \theta_1$$

Assuming $n_0 = 1$

$$\sin \theta_i = n_1 \sin \theta_1$$

Repeating this equation for each boundary interface,

$$\sin \theta_i = n_n \sin \theta_n$$

The refractive index $= \sqrt{\varepsilon_r}$.

Hence, if N is the electron density in the nth layer,

$$\sin \theta_i = \left[1 - \frac{81\,N}{f^2} \right]^{1/2} \sin \theta_n$$

As the electron density increases there will be some layer for which the above relationship cannot be satisfied, i.e. when $\sin \theta_n \geqslant 1$ and the wave no longer passes through the ionosphere but is reflected back to earth. For a given value of incident angle θ_i, this occurs, from the last equation, when N satisfies the relationship

$$N = \frac{f^2}{81} \cos^2 \theta_i$$

The electron density is a function of altitude, so the altitude at which reflection occurs varies with the angle of the incident wave θ_i and the frequency of the signal, f.

There is a very significant variation in the density and height of the ionospheric layers with time of day and season of the year. In fact the prediction of radiowave propagation through the ionosphere is very much akin to the prediction of weather forecasts and it has the same level of success.

It is possible to forecast reasonably well how the ionosphere is likely to behave and therefore to choose those frequencies which will produce the most effective reception. This generally means that through the 24-hour period the transmission frequency has to vary significantly; it would not be unusual for the frequency to have to change by a factor of 2 over a very small period of time to respond to, say, the end of darkness. Nowadays these considerations of ionospheric condition are studied by computer, and graphs of recommended transmission frequency against time of day are plotted.

Apart from the skip distance problem, which can be overcome to some extent by having a broad elevation in the radiated pattern, there is another problem with multiple reflections. Under certain circumstances a wave can be reflected by the ionosphere and then reflected once at the ground and again at one of the ionospheric layers so that a multipath signal is eventually received. In that case interference will result between the various signals.

A useful discussion on the general approach to coping with ionospheric problems is given in reference 7.

7.22.2 Line of slight

At frequencies above the VHF range the most important propagation mechanism is line of sight. At these frequencies the ionosphere ceases to refract the wave which goes straight through the various layers; and the groundwave is attenuated very rapidly. This means that from a practical point of view transmitter and receiver must be in visible contact, although there will be a small degree of defraction around objects at the lower frequencies. This results in relatively poor over-the-horizon communication and it raises a problem for coverage in the higher television broadcast bands where transmitter and customers are in hilly regions.

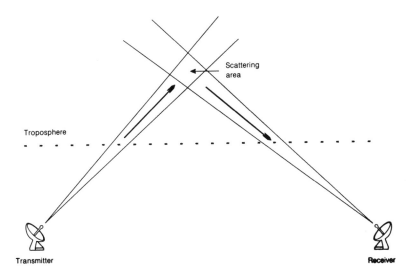

Fig. 7.28 Tropospheric scatter.

7.22.3 Tropospheric scatter

At frequencies above about 400 MHz there is significant scatter in the troposphere (below the ionosphere). This provides the means for extending the range of a communication link, as shown in Fig. 7.28. The signal level may be low, and the link is subject to fading due to variations in the tropospheric height, but for some applications it is the most practical method of providing a high capacity communications channel.

REFERENCES

1. Jordan, E. C., *Electromagnetic Waves and Radiating Systems*, Prentice-Hall, London, 1968.
2. Kraus, J. D., *Antennas*, McGraw-Hill, Maidenhead, 1950.
3. Jasik, H., *Antenna Engineering Handbook*, McGraw-Hill, Maidenhead, 1961.
4. Fradin, A. Z., *Microwave Antennas*, Pergamon, Oxford, 1961.
5. Weeks, W. L., *Antennas Engineering*, McGraw-Hill, Maidenhead, 1968.
6. Rudge, A. W., Olver, A. D., Milne, K. and Knight, P., *The Handbook of Antenna Design* (Vol. I), Peter Peregrinus, London, 1982.
7. Rudge, A. W., Olver, A. D., Milne, K. and Knight P., *The Handbook of Antenna Design* (Vol. II), Peter Peregrinus, London, 1983.
8. Pozar, D. M., Microstrip Antennas, *Proc. IEEE*, Jan. 1992, 79–91.

APPENDIX 7.1   FIELD COMPONENTS PRODUCED BY A SHORT CURRENT ELEMENT

In Section 7.3 we used Eqn (7.5) to represent the relationship between the retarded vector potential and the magnetic field $\bar{H}$, viz.

$$\nabla \times \bar{A} = \bar{H} \tag{7.5}$$

and the components of $\bar{A}$ for the current filament shown in Fig. 7.4 are

$$
\left.
\begin{aligned}
A_r &= A_z \cos \theta = \frac{Il}{4\pi r} \sin \omega \left[t - (r/c)\right] \cos \theta \\[2mm]
A_\theta &= -A_z \sin \theta = -\frac{Il}{4\pi r} \sin \omega \left[t - (r/c)\right] \sin \theta \\[2mm]
A_\phi &= 0
\end{aligned}
\right\} \tag{7.6}
$$

In spherical coordinates, $\nabla \times \bar{A}$ is expressed as

$$
\nabla \times \bar{A} = \bar{a}_r \frac{1}{r \sin \theta}\left[\frac{\partial}{\partial \theta}(A_\phi \sin \theta) - \frac{\partial A_\theta}{\partial \phi}\right] + \bar{a}_\theta \frac{1}{r}\left[\frac{1}{\sin \theta}\frac{\partial A_r}{\partial \phi} - \frac{\partial}{\partial r}(rA_\phi)\right]
$$
$$
+ \bar{a}_\phi \frac{1}{r}\left[\frac{\partial}{\partial r}(rA_\theta) - \frac{\partial A_r}{\partial \theta}\right] \tag{7.30}
$$

where $\bar{a}_r, \bar{a}_\theta, \bar{a}_\phi$ are the unit vectors in the r, θ and ϕ directions, respectively.

Substituting from Eqn (7.6) into Eqn (7.30), the $\bar{a}_r$ and $\bar{a}_\theta$ components are zero, leaving

$$\nabla \times \bar{A} = \bar{a}_\phi \frac{1}{r} \left[\frac{\partial}{\partial r} \left(-\frac{Il}{4\pi} \sin \omega \left[t - (r/c) \right] \sin \theta \right) \right.$$

$$\left. - \frac{\partial}{\partial \theta} \left(\frac{Il}{4\pi r} \sin \omega \left[t - (r/c) \right] \cos \theta \right) \right]$$

$$= \bar{a}_\phi \frac{1}{r} \left[\frac{Il}{4\pi} \frac{\omega}{c} \cos \omega \left[t - (r/c) \right] \sin \theta + \frac{Il}{4\pi r} \sin \omega \left[t - (r/c) \right] \sin \theta \right] \qquad (7.31)$$

and, knowing that $\bar{H} = \bar{a}_r H_r + \bar{a}_\theta H_\theta + \bar{a}_\phi H_\phi$, we have, from Eqns (7.5) and (7.31),

$$H_\phi = \frac{Il}{4\pi} \sin \theta \left[\frac{\omega}{cr} \cos \omega \left[t - (r/c) \right] + \frac{1}{r^2} \sin \omega \left[t - (r/c) \right] \right] \qquad (7.32)$$

For the space outside the current element, the relationship between the H field and the E field is given by one of Maxwell's equations [Eqn. (9.10) with $J = 0$], i.e.

$$\nabla \times \bar{H} = \varepsilon \frac{\partial \bar{E}}{\partial r} = \varepsilon \left[\bar{a}_r \frac{\partial E_r}{\partial t} + \bar{a}_\theta \frac{\partial E_\theta}{\partial t} + \bar{a}_\phi \frac{\partial E_\phi}{\partial t} \right] \qquad (7.33)$$

Thus, if we expand the left-hand side of this equation, then with the help of Eqn (7.32) we can find the electric field components. Since there is only one component to the $\bar{H}$ field, H_ϕ, we have by analogy with Eqn (7.30)

$$\nabla \times \bar{H} = \bar{a}_r \frac{1}{r \sin \theta} \frac{\partial}{\partial \theta} (H_\phi \sin \theta) - \bar{a}_\theta \frac{1}{r} \cdot \frac{\partial}{\partial r} (r H_\phi)$$

Substituting from Eqn (7.32), this becomes

$$\nabla \times \bar{H} = \bar{a}_r \frac{1}{r \sin \theta} \frac{\partial}{\partial \theta} \left\{ \frac{Il \sin^2 \theta}{4\pi} \left[\frac{\omega}{cr} \cos \omega \left[t - (r/c) \right] + \frac{1}{r^2} \sin \omega \left[t - (r/c) \right] \right] \right\}$$

$$- \bar{a}_\theta \frac{1}{r} \frac{\partial}{\partial r} \left\{ \frac{Il \sin \theta}{4\pi} \left[\frac{\omega}{c} \cos \omega \left[t - (r/c) \right] + \frac{1}{r} \sin \omega \left[t - (r/c) \right] \right] \right\}$$

After performing the differentiations, and rearranging the terms,

$$\nabla \times \bar{H} = \bar{a}_r \frac{Il \cos \theta}{2\pi} \left[\frac{\omega}{cr^2} \cos \omega \left[t - (r/c) \right] + \frac{1}{r^3} \sin \omega \left[t - (r/c) \right] \right]$$

$$+ \bar{a}_\theta \frac{Il \sin \theta}{4\pi} \left[-\frac{\omega^2}{c^2 r} \sin \omega \left[t - (r/c) \right] + \frac{\omega}{cr^2} \cos \omega \left[t - (r/c) \right] \right.$$

$$\left. + \frac{1}{r^3} \sin \omega \left[t - (r/c) \right] \right] \qquad (7.34)$$

Equating the $\bar{a}_\theta$ from the right-hand sides of Eqns (7.33) and (7.34),

$$E_\theta = \frac{1}{\varepsilon} \frac{Il \sin \theta}{4\pi r} \int \left[-\frac{\omega^2}{c^2} \sin \omega \left[t - (r/c) \right] + \frac{\omega}{cr} \cos \omega \left[t - (r/c) \right] \right.$$

$$\left. + \frac{1}{r^2} \sin \omega \left[t - (r/c) \right] \right] dt$$

which, after integrating, becomes

$$E_\theta = \frac{Il\sin\theta}{4\pi\varepsilon}\left[\frac{\omega}{rc^2}\cos\omega\left[t-(r/c)\right]+\frac{1}{cr^2}\sin\omega\left[t-(r/c)\right]\right.$$
$$\left. -\frac{1}{\omega r^3}\cos\omega\left[t-(r/c)\right]\right] \qquad (7.35)$$

By a similar argument, the $\bar{a}_r$ terms on the right-hand sides of Eqns (7.33) and (7.34) are used to find the radial component of the E field:

$$E_r = \frac{Il\cos\theta}{2\pi\varepsilon}\left[\frac{1}{cr^2}\sin\omega\left[t-(r/c)\right]-\frac{1}{\omega r^3}\cos\omega\left[t-(r/c)\right]\right] \qquad (7.36)$$

We have shown, therefore, that the non-zero field components in the space surrounding the current element are

$$H_\phi = \frac{Il\sin\theta}{4\pi}\left[\frac{\omega}{cr}\cos\omega\left[t-(r/c)\right]+\frac{1}{r^2}\sin\omega\left[t-(r/c)\right]\right] \qquad (7.37)$$

$$E_\theta = \frac{Il\sin\theta}{4\pi\varepsilon}\left[\frac{\omega}{c^2r}\cos\omega\left[t-(r/c)\right]+\frac{1}{cr^2}\sin\omega\left[t-(r/c)\right]\right.$$
$$\left. -\frac{1}{\omega r^3}\cos\omega\left[t-(r/c)\right]\right] \qquad (7.38)$$

$$E_r = \frac{Il\cos\theta}{2\pi\varepsilon}\left[\frac{1}{cr^2}\sin\omega\left[t-(r/c)\right]-\frac{1}{\omega r^3}\cos\omega\left[t-(r/c)\right]\right] \qquad (7.39)$$

corresponding with Eqns (7.7)–(7.9) in Section 7.3.

By substituting $\omega = 2\pi f, c = f\lambda$ and $\eta = \sqrt{(\mu/\varepsilon)}$ (where η is the free-space impedance), these equations can be expressed in a form which explicitly involves λ:

$$H_\phi = \frac{Il\sin\theta}{2\lambda}\left[\frac{1}{r}\cos\omega\left[t-(r/c)\right]+\frac{\lambda}{2\pi r^2}\sin\omega\left[t-(r/c)\right]\right] \qquad (7.40)$$

$$E_\theta = \frac{\eta Il\sin\theta}{2\lambda}\left[\frac{1}{r}\cos\omega\left[t-(r/c)\right]+\frac{\lambda}{2\pi r^2}\sin\omega\left[t-(r/c)\right]\right.$$
$$\left. -\frac{\lambda^2}{4\pi^2r^3}\cos\omega\left[t-(r/c)\right]\right] \qquad (7.41)$$

$$E_r = \frac{\eta Il\cos\theta}{\lambda}\left[\frac{\lambda}{2\pi r^2}\sin\omega\left[t-(r/c)\right]-\frac{\lambda^2}{4\pi^2r^3}\cos\omega\left[t-(r/c)\right]\right] \qquad (7.42)$$

PROBLEMS

7.1 Draw the polar diagram produced by two isotropic antennas, which are fed with equal currents if:

(a) the separation between the antennas is (i) 0.25λ; (ii) 0.7λ; (iii) 1.1λ;
(b) the phase difference between the currents is (i) $0°$; (ii) $30°$; (iii) $90°$.

7.2 Two isotropic elements, 0.4λ apart, are fed with currents of equal magnitude and a phase difference of 90°. Find the number of nulls in the polar diagram, and their direction, and the magnitude of the field normal to the line joining the elements.

Answer: 2 nulls, $\pm 51.2°$, 0.7.

7.3 Find the position of the main beam, relative to the line of the array, if five elements, 0.5λ apart, are fed with currents of equal magnitude and phase.

Answer: Main beam normal to the line of elements.

7.4 Repeat problem 7.3 for separations between the elements of (i) 0.25λ and (ii) 0.625λ, and sketch the radiation pattern in all three cases, noting the effect of increasing the spacing between the elements.

7.5 Plot the radiation pattern for an array of ten identical isotropic sources in line, and spaced at intervals of 0.375λ, if the phase difference between each source is $-135°$.

7.6 Use pattern multiplication to find the polar diagram of the array of isotropic radiators shown, assuming that the current magnitude is the same in each radiator.

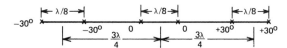

7.7 Use pattern multiplication to find the radiation pattern of an array of three isotropic elements in line, spaced at half-wavelength intervals, if the currents are in phase, but the magnitude of the current in the middle element is twice that in the outer ones.

7.8 Sketch the polar diagram of a six-element linear array in which the elements are fed with currents of equal magnitude, and a constant phase difference of 30°. The distance between adjacent elements is 0.6λ.

Compare this polar diagram with that produced by the array shown:

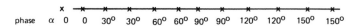

Spacing between adjacent elements is 0.3λ.

8 Active microwave devices

We are considering here systems that operate at frequencies above about 1 GHz, covering the bands used for industrial and domestic heating, microwave communications, satellite communications and radar.

A notable feature of this range of frequencies is that some of the active devices are vacuum tubes. The power levels required for many systems cannot be achieved with solid-state techniques and this has meant a continuous development and refinement of the vacuum tubes that were first introduced many years ago.

Whether or not a solid-state device can be used in a particular application will depend almost entirely on the power levels involved. If these levels do not exceed 100 W, solid-state will be a natural choice. For higher powers, however, vacuum tubes must be used; which type will depend on other considerations such as bandwidth and noise level.

Here we will not look at a comprehensive range of devices, but we will study the operation of the most common, starting with the oldest of the vacuum tubes currently in use, the klystron.

8.1 KLYSTRON

The origins of this tube are not clear. In the late 1930s, workers in several countries wrote about the possibility of using velocity modulation to generate microwaves; the space-charge modulation approach, which formed the basis of operation of lower frequency valves such as the triode, could not operate at such high frequencies. Which group should be credited with actually making the first prototype klystron is not known. The onset of war veiled such things in secrecy. However, we do know that much early work was done in the USA and in the UK, where there was intense pressure to develop satisfactory oscillators and amplifiers for use in the new detection system called radar.

8.2 TWO-CAVITY KLYSTRON

The basic form of the klystron amplifier is shown in Fig. 8.1. As we can see, it consists of a cathode, electron gun, two cavities and a collector. Electrons are generated at the cathode, directed into a fine stream along the axis of the tube by the electron gun, and, after passing through two cavities, they leave the

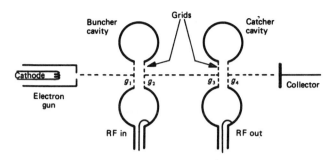

Fig. 8.1 Diagram of a two-cavity klystron.

tube via the collector. The cavity nearer to the cathode is called the buncher, and the other is called the catcher. The microwave signal to be amplified is coupled into the buncher cavity by either a waveguide or a coaxial loop. A similar device is used to abstract the amplified signal from the catcher cavity. The shape of the cavities will depend on the particular design; it could be inside the vacuum tube surrounding the electron gun, electron beam and collector, with the coupling by waveguide designed to launch the signal into the cavity, or it could be partly outside the tube, which would facilitate easier coupling of the microwaves, but would also require glass-to-metal seals between the cavities and the vacuum tube. Whatever the method used, the basic shape of the cavities is toroidal with two fine mesh grids across the axis to allow the electron beam to penetrate. The grids contain the cavity fields between them. There are some klystrons in which grids are not used; then there is no impairment to the passage of the electrons, but there are fringe fields around the axial gaps in the cavities.

The term 'velocity modulation' describes the action of the microwave signal on the electron beam. In our discussion we are talking about the steady-state operation of the valve; that is, the way in which the valve works after it has warmed up and any transient effects have died away.

In all microwave tubes there is a transfer-of-energy mechanism that we will accept without attempting to explain how it takes place. If an electron passes through an electric field it experiences a force that causes it to change velocity. If the electron is speeded up, its kinetic energy is increased, and that increase must be taken from the field. If the electron is decelerated it loses energy, and. the field gains an equivalent amount, so that energy in the system is conserved.

Now we consider the passage of an electron between the grids of a cavity (Fig. 8.2). In the steady state, the potential across the grids will change at the microwave frequency of the field within the cavity itself. On one half-cycle [Fig. 8.2(a)] grid 1 will be positive with respect to grid 2. The electron passing through the grids in the direction shown will experience a decelerating force and will therefore slow down, giving up energy to the field in the cavity. The degree of deceleration will depend on the field across the grids at the time of the passage of the electron. We assume in our discussion that the grids are close together and the field remains constant during the transit of the

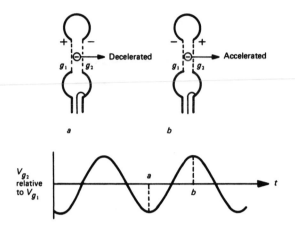

Fig. 8.2 Effect of grids on the electron beam in a two-cavity klystron.

electron. On the other half-cycle [Fig. 8.2(b)] the electron sees an accelerating force and is therefore speeded up, absorbing energy from the cavity field. When there is no potential difference between the grids the electron will pass through with velocity unchanged.

To clarify this velocity modulation mechanism further we can represent the movement of electrons, relative to the microwave field across the cavity grids, on a distance–time diagram, sometimes called the Applegate diagram after an early American worker on klystron characteristics (Fig. 8.3).

The sinusoid at the bottom of the diagram represents the fields across the grids. The lines above it represent the passage of an electron; the angle of each line is a measure of its velocity. Consequently, an electron passing through the grids when the field is maximum accelerating (point A) will have a larger slope (further distance in a given time) than one that leaves when there is no potential difference between the grids (point B); and that in turn has a larger slope than an electron suffering maximum deceleration (point C). As we might

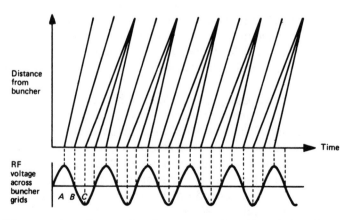

Fig. 8.3 Distance–time diagram for electrons passing through buncher.

expect the faster electrons catch up the slower ones, and bunches of electrons are formed. The diagram shows that these bunches form around the electrons that pass through with unchanged velocity as the field goes from decelerating to accelerating.

In the two-cavity klystron it is the function of the first cavity to create these bunches – hence its name. The cavity is tuned, by either adjusting a screw in the wall or deforming the wall itself in some way, to the incoming microwave frequency. When the electron stream, which we assume to be approaching the buncher at uniform velocity, passes through the grids, the velocity modulation causes bunches to form at some point along the axis of the tube in the direction of the collector. The catcher is placed at this point, allowing the tight bunches to arrive between the grids. The distance between the buncher and catcher will be fixed once the valve is constructed. However, the distance taken to form the bunches will depend on their initial velocity, and this allows the collector voltage to control the bunching distance. In the steady state, the bunches arrive between the catcher grids when grid 3 is positive with respect to grid 4 (Fig. 8.1). Then the bunch is decelerated, and the cavity field absorbs the loss in kinetic energy, thus enhancing the strength of the field in the catcher cavity.

We can see from Fig. 8.3 that, although bunches are formed, if there is a continuous stream of electrons from the cathode, not all electrons are involved in the bunching process. Those that are not will not add to the catcher field and may in fact take energy from it. This mechanism reduces the efficiency of the valve.

The electron gun, consisting of a complex arrangement of electrodes, is designed to focus the electron stream into a filament along the axis of the tube. There will be some blurring at the edges of the beam, because of the space-charge forces that tend to spread the beam out.

The collector is designed to perform two functions. It must collect the electrons from the beam, without allowing any reflections or secondary emissions, and it must conduct away the heat generated by the impact of the electron beam. The first is achieved by shaping the collector in such a way that any secondary electrons do not pass back down the tube. The second requires that either a large structure of good conducting material is used to allow natural convection cooling, or that forced-air blowing be included in the overall system design.

It is possible to feed some of the catcher field back into the buncher by providing a suitable coupling mechanism between them, and so create a two-cavity klystron oscillator, but it is not much used in practice.

8.3 REFLEX KLYSTRON

The commonest klystron oscillator, particularly for laboratory work or for use as a local oscillator, is the reflex klystron. It uses only one cavity. A schematic representation is shown in Fig. 8.4. An electron gun directs electrons from the cathode through the grids of the cavity. The velocity of the electron beam is determined by the resonator – cathode voltage. As the beam

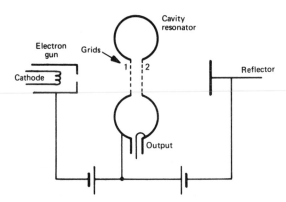

Fig. 8.4 Diagram of a reflex klystron.

passes through the resonator cavity grids, it is velocity-modulated, as we discussed in the last section with the aid of Fig. 8.2. Passing from the resonator grids the beam travels towards the reflector which is at a large negative potential. The beam is therefore in a retarding field, which it is eventually unable to resist, and it is forced to reverse direction and travel back down the tube towards the cathode. It therefore passes through the grids a second time and, on emerging from the grids, the beam is collected on the body of the resonator. The technical problem of providing adequate heat dissipation limits the power output of this tube.

The influence of the field in the resonator on the electron beam is similar to that of the buncher and catcher in the two-cavity klystron. The modulated beam travelling through the cavity for the first time gradually breaks up. The faster electrons travel further towards the reflector, before having to turn back, than those electrons that have already been retarded by the action of the resonator field. The depth of penetration of the electrons into the reflector region is governed by the reflector voltage. To make the valve oscillate, the returning beam must arrive back between the grids in such a way

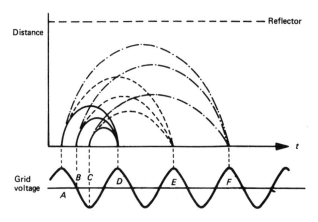

Fig. 8.5 Distance–time diagram for reflex klystron.

that (a) the bunches are completely formed, and (b) the field is such that it provides maximum retardation, i.e. grid 1 is maximum negative with respect to grid 2. We can see this condition by referring to Fig. 8.5, which shows a distance–time diagram for the reflex klystron. The bunches form around electrons that pass through the grids for the first time as the field is moving from accelerating to decelerating (point B), and to achieve maximum transfer of energy from the beam to the resonator field the bunch must arrive back between the grids when the field is at D, E, F, etc. The transit time for the electrons in the resonator–reflector region, to give maximum output, is $(n - \frac{1}{4})T$, where T is one cycle of the field and $n = 1, 2, 3 \dots$.

Evidently, from the diagram, point D is at $n = 1$, E at $n = 2$, etc. Two other conditions are of interest. If the bunches arrive back between the grids when grid 1 is positive with respect to grid 2, they will be in an accelerating field and will therefore take energy from the field in the cavity, and there will be no output. Second, the bunches could arrive between the grids when there is a retarding field, but it is not at a maximum. Then the amount of energy taken from the bunches will be reduced, and the output from the klystron will fall.

When using a reflex klystron it is normal practice to adjust the resonator and reflector voltages to achieve maximum output. The frequency can normally be varied by a tuning screw, which slightly deforms the shape of the resonator.

The variation of output power with reflector voltage is shown in Fig. 8.6. The maxima are clearly defined. Each hump is called a mode and, as we can see, there may be several. In general, the amplitude of successive maxima increases as the reflector voltage increases; thus the largest output is achieved in mode 1, when $n = 1$ in the above formula, corresponding to a high voltage on the reflector. Occasionally, however, mode 1 is not the largest; second-order effects, such as multiple transits through the grids, reduce its level below that of mode 2.

Although the resonant frequency of the cavity itself is determined entirely by its geometry and size, when the valve is operating other factors affect the output frequency. There are three main components that combine to form the total impedance of the system; in their simplest form they can be considered as three parallel admittances, as shown in Fig. 8.7.

Y_e is the admittance of the electron beam, Y_c is the admittance of the cavity, and Y_L is the load, or output, admittance. Variation of the reflector or

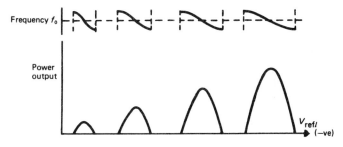

Fig. 8.6 Output characteristics of reflex klystron.

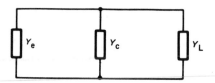

Fig. 8.7 Simplified equivalent circuit for reflex klystron.

resonator voltage will change the bunching and velocity of the electron beam, thus varying the beam current and hence Y_e. The output frequency will therefore be varied. Figure 8.6 shows a typical example of the variation of frequency with reflector voltage. We can see that the output frequency varies across each mode; the frequency is the same at each mode maximum, but the frequency range of the variation across the mode may not be constant. This frequency dependence on the reflector voltage is often called electronic tuning, and it is this characteristic that made the reflex klystron attractive as a local oscillator in a radar receiver. By means of an automatic frequency-control circuit, the frequency of the received microwave signal was compared with that from the klystron, and if the difference was not equal to the required intermediate frequency (typically 45 or 60 MHz) the reflector voltage could be adjusted to bring the klystron output frequency to an appropriate value.

We have already noted that the cavity can be tuned mechanically by adjusting a screw or deforming a flexible wall. This mechanical change results in a change in Y_c, the cavity admittance. Since Y_L is included in the equivalent admittance of the tube, the load can have an effect on the output; if the load changes, Y_L will change, and hence the output frequency will alter. This effect can cause difficulties in systems where the load impedance is not constant, and to overcome it some form of device, such as an isolator, is used to block any reflections coming back into the generator.

The reflex klystron can exhibit a hysteresis effect, which sometimes causes problems in applications where large frequency variations are required. Figure 8.8 shows two representations of this effect. In Fig. 8.8(a) we see that as the reflector voltage is increased the output follows the normal mode shape, but as the reflector voltage is decreased the characteristic follows a different path. Figure 8.8(b) shows how hysteresis affects the curve of resonator voltage against reflector voltage for a fixed output power. Instead of being a single curve it is double valued. The characteristic is

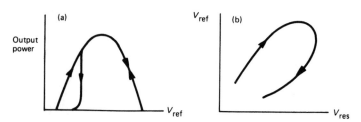

Fig. 8.8 Hysteresis effect (a) across a mode; (b) constant power curve.

produced by following one side of the curve, by increasing both resonator and reflector voltages, and then following the other side as those voltages are decreased.

Hysteresis is caused by second-order effects that we have not taken into account.

8.4 MAGNETRON

The reflex klystron is not able to deliver large quantities of power. For applications such as industrial heating, domestic cooking and radar, from 1 kW up to several megawatts of pulsed power may be required. To generate the high powers required for radar and microwave heating a different approach from that of the reflex klystron is necessary. The cavity magnetron developed by Randell and Boot in 1940 is the most common means of generating high powers in the low frequencies of the microwave range. In fact, as has been pointed out by Mr Selby Lounds, the magnetron principle was invented some 20 years earlier, being the subject of a patent by A W Hull of the General Electric Company of America in 1921, and Yagi and his group were working on similar devices in the 1920s. It was the shaping of the anode block into a series of cavities that changed a mechanism for generating low powers into one that could produce several megawatts. Although its design has been refined since then, there have been very few significant changes from the crude prototypes that demonstrated that the first magnetrons would work.

Figure 8.9 shows the essential features of the valve. It has five major components:

(i) an axial magnetic field, usually produced by a permanent magnet;
(ii) a cathode that is heated to generate electrons;
(iii) an anode block that has several slots cut into it (slow-wave structure);
(iv) a coupling device, either a loop or a hole, to couple energy from a slot in the anode block to external circuitry;
(v) the space between the cathode and the anode.

The operation of the magnetron is dependent on the interactions between

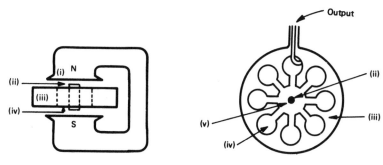

Fig. 8.9 Diagram of magnetron.

the electrons travelling in the anode–cathode space and the static and RF electromagnetic fields in that region. The analysis is complex and will not be considered here.

We will consider briefly the action of the dc fields on electrons leaving the cathode, and we shall also look at the anode block and consider it as a slow-wave structure, whose purpose is to slow down a wave circulating in the cathode–anode space to a velocity approximately the same as that of the electrons moving in that space. Energy can then be transferred from the electron beam to the electromagnetic wave. The frequency of the output depends on the design of the slow-wave structure, as we shall see.

Between the anode and the cathode there is a strong electric field, produced by a high negative dc voltage on the cathode. In the absence of a magnetic field, an electron leaving the cathode moves radially to the anode (path 1 in Fig. 8.10). In this diagram we assume that the anode block has no indentations. An axial magnetic field applied to the structure will create an additional, rotational force on an electron in the anode–cathode space. The force is a function of both the electron velocity (determined by the anode–cathode voltage) and the strength of the magnetic field. The combination of radial and rotational forces on the electron can produce different trajectories, as shown in Fig. 8.10. If the electron velocity is not up to some critical value, the electron will follow a curved path on to the anode – path 2 in the diagram. If the velocity of the electron is very high, the rotational force of the magnetic field will draw the electron back into the cathode (path 4). Between these two extremes an electron, by a suitable combination of fields, will follow a curved trajectory almost to the anode, and then follow a wide loop back to the cathode (path 3). These paths are those that would occur in the concentric arrangement shown in Fig. 8.10. However, the function of the magnetron is to produce microwave oscillations of an electromagnetic field. To do that, the field must be able to exist inside the tube, and it must also be capable of interacting with the electron beam, from which it will receive its energy. The basic mechanism involved is of an electromagnetic wave travelling around the space between cathode and anode and, during its progress, absorbing energy from the electron cloud in that space. For satisfactory interaction between electrons and the beam to occur, the electron velocity and the electromagnetic field travelling-wave velocity must be substantially the same, as we noted earlier. The slow-wave structure used to support a travelling

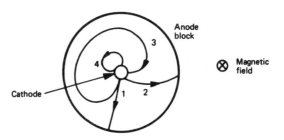

Fig. 8.10 Electron-paths in anode-cathode space.

wave, whose phase velocity is about the same as that of the electron beam (i.e. about one tenth of the velocity of light), consists of a series of cavities cut into the anode block. The simplest shape is an even number of circular cavities evenly distributed around the block, and connected to the cathode–anode space by a short coupling slot. The cavities are as similar as they can be made, and we can assume that they resonate at identical frequencies. They behave in the same way as the cylindrical cavity described in Section 9.21. Normally they would be operated in the fundamental TE_{101} mode.

The fields at the coupling holes will fringe into the cathode–anode space and influence any electron passing by. If the polarity of the field is such that an electron is accelerated it will take up energy from the cavity. The consequence of the increase in electron velocity is to move the electron back towards the cathode, and there it will create secondary electrons. (In the design of the cathode, this secondary emission, due to the return of high-velocity primary electrons, is a significant source of electrons, and allows the valve to run at a lower cathode temperature than would otherwise be possible.)

On the other hand, if the polarity of the field is in the opposite direction, the electrons passing through it will be slowed down and lose energy to the field. This is the condition required for successful operation. The loss of energy causes the electron to move into a wider orbit, and as it leaves the influence of that cavity it is affected by the field from the next. If by the time the electron has moved from the field due to the first cavity to that due to the second, the field polarities have changed, then it will be decelerated by that field, move out into a wider orbit, and continue to the next cavity where the field will again be a retarding one, and so on. Eventually the electron orbit will take it on to the anode. The trajectory just described is shown in Fig. 8.11.

In fact, the structure of the anode block can support a large number of travelling wave modes. The one we have described, in which the fields at alternate slots are 180° out of phase, is called the π mode. It is the one most commonly used in practice. Because a large number of modes are possible there is an inherent instability as the travelling wave can slip from one mode to another. Early in the development of the magnetron it was appreciated that better stability could be achieved if the π relationship could be maintained. A

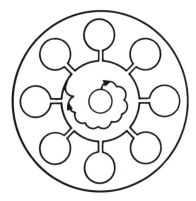

Fig. 8.11 Path of electron undergoing deceleration at successive slots.

Fig. 8.12 Strapping between alternate slots.

simple, but effective, method was 'strapping', in which alternate 'pole pieces' are interconnected by metal straps over the anode block, as shown in Fig. 8.12. This device was very successful in reducing the valve instabilities.

The coupling loop or waveguide aperture used to take energy from the valve (Fig. 8.9) can be placed in any one of the cavities. Work performed by some of the original designers showed that there was no advantage in coupling from more than one cavity; that chosen will depend on which is the most convenient mechanically.

Many different configurations have been used for the anode block, and some are shown in Fig. 8.13. Of particular significance is the 'rising sun' arrangement of Fig. 8.13(a), which is designed for use at high frequencies, where the slot and cavity shape are more difficult to manufacture. The alternate short and long slots support field patterns that stabilize the π mode without the need to use strapping.

8.5 TRAVELLING-WAVE TUBE (TWT)

The klystron amplifier has the drawback of being a narrowband device. For some applications, particularly in communications, wideband operation is essential but the resonant cavities of the klystron limit its response to a few megahertz. The purpose of the klystron cavities is to provide an RF field that can velocity-modulate an electron beam, in the case of the buncher, or respond to such modulation in the case of the catcher. However, Kompfner[1] reported that another method of allowing the RF field to interact with the electron beam could be used without relying on resonant cavities. He introduced the travelling-wave tube in which the electromagnetic signal to be amplified is transmitted close to an electron beam for a distance of several wavelengths; their proximity allows an interaction to take place in which the beam loses energy to the RF wave. We have noted several times in the preceding sections that, before a constructive interaction can occur, the electron beam must be travelling slightly faster than the electromagnetic wave. This means that, for the TWT to function as an amplifier, the wave must be slowed down to about the same velocity as the beam, which for normal operating conditions is about one tenth of the velocity of light. The wave velocity of interest is that along the axis of the tube. To achieve this reduction, a slow-wave structure is used. For low-power tubes, the slow-wave structure is a helix. A diagram of a helix TWT is shown in Fig. 8.14. The axial velocity component of the wave propagating along the helix is $c \sin \alpha$, where α is the helix

Fig. 8.13 Various anode block shapes. (a) Rising sun, (b) hole and slot, (c) slot.

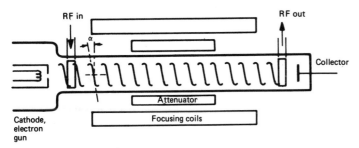

Fig. 8.14 Diagram of travelling-wave tube.

pitch and c is the velocity of light. For small α this velocity is approximately $c\alpha$, where α is in radians.

Other components of the TWT can be seen in the diagram. The electron beam is generated by a cathode, and focused by an electron gun. Additional focusing is necessary to confine the beam to the axis over the length of the tube, so it is provided by an axial magnetic field, produced by a coil surrounding the tube. The beam passes through the helix and then strikes a collector, where it is absorbed. The collector is a robust piece of copper, shaped to reduce secondary emission, and attached to a large heat sink.

The helix is made from fine wire and supported by ceramic rods. Its diameter is substantially greater than that of the beam. At each end of the helix some coupling device must be used to interconnect the electromagnetic wave with the external circuitry.

There are two interrelated mechanisms involved in the operation of the TWT:

(i) the creation of travelling waves on the electron beam, due to the action of the axial field produced by the signal on the helix;
(ii) the effect of these electron beam travelling waves on the propagation characteristics of the helix.

The slow-wave on the helix produces an axial electric field that, for a single-frequency signal, is sinusoidal in the z direction. Some sections of the beam will be accelerated by this axial field, and some decelerated. The natural charge repulsion inside the beam will cause slight bunching to occur, indicative of velocity modulation as we discussed in earlier sections. Figure 8.15 shows schematically the relationship between the axial field and the beam. The diagram represents one instant of time. However, the electric field will travel along the helix, and that induces two travelling waves on the beam. A theoretical description of this mechanism of setting up travelling waves on the beam by slight perturbations of the electron flow is given in several texts, e.g. Atwater[2] and Collin.[3]

The existence of the two travelling waves on the beam has an influence on the characteristics of the slow-wave structure. In Fig. 8.16 the beam current, and a transmission line representation of the slow-wave structure, are shown. There will also be a small leakage current δi between the beam and the structure. This leakage current, which is directly proportional to the current

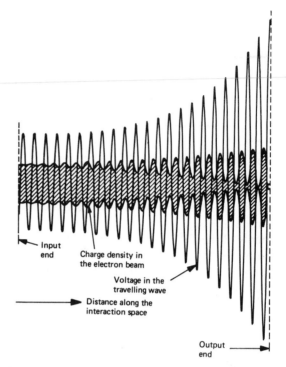

Fig. 8.15 Transfer of energy from beam to signal. Reproduced from Reich, Shalnik, Ordung and Kraus (1957), *Microwave Principles*, Van Nostrand Reinhold, New York, by permission.

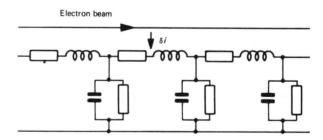

Fig. 8.16 Transmission line representation of travelling-wave tube.

on the beam, will contain components of the two beam travelling waves. The combination of the normal transmission line travelling waves on the helix when no beam is present and the travelling waves on the electron beam, is to produce four travelling waves on the helix in the presence of the beam. One is a backward wave and the other three are forward waves. The backward wave, if allowed to travel to the input coupler, would be reflected there and then travel forward, thus causing interference with the forward waves and possibly generating oscillations. An attenuator (or an isolator) is placed at some point along the tube to prevent such interference. Of the three forward waves, one

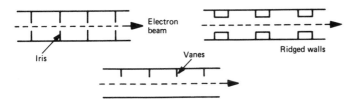

Fig. 8.17 Waveguide slow-wave structures.

has no amplitude variation (the propagation constant is imaginary), another is attenuated, but the third grows exponentially with z, the distance from the input end of the helix. It is this last wave that gives the amplification required. The actual gain achieved will depend on how efficiently the energy in the beam can be transferred to the field on the helix. The growth wave is continuously amplified as it passes along the tube, and at the output end it is coupled to the external circuit.

Typically, a helix type TWT will operate over an octave bandwidth, and have a gain of 30 dB or more. Uniform gain across the band may be essential for many applications, and this will limit the bandwidth available.

The helix slow-wave structure is convenient, and easy to manufacture, but it can be used only for low-power tubes. The description above assumes that the electron beam consists of electrons moving in straight lines from cathode to anode. In practice the electron flow is much more complicated because of the mutual repulsion between the electrons, and this causes the beam to be blurred at the edges. For high power, the electron beam is of large diameter and it could easily damage the helix if it came in contact with it. The helix itself is not able to withstand bombardment by stray electrons; hence other structures are used in high-power systems. Figure 8.17 shows examples of some slow-wave structures based on waveguide propagation.

One of the most important civil applications of the TWT is as a high power amplifier (HPA) in a satellite transponder (see Section 14.4).

8.6 SOLID-STATE DEVICES

The vacuum tubes described earlier, although capable of operating at high power levels, suffer from several major disadvantages; they require stable, high voltage, power supplies; they are mechanically fragile, and they have relatively short lives. Solid-state devices, on the other hand, are extremely robust, have long working lives, and operate at a few tens of volts. They are inherently smaller and more convenient to use, and cheaper to buy. Consequently they are used for all applications not requiring a continuous power greater than several tens of watts. As has happened at lower frequencies, there have been several different devices proposed and developed, but here we shall look only at the three most important; the Gunn diode, the IMPATT diode and the GaAsFET.

8.7 GUNN DIODE

The excitement following the invention of the transistor in 1948 led to an upsurge in active device research for microwave applications. Although many modifications were made to bipolar transistor designs, in an attempt to make them suitable for operation at microwave frequencies, there were clearly some intrinsic limitations that prevented lower-frequency techniques from being applied successfully. From 1961 to 1963, however, there was a major research breakthrough. In 1961 Ridley and Watkins,[4] and in 1962 Hilsum,[5] wrote theoretical papers predicting that, given certain conditions that are satisfied by a few compounds, microwave oscillations could be produced directly from a thin sample of such a material. In 1963, Gunn,[6] while examining the noise properties of a slice of GaAs, observed a regular pulse of current from his sample when the applied voltage exceeded some level; the Gunn device was discovered. Gunn followed up this work with a set of very precise and difficult measurements in which he investigated the variations of electric field across the device when it was operating in an oscillatory mode, and as a result he confirmed many of the earlier theoretical predictions.

We can see the requirements of a suitable material from the energy level diagram of GaAs shown in Fig. 8.18. The conduction band is separated from the valence band by what is called the 'forbidden gap', which is 1.4 eV wide. When there is no field across the diode (so called because it has two terminals, a cathode and an anode, and not because it exhibits any rectifying properties) the free conduction electrons reside in the lowest conduction state, shown as the lower valley in the diagram. As the electric field across the device increases, the drift velocity also increases; the straight line OA in Fig. 8.19 shows the relationship between them. The gradient of OA is known as the electron mobility in the lower valley, say μ_L. As the electric field is increased, it eventually achieves a value E_{Th} at which the linear relationship between electron velocity and electric field breaks down, and the velocity begins to decrease, along CD. There is now a negative gradient until the field reaches E_v at point D, after which the velocity again increases linearly with electric field. Beyond D the characteristic follows the line OB which represents an electron mobility of μ_u. Clearly, μ_u is smaller than μ_L.

Along DB the conduction electrons, which were in the lower valley, have now moved to the upper valley where their mobility is reduced, and their

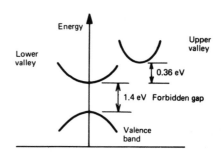

Fig. 8.18 Energy-level diagram for GaAs.

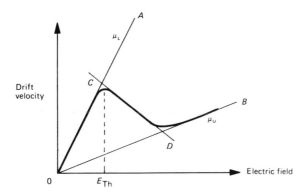

Fig. 8.19 Variation of drift velocity with electric field in a Gunn device.

mass is increased. Between C and D the transition over the 0.36 eV gap between the valleys is taking place. To avoid breakdown of the diode material, which must have suitable conduction band energy levels, the difference between them must be much smaller than the forbidden gap. There are some III–V compounds that satisfy this requirement. GaAs is the best understood because, since the Gunn experiments, it has been used widely as a diode material. Another compound of interest is InP which may eventually prove to be even more appropriate. Intensive work is being done in several countries to overcome some of the technical problems associated with its production.

The distribution of electric field across the diode is constant until E_{Th} is exceeded. Thereafter the diode is operating in the unstable CD region. Gunn showed that in this region a domain of high-intensity field and low-mobility electrons forms at the cathode, and gradually drifts across to the anode, where it leaves the device and produces a pulse of current in the external circuit. The variation of output current with time is shown in Fig. 8.20. Once a domain forms at the cathode no further domain can be sustained, so there is only one domain present at any time. Some properties of the cathode region ensure that the domain begins there consistently, and therefore the periodicity of the pulses is determined by the transit time of the domain across the device, which in turn is related to the device thickness. For operation in the range 5–10 GHz the material would be about 10 μm across. To sustain oscillations in a circuit, the circuit resistance and the device resistance must cancel,

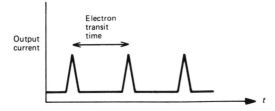

Fig. 8.20 Output pulses from Gunn device..

implying that at the frequency of interest the device resistance must be negative. That this is so can be seen from Fig. 8.19. The electron drift velocity is related directly to current flow through the device, and the electric field is proportional to the terminal voltage; hence the line CD represents a negative resistance. In practice it will have a value of no more than $10\,\Omega$. Apart from this negative resistance, there will be reactive components associated with the device, its leads and the packaging, all of which will affect the operating frequency. A simplified equivalent circuit is shown in Fig. 8.21.

The mode of operation described above, in which the applied voltage is maintained above E_{Th}, is called the transit-time mode. To produce oscilla-

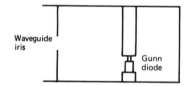

Fig. 8.21 Simplified equivalent circuit of Gunn device.

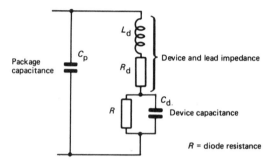

Fig. 8.22 Waveguide mounting for Gunn device.

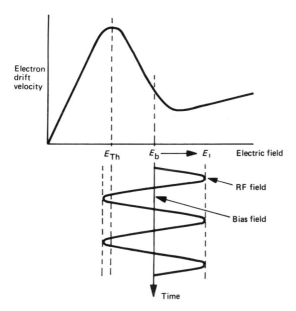

Fig. 8.23 Biasing for the LSA operation mode.

tions, the diode is usually placed in a resonant microwave circuit which acts as a high Q tuned system, giving a sinusoidal output. A schematic of a mounting for the Gunn diode is shown in Fig. 8.22. The electrodes are produced by crystal growth on to a conducting substrate, or by ion implantation. One of the factors that limits the power obtainable is the heat dissipation that can be achieved. To maximize the conductivity of heat from the device a large integral heat sink, forming part of the substrate, is used.

In the transit-time mode we saw that the natural oscillating frequency is determined by the length of the sample: the thinner the sample, the shorter the transit time, and hence the higher the frequency. However, the transit time mode is not the only one that is possible.[7] By applying a dc bias which takes the electric field just below E_{Th}, the diode can be made to switch into the negative resistance region as the alternating voltage takes the field value above E_{Th} (Fig. 8.23). Then the frequency of oscillation can be determined entirely by the external circuit in which the device is mounted. This mode of operation is known as the limited space-charge accumulation (LSA) mode.[8,9]

8.8 IMPATT DIODE

The IMPATT diode takes its name from IMPact Avalanche and Transit Time diode. It is one of several microwave devices based on the avalanche growth of carriers through a diode junction placed in a very high electric field. Sometimes called the Read diode after the man who proposed it, this device was first produced experimentally in 1965 by Johnston, de Loach and Cohen.[10] Since then IMPATT oscillators have been made for frequencies throughout the microwave spectrum, from 300 MHz to 300 GHz. They are more effficient (typical value about 15%), and can produce significantly more power than the Gunn device, but at the expense of greater noise and higher power supply voltage.

As the name implies, there are two factors that are particularly important in the operation of IMPATT diodes. First is avalanche multiplication. This phenomenon can occur at electric fields above about 10^5 V/cm. Carriers receive sufficient energy to produce electron–hole pairs by impact ionization. The ionization rate is strongly, but not linearly, related to the electric field; a five-fold increase in ionization rate can be produced by doubling the electric field.

The other important factor is the finite carrier drift velocity, and its variation with electric field strength. The electric fields used in IMPATT devices are much higher than the threshold fields discussed in the previous section on Gunn devices. Operating fields are above 10^5 V/cm and at such levels the electron drift velocity variation with electric field suffers a saturation effect and remains substantially constant. The value of the drift velocity depends particularly on the device material; 6×10^6 cm/s and 8×10^6 cm/s are typical for Si and GaAs, respectively, when used at an operating temperature of about 200 °C.

The basic form of the IMPATT is shown in Fig. 8.24; electrons are released at the p–n junction and then drift through an intrinsic n region. There are two significant widths associated with the device; W_A, the avalanche region, in which ionization takes place and a concentration of electrons is formed, and W_D, the drift region through which the electron bunch moves at velocity v_s.

The field across the p^+–n junction is well in excess of the breakdown value, with the maximum field E_p being approximately given by

$$E_p = \frac{q}{\varepsilon} N W$$

where W is the depletion width $= W_A + W_D$, N is the doping concentration in the n-type material, q is the electron charge and ε is the permittivity of the material.

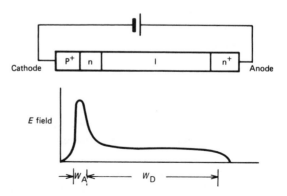

Fig. 8.24 IMPACT diode structure.

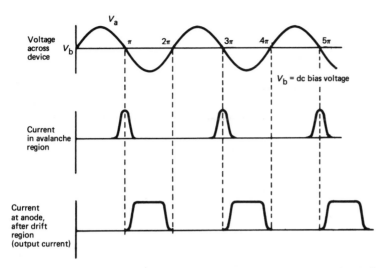

Fig. 8.25 Delay between output current and applied voltage in IMPATT diode.

When an IMPATT is used in an oscillator circuit it is reverse biased by a fixed voltage V_b, and superimposed on V_b is the voltage due to the microwave field in the oscillator, V_a. If V_b is chosen such as to just support avalanche breakdown, we can see from Fig. 8.25 how the diode sustains oscillations in the circuit.

The ionization mechanism has an inherent inertia; consequently the ionization rate does not respond instantaneously to changes in junction voltage. As V_a increases, the ionization rate increases, and continues to increase when V_a falls, until V_a reaches zero. The ionization then decreases also. The resulting electron density is shown as a function of time, relative to V_a in Fig. 8.25. The avalanche current peak is approximately 90° behind the maximum of V_a. The electrons leave the avalanche region after distance W_A and enter the drift region which they traverse at velocity v_s. The current at the terminals of the IMPATT, in response to V_a, is shown by the bottom curve of Fig. 8.25. The fundamental will be approximately 180° behind the V_a maximum. The angular displacement through the drift region is θ where

$$\theta = \omega \frac{W_D}{v_s} \qquad (8.1)$$

For complete antiphase between V_a and the output current, the oscillation frequency will be such as to make $\theta = \pi$. This occurs at the Read frequency, f_r, where

$$f_r = \frac{v_s}{2W} \qquad (8.2)$$

The width of a Si-based device, which has $v_s \simeq 6 \times 10^6$ cm/s for an operating frequency of 10 GHz, would therefore be about 3 µm.

Read's equation

The theoretical basis for the IMPATT device was provided by Read in 1958. Following his work, several refinements have been made to his analysis to account for the manner of realizing the device. The variation of avalanche current density J_a with time is given by the Read equation:

$$\frac{dJ_a(t)}{dt} = \frac{3J_a(t)}{\tau_a}\left[\int_0^W \alpha(E)\,dx - 1\right] \tag{8.3}$$

where x is the distance through the device from the junction and α is the ionization coefficient, which is the probability that a carrier will experience an ionizing collision in a unit length. τ_a is the time taken for an electron travelling at the saturation velocity v_s to travel over the avalanche region, i.e.

$$\tau_a = W_A/v_s$$

Associated with the avalanche region is an inductance L_A and a capacitance C_A:

$$L_A = \frac{W_A}{AJ_{DC}k} \tag{8.4}$$

$$C_A = \frac{\varepsilon A}{W_A} \tag{8.5}$$

A is the device area, J_{DC} the mean current density and k is a factor relating the characteristics of the semiconductor material to the ionization coefficient, at maximum field:

$$k = \frac{3}{\tau_a}\frac{\varepsilon}{qN}\alpha(E_p)$$

From Eqns (8.4) and (8.5) the resonant frequency of the avalanche region is f_{ra} where

$$f_{ra}^2 = \frac{1}{(2\pi)^2 L_A C_A} = \frac{J_{DC}k}{\varepsilon} \tag{8.6}$$

At frequencies above f_{ra} the device resistance becomes negative, and the reactance changes from inductive to capacitive.

Power frequency limitations

There is a maximum voltage V_m, which may be applied to the diode, corresponding to the maximum field E_m which the diode can withstand without breakdown over the depletion layer W. When the electron velocity is at its saturation level v_s, the maximum current I_m is

$$I_m = \frac{E_m \varepsilon v_s A}{W} \tag{8.7}$$

where A is the cross-sectional area of the device and ε the permittivity of the semiconductor material.

The maximum power input

$$P_m = I_m V_m \tag{8.8}$$

$$= \frac{E_m \varepsilon v_s A}{W} E_m W$$

$$= E_m^2 \varepsilon v_s A$$

The depletion layer capacitance $C = \varepsilon A / W$. Therefore

$$P_m = \frac{E_m^2 v_s^2}{\Delta \pi f^2 X_c} \tag{8.9}$$

where

$$X_c = \frac{1}{2\pi f C} \quad \text{and} \quad f = \frac{v_s}{2W}$$

Hence if X_c is limited, P_m is proportional to $1/f^2$ in the millimetre wave range.

Thermal limitation

The major limiting factor other than the intrinsic limit implied above is the amount of heat that can be conducted away from the junction. The dissipation of heat can be improved by mounting the device on a high conductivity material such as copper or diamond. Most of the heat is generated in the avalanche region where, for GaAs or Si, the operating temperature may be about 200 °C. The thermal conductivity of the diode depends on the thermal conductivity of the semiconductor material, the thermal resistance between the semiconductor and the heat sink, and the thermal conductivity of the sink material. For a particular system, the power dissipation decreases linearly with frequency.

8.9 FIELD-EFFECT TRANSISTORS

Probably the most promising type of microwave solid-state device is the field-effect transistor in one or other of its many forms. Based on GaAs, it is used for various purposes: high power generation, low noise and high frequency amplification in satellite and microwave terrestrial links. GaAs has a higher electron mobility than Si, and therefore a shorter transit time, allowing better high-frequency performance.

The basic form of the device is shown in Fig. 8.26. The buffer layer isolates the n-type active layer from the semi-insulating substrate. The active layer, which is less than half a micrometre deep, has a doping level of some 10^{17} donors/cm^3. Source and drain ohmic contacts are placed either directly on to the active layer, or on to another, more heavily doped, layer in the contact area. Electrons travel from the source to the drain, which is positive relative to the source, via the active layer. The path available for the electrons is called the channel. With this arrangement the current from source to drain, I_{DS},

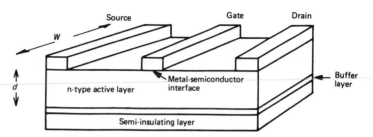

Fig. 8.26 Microwave FET.

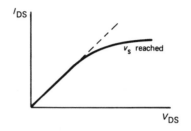

Fig. 8.27 Source–drain current–voltage characteristic for FET.

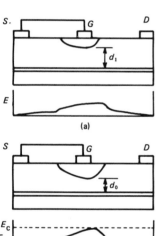

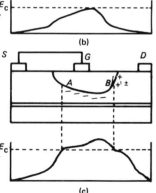

Fig. 8.28 Effect of applied voltage on current channel in a FET

increases as the applied voltage V_{DS} increases. The relationship between I_{DS} and V_{DS} is linear at low voltage as the electron velocity v varies linearly with the electric field produced by V_{DS}. As the voltage increases, v increases less quickly than the electric field, and eventually when a critical value of E is reached, $E_c \simeq 3 \times 10^3$ V/cm, the electron drift velocity saturates, at v_s. This characteristic is shown in Fig. 8.27.

The current I_{DS} is given by

$$I_{DS} = wdvqn \tag{8.10}$$

where wd = width × depth of the current channel, and $v.q.n.$ = velocity, unit charge and density of the electron flow, respectively.

In the case just discussed, where the field just reaches E_c, the minimum depth of the channel is d_0. Clearly I_{DS} varies directly with v and therefore reaches a saturation value when $v = v_s$.

By introducing an electrode (the gate) between source and drain, an additional control on I_{DS} is possible. The gate, which is reverse biased, creates a depletion region in the adjacent active layer, which has a depth proportional to the gate voltage. This depletion region has no conduction electrons, and therefore the depth of the current channel is now reduced to d_1, shown in Fig. 8.28(a).

For low values of V_{GS} the current I_{DS} varies linearly with V_{DS}, and, as V_{DS} increases, eventually saturates. As we would expect, the saturation current and the value of V_{DS} at which it is reached decrease as VGS becomes more negative. A typical family of curves is shown in Fig. 8.29.

When V_{GS} is zero the gate still has an effect: the field beneath the gate produces a depletion layer. The reduction in channel depth produced by the

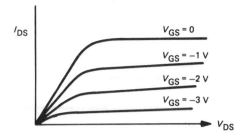

Fig. 8.29 Variation of output current with applied voltage.

depletion layer induces an increase in electron drift velocity as the field beneath the active layer increases [Fig. 8.28(b)]. When the field reaches the critical value E_c the velocity saturates, the depth of the channel is d_0, and the current through the device is given by

$$I_{DS} = wqnv_sd_0 \tag{8.11}$$

Any further increase in V_{DS} deepens the depletion layer, hence reducing d still further. For I_{DS} to remain constant in the area of saturated velocity, the electron density increases. Figure 8.28(c) shows the effect of this deeper depletion layer. The point at which the field becomes equal to E_c moves towards the source, and therefore is reduced in voltage (point A), producing a deeper channel than d_0, and therefore a larger current. This suggests that the current characteristic, after the saturation effect in Fig. 8.28(c), increases again with V_{DS}. GaAs demonstrates such a gentle positive slope beyond the initial saturation.

Another feature of the depletion layer on the drain side of A, in Fig. 8.28(c), is that an accumulation of electrons develops over the distance AB where the field is greater than E_c; point B is that position where the channel depth reaches the value it had at A. The value of n increases to maintain I_{DS} since d is reduced, and the electron velocity is saturated. Beyond B, a positive charge layer is formed which helps to maintain constant I_{DS}.

The bottom edge of the depletion layer is not parallel to the surface of the device, but is deeper on the drain side. This is due to the voltage gradient over the width of the gate producing a wedge-shaped layer, with rounded edges. As we can see from the diagrams, the precise shape varies with V_{DS}.

This qualitative description shows the effect of the gate electrode when it is connected directly to the source. By its influence, the I_{DS}/V_{DS} characteristic is modified. However, the device is designed to be used as a transistor with a control voltage on the gate. As we noted in Fig. 8.29, by negatively biasing the gate the saturation effect occurs at a lower V_{DS} than for $V_{GS} = 0$. The reverse biased gate junction deepens the depletion layer and hence restricts I_{DS}. Eventually the depletion layer reaches the buffer layer and the channel is closed. The gate voltage at which this occurs, known as the pinch-off voltage V_p, is determined by the electrical properties, and the depth, of the active

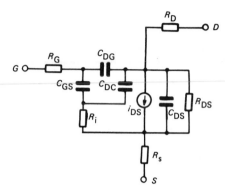

Fig. 8.30 FET equivalent circuit.

layer:

$$V_p = \frac{qnd^2}{2\varepsilon\varepsilon_0}$$ (8.12)

where q, n and d are as we defined them earlier, and ε, ε_0 are the relative permittivity of GaAs and the permittivity of free space, respectively.

A lumped component equivalent circuit for the device is shown in Fig. 8.30. From this circuit, the maximum frequency of oscillation is

$$f_m = \frac{g_m}{4\pi(C_{GS} + C_{DG})}\left[\frac{R_i + R_s + R_G}{R_{DS}} + \frac{g_m}{(C_{GS} + C_{DG})}C_{DG}R_G\right]^{1/2}$$ (8.13)

The performance of the device is related to the bias circuit used to establish and maintain the dc gate voltage; variations in this voltage induce noise and therefore low values of source and gate resistance are desirable. To minimize the variation in bias point, a two-source biasing configuration is preferred, such as that shown in Fig. 8.31.

This brief description refers to the MESFET (metal semiconductor field-effect transistor) which has a metal gate (Schottky) contact. It has a very low noise figure, and is capable of operating at high millimetre wave frequencies. Many other FETs exist, or are being developed; by varying the geometry of the device and the nature of the gate, particular features are enhanced. For further information, see Lamming[11] and Ha.[12]

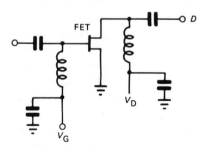

Fig. 8.31 Biasing network for an FET.

8.10 CONCLUSIONS

FET devices are the most common of the semiconductor devices currently in use. There are various forms, and several versions of each. One of the most promising recent developments has been the HEMT (high electron mobility transistor), which, although invented in the 1970s has now generated considerable interest. Previously it offered increased gain when operated at low temperatures, being dependent on a thin layer of AlGaAs for which the mobility increased considerably as the temperature was reduced to 77 K. New combinations of III–V materials have been used, based on InAs, and these promise significant gain improvements at room temperatures.

As we noted either, there are many applications that require power levels above those which can be achieved from solid-state devices. Consequently, there continues to be significant development in microwave vacuum tube design, in an attempt to improve the efficiency and performance of those tubes that have been established for nearly half a century, and to discover entirely new ways of generating high power.

REFERENCES

1. Kompfner, R., "The travelling-wave tube", *Wireless World*, **52**, 349 (1946).
2. Atwater, H. A., *Introduction to Microwave Theory*, McGraw-Hill, Maidenhead, 1962.
3. Collin, R. E., *Foundation for Microwave Engineering*, McGraw-Hill, Maidenhead, 1966.
4. Ridley, B. K. and Watkins, T. B., "The possibility of negative resistance effects in semiconductors", *Proceedings of the Physical Society*, **78**, 293–304 (Aug 1961).
5. Hilsum, C., "Transferred electron amplifiers and oscillators", *Proceedings of the Institute of Radio Engineers*, **50**(2), 185–9 (Feb 1962).
6. Gunn, J. B., "Microwave oscillations of current in III–V semiconductors", *Solid-state Communications*, **1**, 88–91 (Sept 1963).
7. Pattison, J., "Active microwave devices", *Microwave Solid-state Devices and Applications* (ed. Morgan, D. V. and Howe, M. J.), Pergamon, Oxford, 1980, Chapter 2.
8. Hobson, G. S., *The Gunn Effect*, Clarendon, Oxford, 1974.
9. Bosch, B. G. and Engelmann, R. W., *Gunn-effect Electronics*, Pitman, London, 1975.
10. Johnston, R. L., de Loach, B. C. and Cohen, B. G., "A silicon diode microwave oscillator", *Bell System Technical Journal*, **44**, 2 (Feb 1965).
11. Lamming, J., "Active microwave devices – FET's and BJT's", *Microwave Solid-state Devices and Applications* (ed. Morgan, D. V. and Howe, M. J.), Pergamon, Oxford, 1980, Chapter 4.
12. Ha, Tri T., *Solid-state Microwave Amplifier Design*, Wiley, Chichester, 1981.

9 Passive microwave devices

In the last chapter we considered some common microwave active devices. When microwave energy is generated it is used by either radiating it into space via an antenna or passing it along a transmission system. In either case some circuitry will be involved at both the transmitting and receiving ends. This circuitry, and the transmission system, have features that are special to signals of microwave frequencies. In this chapter we will consider first the commonest microwave component, the rectangular waveguide, and go on to look at several of the passive devices that comprise microwave circuitry.

9.1 WAVEGUIDES

Waveguides are distinguished by having only one guiding surface. They are of two main types: (a) metal tubes of any cross-section, and (b) dielectric rods. In the first case the wave travels down the inside of the tube and there is no leakage to the outside. In the second, the wave is propagated over the outer surface of the rod and there is a considerable field in the surrounding medium.

In this chapter we shall be looking at the first type only.

For best transmission, tubular waveguides should be made of a low-loss, conducting material. If runs are short, and cost is important, aluminium or brass will be used, but good quality waveguides having low attenuation are made from copper which, for some applications where low loss is paramount, might be plated with silver or gold. The energy is propagated along the inside of the tube so the inner surface should be smooth and the cross-section must be uniform throughout the length of the guide.

9.2 RECTANGULAR WAVEGUIDE

Power is transmitted down a waveguide by electric and magnetic fields which travel in a manner similar to that of travelling voltage waves in a transmission line.

Figure 9.1 shows a rectangular waveguide of internal cross-section $a \times b$, in relation to a right-handed system of coordinates. This means that the broad face of the guide is in the x direction, the narrow face in the y direction, and the wave is propagated in the z direction.

The wave inside the guide is described in terms of a field pattern. Figure 9.2 shows the pattern for the fundamental rectangular waveguide mode, TE_{10}

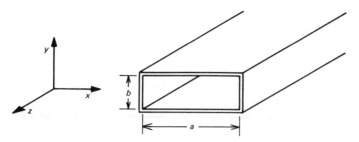

Fig. 9.1 Rectangular waveguide dimensions.

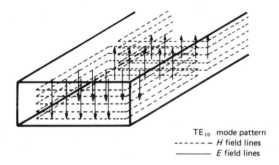

TE$_{10}$ mode pattern
- - - - - - *H* field lines
———— *E* field lines

Fig. 9.2 Field patterns of TE$_{10}$ rectangular waveguide mode.

(defined below). The way in which this field pattern is set up can be considered from two points of view, which will be considered in turn in the following sections.

9.3 INTERFERENCE OF TWO PLANE WAVES

In Figure 9.3 two uniform plane waves of identical amplitude and phase are shown travelling in different directions. In each, the electric field vector is normal to the plane of the paper. As the waves travel across each other there will be constructive and destructive intereference, resulting in a field pattern that is the algebraic sum of the component fields at each point. Where a maximum of one of the waves crosses a maximum of the other there will be a maximum in the resultant field, and conversely there will be a minimum where the minima of the two component fields are coincident. Examples of these points are at *E* for the minimum and *F* for the maximum in Fig. 9.3.

At points where there is a maximum of one field coincident with a minimum of the other there will be zero resultant field. In Fig. 9.3 a line is drawn joining such points, shown as *AA*. Since there is zero electric field along this line, a conducting surface could be inserted parallel to the *E* field lines along *AA* without perturbing the field. Similarly, a conducting surface could be placed along *BB*. These two conducting planes could thus form the side

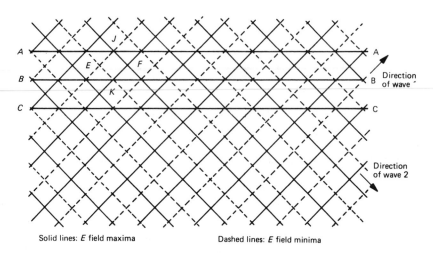

Solid lines: *E* field maxima   Dashed lines: *E* field minima

Fig. 9.3 Waveguide field pattern from two plane waves.

walls of a rectangular waveguide. Top and bottom conducting walls, parallel with the plane of the paper, can be used without altering the field pattern since the *E* field lines will meet them normally.

The resultant field pattern between *AA* and *BB* in Fig. 9.3 is therefore a representation of one that can exist inside a rectangular waveguide. As we shall see later the pattern between *AA* and *BB* represents the fundamental rectangular waveguide mode. If, instead of the side walls being along *AA* and *BB*, they were placed along *AA* and *CC*, a different pattern, representing a second-order mode, would exist. The question of modes is discussed in Section 9.7.

9.4 CUT-OFF WAVELENGTH

The pattern shown in Fig. 9.3 between lines *AA* and *BB* can be used to quantify the cut-off wavelength in terms of the free space wavelength and the guide width, i.e. the distance between *AA* and *BB*. An expanded representation of the pattern between a maximum and a minimum is given in Fig. 9.4. We can see that:

λ_g = wavelength of the field pattern inside the guide, i.e. the distance between two adjacent points of similar phase in the direction of propagation

λ_0 = wavelength of the wave in the medium filling the guide, say air

$$= \frac{\text{velocity of light in air}}{\text{frequency of the wave}} = \frac{c}{f}$$

From the diagram, if the angle between the direction of propagation of the two component waves is 2α (i.e. each wave propagates at an angle α to the axis

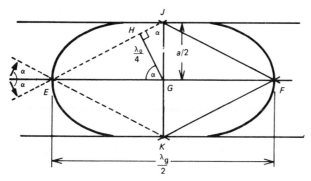

Fig. 9.4 Details of field pattern.

of the guide),

$$\sin\alpha = \frac{\lambda_0}{2}\frac{1}{a} \quad \text{and} \quad \cos\alpha = \frac{\lambda_0}{\lambda_g}$$

Using the identity $\cos^2\alpha + \sin^2\alpha = 1$

$$\frac{\lambda_0^2}{(2a)^2} + \frac{\lambda_0^2}{\lambda_g^2} = 1$$

or

$$\frac{1}{\lambda_0^2} = \frac{1}{\lambda_g^2} + \frac{1}{(2a)^2} \tag{9.1}$$

where a is the width of the guide. $2a$ is called the cut-off wavelength λ_c for the TE$_{10}$ mode; hence Eqn (9.1) becomes

$$\frac{1}{\lambda_0^2} = \frac{1}{\lambda_g^2} + \frac{1}{\lambda_c^2} \tag{9.2}$$

The cut-off wavelength is so called because it is the maximum value that λ_0 can have. If λ_0 becomes greater than λ_c the wave will not propagate down the guide. Thus the waveguide behaves as a high-pass filter: only waves of frequency greater than the cut-off value $f_c(= c/\lambda_c)$ can be carried. This aspect of waveguide behaviour will be considered in further detail in later sections.

9.5 PHASE VELOCITY

The rate at which a particular point in the field pattern travels down the guide is known as the phase velocity. From Fig. 9.4 consider the movement of point E towards F. In order to maintain the phase of E constant it must arrive at G at the same time as point H on one of the wavefronts. H is moving with velocity c, and covers a distance $\lambda_0/4$, whereas E travels $\lambda_g/4$ in the same time.

Therefore the velocity of E, i.e. the phase velocity v_{ph}, is given by

$$v_{ph} = c \frac{\lambda_g}{\lambda_0} \tag{9.3}$$

9.6 GROUP VELOCITY

It would violate the theory of relativity if energy were transported down the guide at the phase velocity, and in fact that does not happen. The velocity of energy movement is known as the group velocity, and in terms of our plane wave model in Fig. 9.4 it is equal to the component of the plane wave velocity in the z direction. The plane wave travels at velocity c in air so the component in the z direction is

$$v_g = c \cos \alpha = c \frac{\lambda_0}{\lambda_g} \tag{9.4}$$

It is worth noting from the last two equations that

$$v_{ph} \cdot v_g = c^2 \tag{9.5}$$

9.7 RECTANGULAR WAVEGUIDE MODES

The above discussion has been concerned with a particular field pattern inside the waveguide – that given by Figs. 9.2 and 9.3 and the section between AA and BB of Fig. 9.4. As we noted earlier, it is the TE_{10} mode, often called the fundamental rectangular waveguide mode because it has the lowest cut-off frequency.

In the designation TE_{10} the letters stand for transverse electric, indicating that the field pattern has no electric component in the z direction, it being contained entirely in the $x-y$, transverse, plane. The numerals 1, 0 are used to indicate the number of half-sinusoidal variations of the E field in the x and y directions, respectively. For any value of y the electric field goes through one half-sinusoidal variation with x, i.e. from zero at one wall, through a maximum in the guide-centre, to zero at the other wall (Fig. 9.2). For any value of x there is no variation of field with y, i.e. the field is constant in the y direction. Thus TE_{10} designates a transverse electric mode having one half-sinusoidal variation in the x direction (across the broad side of the guide) and no variation in the y direction.

In fact, there are many possible TE modes. The general designation is TE_{mn} where m is the number of half-sinusoidal variations in the x direction, and n the number of similar variations in the y direction. We shall see later that the cut-off wavelength for the TE_{mn} mode is given by

$$\frac{1}{(\lambda_{c_{mn}})^2} = \left(\frac{m}{2a}\right)^2 + \left(\frac{n}{2b}\right)^2 \tag{9.6}$$

where a and b are the dimensions shown in Fig. 9.1.

There is another set of modes possible in a rectangular waveguide; the transverse magnetic, or TM, modes, in which there is no component of magnetic field in the direction of propagation. The general mode is designated TM_{mn}, and it too has a cut-off wavelength given by Eqn (9.6). The suffices mn have the same meaning as in TE_{mn}.

Unlike coaxial transmission line, there is no TEM mode in a waveguide: a mode cannot exist in which both the electric and magnetic vectors have no component in the z direction.

9.8 FIELD THEORY OF PROPAGATION ALONG A RECTANGULAR WAVEGUIDE

Although the above discussion, in which the fields inside the waveguide are considered to result from the intereference of two plane waves, gives a very useful insight into the mechanism of waveguide propagation, it has some disadvantages. In particular, it does not allow an evaluation of the field components inside the guide. A deeper appreciation can be obtained by using Maxwell's equations. These equations state in generalized form the relationship that must exist between the electric and magnetic fields in any medium, provided the characteristics of the medium and the initial conditions of the fields are known. It would be inappropriate to develop Maxwell's equations here, but several texts discuss their derivation in some detail, and reference should be made to one of these, e.g. Bleaney and Bleaney[1] or Paul and Nasar,[2] for background information.

We start by stating the equations in generalized vector form:

$$\nabla \cdot \bar{E} = \frac{\rho}{\varepsilon} \tag{9.7}$$

$$\nabla \cdot \bar{B} = 0 \tag{9.8}$$

$$\nabla \times \bar{E} = -\frac{\partial \bar{B}}{\partial t} \tag{9.9}$$

$$\nabla \times \bar{H} = \bar{J} + \varepsilon \frac{\partial \bar{E}}{\partial t} \tag{9.10}$$

$$\bar{B} = \mu \bar{H} \tag{9.11}$$

$\bar{E}$ and $\bar{H}$ are the electric and magnetic field vectors, respectively, $\bar{B}$ is the magnetic flux density, and $\bar{J}$ is the conduction current density. All four vectors are related to the space inside the guide; for air the value of $\bar{J}$ is zero.

The other symbols represent the electric characteristics of the material filling the guide. Normally the charge density (ρ) is zero, and for air the permittivity (ε_0) and the permeability (μ_0) are those for free space, i.e.

$$\varepsilon_0 = \frac{1}{36\pi \times 10^9} \quad \text{farad/m}$$

$$\mu_0 = 4\pi \times 10^7 \quad \text{H/m}$$

Although the following analysis is lengthy, it allows us to build up an awareness of an analytical description of waveguide propagation. By considering the argument in some detail, rather than merely isolating the important equations, we can appreciate more fully the underlying theory, and that will be of value in discussing several of the devices considered later in the chapter.

If we assume immediately that there is a sinusoidal signal being transmitted down the guide, then the time derivatives in Eqns (9.9) and (9.10) can be replaced by $j\omega$, and they become, when $\bar{B}$ is replaced by μH,

$$\nabla \times \bar{E} = -j\omega\mu\bar{H} \tag{9.12}$$

$$\nabla \times \bar{H} = j\omega\varepsilon\bar{E} \tag{9.13}$$

and, since $\rho = 0$,

$$\nabla \cdot \bar{E} = 0 \tag{9.14}$$

Now we must solve these equations for waveguide having a rectangular cross-section.

Taking the curl of Eqn (9.12), and substituting from Eqn (9.13),

$$\nabla \times \nabla \times \bar{E} = \omega^2 \varepsilon\mu\bar{E} \tag{9.15}$$

Using Eqn (9.15) in the vector identity,

$$\nabla \times \nabla \times \bar{E} = -\nabla^2\bar{E} + \nabla(\nabla \cdot \bar{E})$$

and combining the result with Eqn (9.14), we have

$$\nabla^2\bar{E} + \mu\varepsilon\omega^2\bar{E} = 0 \tag{9.16}$$

This is a vector representation of the E field wave equation, and to solve it for the waveguide it must be put into an appropriate coordinate system. In rectangular coordinates

$$\nabla^2\bar{E} = \bar{x}\nabla^2 E_x + \bar{y}\nabla^2 E_y + \bar{z}\nabla^2 E_z \tag{9.17}$$

From Eqns (9.16) and (9.17) there are three equations to be solved, one for each coordinate direction:

$$\nabla^2 E_x = -\mu\varepsilon\omega^2 E_x \tag{9.18}$$

$$\nabla^2 E_y = -\mu\varepsilon\omega^2 E_y \tag{9.19}$$

$$\nabla^2 E_z = -\mu\varepsilon\omega^2 E_z \tag{9.20}$$

By developing Eqns (9.16) and (9.17) in terms of $\bar{H}$ rather than $\bar{E}$, a similar set of equations is produced for the $\bar{H}$ components. The six equations resulting are related by the boundary conditions at the waveguide walls, and these conditions must be used when the equations are solved.

We will now solve Eqn (9.19), which can be expanded to

$$\frac{\partial^2 E_y}{\partial x^2} + \frac{\partial^2 E_y}{\partial y^2} + \frac{\partial^2 E_y}{\partial z^2} = -\mu\varepsilon\omega^2 E_y \tag{9.21}$$

In principle, E_y may vary with x, y and z, and variation in one direction will be independent of variation in another, so the technique of separation of variables can be used.

Let the solution of E_y be

$$E_y = XYZ \tag{9.22}$$

where XYZ are functions of only x, y, and z, respectively.

Then, substituting Eqn (9.22) into Eqn (9.21),

$$\frac{\partial^2 X}{\partial x^2} YZ + X \frac{\partial^2 Y}{\partial y^2} Z + XY \frac{\partial^2 Z}{\partial z^2} = -\mu\varepsilon\omega^2 XYZ$$

or

$$\frac{1}{X} \frac{\partial^2 X}{\partial x^2} + \frac{1}{Y} \frac{\partial^2 Y}{\partial y^2} + \frac{1}{Z} \frac{\partial^2 Z}{\partial z^2} = -\mu\varepsilon\omega^2 \tag{9.23}$$

At any specific frequency the right-hand side of Eqn (9.23) is constant, so for this expression to be valid each of the terms on the left-hand side must also be constant; hence we can write

$$\frac{1}{X} \frac{\partial^2 X}{\partial x^2} = k_x^2 \tag{9.24}$$

$$\frac{1}{Y} \frac{\partial^2 Y}{\partial y^2} = k_y^2 \tag{9.25}$$

$$\frac{1}{Z} \frac{\partial^2 Z}{\partial z^2} = k_z^2 \tag{9.26}$$

The significance of k_x, k_y and k_z will be clear later.

Introducing $k^2 = \mu\varepsilon\omega^2$, Eqn (9.23) gives

$$k_x^2 + k_y^2 + k_z^2 = -k^2 = -\mu\varepsilon\omega^2 \tag{9.27}$$

We can distinguish between the transverse and z direction components; letting

$$k_x^2 + k_y^2 = -k_c^2 \tag{9.28}$$

we have

$$k_z^2 = -k^2 + k_c^2 \tag{9.29}$$

or

$$k_z = \pm j \sqrt{(k^2 - k_c^2)} \tag{9.30}$$

The solution of Eqn (9.26), which gives the variation in the z direction, is, by analogy with voltage waves on a transmission line (following Eqn (6.11))

$$Z = \exp(\pm k_z z) \tag{9.31}$$

and from Eqn (9.30), when $k^2 > k_c^2$,

$$k_z = \pm j\beta_g \tag{9.32}$$

where

$$\beta_g = \sqrt{(k^2 - k_c^2)} \tag{9.33}$$

hence

$$Z = \exp(\pm j\beta_g z) \tag{9.34}$$

Then E_y is given by

$$E_y = XY \exp[j(\omega t \pm \beta_g z)] \tag{9.35}$$

where here the time dependence implicit in assuming a sinusoidal signal is included in the term $\exp(j\omega t)$.

Equation (9.35) represents two travelling waves, one in the forward and one in the backward direction, in a similar way to the waves on a transmission line (see Section 6.1).

This method of solution could be used to develop expressions for the other components in the E and H fields.

For normal propagation $k^2 > k_c^2$, making β_g real and k_z imaginary, and therefore $\exp(\pm j\beta_g z)$ is a phase term. However, if $k_c^2 > k^2$, β_g is imaginary and k_z is real, indicating attenuation. In that case there is rapid attenuation. The transition between propagation and attenuation occurs at that value of $\omega(= \omega_c)$ for which

$$k_c^2 = k^2$$

i.e.

$$k_c^2 = \mu\varepsilon\omega_c^2$$

or

$$\omega_c = k_c/\sqrt{(\mu\varepsilon)} \tag{9.36}$$

k_c has dimensions of 1/length and it can be defined by

$$k_c = 2\pi/\lambda_c \tag{9.37}$$

We can rewrite Eqn (9.29), i.e.

$$k_z^2 = -k^2 + k_c^2$$

as

$$-\left(\frac{2\pi}{\lambda_g}\right)^2 = -\left(\frac{2\pi}{\lambda_0}\right)^2 + \left(\frac{2\pi}{\lambda_c}\right)^2$$

or, rearranging terms,

$$\frac{1}{\lambda_0^2} = \frac{1}{\lambda_g^2} + \frac{1}{\lambda_c^2} \tag{9.2}$$

where

$$\lambda_g = \frac{2\pi}{\beta_g} \quad \text{(cf. Eqn (6.18))}$$

and, using $\omega = 2\pi f_0 = c/\lambda_0$ in Eqn (9.27)

$$\lambda_0 = \frac{2\pi}{k} = \frac{2\pi}{\mu\varepsilon\omega^2}$$

These wavelengths λ_0, λ_g and λ_c have been discussed in Section 9.4. Here we are not yet able to specify λ_c, because we have not considered a particular mode, but that will be done shortly.

9.9 TRANSVERSE ELECTRIC (TE) MODES

In the last section we derived some general expressions for modes in a rectangular waveguide, but we cannot go further without specifying the class of mode in which we are interested. As we saw in Section 9.7 there are two classes, TE and TM. In this section we consider TE modes in a rectangular guide, and we can use our earlier equations to set up the field equations in the waveguide; we then introduce the effect of having no component of the E field in the z direction, thus allowing a solution to Maxwell's equations to be obtained.

Returning to Eqn (9.12),

$$\nabla \times \bar{E} = -j\omega\mu\bar{H}$$

and expressing it in Cartesian coordinates,

$$\bar{x}\left(\frac{\partial E_z}{\partial y} - \frac{\partial E_y}{\partial z}\right) + \bar{y}\left(\frac{\partial E_x}{\partial z} - \frac{\partial E_z}{\partial x}\right) + \bar{z}\left(\frac{\partial E_y}{\partial x} - \frac{\partial E_x}{\partial y}\right)$$

$$= -j\omega\mu(\bar{x}H_x + \bar{y}H_y + \bar{z}H_z) \tag{9.38}$$

For TE modes there is no E field in the z direction, and hence $E_z = 0$. From Eqn (9.31) the dependence of the other fields in the z direction is $\exp(\pm k_z)$, giving $\partial/\partial z = k_z$.

Using these two conditions in Eqn (9.38),

$$\bar{x}(-k_z E_y) + \bar{y}(k_z E_x) + \bar{z}\left(\frac{\partial E_y}{\partial x} - \frac{\partial E_x}{\partial y}\right)$$

$$= -j\omega\mu(\bar{x}H_x + \bar{y}H_y + \bar{z}H_z)$$

Equating x, y and z components,

$$k_z E_y = j\omega\mu H_x \tag{9.39}$$

$$k_z E_x = -j\omega\mu H_y \tag{9.40}$$

$$\frac{\partial E_y}{\partial x} - \frac{\partial E_x}{\partial y} = -j\omega\mu H_z \tag{9.41}$$

By a similar development from Eqn (9.13), additional relationships between the E and H fields can be found, viz.

$$\frac{\partial H_z}{\partial y} - k_z H_y = j\omega\varepsilon E_x \tag{9.42}$$

$$k_z H_x - \frac{\partial H_z}{\partial x} = j\omega\varepsilon E_y \tag{9.43}$$

$$\frac{\partial H_y}{\partial x} - \frac{\partial H_x}{\partial y} = 0 \tag{9.44}$$

To find the field components we can first use these equations to express the transverse field components E_x, E_y, H_x, H_y in terms of the single component in the z direction, H_z.

From Eqns (9.40) and (9.42),

$$\frac{\partial H_z}{\partial y} + \frac{k_z^2}{j\omega\mu} E_x = j\omega\varepsilon E_x$$

$$j\omega\mu \frac{\partial H_z}{\partial y} + k_z^2 E_x = -\omega^2\mu\varepsilon E_x$$

or

$$j\omega\mu \frac{\partial H_z}{\partial y} = -(k_z^2 + \omega^2\mu\varepsilon) E_x$$

i.e.

$$E_x = -\frac{j\omega\mu}{k_z^2 + \omega^2\mu\varepsilon} \frac{\partial H_z}{\partial y}$$

$$= -\frac{j\omega\mu}{k_c^2} \frac{\partial H_z}{\partial y} \tag{9.45}$$

where

$$k_c^2 = k_z^2 + \omega^2\mu\varepsilon$$

Substituting Eqn (9.45) back into Eqn (9.40), and rearranging terms gives

$$H_y = \frac{k_z}{k_c^2} \frac{\partial H_z}{\partial y} \tag{9.46}$$

In a similar way we can find expressions for the other transverse components:

$$E_y = \frac{j\omega\mu}{k_c^2} \frac{\partial H_z}{\partial x} \tag{9.47}$$

and

$$H_x = \frac{k_z \partial H_z}{k_c^2 \partial x} \tag{9.48}$$

Thus in order to find the transverse field components we only need to determine H_z. It can be found as follows.

By analogy with the discussion on E_y in Section 9.8, H_z must be a solution of

$$\nabla^2 H_z + \mu\varepsilon\omega^2 H_z = 0 \tag{9.49}$$

and in the same way it will have the form

$$H_z = XY \exp\left[j(\omega t \pm \beta_g z)\right] \tag{9.50}$$

We remember that X and Y must each satisfy second-order differential equations, so they must each have two solutions. Two satisfactory solutions are

$$X = A \sin jk_x x + B \cos jk_x x \qquad (9.51)$$

$$Y = C \sin jk_y y + D \cos jk_y y \qquad (9.52)$$

Then the full solution for H_z is

$$H_z = (A \sin jk_x x + B \cos jk_x x)(C \sin jk_y y + D \cos jk_y y) \exp [j(\omega t \pm \beta_g z)] \qquad (9.53)$$

The unknowns in this expression (A, B, C, D, k_x, k_y) are found from either the boundary or the initial conditions. The boundary conditions imposed by the walls of the guide are as follows:

(i) there can be no normal component of magnetic field at the guide walls;
(ii) there can be no tangential component of electric field at the guide walls.

Condition (i) means that

$$\frac{\partial H_z}{\partial x} = 0 \quad \text{at} \quad x = 0, a$$

and

$$\frac{\partial H_z}{\partial y} = 0 \quad \text{at} \quad y = 0, b$$

thus making $A = C = 0$ in Eqn (9.53), leaving

$$H_z = H_1 \cos jk_x x \cos jk_y y \exp (\pm j\beta_g z) \qquad (9.54)$$

(omitting, but not forgetting, the time-dependent term $\exp [j\omega t]$), $H_1 = BD$, and is determined by the initial field applied to the guide.
   If we use this value of H_z in Eqn (9.47) we have

$$E_y = \frac{\omega \mu k_x H_1}{k_c^2} \sin jk_x x \cos jk_y y \exp (\pm j\beta_g z)$$

and from condition (ii) $E_y = 0$ at $x = 0, a$, which is satified by making

$$jk_x = m\pi/a \qquad (9.55)$$

By a similar argument, consideration of E_x imposes the relationship

$$jk_y = n\pi/b \qquad (9.56)$$

Then, substituting for jk_x and jk_y from Eqns (9.55) and (9.56) into Eqn (9.54) gives

$$H_z = H_1 \cos \frac{m\pi}{a} x \cos \frac{n\pi}{b} y \exp (\pm j\beta_g z) \qquad (9.57)$$

where m and n can be any integer and indicate the number of half-sinusoidal variations of field in the x and y directions, respectively, i.e. they refer to the TE_{mn} mode.
   By inserting Eqn (9.57) into Eqns (9.45), (9.46) and (9.48), the full set of field

components for the TE_{mn} mode are obtained, i.e.

$$E_x = \frac{j\omega\mu n\pi H_1}{bk_c^2} \cos\frac{m\pi x}{a} \sin\frac{n\pi y}{b} \exp(\pm j\beta_g z) \qquad (9.58)$$

$$E_y = -\frac{j\omega\mu m\pi H_1}{ak_c^2} \sin\frac{m\pi x}{a} \cos\frac{n\pi y}{b} \exp(\pm j\beta_g z) \qquad (9.59)$$

$$E_z = 0 \qquad (9.60)$$

$$H_x = -\frac{j\beta_g m\pi H_1}{ak_c^2} \sin\frac{m\pi x}{a} \cos\frac{n\pi y}{b} \exp(\pm j\beta_g z) \qquad (9.61)$$

$$H_y = -\frac{j\beta_g n\pi H_1}{bk_c^2} \cos\frac{m\pi x}{a} \sin\frac{n\pi y}{b} \exp(\pm j\beta_g z) \qquad (9.62)$$

$$H_z = H_1 \cos\frac{m\pi x}{a} \cos\frac{n\pi y}{b} \exp(\pm j\beta_g z) \qquad (9.63)$$

where, from Eqns (9.27) and (9.28) as before,

$$k_c^2 = \mu\varepsilon\omega^2 + k_z^2$$

From Eqn (9.28)

$$k_c^2 = -(k_x^2 + k_y^2)$$

Substituting for k_x and k_y from Eqns (9.55) and (9.56), respectively,

$$k_c^2 = \left[\left(\frac{m\pi}{a}\right)^2 + \left(\frac{n\pi}{b}\right)^2\right]$$

therefore

$$\mu\varepsilon\omega^2 + k_z^2 = \left(\frac{m\pi}{a}\right)^2 + \left(\frac{n\pi}{b}\right)^2$$

From $\omega = 2\pi f$, $\qquad \omega^2 = \left(\frac{2\pi c}{\lambda_0}\right)^2 \qquad$ and $\qquad c = 1/\sqrt{(\mu\varepsilon)}$

therefore

$$\mu\varepsilon\omega^2 = \left(\frac{2\pi}{\lambda_0}\right)^2$$

and

$$k_z^2 = (j\beta_g)^2 = -\left(\frac{2\pi}{\lambda_g}\right)^2$$

hence

$$\left(\frac{2\pi}{\lambda_0}\right)^2 - \left(\frac{2\pi}{\lambda_g}\right)^2 = \left(\frac{m\pi}{a}\right)^2 + \left(\frac{n\pi}{b}\right)^2$$

Comparing this equation with Eqn (9.6)

$$\left(\frac{2\pi}{\lambda_c}\right)^2 = \left(\frac{m\pi}{a}\right)^2 + \left(\frac{n\pi}{b}\right)^2$$

therefore

$$\frac{1}{\lambda_c^2} = \left(\frac{m}{2a}\right)^2 + \left(\frac{n}{2b}\right)^2 \tag{9.64}$$

This expression, which for the TE_{10} mode gives $\lambda_c = 2a$ as in Section 9.4, allows us to find the cut-off wavelength for any TE mode, given the cross-sectional dimensions of the waveguide. We can see that λ_c will vary with the size of guide, and therefore with its aspect ratio a/b.

9.10   TRANSVERSE MAGNETIC (TM) MODES

As we noted in Section 9.7, the other class of modes in a rectangular waveguide is that in which the magnetic field has no component in the z direction (i.e. $H_z = 0$). Modes of this type are called transverse magnetic because the magnetic field is contained entirely in the xy plane. The general mode designation is TM_{mn} where m and n have the same meaning as in Section 9.7.

By an analysis similar to that used in considering TE modes, the six field components of the TM_{mn} mode are obtained:

$$E_x = \frac{j\beta_g m\pi E_1}{ak_c^2} \cos\frac{m\pi x}{a} \sin\frac{n\pi y}{b} \exp(\pm j\beta_g z) \tag{9.65}$$

$$E_y = \frac{j\beta_g n\pi E_1}{bk_c^2} \sin\frac{m\pi x}{a} \cos\frac{n\pi y}{b} \exp(\pm j\beta_g z) \tag{9.66}$$

$$E_z = E_1 \sin\frac{m\pi x}{a} \sin\frac{n\pi y}{b} \exp(\pm j\beta_g z) \tag{9.67}$$

$$H_x = \frac{j\omega\varepsilon n\pi E_1}{bk_c^2} \sin\frac{m\pi x}{a} \cos\frac{n\pi y}{b} \exp(\pm j\beta_g z) \tag{9.68}$$

$$H_y = -\frac{j\omega\varepsilon m\pi E_1}{ak_c^2} \cos\frac{m\pi x}{a} \sin\frac{n\pi y}{b} \exp(\pm j\beta_g z) \tag{9.69}$$

$$H_z = 0 \tag{9.70}$$

Equations (9.64) and (9.33) for the cut-off wavelength and the phase constant are also valid for TM modes. If we examine Eqn (9.64) again, and express it in terms of cut-off frequency rather than wavelength, it becomes

$$f_c = c\left[\left(\frac{m}{2a}\right)^2 + \left(\frac{n}{2b}\right)^2\right]^{1/2} \tag{9.71}$$

and it is clear that there is a dependence on a and b as well as on m and n. Taking as an example $b = 2a$ we can show the relationship between the cut-off frequencies of several modes on the simple diagram of Fig. 9.5. The cut-off values are given relative to that for the TE_{10} mode, and it can now be seen why it is called the fundamental. Some modes (e.g. TE_{11}, TM_{11}) have the same value of f_c; they are called degenerate because propagation that starts in one mode can easily change (i.e. degenerate) into the other. Such things as

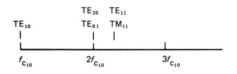

$f_{C_{10}}$ – cut-off frequency for TE_{10} mode

Fig. 9.5 Cut-off frequencies for low-order rectangular modes.

bends, roughness in the guide or poorly fitted flanges can cause this mode conversion and because waveguide detectors are mode dependent it results in a loss of power.

9.11 FIELD EQUATIONS FOR THE FUNDAMENTAL TE_{10} MODE

Putting $m = 1$, $n = 0$ into Eqns (9.58) to (9.63) and adjusting the coefficients, we have the following field components:

$$E_x = 0 \tag{9.72}$$

$$E_y = E_1 \sin \frac{\pi x}{a} \exp(\pm j\beta_g z) \tag{9.73}$$

$$E_z = 0 \tag{9.74}$$

$$H_x = \frac{\beta_g}{\omega\mu} E_1 \sin \frac{\pi x}{a} \exp(\pm j\beta_g z) \tag{9.75}$$

$$H_y = 0 \tag{9.76}$$

$$H_z = \frac{jak_c^2 E_1}{\omega\mu\pi} \cos \frac{\pi x}{a} \exp(\pm j\beta_g z) \tag{9.77}$$

These are the equations of the field patterns considered in Section 9.9 and which are shown in Fig. 9.2. As well as being the fundamental, this is also the simplest rectangular waveguide mode; there is only one half-sinusoidal variation of electric field across the guide in the x direction and none in the y direction. Similar points of phase down the guide are separated by λ_g.

From Eqns (9.72) to (9.77) the field patterns for the TE_{10} mode can be drawn, giving the diagram we discussed earlier (Fig. 9.2). The introduction of a cut-off frequency implies that the phase constant β_g is frequency dependent, unlike the phase constant β discussed in Chapter 6 on transmission lines. The variation of β_g with frequency can be shown on an ω–β diagram. Figure 9.6 shows a typical characteristic for rectangular waveguide with β_g falling to zero (no propagation) when $\omega = \omega_c$, the cut-off frequency.

At high frequencies the ω–β diagram is linear. For comparison, the ω–β characteristic of a transmission line is also shown. The equations relating ω and β are

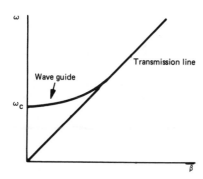

Fig. 9.6 Rectangular waveguide $\omega - \beta$ diagram.

for waveguide $$\omega \sqrt{(\mu\varepsilon)} = \sqrt{(k_c^2 + \beta_g^2)} \qquad (9.78)$$

where, for a given guide, μ, ε and k_c are constant, and

for transmission line $$\omega \sqrt{(\mu\varepsilon)} = \beta \qquad (9.79)$$

9.12 ATTENUATION IN RECTANGULAR WAVEGUIDE

In the above analysis we have assumed implicitly that the walls of the guide are perfect conductors. Although the conductivity may be very high, there will be some loss at the waveguide walls caused by two factors. First, the finite conductivity of the wall material gives rise to ohmic loss and, second, surface roughness, which becomes more important as the frequency increases. The second cause can be reduced at the cost of a high-quality polished finish on the inside wall surface. The ohmic loss is a function of the wall itself and therefore it is intrinsic once the wall material has been chosen.

An expression for the attenuation along a rectangular air-filled waveguide, due to ohmic loss, can be obtained by finding the wall currents and then calculating the power loss into the guide wall. In this way the attenuation is shown[5,7] to be given by

$$\alpha = \frac{R_s}{120\pi b} \frac{2\pi}{\lambda_0 \beta_g} \left[1 + 2\frac{b}{a}\left(\frac{f_c}{f}\right)^2 \right] \quad \text{nepers/m} \qquad (9.80)$$

where R_s is the surface resistivity of the waveguide inside wall.

Figure 9.7 shows how this attenuation varies with frequency. Each curve has a characteristic shape; as the frequency increases, there is a rapid fall in attenuation as the cut-off value is exceeded. Thereafter, the attenuation falls gently to a minimum, a short distance from cut-off, and then gradually increases with frequency.

It can be seen that the value of attenuation varies with the size and aspect ratio of the waveguide cross-section and with the mode number, as well as with the guide conductivity.

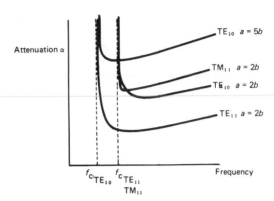

Fig. 9.7 Rectangular waveguide attenuation characteristics.

9.13 EVANESCENT MODES

Although we have discussed the cut-off mechanism as an effect that occurs suddenly as the frequency is reduced, in fact there is some propagation below cut-off, but very little. Below cut-off the propagation coefficient is real, and therefore the wave attenuates exponentially with distance from the source. The modes that are present in the region of the source end of the waveguide when it is operating below cut-off are called evanescent modes.

For transmission purposes, evanescent modes cannot be used, but there are some applications where the rapid attenuation can be useful. Because the forward wave attenuates very rapidly, there is little or no reflection from a short circuit or other junction placed in the guide, and any reflected wave is also heavily attenuated. This means that the reflected wave can be ignored allowing the guide input impedance to appear to be purely reactive, a characteristic that finds application in a number of measuring techniques.

9.14 RECTANGULAR MODE PATTERNS

We discussed the existence of higher-order modes in an earlier section, but the variation of fields in both the x and y directions makes the shapes of the field patterns for the general mn mode difficult to visualize.

In Fig. 9.8 we present patterns for some of the lower-order modes alongside diagrams showing methods of exciting them. Notice that the E and H field lines cross at right-angles, that the magnetic field lines do not terminate on the walls of the guide but form closed loops, and that the positions of maximum field in the cross-section of the guide vary with the mode number.

9.15 CIRCULAR WAVEGUIDES

Although the rectangular cross-section is that most commonly used in waveguides, there are several others that are used from time to time, some of

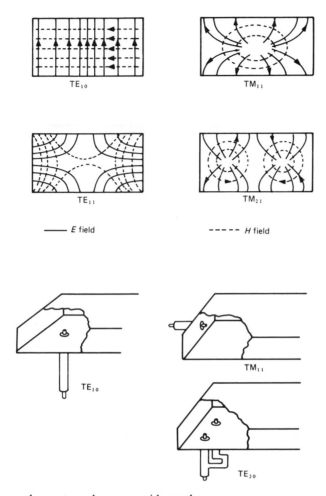

Fig. 9.8 Low-order rectangular waveguide modes.

which are shown in Fig. 9.9. The field patterns can be found by solving Maxwell's equations and applying the appropriate boundary conditions, which usually means that a coordinate system allied to the shape of the cross-section must be introduced. As an example, we shall look briefly at the solution for a circular waveguide, which is used in several special devices that need the mode-pattern symmetry which is characteristic of some of the low-order modes.

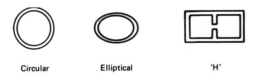

Circular   Elliptical   'H'

Fig. 9.9 Waveguide cross-sections.

It is not necessary to go through all the steps we followed earlier; we can assume that the space inside the guide has the same effect on Maxwell's equations as in the rectangular waveguide case.

Starting with Eqns (9.12) and (9.13),

$$\nabla \times \bar{E} = -j\omega\mu\bar{H} \tag{9.12}$$

$$\nabla \times \bar{H} = j\omega\varepsilon\bar{E} \tag{9.13}$$

we can use a similar argument to that of Section 9.7. Equation (9.12), expressed in cylindrical coordinates, has the form

$$\bar{r}\frac{1}{r}\left[\frac{\partial E_z}{\partial \theta} - \frac{\partial(rE_\theta)}{\partial z}\right] + \bar{\theta}\left[\frac{\partial E_r}{\partial z} - \frac{\partial E_z}{\partial r}\right] + \bar{z}\frac{1}{r}\left[\frac{\partial(rE_\theta)}{\partial r} - \frac{\partial E_r}{\partial \theta}\right]$$

$$= -j\omega\mu[\bar{r}H_r + \bar{\theta}H_\theta + \bar{z}H_z] \tag{9.81}$$

Separating the three components, and equating left and right sides of Eqn (9.81),

$$\frac{1}{r}\left[\frac{\partial E_z}{\partial \theta} + \frac{\partial(rE_\theta)}{\partial z}\right] = -j\omega\mu H_r \tag{9.82}$$

$$\frac{\partial E_r}{\partial z} - \frac{\partial E_z}{\partial r} = -j\omega\mu H_\theta \tag{9.83}$$

$$\frac{1}{r}\left[\frac{\partial(rE_\theta)}{\partial r} - \frac{\partial E_r}{\partial \theta}\right] = -j\omega\mu H_z \tag{9.84}$$

As before, we will assume that the z dependence of the fields is $\exp(k_z)$, and therefore we can replace $\partial/\partial z$ in Eqns (9.82) and (9.83) by k_z. Then

$$\frac{1}{r}\frac{\partial E_z}{\partial \theta} - k_z E_\theta = -j\omega\mu H_r \tag{9.85}$$

$$k_z E_r - \frac{\partial E_z}{\partial r} = -j\omega\mu H_\theta \tag{9.86}$$

$$\frac{1}{r}\left[\frac{\partial(rE_\theta)}{\partial r} - \frac{\partial E_r}{\partial \theta}\right] = -j\omega\mu H_z \tag{9.87}$$

Similarly, starting with Eqn (9.13), the following additional equations are developed:

$$\frac{1}{r}\frac{\partial H_z}{\partial \theta} - k_z H_\theta = j\omega\varepsilon E_r \tag{9.88}$$

$$k_z H_r - \frac{\partial H_z}{\partial r} = j\omega\varepsilon E_\theta \tag{9.89}$$

$$\frac{1}{r}\frac{\partial}{\partial r}(rH_\theta) - \frac{1}{r}\frac{\partial H_r}{\partial \theta} = j\omega\varepsilon E_z \tag{9.90}$$

Again, two families of modes are possible, transverse electric (TE) and transverse magnetic (TM).

9.16 CIRCULAR WAVEGUIDE (TE) MODES

By definition, the z component of electric field is zero, and putting $E_z = 0$ into Eqns (9.85) to (9.86) reduces them to

$$k_z E_\theta = j\omega\mu H_r \tag{9.91}$$

$$k_z E_r = -j\omega\mu H_\theta \tag{9.92}$$

$$\frac{1}{r}\frac{\partial}{\partial r}(rE_\theta) - \frac{1}{r}\frac{\partial E_r}{\partial \theta} = -j\omega\mu H_z \tag{9.93}$$

$$\frac{1}{r}\frac{\partial H_z}{\partial \theta} - k_z H_\theta = j\omega\varepsilon E_r \tag{9.94}$$

$$k_z H_r - \frac{\partial H_z}{\partial r} = j\omega\varepsilon E_\theta \tag{9.95}$$

$$\frac{1}{r}\frac{\partial}{\partial r}(rH_\theta) - \frac{1}{r}\frac{\partial H_r}{\partial \theta} = 0 \tag{9.96}$$

The only z component is H_z, and by algebraic manipulation of the above equations the other components can be obtained in relation to it. For example, from Eqn (9.88),

$$H_\theta = -\frac{1}{k_z}\left[j\omega\varepsilon E_r - \frac{1}{r}\frac{\partial H_z}{\partial \theta}\right]$$

and putting that into Eqn (9.92) gives

$$E_r = \frac{j\omega\mu}{k_z^2}\left[j\omega\varepsilon E_r - \frac{1}{r}\frac{\partial H_z}{\partial \theta}\right]$$

Hence

$$E_r = \frac{-j\omega\mu}{r(k_z^2 + \omega^2\varepsilon\mu)}\frac{\partial H_z}{\partial \theta}$$

or, if k_c is introduced from Eqn (9.45),

$$E_r = \frac{-j\omega\mu}{k_c^2}\frac{1}{r}\frac{\partial H_z}{\partial \theta} \tag{9.97}$$

By a similar sort of argument the other three components are

$$E_\theta = \frac{j\omega\mu}{k_c^2}\frac{\partial H_z}{\partial r} \tag{9.98}$$

$$H_r = \frac{k_z}{k_c^2}\frac{\partial H_z}{\partial r} \tag{9.99}$$

$$H_\theta = \frac{k_z}{k_c^2}\frac{1}{r}\frac{\partial H_z}{\partial \theta} \tag{9.100}$$

Thus if we can find an expression for H_z we can then find the values of the other components.

By analogy with the arguments leading to Eqn (9.20), H_z must satisfy the vector wave equation

$$\nabla^2 H_z = -\omega^2 \varepsilon \mu H_z \qquad (9.101)$$

which has a solution of the form

$$H_z = R\theta Z \qquad (9.102)$$

where R, θ and Z refer to the three cylindrical coordinates of Fig. 9.10. In cylindrical coordinates

$$\nabla^2 H_z = \frac{\partial^2 H_z}{\partial r^2} + \frac{1}{r}\frac{\partial H_z}{\partial r} + \frac{1}{r^2}\frac{\partial^2 H_z}{\partial \theta^2} + \frac{\partial^2 H_z}{\partial z^2} \qquad (9.103)$$

Putting Eqns (9.103) and (9.102) into Eqn (9.101), and remembering that the z component solution is

$$Z = H_z \exp(k_z z) \qquad (9.104)$$

Eqn (9.101) becomes

$$\theta Z \frac{\partial^2 Z}{\partial r^2} + \frac{\theta Z}{r}\frac{\partial R}{\partial r} + \frac{RZ}{r^2}\frac{\partial^2 \theta}{\partial \theta^2} + R\theta k_z^2 Z = k^2 R\theta Z$$

and dividing by $R\theta Z$

$$\frac{1}{R}\frac{\partial^2 R}{\partial r^2} + \frac{1}{Rr}\frac{\partial R}{\partial r} + \frac{1}{\theta r^2}\frac{\partial^2 \theta}{\partial \theta^2} + k_z^2 = -k^2$$

or

$$\frac{1}{R}\frac{\partial^2 R}{\partial r^2} + \frac{1}{r}\frac{1}{R}\frac{\partial R}{\partial r} + \frac{1}{r^2}\frac{1}{\theta}\frac{\partial^2 \theta}{\partial \theta^2} + k_c^2 = 0 \qquad (9.105)$$

where, as before [Eqn (9.29)],

$$k^2 + k_z^2 = k_c^2$$

From this we can establish two independent expressions:

$$\frac{r^2}{R}\frac{\partial^2 R}{\partial r^2} + \frac{r}{R}\frac{\partial R}{\partial r} + k_c^2 = n^2 \qquad (9.106)$$

and

$$\frac{1}{\theta}\frac{\partial^2 \theta}{\partial \theta^2} = -n^2 \qquad (9.107)$$

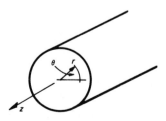

Fig. 9.10 Circular waveguide.

where n^2 is a constant. The solution for θ is straightforward:

$$\frac{\partial^2 \theta}{\partial \theta^2} + n^2 \theta = 0 \tag{9.108}$$

Hence

$$\theta = A_n \sin n\theta + B_n \cos n\theta \tag{9.109}$$

However, the solution for R is less simple. Equation (9.106),

$$\frac{\partial^2 R}{\partial r^2} + \frac{1}{r}\frac{\partial R}{\partial r} + \left(k_c^2 - \frac{n^2}{v^2}\right)R = 0$$

does not have a simple closed-form solution. It appears in several problems and it is called the Bessel equation. Solutions are given in tabular form and designated by the symbols $J_n(k_c r)$ and $N_n(k_c r)$. Thus the solution Eqn (9.105) is written

$$R = C_n J_n(k_c r) + D_n N_n(k_c r) \tag{9.110}$$

$J_n(k_c r)$ is called the nth-order Bessel function of the first kind, of argument $k_c r$, and $N_n(k_c r)$ is the nth-order Bessel function of the second kind.

Combining Eqns (9.104), (9.109), and (9.110) into Eqn (9.102),

$$H_z = [C_n J_n(k_c r) + D_n N_n(k_c r)][A_n \sin n\theta + B_n \cos n\theta]\exp(k_z z) \tag{9.111}$$

The coefficients, C_n, D_n, A_n and B_n are obtained from the boundary conditions or initial fields in the guide.

The nth-order Bessel function of the second kind, $N_n(k_c r)$, increases to infinity as r approaches zero. The effect of having this term in the expression for H_z would therefore be to have a field of infinite magnitude along the axis of the waveguide. This is not possible and hence, on physical grounds, $D_n = 0$, leaving

$$H_z = C_n J_n(k_c r)[A_n \sin n\theta + B_n \cos n\theta]\exp(k_z z) \tag{9.112}$$

The terms giving the variation of field with θ depend on the positioning of the $\theta = 0$ reference. By choosing it correctly, A_n can be made zero without imposing additional constraints on the field. Then

$$H_z = B_n \cos n\theta \cdot C_n J_n(k_c r)\exp(k_z z)$$

or

$$H_z = H_1 \cos n\theta \cdot J_n(k_c r)\exp(k_z z) \tag{9.113}$$

where H_1 is determined by the initial field applied to the guide. This solution is not yet complete: H_z must satisfy the conditions at the internal surface of the waveguide. At the guide wall, radius a, $J_n(k_c a)$ must be zero since H_z must be zero there. In addition, also at the guide walls, radius a, $\partial H_z / \partial r = 0$ using boundary conditions (ii) following Eqn (9.53), which implies that E_θ is zero from Eqn (9.113), [Eqn (9.98)]. For $\partial H_z / \partial r = 0$,

$$\frac{\partial}{\partial r}[H_1 J_n(k_c a)\cos n\theta \exp(\pm j\beta_g z)] = 0$$

giving

$$\frac{\partial}{\partial r} J_n(k_c a) = 0 \tag{9.114}$$

$J_n(k_c a)$ does not have a regular periodicity so there is no simple expression for k_c. Instead, tabulated values of the roots of Eqn (9.114) must be used. These depend on n and they are denoted by p'_{nm}. The prime indicates the roots of the first derivative of $J_n(k_c a)$, n is the Bessel function order, and m is the number of the root or the number of zeros of the field [i.e. of $J'_n(k_c a)$] in the range 0 to a.

So now we have developed a full solution of H_z, i.e.

$$H_z = H_1 J_n\left(\frac{p'_{nm}}{a}, r\right) \cos n\theta \exp\left(\pm j\beta_g z\right) \tag{9.115}$$

Returning to the other field components, Eqn (9.115) can be used in Eqns (9.97), (9.99) and (9.100), giving

$$E_r = -\frac{j\omega\mu}{k_c^2} \frac{1}{r} \frac{\partial}{\partial\theta}\left[H_1 J_n\left(\frac{p'_{nm}}{a}, r\right) \cos n\theta \exp\left(\pm j\beta_g z\right)\right]$$

$$= \frac{j\omega\mu n}{r k_c^2} H_1 j_n\left(\frac{p'_{nm}}{a}, r\right) \sin n\theta \exp\left(\pm j\beta_g z\right) \tag{9.116}$$

$$E_\theta = \frac{j\omega\mu}{k_c^2} \frac{\partial}{\partial r}\left[H_1 J_n\left(\frac{p'_{nm}}{a}, r\right) \cos n\theta \exp\left(\pm j\beta_g z\right)\right]$$

$$= \frac{j\omega\mu}{k_c^2} H_1 \left(\frac{\partial}{\partial r}\left[J_n\left(\frac{p'_{nm}}{a}, r\right)\right]\right) \cos n\theta \exp\left(\pm j\beta_g z\right) \tag{9.117}$$

$$H_r = \frac{k_z}{k_c^2} H_1 \left[\frac{\partial}{\partial r} J_n\left(\frac{p'_{nm}}{a}, r\right)\right] \cos n\theta \exp\left(\pm j\beta_g z\right) \tag{9.118}$$

$$H_\theta = \frac{k_z}{k_c^2} \frac{1}{r} n H_1 J_n\left(\frac{p'_{nm}}{a}, r\right) \sin n\theta \exp\left(\pm j\beta_g z\right) \tag{9.119}$$

As before, we have obtained the transverse field components in terms of the single longitudinal component H_z.

9.17 CIRCULAR MODE CUT-OFF FREQUENCY

The nomenclature used to describe the transverse electric fields in a circular waveguide is similar to that in a rectangular waveguide. TE_{nm} designates the nm mode. The cut-off wavelength in the circular case is given by

$$\lambda_{c_{nm}} = \frac{2\pi a}{p'_{nm}} \tag{9.120}$$

and the cut-off frequency by

$$f_{c_{nm}} = \frac{p'_{nm} c}{2\pi a} \tag{9.121}$$

Table 9.1 First ten TE modes in circular waveguide, and corresponding values of p'_{nm}

Mode	p'_{nm}
TE_{11}	1.841
TE_{21}	3.054
TE_{01}	3.832
TE_{31}	4.201
TE_{41}	5.318
TE_{12}	5.331
TE_{51}	6.416
TE_{22}	6.706
TE_{02}	7.016
TE_{32}	8.015

Values of p'_{nm} are given in Table 9.1. We can see from the table that the lowest cut-off frequency is for the TE_{11} mode, when $p'_{11} = 1.841$. This is called the fundamental or dominant mode. The field patterns across the guide for some of the lowest-order modes are shown in Fig. 9.11.

9.18 ATTENUATION IN CIRCULAR WAVEGUIDE

The waveguide walls have finite conductivity and so give rise to some ohmic loss. The attenuation of a TE_{nm} wave caused by the wall losses is given by

$$\alpha = \frac{R_s}{a\eta}\frac{1}{(1-(f_c/f)^2)^{1/2}}\left[\left(\frac{f_c}{f}\right)^2 + \frac{n^2}{(p'_{nm})^2 - n^2}\right] \qquad (9.122)$$

We can see that $n = 0$ is a special case, for then

$$\alpha = \frac{R_s}{a\eta}\frac{1}{\sqrt{[1-(f_{c01}/f)^2]}} \qquad (9.123)$$

and the variation of α with frequency is negative, i.e. the attenuation decreases as the frequency increases.

Typical curves of TE mode attenuation are shown in Fig. 9.12 and the difference between the TE_{0m} modes and the others is clear.

Another peculiar property is that the attenuation of these modes decreases as a, the guide radius, is increased. These two properties, although appreciated for some time, led to a worldwide interest in the mid-1950s when communication systems were required which could transmit high-frequency carriers over long distances with low loss.[3] The TE_{01} circular mode was

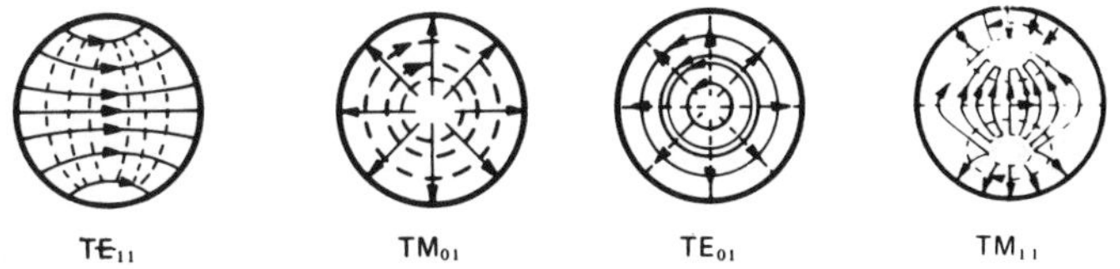

TE₁₁   TM₀₁   TE₀₁   TM₁₁

Fig. 9.11 Low-order circular waveguide modes.

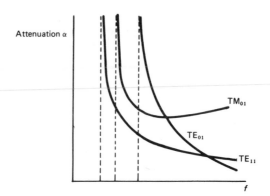

Fig. 9.12 Circular waveguide attenuation characteristics.

chosen, and laboratories in the UK, USA and Japan spent the next 20 years in developing production systems. Several were used successfully. From our discussion there would appear to be few problems, but in practice there are several which produced major difficulties. The TE_{01} mode is not the fundamental, and the waveguide is overmoded because of its large diameter. These factors mean that several modes can propagate down the guide, and at bends and corners, which are essential parts of a realistic system, the TE_{01} mode can be easily converted into other modes, with a consequent loss of power. These other spurious modes may have different propagation velocities from the TE_{01}, and that can have the effect of degrading the shape of the signal if there is a double mode conversion – from TE_{01} to spurious mode at one point along the guide and then back from spurious mode to TE_{01} at a further point. The performance of the guide can be improved by using something more complex than a simple circular cylindrical tube. By putting a helix of fine wire, set in resin, on the inner surface, many of the spurious modes can be filtered out since the helix will attenuate all non-TE_{0m} modes. Of course, the helix adds significantly to the cost of making the guide, but if the spacing between repeaters along the guide can be reduced thereby, a net saving may result. The precision quality of the waveguide requires that it is inserted into some protective jacket and laid in such a way that there are no tight bends. The latter considerations again add to the cost. Until the early 1970s it seemed that, on balance, circular waveguide would be an attractive long-haul transmission medium. Then optical fibres came along, and although for a few years both systems were developed in parallel in the large telecommunications laboratories, it soon became clear that fibre systems (see Chapter 12) could be better, and now their performance is such that the waveguide approach is no longer viable.

9.19 RECTANGULAR CAVITY

Figure 9.13(a) shows a rectangular waveguide cavity, and we can recognize it as a length of waveguide, terminated at each end by a short-circuiting face.

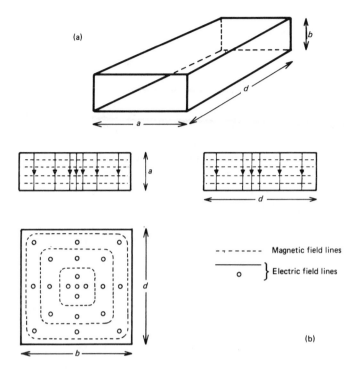

Fig. 9.13 (a) Rectangular waveguide cavity; (b) field pattern for TE_{101} mode.

Waves are reflected back and forth from these end faces. The cavity is highly resonant; oscillations occur over a very narrow band of frequencies centred on the resonant frequency. Tuning is either fixed, when all the cavity dimensions can have just one value, or variable, when one end face is attached to a plunger, allowing the resonant length of the cavity to be varied.

Any of the rectangular waveguide modes of Section 9.7 can exist within the cavity. We shall consider in detail the fundamental TE_{01} mode. The boundary conditions used in the development of the solutions to the field equations within the waveguide also apply in the cavity, which puts an additional constraint on the field pattern. The conducting walls at $z = 0$ and $z = d$ (the end walls) impose the condition that at these walls the tangential electric field, E_y, must be zero. Thus d must be an integral number of half guide-wavelengths long. The lowest order cavity mode is when $d = \lambda_g/2$. Then the E field is zero at the end walls and has one maximum, in the centre of the cavity.

If the width of the guide is a and the free-space wavelength is λ_0, the first resonant length can be found from Eqn (9.2) to be

$$d = \frac{\lambda_g}{2} = \frac{\lambda_0}{2[1 - (\lambda_0/2a)^2]^{1/2}} \tag{9.124}$$

The resonant frequency corresponding to a cavity of width a and length d in this fundamental (TE_{101}) mode is found directly from

$$f_{r_{101}} = c/\lambda_{0_{101}}$$

where

$$\lambda_{0_{101}} = 2ad/(a^2 + d^2)^{1/2} \tag{9.125}$$

Clearly, equations similar to Eqns (9.124) and (9.125) can be derived easily for TE_{10p} modes where p can, in principle, have any integer value in the range $1 \leqslant p \leqslant \infty$.

In the TE_{10p} modes there is no variation of E field with y and hence the height of the cavity b is not involved in the last two equations.

The E field inside the cavity in the TE_{101} mode can be considered as the sum of the incident and reflected waves, i.e.

$$E_y = [E_1 \exp(-j\beta_g z) + E_2 \exp(j\beta_g z)] \sin\left(\frac{\pi x}{a}\right) \tag{9.126}$$

and, by analogy, the H_x and H_z components are

$$H_x = -\frac{\sqrt{[1-(\lambda_0/2a)^2]}}{\eta}[E_1 \exp(-j\beta_g z) - E_2 \exp(j\beta_g z)] \sin(\pi x/a) \tag{9.127}$$

$$H_z = \frac{\lambda_0}{2a\eta}[E_1 \exp(-j\beta_g z) - E_2 \exp(j\beta_g z)] \cos(\pi x/a) \tag{9.128}$$

These equations can be simplified if we apply some of our earlier statements about the properties of the cavity. Assuming that the wall losses are negligible, the magnitude of the incident and reflected waves at the end wall will be equal, although of opposite sign, i.e. $E_1 = -E_2$. Also, because we are using the TE_{101} mode, in which $E_y = 0$ at $z = 0$ and $z = d$, and there is only one E field maximum between the end walls, $\beta_g = \pi/d$. Then Eqn (9.126) can be written

$$E_y = E_1\left[\exp\left(-j\frac{\pi}{d}z\right) - \exp\left(j\frac{\pi}{d}z\right)\right] \sin\frac{\pi x}{a}$$

$$= -2jE_1 \sin\frac{\pi}{d}z \sin\frac{\pi}{a}x \tag{9.129}$$

and using this expression in Eqns (9.127) and (9.128) we have

$$H_x = \frac{2E_1}{\eta}\frac{\lambda}{2d} \sin\frac{\pi x}{a} \cos\frac{\pi z}{d} \tag{9.130}$$

$$H_z = -\frac{\lambda}{2a}\frac{2E_1}{\eta} \cos\frac{\pi x}{a} \sin\frac{\pi z}{d} \tag{9.131}$$

where

$$\eta = \sqrt{(\mu/\varepsilon)}$$

From these three expressions the field patterns inside the cavity can be plotted for the fundamental mode. Figure 9.13(b) shows a representation of these fields.

Higher-order modes

We can extend our discussion of the TE_{101} mode to the more general case of the TE_{mnp} mode where m, n and p represent the number of half-sinusoidal variations of field in the x, y and z directions, respectively. In theory, each parameter can have an infinite number of values.

The rectangular waveguide TE_{mn} mode has an H_z field of the form.

$$H_z = [A \exp(-j\beta_g z) + B \exp(j\beta_g z)] \cos\frac{m\pi}{a} x \cos\frac{n\pi}{b} y \qquad (9.132)$$

As in our discussion on the TE_{101} mode, we can assume that $A = -B$ and that $H_z = 0$ at $z = 0, d$ making $\beta_g = p\pi/d$.

Then Eqn (9.132) can be written

$$H_z = -2jA \sin\frac{p\pi}{d} z \cos\frac{m\pi}{a} x \cos\frac{n\pi}{b} y \qquad (9.133)$$

For the general TE_{mnp} mode the relationship between the transverse field components E_x, E_y, H_x and H_y, are the same as those given by Eqns (9.58) to (9.62) in Section 9.9.

Substituting Eqn (9.133) into those equations gives

$$H_x = \frac{2jA}{k_c^2}\left(\frac{p\pi}{d}\right)\left(\frac{m\pi}{a}\right) \sin\frac{m\pi}{a} x \cos\frac{n\pi}{b} y \cos\frac{p\pi}{d} z \qquad (9.134)$$

$$H_y = \frac{2jA}{k_c^2}\left(\frac{p\pi}{d}\right)\left(\frac{n\pi}{b}\right) \cos\frac{m\pi}{a} x \sin\frac{n\pi}{b} y \cos\frac{p\pi}{d} z \qquad (9.135)$$

$$E_x = \frac{2A\omega\mu}{k_c^2}\left(\frac{n\pi}{b}\right) \cos\frac{m\pi}{a} x \sin\frac{n\pi}{b} y \sin\frac{p\pi}{d} z \qquad (9.136)$$

$$E_y = -\frac{2A\omega\mu}{k_c^2}\left(\frac{m\pi}{a}\right) \sin\frac{m\pi}{a} x \cos\frac{n\pi}{b} y \sin\frac{p\pi}{d} z \qquad (9.137)$$

where as before

$$k_c^2 = \left(\frac{m\pi}{a}\right)^2 + \left(\frac{n\pi}{b}\right)^2 \qquad (9.138)$$

For the TE_{mnp} cavity mode

$$\beta_g = \frac{p\pi}{a} = \left[\left(\frac{2\pi}{\lambda_0}\right)^2 - k_c^2\right]^{1/2} \qquad (9.139)$$

Rearranging Eqns (9.138) and (9.139) we can derive the resonant free-space wavelength for a cavity of dimensions a, b, d operating in the TE_{mnp} mode, viz.

$$\lambda_0 = 2\pi \bigg/ \left[\left(\frac{m\pi}{a}\right)^2 + \left(\frac{n\pi}{b}\right)^2 + \left(\frac{p\pi}{d}\right)^2\right]^{1/2} \qquad (9.140)$$

It is worth noting that, for a particular resonant frequency, the resonant length of the cavity, d, depends also on the cross-sectional dimensions of the waveguide appropriate to the mode of interest [Eqns (9.125) and (9.140)].

One of the main applications of multimode cavities is in microwave heating;[4] for example, in microwave ovens, where the object is to heat the sample uniformly, there is a need for a uniform field over the central space to be occupied by the sample, and this uniformity can be achieved by a combination of high-order modes and the use of a mode-stirrer, which is a metal plate, made to change its position and so vary the resonant length, and therefore the position of field maxima in the oven.

9.20 CAVITY Q – TE$_{101}$ MODE

As in all resonant systems, there are some losses, and eventually the oscillations in the cavity will decay. It is shown in other texts[5,6] that the cavity Q, which is a measure of the sharpness of the resonance curve, can be expressed as a ratio of the energy stored in the cavity to the mean loss at the walls, or

$$Q = \omega_r \frac{W_s}{W_L} = \omega_r \times \frac{\text{energy stored in the cavity}}{\text{average power loss}} \qquad (9.141)$$

where ω_r is the angular resonant frequency of the cavity.

W_s is the energy stored in the fields within the cavity. It is shared between the electric and magnetic fields, and it can therefore be found by calculating the maximum of one, when the other is zero.

The energy in an electric field is $\frac{1}{2}\varepsilon E^2$ integrated over the volume of the cavity:

$$W_s = \frac{\varepsilon}{2} \int_0^d \int_0^b \int_0^a |E_y|^2 \, dx \, dy \, dz \qquad (9.142)$$

since there is only one field component, E_y in the TE$_{101}$ mode.

Substituting for E_y from Eqn (9.129),

$$W_s = \frac{\varepsilon}{2} \int_0^d \int_0^b \int_0^a \frac{\varepsilon}{2} E_0^2 \sin^2 \frac{\pi x}{a} \sin^2 \frac{\pi z}{d} \, dx \, dy \, dz$$

where E_0 depends on the initial field in the cavity.

Remembering to replace $\sin^2 \theta$ by $\frac{1}{2}[1 - \cos 2\theta]$, this expression is integrated and reduces to

$$W_s = \frac{\varepsilon a b d E_0^2}{8} \qquad (9.143)$$

The average power loss at the walls is $\frac{1}{2}I^2 R_s$ where I is the peak wall current induced by the fields in the cavity, and R_s is the resistivity of the walls. The current on any wall is directly related to the tangential magnetic field at the wall surface. This field will vary with position in the cavity and therefore integration is required over each of the wall surfaces. The symmetry of the system means that opposite walls will have identical losses, hence

$$W_L = \frac{R_s}{2} \left\{ 2 \int_0^b \int_0^a |H_x|_{z=0}^2 \, dx \, dy + 2 \int_0^d \int_0^b |H_z|_{x=0}^2 \, dy \, dz \right.$$
$$\left. + 2 \int_0^d \int_0^a (|H_x|^2 + |H_z|^2) \, dx \, dz \right\} \qquad (9.144)$$

i.e.

W_L = loss in (end walls + side walls + top and bottom walls)
If we now develop Eqn (9.144) using

$$H_x = -j\frac{E_0}{\eta}\frac{\lambda_g}{2d}\sin\frac{\pi x}{a}\cos\frac{\pi z}{d}$$

and

$$H_z = j\frac{E_0}{\eta}\frac{\lambda_g}{2a}\cos\frac{\pi x}{a}\sin\frac{\pi z}{d}$$

We have

$$W_L = \frac{R_s \lambda_g^2 E_0^2}{8\eta^2}\left[\frac{ab}{d^2} + \frac{bd}{a^2} + \frac{1}{2}\left(\frac{a}{d} + \frac{d}{a}\right)\right] \qquad (9.145)$$

Then, from Eqns (9.143) and (9.145) and after some rearranging, the Q factor becomes

$$Q = \frac{\pi\eta}{R_s}\left[\frac{2b(a^2 + d^2)^{3/2}}{ad(a^2 + d^2) + 2b(a^3 + d^3)}\right] \qquad (9.146)$$

9.21   CIRCULAR CAVITY

By analogy with the previous section we can see that a length of circular waveguide with short-circuited ends will also form a cavity. The separation of the ends, d in Fig. 9.14(a), determines its resonant frequency; resonance occurs when d is an integral number of half-guide-wavelengths long. For each of the circular waveguide modes (Section 9.17) there is a corresponding family of cavity resonating modes, e.g. the fundamental TE_{11} circular waveguide mode will give rise to the TE_{11l} family of cylindrical cavity modes, where l indicates the number of half-guide-wavelengths in the length of the cavity. The fundamental cavity mode corresponds to $d = \lambda_g/2$, and it is designated the TE_{11} mode. In terms of the dimensions of the cavity, the resonant frequency of the TE_{111} mode is given by

$$f_{111} = \frac{c}{2\pi}\left[\left(\frac{\pi}{d}\right)^2 + \left(\frac{p'_{11}}{a}\right)^2\right]^{1/2}$$

where p'_{11} is the Bessel function coefficient of Section 9.17.

Another mode which has several applications is the TE_{011}. Its field pattern, shown in Fig. 9.14(b), produces no current flow between the cylindrical wall and the cavity ends. This means that the cavity can be constructed to allow the distance d to be varied (in applications where the cavity is to be tuned, as in a wavemeter), without imposing very rigorous constraints on its fabrication; good contact between the wall and the end faces is not essential.

The Q factor for TE modes has a form which is a function of the dimensions of the cavity, the mode number, and the skin depth of the wall material, δ_s.

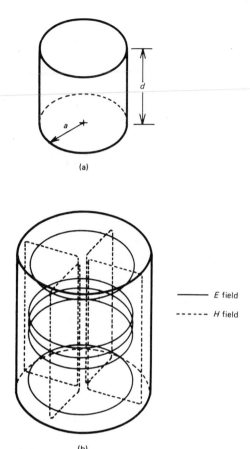

      ——— E field

      - - - - H field

(a)

(b)

Fig. 9.14 (a) Circular waveguide cavity; (b) field pattern for TE_{011} mode.

From Collin,[5] for the TE_{mnl} mode

$$Q = \frac{\lambda_0}{2\pi\delta_s}\left[1 - \left(\frac{n}{p'_{nm}}\right)^2\right]\left[(p'_{nm})^2 + \left(\frac{l\pi a}{d}\right)^2\right]^{3/2} \Bigg/$$

$$\left[(p'_{nm})^2 + \frac{2a}{d}\left(\frac{l\pi a}{d}\right)^2 + \left(1 - \frac{2a}{d}\right)\left(\frac{nl\pi a}{p'_{nm}d}\right)^2\right]$$

where δ_s is related to the conductivity of the walls, σ, by

$$\delta_s = \sqrt{(2/\omega\mu\sigma)}$$

Coupling into the cavity depends on the mode required in the particular application. The aperture between the cavity and the waveguide feed has to be so placed that, if possible, only the mode of interest is generated. Some modes are degenerate and special techniques may be necessary to absorb or avoid those which are unwanted.

    A full discussion on circular waveguide cavities can be found in Harvey[7] and Marcuvitz.[8]

9.22 RECTANGULAR WAVEGUIDE COMPONENTS – TE$_{10}$ MODE

In any practical application, several components are essential to provide a useable microwave system. A typical layout is shown in Fig. 9.15. In this section we will discuss briefly the passive devices shown; generators have been considered separately in Chapter 8.

Nearly all components are designed to operate in a particular mode. In our discussion we will assume that the fundamental TE$_{10}$ mode is being transmitted. The purpose of most passive microwave devices is self-evident from their names, but their operation is not always quite so obvious. In most cases there has been considerable analytical work done to evaluate how these devices work, and their operating performance characteristics. Here we shall not consider any analysis, and if the reader would like to explore this area, reference should be made to the specialist texts available.[5,7]

9.23 WAVEGUIDE – COAXIAL TRANSFORMER

To launch a wave into the guide some mechanism is required that will transform the RF energy on the feed from the generator (often coaxial cable) into microwave energy. Alternatively, and usually at higher frequencies, the generator is mounted directly on to the waveguide and indeed may be integral with it. Examples of a coaxial–waveguide transformer and a directly coupled feed are shown in Fig. 9.16.

In Fig. 9.16(a), energy is coupled from the generator via a simple probe, the centre of a coaxial feed. The probe penetrates into the guide through a broad face, and acts like an aerial. The electric fields induced between the probe and the waveguide propagate in all directions. By placing a short-circuit behind

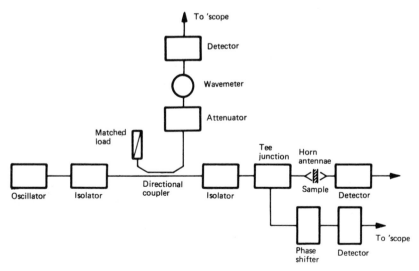

Fig. 9.15 Typical microwave circuit.

Fig. 9.16 Coaxial–waveguide coupler.

the probe, energy reflected from the short is propagated in the forward direction, in phase with that leaving the probe. The degree of coupling between the probe and the guide depends on the depth of penetration of the probe into the guide. The guide itself acts as a filter, which causes the many spurious (often evanescent) modes generated at the probe to be attenuated within a short distance.

Sometimes more rigidity or a greater bandwidth is required at the transformer than can be achieved by a simple probe. A horizontal bar [Fig. 9.16(b)] attached to the probe makes the feed mechanically more robust, and at the same time increases the bandwidth of the device.

In general, these feeds are reciprocal devices that can be used as either transmitters or receivers.

In the above paragraphs we have looked only at electric field probes, but the TE_{10} mode can also be generated by inducing a magnetic field in the guide. This is done by placing a loop through an end wall. The current variation produced in the loop by a signal from the generator induces a magnetic field, and again the filtering action of the guide creates the conditions for propagation of the TE_{10} mode.

9.24 ATTENUATOR

If the amplitude of the wave in the guide is too high, it must be reduced by using an attenuator, which consists of a vane made of a lossy material. The degree of attenuation depends on the strength of field at the vane; by varying the position of the vane, different levels of attenuation are possible. Two common forms of attenuator are shown in Fig. 9.17. In Fig. 9.17(a) the vane is attached to a micrometer through the sidewall of the guide. Attenuation is a minimum when the micrometer is fully extended, and the vane is against the sidewalls of the waveguide, where the electric field is at a minimum. Maximum attenuation occurs when the micrometer is screwed in and the vane lies along the axis of the guide in a position of maximum electric field. Good quality attenuators are tested in the manufacturer's laboratory and a calibration chart is issued to relate the reading on the micrometer screw to the attenuation introduced by the vane. For some applications, an attenuator is used to reduce the level of the signal in the guide to a workable value, and in that case a simple ungraduated screw will suffice. In order to avoid reflections the ends of the attenuator vane may be either stepped or tapered. The

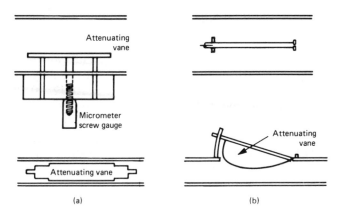

Fig. 9.17 Waveguide attenuators.

attenuator in Fig. 9.17(b) shows a different type of mechanism – often used in the high-frequency microwave, or millimetre, range. The vane has a cycloid profile and it is introduced into a slot along the centre of a broad face of the guide. The degree of attenuation increases with the depth of penetration into the waveguide, achieving a maximum when the vane is fully inserted.

9.25 DIRECTIONAL COUPLER

In many systems a proportion of the field is taken from the main waveguide to be measured or sampled in some way. A directional coupler allows a known fraction of the power in the primary guide to be filtered off. It consists of a second waveguide, contiguous with the first, and coupled to it by one or more holes, which may be circular, rectangular or cruciform in shape. The degree of coupling depends on the size of the hole and its position relative to the axis of the waveguide, if it is in the broad face. A double (four-port) coupler is often used in which energy is passing from port 1 to port 2 in the primary guide. Port 3 couples out a fraction of the forward wave and port 4 couples out a similar proportion of the backward wave. In this type of coupler (Fig. 9.18) the coupling holes are separated by $\lambda_g/4$. Hence, part of the forward wave from port 1 to port 2 is coupled through hole A and an equal part through hole B. The coupled fields add and are detected at port 3. Power from the forward wave coupled through hole B and travelling towards port 4 is

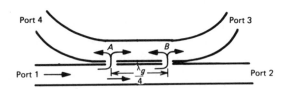

Fig. 9.18 Directional coupler.

cancelled by the power from hole A travelling in that direction. The effect of the coupling holes is given by two parameters:

(i) Coupling factor. With matched loads at ports 2 and 3, any power at port 4 is due to that coupled from the primary guide. The coupling factor is given by $10 \log (P_1/P_4)$ dB. The typical values are 3, 10, 12, 20 dB.
(ii) Directivity. The coupling holes will themselves produce reflections, and there will not be total isolation between ports 3 and 4. The directivity measures the ratio of the power in these ports for a matched load at port 2, i.e. directivity $= 10 \log (P_4/P_3)$ dB and this value would not be expected to be less than 30 dB.

9.26 TEE JUNCTIONS

Rectangular waveguides can be joined by using simple tee devices. Three commonly used, fundamental mode, junctions are shown in Fig. 9.19. To distinguish between junctions into the broad or narrow side of the guide, the first [Fig. 9.19(a)] is called an E plane tee, and the second [Fig. 9.19(b)] is called an H plane tee. Apart from the difference in geometry, these tees have different characteristics. In the E plane tee a wave entering port 1 divides equally into ports 2 and 3, but in phase opposition, so that at a given distance from the junction the signal out of port 2 is 180° out of phase with that from port 3. The H plane junction does not exhibit such phase reversal. The E field change across the E and H plane junctions is shown in the diagram, and explains the difference between them.

The third tee is a four-port, known as a hybrid, since it combines an E and an H plane junction. It has the property that a wave entering one port divides into the two adjacent ports, but does not appear at the opposite port. This characteristic allows the junction to be used in any systems in which two signals are to be processed, but must be isolated. For example, as we see in Fig. 9.19(d), this device (sometimes called a magic-tee) can be used to isolate the receiver from the transmitter, in a radar system that has a common transmit–receive aerial. The hybrid-tee has the disadvantage that half of the power is lost – in this example half of the received power goes to the receiver, but the other half is lost in the transmitter, and in a radar system that would degrade the sensitivity of the system considerably.

9.27 WAVEMETER

Traditionally, frequency has been measured by first measuring the wavelength of the signal in the waveguide. This can be done easily by using a wavemeter that is a waveguide cavity having one end wall attached to a piston. The position of the piston can be determined from the micrometer gauge to which it is attached. Energy from the waveguide is coupled into the cavity via small coupling holes that, in the case of a cylindrical cavity, would be via the fixed

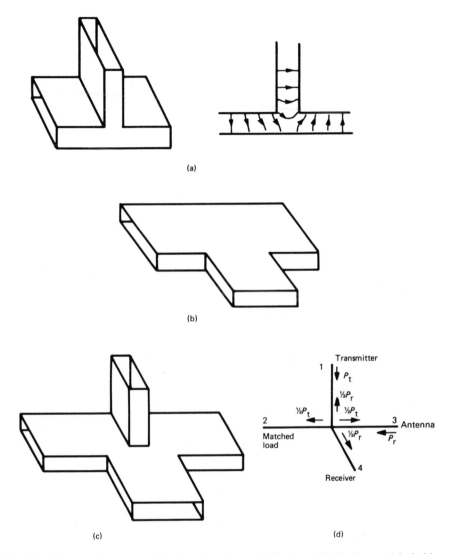

Fig. 9.19 Rectangular waveguide Tee junctions: (a) *E* plane; (b) *H* plane; (c) hybrid.

end wall, or for a rectangular cavity via a broad face of the cavity guide [Figs. 9.20(a) and (b)]. Connection into the cavity can be one-port (absorption), or two-port (transmission). The single port wavemeter absorbs energy from the waveguide when it is adjusted to a resonant frequency, and consequently the output from a detector at *D* in Fig. 9.20(a) will dip sharply at resonance. Conversely, in a transmission-type cavity, energy passes through the cavity most effectively at resonance; hence the detector in Fig. 9.20(b) will detect a surge of current at that frequency. In either case, the *Q* of the wavemeter will normally exceed 3000, and for high-quality wavemeters it may be greater than 10 000.

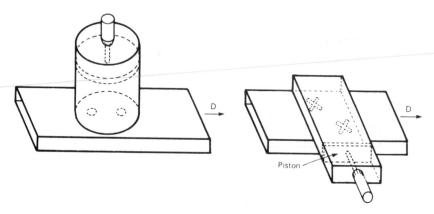

Fig. 9.20 Wavemeters: (a) cylindrical cavity; (b) rectangular cavity.

9.28 MATCHED LOAD

There are several positions in a microwave circuit, e.g. at one side of a directional coupler, beyond a wavemeter, or at the opposite end of a sample to be measured, when a matched termination is required. A termination provides a match if it does not induce a reflected wave. Hence most matched loads operate on the principle of completely absorbing incident energy so that no reflections can take place. Figure 9.21 shows a typical matched load. The absorbing material is wood, or some other lossy dielectric, and it is shaped to offer a gradual taper to the incident wave, and therefore no reflecting surface. To further reduce the possibility of reflection, the ends of the legs of the double wedge are displaced by $\lambda_g/4$, so that any reflections from these ends cancel. The dielectric insert is several wavelengths long to ensure that any energy reflected from the distant end is absorbed on the return path.

9.29 SHORT CIRCUIT

In Section 6.10 the use of stubs in transmission lines was discussed and there it was shown that the input impedance of a short-circuited stub can be altered by varying the stub length. A short circuit can be used in the same way in microwave systems. By attaching a short circuit to a plunger, the length of the

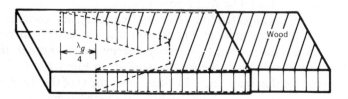

Fig. 9.21 Matched load.

short-circuited waveguide can be adjusted to provide the required input impedance. A typical short circuit is shown in Fig. 9.22.

9.30 ISOLATOR

Here we are not able to develop the theoretical basis for ferrites; we will acknowledge that a ferrite is a material that, when subjected to a magnetic field, will cause a polarized electromagnetic wave passing through it to rotate. The direction of rotation is not dependent on the direction of propagation of the wave. Hence a device can be constructed such that in one direction the rotation assists in the transmission of the wave, whereas in the opposite direction the rotation blocks propagation. Reference to Fig. 9.23 will explain how this can happen. Let us assume that the ferrite and magnetic field are so chosen that the electric field vector is rotated clockwise by $45°$. By using a $45°$ anticlockwise twist in the waveguide input before the ferrite, the rotation by the ferrite restores the wave to its original polarization and it can leave the output guide in the normal way. For a wave travelling in the opposite direction, however, the rotation through the ferrite is still clockwise, but the rotation through the $45°$ twist is clockwise also, resulting in the wave approaching the input guide horizontally polarized, and thus unable to pass

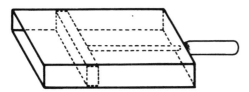

Fig. 9.22 Variable position short circuit.

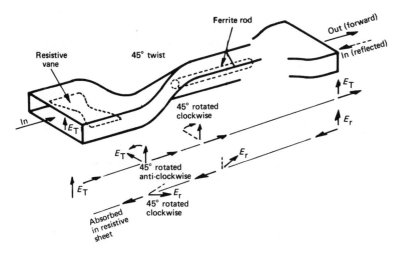

Fig. 9.23 Principles of operation of isolator.

through the system. There would be a reflection in such a case, but the resistive vane absorbs the energy in this backward wave and prevents reflections occurring. The loss in the forward direction is very small, whereas that in the backward direction might be as high as 30 dB. This device is used chiefly to isolate a generator from a system, to prevent mismatches in the system from causing reflections which might interfere with the performance of the generator.

Another type of isolator is shown in Fig. 9.24. Here the magnetic field is across the guide, produced by a substantial permanent magnet.

9.31 MICROSTRIP

The development of microwave integrated circuits, involving substrate-based sources, detectors and passive components, has necessitated the development

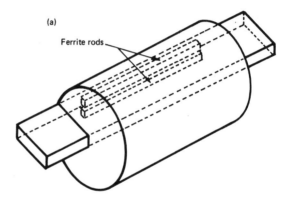

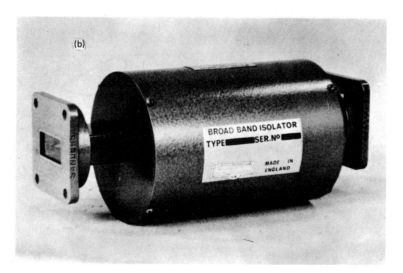

Fig. 9.24 Ferrite isolator. Reproduced from Marconi catalogue by permission of Sanders Division Marconi Instruments Ltd.

of waveguiding techniques suitable for integrated circuit applications. Some of the structures investigated for guiding waves on substrates are shown in Fig. 9.25. The most commonly used is the microstrip line and that is the structure that we will consider here. The dimensions of interest are indicated in Fig. 9.26.

The ground plane and the strip are of good conducting material such as copper, while the substrate is of a dielectric material with low loss and high permittivity, typically $\varepsilon_r \approx 10$.

We noted earlier, in Section 9.7, that in pipe waveguides, the operating modes are TE and TM. In microstrip, the modes are extremely complex, but when the dielectric has a high permittivity they are almost TEM. The complication arises from the fact that the propagation is not entirely through the dielectric material. Figure 9.27 shows the E field pattern about the line and it is clear that while most of the field is in the substrate some of it is not. The analysis is therefore difficult.

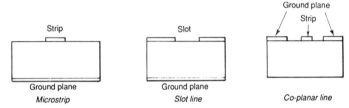

Fig. 9.25 Microwave integrated circuit structures.

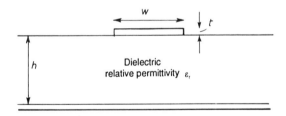

Fig. 9.26 Microstrip dimensions.

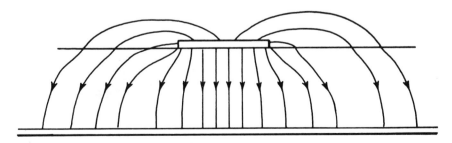

Fig. 9.27 E-field distribution in microstrip.

Here we will not develop the underlying theory of propagation on a microstrip system; if the reader wishes to do so, excellent treatments are found in references 9 and 10.

If the energy propagated along the microstrip were confined entirely to the dielectric, the analysis would be relatively straightforward, consisting of calculating the velocity along the microstrip from

$$v = c/\sqrt{\varepsilon_r}$$

and from that would follow good approximations for the other parameters of interest, in particular the capacitance and characteristic impedance.

Given that only a small amount of the energy propagates through air, the line can be modelled by a system in which all the energy travels in the substrate, but for which the dielectric permittivity is modified from its real value to an effective value ε_e. The difficulty is then to devise an expression ε_e in terms of w, h, and ε_r which can be verified by experiment. Many workers have tackled the problem of providing a suitable expression for ε_e and in the literature there is a variety of different equations available. One such equation is the following for the characteristic impedance, Z_0:

$$Z_0 = \frac{42.4}{\sqrt{(\varepsilon_r + 1)}} \ln\left\{ 1 + \left(\frac{4h}{w}\right)\left[\left(\frac{14 + 8/\varepsilon_r}{11}\right)\left(\frac{4h}{w}\right) \right.\right.$$
$$\left.\left. + \sqrt{\left(\left(\frac{14 + 8/\varepsilon_r}{11}\right)^2 \left(\frac{4h}{w}\right)^2 + \left(\frac{1 + 1/\varepsilon_r}{2}\right)\pi^2\right)} \right] \right\} \quad (9.147)$$

Clearly Eqn (9.147) is an empirical relationship. From an analytical point of view it gives an expression for Z_0 which can be used to derive other parameters. However, a designer is usually required to determine the dimensions of the strip line given specified values for ε_r and Z_0. To do that Eqn (9.147) would be rearranged with w/h as a subject. The equations used for ε_r can vary with the value of w/h and therefore care must be taken in their application. Reference 11 gives one of the clearest expositions of the basic features of microstrip design.

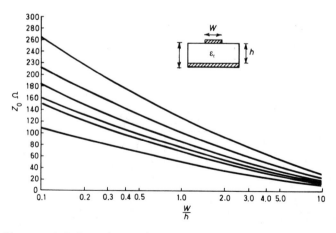

Fig. 9.28 Characteristic impedance of microstrip line.

In order to design a microstrip line it is necessary usually to use design curves of the type shown in Fig. 9.28. There it can be seen that the characteristic impedance of the microstrip is portrayed as a function of the dimensions w and h for a variety of values of dielectric permittivity. Hence the curves can be used either to find Z_0 for a particular strip line or to determine the ratio of w/h once the required characteristic impedance is known.

REFERENCES

1. Bleaney, B. I. and Bleaney, B., *Electricity and Magnetism*, Oxford University Press, Oxford, 1978.
2. Paul, C. R. and Nasar, S. A., *Introduction to Electromagnetic Fields*, McGraw-Hill, Maidenhead, 1982.
3. Karbowiak, A. E., *Trunk Waveguide Communication*, Chapman & Hall, London, 1965.
4. Okress, E. C., *Microwave Power Engineering*, Vol. 2, Academic Press, London, 1968.
5. Collin, R. E., *Foundation of Microwave Engineering*, McGraw-Hill, Maidenhead, 1966.
6. Atwater, H. A., *Introduction to Microwave Theory*, McGraw-Hill, Maidenhead, 1962.
7. Harvey, A. F., *Microwave Engineering*, Academic Press, London, 1963.
8. Marcuvitz, M., *Waveguide Handbook*, McGraw-Hill, Maidenhead, 1951.
9. Gupta, H. C. and Sing, A., *Microwave Integrated Circuits*, Chapter 3, Wiley Eastern, 1974.
10. Edwards, T. C., *Foundations for Microstrip Circuit Design*, Wiley, Chichester, 1981.
11. Combes, P. F., Graffeuil, J. and Santerean, J. F., *Microwave Components, Devices and Active Circuits*, Wiley, 1987.

PROBLEMS

9.1 Find the values of $\lambda_0, \lambda_g, \lambda_c, v_{ph}$ and v_g for a rectangular waveguide of inner cross-section 2.3 cm × 1.0 cm when it carries a signal of 10 GHz in the TE_{10} mode.

Answer: 3 cm, 3.96 cm, 4.6 cm, 3.96×10^8 m/s, 2.27×10^8 m/s.

9.2 A rectangular waveguide of cross-section $a \times b$ is filled with a dielectric material of permittivity $\varepsilon = k\varepsilon_0$. Show that for the TE_{10} mode the guide wavelength is given by

$$\lambda_g = \frac{\lambda_0}{[k - (\lambda_0/\lambda_c)^2]^{1/2}}$$

9.3 (a) Show that the superposition of two rectangular TE_{10} modes of equal amplitude and opposite direction of travel (as in a short-circuited waveguide) produces standing wave maxima and minima of field along the guide.
(b) Show also that the E field and H field maxima are displaced along the guide by $\lambda_g/4$.

9.4 Calculate the minimum value of a, in a rectangular waveguide of width a, operating in the TE_{10} mode over the frequency range from 10 to 11.5 GHz if the variation in v_g is not to exceed 20% of its value at 10 GHz.

Answer: 1.87 cm.

9.5 A TE_{10} signal at 10 GHz is propagating down a waveguide of width $a = 2$ cm. What is the change in phase velocity if the width is increased to 2.4 cm?

Answer: 0.70×10^8 m/s.

9.6 Find the distance between two adjacent wave minima in a 2.3 cm × 1.0 cm rectangular air-filled waveguide which is propagating in the TE_{10} mode at the frequency at which the TE_{21} mode could just begin to propagate.

Answer: 0.799 cm.

9.7 Sketch the electric and magnetic field patterns in both the cross-sectional and longitudinal planes of a rectangular waveguide operating in the TE_{20} mode. From the model of two interfering plane waves, find the cut-off wavelength for that mode in terms of the cross-sectional dimensions $a \times b (\lambda_c = a)$.

9.8 An air-filled rectangular waveguide has a TE_{10} mode cut-off frequency of 8 GHz. The signal frequency is 15 GHz, and the ratio $a/b = 2$. If the guide is now filled with perspex of relative permittivity 2.5, find the guide wavelength and phase velocity of the signal in the guide.

Would you expect higher order modes to be propagated in the dielectric-filled guide?

Answer: 1.34 cm, 2.02×10^8 m/s; yes.

9.9 A wave of frequency 9.95 GHz is travelling in the TE_{10} mode down a rectangular air-filled waveguide of width 1.5 cm. How far along the guide will the wave travel before it is attenuated to $1/e$ of its initial values?

Answer: 5.15 cm.

Telephony $\boxed{10}$

The telephone system is the largest integrated system in the world. Individual subscribers, separated by half the globe, can, in many cases automatically, make contact using this system. There is very rapid progress towards a complete direct-dialling international network, and it now seems very likely that the development of microcircuits, telecommunications satellites and new transmission media will allow a telephone link to be established to an individual in the remotest part of the earth via his own radio transceiver, identified by his unique world telephone number and accessed via satellite.

As a subject, telephony is usually restricted to considerations of the way in which subscribers are inter-connected, and an analysis of the traffic they generate. Traditionally, voice traffic, limited to a 3.4 kHz bandwidth, is assumed to be the nature of the signal sent over the telephone system, and we shall take that view in this chapter. However, the rapid growth of data traffic, generated by offices, banks, computers, airlines, etc., and the introduction of digital telephone systems in many countries, have raised the question of putting data over the telephone network and thus providing an integrated service. We will take a brief look at this area at the end of the chapter.

Telephony introduces several new concepts that appear strange at first, and tele-traffic theory (the study of telephone traffic and its relation to the design of switching and network systems) uses probabilistic models that need to be studied carefully in order to be understood, although the level of difficulty is well within the scope of an undergraduate statistics course, which most engineers follow.

We start with a section on traffic theory to establish a mathematical foundation. Systems were developed without the use of such theory, and indeed a mathematical analysis for many aspects of modern computer-controlled exchanges has not yet been developed, but if analysis can be applied it allows designers to make full use of their system, or an administration to predict how a manufacturer's equipment will perform.

10.1 TELEPHONE TRAFFIC

The use made of a telephone exchange, or a group of trunk circuits, is determined by both the rate at which calls arrive and the length of time for which they are held. The term 'traffic' takes these two quantities into account. Two units of traffic are employed: the CCS (hundred-call seconds) which is used principally in the USA; and the erlang which is used in Europe and other parts of the world.

If a circuit carries one call continuously for one hour it is said to carry one erlang (= 36 CCS) of traffic. The erlang, named after the Danish pioneer of tele-traffic theory, A.K. Erlang,[1] is expressed in mathematical terms as

$$A = \lambda s \qquad (10.1)$$

where A = traffic in erlangs, λ = mean call arrival rate (= calls per unit time) and s = mean call holding time measured in the same time units as λ.

10.2 INTER-ARRIVAL AND CALL HOLDING TIMES

Usually the quantities λ and s in Eqn (10.1) are mean values because in practice calls arrive at random and they last for a random length of time. The associated inter-arrival and holding time distributions could be determined by taking measurements over a long period, and building up a pattern for these parameters. However, apart from being a lengthy process, this approach has the disadvantage that the results would be accurate only for the time and place of the measurements; it would not necessarily follow that such distributions were applicable in general. To have some idea of the shape of the inter-arrival time and holding time distributions is, however, most important from the traffic theorist's point of view, and if they can be approximated by well-understood mathematical models, the development of a theoretical foundation to the subject can follow.

Over the years there have been many series of measurements on traffic from subscribers, and there is widespread agreement that the negative exponential distribution, with an appropriate choice of mean value, is satisfactory for both parameters. Figure 10.1 compares the negative exponential holding time curve with the results of a typical measurement. We notice that the ends of the distribution depart from reality, but, taking into account some of its attractive features, it is sufficiently accurate to stand as a valuable theoretical model.

The negative exponential distribution can be handled with comparative ease, but the property that is most appealing is that it is 'without memory', i.e. the probability that a call arrives in an interval dt is not dependent on the time

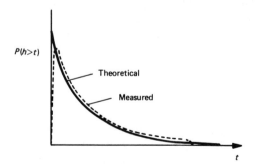

Fig. 10.1 Call holding time distribution.

that has elapsed since the last call or, for the holding time distribution, the probability that a call releases in a given interval dt is not related to the time for which the call has been in progress. Some thought will satisfy you that analysis would be significantly complicated if the history immediately before dt had to be taken into account.

The numerical values used for the mean inter-arrival time and the mean holding time depend on the application. The mean inter-arrival time is the inverse of the mean calling intensity, and therefore depends on the level of offered traffic. For the mean holding time, telephone administrations assume some reasonable value, typically 120 or 180 seconds. Calls arriving with a negative exponential inter-arrival time distribution are said to be Poissonian. A formal discussion of the Poissonian distribution is given in Section 13.3.

10.3 TRAFFIC VARIATION

The amount of telephone traffic passing through a particular exchange depends on several factors: the nature of the subscribers on the exchange (business, residential or mixed), the time of day, the month of the year, the occurrence of holidays and catastrophies, and the tariff in operation.

Figure 10.2 gives a typical distribution of traffic on a weekday in an exchange with a preponderance of business subscribers.

We see from the diagram that the night-time traffic is low, that there is a morning peak between 9.30 am and noon and that there is an afternoon peak between 2.00 pm and 4.30 pm. There is a slight increase in traffic between 6.00 pm and 10.00 pm owing to cheap-rate trunk calls from the residential subscribers. The heavy afternoon peak is typical of a system that offers the business subscriber a financial incentive for leaving trunk calls to the afternoon – by charging more for morning calls. The difference between the morning and afternoon peaks could be reduced by lowering the tariff differential. From the point of view of efficient use of the system, the distribution of traffic should be as uniform as possible throughout the day, and one of the functions of having a variable tariff scheme is to induce subscribers to use parts of the day that would not be their first choice.

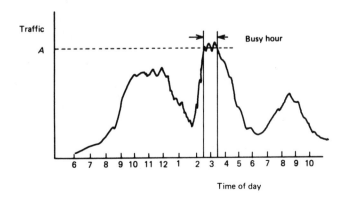

Fig. 10.2 Typical daily traffic distribution

10.4 BUSY HOUR

In the last two sections we have discussed the variation of traffic and its dependence on several factors. However, the designer must be able to use some traffic value against which he can design his exchange or determine the number of trunk circuits required. The traffic level is an average, taken over several days, and over the busiest period. The period is usually one hour, and the average traffic over that hour is called the 'busy hour' traffic. In Fig. 10.2 the busy hour is 2.30 pm to 3.30 pm, and the mean traffic over that period, A, would be used as the value of the busy hour traffic. In general, when traffic levels are quoted for routes or switching systems, it is implied that they are busy hour traffics. For example, if a circuit is said to carry 0.6 erlangs, it will be busy, on average, for 0.6 h (36 min) during the busy hour (not necessarily consecutive periods of time).

10.5 LOST-CALLS-CLEARED SYSTEMS

There are two major types of system used in telephony – lost-calls-cleared, and delay; although in a delay system a waiting subscriber may become impatient and leave, or the system may run out of holding positions. In a lost-calls-cleared system, if a call arrives to find there is no free path available to its destination it is cleared down. Once the system has tried to set up the call and has been unable to do so, there is no mechanism for putting the call in a queue to await a free path. Most of the theory developed in this chapter will concern lost-calls-cleared systems.

10.6 EQUATIONS OF STATE

Calls arrive at the network and depart from it at random, and the number of calls in progress, that is the number of devices busy, will vary in a random fashion. In the following we use N to be the maximum number of devices in the system, and i to indicate the number of calls in progress. Clearly, i is equal to the number of busy devices, and it is often called the state of the network. When $i = N$ the network is full and we use the term 'congested' to describe it; any further calls arriving when the network is in state N will be unable to find a free device.

In developing a mathematical model to allow us to analyse the properties of the system (the most important being the probability that the network is congested), we make the following assumptions:

(i) the calls are independent;
(ii) the rate of call arrivals when the network is in state i is λ_i;
(iii) the rate of call departure when the network is in state i is μ_i;
(iv) only one event (departure or arrival) can occur at a particular instant of time.

One common method of representing the states of a system, and the transitions between them, is by the state transition diagram of Fig. 10.3. There are

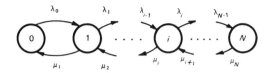

Fig. 10.3 State rate transition diagram.

$N+1$ states, going in unit steps from no calls in progress (state 0) to a congested system (state N), i.e.

$$0 \leqslant i \leqslant N \quad i = 0, 1, 2 \ldots$$

The parameters λ_i and μ_i are the rates of call arrival and departure, respectively, when the network is in state i. As we saw earlier, λ_i is related directly to the traffic offered to the system, and μ_i is determined by the nature of that traffic. Our objective is to develop an expression for the probability that the system is in state i, i.e. we want to find the state probabilities.

The probability that the system is in state i at some time $t + dt$ is equal to the sum of the following:

(i) the probability that the system was in state i at time t and no call arrived or departed in time dt;
(ii) the probability that the system was in state $i - 1$ at time t and a call arrived in time dt;
(iii) the probability that the system was in state $i + 1$ at time t and a call released during time dt.

As before, we are assuming here that no more than one transition can take place in the interval dt.

In symbols we can write this as

$$[i]_{t+dt} = [i]_t(1 - \lambda_i dt - \mu_i dt) + [i+1]_t(\mu_{i+1} dt) + [i-1]_t(\lambda_{i-1} dt) \quad (10.2)$$

where $[x] \equiv$ probability that the network is in state x.

Rearranging this expression

$$\frac{[i]_{t+dt} - [i]_t}{dt} = -(\lambda_i + \mu_i)[i]_t + \mu_{i+1}[i+1]_t + \lambda_{i-1}[i-1]_t \quad i = 1, 2, 3, \ldots$$
$$(10.3)$$

$i = 0$ is a special case, and

$$\mu_0 = 0 \text{ (no departure from state 0)}$$
$$\lambda_{-1} = 0 \text{ (state 1 does not exist)}$$

Then Eqn (10.3) becomes

$$\frac{[0]_{t+dt} - [0]_t}{dt} = -\lambda_0[0]_t + \mu_1[1]_t \quad (10.4)$$

As $dt \to 0$ Eqns (10.4) and (10.3) can be written in differential form:

$$\frac{d[0]_t}{dt} = -\lambda_0[0]_t + \mu_1[1]_t \quad (10.5)$$

and from Eqn 10.3

$$\frac{d[i]_t}{dt} = -(\lambda_i + \mu_i)[i]_t + \mu_{i+1}[i+1]_t + \lambda_{i-1}[i-1]_t \qquad (10.6)$$

10.7 STATISTICAL EQUILIBRIUM

In section 10.4 we discussed busy hour traffic. If the mean traffic over that period is used for design purposes we can assume that there will be no long-term change in the state probabilities with time. Therefore we can assume that during the busy hour the time dependence of the state probabilities is zero, and $d[i]/dt = 0$.

Applying this condition to Eqns (10.5) and (10.6) gives

$$\lambda_0[0] = \mu_1[1] \qquad (10.7)$$

$$(\lambda_1 + \mu_1)[1] = \lambda_0[0] + \mu_2[2] \quad \text{etc.}$$

In general

$$(\lambda_i + \mu_1)[i] = \lambda_{i-1}[i-1] + \mu_{i+1}[i+1] \qquad (10.8)$$

Substituting successive equations gives

$$\lambda_1[1] = \mu_2[2]$$

$$\lambda_2[2] = \mu_3[3] \quad \text{etc.}$$

The general expression becomes

$$\lambda_{i-1}[i-1] = \mu_i[i] \qquad (10.9)$$

or, in words,

(probability of being in state $i-1$) × (rate of call arrivals in state $i-1$)
    = (probability of being in state i) × (rate of call departure in state i)

In the long run, then, we can say that the number of calls leaving the system is equal to the number of calls arriving. If this were not so, as time passed the state of the system would gradually creep towards 0 or N and stay there.

Statistical equilibrium is a concept that lies at the heart of the models we are going to develop in the next few pages.

10.8 STATE PROBABILITY

We have made considerable use of the notion of state probability $[i]$ in the last section, and it is worthwhile remembering, before developing the theory further, that this probability is equivalent to the proportion of the busy hour that the system is in state i.

In Fig. 10.4 the call arrivals and departures are plotted over the busy hour for a system having $N = 5$. This diagram shows the way in which the system changes state; note that only one transition can occur at any instant.

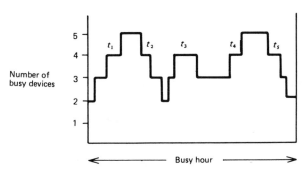

Fig. 10.4 Changes of state for a 5-device system.

Taking state 4 as an example:

$$[4] = \frac{t_1 + t_2 + t_3 + t_4 + t_5}{3600} = \frac{\sum \text{Times spent in state 4}}{\text{Busy hour}}$$

which is the fraction of the busy hour during which exactly four devices are busy.

We can also see from that diagram that

$$[0] + [1] + [2] + [3] + [4] + [5] = 1$$

In general, if there are N possible states

$$\sum_{i=0}^{N} [i] = 1 \tag{10.10}$$

Although this may appear to be rather obvious, it is a most important equation, as we shall see.

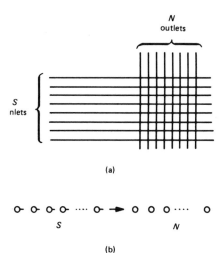

Fig. 10.5 Full availability system: (a) switch representation; (b) theoretical model.

10.9 FULL-AVAILABILITY MODELS

A full-availability system is one in which there are S sources of traffic and N outlets and any free source can seize any free outlet, regardless of the state of the system. The simplest example is a matrix switch with $S \times N$ cross-points as shown in Fig. 10.5(a). A straightforward representation, which we shall use, is shown in Fig. 10.5(b).

In order to evaluate $[i]$ for a full-availability group we need to have information about the offered traffic and how it depends on the state of the system; that in turn depends on the size of S in comparison with N.

10.10 ERLANG DISTRIBUTION: $S \gg N$

Constant traffic

A N

Fig. 10.6 Erlang full-availability model.

If the number of sources is very much greater than the number of devices, it can be assumed that, no matter how many devices are busy, the rate of call arrivals will be constant. We can therefore ignore the number of sources, and our model becomes that shown in Fig. 10.6. A erlangs are offered to N devices.

Assuming the system is in statistical equilibrium, Eqn (10.9) holds. In this case the rate of call arrivals is constant, and, using Eqn (10.1),

$$\lambda_i = \lambda = \frac{A}{s} = \frac{\text{Mean offered traffic}}{\text{Mean holding time}}$$

and the rate of call departures $= \dfrac{\text{Number of calls in progress}}{\text{Mean holding time}}$

i.e.

$$\mu_i = \frac{i}{s}$$

Then from Eqn (10.9)

$$[i]\frac{i}{s} = [i-1]\frac{A}{s}$$

Putting in values for i:

$$[1] = A[0]$$

$$[2] = \frac{A^2}{2}[0]$$

$$\vdots$$

$$[i] = \frac{A^i}{i!}[0] \tag{10.11}$$

$$\vdots$$

$$[N] = \frac{A^N}{N!}[0]$$

Using Eqn (10.10)

$$[0] + A[0] + \cdots + \frac{A^N}{N!}[0] = 1$$

Hence

$$[0] = \frac{1}{\displaystyle\sum_{v=0}^{N} \frac{A^v}{v!}}$$

and using this in Eqn (10.11), the probability of being in state i is

$$[i] = \frac{A^i}{i!} \bigg/ \sum_{v=0}^{N} \frac{A^v}{v!} \tag{10.12}$$

Thus we can find $[i]$ for all values of i if we know A and N.

Equation (10.12) is called the probability distribution of the states of the system, and it is a truncated form of the Poisson distribution (Section 13.3).

10.11 TIME CONGESTION

Time congestion E is defined as the proportion of the busy hour for which the network is in state N, i.e.

$$\text{Time congestion } E = [N]$$

It follows directly from equation (10.12) that

$$E = [N] = \frac{A^N}{N!} \bigg/ \sum_{v=0}^{N} \frac{A^v}{v!} = E_N(A) \tag{10.13}$$

This is known as Erlang's loss formula. It has been tabulated extensively and it can be evaluated simply on a computer by using the iteration

$$E_i(A) = \frac{A E_{i-1}(A)}{i + A E_{i-1}(A)} \tag{10.14}$$

and the initial condition that $E_0(A) = 1$.

10.12 CALL CONGESTION

Another type of congestion is call congestion, B. It is the probability that a call arrives to find the system fully occupied, i.e.

$$B = \frac{\text{Number of calls arriving when the system is in state } N}{\text{Total number of calls arriving in the busy hour}}$$

In the Erlang case,

$$B = \lambda \cdot [N] \bigg/ \sum_{i=0}^{N} \lambda \cdot [i]$$

and since λ is constant

$$B = [N] \Big/ \sum_{i=0}^{N} [i]$$

Again applying Eqn (10.10), we have

$$B = [N]$$

Thus for the Erlang case

call congestion = time congestion

This model is used extensively in practice. It is applied whenever the assumption can be made that the calling rate is independent of the number of calls in progress.

10.13   BERNOULLI DISTRIBUTION: $S \leqslant N$

If the number of sources is less than the number of devices, a different model has to be used becase we cannot assume that the traffic intensity is indepenent of the number of calls in progress. The model must now take account of the number of sources (Fig. 10.7). The simplest way of taking account of the state of the system is to relate the traffic offered to the traffic offered per free source and the number of free sources available, i.e.

$$\lambda_i = (S - i)\alpha/s \tag{10.15}$$

where α is the traffic per free source, $S - i$ is the number of free sources and, as before, the mean holding time is s.

Applying the statistical equilibrium Eqn (10.9), once more

$$\lambda_{i-1}[i-1] = \mu_i[i]$$

$\mu_i = i/s$ as in the Erlang case, and if this is used, along with λ_{i-1} from Eqn (10.15), we have

$$(S - i + 1)[i-1]\frac{\alpha}{s} = \frac{i}{s}[i]$$

or

$$[i] = \frac{S - i + 1}{i}\alpha[i-1]$$

giving

$$[1] = S \cdot \alpha \cdot [0]$$

$$[2] = \frac{S-1}{2} \cdot \alpha \cdot [1] = \frac{S(S-1)\alpha^2}{2}[0] \quad \text{etc.}$$

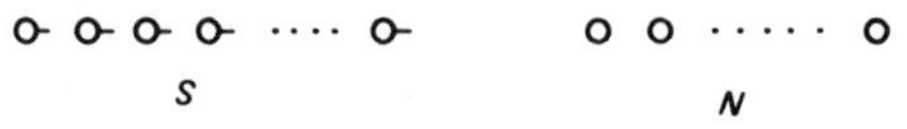

Fig. 10.7 Limited source full-availability model.

and generally

$$[i] = \frac{S!}{(S-i)!i!} \alpha^i[0] = \binom{S}{i} \alpha^i[0]$$

Summing over all i and applying Eqn (10.10),

$$\sum_{i=0}^{S} [i] = 1 = \left[1 + S\alpha + \frac{S(S-1)\alpha^2}{2} + \frac{S(S-1)(S-2)\alpha^3}{3!} + \cdots + \alpha^S \right][0]$$

(Note that the summation is from $0 \leqslant i \leqslant S$ since $S \leqslant N$.)

$$[0] = \frac{1}{1 + S\alpha + S(S-1)\frac{\alpha^2}{2!} + \cdots + \alpha^S} = \frac{1}{(1+\alpha)^S}$$

and the general term can now be written

$$[i] = \binom{S}{i} \alpha^i \frac{1}{(1+\alpha)^S}$$

$$= \binom{S}{i} \left(\frac{\alpha}{1+\alpha}\right)^i \left(1 - \frac{\alpha}{1+\alpha}\right)^{S-i}$$

or

$$[i] = \binom{S}{i} a^i (1-a)^{S-i} \tag{10.16}$$

where

$$a = \frac{\alpha}{1+\alpha}$$

If $S < N$ there will be no time congestion because the state N is never reached; there will always be at least $N - S$ free devices.

If $S = N$ then the time congestion is

$$[N] = a^S = a^N \tag{10.17}$$

But the call congestion will be zero because there are no calls arriving when the network is in state N, all the sources being busy.

The quantity a in Eqn (10.16) is the carried traffic per device. Note that Eqn (10.16) is the classical form of the Bernoulli distribution.

10.14 ENGSET DISTRIBUTION: $S > N$

This case models the situation in which the number of sources is greater than the number of devices, but not so large that the traffic offered is constant. As in the Bernoulli case we shall assume here that the traffic offered is related directly to the number of free sources, allowing us to start with the same state equation, viz.

$$(S - i + 1)[i-1]\frac{\alpha}{S} = \frac{i}{S}[i] \tag{10.18}$$

from which

$$[1] = S\alpha[0]$$

$$[2] = \frac{S(S-1)\alpha^2}{2}[0] \quad \text{etc.}$$

and generally

$$[i] = \binom{S}{i}\alpha^i[0] \tag{10.19}$$

We can now sum $[i]$ over all states $0 \leqslant i \leqslant N$ and apply Eqn (10.10), i.e.

$$\sum_{i=0}^{N}\binom{S}{i}\alpha^i[0] = 1$$

or, changing the variable to avoid confusion later,

$$[0] = 1 \bigg/ \sum_{v=0}^{N}\binom{S}{v}\alpha^v$$

Then, inserting this expression in Eqn (10.19) gives

$$[i] = \binom{S}{i}\alpha^i \bigg/ \sum_{v=0}^{N}\binom{S}{v}\alpha^v \tag{10.20}$$

10.14.1 Time congestion

From Eqn (10.20)

$$[N] = \binom{S}{N}\alpha^N \bigg/ \sum_{v=0}^{N}\binom{S}{v}\alpha^v \tag{10.21}$$

which is denoted by $E(N, S, \alpha)$. Values of E for given N, S and α can either be calculated from Eqn (10.21) or be obtained from tabulated results.

10.14.2 Call congestion

From the previous defintion

$$B = (S-N)\frac{\alpha}{S}[N] \bigg/ \sum_{i=0}^{N}[i](S-i)\frac{\alpha}{S}$$

Substituting from Eqn (10.21) for $[N]$ and rearranging gives

$$B = \binom{S-1}{N}\alpha^N \bigg/ \sum_{v=0}^{N}\binom{S-1}{v}\alpha^v$$

$$= E(N, S-1, \alpha) \tag{10.22}$$

10.15 PROBABILITY OF OCCUPANCY OF PARTICULAR DEVICES

We have just developed the three most important full-availability models, and in each case the general term has been $[i]$, the probability that any i devices are busy.

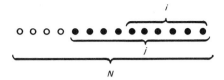

Fig. 10.8 Model to find $H(j)$.

There are some applications where a distinction must be made between the two probabilities of i busy devices:

(i) that any i devices are busy, denoted by $G(i)$;
(ii) that a particular i devices are busy, denoted by $\bar{H}(i)$.

The first is what we have calculated so far. To find the second, $H(i)$, we consider the situation shown in Fig. 10.8.

The problem is, given that j out of N devices are busy, what is the probability, $H(i)$, that a particular i are busy?

We can argue as follows:

Given that j devices in total are busy, there are $j-i$ busy devices not included in our i, and these $j-i$ can be distributed over the $N-i$ positions available for them in $\binom{N-i}{j-i}$ ways. Then

$$\frac{\text{Number of ways the particular } i \text{ are busy}}{\text{Total number of ways of distributing } j \text{ over } N} = \frac{\binom{N-i}{j-i}}{\binom{N}{j}}$$

Hence

$$H(i) = \sum_{j=i}^{N}\left[\binom{N-i}{j-i}\Big/\binom{N}{j}\right]P(j) \tag{10.23}$$

noting that the summation can only be over the range for which $j \geqslant i$. The value of $H(i)$ depends of course on the quantity $P(j)$ which, in the terms introduced at the beginning of this section, is equivalent to $G(j)$. Applying the models of Section 10.2, there are three cases.

10.15.1 Erlang

$$H(i) = \sum_{j=i}^{N}\left[\binom{N-i}{j-i}\Big/\binom{N}{j}\right]P(j)$$

$$= \sum_{j=i}^{N}\left[\binom{N-i}{j-i}\Big/\binom{N}{j}\right]\cdot\left[\frac{A^j}{j!}\Big/\sum_{v=0}^{N}\frac{A^v}{v!}\right] \tag{10.24}$$

from Eqn (10.13)

$$= \sum_{j=i}^{N}\frac{(N-i)!}{(j-i)!(N-j)!}\cdot\frac{j!(N-j)!}{N!}\cdot\left[\frac{A^j}{j!}\Big/\sum_{v=0}^{N}\frac{A^v}{v!}\right]$$

$$= \left[A^N \Big/ N! \sum_{v=0}^{N} \frac{A^v}{v!} \right] \cdot \frac{(N-i)!}{A^N} \sum_{j=i}^{N} \frac{A^j}{(j-i)!}$$

$$= \left[\frac{A^N}{N!} \Big/ \sum_{v=0}^{N} \frac{A^v}{v!} \right] \cdot \frac{(N-i)!}{A^N} \left[\frac{A^i}{1} + \frac{A^{i+1}}{1} + \cdots + \frac{A^N}{(N-i)!} \right]$$

$$= E_N(A) \cdot \frac{A^i (N-i)!}{A^N} \left[1 + A + \cdots + \frac{A^{N-i}}{(N-i)!} \right]$$

$$= \frac{E_N(A)}{E_{N-i}(A)} \tag{10.25}$$

Showing that $H(i)$ in the Erlang case can be found from two Erlang functions.

10.15.2 Bernoulli

In this case

$$P(j) = \binom{N}{j} a^j (1-a)^{N-j}$$

and

$$H(i) = \sum_{j=i}^{S} \left[\binom{N-i}{j-i} \Big/ \binom{N}{j} \right] \binom{N}{j} a^j (1-a)^{N-j}$$

Expanding and rearranging terms, this reduces to

$$H(i) = \frac{\binom{S}{i}}{\binom{N}{i}} a^i \quad \text{for} \quad S < N \tag{10.26}$$

In the special case when $S = N$, which is common in practice,

$$H(i) = a^i \tag{10.27}$$

10.15.3 Engset

From the earlier analysis

$$P(j) = \binom{S}{j} \alpha^j \Big/ \sum_{v=0}^{N} \binom{S}{v} \alpha^v$$

so

$$H(i) = \sum_{j=i}^{N} \left[\binom{N-i}{j-i} \Big/ \binom{N}{j} \right] \cdot \left[\binom{S}{j} \alpha^j \Big/ \sum_{v=0}^{N} \binom{S}{v} \alpha^v \right] \tag{10.28}$$

which becomes

$$H(i) = \frac{E(N, S, \alpha)}{E(N-i, S-i, \alpha)} \tag{10.29}$$

Summarizing the results obtained for the full availability models considered above, we can tabulate the formulae we have derived as shown in Table 10.1.

Table 10.1

	$G(i)$	$H(i)$
Erlang	$E_i(A) = (A^i/i!) \bigg/ \sum_{v=0}^{N} (A^v/v!)$	$\dfrac{E_N(A)}{E_{N-i}(A)}$
Bernoulli	$\dbinom{N}{i} a^i (1-a)^{N-i}$	$\dfrac{\dbinom{S}{i}}{\dbinom{N}{i}} a^i \quad (a^i \text{ for } S=N)$
Engset	$E(N,S,\alpha) = \dbinom{S}{i} a^i \bigg/ \sum_{v=0}^{N} \dbinom{S}{v} \alpha^v$	$\dfrac{E(N,S,\alpha)}{E(N-i,S-i,\alpha)}$

10.16 LINK SYSTEMS

A link system is a multi-stage switching network that consists of a set of first-stage switches, connected to a set of second-stage switches, which in turn are connected to a set of third-stage switches, and so on. The number of stages of switching used depends on the purpose of the system. We will consider first a two-stage system as shown in Fig. 10.9(a). There we see that there are four inlets and five outlets on each of the six first-stage switches, and six inlets and six outlets on each of the five second-stage switches.

The connections, or links between the two switching stages are shown in Fig. 10.9(b). We see that the diagram is very messy, and even if a link is drawn only when there is a call in progress along it, the lines indicating a particular call will be difficult to trace when the number of calls is high. A different type of representation suits our purposes better, i.e. one that was first introduced by workers at the Swedish company L.M. Ericsson, and is called a chicken diagram. The chicken diagram representation of the network in Fig. 10.9(a) is shown in Fig. 10.9(c). Now the link connections are not shown, but are implied by the diagram. A column on the A matrix represents the inlets of a first-stage switch; the corresponding column on the B matrix represents the outlets of the first-stage switch. A row on the B matrix corresponds to the inlets on a second-stage switch, and a row on the C matrix corresponds to the outlets on the same second-stage switch. Calls can be traced through the network by placing a number on the appropriate place in each matrix. An example of the network representation, and a chicken diagram representation corresponding to it are shown in Fig. 10.10. In addition, several calls have been placed in the system, and the method of representing them is shown clearly in each diagram. The reader should study these diagrams carefully until the relationship between them becomes clear; the network is discussed further in the next section.

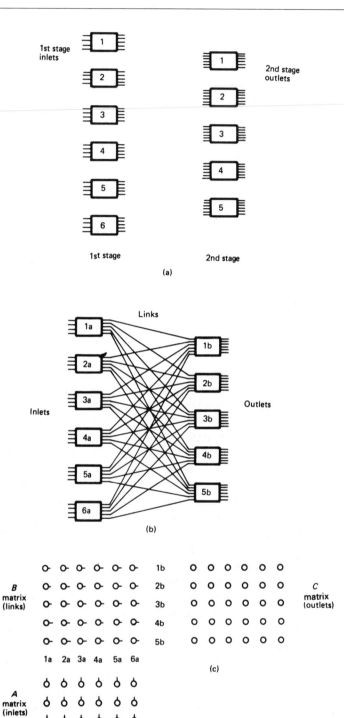

Fig. 10.9 Two-stage links system; (a) switches; (b) complete with links; (c) chicken diagram equivalent.

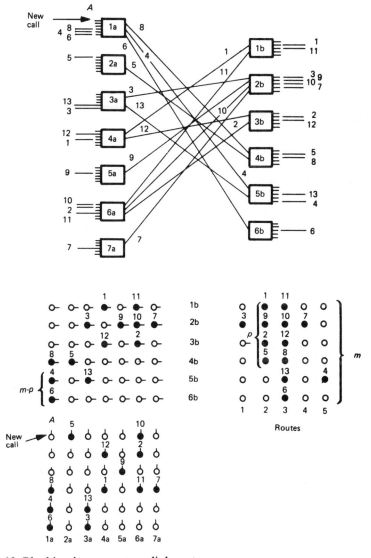

Fig. 10.10 Blocking in a two-stage link system.

10.17 PROBABILITY OF BLOCKING

In order to determine the equipment requirement of the system or, putting that in another way, to find out if the system will perform as it should, some measure of performance is required. That most often used is the probability of blocking, which is the proportion of calls, in the long term, that are rejected. If we consider the general case of a two-stage link system in which the corresponding outlets on the second-stage switches form a route (which is a group of circuits going to the same destination), as indicated by the chicken diagram in Fig. 10.10(b), we can see that there are two ways in which a call may be blocked:

(i) there may be no free route circuit;
(ii) there may be no path available between the inlet carrying the incoming call and the free route circuits.

The diagram shows both of these conditions; route 3 has no free circuits available, whereas if inlet A on first-stage switch 1 is wanting to be connected to route 2, on which there are free route circuits, it cannot be, because there is no available link.

If we now look at this particular connection between inlet A and route 2 further, we can develop from it an analytical approach to the probability of blocking. The part of the network involved in the connection from A to route 2 is shown in Fig. 10.11.

There are m links to a route from any first-stage switch, since any route has one circuit on each of the second-stage switches. Thus a route consists of m outlet circuits, one on each second-stage switch.

Let us assume that there are p busy circuits on the route, then a call will not be able to be connected if, at the same time, the particular $m - p$ links, having access to the free $m - p$ route circuits, are busy, as they are in the diagram.

The first probability, that there are any p route circuits busy, we denote by $G(p)$, and the second probability, that $m - p$ particular devices are busy is denoted by $H(m - p)$. The probability of blocking, that is the probability that a call cannot be connected, is the joint probability of these two events, summed over all possible values of the variable p. Denoting the probability of blocking by E, we have then that

$$E = \sum_{p=0}^{m} G(p) H(m - p) \qquad (10.30)$$

This apparently simple expression, known as the **Jacobaeus equation**, after its development by Christian Jacobaeus of L.M. Ericsson[2] in 1950, has been the cornerstone of blocking calculations on link systems for many years. There are some assumptions implicit in its form which become important, and indeed limiting, under certain circumstances: the most significant is that the

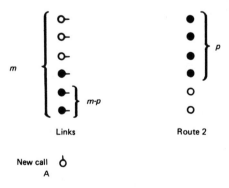

Fig. 10.11 Paths from a particular inlet to a route.

two probabilities $H(m-p)$ and $G(p)$ are independent. Provided that the traffic level in the network is low, this independence assumption can be made without concern, but at higher traffic levels the analysis must be modified to take the conditional dependence of $G(p)$ on $H(m-p)$ into account.

The evaluation of Eqn (10.30) depends on the probability distributions appropriate to $G(p)$ and $H(m-p)$. In most cases one of those summarized in Table 10.1 is chosen. Often $G(p)$ will be Erlang because the number of route circuits (m) will be much smaller than the number of potential traffic sources (all the inlets on the first stage). The distribution used for $H(m-p)$ will depend entirely on the relationship between the inlets and outlets of the first-stage switches. An example will demonstrate how the Jacobaeus equation can be applied to a two-stage system.

EXAMPLE: In a two-stage link system the outlet circuits are combined into routes – each route has one circuit from each outlet switch. Using a chicken diagram representation, find the internal link congestion if the total traffic offered to the network is 42 erlangs, and if the outlets are formed into routes, as explained in the first sentence. Assume that there is equal demand for each route.

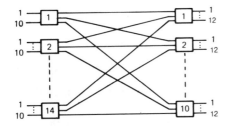

The chicken diagram representation is

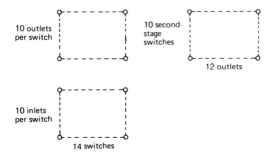

There are 12 routes, and 42 erlangs of total offered traffic; therefore

each route is offered $42/12 = 3.5$ erlangs

The Jacobaeus equation gives the probability of blocking as

$$\sum_{p=0}^{m} G(p)\,H(m-p)$$

where $G(p)$ = probability the p devices are busy on the route and $H(m-p)$ = probability that $m-p$ particular links are busy. Since there are 140 inlets to the network, and each route consists of 10 devices, we can assume an Erlang distribution on the route, i.e.

$$G(p) = \frac{A^p}{p!} \bigg/ \sum_{v=0}^{m} \frac{A^v}{v!}$$

There is the same number of inlets as outlets on a first-stage switch, and therefore the Bernoulli distribution is appropriate for $H(m-p)$, giving, from Table 10.1,

$$H(m-p) = a^{m-p}$$

where a = traffic carried per link.

There are 140 links in total; therefore, assuming the congestion to be small,

$$a = 42/140 = 0.3 \text{ erlangs}$$

Hence the total blocking, E, is

$$E = \sum_{p=0}^{10} \frac{(3.5)^p/p!}{\sum\limits_{v=0}^{10} 3.5^v/v!} (0.3)^{10-p}$$

Evaluating this expression for each value of p, and summing, gives the total

$$E = 0.0082$$

The internal link congestion = total congestion − route congestion.

In any route there are 10 circuits, and we calculated earlier that the traffic offered per route is 3.5 erlangs. Therefore the route congestion = $E_{10}(3.5) = 0.0023$. Hence internal congestion = $0.0082 - 0.0023 = 0.0059$.

Although we shall not do more here, the Jacobeaus approach can be used for a wide variety of link system structures, but they do not involve any change in the principles we have developed.

10.18 DELAY SYSTEM

In the foregoing sections we dealt exclusively with lost-calls-cleared systems in which a call is lost if it arrives when all the devices are occupied. In such systems there is no mechanism for allowing calls to wait until a free device becomes available.

If queueing is allowed, however, the call can be delayed, or held within the system until it has access to a free device. In this section we shall briefly examine the simplest delay model. It has the following attributes:

  (i) calls arrive at random with constant mean traffic and negative-exponential distribution of inter-arrival times (Poisson offered traffic);
 (ii) the holding times follow a negative-exponential distribution;
(iii) there are n devices (servers);
(iv) the queue can accept an infinite number of waiting calls;

(v) the service discipline is first in–first out, i.e. calls are passed to devices in order of arrival.

Queues satisfying (i), (ii) and (iii) are usually classified as $M/M/n$ in books on queueing theory, after the nomenclature of Kendall[3] who proposed a systematic classification of queues.

The queue we will now study is idealized through the assumption (iv), which could not be achieved in practice, but there are many real situations which are sufficiently close to it to make the analysis applicable.

If, as before, the mean calling rate and the mean departure rate at any instant are given by λ_i and μ_i, where i is the number of calls is the system, we can draw a transition diagram as shown in Fig. 10.12.

If we compare this diagram with that of Fig. 10.3 we can see two differences. In the delay system there is an infinite number of states, and the departure coefficient μ_i is dependent on the number of calls in progress when $i \leqslant n$, and thereafter it is equal to μ_n for all i.

We can use the same approach as we did in the analysis of the Erlang loss formula, since here too we have assumed that the mean traffic offered is constant.

Starting with the relevant birth and death equation, assuming that we have statistical equilibrium (which in a queueing system requires that the mean traffic offered per device is less than unity, i.e. $A/n < 1$), we have

$$\mu_i[i] = \frac{A}{s}[i-1]$$

and

$$\mu_i = \frac{i}{s}$$

Therefore

$$[i] = \frac{A}{i}[i-1]$$

Putting in values

$$[1] = A[0]$$

$$[2] = \frac{A}{2}[1] = \frac{A^2}{2}[0]$$

$$[3] = \frac{A}{3}[2] = \frac{A^3}{3!}[0]$$

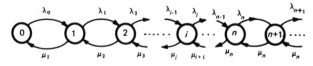

Fig. 10.12 State transition diagram for a delay system.

etc. to

$$[n-1] = \frac{A^{n-1}}{(n-1)!}[0]$$

$$[n] = \frac{A^n}{n!}[0] = \frac{A}{n}[n-1]$$

For $i > n$ the factor A/n will remain constant, so we have

$$[n+1] = \frac{A}{n}[n]$$

$$[n+2] = \left(\frac{A}{n}\right)^2 [n]$$

and in general

$$[i] = \left(\frac{A}{n}\right)^{i-n} [n] \qquad (10.31)$$

Again we use the normalizing condition that the sum of all state probabilities is unity:

$$\sum_{i=0}^{\infty} [i] = 1 = [0] \left\{ \sum_{i=0}^{n-1} \frac{A^i}{i!} + \sum_{k=0}^{\infty} \left(\frac{A}{n}\right)^k \frac{A^n}{n!} \right\}$$

where the variable k is used instead of i in the range $i \geq n$.

$\sum_{k=0}^{\infty} (A/n)^k$ is the infinite sum of a geometric series of ratio A/n, which has the value $1/[1-(A/n)]$.

Therefore

$$1 = [0] \left\{ \sum_{i=0}^{n-1} \frac{A^i}{i!} + \frac{A^n}{n!} \cdot \frac{1}{1-(A/n)} \right\}$$

or

$$[0] = 1 \left/ \left[\sum_{i=0}^{n-1} \frac{A^i}{i!} + \frac{A^n}{n!} \frac{n}{n-A} \right] \right. \qquad (10.32)$$

The probability that a call is delayed is the same as the probability that all devices are busy, which is the sum of the state probabilities $[i]$ in the range $i \geq n$, i.e.

$$\text{Prob (call is delayed)} = \sum_{i=n}^{\infty} [i]$$

$$= \sum_{i=n}^{\infty} \left(\frac{A}{n}\right)^{i-n} \frac{A^n}{n!}[0]$$

$$= \frac{A^n}{n!} \frac{n}{n-A}[0]$$

$$= \frac{(A^n/n!)\, n/(n-A)}{\sum_{i=0}^{n-1} A^i/j! + (A^n/n!)\cdot n/(n-A)}$$

This expression, symbolized by $E_{2,n}(A)$, is known as the Erlang delay formula. For purposes of computation it can be expressed in terms of the Erlang loss formula as

$$E_{2,n}(A) = \frac{nE_n(A)}{n - A + AE_n(A)} \tag{10.33}$$

where $E_n(A)$ is given by Eqn. (10.13).

In queueing systems, several parameters other than the probability that a call is delayed may be of interest, e.g. the mean queue length, the mean delay of all calls, the mean delay of delayed calls, maximum delay, and so on. Further treatment of queues related to telephony can be found in Cooper[3] and Fry.[4]

10.18.1 M/M/1 queue

Modern systems depend on software and processes for their control and management. In many of these, and in packet switched systems, buffers are used to queue tasks awaiting processing. All buffered systems can be analysed by using queueing theory, which is a set of appropriate mathematical models from which such metrics as delay time can be determined. The following analysis, expressed in terms of telephone calls to make it consistent with our discussion, is applicable to any analogous queueing system.

Earlier we determined the probability that a call is delayed. Here we will use a simple model to find the mean delay of a call in a queue with an infinite buffer space. With no end to the size of the buffer, it is not possible for calls to be lost due to blocking; although they may disappear through impatience.

In addition to having an infinite length queue we also reduce the number of servers to one, hence we are analysing the M/M/1 queue.

We noted earlier that the probability that the system is in state i is given by

$$\mu_i[i] = \frac{A}{s}[i-1] \tag{10.34}$$

Since there is only one server, μ is fixed and independent of i. In addition, we can use Eqn (10.1) to express A in terms of λ. Thus,

$$\mu_i = \mu \quad \text{and} \quad A = \lambda.s$$

therefore

$$\mu[i] = \frac{\lambda.s}{s}[i-1] \tag{10.35}$$

Putting in values

$$\left.\begin{array}{l} [1] = \lambda/\mu\,[0] \\[2mm] [2] = (\lambda/\mu)^2\,[0] \end{array}\right\} \tag{10.36}$$

and so on to

$$[i] = (\lambda/\mu)^i\,[0] \tag{10.37}$$

We assume here that the queue can grow indefinitely, since $0 < i < \infty$. The normalization equation therefore becomes:

$$\sum_{i=0}^{\infty} [i] = 1$$

i.e.

$$\sum_{i=0}^{\infty} (\lambda/\mu)^i [0] = 1$$

giving

$$[0] = \frac{1}{\sum_{i=0}^{\infty} (\lambda/\mu)^i}$$

hence

$$[i] = \frac{(\lambda/\mu)^i}{\sum_{i=0}^{\infty} (\lambda/\mu)^i} \tag{10.38}$$

We know that λ must be less than μ to avoid the queue growing without limit. The quantity λ/μ is known as the utilization ρ. Using this form,

$$[i] = \frac{\rho^i}{\sum_{i=0}^{\infty} \rho^i}$$

The denominator is the infinite series representation of $(1 - \rho)^{-1}$; hence

$$[i] = \rho^i (1 - \rho)$$

where $[i]$ is the probability that there are i customers in the system at any time.

The mean number of customers in the system, N, is given by the weighted sum of the state probabilities:

$$N = \sum_{i=0}^{\infty} i[i]$$

$$= \sum_{i=0}^{\infty} i . \rho^i (1 - \rho)$$

$$= \frac{\rho(1 - \rho)}{(1 - \rho)^2}$$

$$= \frac{\rho}{1 - \rho}$$

or

$$N = \frac{\lambda}{\mu - \lambda} \tag{10.39}$$

The mean time a customer spends in the system is the average time for delayed and undelayed customers combined, and is the sum of the queueing and service times. This sojourn time we will denote by T. It can be related to N

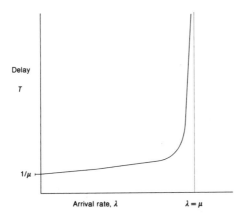

Fig. 10.13 Delay in an M/M/1 queue.

and λ by Little's Law:

$$\lambda T = N \qquad\qquad (10.40)$$

Therefore

$$T = N/\lambda = (\mu - \lambda)^{-1}$$

Using this expression, we can see from Fig. 10.13 that if the mean service time μ is fixed, then T is a non-linear function of λ. We can also see that the minimum mean sojourn time is the mean service time $1/\mu$ and that as $\lambda \to \mu$ the delay becomes very large indeed. This last condition is sometimes called the point of instability of the queue.

Although the M/M/1 is the simplest of all queues to analyse, and although the analysis depends on the assumption of infinite buffer space, it gives a very good insight into the behaviour of queues in general.

10.19 SIMULATION

Modern systems involve a considerable amount of control by microprocessors dedicated to a particular small task and operating on an individual bit or word basis. The sequence of tasks involved in passing a call through the system can be represented as a series of queues, and at each queue there is the possibility of delay. If the traffic is very high, the total delay may become so great that the processors cannot cope and the throughput falls dramatically. Tele-traffic theory has to provide the tools to predict this degradation in performance. However, sufficient analytical power is not available in queueing theory. It provides the facility to study fairly simple problems, but when there are several queues interacting the models can become too difficult to analyse. Resort must then be made to simulation. A computer program is written to represent the behaviour of the system to be studied and it is used in conjunction with some form of simulation package which generates calls in a prescribed way (e.g. with negative exponential inter-arrival and call holding time distributions), provides random number streams and organizes the simulation of the system operation. Simulation is a most powerful tool and it can be used for any system. It relies on:

(i) sufficient computer power;
(ii) a satisfactory software model of the system being studied;
(iii) the correct choice of the behaviour of the traffic offered;
(iv) a proper evaluation of which results to take;
(v) sufficient length and frequency of runs to account for the effects of correlation.

The last condition is one that must be in mind when a simulation is being organized; correlation can produce errors in results if it is not taken into account. In particular, the urge to reduce cost by making runs short should not inhibit the need for long runs to make the results reliable.

Simulation also has disadvantages. It is costly because it requires long runs and therefore considerable computer time, it is time consuming for the analyst because the programs to model the system being studied are often very complex, and the results are not of general applicability since they are related to a particular set of initial conditions such as system dimensions and traffic level.

Until recently, simulation programs were written on the assumption that the traffic offered to the network would be telephony, the representation of which is well understood, as describe earlier. Thus the creation of traffic models was straightforward. Now the interest is in developing systems that can handle many types of traffic, some of which will not be well understood at all. In addition, it used to be assumed that the traffic would be more or less balanced across the network. If the network was a switch then the traffic on each inlet would be the same, and if the network was a representation of the interconnection of exchanges, then the offered traffic at the nodes was assumed to be uniform. This again can no longer be justified. There are two questions here. First, what is an appropriate distribution for each of the traffic types that might be carried, and second, how can simulations be compared when it may not be clear what variation in traffic load is assumed across the network?

There is a general move towards using proprietary software rather than custom-built simulation packages. The advantage is that the proprietary packages are usually well written and well supported, giving a high degree of reliability. The mechanisms for obtaining results are also well developed and simple to use. The disadvantages are that in general such packages are slow, and to some degree inflexible; they do not give the user complete control. Often, experienced users will combine some element of their own programming into the system.

One further difficulty with simulation as a methodology is that it is easy to simulate the model rather than the system. This is of particular concern when using simulation to validate analytical results. If the simulation merely parallels the analysis, creating a representation of the same model, it will not be surprising if the results agree very closely; indeed it would be a cause for some concern if they did not.

10.20 SWITCHING AND SIGNALLING

When a subscriber makes a telephone call a series of events takes place which have the following sequence, illustrated in Fig. 10.14.

1. Calling subscriber lifts handset.
2. Exchange detects demand for call.
3. Check is made to find free equipment in the exchange.
4. When equipment becomes free, dial tone is returned to the caller. If

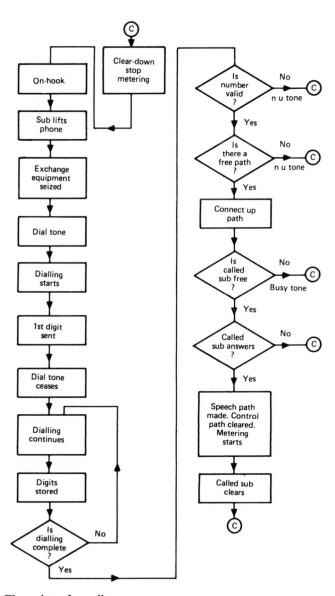

Fig. 10.14 Flow-chart for call set-up.

equipment is not immediately available at the exchange there may be a long dial-tone delay.

5. Subscriber dials digits.
6. Exchange interprets the digits and routes call to its destination.
7. Check is made to determine whether or not the called subscriber is free.
8. If called subscriber is busy, busy tone is returned to caller.
9. If called subscriber is free, ringing tone is returned to caller and ring current is sent forward to ring called subscriber's bell.
10. If called subscriber is unavailable, caller hangs up and equipment clears down.
11. If called subscriber answers, ring tone and ring current cease.
12. Speech path is set up and metering commences.
13. When conversation ceases called subscriber hangs and metering ceases.
14. Calling subscriber hangs up and system clears down.

Note that if an unallowed number is dialled, or service to the number dialled has been discontinued, number-unobtainable tone is returned to the caller.

From the above description we can see that the process of signalling has at least three functions:

(i) To indicate the state of the call to each subscriber by

    (a) dial tone,
    (b) ring tone and ring current,
    (c) busy tone or number-unobtainable tone.

(ii) To tell system what to do next, by indicating the path for the call, for example.
(iii) To initiate a billing procedure – usually by tripping the calling subscriber's meter at the correct charging rate – to enable the administration to gather the revenue needed to provide the service.

In the early days of telephony when calls were handled by an operator, these signalling functions were done manually. Indeed in small communities having up to 200 lines, or so, the operator provided an automatic call-transfer facility of some sophistication: the more interest the operator had in local affairs the better that part of the service could be!

The flow-chart of Fig. 10.14 applies particularly to the setting up of calls that share the same exchange. Although most calls are still of that type, in many large cities an increasing number of calls involve inter-exchange working, and the use of direct-dialling systems has created a greater demand for trunk calls. Inter-exchange signalling is necessary to allow the calling subscriber's local exchange to keep control of the call, and later we shall look in some detail at the signalling systems used for the purpose. Evidently, inter-exchange systems are not restricted in speed as are the subscriber's signalling systems by the dialling process, and they are not required to produce the tones which the local exchanges must produce to keep the subscriber informed of the progress of the call.

10.21   TYPES OF SIGNALLING SYSTEM

We shall not look at any of the historical systems, but we shall look at those which are in use. They fall into four main categories:

  (i) loop disconnect – dc signalling;
 (ii) multi-frequency – ac signalling;
(iii) voice frequency – ac signalling;
(iv) common channel signalling.

10.21.1   Loop disconnect signalling

Until recently, the universal means at the subscriber's disposal for indicating the number he wished to call was the telephone dial, and it is still present on many handsets. It was developed to its present state many years ago, but although robust, it has a major disadvantage: it operates very slowly by the standards of modern electronics, and its slow motion places a definite limit on the speed at which signals can be sent to the exchange.

When the dial is rotated the finger hole corresponding to the required digit is pulled round to the finger guard and then the dial is released. A governor inside the dial causes it to rotate back automatically at a fixed speed, causing a series of pulses to be sent down the subscriber's line. After the dial has returned to its rest position the next digit is dialled. The time between the last of the pulses for one digit, and the first pulse of the next is called the inter-digit pause. It is this pause that allows the exchange to recognize the end of a digit. A typical sequence, with average operating times (nominally the pulses are sent at a rate of 10 per second) is shown in Fig. 10.15. Although the dials are designed to operate at a nominal 10 pulses per second, the exchange equipment will still function reliably if the pulse speed is between 7 and 12 pulses per second and the break is between 63 and 72% of the pulse period.

10.21.2   Multi-frequency (mf) signalling

Modern handsets are fitted with key-pads instead of dials to facilitate a much more rapid transfer of signals between handset and exchange. Generally the keypads send out frequencies instead of pulses to represent a digit. Most systems use two frequencies to represent a particular digit. Whether or not the much increased speed of mf signalling can be realized depends on the local exchange. An mf key-phone working into a Strowger-type exchange gives no overall improvement, it just changes the position in the process at which the

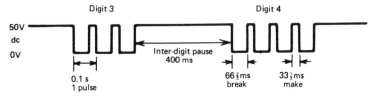

Fig. 10.15 Ideal output from telephone dial.

caller must wait: when a dial is used there is a long inter-digit pause and short post-dial delay, whereas in an mf system there is little inter-digit pause because the numbers can be keyed in very quickly, but if the exchange equipment operates at 10 pulses/s the mf signal must be converted into normal digits producing a lengthy post-dial delay.

However, in modern exchanges the equipment will respond directly to the mf signals and the call can be set up very quickly.

10.21.3 Voice-frequency (vf) signalling

The use of dc signalling is limited by several factors. In particular, over long lines the variation of equipment performance, and the effect of the characteristics of the lines used to transmit between exchanges, degrade the pulse shape of the dialled digits to such an extent that errors can be caused. In addition, the receiving equipment requires a higher voltage than would be necessary if the system used alternating current signals. Apart from these considerations multi-channel working makes dc signalling inappropriate.

The normal telephone channel occupies a bandwidth of 300–3400 Hz, from the range of 0–4000 Hz which it is allocated. If ac signalling is to be used it must operate at frequencies within this range, and it is therefore known as voice-frequency signalling. The frequencies used can be either inside the normal speech band (in-band signalling) or outside that band and within the 0–4000 Hz range (out-of-band signalling). As the frequencies used for vf signalling are the same as those used in speech, special care must be taken to ensure that the two functions do not interfere. By understanding something of the characteristics of speech it is possible to specify signalling systems that will not be operated erroneously by speech frequencies. Since signalling is done by tones within the baseband of the telephone channel it is possible to use vf signalling when the channels are multiplexed on to a common carrier – either line or radio.

10.21.4 Common channel signalling

An alternative to a signalling system being associated individually with a speech path is to couple the signals for a large number of calls together and send them on a separate common signalling channel. This method of signalling is particularly useful for inter-exchange working, when the signals will be sent over a data-type high-speed link, and in PCM systems. In the latter, one channel is dedicated to signalling in a 32 channel system. The signals must be coded so that they can be related to a specific speech channel. There is an inherent security problem in common channel signalling, so provision should be made for a second signalling channel to be used in the event of failure, and an error-detection facility must be provided.

10.22 VOICE-FREQUENCY SIGNALLING

As noted in Section 10.21, vf signalling can be either in-band or out-of-band (although some writers refer to in-band as vf signalling, and treat out-of-band separately).

10.22.1   In-band signalling

Since the signalling frequency is within the 300–3400 Hz speech bandwidth there are obvious problems associated with this signalling method. It cannot be operated during speech and the equipment must be able to distinguish between a speech pattern and a signal. There are two parameters available for variation: the signal frequency and the signal recognition time. Other considerations that will assist in distinguishing between speech and signal are:

(i) speech at the signal frequency is accompanied by other frequencies;
(ii) more than one signal frequency could be used;
(iii) the signals could be coded bursts of the signal frequency.

10.22.2   Choice of frequency

This is related to the frequency characteristics of speech. As shown in Fig. 10.16, the energy in English is predominant at lower frequencies (maximum at 500 Hz or thereabouts) and it falls gradually over the rest of the band. This suggests that a high signal frequency should be used to reduce the possibility of imitation by speech frequencies. However, there are other considerations that suggest that frequencies at the upper end of the band should not be used. At the higher frequencies there is an increase in crosstalk, so low-level transmission is necessary. If the amplitude is low the receiver must be very sensitive, raising the prospect of imitation signalling by low-level speech. The channel characteristics may vary at the upper end of the band on different links so that on some older links the cut-off may be below 3400 Hz. In addition, the variation of amplitude with frequency is quite marked at the high frequencies so any change in signal frequency will result in a significant change in signal amplitude. Between these considerations some compromise is necessary and in practice the frequency chosen lies in the range 2040–3000 Hz.

10.22.3   Signal duration

By delaying the recognition of the signal until it has persisted for some time the chance of signal imitation is reduced significantly. Using a guard circuit (see below) and a 40 ms recognition delay the probability of the signal receiver responding to a speech frequency is reduced to a low level. Figure 10.17(a) shows the degree of improvement that may be possible as the recognition time is increased.

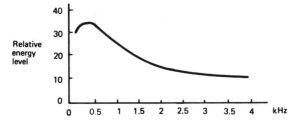

Fig. 10.16 Variation of energy with frequency in English speech.

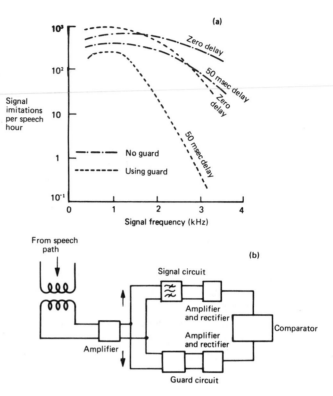

Fig. 10.17 (a) Effect of guard circuits, with and without delay; (b) block diagram of system using guard circuit.

10.22.4 Guard circuit

As can be seen in Fig. 10.17(a) the effect of signal recognition delay is greatly enhanced if a guard circuit is used in the receiver to increase the rejection of imitation signals. Figure 10.17(b) shows the basic features of such a system. The circuit shown there forms part of the vf receiver shown in Fig. 10.18.

The energy coupled from the speech channel is amplified and then passes along two paths:

(i) through a bandpass filter tuned to the signal frequency, forming part of the signal circuit;
(ii) through the guard circuit which includes a bandstop filter allowing all but the signal frequency to pass.

The outputs from (i) and (ii) are compared. When a signal arrives the signal circuit responds strongly and the guard circuit weakly, leading to a strong positive signal being detected by the signal-detector circuit. When speech is present, any imitation signals passing through the signal circuit are attenuated by the strong signal passing through the guard circuit, thus reducing the chance of spurious imitation.

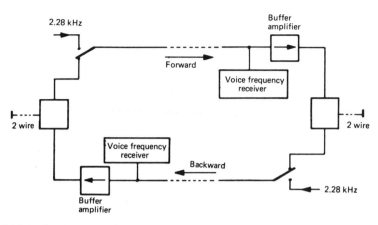

Fig. 10.18 Voice-frequency signalling system.

10.22.5 Basic layout of the system

Two-way working is achieved by combining two forward wires and two backward wires into a four-wire system (Fig. 10.18). The tone generators are switched in as required and the line is monitored constantly by the vf receivers for the presence of a signal.

Echo suppression can be used if required.

When employed in a multi-link connection, signalling is generally done on a link-by-link basis and it is important that signals from one link do not leak over to the next. To prevent such leakage, line splitting is used. There are various methods of performing the split; a common one is the buffer amplifier which, on the signal being detected by the vf receiver, is biased below cut-off, which effectively breaks the line after the receiver while the signal is in progress. The buffer also protects the receiver from local-end interference. Because the line splitting is initiated by the receiver, it is necessary to reduce the chance of signal imitation, so recognition delay is also used at this point. The receiver on detecting a signal frequency, waits for 2 ms before initiating the line split. During that 2 ms there will be leakage into the next link, but it will not be a problem as it is much less than the normal recognition delay of 40 ms used in the receiver.

Line splitting may also be inserted at the transmit end of each direction to prevent switching transients passing down the line at the same time as the signal.

10.22.6 Types of vf signalling

There are two basic types, pulse and continuous, and the recognition of a particular signal can be based on:

 (i) the signal direction;
(ii) the position of the signal in a sequence;

(iii) the frequency content;
(iv) its length.

Pulse signalling

The recognition of a signal is determined from its length and its sequence. The following points can be made about pulse signalling:

(i) It has a higher signal repertoire than continuous signalling.
(ii) It can be transmitted at a higher voltage level and therefore provides a better SNR.
(iii) It is less influenced by interference.
(iv) It complicates the dc/ac and ac/dc conversions because the pulses have to be carefully timed.
(v) It requires a memory facility at the receiver for pulse recognition.

EXAMPLE: The UK AC9 signalling system is typical of that used in many countries. The signal frequency is 2280 Hz and the pulses used are shown in Table 10.2.

The calling subscriber's clear signal should initiate the backward release-guard signal. If the release guard is not received, the outgoing end sends seizure followed by forward clear until release guard is received.

A continuous tone busies the sending end relay set (busy back signal) and raises an alarm if it continues for several minutes.

The release-guard facility ensures that calls are not passed down to the receiving end if there is a fault. When a fault occurs and the forward-clear signal is sent, the outgoing equipment is guarded from accepting another call until the release-guard signal is received. The repeated use of the seizure and calling-subscriber-clear signal will initiate the release-guard signal once the fault has cleared.

Note that the backward answer signal and the called-subscriber-clear signal have the same pulse length, but they are distinguishable by their place in the sequence.

Metering is stopped by either the forward-clear or backward-clear signals.

Continuous signalling

There are two types of continuous vf signalling: (i) compelled and (ii) two-state non-compelled. Both are used, (i) in CCITT system 5 and (ii) in Bell SF. In (i) a

Table 10.2

Signal	Duration (ms)
Seizure (forward)	65
Dial pulsing (forward) (loop disconnect)	57
Called subscriber answers (backward)	250
Called subscriber clears (backward)	250
Calling subscriber clears (forward)	900
Release-guard (backward)	1000
Busy (backward)	Continuous

signal ceases when an acknowledgement is received, whereas in (ii) the signal information is carried by the change in state, and acknowledgement is not used.

In terms of reliability (i) is preferable to pulse, which is preferable to (ii), but (i) is much slower than (ii) because it depends on acknowledgement signals, and hence on propagation time. This slowness of (i) means that it is used only in special applications. In the second continuous mode, (ii), the tone is on for much of the non-speech time, which can lead to overload on the transmission system if the signal level is not limited.

10.23   OUT-BAND SIGNALLING

This term is generally applied to a signalling system in which the signal frequency is in the range 3400–4000 Hz. The CCITT recommended frequency is 3825 Hz, although 3700 and 3850 Hz are also used. This method is applicable only to carrier systems because the equipment used for baseband transmission may attenuate the signal frequency.

Compared with in-band signalling there are two advantages:

(i) There is no need to take steps to avoid signal imitation from speech-guard circuits and line splits are not required.
(ii) Signals and speech can be transmitted simultaneously.

The general layout of an out-band signalling system is shown in Fig. 10.19.

Signalling across the switching stage is usually dc, hence out-band signalling is link by link with ac/dc, dc/ac converters at each end. Filtering is used as

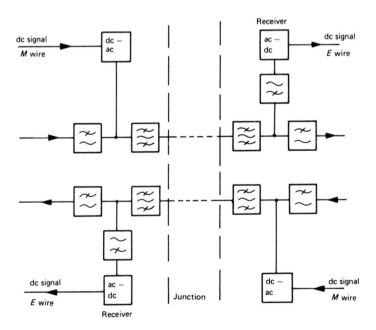

Fig. 10.19 Transmitter and receiver for out-band signalling.

Table 10.3

Signal	Tone	
	Forward	Backward
Idle	Off	Off
Seizure	On	Off
Decadic address	Off (pulse breaks)	Off
Release	Off	On when clear forward first
Answer	On	On
Clear Back	On	Off
Release guard	Off	On
Blocking	Off	On

shown to ensure that there are no spurious frequencies and to isolate the signal before the receiver.

This type of signalling has the advantage of simplicity and consequent cheapness. It can be either two-stage continuous or pulse. There are inherent difficulties in each mode; continuous cannot be at too high a level because of the risk of overload, so the receiver needs to be sensitive, whereas pulse requires more sophisticated circuitry to perform the memory function. On balance, the continuous mode is preferred because of its simplicity (compared with in-band signalling where the pulse mode has advantages).

Out-band signalling is more attractive than in-band, but cannot always be applied without considerable expense to existing transmission plant; consequently it is commonly used in new fdm systems, but vf is the most common on existing systems.

There are two modes of continuous signalling, tone-off idle and tone-on idle, and there is little to choose between them for many applications, although tone-on idle would be preferred if the system were in general use because tone-off idle would tend to overload the transmission system.

A typical signal code for a tone-off idle system is shown in Table 10.3.

10.24 SWITCHING NETWORKS

In switching systems the number of cross-points is a good measure of cost, and part of the effort involved in switch design is concerned with reducing the number.

An idea of the way in which the number of cross-points can be reduced, and the effect of that reduction on network performance, can be obtained from the following simple approach.

If there are N lines into and N lines from a single switch the number of cross-points required is N^2, e.g. the matrix shown in Fig. 10.20 requires 10 000 cross-points.

This large number of cross-points allows any free inlet to be connected to any free outlet regardless of the connections made between other inlets and outlets – a full availability lossless system.

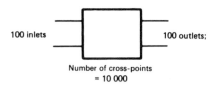

Fig. 10.20 Full-availability switch.

However, each inlet, if it is a subscriber's line, will carry a very small load, probably less than 0.1 erlangs, so it will be busy for only 6 min on average during the busy hour. This means that there will be fewer than 100 calls in progress simultaneously and therefore the system could operate with many fewer cross-points.

As a first step towards reducing the size of the network the matrix can be split into two parts; the first part having 100 inlets and 25 outlets and the second part having 25 inlets and 100 outlets [Fig. 10.21(a)]. This reduces the number of cross-points by 50%.

A further saving can be made if the inlet and outlets switches are further subdivided, as in Fig. 10.21(b). However, this last division raises serious problems. Each inlet switch has access to only one outlet switch so, for example, inlet switch 3 cannot route a call to outlet switch 5, etc. One way round this diffficulty would-be to interconnect the links between the inlet and outlet switches in the manner shown in Fig. 10.22, but that itself has the drawback that only one of the calls from an inlet switch can be connected with a particular outlet switch.

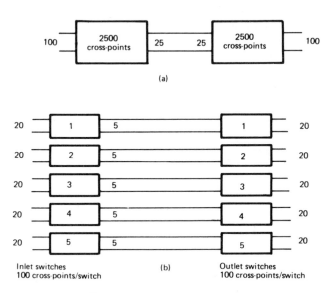

Fig. 10.21 (a) Cross-point reduction by vertical partitioning, (b) Cross-point reduction by vertical and horizontal partitioning.

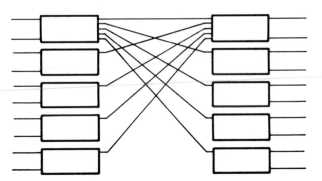

Fig. 10.22 Diagonal link connections to increase availability.

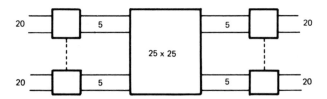

Fig. 10.23 Three-stage link system.

It seems that the reduction in availability has been too great and that more cross-points need to be used to make the system less restrictive. A middle stage can be included to make the distribution of paths between the inlet and outlet stages more satisfactory. This distributor stage will, in this example, have 25×25 cross-points as shown in Fig. 10.23.

In the same way as was done with the inlet and outlet stages, the distributor stage can be subdivided into five 5×5 switches and the link pattern between first and second stages, and second and third stages arranged to provide five paths through the network from any first stage (inlet) switch to any third stage (outlet) switch (Fig. 10.24).

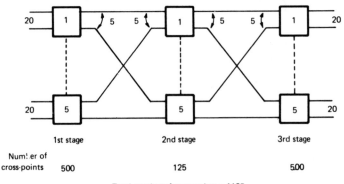

Fig. 10.24 Cross-point reduction on middle stage. Total number of cross-points = 1125.

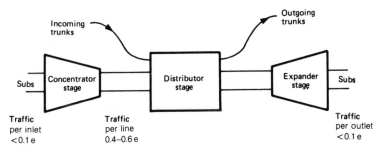

Fig. 10.25 Functional diagram of three-stage link system.

Clearly there has been a very large reduction in the number of cross-points. The chance that a call is lost because of the reduced availability is a question that can be answered by calculating the internal blocking of the network at given traffic levels, using the Jacobaeus equation, or some other suitable method.

Considering once again the three stages of the network in Fig. 10.24, we can see that the first stage has many more inlets than outlets (100 and 25, respectively), the third stage has many more outlets than inlets, and the second stage has an equal number of inlets and outlets. Figure 10.25 represents this aspect of the three stages diagrammatically. For obvious reasons the stages are named:

First stage   many more inlets than outlets – concentrator
Second stage   same number of inlets and outlets – distributor
Third stage   many more outlets than inlets – expander

This division into concentration, distribution and expansion stages typifies many switching systems. Subscribers' lines, with very low traffic per line, are concentrated to high-usage circuits operating at about 0.4–0.6 erlangs, and the distribution stage carries that traffic. Subscribers' lines are attached to the outlets of the system so there must be an expansion stage between the distribution stage and the subscribers.

Trunk circuits, which are designed for high traffic, come directly to, and go directly from, the distribution stage, as shown.

10.25 BASIC ANALOGUE SWITCHING SYSTEMS

Switching systems can be classified in several different ways, but here we shall restrict our interest to separating them on the basis of their method of control. Two methods are in common use: (i) step-by-step; or (ii) centralized control.

10.26 STEP-BY-STEP

In step-by-step systems, the control path and speech path are the same. As each digit is dialled, the connection is made to one further stage, until the final selection when two digits are required.

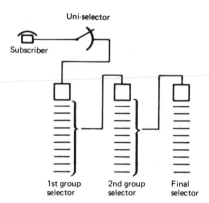

Fig. 10.26 Diagrammatic representation of a step-by-step system.

Strowger-type systems operate in this way. As an example consider a local exchange call on a four-digit number system (Fig. 10.26).

When the caller goes off-hook, the line circuit connects his line to a free first numeral selector (group selector). This connection may take some time, and not until the first selector is seized is dial tone returned to the caller. The dial-tone delay is a measure of system capacity from the subscriber's point of view, and most administrations specify a maximum value for this time. If it is too long, the subscriber feels that the system is not responding to his demand for service and he may hang-up.

Having seized the first selector the control waits for the first digit to be dialled; on its receipt the first selector wipers are racked up to the corresponding level and hunt round for a free outlet to a second selector. The second digit causes the second selector to go through a similar process and seize a final selector. The final selector has subscribers' line circuits attached to it, and so each of its hundred contacts represents a different subscriber. Consequently it needs two digits for its operation.

If at some stage during the setting up of the call, after the dial tone has been received a free selector is not available, a busy tone is returned to the caller; the wipers, having hunted over all the outlets to find a free selector at the next stage, without success, automatically return to their home position. In this method of control, part of the path is set up and then the system waits for more information before extending the path a little further and waiting again, thus setting up the call in a step-by-step manner.

10.27 COMMON CONTROL

The step-by-step system described earlier operates on the principle that the control of the call follows the same path as the speech circuit. This means that all the control equipment is provided on a per-call basis. However, some of it need not be, and savings can be made by separating some of the control from the speech path. It can then be used only as required by a call, and can then

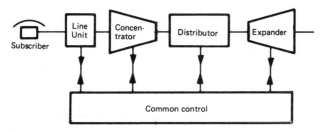

Fig. 10.27 Use of common control.

become available for other calls. For example, the register–translator, which is only required during the routing, or path selection process, could be placed in a common-control area, used to set up a path between caller and called, and then released for use in setting up another call. Figure 10.27 shows a diagram of a common-control system.

The switching block will not be of the Strowger-type, but either a cross-bar switch or a reed-relay system. A cross-bar switch consists of a matrix of horizontal and vertical conductors that can be made to interconnect at any required cross-point by the vertical bar trapping a metal finger attached to the horizontal bar when the latter is tilted by a relay operated by the marker-circuit in the common-control. The reed-relay switch is an 'electronic' cross-point consisting of two contacts inside an evacuated envelope and surrounded by an inductive coil. When the coil is energized it induces a magnetic field which forces the contacts together. The reed-relay switch is also of the matrix type with horizontal and vertical wires, the interconnection again being controlled by the marker.

Although systems differ in their arrangements and in the names used to identify particular units, Fig. 10.28 shows the basic form of a common-control exchange.

Subscribers are attached directly to a subscriber's line unit which recognizes a request-for-call condition. When the calling subscriber goes off-hook, the line unit indicates the request to the common control that seeks a free outlet from the first switching stage, which is a concentrator, and returns dial

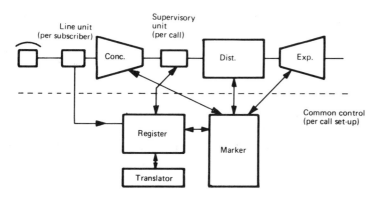

Fig. 10.28 Centralized control exchange.

tone to the caller. Each outlet from this first stage of switching is connected to a call-monitoring (supervisory) unit that maintains an awareness of the call for metering purposes, and reacts to a clear-down signal. The calling subscriber dials and the digits are stored in a register in the common control. If translation is required, the register passes the digits to a translator when one becomes available, and the translation is sent to the marker. The marker examines the switching units for a free path to the called outlet and tests the outlet to find out if it is free. If it is not, busy tone is returned to the caller. If it is free, a path through the switches is set up, ring-current is sent to the called subscriber and ring-tone is sent back to the caller. The common control is released and the progress of the call is noted by the call-monitoring unit.

There will usually be fewer translators provided than registers and many fewer registers than call monitoring units, the provision being proportional to the relative holding times of these elements except that in an exchange which requires only one translator, two may be provided for security reasons.

There is considerable saving in expense if the most provided control equipment, the subscriber's line unit, is made as simple as possible, and more complex features of the control are placed in those units that are required on a per set-up basis only, like the registers and markers.

10.28 MULTI-FREQUENCY SIGNALLING

The rather tedious procedure of dialling long numbers can be alleviated by using a key-pad, which allows the caller to press the required number instead of operating the dial (Fig. 10.29). This method reduces the dialling time considerably, and since the signalling is no longer at 10 pulses/s it too can be speeded up. A digit is signalled by a unique combination of two frequencies, one from each of the rows in Fig. 10.29. Detection at the exchange is carried

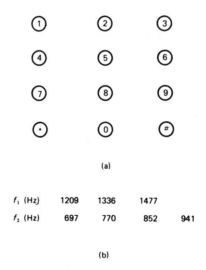

(a)

f_1 (Hz)	1209	1336	1477	
f_2 (Hz)	697	770	852	941

(b)

Fig. 10.29 (a) Key-pad arrangement; (b) frequencies used in key-pad.

out by using frequency filters. The frequencies are within the voiceband and care must be taken to reduce the risk of speech imitation. As we saw in our discussion on in-band vf signalling, there are two approaches; one involves a guard circuit and the other, signal recognition delay. Both of these methods are used.

Another benefit of the key-pad, which will be exploited in digital systems, is that there are more combinations of frequencies than the ten required for digits. If there are 16 possible combinations – one frequency from four, twice – there are five potential combinations that could used to signal other information to the exchange. Some systems have only two additional unused combinations because the pair of frequencies are chosen from a group of four and a group of three, giving 12 in all.

The development of common control based on computer operation removed some of the constraints on inter-exchange signalling. Very rapid reception, detection and processing were possible, and for signalling between processors care need not be taken to counter speech imitation. CCITT No. 4 was developed as a fast-inter-exchange mf signalling method based on transmitting signals as a combination of two from five or two from six, depending on the repertoire required.

10.29 THE TELEPHONE NETWORK

The telephone network has developed dramatically in recent years so that it is now possible to make calls automatically between subscribers separated by thousands of miles. Long-distance calls pass through several stages of switching and several possible transmission links before reaching their destination, and to make such calls possible many facets of telecommunications must be integrated and reasonable compromises made by systems designers. In practice, there are very few occasions when a telecommunications administration has the opportunity to design a network from the beginning. Usually some sort of network exists already and there is a demand for increased service: more lines, better performance, more facilities, more comprehensive direct dialling, etc. This demand has to be satisfied by grafting on to the existing system new switching and transmission equipment that may be totally different in type and in its principles of operation. Recent years have seen two revolutionary developments: the integrated circuit leading to cheap, reliable, small and very powerful digital systems; and the optical fibre, which is much better in many respects than conventional transmission media. These two have led to an entirely new approach, and enormously enhanced facilities for future systems. One of the major planning problems is how to graft such new systems on to existing equipment. We shall return to this point later.

Although the structure of telephone networks has developed piecemeal as demand has increased, it still has some identifiable form. The passage of a call through a national network can be represented by the multi-level diagram shown in Fig. 10.30. Before going further, some comment is necessary on the terminology to be used. The words used in North America and the UK differ

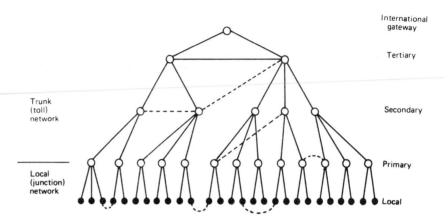

Fig. 10.30 Hierarchical routing system.

Table 10.4 Terminology

CCITT	USA	UK
Primary centre	Toll centre	Group switching centre
Secondary centre	Primary centre	District switching centre
Tertiary centre	Sectional centre	Main switching centre
Trunk exchange	Toll office	Trunk exchange
Trunk network	Toll network	Trunk network
Trunk circuit	Trunk	Trunk (circuit)
Local exchange	End (central) office	Local exchange
Junction circuit	Inter-office trunk	Junction

significantly in their meaning. Here we will refer to the various levels of switching by using the CCITT recommended terms. Table 10.4 relates them to those commonly used in North America and the UK.

From Fig. 10.30 we can see that there are several levels of switching that combine to form the complete network. It is usual to think of the systems in two parts. The first is the junction network, serving the subscriber and consisting of the link from subscriber to local exchange, from local exchange to primary switching centre, and back to the called subscriber via another local exchange. The second part of the system is the trunk network, which is concerned only with calls passing at primary centre level and above. Thus the primary centre is associated with both parts of the network.

The structure shown by the solid lines represents the basic hierarchical network of the system, and telephone calls routed along these links are said to be travelling on a backbone or final route. When there is much traffic between exchanges, either belonging to the same cluster, or at different levels, direct routes may be installed to save a stage of switching. Such routes are shown as broken lines in Fig. 10.30. These direct routes are dimensioned to a different grade of service from the final routes. Because a call finding all the direct-route circuits busy can be placed on the backbone route, the direct route can be dimensioned to a higher grade of service than the final route, and thus work more efficiently.

The number of exchanges at the various levels depends on several factors: the physical extent of the network, the number of subscribers, the amount of traffic, the forecast growth and the transmission methods used. Beyond the top level of the national system is a layer that gives access to the international network. This layer may consist of one or more international (usually called gateway) exchanges.

10.30 NUMBERING SCHEMES

In modern systems, the numbering scheme used by a telephone administration to allocate subscribers' numbers has an underlying plan, and there are a few constraints on the development of the plan that must be taken into account:

(i) It must provide each subscriber with a unique number within the national network.
(ii) The allocation to areas must be able to meet forecast growth for several decades.
(iii) The number of digits should not exceed that recommended by CCITT.

In principle, (i) is easy to satisfy if (ii) and (iii) have been met. The length of the number recommended by CCITT is $11 - n$ where n is the country code (see below). If, for example, n is 2, then the national number should not exceed nine digits in length. These digits are used to denote the subscriber's number on the local exchange, the exchange within a given area, and the area within the national numbering scheme. In many local exchanges there is a maximum capacity of 10 000 lines, thus the last four digits of the national number are allocated to the subscriber's number in the exchange. Of the remaining five digits in our example, the first two would denote the area and the remaining three the exchange within the area. Thus the number has the form shown in Fig. 10.31.

The division of digits between area exchange and subscriber's code may not be the same as that shown – but all three components exist in every number.

For automatic long-distance dialling a prefix is necessary to indicate to the exchange equipment that a trunk call is being made. In many countries a '0' is used, but any other digit would do. In calculating the length of the national number the prefix is not included.

Fig. 10.31 National telephone number.

10.30.1 International number

In the introduction to this chapter we imagined a future when person-to-person calls could be made very easily via individual instruments and satellite links. For such availability of connections to be possible each subscriber in the world must have a unique number. To achieve that, some agreement between countries is essential; each country must be identified by a number different from that of all other countries. That is achieved by agreement through CCITT on the way in which these codes, called country codes, are allocated.

The first digit of the country code is the zone code; the world is divided up into nine zones and each country belongs to a zone. The relationship between zone number and geographical area is shown in Table 10.5.

In all but two of the zones, one or two digits are added to the zone number to produce the country code. For example, Brazil has a zone number 5 and an additional country code digit 5 to give its country code 55. Brunei is in zone 6 and two further digits are added to give a country code 673.

The two exceptions are zones 1 (North America and the Caribbean) and 7 (USSR). Throughout each of these zones there is a linked numbering scheme that means, for example, that no subscriber in Canada has the same national number as a subscriber in the USA. Consequently, to connect to anyone in zone 1 the digit 1 is followed by the national number. A similar situation exists in the USSR. Europe is at the other extreme; there are many countries with large national networks that have nine digits in their national numbers. For these, a two-digit country code is required, and that can only be achieved by having two zone numbers allocated to Europe.

The division of the world into the zones shown in Table 10.5 is intended to be satisfactory until early in the next century, but clearly, as some large countries develop their telephone networks, some adjustment will be necessary at some future time.

The national and international numbering schemes we have discussed above are the simplest. However, in several parts of the world there are small exceptions, particularly in regard to local calls. In the scheme where the national number is used for all calls within a country, it can lead to irritation on the part of the subscriber and long set-up times for the exchange equipment. Consequently, in many countries local calls use a shorter code. For calls within the same area, the area code is omitted, and for small single-exchange areas no exchange code is used for own-exchange calls.

Table 10.5 Zone numbers

1	North America	6	Australasia
2	Africa	7	USSR
3	Europe	8	Eastern Asia
4	Europe	9	Far East and Middle East
5	South America	0	Spare

Coupled with this last arrangement will be a very short code for calls to adjacent exchanges; these arrangements are particularly well suited to rural areas. The disadvantage of short codes is that they change with the location of the calling subscriber, and therefore a short code directory must be available in each exchange area.

The above description relates to the current practice of allocating numbers to premises. In countries where the telephone network is a public utility, the creation of a numbering plan and the allocation of numbers are the responsibilities of the network operator, the PTT. However, in a privatized, non-monopolistic environment the responsibility for numbering will fall to some other organization. It could be the regulator, or a government ministry, but it is not feasible to allow one operator, among several, to have control of numbering since that would create an intolerable degree of market advantage.

As the number of operators, services and service providers increases there is a substantial demand for the allocation of more numbers. For example, the widespread use of facsimile, dialled up in the same way as a telephone call, has placed a considerable pressure on numbering plans.

There is likely to be an even more significant change in the near future. In anticipation of a general use of mobile systems, plans have been developed to relate numbers to people, rather than to premises. A personal numbering scheme will probably be introduced so that access can be made via any handset, identification of the user being established by using a smart card or similar electronic code that can be entered by the user; this mechanism would also ensure correct billing.

10.31 ROUTING CALLS

The early type of switching equipment, called step-by-step or Strowger, operated by using the dialled pulses to move the selectors to the position corresponding to the digit dialled. In many ways this was an excellent system, but one major disadvantage was that it allowed no flexibility in the way calls were routed – the route was predetermined by the dialled digits. Although some systems were modified to overcome this problem it was not until common-control equipment became widely used that the path a call took between calling and called subscribers could be chosen to allow the most efficient use of the available capacity in the system. The function of the common control in the routing process was to store the dialled digits in a register and then translate them into routing digits which would indicate to the switching system the path to take through the network. This register–translator combination is essential to automatic trunk and international dialling schemes; it allows the telephone administration to manage the system efficiently by changing routes as circumstances alter without having to change subscribers' numbers. This therefore separates the subscriber from the system. The subscriber dials the national number from any location and the register – translator automatically selects an appropriate route.

10.32   DIGITAL SYSTEMS

The telephone network worldwide is essentially digital, having developed from being predominantly analogue in the 1970s. The change was brought about by a combination of technological and theoretical improvements, and it gave rise to the notion that by using a digital representation of voice, the same system could be used to carry data.

The first mainstream digital exchanges were installed in the late 1970s, and they will be superceded by the new systems, based on packet rather than circuit switching, that are being introduced on an experimental basis in the mid-1990s. They are the initial steps towards a high-speed, multi-media network that will carry various forms of voice, data and video traffic.

In this chapter we will concentrate on the switching and transmission used in the public switched telephone network (PSTN), including the integrated services digital network for voice and data, known as ISDN. However, the concluding sections will introduce the intentions and principles underlying the new broadband network, B-ISDN that will depend on the synchronous digital hierarchy, SDH, and use a high-speed packet format, called asynchronous transfer mode, ATM.

At the outset it is worth emphasizing the distinction between ISDN and B-ISDN. There are several differences, but in the context of this chapter the most imporant is that ISDN is a natural extension of the fixed channel, circuit-switched, 64 kb/s telephone network, whereas B-ISDN uses the totally different technology of high-speed packet switching.

10.33   PCM TELEPHONY

In Chapter 3 the principles of PCM were discussed so the details will not be repeated here. However, it is worth reiterating why the building block for the telephone network is a 64 kb/s channel. The basic system is shown in Fig. 10.32. The analogue output from the microphone in the handset is filtered to ensure that it is band-limited to the range 300–3400 Hz. The filter output is sampled at 8 kHz/s, giving 125 µs between samples. Later, we will refer to this sample separation time (or frame time as it is usually called) as one

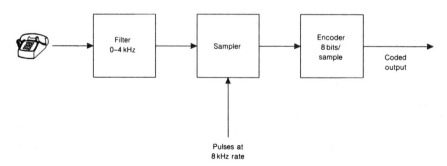

Fig. 10.32

of the parameters that limits the flexibility of the network; it is essential that samples from a particular telephone arrive at precisely 125 µs intervals.

Returning to the PCM system, a sample is changed from an amplitude into a series of 8 bits in an encoder. Thus the bit rate used to carry the signal representing the speech sample is 8000×8, or 64 kb/s. For the standard PSTN this channel bit rate is fixed.

There are two transmission systems used world-wide, one with a 24 channel format and the other based on 32 channels. The former is used in North America and Japan, and the latter in the rest of the world. The larger format operates with a frame of 32 timeslots, each one equivalent to a speech channel, although only 30 of the slots carry speech. The other two are used for sychronisation and signalling as described below. All 32 timeslots must fit into the 125 µs frame. A timeslot is therefore approximately 3.9 µs long. Sixteen frames are combined to form a multiframe which takes 2 ms to transmit.

The network uses the signalling information to ensure that the speech channels are sent to the correct destination and that release, cleardown and metering are carried out at the appropriate times.

Two modes of signalling are used in digital systems: channel associated and common channel. The second, which relates to digital links between exchanges, will be described later. Channel associated signalling is used on the PCM transmission path when traffic is arriving from different sources such as analogue connections, or a multiplexor. In this case, signals are associated with each channel by allocating in each multiframe one signalling half-word in timeslot 16 to each channel; thus the 8 bit signalling slot contains signalling information of 4 bits for two channels in each frame.

Figure 10.33 shows that in all but the first frame (frame 0), timeslot 16 is used to carry signalling. Hence in frame 1, timeslot 16 carries the signals related to speech channels 1 and 16, in frame 2 it carries those for channels 2 and 17, and so on until frame 15 when it has signals for channels 15 and 30. The next frame is the first of the following multiframe and the sequence is repeated.

Timeslot 0 in each frame, and timeslot 16 in frame 0 carry synchronization and alignment words to ensure that the transmission and reception of the system are in step.

When common channel signalling is used the above arrangement is not relevant. Instead, the capacity of timeslot 16 is made available to the signalling packets as required, and it is likely that some of the signalling information is not related to the traffic being carried on the voice channels.

The detailed relationship between the multiframe and the timeslots is shown in Fig. 10.33.

Digital exchanges consist of the basic components shown in Fig. 10.39. Most of the control of the system is handled by microprocessor devices, singly or in clusters, which are driven by software. Here we are not able to discuss the huge new field of telecommunications software engineering, but it provides the most important challenge in modern system design. The software must be efficient, reliable, secure, understandable and well documented. In theory it affords a degree of flexibility in the operation of the system which is

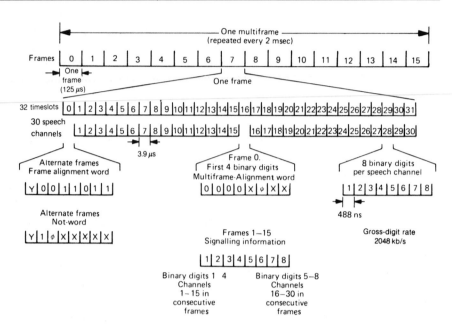

Fig. 10.33 32-frame PCM multiplex frame arrangement: *x*, digits not allocated to any particular function and set to state *one*; *y*, reserved for international use (normally set to state *one*); 0, digits normally *zero* but changed to *one* when loss of frame alignment occurs and/or system-fail alarm occurs (timeslot 0 only) or when loss of multiframe alignment occurs (timeslot 16 only).

much higher than is possible in hard-wire control. However, such is the complexity of large telephone networks that software development presents the most difficult problems to designers, for it must last for tens of years and although made up of very long programs, it must be able to cope with dramatic changes in hardware technology.

In analogue exchanges the usual figure of merit used is the grade of service, or probability of blocking, but in digital switches the blocking is virtually zero. In these systems the major problems concern delay in the processing of calls caused by the processor units becoming overloaded. The analysis of such problems is very difficult and if simple queueing theory models do not apply resort must be made to computer simulation. One of the difficult tasks of the software engineer is to produce a satisfactory compromise between short efficient programs and those that are longer, more complex and more reliable and secure.

10.34 DIGITAL TRANSMISSION HIERARCHY

In the last few paragraphs we have discussed PCM transmission at baseband. In practice, in order to provide the required capacity, and to exploit the bandwidth available on cable or fibre, channels must be put together in an orderly manner. The basic multiplexed element is the 2 Mb/s multiframe as

	24 channel	32 channel
Primary rate	1544 kb/s	2048 kb/s
Secondary rate	6312 kb/s	8448 kb/s
Tertiary rate	44736 kb/s	34368 kb/s
Quaternary rate	139264 kb/s	139264 kb/s

Fig. 10.34 Digital transmission hierarchies.

discussed earlier. Multiframes can be multiplexed into superframes, and superframes into hyperframes, and this sequence can continue until there is sufficient capacity, or the bandwidth of the transmission medium is exceeded. Figure 10.34 shows the hierarchies used in telephone networks based on 24 and 32 channel frames.

10.35 COMMON-CHANNEL SIGNALLING (CCS)

End-to-end digitization, processor control of exchanges, the demand for a larger range of facilities and services, and the need for automatic, high-speed network management, all place heavy demands on the signalling arrangements, and they cannot be met by the channel-associated signalling mechanisms discussed above. A flexible, digital approach is required that has the capacity for a large repertoire of signals and that can handle a variety of traffic types. By general agreement the system used is that recommended by CCITT as Signalling System No. 7. We shall refer to it by the shorter title SS7 for convenience.

The flexibility of SS7 derives from three factors; the dis-association of the signalling from the speech channels, the consequent opportunity to have a logically separate signalling network, overlaying that for transmission, and a packet-based message transfer protocol, allowing the channel to be used for information, management and control, as well as for signalling for individual calls.

SS7 is used to send information between exchanges. In principle it is comparatively simple, but its implementation consists of a large amount of detail, and there are small but significant differences between the various systems in use, making it necessary to introduce interface circuitry between national implementations in order to facilitate international connections.

10.35.1 Overlay network

Although in practice the physical implementation of the signalling network is usually co-incident with the user network, it does not have to be so. It is quite feasible for the signalling network to be entirely detached from the traffic network. We can represent this detachment by considering the logical position. The logical representation of the two networks is as shown in Fig. 10.35. Note that the signalling network is connected to a digital exchange via a signalling point (SP). Because the signalling network is logically separate, there can be signalling transit units. Such units are known as signalling transfer points (STPs) and they act as intermediate nodes.

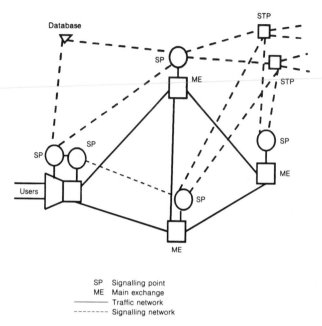

SP   Signalling point
ME   Main exchange
———— Traffic network
------- Signalling network

Fig. 10.35 SS7 non-associated overlay network.

10.35.2   Reference model

Knowing that signalling information is sent in the form of packets, and that SPs and STPs are the nodes of a signalling network, it is not surprising that the protocols used in SS7 can be expressed as a multi-layer architecture. The reference model used to describe SS7 pre-dates the OSI seven-layer model, but the two can be related. Traditionally SS7 is described by a four-layer model, as shown in Fig. 10.36.

Layer 1 is the physical layer. It specifies the interface conditions and functions across the boundary to layer 2, making SS7 independent of the specific physical link medium. In digital systems operating in standard PCM format the physical layer is provided by the signalling channel, TS 16, which may be transmitted over a wide variety of physical media including coaxial cable and optical fibre.

Layer 2 is the signalling link layer. The packet to be sent over the physical layer is given some form, and incorporates codes to ensure that it can be transmitted to the correct destination with a high probability that it will be error free. Source and destination codes, and error detection and correction codes are incorporated, along with codes that allow flow control mechanisms to be used to restrict the transmission of signalling packets to a rate that can be handled by the receiver.

At this level there are three types of packet known as

MSU   Message signalling unit
LSSU   Link status signalling unit
FISU   Fill-in signalling unit

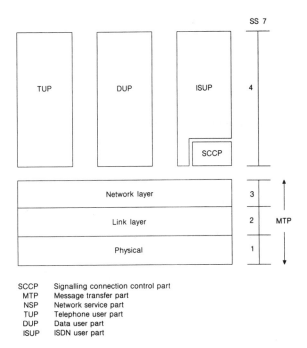

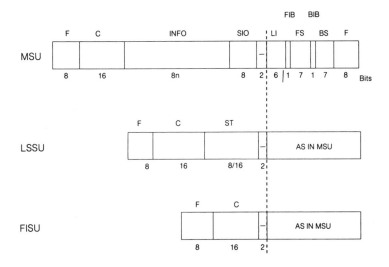

Fig. 10.36 SS7 reference model.

Fig. 10.37 SS7 layer 2 data units.

The frame format for each of these units is shown in Fig. 10.37, where it can be seen that much of the overhead information is common to all three types.

The functions of these frames are as follows:

MSU Transports higher-level information
LSSU Sets up the link and is used in flow control

FISU   Is transmitted when no other information is being sent in order to maintain the link; i.e. it ensures that the receiver is correctly set up, and that it does not clear down the connection too early.

The framing structure of the MSU is as follows:

Flag	indicates the start and finish of the frame and it has the fixed form 01111110
FS, FIB	
BS, BIB	used in the error control mechanism
LI	indicates the length of the frame from C to LI, and the type of SU
C	uses a standard CCITT check sequence
SIO, INFO	carry information from higher levels
ST	carries information on the condition of the link and its alignment status.

Layer 2 functionality is sufficient for passing control information across a single link between exchanges. The information field is not interpreted at this level, so it is passed unprocessed from level 1 to level 2.

Layer 3 is the signalling network layer. As in data networks, layer 3 is concerned with network functions, in particular with providing routing information for the STPs to ensure that the packets are transmitted efficiently through the network. When there is more than one link between source and destination signalling point the layer 3 functions are required.

Layers 1–3 constitute the message transfer part (MTP).

Layer 4 in the SS7 model depends on the application. It is known as the user part, in general, but there are several such parts: The telephone user part (TUP), data user part (DUP), and ISDN user part (ISUP) are the three of present interest, but as new services are introduced, additional user parts may be defined.

The user parts were designed with a functionality that cannot be mapped easily onto the OSI reference model. This raises problems for future developments in an open system environment, particularly for user applications that are not based on circuit switching, such as network management.

The telephone, data and ISDN user parts are unlikely to be redefined. They operate directly into the message transfer part of the SS7 model and their overall structure is established.

It is perhaps worth emphasizing that from the layered model perspective the upper layer users, normally thought of in data networks as the users of the system, are the control processors in the exchanges at the ends of the signalling link path. To benefit fully from the scope of SS7 these processors use all the upper layers and can therefore be regarded as the users of the system.

The early specification of SS7 not only treated the upper layers as one set of functions, but chose a functionality for layer 3 of the MTP that is slightly different from layer 3 in the OSI model. To map the SS7 model onto the 3/4 interface of OSI requires an additional sublayer, called the signalling connection control part (SCCP). By adding this sublayer to the top of MTP, new

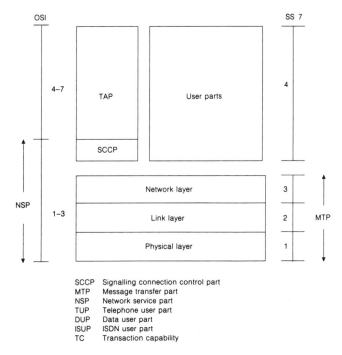

Fig. 10.38 SS7/OSI reference model.

user parts can be developed in an OSI framework. The upper layer, 4 in SS7, is given more structure too by subdividing it into transaction capabilities sublayers.

In Fig. 10.38 the complete layered structure of SS7 is given, with the associated OSI layers indicated for comparison.

The universal acceptance of SS7 as the digital signalling system for the telephone network has enhanced considerably the global operation of the network. It is essential that there is a common understanding and interpretation of the recommendations. In practice there are three ways in which this requirement proves to be difficult. The recommendations may be interpreted ambiguously. This may be because they are ambiguous in the obvious sense; not sufficiently explicit and therefore leave scope for different, equally valid, interpretations. The ambiguity may also arise, however, because different developers interpret the same terminology in different ways; the writers of the recommendations having been wrong in assuming that their semantics had only a single meaning.

Secondly, there may be some implementation decisions that in practice do not produce the precise functionality at an interface as set out in the recommendations, leading to an inability of pieces of equipment to interwork.

Finally, there is a need, for various reasons, to include options within the recommendations, for example when a specific feature is required by one operator but not by others. The provision of options must be made rarely since, unless the effects on interworking can be comprehensively evaluated, unforeseen difficulties will occur.

The mutual self-interest of having a universally recognized signalling system has lead the interested parties to become committed to establishing an efficient mechanism, through CCITT, for iterating towards a well-defined system by reporting problems, and suggesting solutions.

10.36   DIGITAL SWITCHING

The digital switch can have many structural forms depending on the application, the number of connections required, and the technology used. The system shown schematically in Fig. 10.39 is a local exchange and it shows that there are two types of switch involved. A subscriber switch to act as a concentrator, and a central switch which has a distribution function.

The architecture of the subscriber switch will depend to a large extent on the number of subscribers to be attached to the exchange. If a small rural exchange is being connected the switch may act as no more than a multiplexor, being the mechanism whereby up to 30 speech channels are time division multiplexed on to a single PCM carrier. In such a case, since a group of 30 channels is the smallest PCM link available, it may be sensible to allow all subscriber lines full availability access. However if the number of subscribers is somewhat larger a concentrator will be necessary. As in analogue systems, a concentrator provides significantly fewer lines on the outlet side than are attached at the input. The fact that the inlets are subscribers indicates that generally the mean traffic per inlet line will be less than 0.1 erlang thus a very low loss probability can be achieved even if the number of outlets is no more than a fraction of the number of inlets.

For digital systems the interworking between the digital switch (exchange) and the digital transmission link (PCM line) is straightforward, using open system signalling SS7. However, if the switch has to carry traffic originating on an analogue line then the signalling will be one of many possible analogue signalling protocols, such as those described earlier. In that case interworking is difficult, and special interface units are required to ensure smooth oper-

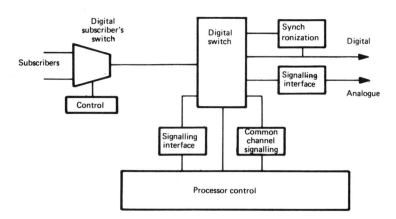

Fig. 10.39 Digital telephone exchange.

ation. These interface circuits can be exteremely expensive, adding considerably to the cost of the switch. In practice, since there are many types of analogue signalling, there may be a need for more than one interface in a particular installation. Because of the variations in the detail of signalling system protocols it is highly likely that interfaces have to be developed on a one-off basis; often, a general design to interface SS7 to each analogue signalling system will not be a feasible solution.

10.36.1 Time switching and space switching

The last section dealt with the notion that subscribers lines could be connected to PCM links via a concentrator or a multiplexor. However, that is not the only function of digital switching. It is often used to interconnect a calling subscriber on an incoming line with a called subscriber on an outgoing line, or as a distribution stage as discussed earlier.

We saw in Section 10.33 that the channels on a PCM link are structured into a frame and multiframe arrangement so that any attempt to interconnect calling and called subscribers will require a timeslot adjustment; see Fig. 10.40 for an example. Five subscribers on the inlet side of a multiplexor are able to call subscribers on the output side; if, for example, subscriber C on inlet timeslot 3 wants to be connected to subscriber V on outlet timeslot 4, the inlet sample must be delayed in the switch for one timslot before being sent out on timeslot 4 for the connection. The type of switch designed to provide the appropriate delay is called a time switch. The sample from the inlet channel is stored in a buffer and then read out when the appropriate output timeslot arrives. Figure 10.41 illustrates the way in which this is done. Synchronization is obtained from the frame and multiframe alignment signals in the PCM format (Section 10.33). Each input timeslot word is stored in a buffer. A control store holds information on the time at which each sample has to be read out. At the appropriate time this control store connects the data buffer to the output line. The control store instructions are derived from a central processing unit which responds to the call request from subscribers.

Time switching is clearly not enough if there is more than one incoming and outgoing PCM line. Any channel on an inlet line must be able to obtain access to any channel on any outgoing line and to do this, space switching is necessary to provide the linking between the particular inlet and outlet lines.

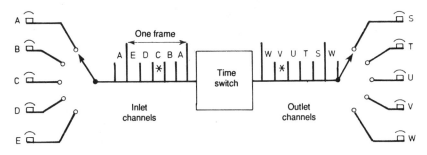

Fig. 10.40 Time switching.

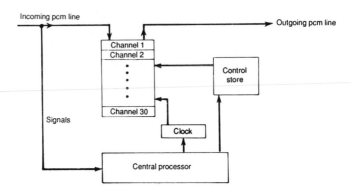

Fig. 10.41 Buffer delay for time switching.

For complete flexibility in the operation of the digital switch some combination of time and space switching is required. Although it is possible to achieve full switching with just one time switch (T) in combination with one space switch (S), such a switch would probably suffer internal blocking. More commonly a three-stage arrangement is used, either STS or TST. As an example we will consider the TST configuration, shown in Fig. 10.42. Each inlet time switch has one four-channel inlet line. A frame has four timeslots, and on the outlet side of the system there are three outlet time switches each connected to a PCM line of four channels. A, B and C are the inlet lines and P, Q and R are the outlet lines. Assume the connections required are as follows:

Inlet Channel	A1	A2	A3	A4	B1	B2	B3	B4	C1	C2	C3	C4
Outlet Channel	R4	R3	Q4	R1	P3	P4	P2	Q2	R2	Q1	P1	Q3
Delay (timeslots)	3	1	1	1	2	2	3	2	1	3	2	3

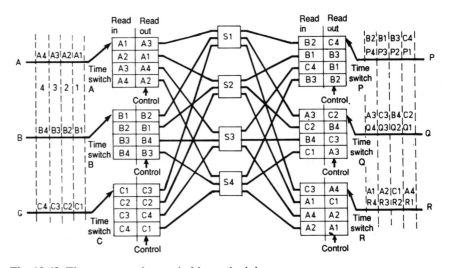

Fig. 10.42 Time–space–time switching principles.

Also shown is the number of timeslots delay that are necessary to make the input/output connections required. Some of these connections require a delay which crosses the frame boundary, e.g. to connect B3 to P2 requires a delay of three timeslots, meaning that a sample which arrives in timeslot 3 has to be held until timeslot 4 and timeslot 1 of the next frame, have passed; the sample is then released into timeslot 2 of the new frame to make the connection.

On the inlet side of the first-stage time switches, the samples in the speech channels are stored in cyclic order – as shown by the read-in column of the inlet time switches. Similarly the channels are read out from the outlet time switches in cyclic order, from timeslot 1 to timeslot 4 as shown. Consequently, any time delay required must be produced by the combined delay between the two time switching stages.

Consider the call in channel A1. It is to be connected to channel R4, requiring a delay of three timeslots between the input and output of the system. From Fig. 10.42 we can see that A1 is read into timeslot 1 and out in timeslot 2. (The choice of which timeslot to read it out is somewhat arbitrary, being determined to some extent by the calls already set up in the system.) A1 is switched in the second space switch to time switch R and is read into that switch in timeslot 2. The additional delay of two timeslots is provided in time switch R since A1 is stored until timeslot 4 arrives, when it is read out to line R as required.

In the above description we noted that Al was switched in timeslot 2 by space switch S2, and was read into time switch R in the same timeslot. Generally there is a logical space switch for each timeslot; space switch 1 for timeslot 1, space switch 2 for timeslot 2 and so on. Thus there can be no change of timeslot across a space switch.

In reality a physically separate space switch is not required for each timeslot. Since the timeslot switching arrangements occur sequentially it is possible to use one space switch for all the switching required since the cross-connections inside the switch can be changed at the end of each timeslot. Thus, provided the switching speed is sufficiently high, a single space switch will suffice. This does not alter Fig. 10.42 in which the space switch was shown as a separate entity for each timeslot; that diagram gives a correct notional impression of the way in which the switch operates.

A more general TST switch of the type found in a distribution stage is shown in Fig. 10.43. The time switch is split into several units, each having M PCM links of L channels. Consequently, if the time switch is non-blocking it will have an outlet highway of $N = ML$ time slots. The space switch is square with R inlet highways and R outlet highways. The purpose of the TST unit is to allow a particular call, which occupies a specific channel into one of the time switches, to be connected to a particular outlet channel. Basically the switching is between highways on either side of the space switch. Each highway has N timeslots and in order for a particular call to be connected say from $H1$ to $H3$ it must find a timeslot which is free in both highways. This slot may not be the same as the required incoming and outgoing slots for the call, and so some time delay, provided by the time switches, is necessary. The method used to·produce the delay again depends on the technology em-

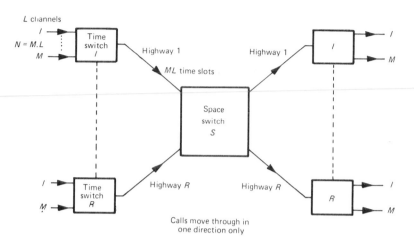

Fig. 10.43 Network representation of a digital switch.

ployed, but the delay may be from 1 to 31 timeslots depending on the relative positions of the timeslots in and out of the unit, and the chosen free timeslot in the space switch.

To understand the behaviour of the TST switch in terms of the link systems considered earlier it is important to appreciate that for each timeslot the interconnections in the space switch will be different; at each timeslot there will be a different set of calls in progress and the connections between the highways will last for only one timeslot period then new connections will be established. This can be represented by having N space switches (Fig. 10.44), one for each timeslot.

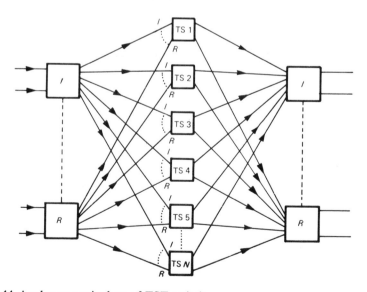

Fig. 10.44 Analogue equivalent of TST switch.

Whether or not blocking occurs in the TST unit depends entirely on the dimensions of the space switch, and since in modern systems switches are comparatively inexpensive they are usually large enough to make blocking negligible. For total non-blocking there must be at least as many outlets as inlets on the time switches, and the space switch highways must have $2N-1$ timeslots, where N is the number of timeslots in a link to a time switch.

Digital switches are uni-directional, and that implies that two paths are required to connect two channels X and Y, one for conversation from X to Y and the other for conversation from Y to X. To reduce the control process, the X to Y slot is chosen according to whatever rules are used by the designer, and the Y and X interconnection is allocated a fixed number of timeslots from it, e.g. one, or half a frame. By this method, if the X to Y connection is available, the Y to X must also be free.

The digital switch just described is situated in a main exchange, forming part of the trunk network. Interconnections between exchanges are made, for the information paths, via PCM links. However, signalling is carried over a common channel, using signalling system CCITT No. 7 as described in Section 10.35. The inter-exchange signals will be concerned not only with setting up calls between exchanges, but with accounting, administration fault diagnosis and maintenance. As described earlier, the CCITT No. 7 system transmits signals as messages that can have variable length. Each message is preceded by labels that identify its originating and destination exchanges, the type of message (call handling, fault, etc.) and includes error-detection and acknowledgement bits. If an error is detected in a message, that and all subsequent messages are retransmitted to ensure that the sequence received at the far end is in the correct order.

The error rate has an important bearing on the capacity of the signalling channel; retransmission of incorrectly received messages obviously takes up time that could be used for new signals and consequently slows down the overall process. The capacity is specified in terms of number of messages per busy hour, given that the delay from end to end is not greater than some predetermined value.

On the subscriber side of the digital switch there will be a local unit of some description. For large areas a local digital exchange would be used with mf signalling from subscriber to exchange where conversion to a PCM format would take place before concentration through a digital switch. Alternatively for very small units no exchange facility would be available, but a simple digital concentrator would be used to take in the analogue channels, convert them to PCM and multiplex them on to a single highway to the nearest local exchange. Calls between subscribers on the same concentrator would then have to pass through the local exchange. Signalling in these PCM links would be on TS16 and the control, software and firmware at the main exchange would convert it to common channel if a trunk call were required.

The introduction of digital systems is rarely a starting point for the telephone system. Usually a system exists and the digital equipment has to be grafted on to it. Whatever method is used, interworking between the old and new systems is required, and one area of difficulty is the interfacing of various analogue signalling systems with the new equipment designed to operate on

TS16 or common channel. This interfacing can be a severe problem if there are many existing signalling schemes in a particular network, and the development of satisfactory units can add considerably to system costs.

10.37 INTEGRATED SERVICES DIGITAL NETWORK (ISDN)

Alongside the digitization of the telephone network there were parallel developments concerning data communications. Coming from a main-frame computer background, and hence based on a data processing environment, the transport of data was determined by the need to transfer large quantities of information between different sites or institutions within a market sector. This lead to the establishment of large private data networks, using modems over that part of the system linked by the PSTN. Once computing developed into a more distributed, ubiquitous environment, and workstations and PCs replaced mainframes for many uses, the attraction of a common communications network instead of separate networks for data and voice was obvious.

Since the telephony network had by far the larger penetration, and because it was going digital, the obvious step was to modify it to be able to carry digital data as well as voice. In principle this would provide the means to simplify the user's system, enabling one desktop unit to handle both voice and data traffic simultaneously.

The approach used was to establish a set of standards that would be generally acceptable and; based on the 64 kb/s channel, offer the facility to carry either voice or data. Thus, by coupling together a number of these channels, both types of traffic could be exchanged simultaneously.

There are two types of ISDN connection available. The smaller is known as basic rate in which the network allocates three channels to the link; of the total bandwidth of 192 kb/s thereby allocated, two 64 kb/s channels and one 16 kb/s channel are available to the user. The 64 kb/s channels are known as B channels, whereas the 16 kb/s channel is known as a D channel.

The larger system (known as primary rate ISDN) takes up a full multiframe and offers to the user 30 B channels and one D channel. In this case the D channel has a bandwidth of 64 kb/s.

In each case the D channel carries signalling which, between exchanges, is based on SS 7. If there is spare capacity on the D channel it can be used for low-speed data such as information for process monitoring.

10.38 ISDN TERMINAL EQUIPMENT REFERENCE MODEL

It is not possible to examine ISDN in detail here. It is specified by the extensive CCITT Recommendations in the I-series. In general, these Recommendations have been agreed by most operators, manufacturers and service providers, but because of their complexity, and because at one or two points they leave scope for different interpretations on detail, there are in fact small differences between the systems being implemented in different countries. This leads to interworking problems that have to be resolved through the use of interface circuitry. With time, the differences will be reduced.

In what follows we will disregard any of these small variations, and describe the main features and functions of ISDN.

In developing the Recommendations, CCITT was concerned not to influence more than necessary the way in which ISDN systems would be implemented. Its Recommendations are therefore based on functions, interface reference points and frame structures, rather than on the circuitry and software that might be used in a particular system. This has the advantage that ISDN can be discussed without reference to the technology to be used, but it has the disadvantage that at times the description is apparently vague and difficult to relate to a physical system.

In its simplest form, the reference model is shown in Fig. 10.45.

10.38.1 User-network interface functions

The functions of the boxes in Fig. 10.45 are described in CCITT Recommendations in general terms. The terminal equipment that can be connected to ISDN is of two types, TE1 and TE2. TE1 terminals are those that satisfy the ISDN Recommendations, and can therefore be connected directly to the network, whereas TE2 terminals have interfaces at the R reference point that correspond to a non-ISDN (though CCITT) standard. These terminals are connected to the network via a terminal adapter which provides the necessary standards interface.

NT1 and NT2 are the network units that terminate the operator's network and interface with the terminals:

NT1 provides physical layer functions, isolating the user from the loop to the local exchange. Some maintenance functions may be provided in NT1 and it forms the connection interface into the user's premises; perhaps offering a multi-drop connection from several terminal types (e.g. telephone, data source) on the user's premises to one controller, or indeed combinations of connections between these internal networks.

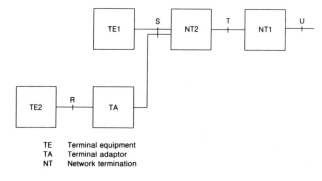

TE	Terminal equipment
TA	Terminal adaptor
NT	Network termination

Fig. 10.45 ISDN terminal reference model.

NT2 can provide a variety of functions; switching, services within the user premises, multi-point connections, and the functionality of the first three OSI layers. Hence, NT2 could represent a PABX or a LAN or a terminal controller, or indeed combinations of connections between these networks.

In a particular installation there are several possible arrangements for NT1 and NT2. NT1 may be missing, or both NT1 and NT2 functions may be integrated. The recommendations contain examples of such combinations, as shown in Fig. 10.46.

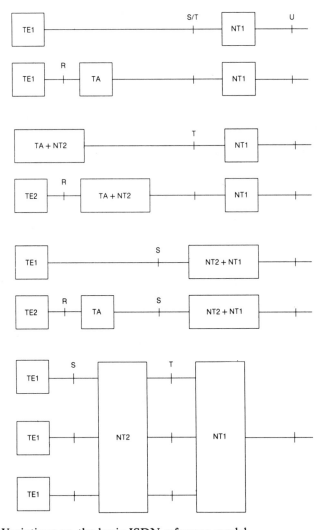

Fig. 10.46 Variations on the basic ISDN reference model.

10.38.2 Reference points

The reference points identify particular positions in the system against which the various parties involved can identify the functions of the blocks discussed above.

The S reference point, separating the terminal equipment from the network termination, provides, outside the US, the demarcation between the operator's equipment and that of the user.

The T reference point provides the interface between the two network termination functions, NT1 and NT2 described earlier.

We have assumed that the interface between the user and the network appears at point T. However, in the USA, in order to satisfy the intention of the regulatory arrangements, that attempt to minimize the part of the network that is regulated, the network access point is considered to be U. In that case, NT1 and NT2 are considered to be part of the user's equipment and the network is responsible for all functions to the right of NT1.

10.39 FRAMING

The three access channels in the basic rate service are multiplexed onto one line. In addition to the two B channels and the D channel, synchronization, framing and some management function information, are sent outside these channels and it is this overhead that uses the additional 48 kb/s capacity available in the transmission rate of 192 kb/s on the basic rate system.

Figure 10.47 shows the framing arrangement at the physical layer on the line between the exchange and the NT1 interface. It consists of 48 bits transmitted in 250 µs, corresponding to a bit rate of 192 kb/s. Examination of the frame shows that the B channels each have 16 bits, and the D channel 4 bits, equivalent to 64 kb/s and 16 kb/s, respectively. The remaining bits ensure that the AMI transmission (Section 3.8), works correctly, and that framing is synchronized at each end of the link.

In the primary rate system a complete 64 kb/s channel is available to carry overhead bits.

10.40 SIGNALLING CHANNEL PROTOCOLS

ISUP – Integrated services user part

In Section 10.36 we discussed user parts as the upper layers of the SS7 architecture, with the MTP layers below them. The user parts DUP, TUP and ISUP are related to circuit switching, being concerned with the automatic setting up, clearing down and management of calls.

The ISUP is more flexible than the other two, and will eventually supercede them if ISDN becomes the normal type of connection. It differs from DUP and TUP in that it relies more on the use of optional fields, thus providing flexibility, but at the cost of more processing at the end exchanges.

The details of the codes and protocols used to establish and monitor

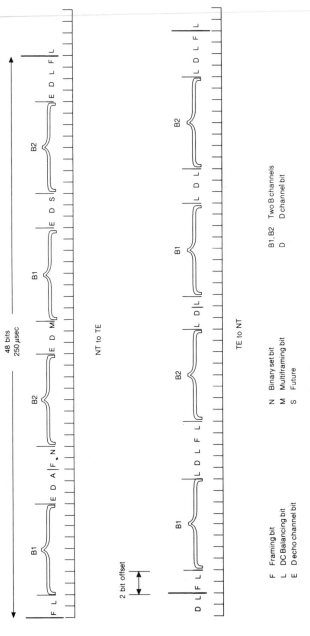

Fig. 10.47 Bit pattern for basic ISDN transmission.

Fig. 10.48 ISUP, frame structure.

connections in normal and abnormal conditions is to be found in the appropriate CCITT Recommendations, but essentially the ISUP frame, that fits into the MTP frame, has the form shown in Fig. 10.48.

CIC is a circuit identification code linked to the end-to-end virtual circuit being established. The message type code is a necessary prelude to the other fields, indicating the function and structure of the message being sent. Similarly, the fixed and variable segments of the mandatory fields are essential, and related to the message type field. The optional field can be of any length up to 131 octets, accounting for more than a dozen parameters. There is a trade-off necessary between the complexity involved with the field and the flexibility that a wide repertoire presents.

10.40.1 Local access D channel

The information sent over the D channel is carried on an HDLC type protocol known as LAP-D (channel D link access protocol). Two mechanisms are available: (a) acknowledged information transfer; and (b) unacknowledged information transfer. The difference being that in the first case the protocol allows for the identification and rejection of erroneous frames, and provides error and flow control facilities. Frames have sequence numbers to identify order. In the unacknowledged mechanism there is no error or flow control facility and hence there is no need to number frames. Erroneous frames are discarded, but there is no attempt made to correct errors or initiate retransmission.

The frame format is shown in Fig. 10.49. The significant difference between LAP-D and HDLC is the inclusion in LAP-D of the terminal endpoint identifier (TEI) and the services access point identifier (SAPI) fields. The TEI

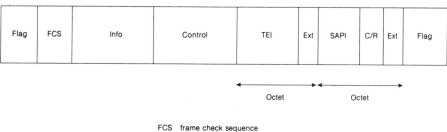

Fig. 10.49 LAP-D frame structure.

field is used to identify the terminal equipment in the connection, since one ISDN line may be accessing several terminals. The SAPI field indicates the signalling point to which the message is addressed in the particular connection.

10.41 ISDN SERVICES

Two distinct groups of services are identified in the Recommendations. These are known as:

(1) Bearer services
(2) Teleservices

In general bearer services are associated with, and limited to, those services that can be offered over the first three protocol layers. Hence the name bearer.

Teleservices are more complex and in principle involve the higher layer levels of the OSI architecture.

With reference to Fig. 10.46, bearer services access the network at the NT1 or NT2 points, whereas teleservices, using the non-communication layers, access the network via either TE1 or TE2. Because the bearer services are limited to the first three layers, they are described within the recommendations (Fig. 10.50) in terms of communication channel characteristics. Any service that fits into one of the categories of bearer services may be transmitted in this class.

Teleservices, also listed in Fig. 10.50, are described more clearly in terms of the application. Teleservices will be extended as more services become available, but at present they include the voice, text and image services in common use, and which will be available over the ISDN network. Interactive video will be accommodated, and voice storage and message retrieval services will be added; these are already available as separate services on some telephone networks.

Essentially the teleservice function is used to provide a high level exchange of information efficiently and reliably, incorporating the communications facilities of the lower levels with the intelligence available in a variety of

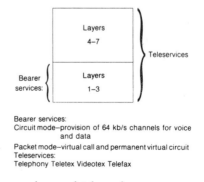

Bearer services:
Circuit mode–provision of 64 kb/s channels for voice and data
Packet mode–virtual call and permanent virtual circuit
Teleservices:
Telephony Teletex Videotex Telefax

Fig. 10.50 ISDN bearer services and teleservices.

computer-based terminals that add the high-level functionality. From the user's point of view, teleservices make the communications link transparent, and provide a convenient means of establishing, maintaining and clearing down a connection that transfers information in the format required by the user, between terminals that are compatible, and offers a range of input and output presentations.

10.42 BROADBAND COMMUNICATIONS

There is an expectation in the communications industry that video traffic on the network will gradually increase to form the bulk of the information carried. If that happens the 64 kb/s network will not be able to continue to provide a sufficiently high quality of service. Even if several concatenated channels could be allocated readily, the present network would be inadequate because of intrinsic limitations that would create insurmountable problems.

In the late 1980s, the need to circumvent these restrictions for future communications was addressed. The result was a proposal by AT&T of a system called SONET which, after some modification, was adopted by CCITT under the name synchronous digital hierarchy, or SDH for short. At first the title seems strange, but it refers to the main feature of the technique which is to provide a simple method of multiplexing and demultiplexing information streams of various bit rates; in contrast to the existing system which is now referred to as the plesiochronous network, or PDH.

Ultimately the large variety of information streams – voice, many forms of data, and various classes of video – will be carried over a packet-based network known as asynchronous transfer mode, or ATM. SDH is a transport mechanism which can accommodate the existing network, but which is also capable of carrying ATM packets. Thus SDH not only offers distinct advantages over the plesiochronous network, but it also avoids the effects of obsolescence something that is essential in any feasible replacement system.

Generically, the future high-speed broadband network is referred to as B-ISDN (broadband-ISDN) but it is almost certain that its implementation will be in the form of ATM, so we will restrict our attention to that.

Communications networks are pluralistic in several ways. Traditionally, in each nation state there was one public service operator, the national Ministry for Posts, Telegraph and Telecommunications (the PTT), and one type of traffic, telephony. Now there are in some countries several operators, private rather than public, and many types of traffic. The complexity of networks is increasing rapidly, for in addition to a variety of operators and sources, there are many new services and features being made available. In addition there are increasing demands for complex automatic network management facilities. These all place a considerable burden on the software that controls and supports the network, and on the signalling system that provides the means of communicating between exchanges.

In this chapter we will examine SDH and ATM in some detail, and put them in the context of current developments.

10.42.1 Limitations of the present network

The plesiochronous network has served us well since the advent of digital switching in the 1970s. However, as the demand for higher and higher bit rates occurs, the limitations of the system become apparent. The nature of the PCM multiframe structure is that the timeslots and frames have to be maintained in synchronism by constant reference between the exchange and network clocks. However, while superficially that constant comparison should provide a high degree of synchronism, in fact there can be considerable drift because of different propagation requirements and characteristics over different links. This drift has to be corrected by inserting bits in order to adjust the time in a bit stream so that it regains network synchronization.

The process of adding bits to adjust the position of the start of a word or frame is called bit-stuffing. Although this mechanism avoids slippage at the transmitter end, it causes other serious problems with regard to demultiplexing. For many applications it is desirable to be able to extract one of the component traffic streams without having to demultiplex all of the streams that make up the high-speed channel. In a plesiochronous system that is not possible because the individual bit streams cannot be identified easily, due to the effect of the bit-stuffing.

Figure 10.51 shows the stages in a PDH multiplexor. In order to isolate one 2 Mb/s stream it is necessary to completely demultiplex the hierarchy and then multiplex it up again. For POTS (plain old telephone system, a term used to describe the network prior to the development of ISDN) this clumsy structure did not cause much of a problem, but as various traffic types are linked on to the network there is a frequent need to be able to isolate a particular channel, or to insert a new one along the transmission path. The expense of having multiplexing equipment at each access point is prohibitive in terms of both complexity and cost.

SDH offers a cheaper, more convenient and more flexible approach.

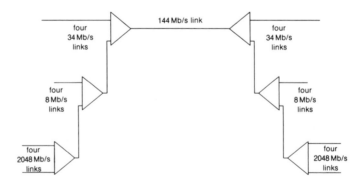

Hierarchy shown is based on 2048 Mb/s lines
i.e. on 32 channel 64 kb/s pcm multiframe

Fig. 10.51 Plesiochronous hierarchy for 2 Mb/s systems.

10.43   SYNCHRONOUS DIGITAL HIERARCHY (SDH)

Built around a basic 125 µs frame, SDH can support both 64 kb/s PCM channels and asynchronous ATM cells. Thus it is attractive as a link between POTS and broadband multi-media networks.

The 125 µs frame is known as STM-1 and consists of 2430 bytes. In discussing the operation of SDH, and its versatility, the frame is understood most easily by representing it as a rectangular matrix having 270 columns and 9 rows, Fig. 10.52. The frame payload is restricted to 261×9 bytes, the first nine columns being an overhead, used for control. The payload area is called a virtual container, VC. Both synchronous and asynchronous working can be used; in synchronous mode the frame in column 1 of row 1 is a start of frame byte, thus providing a reference for all other bytes.

The overhead part of the frame is used to indicate the 'address' of each byte. Given 2430 bytes in 125 µs, the information rate is 19 440 000 bytes/s, or 155.52 Mb/s.

The payload available to the user is reduced by the nine-column overhead noted earlier, thus the channel capacity of an STM-1 frame is $155.52 \times 261/270 = 150.34$ Mb/s. This payload area is called a virtual container level 4, or VC-4 to distinguish it from smaller container sub-sets.

The virtual container gives a high degree of versatility. It can be subdivided into tributary units, and the timing of the start of the VC can be made flexible. By using a pointer within the STM-1 overhead field the start of the VC-4 can be indicated, and provided that this pointer is updated, VC-4 can float within the STM-1 frame. Indeed, the VC-4 does not have to be completely contained within one STM-l but can spread over a frame boundary, as shown in Fig. 10.53. By using this facility the timing between the STM-l frame and the VC-4 can be adjusted to accommodate transmission delay at various points in the network.

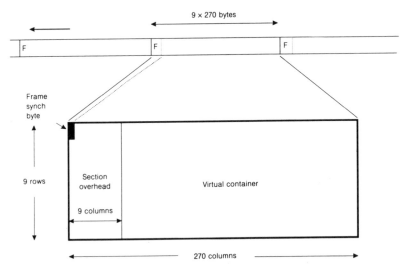

Fig. 10.52 SDH frame format.

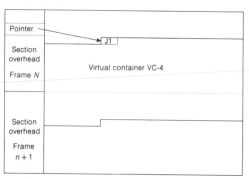

J1 is the first byte of VC-4 and its location is indicated by
the pointer in the section overhead of frame *n*

Fig. 10.53

The pointers can also be used to provide timing adjustment at a synchronizing element to modify the timing of several incoming SDH links so that they are properly aligned at the output.

Although the container VC-4 has a nominal capacity of 150.34 Mb/s, it is not all available to the user. VC-4 remains intact from transmitting node to receiving node, which may be separated by intermediate nodes. To ensure the integrity of the transmission over the whole path, a path overhead is included in VC-4, occupying the first byte of each row. Thus the payload capacity of VC-4 is reduced to 149.76 Mb/s. Evidently VC-4 can accomodate without difficulty the current multiplex rate of 139.26 Mb/s (nominal 140 Mb/s). Thus VC-4 is designed to accept input tributaries at this rate, which corresponds to that required to transmit broadcast quality television.

We discussed earlier that one of the benefits of SDH is the ease with which, compared to the plesiochronous network, tributaries of different bit rate can be multiplexed into or out of the system. This is due to the virtual circuit mechanism.

We have just discussed the 140 Mb/s tributary. Another CEPT bit rate that is well used is 2.048 Mb/s. This is included in the STM-1 frame as a set of 63 tributaries, each of 4 columns width. Such tributary units (called TU12) provide a rate of $4 \times 9 \times 8000 \times 8 = 2.304$ Mb/s.

Other lower bit rates are accommodated by using appropriate TUs to fill the basic VC-4 container. The North American system requires some different sizes of TU from the European, and these are given specific labels: e.g., TU11 occupies three columns and therefore represents 1.728 Mb/s, sufficient to support the DS1 rate of 1.544 Mb/s, and TU2 is 12 columns wide, giving a sub container of 6.912 Mb/s, sufficient for the North American DS2 rate. These TUs are represented diagrammatically in Fig. 10.54. By using the pointing system in the frame overhead, particular channels can be identified, and then abstracted or inserted into the appropriate TU.

Similarly, the 34 Mb/s nominal bit rate is accommodated in a tributary unit that occupies 36 columns; hence seven such TUs can be placed in one VC-4. For each of these bit rates there is a difference between that required by the

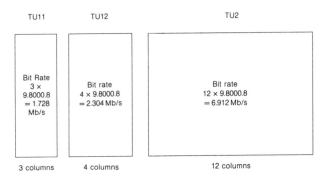

Fig. 10.54 Transmission units for an STM-1 frame.

tributary traffic and that available within the TU in the frame. The difference is made up by adding stuffing bits to the tributary traffic to fill the TU. Additional stuffing bits are needed to completely fill the VC-4 subframe when all of the TUs are in place.

Figure 10.55 shows the functional requirements of the assembler and dis-assembler. To the tributary traffic, overheads to control and manage the connection are added, and the bits are stuffed into the stream to fill the VC-4. At the dis-assembler the reverse procedure occurs. As far as the payload of the STM-1 is concerned, the VC-4, it is transmitted intact across the network, whereas the overhead bits on the STM-1 frame are interpreted and adjusted at each node.

STM-1 is the basic SDH unit, but as we have seen, it has a maximum nominal capacity of 140 Mb/s. Higher rates will be required for some applications, and in principle $n \times$ STM-1 can be provided by byte interleaving n STM-1 frames. In practice only two have been defined, STM-4 and STM-16. An STM-4 frame consists of 4 byte-interleaved STM-1 frames, as shown in Fig. 10.56. The result is a frame having an overhead field of 36×9 bytes, and a payload of 1044×9 bytes. The payload is therefore equivalent to $1044 \times 9 \times 8000 \times 8 = 601.344$ Mb/s.

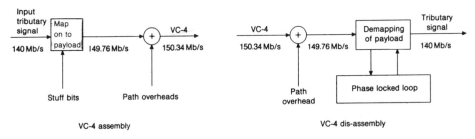

Fig. 10.55 Functional schematic for an SDH assembler/dis-assembler.

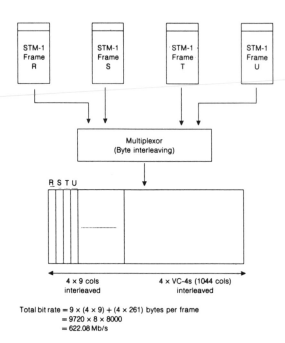

Fig. 10.56 STM-4 frame formed by interleaving four STM-1 frames.

10.44 ASYNCHRONOUS TRANSFER MODE (ATM)

It is clear that the 64 kb/s timeslotted PCM transmission system is ideal for voice traffic, but severely restricted if other sources are to be attached to the network. There are two problems with it. First, the timeslot–channel mapping is totally inflexible, and when there is the possibility of a wide range of traffic speeds this is a significant limitation. Secondly, while in principle it is possible for channels to be concatenated to provide for broader bandwidth requirements, in practice the switching and control mechanisms do not make such merging a practical proposition if the service must be available more or less on demand.

To provide for multiple bandwidths an entirely different system must be developed. One that is backward compatible with POTS, but which offers the means of carrying synchronous and asynchronous traffic, from low-speed data to high-definition video.

ATM is one mechanism that will meet that objective, and it is most likely to be the one which will be adopted worldwide. There is one transport mechanism, and several potential types of traffic so it is not surprising that ATM, like any other contenders, is a compromise; offering a reasonable means of transporting most types of traffic. It will not be quite as good as PCM for voice, and perhaps not as good as X25 for data, particularly in terms of error rates and error control, but it will be good enough for both, and certainly better than PCM or X25 as a means of carrying voice, data and video on the same network.

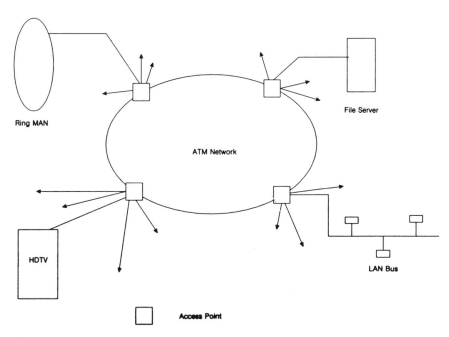

Fig. 10.57 ATM network schematic.

The layout of the network is shown in Fig. 10.57. Because the ATM network itself will have high capacity links, it is likely that the access points will be connected to some concentrating or capacity sharing system such as a LAN, MAN or PBX, or to terminals requiring direct high-capacity access such as large file servers, or high-definition television. The access point will have various functions: mulitplexing, flow control, admission control, and policing, all from the network's point of view, and suitable interfacing software to allow interaction with the access network.

Transport over the communications subnet is via short packets, known as cells. The size of the cell is chosen to give a reasonable compromise that produces a satisfactory performance for all types of traffic.

The transport mechanism is similar to virtual circuit operation in a packet switched network. At set-up the connection is established from end-to-end and that path is maintained for the duration of the exchange of information. The cell, like all packets, consists of two parts, the payload field that contains the information to be sent, and the header that carries mainly addressing and error control. The cell is shown in Fig. 10.58. The information field is 48 bytes long and the header is 5 bytes. This fits exactly the size of the packet in DQDB (see section 13.10).

The detailed structure of the header is also shown in Fig. 10.58. At present some of the fields are not yet completely defined in the CCITT Recommendations, or their purpose has yet to be fully examined. The VCI (virtual channel identifier) and the VPI (virtual path identifier) are both used to determine the route taken by the cell from end to end. The VPI is semi-permanent. It indicates the path to be taken between a paticular source–destination pair,

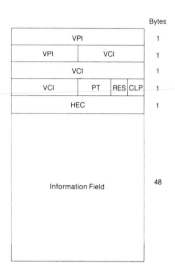

Fig. 10.58 ATM cell structure.

and this is fixed in the short term. The VCI indicates the particular channel to be used by the cell on a specific link between two nodes. At a node the incoming VCI is translated to that appropriate for the required output link. One approach is to have a look-up table at the node. The table is updated from time to time by means of a signalling channel.

The error control field is used for the header only. It ensures that the addressing is unlikely to contain errors. Unlike in X25 there is no link-by-link error control on the information field. Voice traffic is not very sensitive to cell loss and it has substantial inherent redundancy so error control is unnecessary. For data however, error control is essential since data has no inherent redundance, so any errors have to be corrected. In X25 this is done at each node traversed. That procedure is slow and not suitable for high-speed broadband applications. In ATM error control is done on an end-to-end basis. All the information is transferred to the destination address and if there are some bits in error the whole message, or that part that has been corrupted, is retransmitted. This reduces to a minimum the time used in processing at the nodes, at the cost of having to wait rather a long time for errors to be corrected. In practice, for many applications, error control is the concern and responsibility of the end user, and occasionally the requirements imposed by the user are very stringent, making it wasteful if the network itself were also to provide protection.

The cell is carried by a broadband transmission medium, probably optical fibre. The cells travel in a continuous stream. If the cell is occupied the destination address is clear. When there is no information wanting to transmit the system sends dummy cells to ensure that propagation is continuous. In theory it would be possible for the link between two nodes to be completely filled with continuous live cells, but in practice a utilization of about 80% would be the maximum that could be realistically used.

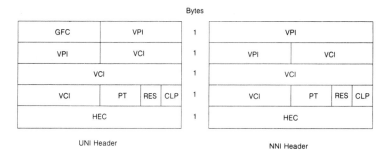

Fig. 10.59 UNI and NNI headers for ATM cells.

The payload type field is used to describe in coded form the type of information being carried, and in particular it indicates when a cell is carrying signalling or management information.

The priority bit provides a two-level priority facility. In default value the cell is liable to loss if buffer space is exceeded, whereas in priority mode the cell is processed in preference to those with no priority status.

There is a slight difference between the headers of cells working over the user/network interface (UNI), and those over the network/network interface (NNI), as can be seen in Fig. 10.59.

10.44.1 Statistical multiplexing

Above we have considered the salient features of ATM as a transport mechanism for a broadband communications network. The reason for starting with the network is to establish in general terms how it works, and now we can look at it more from the perspective of the user and the services it offers

Remember that the justification for changing from the current time division multiplexed system to the packet-based ATM is that there is a need to be able to carry a large number of different types of traffic. That being so, how is that traffic put on to the ATM network in such a way that it can be granted sufficient bandwidth in a flexible manner without interfering with other traffic, either already on the network, or likely to use it in future?

The link between traffic sources and the network is a multiplexor. On the user side there are lines to the potential sources, and on the network side there is a very high bit rate link into the network itself, Fig. 10.60. Putting that in another way, the multiplexor can be seen as a resource allocator. It controls a large bandwidth link and has the function of allocating access to that link in as efficiently as possible. The objective is to give each traffic type the bandwidth it needs, without being wasteful and thereby reducing the number of sources that can be served.

Because the cell payload size is very small, a particular message will have to be accommodated by sending it in many cells. Clearly a high-definition television message will require many more cells than a low-speed data

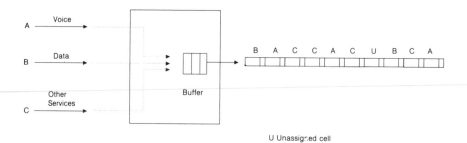

Fig. 10.60 Statistical multiplexor.

message. The statistical multiplexor works in such a way that, provided the collective mean demand of the traffic sources does not exceed the mean capacity of its outlet link, all the traffic can be handled automatically, each being given, in the long term, the bandwidth it needs.

As messages arrive at the multiplexor they are structured into cell-sized packets and queued in a first-come-first-served buffer. The outlet of the buffer goes directly to the outlet link of the multiplexor. Thus there will be many more packets from high-speed sources than from those operating at low bit rates. Consequently, in the outlet stream there will be many more cells from the high-bandwidth source.

Although the outlet provides enough capacity for all possible inlet sources, there is an implicit variation in the delay experienced by different cells. For example, if voice traffic is being transmitted, the number of cells between adjacent voice channel cells in the output stream will not be uniform because of the high probability that consecutive samples will have to be accommodated along with traffic from other sources. The statistical nature of the arriving traffic streams will produce variation in the arrival time of the voice stream cells, unlike in time division multiplex systems where the arrival of voice samples is at precise intervals of 125 µs. This variation in arrival times can be particularly troublesome for voice traffic since its efficient demodulation depends on this precision of timing. In ATM steps are taken to ensure that the delay variation does not exceed specified limits. Cells from the multiplexor are contiguous. They may be from any source, or from none. If the buffer is empty an unassigned cell is transmitted. The buffer is dimensioned so that the probability that a cell is lost due to the buffer being full is less than one in 10^{-9}

As we might expect in a data-communication type of mechanism the description of the system can be consistently described in terms of the OSI reference model. Figure 10.61 shows that which applies to ATM. The upper layers are the responsibility of the user or service provider. The network provides the physical layer, probably optical fibre, and the ATM layer which contains the mechanisms described above. On top of the ATM layer there is a higher level 2 sublayer known as the adaptation layer (AAL). This provides the interface between the ATM layer and the higher layer functions that may be user-related, signalling or management services.

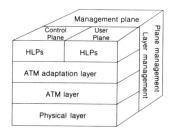

HLPs Higher layer protocols

Fig. 10.61 Layered model for ATM.

The AAL is responsible for mapping the protocol data units (PDUs) of the higher layer functions into the ATM cell information fields, and vice versa, thus performing segmentation and reassembly of the PDUs. The AAL has a different function for each class of service that might be transported over the ATM network. Earlier we considered in very general terms the variety of sources that could generate traffic, such as low-speed data, voice, and video. The network must also carry management and control traffic and that will have particular characteristics.

The CCITT Recommendation considers four classes of traffic. In essence, traffic may require source and destination to be synchronized in order that information can be correctly reconstructed at the receiver; such traffic is sometimes called real time. Data services, in general do not require source and destination users to be aware of any time relationship. Although bit-rate requirements can vary, the actual rate is not significant in determining the class of traffic. More important is the constancy of the rate. Traffic is divided into two groups, that with constant bit rate, and that in which it varies. Finally, most traffic is transported in connection mode, but in principle there are connection-mode and connectionless mode services.

Although if each of the above variables is considered there are eight possible combinations (from three binary conditions) the CCITT Recommendation considers only four classes as shown in Fig. 10.62, known as Class A, B, C and D.

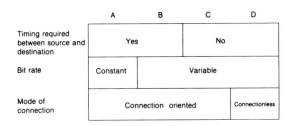

Fig. 10.62 ATM service classes.

Class A traffic is typically voice, or fixed bit rate video. In both there is a synchronization required between source and destination, requiring timing information to be sent over the virtual circuit, and the bit rate used is fixed and non-varying.

Class B is similar to Class A except that there is variation in the bit rate. Some modulation techniques used in audio or video systems result in variation of bit rate as a means of conserving bandwidth; high bit rates occur when there is a strong change in the source audio or video signal.

Class C traffic is not time sensitive, and the bit rate may vary.

In Classes A, B, and C the mode is connection oriented.

Class D traffic is connectionless and, as in Class C, the bit rate is variable.

10.44.2 Traffic management

The large variety of traffic rates and distributions that users may generate introduce entirely new problems for the management and control of broadband networks compared with POTS. There are four main functions:

acceptance
flow control
policing
congestion limitation.

Call acceptance is the mechanism whereby the characteristics of the traffic to be transmitted are relayed to the manager, and a decision taken on whether or not there is sufficient capacity to accept the traffic.

Flow control is related to call acceptance and refers to the management of the network in such a way that if traffic is accepted, it is guaranteed a high probability of successful transmission.

When a call is accepted for transmission a contract is established whereby the user agrees to supply traffic with characteristics within some stated limits, and the network agrees to transmit that traffic. Policing is the fuction of monitoring the traffic actually sent to ensure that the user does not break the contract by sending traffic that falls outside the agreed specification. If the traffic is not policed a user could send more traffic than agreed and therefore defraud the operator, or create difficulties by interfering with other users.

Congestion control concerns re-routing traffic if there is some bottleneck within the network. Normally it would consist of re-routing traffic to avoid the obstacle.

Source traffic is described in terms of its probability distribution function. Normally that function will not be known and therefore some approximation is used. The normal approach is to consider only the first two moments of the distribution, the mean and the variance. This is not ideal for some applications, and the peak bit rate is also used as an important metric.

The traffic management mechanisms raise central questions about the application and implementation of ATM. The principles discussed earlier are straightforward and understandable. However, because of the versatility of the network in being able to carry any type of traffic, there are serious questions still to be resolved about the most appropriate way of carrying out these management functions. The problems all revolve around the difficulty of defining traffic. Experience with telephony has shown that to describe its traffic as having a negative exponential holding time and a negative exponential inter-arrival time is entirely consistent with observations over a long period of time. The traffic arriving from a number of telephone handsets into an exchange can be accurately described as Poissonian, as discussed at the beginning of this chapter.

In ATM, however, things are not so simple. The complexity and range of traffic types creates a serious tension between the need to characterize them accurately in terms of probability distributions, in order that the system can be designed, and can operate efficiently, while at the same time there is a need to make the characterization as simple as possible, so that the resulting analysis and control mechanisms do not become too complicated or expensive.

It is not surprising that these issues have proved to be a rich source of research projects as a large variety of techniques are suggested and evaluated. As is usually the case in complex activities, there is no single best way. Some mechanisms are simple and robust, but relatively unsophisticated, while others can be made sensitive to changes in the traffic, but lack practical attraction. However, there will shortly come a time when either the standards bodies or the industry will agree on particular techniques and further discussion will be to a large extent academic.

There are a few parameters that are used to characterize performance that are relatively uncontroversial. How important each of them are will depend on the traffic type. Cell delay, variation in delay and cell loss probability are the most important.

10.45 CONCLUSION

There are periods of significant change and periods of consolidation in the history of many systems. In communications, significant change has been driven by a combination of technical development and market need. The relatively recent developments of optical fibre transmission and VLSI devices made possible a large number of improvements that could not have happened otherwise. The significant system changes have been from analogue to digital working, the use of cellular and satellite communications, the enormous capacity of optical links, and the power and versatility of software control.

We are rapidly moving towards a stage when the traditional constraints on telecommunications systems are removed. Very large switches are now commonplace, extremely high bandwidth is easily provided, and there are sophisticated services and facilities readily set up. The next clear step is the introduction of multi-media networks. Data communications and telephony are moving closer together, via ISDN and LANs, and soon video will be transported over the same network. The development phase of B-ISDN is giving way to practical devices, and networks will become increasingly packetized.

Eventually the communications network will be capable of carrying all types of traffic of whatever speed. The problems for systems designers will then be related to the automatic, effective and efficient management of a network that carries huge amounts of traffic, that has to provide complex features, and that is an integral part of the economies of companies and countries.

REFERENCES

1. Brockmeyer, B., Halstron, H. L. and Jensen, A., *The Life and Works of A. K. Erlang*, Copenhagen Telephone Company, 1943.
2. Jacobaeus, C., 'A study on congestion in link systems', *Ericsson Technics,* No. 48, 1950.
3. Cooper, R. B., *Introduction to Queueing Theory*, Edward Arnold, London, 1981, Chapter 5.
4. Fry, T. C., *Probability and its Engineering Uses*, Van Nostrand Reinhold, Wokingham, 1965.
5. Fishman, G.S., *Principles of Discrete Event Simulation*, Wiley, Chichester, 1978.
6. Flood, J. E., *Telecommunication Networks*, Peter Perigrinus, London, 1975.
7. Bear, D., *Telecommunication Traffic Engineering*, Peter Perigrinus, London, 1976.
8. Hills, M. T., *Telecommunication Switching Principles*, George Allen & Unwin, London, 1979.

PROBLEMS

10.1 A traffic-recording machine takes measurements of the number of busy devices in a group every three minutes during the busy hour. If the sum of the devices busy over that period is 600, what is the value of traffic carried?

Answer: 30 erlangs.

10.2 A loss-system full availability group consists of five devices. If the mean call holding time is 180 seconds, and the call intensity is 80 calls/hour, what is the mean load per device?

Answer: 0.8 erlangs.

10.3 In a particular system, it was found that during the busy hour, the average number of calls in progress simultaneously in a certain full availability group of circuits was 15. All circuits were busy for a total of 30 seconds during the busy hour. Calculate the traffic offered to the group.

Answer: 15.13 erlangs.

10.4 A group of eight circuits is offered 6 erlangs of traffic. Find the time congestion of the group, and calculate how much traffic is lost.

Answer: 0.122; 0.73 erlangs.

10.5 The overflow traffic from the eight circuits in Problem 10.4 is added to a ninth circuit. What traffic will it carry?

Answer: 0.42 erlangs.

10.6 A system of six telephones has full availability access to six devices. Find the probability that 1, 2, ..., 6 devices are busy. What is (a) the call, and (b) the time congestion of the system if the carried traffic is 2.4 erlangs?

Answer: (a) 0; (b) 0.004.

10.7 (a) Two erlangs of traffic are fed to three devices. What is the congestion, and how much traffic is lost?
(b) Two erlangs of traffic are fed to one device. The overflow is fed to a second device, and the overflow from that to a third. What is the overall congestion, and is the value of the traffic lost the same as in (a)?

Answer: (a) 0.2105, 0.421 erlangs; (b) 0.165, No.

10.8 Show that, if the assumption of statistical equilibrium is valid, the probability of a system being in state i is given in terms of the probability that it is in state 0 by

$$[i] = \frac{\prod\limits_{j=0}^{i-1} \lambda_j}{\prod\limits_{j=0}^{i} \mu_j} [0]$$

10.9 The state transition diagram below represents a system with an infinite number of devices subjected to calls arriving at random with fixed mean arrival rate, λ.

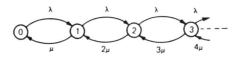

If $\lambda/\mu = A$, the mean offered traffic, use the birth and death equations, and assume statistical equilibrium, to show that

$$[i] = \frac{A^i \exp(-A)}{i!}$$

10.10 An infinite number of sources feed 5 erlangs of traffic into eight devices. Find the probability of the network being in each of its possible states, i, and check that $\sum_i [i] = 1$. Plot $[i]$ against i.

Answer: 0.0072, 0.0362, 0.0904, 0.1506, 0.1883, 0.1883, 0.1569, 0.1121, 0.0700.

10.11 Repeat question 10 for (i) 2 erlangs, and (ii) 7 erlangs of traffic, noting the variation in both the congestion and the distribution of $[i]$, with increasing traffic.

Answer: (i) 0.1354, 0.2707, 0.2707, 0.1805, 0.0902, 0.0361, 0.0120, 0.0035, 0.0009;
(ii) 0.0013, 0.0088, 0.0306, 0.0715, 0.1251, 0.1751, 0.2044, 0.2044, 0.1788.

10.12 A telephone route of n circuits has to carry a normal load of 3 erlangs. If the grade of service must not exceed 0.03, what is the smallest value that n can have? In an emergency, there is a 20% increase in offered traffic. What will be the grade of service in this overload condition?

Answer: $n = 7, 0.0438$.

10.13 Draw the chicken diagram equivalent of the two-stage link network shown below, and calculate the internal blocking.

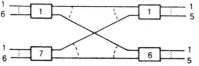

Corresponding outlets from second-stage switches form a route.

Traffic offered per route = 2.5 erlangs
Traffic carried per inlet switch = 1.8 erlangs

Answer: 0.0411.

10.14 A time–space digital switching configuration is shown. What are the functions of the time and space elements in relation to the speech channels on the PCM inlets and outlets? Draw the analogue equivalent of this switch.

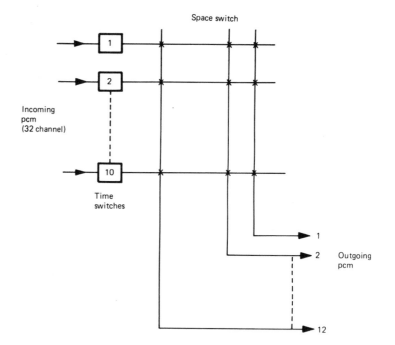

Television systems 11

11.1 INTRODUCTION

This chapter is concerned primarily with the fundamentals of domestic broadcast television systems which now includes direct broadcast satellite systems. There are, of course, many specialized closed-circuit television systems, but these are all based upon the same principles as the broadcast system and will not therefore be considered separately.

Television has progressed over a period of about 70 years from John Logie Baird's mechanical scan system to the modern highly sophisticated colour systems that can transmit both entertainment programmes and teledata information. During this period the quality of sound and visual image reproduction has increased steadily and digital techniques, in particular, have made rapid inroads in what has been basically an analogue technology. The basic principles governing the transmission of visual information by electrical means have, however, remained unchanged. These principles rely upon a property of the human eye/brain combination known as 'persistence of vision'. This means that under certain conditions, the brain is unable to differentiate between a moving image focused on the retina of the eye and a rapid sequence of still images. Television is specifically designed for the human eye/brain combination, and it is appropriate to begin by considering the subjective response of this combination to incident light energy.

11.2 MEASUREMENT OF LIGHT AND THE RESPONSE OF THE EYE

Light is a form of electromagnetic radiation and the power radiated from any source is measured objectively in watts. Light energy is visible only over a very restricted range of wavelengths, from 4×10^{-9} m to 7×10^{-9} m. The subjective sense of 'intensity' varies with the wavelength of incident light, where in this context intensity means the impression of an observer of whether a light source is bright or dim. The response of the eye varies between individuals but the variation in this response has been found to be small enough to allow a 'standard observer response' to be defined. This standard response is shown in Fig. 11.1.

A black body radiator radiates equal energy at all wavelengths. The luminous intensity of such a source is obtained by weighting the uniform spectral density of the source by the response of the standard observer. The resulting subjective unit is the candela and is defined as 1/60th of the luminous intensity per cm² of a black body radiator maintained at a

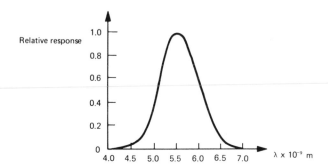

Fig. 11.1 The standard observer response.

temperature of 2042 K. The lumen is the amount of light passing through unit area at unit distance from a point source of 1 candela, i.e. the total emission from a source of 1 candela is 4π lumens. The lumen is the unit of luminous flux and, unlike the power output from a light source, depends upon the response of the standard observer.

It is clear from Fig 11.1 that two light sources of equal power but different colour do not necessarily have the same luminous flux. Two light sources of different colour that appear equally bright to the standard observer do have the same value of luminous flux measured in lumens.

The significance of the last statement will become clear when colour theory is considered in Section 11.8. As far as monochrome television is concerned the response of the camera when viewing a coloured scene should closely resemble the response of the standard observer, i.e. the electrical output should be proportional to the brightness of the scene measured in lumens. Such an output signal is termed the luminance signal.

11.3   THE MONOCHROME TELEVISION WAVEFORM

In Europe broadcast television pictures are transmitted using the 625 line standard (it should be noted however that vigorous efforts are being made to define a universal European high-definition standard). The 625 line signal waveform will be considered as the means of explaining the transmission of a three-dimensional vision signal (i.e. one that is a function of time and both horizontal and vertical position) over a one-dimensional channel (i.e. a channel in which voltage is a function of time). The three-dimensional visual image to be transmitted is focused onto a photosensitive (in the electrical sense) surface by an optical lens system. The photosensitive surface may be considered as a large number of separate photoelectric transducers. Each of these transducers produces an electrical output proportional to the intensity of the image that falls upon it. The original image is thus decomposed into a large number of picture elements known as pixels (or pels). The electrical output corresponding to each pixel is one-dimensional since it is a function only of time. The three-dimensional signal, which is the combined output of all pixels, is actually transmitted as a rapid sequence of one-dimensional outputs. The original optical image is then reconstructed by displaying each pixel in its correct spatial position.

This image is produced, in a domestic television receiver, by modulating the beam current of a cathode ray tube (CRT). The screen of such a tube is covered with a photo-emissive layer and will have the same number of pixels as the photosensitive plate in the image capture system. The image is reproduced by effectively connecting each pixel in the image capture system to the corresponding pixel in the CRT in turn. This is achieved by scanning the photo-emissive screen in the CRT with an electronic beam synchronized to the scanning beam in the image capture system. Sequential scanning of the image capture system and CRT are shown diagramatically in Fig. 11.2. The electron beam is deflected in both vertical and horizontal directions. The scan begins at point *A* in Fig. 11.2 and proceeds to point *B*. During flyback, which occurs very rapidly, no information is transmitted. The next line is scanned from *C* to *D* and the process is repeated until point *F* is reached. At this point flyback occurs to point *A* and the scanning of the complete picture is repeated.

The above explanation is based on the assumption that both image capture and display systems are based on vacuum tubes with scanning electron beams. In many personal computer systems, for example, the CRT has been replaced by a liquid crystal display in which the individual pixels are addressed by a switching waveform generated from a digital circuit and such displays are also employed in miniature TV receivers. It is also commonplace for image capture devices to be based on solid-state technology (see Section 11.18) and in such cases the scanning electron beam is replaced by a similar switching waveform. The scanning electron beams are thus replaced by synchronized switching waveforms, but the scanning process illustrated in Fig. 11.2 is still relevant to the production of television images.

It can be seen from Fig. 11.2 that each picture scan takes a finite time. If this time is longer than the duration of persistence of vision the eye will perceive flicker on the picture. If the scan time is very short the number of pixels/second and hence the signal bandwidth will be large. The optimum scan rate is therefore one that is just fast enough to avoid flicker. The picture frequency at which flicker occurs is related to the luminance of the picture being viewed. For television systems, flicker is avoided when the picture rate is as low as 25 complete scans/second. This is made possible by use of interlaced scanning, which is shown in Fig. 11.3.

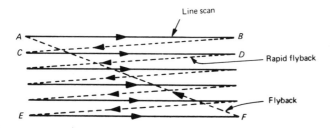

Fig. 11.2 Sequential scanning.

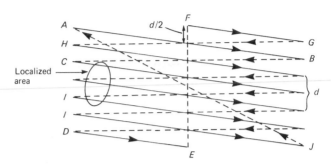

Fig. 11.3 Interlaced scanning.

The scan now occurs in two fields, the odd field starts at A and the line is scanned until point B. Flyback then occurs from B to C and so on. It will be seen from Fig. 11.3 that the last line is scanned from D to E, i.e. field flyback occurs from E to F. (In fact this flyback takes the equivalent of several line scan periods so will not be a straight line as shown in Fig. 11.3.) The even field starts half-way across the picture at F and the scan of the first line finishes at G, flyback occurring from G to H. The last line of the even field is scanned from I to J, flyback then occurs to point A, which is the start of the next odd field.

Interlaced scanning avoids flicker because although each complete picture is scanned 25 times/second, a localized area of the picture is scanned by both odd and even fields. Thus each such area is apparently scanned 50 times/second. This deception relies on the fact that the variation in brightness in a vertical direction usually occurs gradually. If there is a significant difference in the brightness between one line and the next, some localized flicker will be observed. It should be noted that two field frequencies are in common use i.e. 50 fields/second or 60 fields/second depending on the local power supply frequency.

It has been pointed out that it is necessary to synchronize the scans in the image capture system and the display device. To achieve this, a line-synchronizing pulse is transmitted at the end of each line and a field-synchronizing pulse is transmitted at the end of each field. The synchronizing pulses trigger the flyback circuits in the television receiver. The transmitted video signal thus has distinct components, i.e. the picture signal that represents the variation in brightness of each line and the synchronizing pulses that are transmitted below black level. The composite waveform is shown in Fig. 11.4.

The 625 line transmissions in the UK use negative amplitude modulation of the vision carrier, i.e. an increase in picture brightness produces a decrease in signal level. A portion of the envelope of the modulated signal is shown in Fig. 11.5. The advantage of negative modulation is considered to be that the black spots produced on the screen by some forms of ignition interference are less objectionable than the white spots that would be produced in the same circumstances if positive modulation were used. The ratio WB/BS shown in Fig. 11.5 is known as the picture/sync ratio and has a value of 7/3. This is a compromise figure that produces a reasonable picture quality and adequate

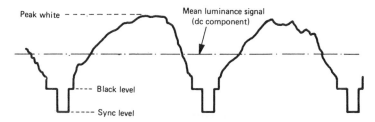

Fig. 11.4 Brightness and synchronizing information.

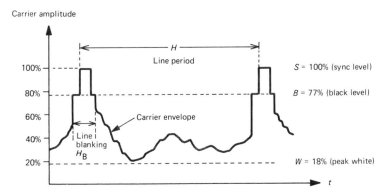

Fig. 11.5 Transmitted carrier envelope.

synchronization in poor SNR conditions. Figure 11.5 shows that a portion of each line before, and just after, each line synchronizing pulse is at black level. This allows the line flyback to be suppressed and hence no visual output is produced during this interval. A similar provision is made at the end of each field, but in this case the duration of black level extends for 25 line periods. This means that in one complete picture (two fields) there are 50 lines which are suppressed. These lines are actually used for various forms of data transmission which is considered in Section 11.8.

11.4 BANDWIDTH OF A TELEVISION WAVEFORM

The bandwidth of a television waveform is directly related to the number of pixels transmitted per second. In each 625 line picture 25 lines are blanked off at the end of each field and the number of lines seen by a viewer is thus $(L - L_B)$ where $L = 625$ and $L_B = 50$. The vertical resolution of the picture is equal to the number of lines seen. Assuming equal horizontal and vertical resolution the number of pixels per line is $A (L - L_B)$ where $A(= 4/3)$ is the picture aspect ratio (width/height). Since each line is blanked out for a period of H_B, the number of pixels transmitted per second is

$$P_s = \frac{A(L - L_B)}{H - H_B} \tag{11.1}$$

For a 625 line system the line period $H = 64\,\mu s$ and $H_B = 12.05\,\mu s$ giving a value of $P_s = 14.76 \times 10^6$ pixels/second. The maximum signal bandwidth occurs when adjacent pixels alternate between black level and peak white. The video signal is then represented by a square wave as shown in Fig. 11.6. It can be seen from this figure that it is possible to transmit adjacent black and peak white levels by a sine wave with a period equal to the period of two pixels. The bandwidth required is then $1/T$ which, in the 625 line system, is 7.38 MHz. It is clear from Fig. 11.6 that if the transmission bandwidth is restricted to this figure then the sharp edges of the square wave will be lost. This is virtually undetectable in practice because of the finite size of the scanning beam, which does not allow detail of this resolution to be displayed in any case. Statistically speaking, very few pictures require the degree of resolution allowed by a bandwidth of 7.38 MHz. Use is made of this fact to reduce the signal bandwidth even further. The actual bandwidth allowed for 625 line transmissions is $K \times 7.38$ MHz. The constant K is known as the Kell factor. The value of K depends upon several elements, e.g. the width of each line on the CRT and the low-pass filtering effect of the eye (which itself depends upon how close the viewer is to the CRT). The value of K used in Great Britain is 0.73, which gives an acceptable subjective picture quality and results in a video bandwidth of 5.5 MHz.

11.5 CHOICE OF NUMBER OF LINES

There have been various standards in operation throughout the world, but this has now been essentially reduced to two, 525 lines in North America, Greenland and Japan and 625 lines elsewhere. All the systems that have been used have one feature in common, the use of an odd number of lines. The reason for this can be explained by reference to Fig. 11.3. If each picture has $2n + 1$ lines, then each field will have $n + \frac{1}{2}$ lines. If it is assumed that the distance between adjacent lines of the same field is d the odd field scan finishes halfway across the last line. This means that field flyback (if assumed instantaneous) would cause the even field to start halfway across the first line and would therefore be a distance of $d/2$ above the first line of the previous field. Interlacing is thus achieved automatically with the same field deflection on both odd and even fields.

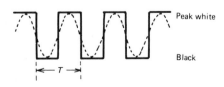

Fig. 11.6 Television picture bandwidth.

11.6   SYNCHRONIZING PULSES

Pulses are required to trigger both line and field timebase generators in the
receiver. In early receiver designs the transmitted synchronizing pulses were
used directly and this imposed a number of requirements on these pulses. The
detail of the line synchronizing pulse is shown in Fig. 11.7. The line flyback is
triggered by the leading edge of the synchronizing pulse. To avoid a ragged
picture the triggering must be fairly precise. This means that the rise time of

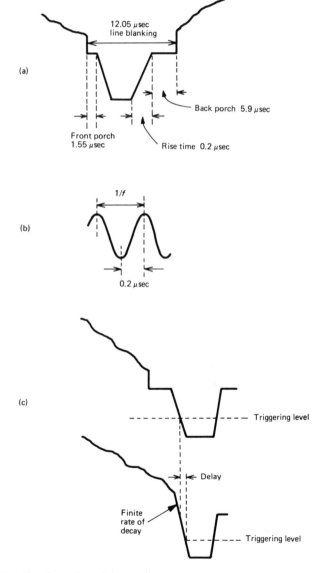

Fig. 11.7 Details of synchronizing pulses.

the pulse should be as short as possible and trigger level should not be affected by the level of the luminance signal just before the synchronizing pulse. In the 625 line system a rise time of 0.2 μs is specified and it may be seen, by reference to Fig. 11.7(b), that this requires a signal bandwidth exceeding $f = 2.5$ MHz. This is well within the allotted bandwidth of 5.5 MHz.

It will be seen from Fig. 11.7(a) that the video signal is reduced to black level for a short interval before and after the line sync pulse; these intervals are known as the front and back porch, respectively. The function of the front porch is illustrated in Fig. 11.7(c). The luminance signal can have any value between peak white and black level at the end of each line. In the absence of the front porch the time taken for the video signal to drop to the triggering level would vary with the amplitude of the luminance signal just before the leading edge of the synchronizing pulse. The timing of the flyback would then vary from line to line giving a ragged picture. The front porch ensures that each line is reduced to black level before the leading edge of the synchronizing pulse. The flyback then occurs at the same instant at the end of each line.

The back porch has two functions – it ensures that the beam is blanked off during line flyback and also provides a convenient reference for restoring the zero frequency component to the signal which has been passed through ac coupled amplifiers.

The field synchronizing pulses occur at the end of each field and their purpose is to trigger the vertical deflection system. It is important that precise triggering is affected to ensure correct interlace of the odd and even fields. The field synchronizing pulses must obviously be distinguishable from the line synchronizing pulses. The picture sync ratio is fixed as 7/3, and hence to distinguish the two types of pulses the field synchronizing pulses are made much wider than the line synchronizing pulses. These pulses have, in fact, a duration of approximately 2.5 line periods.

Modern receivers do not use the transmitted synchronizing pulses directly but use a form of line synchronization known as flywheel synchronization. Instead of using individual sync pulses to trigger each line scan the frequency and phase of locally generated line synchronizing pulses are compared with those present on the incoming video waveform in a phase detector. Any phase error is averaged and fed back to a voltage-controlled oscillator which pulls the two streams of synchronizing pulses into phase coincidence. This averaging effect minimizes the effect of noise and interference on the incoming sync pulses and produces superior picture quality. It is therefore no longer strictly necessary to provide line synchronization pulses during field flyback for modern TV receivers.

Figure 11.8 shows the detail of the field and equalizing pulses transmitted in the UK. The equalizing pulses also relate to early TV designs in which line and field pulses were separated using an integrator. (Integrators produced a higher output for wide field pulses than for narrow line pulses). The equalizing pulses were inserted to ensure that the residual output of the integrator was the same on odd and even fields, thereby ensuring accurate interlacing. In modern receivers line and field pulses are separated in a single IC which also accommodates flywheel synchronization. In such devices simple counter circuits produce the field synchronizing pulses at the correct instants.

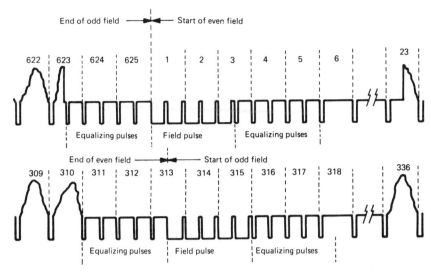

Fig. 11.8 Field sync period in UK 625 line transmissions.

11.7   THE TELEVISION RECEIVER

The domestic television receiver is required to receive signals over a wide bandwidth and is based on the superheterodyne principle. The television receiver is considerably more complicated than a broadcast radio receiver because the former is required to reproduce a video signal, synchronized scanning waveforms for the CRT, and a sound signal. The block diagram of a typical monochrome receiver is given in Fig. 11.9.

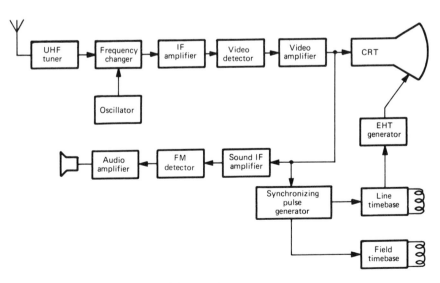

Fig. 11.9 Superheterodyne television receiver.

The British 625 line system transmits the video signal using amplitude modulation of the vision carrier and frequency modulation of a sound carrier 6 MHz above the vision carrier. (The original FM sound carrier is now accompanied by an additional carrier 6.552 MHz above the video carrier which is used for digital transmission of stereophonic sound. The coding scheme used is known as NICAM, which is covered in detail in Section 3.6.) The local oscillator in the receiver operates at 39.5 MHz above the vision carrier, the difference frequency being used as the vision IF. The relative positions of sound and vision carriers will of course be reversed in the IF signal. The spectrum of the IF signal and the frequency response of the IF amplifier are shown in Fig. 11.10.

The asymmetric response of the IF amplifier around the frequency of 39.5 MHz is required to compensate for the vestigial sideband transmission of the vision signal. The detection of VSB signals is discussed fully in Section 2.12. The IF amplifier response is 30 dB down at the FM sound carrier frequency. This is required because both sound and vision carriers are applied

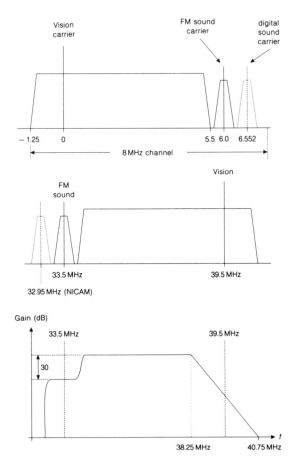

Fig. 11.10 Signal spectra: (a) transmitted signal spectrum; (b) spectrum of 625 IF signal; (c) IF amplifier response.

to the vision detector, which then produces sum and difference frequencies between these two carriers. The difference frequency of 6 MHz is used as the sound IF and this is known as the 'intercarrier sound' principle. This component will be modulated in both amplitude and frequency. The 30 dB drop in the IF amplifier response at the sound carrier frequency ensures that the amplitude of the sound carrier is much less than the vision carrier. The resulting depth of amplitude modulation of the 6 MHz component is then consequently small and easily removed by limiting. (Note that in receivers designed to accommodate NICAM the digital sound carrier is extracted from the mixer output by a filter with a centre frequency of 32.95 MHz.)

Clearly the intercarrier sound principle requires both sound and vision carriers to be present at all times. This requirement is met by ensuring (see Fig. 11.5) that the vision carrier has a minimum amplitude of not less than 18% of its peak value. The demodulated vision signal is the drive signal for the cathode ray display tube. This is a thermionic device, in which the intensity of the electron beam emitted from the cathode is a function of the cathode-to-grid voltage. The electron beam is electronically focused on to a screen coated with a photo-emissive phosphor, which produces the displayed image. The light output of a CRT is not a linear function of the cathode-to-grid voltage, but is equal to this voltage raised to power γ. This means that the output voltage E_Y produced by a television camera would not produce an acceptable image on a CRT. The non-linearity of the display device is equalized by pre-distorting the image capture system output voltage. The transmitted gamma corrected signal is $E_Y^{1/\gamma}$ and has an important but unwanted effect on colour images, as will be considered in Section 11.13.

In addition to gamma correction, the vision signal must be further modified before it is applied to the cathode and grid terminals of the CRT. The dc component of the vision signal, which represents the average brightness of a picture, is removed when the vision signal is transmitted via ac-coupled amplifiers. The dc component also affects the synchronization of line and field timebase circuits, as illustrated in Fig. 11.11 and must be restored to the vision signal. A full discussion of the operation of dc restoration circuits is given by Patchett.[1] After dc restoration the complete vision signal, including synchronizing pulses (which are actually below black level), is applied to the control grid of CRT.

The synchronizing pulses are separated from the picture information and are then used to trigger the line and field scan circuits. The scanning process in television tubes is produced using magnetic rather than electrostatic deflection, the electron beam being deflected at right angles to the magnetic field direction. There are several advantages associated with magnetic deflection systems, the main one being that the deflecting force is proportional to the electron beam velocity and thus increases as the anode accelerating voltage increases. Electrostatic systems, which are commonly used for oscilloscope applications, produce a force that is independent of beam velocity. This means that a high accelerating potential, required to give a bright image, would be accompanied by a small deflection because individual electrons would be influenced by the deflecting force for a shorter interval. The current in the deflection coils has a sawtooth waveform, and during flyback the

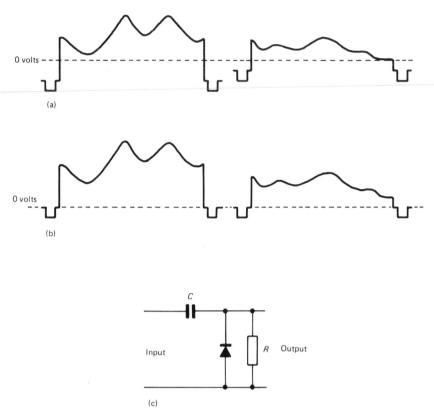

Fig. 11.11 DC restoration: (a) before dc restoration (sync level varies with picture brightness); (b) after dc restoration (sync level fixed); (c) dc restoring circuit.

current changes very rapidly. This produces a considerable induced voltage, which is stepped up by a transformer action and used to form the anode accelerating voltage. A typical accelerating voltage for a colour tube is 15 kV.

Colour television transmission uses the same standards as monochrome transmissions. There are, however, several specific requirements, which must be met to produce acceptable colour images, and in order to understand these requirements it is first necessary to consider the properties of coloured light.

11.8 COLORIMETRY

The colours produced by the display tube in a colour television receiver depend upon the principle of additive colour mixing. This is based on the postulation that a large range of the subjective sensation of colour may be produced by adding certain primary colours in different proportions. This principle is quite different from the principle of the subtractive colour mixing used by artists to create different colours by mixing pigments. A wide range of primary colours can be chosen for the additive colour process, but the primary colours used in television transmission are red, green and blue.

Examples of some colours produced by these primaries are:

$$\text{red} + \text{green} = \text{yellow}$$
$$\text{red} + \text{blue} \ = \text{magenta}$$
$$\text{blue} + \text{green} = \text{cyan}$$
$$\text{red} + \text{blue} + \text{green} = \text{white}$$

Coloured light is usually described in terms of hue (the actual colour), saturation (the dilution of the colour with white light) and luminance (the brightness). The response of the eye is additive, i.e. the luminance of a colour produced by adding three primary colours is the sum of the luminance of the individual primary colours. This algebraic relationship is referred to as Grassman's Law.

White light can be defined in several ways; one convenient definition is equal energy white, in which the energy at all wavelengths has equal brightness because, as is apparent from Fig. 11.1, the sensation of brightness varies with wavelength. Using the response of the standard observer, equal energy white can be expressed in terms of the three primaries, R, G, B as

$$1 \text{ lumen } W = 0.3 \text{ lumen } R + 0.59 \text{ lumen } G + 0.11 \text{ lumen } B \qquad (11.2)$$

In colour television trichromatic units (T units) are used to simplify Eqn (11.2). White light is then said to be composed of equal quantities of red, green and blue light when the latter are expressed in T units:

$$1 \text{ lumen } W = 1T(R) + 1T(G) + 1T(B) \qquad (11.3)$$

where $1T$ unit of red $= 0.3$ lumen, $1T$ unit of green $= 0.59$ lumen, $1T$ unit of blue $= 0.11$ lumen.

In order for Eqn (11.3) to balance, evidently 1 lumen of white $= 3T$ units. Equation (11.3) can be adapted to represent any colour in terms of R, G and B:

$$1T(C) = x\,T(R) + y\,T(G) + z\,T(B) \qquad (11.4)$$

The coefficients x, y and z in Eqn (11.4) are the trichromatic coefficients of the colour C and it is evident that since this equation obeys Grassman's Law then $x + y + z = 1$. This means that the colour C is actually defined in terms of two trichromatic coefficients. If x and y are known then z can be obtained from $z = 1 - (x + y)$. Colour is often represented graphically in the form of a colour triangle as shown in Fig. 11.12, in which white is defined by the point $x = 0.33$ and $y = 0.33$ (hence $z = 0.33$). The point $x = 0, y = 0$, on the other hand, represents saturated blue as in this case $z = 1$. The colour triangle therefore represents both the hue (the actual colour) and the saturation (dilution with white light), for example the point $x = y = 0.25$ represents 50% desaturated blue as $z = 0.5$ is the dominant component. The hue and saturation of any colour which can be produced using RGB primaries lies within the colour triangle of Fig. 11.12, but the range of colours produced in a colour television display system will ultimately depend upon the phosphors used in the CRT. Red emitting phosphors are based on a rare earth material yttrium/oxysulphide/europium, green emitting phosphors are based on zinc sulphide mixed with copper and aluminium, blue emitting phosphors are based on zinc sulphide mixed with silver.

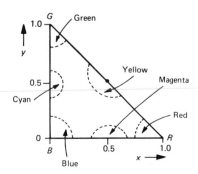

Fig. 11.12 Graphical representation of hue and saturation.

In reality no three primaries exist that can produce the full visible range, and to overcome this problem a universal colour triangle based on hypothetical super-saturated primaries has been defined. All visible colours can then be defined in terms of these hypothetical primaries. This colour triangle is shown in Fig. 11.13.

The primaries used in display tubes in the UK lie within the visible spectrum which can be produced using the hypothetical primaries. The theoretical range of colours that can be produced by a colour television display tube then lies within the dotted triangle shown in Fig. 11.13. It should be noted that although the colour receiver can produce only a limited range of colour, this range is considerably greater than that of high-quality colour film. Equal energy white has the coordinates $x = 0.33$, $y = 0.33$, but the white used for television is slightly different and has the coordinates $x = 0.313$ and

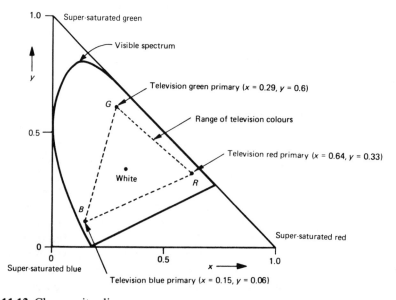

Fig. 11.13 Chromacity diagram.

$y = 0.329$, which is known as illuminant D. This in no way affects the validity of Eqn (11.2).

11.9 REQUIREMENTS OF A COLOUR TELEVISION SYSTEM

It may be concluded, from Eqn (11.4), that in order to transmit colour pictures it is necessary to split the light from the scene being viewed into its red, green and blue components, which are then recombined in the receiver. The bandwidth of the three signal components will be comparable to the figure of 5.5 MHz derived for monochrome signals in Section 11.4. Apart from the excessive bandwidth that such a transmission would require, it would not be compatible with monochrome receivers, i.e. the monochrome receiver would not be able to produce an acceptable black and white image. The requirement of compatibility is an important one and has dominated the development of terrestrial colour television system. However, compatibility is much less of an issue in satellite broadcast systems and this has led to the development of superior colour systems.

This chapter deals primarily with the PAL system which is used in Western Europe (except France and Greece), Australasia and some South American Countries. PAL is based upon the American National Television Standards Committee (NTSC) system, but there are several important differences that will be highlighted at the appropriate point in the text. There are also some minor differences in the PAL standard used in various countries, for example in Australia the sound carrier is 5.5 MHz above the vision carrier. In addition to PAL a brief description of the SECAM system which is used in France, Greece, Cyprus and Eastern Europe is given in Section 11.16, and a brief description of the MAC system, used for some satellite direct broadcast transmissions, is given in Section 11.17.

11.10 THE PAL COLOUR TELEVISION SYSTEM

The basic concept used in all three terrestrial systems mentioned in the previous section is that the colour signal is split into two distinct components. The luminance signal carries the brightness information and is identical to the monochrome signal described earlier. The chrominance signal transmits the colour information (i.e. hue and saturation). Splitting the transmitted information into these categories provides the necessary compatibility because the luminance signal produces an acceptable image in monochrome receivers.

In a monochrome television system, the camera response closely approximates the response of the standard observer. In a colour receiver the red, green and blue signals should approximate to the energy in the primary colours. This is because the eye actually sees the colour image and performs its own relative attenuation on this image, hence when displaying white light red, green and blue signals should be equal in magnitude. The luminance signal can be derived in two ways. One method is to use a standard monochrome camera tube; the other method is to combine the outputs of three separate camera tubes, each tube producing an output corresponding to a different primary colour. When viewing white light the outputs of these three cameras E_R, E_G and E_B are adjusted to be equal. The camera output voltages are therefore interpreted directly in trichromatic units. The luminance signal is derived by combining these outputs according to Eqn (11.2).

$$E_Y = xE_R + yE_G + zE_B \qquad (11.5)$$

where $x = 0.3, y = 0.59$ and $z = 0.11$.

The colour receiver requires a separate knowledge of the E_R, E_G and E_B signals. Ideally the chrominance signal should not transmit any brightness information, as this is already transmitted in the luminance signal. If the effects of gamma correction are ignored it may be shown that this condition is satisfied when the chrominance signal is transmitted in the form of colour-difference signals. The colour-difference signals are:

$$(E_R - E_Y)(E_G - E_Y)(E_B - E_Y)$$

where E_Y is the luminance signal defined by Eqn (11.5)

i.e.

$$\begin{aligned}
(E_R - E_Y) &= (1 - x)E_R - yE_G - zE_B \\
(E_G - E_Y) &= (1 - y)E_G - xE_R - zE_B \\
(E_B - E_Y) &= (1 - z)E_B - xE_R - yE_G
\end{aligned} \qquad (11.6)$$

When transmitting a black and white picture $E_R = E_G = E_B$ and since $x + y + z = 1$, then

$$(E_R - E_Y) = (1 - x - y - z)E_B = 0$$

This result is true for the other colour-difference signals also, indicating that these signals do not contain any brightness information. The original E_R, E_G and E_B can, of course, be reproduced at the receiver simply by adding the luminance signal to the colour-difference signals in turn.

The chrominance signal is in fact completely defined by any two of the colour-difference signals. This can be shown as follows, from Eqn (11.5):

$$E_G = \frac{1}{y} E_Y - \frac{x}{y} E_R - \frac{z}{y} E_B$$

Therefore

$$(E_G - E_Y) = \frac{1 - y}{y} E_Y - \frac{x}{y} E_R - \frac{z}{y} E_B$$

but $(1 - y) = x + z$. Hence

$$(E_G - E_Y) = -\frac{x}{y}(E_R - E_Y) - \frac{y}{z}(E_B - E_Y) \qquad (11.7)$$

The green colour-difference signal can thus be derived from the other two colour-difference signals and need not be transmitted. In fact any one of the colour-difference signals could be reproduced from the other two. The values of $(E_R - E_Y)(E_G - E_Y)$ and $(E_B - E_Y)$ may be calculated for a range of transmitted colours and such calculations reveal that the mean square value of the green colour difference is significantly less than the mean square values of the other colour difference signals. The SNR performance will therefore be greatest when the $(E_R - E_Y)$ and $(E_B - E_Y)$ signals are transmitted.

In addition to providing compatibility, the colour television signal must be confined to the same bandwidth as a monochrome signal, otherwise adjacent channel interference would be a serious problem. The question then has to be answered as to how it is possible to fit the chrominance signal into a bandwidth which is already fully occupied by the luminance signal. Fourier analysis of the luminance signal shows that it does not completely occupy the bandwidth allocated to it. The scanning process may be regarded as a sampling operation, and the spectrum produced will be composed of the sum and the difference frequencies centred on harmonics of the sampling frequency (in this case the line frequency). Using the sampling analogy, and noting that the sampling frequency is much larger than the maximum frequency of the signal being sampled, it is clear that there will be periodic gaps in the spectrum of the sampled signal. This is the case with the luminance signal, which has a spectrum of the form shown in Fig. 11.14.

The chrominance signals will have a similar spectrum; the actual amplitude of the components will of course be different. If it assumed that the red and blue colour-difference signals can be combined, it is apparent from Fig. 11.14 that the chrominance spectrum can be slotted into the gaps in the luminance spectrum. To achieve this it necessary to shift the spectrum of the combined chrominance by modulating a sub-carrier of frequency equal to $\frac{1}{2} n f_L$ where n is an odd integer.

The effect of such a modulated sub-carrier on a monochrome receiver is shown in Fig. 11.15. This component will produce an unwanted modulation of the CRT in the receiver. As the chrominance sub-carrier is an odd multiple of the half-line frequency the unwanted modulation on any line should cancel

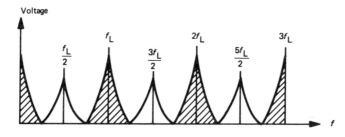

Fig. 11.14 Amplitude spectrum of a luminance signal.

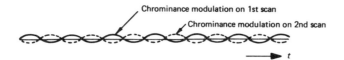

Fig. 11.15 Effect of chrominance signal on a monochrome receiver.

when that line is scanned a second time. In fact this cancellation is not complete. There are several reasons for this, one of the more important being that the light output from a CRT is not linearly related to the grid–cathode voltage (i.e. the decrease in brightness of negative half-cycles of modulation is not equal to the increase in brightness on positive half-cycles of modulation). The result is that a very fine dot pattern is produced on the CRT of a monochrome receiver. This is only visible if the picture is viewed closely and it is usually ignored.

11.11 TRANSMISSION OF THE CHROMINANCE SIGNALS

In the previous section it was assumed that the red and blue colour-difference signals were combined to form a single chrominance signal, the spectrum of which could be fitted into the gaps in the luminance spectrum. The technique used to combine the colour-difference signals uses quadrature amplitude modulation of the chrominance sub-carrier and is similar to the QAM described in Section 3.20. The resulting QAM in this case is given by Eq (11.8)

$$h_c(t) = h_R(t) \cos(2\pi f_{sub}t) + h_B(t) \sin(2\pi f_{sub}t) \tag{11.8}$$

where $h_R(t)$ and $h_B(t)$ represent the red and blue colur-difference signals, respectively. The composite signal may be conveniently slotted into the gaps in the luminance spectrum. Each of the two colour-difference signals is obtained from the chrominance signal at the receiver by coherent detection, which requires the local generation of $\cos(2\pi f_{sub}t)$ and $\sin(2\pi f_{sub}t)$. The outputs of the coherent detectors are:

$$h_c(t) \cos(2\pi f_{sub}t) = \tfrac{1}{2} h_R(t) [1 + \cos(4\pi f_{sub}t)] + \tfrac{1}{2} h_B(t) \sin(4\pi f_{sub}t) \tag{11.9}$$

and

$$h_c(t) \sin(2\pi f_{sub}t) = \tfrac{1}{2} h_B(t) [1 - \cos(4\pi f_{sub}t)] + \tfrac{1}{2} h_R(t) \sin(4\pi f_{sub}t) \tag{11.10}$$

It is clear from these equations that $h_R(t)$ and $h_B(t)$ may be readily obtained by filtering.

Coherent detectors require the locally generated components to be in phase with the suppressed chrominance sub-carriers. The maximum phase error that can be tolerated is in fact $\pm 5°$. The local oscillator thus requires synchronization which is accomplished by transmitting 10 cycles of sub-carrier tone on the back porch of each line synchronizing pulse. The tone is used to synchronize the local oscillators. The detection process is illustrated in Fig. 11.16.

The QAM process is effectively the addition of two double sideband suppressed carriers in phase quadrature. The phasor diagram for the QAM signal may be derived from the phasor diagram of two DSB-AM carriers in phase quadrature as in Fig. 11.17. When the carriers are suppressed the resultant is a single modulated component that varies both in amplitude and phase. It may be shown that the phase of this component represents the transmitted colour (i.e. the hue) and the amplitude of the component

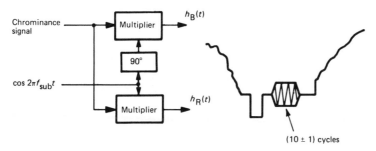

Fig. 11.16 Recovery of colour-difference signals.

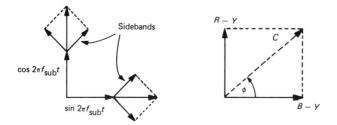

Fig. 11.17 Phasor representation of chrominance sub-carrier.

represents the saturation. The amplitude of the resultant phasor C is proportional to $\sqrt{[(E_R - E_Y)^2 + (E_B - E_Y)^2]}$ and the phase ϕ is proportional to $\tan^{-1}\{(E_R - E_Y)/(E_B - E_Y)\}$. It is convenient to represent various hue and saturation combinations in tabular form. In this context the percentage saturation of any colour is defined in terms of the RGB primaries as

$$S = \frac{\text{maximum amplitude} - \text{minimum amplitude}}{\text{maximum amplitude}} \times 100 \quad (11.11)$$

The hue and saturation for three representative colours are given in Table 11.1 together with the resulting amplitude and phase of the chrominance

Table 11.1 PAL chrominance sub-carrier amplitude and phase

Colour (Hue)	Saturation (%)	E_R	E_G	E_B	E_Y	$E_R - E_Y$	$E_B - E_Y$	C	ϕ
Red	100	1.00	0.00	0.00	0.30	0.70	−0.30	0.76	113.19
Red	50	1.00	0.50	0.50	0.65	0.35	−0.15	0.38	113.19
Red	25	1.00	0.75	0.75	0.83	0.18	−0.08	0.19	113.19
Yellow	100	1.00	1.00	0.00	0.89	0.11	−0.89	0.89	172.95
Yellow	50	1.00	1.00	0.50	0.95	0.06	−0.45	0.45	172.95
Yellow	25	1.00	1.00	0.75	0.98	0.03	−0.22	0.22	172.95
Magenta	100	1.00	0.00	1.00	0.41	0.59	0.59	0.83	45.00
Magenta	50	1.00	0.50	1.00	0.71	0.30	0.30	0.42	45.00
Magenta	25	1.00	0.75	1.00	0.83	0.15	0.15	0.21	45.00

sub-carrier. The last two columns in the table indicate that the phase of the sub-carrier represents the colour transmitted (i.e. the hue) and the amplitude of the sub-carrier represents the saturation. This being so, it is clear that any phase error in the chrominance sub-carrier which may occur during the transmission will thus produce an incorrect hue at the receiver. The PAL system is specifically designed to compensate for this type of phase error and is a development of the NTSC system, which is described in Section 11.14. The sub-carrier frequency used in the British system is 4.433 618 75 MHz, which is equal to $(283.5 + 0.25) \times$ line frequency $+ 25$ Hz. This differs from the odd multiple of the half-line frequency specified earlier (which is the value used in the NTSC system). The reason for the difference is related to the effect of the PAL compensation for sub-carrier phase error. If a sub-carrier frequency of $\frac{1}{2}nf_L$ is used with PAL an objectionable dot pattern is found to 'crawl' across the display. This is avoided by modifying the sub-carrier frequency as indicated.

11.12 THE TRANSMITTED PAL SIGNAL

The bandwidth of the luminance (i.e. monochrome) signal is fixed at approximately 5.5 MHz. In order to fit the modulated sub-carrier into this bandwidth it is necessary to restrict the sidebands to a bandwidth of 1 MHz. The resolution of the chrominance signal is therefore considerably less than the luminance signal. This is not usually detectable by the eye which is insensitive to high-definition colour. The combined luminance and modulated chrominance signal as shown in Fig. 11.18 is then used to modulate the main vision carrier.

This combined signal is less than ideal for several reasons, for instance in the monochrome receiver there is no provision for removing the chrominance sub-carrier and this signal will consequently be applied to the CRT along with the luminance signal. In the colour receiver it is not possible to separate the luminance and chrominance signals completely which gives rise to a form of distortion known as cross-colour distortion. Before discussing these effects in detail it is necessary to consider the significance of gamma correction in the case of colour television.

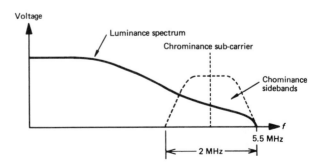

Fig. 11.18 PAL combined lumninance and chrominance signals.

11.13 GAMMA CORRECTION

Most commonly used image capture systems produce an output voltage proportional to the light input and are said to have a gamma value = 1. However, it has already been indicated in Section 11.7 that the light output of a CRT = (grid to cathode voltage)$^\gamma$, where γ has a numerical value of 2.2. To compensate for the non-linearities of the CRT the transmitted luminance signal should be $E_Y^{1/\gamma}$, i.e.

$$E_Y^{1/\gamma} = (0.3E_R + 0.59E_G + 0.11E_B)^{1/\gamma} \qquad (11.12)$$

In a colour system each of the separate colour signals is individually gamma corrected, the resulting luminance signal being

$$E_Y^{1/\gamma} = (0.3E_R)^{1/\gamma} + (0.59E_Y)^{1/\gamma} + (0.11E_B)^{1/\gamma} \qquad (11.13)$$

The actual light output produced at the CRT is obtained by raising Eqns (11.12) and (11.13) to the power γ. Equation (11.12) produces the correct luminance value for all values of $E_R \, E_G \, E_B$ but Eqn (11.13) produces the correct luminance value only when $E_R = E_G = E_B$. This occurs only on black and white scenes. For coloured scenes the light output produced by Eqn (11.13) is less than the light output produced by Eqn (11.12). This means that intensely coloured parts of a picture will be reproduced on a monochrome receiver with a lower luminance than the correct value. This error is reduced, to some extent, by the presence of the chrominance sub-carrier which will have a large amplitude on saturated colours, and no attempt is made to remove this component in monochrome receivers. This is regarded as an acceptable compromise from the compatibility point of view.

The situation in a colour receiver is quite different. The individual gamma-corrected colour signals are obtained and applied simultaneously to the CRT. The light output obtained by raising these individual signals to the power γ is therefore correct, i.e.

$$(E_R^{1/\gamma})^\gamma = E_R$$

The consequence of this is that if the light output is correct in the colour receiver then part of the luminance information in the gamma-corrected colour signals must be contained in the chrominance signal, which is not used in the monochrome receiver. Therefore the gamma-corrected signals do not provide the required monochrome compatibility.

The signal which is used to modulate the vision carrier is E_Y' + chrominance sub-carrier (which is equivalent to a sinusoidal component of varying amplitude and phase). In monochrome transmissions the maximum depth of modulation of the vision carrier occurs on peak white. In colour transmissions maximum modulation occurs on bright yellow, the total video signal amplitude being 1.78 times the peak white value. The next peak occurs at bright cyan, the total amplitude being 1.46 times the peak white value. It is clear from Fig. 11.5 that overmodulation of up to 78% of the vision carrier could result. This overmodulation is reduced to a practically acceptable value of 33% which only occurs rarely. Considering the transmission of bright yellow and taking $\gamma = 2.2$ the values of the luminance and chrominance components are:

$$E'_Y = 0.92 \qquad (E_R^{1/\gamma} - E'_Y) = 0.08 \qquad (E_B^{1/\gamma} - E'_Y) = -0.67$$

For bright cyan the figures are:

$$E'_Y = 0.78 \qquad (E_R^{1/\gamma} - E'_Y) = -0.53 \qquad (E_B^{1/\gamma} - E'_Y) = 0.22$$

Multiplying factors m and n are chosen for the gamma-corrected colour-difference signals to restrict total amplitude of the vision carrier to 1.33, i.e.

$$E'_Y + \{n^2(E_R^{1/\gamma} - E'_Y)^2 + m^2(E_B^{1/\gamma} - E'_Y)^2\}^{1/2} = 1.33 \qquad (11.14)$$

The numerical equations for bright yellow and cyan are

$$0.92 + \{(0.08n^2) + (0.67m)^2\}^{1/2} = 1.33$$

$$0.78 + \{(0.53n)^2 + (0.22m)^2\}^{1/2} = 1.33$$

Solving these equations gives $m = 0.877$ and $n = 0.493$. The γ-corrected colour-difference signals transmitted in the PAL system are called U and V signals and are given by

$$U = 0.493\,(E_B^{1/\gamma} - E'_Y)$$

$$V = 0.877\,(E_R^{1/\gamma} - E'_Y) \qquad (11.15)$$

The original components $E_R^{1/\gamma}$, $E_G^{1/\gamma}$ and $E_B^{1/\gamma}$ may then be reproduced in the receiver for applying separately to the CRT.

It is common practice to use a colour bar test signal known as the CVBS (chroma, video, blanking and syncs) in setting up television receivers and such a signal clearly demonstrates the concept of the allowable overmodulation

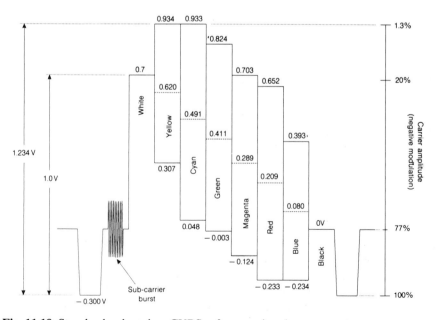

Fig. 11.19 Standard colour bar CVBS reference signal.

outlined above. This standard signal has a nominal value of 1 V peak-to-peak and is the reference signal at the video input and output sockets of cameras, video recorders, monitors and so on. The reference signal is shown in Fig. 11.19. It may be seen from this figure that peak white has a value of 0.7 V, bright yellow has a maximum value of 0.934 V (1.334 × peak white) and bright cyan has a maximum value of 0.933 V (1.332 × peak white).

11.14 THE NTSC CHROMINANCE SIGNAL

The NTSC chrominance signal differs significantly from the PAL equivalent. The NTSC system transmits I and Q signals instead of colour-difference signals. The I and Q signals and their relationship to colour difference signals are shown in Fig. 11.20. The rationale for the use of I and Q signals is based on the observation that the human eye is most sensitive to colour detail in orange and cyan hues and is least sensitive to colour detail in green and magenta hues. This means that the Q signal may be transmitted in a relatively narrow bandwidth, double sideband transmission being employed. The I signal is transmitted in a much larger bandwidth with vestigial sideband transmission.

The I and Q signals may be fully separated at the receiver over the bandwidth of the Q signal, using normal coherent detection with quadrature carriers, as two sidebands are present for both signals. Compensation is necessary for the I signal for the frequencies present in one sideband only, as is usual for vestigial sideband detection.

Restricting the bandwidth of the Q signal in this way means that the highest possible chrominance sub-carrier frequency may be used which produces the finest possible dot pattern on monchrome receivers. In colour receivers the high chrominance sub-carrier frequency means that spurious colour effects

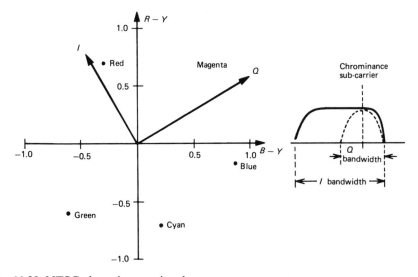

Fig. 11.20 NTSC chrominance signals.

due to adjacent luminance components (which are not fully removed by the sub-carrier detection circuits) are minimized because the luminance energy decreases at the higher video frequencies. This is also true for PAL and is illustrated in Fig. 11.18.

The I and Q signals are derived from the red and blue colour-difference signals according to the relationship specified in Eqn (11.16). These colour-difference signals, and subsequently the colour signals themselves, may be derived from the I and Q signals at the receiver.

$$I = 0.74\,(E_R - E_Y) - 0.27\,(E_B - E_Y)$$

$$(11.16)$$

$$Q = 0.48\,(E_R - E_Y) + 0.41\,(E_B - E_Y)$$

11.15 THE PAL RECEIVER

The PAL system is a development of the NTSC system, in which special provision is made to correct any phase errors that may occur in the chrominance sub-carrier. Such phase errors are mainly due to incorrect synchronization of the local oscillator and sub-carrier generator in the television receiver or to differential phase distortion in the transmission system. It is clear from Table 11.1 that the phase of the chrominance sub-carrier is related to the hue of the image and any phase error will produce incorrect colours. The PAL system compensates for this by reversing the phase of the V signal on alternate lines (PAL in fact means 'Phase Alternation, line by Line'). In the receiver the correct phase of the V signal on alternate lines is restored by averaging the phase error over a period of two lines.

A sub-carrier that has a phase error of θ is illustrated in Fig. 11.21. When the V signal is inverted on alternate lines by the receiver the phase error is also inverted (i.e. a positive phase shift of θ on the received signal becomes a negative phase shift of θ after the phase of the V signal is inverted). If the chrominance information on two adjacent lines is the same the average phase error over this period is zero. In practice the chrominance signal on adjacent lines will not be identical and complete cancellation of phase error will not result.

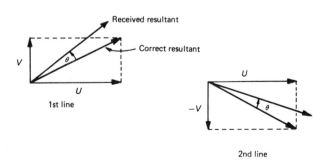

Fig. 11.21 Alternate line inversion of the V signal.

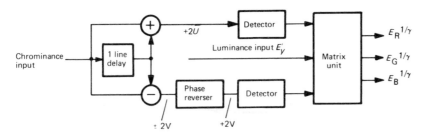

Fig. 11.22 Demodulation of the PAL chrominance sub-carrier.

The averaging is carried out in most modern receivers by adding and subtracting the chrominance signal from its value on the previous line. This requires a delay of 64 μs, which is easily produced using a charge coupled delay line similar to the analogue shift registers shown in Fig. 11.28. A block diagram of the phase error averaging procedure is shown in Fig. 11.22. The matrix unit combines the U and V signals in the required ratios to produce the three gamma-corrected colour-difference signals. These signals are added to the luminance signal to produce the gamma-corrected primary colour waveforms which are fed to the tri-colour display device.

It is clear from the discussion of the PAL system that many compromises are made at various stages of production, transmission and detection of the composite video signal. Non-ideal circuit components invariably add further distortion to the eventual colour image produced. Most of the compromises which occur within the PAL system have been dictated by the requirement of compatibility with monochrome receivers. Ironically the monochrome receiver is now something of a rarity but the limitations due to the requirements of this compatibility remain. In satellite television systems monochrome compatibility was not an issue and much higher quality colour images are possible as a result. Before discussing satellite systems a brief consideration of a third widely used terrestrial system known as SECAM will be given. As with NTSC and PAL this system has also been constrained by the requirements of monochrome compatibility.

11.16 THE SECAM SYSTEM

The *Séquential Couleur à Mémoire* (SECAM) system differs from PAL and NTSC, which both transmit two colour-difference signals simultaneously, by transmitting two colour-difference signals sequentially. The red and blue colour-difference signals are transmitted separately on alternate lines using frequency modulation of the chrominance sub-carrier. There is no local chrominance sub-carrier required in this system, and hence phase distortion is not a problem. However both colour-difference signals are required at the receiver at the same time. SECAM meets this requirement by delaying each colour-difference signal by 1 line interval and using the delayed signal on the next line. Hence each transmitted colour-difference signal is used on two adjacent lines. This obviously halves the colour definition, but this is virtually

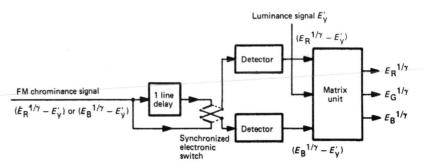

Fig. 11.23 The SECAM decoder.

undetected since the luminance signal transmits the full definition. A block diagram of the SECAM action is given in Fig. 11.23.

The inherent simplicity of the SECAM system suggested by the block diagram of Fig. 11.23 is not in fact realized in practice. The frequency modulated chrominance sub-carrier produces a visible dot pattern on both monochrome and colour receivers which varies with the frequency deviation. Several signal-processing techniques are used to reduce this effect, which consequently increases the complexity of the receiver. Discussion of these techniques is beyond the scope of this text, and the reader is referred to Sims[2] for further information.

11.17 SATELLITE TELEVISION

Satellite communications are covered in detail in Chapter 14. In this section consideration will be confined to the specific issues associated with the transmission of television signals. Direct broadcasting from satellites, in geo-synchronous orbit, to individal homes became established during 1989 with the advent of the Astra 1A satellite. In order to be geo-synchronous, satellites are required to orbit at an altitude of approximately 35 800 km above the equator with a speed of 11 000 km/h, the orbital path is known as the Clarke Belt, after the man who first suggested using satellites for communications (see Section 14.1). Since the geo-synchronous orbital path is fixed it is necessary to regulate the number of satellites permitted and individual satellites are allocated a slot in the orbital path by the World Administrative Radio Conference (WARC). The main satellites visible from Europe are shown in Fig. 11.24.

Satellites are equipped with a number of transponders, receiving programmes and signalling information from an earth station on one frequency and transmitting the programmes back to earth on a different frequency, the frequencies used for the up and down links are in the range 10.95 GHz to 14.5 GHz.

Direct broadcast satellites are usually equipped with up to 10 transponders each of which has a transmitter power of about 100 W. Satellite TV transmissions therefore use frequency modulation to capitalize on the higher

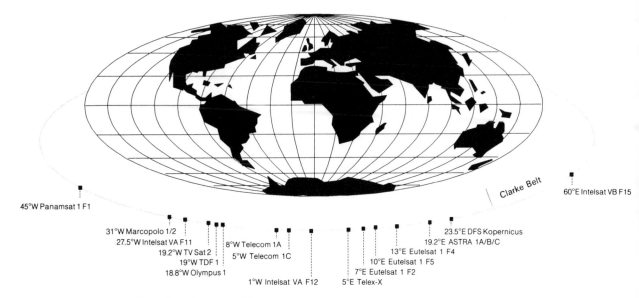

Fig. 11.24 European direct broadcast satellites.

immunity of such transmissions to noise and interference (see Section 4.9). Satellite channels are typically 27 MHz wide and often use the same transmission standards as terrestrial systems in the countries for which they are intended (i.e. PAL, NTSC etc.). These standards are far from ideal as they were designed specifically for amplitude modulation of the vision carrier. However, they do preserve compatibility which reduces the cost of receiving equipment. However, some satellites use an entirely different form of transmission known as multiplexed analogue components system or MAC.

11.17.1 The MAC standard

The terrestrial systems covered earlier in this chapter all exhibit significant compromises which are necessary to maintain compatibility with monochrome transmissions. For example in the PAL system it is not possible to completely separate the chrominance and luminance signals and this produces a form of distortion, in areas of the picture with fine definition, known as cross-colour. This appears in the form of spurious blue/yellow and red/green herringbone patterns. The MAC system is one of several new coding schemes that have been developed for broadcast systems which do not sacrifice quality for the sake of monochrome compatibility and is currently used on the Marcopolo satellite, positioned at 31°W relative to the Greenwich meridian.

The main attribute of the MAC signal is that the chrominance and luminance information are separated completely during transmission so that much higher quality images may be reproduced at the receiver. In addition MAC systems give good quality reception at a signal-to-noise ratio of 11 dB, compared to a figure of about 40 dB for amplitude modulated systems.

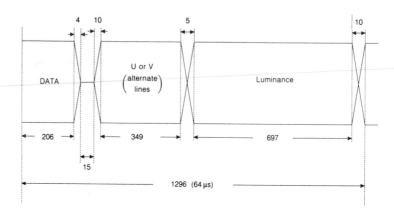

Fig. 11.25 MAC TV line format.

Current MAC transmissions are based on a 625 line standard and therefore are required to maintain the 64 μs line period. In the MAC system this interval is divided into four distinct slots (data, clamp, chrominance and luminance) and time compression techniques are used to accommodate both luminance and chrominance components. The line format for the MAC TV signal in Fig. 11.25.

In this figure the 64 μs line period is divided up into 1296 clock intervals of which 206 intervals are used for data transmission, 15 intervals are used for clamping purposes, 349 intervals are used for the analogue chrominance signal and 697 intervals are used for the analogue luminance signal. The remaining clock intervals are distributed between the necessary ramp-up and ramp-down periods between the four slots. The luminance signal is compressed in the ratio 3:2 and the chrominance signal in the ratio 3:1, a technique for achieving the required time compression is described in Section 11.18 under solid-state image capture devices. The time-compressed luminance and chrominance signals are used to frequency modulate the carrier. The uncompressed luminance signal in MAC has effectively the same bandwidth as in the PAL system. The uncompressed MAC chrominance signal has a bandwidth of 2.8 MHz compared with the 1 MHz figure for PAL.

The MAC digital data signal uses the same carrier frequency as the vision signal but the transmission mode is by PSK. The digit rate is therefore 20.25 Mb/s which fits within the allocated channel bandwidth of 27 MHz. The 206 bit data slot contains one run-in bit, a 6 bit word for line sync, 198 bits of data and one spare bit. The data bits are used to provide several high-quality voice channels, with multi-lingual options and many pages of telextext. The effective data rate is therefore 198 bits for each line period which is 3.09 Mb/s.

The satellite transmission reaches the receiver dish at microwave frequencies (12 GHz for example) and it is necessary to down-convert this frequency to a value suitable for transmission over a coaxial cable to the MAC decoder. The initial amplification and down-conversion to a first IF (usually between 950 MHz and 1.7 GHz) takes place at the point of reception (i.e. at the dish) in

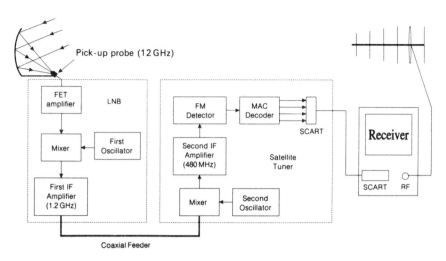

Fig. 11.26 Satellite television receiver system.

a unit known as the low-noise block (LNB). The LNB has a noise figure between 1 dB and 1.6 dB with an overall gain of in the region of 60 dB. The output of the LNB is fed by coaxial cable to a satellite tuner which performs a second down-conversion to an IF of about 480 MHz. The coaxial feeder is also used to supply power to the LNB (typically 200 mA at 15 V). The signal at 480 MHz is then fed to a frequency demodulator and the original luminance and chrominance signals are produced in a MAC decoder. In order to achieve the full potential of MAC transmissions the final output of the decoder can be made available as *RGB* signals and the majority of modern TV receivers on sale in Europe can accept such signals via a EURO-AV (SCART) socket, which means that such a receiver can accommodate both satellite and terrestrial transmissions.

The satellite decoder system actually performs many ancillary functions. For example many broadcasts are scrambled and encrypted to prevent unauthorized reception. Authorized decoders often contain a 'smart card' which may be purchased from the programme company and will facilitate descrambling and decryption enabling normal reception for a fixed period. A typical satellite receiving system is shown in Fig. 11.26. The MAC system is able to accommodate various enhancements, in particular an increase in definition provided by the 1250 line high defintion TV standard with an aspect ratio of 16:9. Using digital techniques this can provide compatibility with the 625 line standard with an aspect ratio of 4:3.

11.18 IMAGE CAPTURE AND DISPLAY SYSTEMS

Image capture systems are the electronic components of the television camera which transform incident light energy into electric signals. The television camera also consists of complex optical systems which are outside the scope of this chapter and will not therefore be covered. There are essentially two forms of image capture systems in widespread use. One is based on a photoconductive vacuum tube, known as the vidicon, and the other form is based on charged-coupled solid-state electronic devices.

11.18.1 The vidicon tube

The construction of the basic vidicon tube is shown in Fig. 11.27. Incident light is focused on to a target disc of continuous photoconductive semiconductor material. The resistance of this material is inversely proportional to the intensity of incident illumination. The photo conductive target is scanned by a low-velocity electron beam, emitted from the cathode, which results in the rear surface of the target being stabilized at approximately cathode potential. The vidicon has electrostatic focusing supplemented by a magnetic focusing coil which is coaxial with the tube. The effect of these two focusing arrangements is to cause individual electrons in the scanning beam to move in a spiral path which coincides with the paths of other electrons at regular distances from the cathode. The focusing arrangements are adjusted so that one of these points of convergence coincides with the target and this produces a scanning spot size diameter in the region of 20 micron.

The front surface, or signal plate, is held at a potential of about 20 V, which produces a current flow through the resistance R_L. When there is no illumination this current is of the order of 20 nA and is known as the dark current. When an image is focused onto the target disc the conductivity of the disc rises in proportion to the intensity of illumination. This allows an electric charge to build up on the rear surface of the target disc and, between scans, the target disc gradually acquires a positive voltage relative to the cathode. The scanning beam deposits sufficient electrons to neutralize this charge and, in so doing, generates a varying current in R_L (typically 200 nA). Since charge is conducted to the rear surface of the target disc over the whole of the interval between scans the vidicon is very sensitive.

However, the vidicon suffers from the two problems of dark current and long persistence (or lag). The dark current tends to produce shading and noise on the low intensity parts of reproduced images. Image persistence arises from the fact that the scanning electron beam does not fully discharge the target plate and this problem is particularly noticeable when levels of

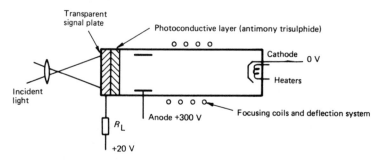

Fig. 11.27 The vidicon tube.

illumination are low. A number of techniques are available to reduce these problems, one of which is the excitation of the face plate by low-level red light generated within the camera assembly.

The vidicon tube described is essentially a monochrome device and will produce a luminance signal. For colour transmission it is necessary to generate the individual RGB signals. This is achieved by using a matrixed face plate in which the front glass surface is covered with thin vertical stripes of RGB colour filter. The vidicon target is similarly divided into vertical strips each one precisely aligned with the corresponding colour filter. All strips on the target corresponding to a primary colour are connected together and brought out to a separate load resistor. Hence by this means it is possible to derive the individual RGB signals from a single tube. These signals are fed to a matrix unit which can produce the gamma-corrected luminance and chrominance signals.

There are several variations on this theme, for example some amateur equipment uses green, cyan and clear filter strips. The target strips behind the green filter produce a G signal, the strips behind the cyan filter produce a $B + G$ signal and the strips behind the clear filter produce a $R + B + G$ signal. It is therefore possible to generate the individual RGB signals by means of a simple matrix unit.

11.18.2 Solid-state image capture devices

Solid-state devices for image capture first appeared in amateur and professional equipment around about 1985. In such devices the photosensitive surface is not continuous but arranged as many thousands of separate silicon photodiodes arranged in horizontal rows equivalent to the lines in a television picture. Each photodiode is therefore equivalent to one pixel and during the 20 ms field period builds up a charge proportional to the light falling on it. Each photodiode is connected to the input of one cell of an analogue shift register, known as a charge-coupled (or bucket brigade) device, by a MOS-FET which is normally OFF. Analogue voltages may be shifted through these devices in the form of the charge on a capacitor. The shift registers are arranged so that adjacent parallel inputs are connected to adjacent photodiodes in a vertical direction as shown in Fig. 11.28.

At the end of each field a transfer pulse is applied to the gate of each of the MOSFETs which causes the charge on each photodiode to be transferred to the appropriate input of the vertical analogue shift register (the number of vertical shift registers is equal to the number of pixels on each television line). Therefore at the end of each field the charge on each pixel is transferred to a cell of one of the vertical analogue shift registers. The complete images is thus stored in the shift registers. The last cell in each vertical analogue shift register is connected to the parallel input of a horizontal analogue shift register. Shift pulses are applied to each vertical shift register simultaneously causing the contents of each cell in the vertical registers to be shifted to the next cell above. The charge in the topmost cell of each vertical register is shifted into one of the cells of the horizontal register.

At this point the cells of the horizontal shift register contain the charge

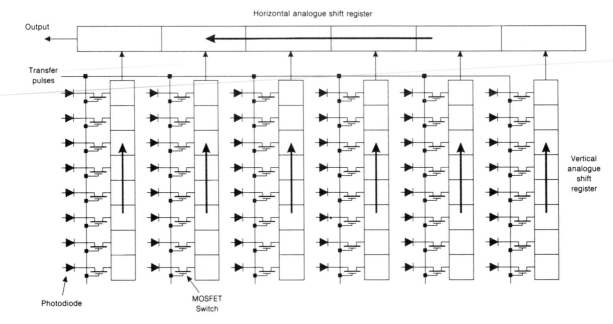

Fig. 11.28 Charge-coupled analogue shift registers in a solid-state image capture device.

values of the pixels of one complete horizontal line. The complete contents of the horizontal register are shifted out serially in an interval corresponding to the line scan interval and, after appropriate filtering, form the analogue video waveform corresponding to one line. When this shift operation is complete the next shift pulse is applied to the vertical registers, thereby loading the horizontal register with the charge values of the pixels of the next complete horizontal line, and so on. The horizontal shift pulse waveform has a frequency of approximately 14.76 MHz, which corresponds to the number of pixels per second given by Eqn (11.1).

It may be noted here that the horizontal shift register may be simply employed to perform time compression of the luminance signal as is required in the MAC transmission system described in Section 11.17. A time compression of 3/2 would be achieved by clocking out the contents of horizontal shift register (one line) at 1.5×14.76 MHz.

Clearly there is no scanning waveform or deflection system required with this type of device and the clock and drive pulses are produced in a timing/divider integrated circuit driven by a precision crystal oscillator. The shift pulses required for the analogue shift registers are more complex than the simple diagram of Fig. 11.28 would suggest. In practice a four-phase switching waveform is required. Techniques for generating colour signals from charge-coupled devices are based on similar principles to those of the vidicon tube.

11.18.3 Display devices

Display devices in colour television systems are largely based on cathode ray tubes, but intense research is ongoing to perfect a large screen solid-state equivalent. This section will be confined to current practice and the basic construction of the colour display tube is as shown in Fig. 11.29.

The colour display tube is required to produce three separate images in the primary colours. The formation of a single colour image is then dependent on the averaging effect of the eye. The electron gun in the colour display tube contains three separate cathodes in line abreast formation. Each cathode produces an electron beam and the three beams are deflected magnetically to form the usual scanning action, but each beam is arranged to strike only the screen phosphor corresponding to its colour. These phosphors are arranged in vertical strips on the screen of the tube each strip producing either red, green or blue light when excited by the electron beam. To ensure that the output from each cathode strikes only its own colour phosphor a mask is placed about 12 mm in front of the screen. This mask is composed of elongated slots which allow the passage of electrons in the scanning beams. The slots are positioned so that electrons passing through strike only the appropriate phosphor and a colour picture is therefore produced.

The mask is clearly a major source of inefficiency in the tube as only about 20% of the incident beams actually reach the phosphor. This means that the mask itself absorbs considerable energy and, in so doing, heats up. Special arrangments are made so that expansion of the mask does not result in inaccuracy when the electron beams strike the screen. Special mounts are incorporated within the tube so that the mask expands in an axial direction.

A variant on the display tube described above is known as the Trinitron. The main difference is that the shadow mask is composed of slots which run the complete height of the tube (rather than the elongated variety shown in

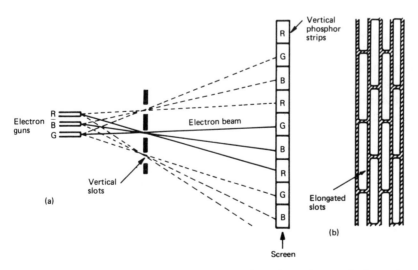

Fig. 11.29 Precision in-line display tube: (a) plan view of tube layout; (b) front view of mask.

Fig. 11.29). There are also some differences in the electron gun assembly and focusing arrangements which allow a smaller spot size than is possible with the arrangement of Fig. 11.29. However, the operating principles of the in-line display tube and the trinitron are broadly similar.

11.19 TELETEXT TRANSMISSION

There are essentially two forms of data transmission for television receivers in the UK known as teletext and viewdata. Teletext is transmitted directly by the broadcast companies and viewdata is transmitted over the switched public telephone network (and therefore requires a modem). Telext has been developed by individual broadcast companies to a common standard and it is possible to display many hundreds of pages of information on each network. It was noted in Section 11.4 that 25 lines on each field are blanked out to allow for the field flyback to return the scan to the top of the picture. Some of these lines (2.5 per field) are used to transmit field synchronizing and equalization pulses and receiver circuits are adjusted so that the remaining blank lines do not appear on the screen. These remaining lines are therefore available for the transmission of data pulses that may occur above the black level. The teletext specification allows for 16 lines on each field to be used for data transmission logic 0 being equivalent to black level and logic 1 being equivalent to 66% of peak white level. These data signals are undetected on a standard receiver but can be separated from the normal video signal in specially equipped receivers.

Each page of teletext information contains up to 24 rows of text with 40 characters per row. Each character is represented by a 7 bit international code with an odd parity check (7 information bits + 1 parity bit ⇒ 8 bit byte). The odd parity has an additional receiver synchronization function when all 7 bits of the standard code have the same value. Unlike the asynchronous data transmission, described in Chapter 4, teletext transmission is synchronous and is therefore more efficient because start and stop bits are not required. Synchronizing information is required, however, and this is transmitted during the first five bytes of each row. This means that each row contains a total of 45 eight bit bytes.

The format of each line is shown in Fig. 11.30. The bits in each byte are identified by sampling the data at the centre of each bit interval, the sampler

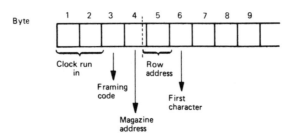

Fig. 11.30 Format of a telext row.

being driven by a locally generated clock. This means that each row contains 360 bits, these bits being transmitted during one line scan interval using NRZ pulses. The bit rate used is 444 × line frequency or 6.9375 Mb/s which means that 360 bits occupy 51.89 µs which allows 12.1 µs for line blanking purposes. The television receiver has a local clock running at this frequency which is synchronized with the incoming data stream by the first two bytes of each line. These bytes contain alternate 1s and 0s at the clock frequency and produce what is known as **clock run in**.

Once the local clock is synchronized it is necessary to detect the beginning of each individual 8 bit byte in order that correct decoding can take place. This is achieved by transmitting a special framing code 11 100100 as byte 3. The data stream is used as a serial input to an 8 bit shift register. When the register contains the framing code a flag is set indicating that the next bit will be the first bit of byte 4, i.e. byte synchronization is achieved. Rows 4 and 5 contain a row address (see later), the first character byte is byte 6 and the first step in decoding a character is to check the parity. A circuit that will produce an 8 bit parity check is shown in Fig. 5.9 and is composed of seven exclusive-OR gates. If a parity check fails (i.e. if the output of this circuit is a binary 0) the character is not decoded, but is replaced by the code for a blank.

Each page of text is composed of 24 rows, hence it is necessary for the decoder to be able to select individual pages and once a page has been selected, to assemble the rows of the page in the correct order. The rows within each page are recognized by a row address code that is transmitted in bytes 4 and 5 of each row. Since there are 24 rows in each page, a minimum of 5 bits is required to specify the number of row addresses. It is necessary at this point to consider the effect of occasional errors. An error in a character byte can be tolerated simply by omitting the character when this occurs. However, an error in the bytes containing the row address is much more serious, as this can cause a row to be wrongly placed within a page. To reduce the probability of such errors occurring, the page and row address are error-protected using a Hamming code. Hamming codes and error detection and correction are discussed in detail in Chapter 5. The code used for row addressing has a Hamming distance of 4, which means that it can correct a single error and can detect a double error.

To provide a Hamming distance of 4 each address bit is accompanied by a parity check bit, hence a total of 10 bits (address + parity) is required. Eight of these bits are transmitted in byte 5 of each row and the other two in byte 4. The remainder of byte 4 is used for a Hamming-coded magazine address which forms part of the page identification. Both row and magazine address codes are transmitted least significant bit first. The format is shown in Fig. 11.31.

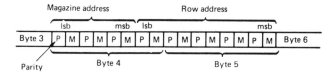

Fig. 11.31 Format of magazine and row address bits.

The function of the row address is to direct the following 40 bytes of text data to the appropriate locations in the page memory. Before this is done the required page must be selected. The top row of each page is called the header row and has the row address 00000. This row contains only 32 text characters instead of the usual 40. The first 8 bytes of text data in the header row carry a page number code, a time code and a control code. These eight bytes are Hamming coded in the same way as bytes 4 and 5 of the other lines. The text in the header row, with the exception of the page number display, is the same for each page.

The page address code (tens and units) is transmitted as two 4 bit binary coded decimal (BCD) numbers with the appropriate Hamming error coding. BCD is used because this can be compared directly with the page selection number entered by the viewer from the decimal keyboard of his/her remote control unit. The page address code occupies bytes 6 (units) and 7 (tens) of the header row. The next four bytes are used for transmitting the minutes (units and tens) and hours (units and tens) of the time code, the Hamming coded BCD format being used for these bytes also. The next two bytes are used to transmit control information.

11.19.1 Page selection

The page is selected by keying in the required three digit number on the viewer's key-pad. Each page is identified by a combination of the page code in the header row and the magazine address code that is transmitted as part of byte 4 of every row. The magazine row contains three digits that are Hamming coded and can thus have any value between 0 and 7. This code is used as the 'hundreds' of the page number identification. The page code is selected from the header row and the magazine code is selected from each transmitted row of text. When a complete match occurs, the following data is written into the page memory. The row address codes are used at this stage to select the appropriate locations within the page memory. When another header row is detected (row address 000000) the page address code are again compared. If there is no match the following data is ignored. In this way only the data from the requested page is transferred to memory. A block diagram of the page selection hardware is shown in Fig. 11.32. If an uncorrected error

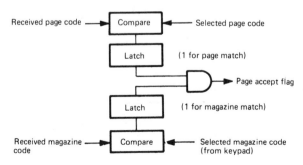

Fig. 11.32 Page selection schematic.

occurs in either the page or row address codes the following data is ignored until the requested page header is repeated.

Normally all header rows are displayed until the requested page header is received. This avoids a blank screen that could occur for up to 30 seconds in an average-sized magazine. In this case the only part of the header that changes on the display is the page number and the time. Full pages are transmitted at the rate of approximately four per second, which means that 25 seconds are required to cycle through 100 pages. This can be inefficient when several blank rows within a page exist. To increase the efficiency of transmission, the page memory is completely filled with the code for 'space' when a new page code is keyed in. Blank lines within a page can then be omitted, i.e. gaps can occur in the row address. Filling the page memory with blanks prevents the row of a previous page from being displayed when the requested page has rows omitted. This technique increases the speed of transmission by up to 25%.

If there are several pages dealing with a common subject, they are sent out in sequence using the same page number. Each new page is transmitted after a delay of about 1 minute, which is sufficient time for the viewer to read the displayed text. The whole sequence of 'self-changing' pages is then repeated continuously.

11.19.2 The page memory

In order to produce the illusion of a fixed image the data representing each page must be scanned 50 times per second. Hence there is a requirement for a complete page memory that can then be scanned sequentially at the required rate. Since each page has 24 rows of 40 characters and each character has a 7

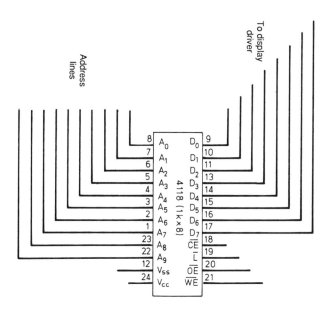

Fig. 11.33 Page memory based on a 1K × 8 bit RAM.

bit code the minimum storage requirement is 6720 bits per page. This may be readily achieved by use of a 1024 × 8 bit random access memory (RAM) as shown in Fig. 11.33.

11.19.3 Character display

The character set in a teletext display is produced using a 6 × 10 dot matrix. The format allows both upper and lower case letters and special purpose symbols to be displayed. The vertical dot resolution is made equal to the line spacing, which means that ten scanned lines are required for one row of text. A typical example is shown in Fig. 11.34.

Each character is in fact represented by a 5 × 9 dot matrix, column 6 being reserved for character separation and line 10 being used for row separation. The character patterns are stored permanently in a teletext read only memory (TROM). The scan proceeds on a line-by-line basis. During each line scan any character will have a corresponding 5 dot code. Hence the TROM must store a series of 5 bit numbers corresponding to the data code. The TROM is addressed by a combination of the character code (7 bits) and the line number (4 bits are required for one to ten lines), the appropriate 5 dot code is then stored at each 11 bit address.

As each line is scanned, the character codes in the particular row of text being displayed are placed upon the address lines of the TROM in sequence together with the 4 digit line scan code. This produces the correct sequence of 5 dot codes on the data lines of the TROM. Smoother characters than the one illustrated can be generated by producing slightly different dot codes on alternate scans. This is known as character rounding.[3]

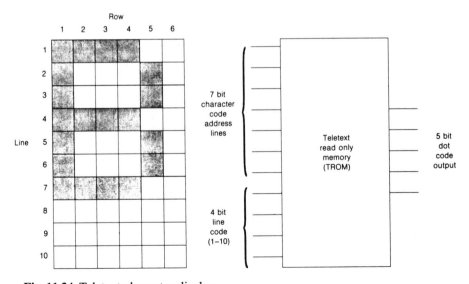

Fig. 11.34 Teletext character display.

11.19.4 Graphics

The teletext display is not restricted to text but can also operate in a graphics mode. This mode is used for the display of simple diagrams and extra large characters. The display in the graphics mode is also divided into a dot matrix, but in this case a 2×6 rather than a 6×10 matrix is used. The graphics mode is indicated by a control code and when in this mode each 7 bit character is interpreted directly as a graphics symbol. The graphics dot matrix and the corresponding bit number is shown in Fig. 11.35. The total number of graphics symbols which can be defined is $2^6 = 64$ symbols.

B1	B4
B2	B5
B3	B6

Fig. 11.35 Graphics dot matrix.

There are many other features of teletext, such as mixing of text and graphics, colour of display, super position of teletext on normal programme pictures, etc. that it is not possible to cover in this volume; the interested reader is referred to Money.[3]

11.20 VIEWDATA

This is a generic term for systems which retrieve and display computer-based information and interactive services using the public switched telephone network and a television receiver or monitor. The main difference between teletext and viewdata is that two way communication is provided between the user and database, which requires a modem to connect the television receiver with the telephone network. This makes possible such services as electronic shopping, home banking etc. The other significant difference is that viewdata services (such as Prestel, which is offered by British Telecom) operate on a menu-driven principle in which the operator selects an item in a series of branches from intermediate menus. The display format of viewdata systems uses the same standard as teletext and thus a specially adapted receiver can be used for both services. A typical set-up is shown in Fig. 11.36.

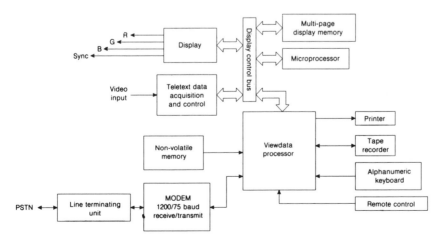

Fig. 11.36 Basic viewdata system.

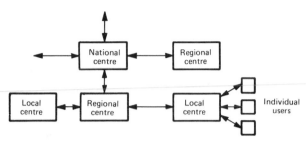

Fig. 11.37 UK national viewdata network.

In the viewdata system only one page is transmitted in response to the page number keyed in by the user. The arrangement of the British national viewdata network is shown in Fig. 11.37. Each of the local centres contains a computer with about 5×10^4 pages of information. Most of the information accessed by the average user will be stored at the local centre, hence the pattern of telephone usage will be on a local call basis. The regional centre is the next stage in the hierarchy. If the requested information is not stored at the local centre it will be automatically transferred from the regional centre over a high-speed data link. The regional centres are connected by high-speed links to a national centre. By this means any user is able to access local information from any one of the regional or local centres. The capabilities of the system in terms of the amount of information are vast, and can, of course, be extended to an international centre at a future date.

A second difference between teletext and viewdata is that the viewdata system uses a cursor instead of page and row address. The position of the cursor is, in effect, controlled by the contents of two counters. One counter represents the position of a character within a row, the other represents the position of a row within a page. The viewdata system uses a similar 7 bit code to teletext for character transmission plus a parity check bit. The viewdata transmission uses an even parity in contrast to teletext which uses odd parity. There are thus $2^7 = 128$ possible characters that can be transmitted, several of which are used for cursor control. For example the code 0001100 (form feed) causes the cursor to return to the top left-hand corner of the screen (i.e. character and row counters are set to zero) and the page memory is filled with blanks. As symbols are received the cursor is moved along the top line (i.e. the character counter is reset and the row counter is incremented by 1. The location address in the page memory RAM is formed from a combination of character and row counter. The cursor can be placed at any point on the screen by using cursor controls, this avoids transmitting blank rows and blanks within a row. For instance the code 0001101 (carriage return) causes the cursor to return to the beginning of the current line (resets character counter to zero) and the code row counter).

11.20.1 Line signals

Data is transmitted to the subscriber using frequency shift keying with asynchronous transmission as described in Section 3.8. The data rate used is 1200 baud with binary 1 being transmitted as a 1300 Hz tone and binary 0 as a

2100 Hz tone. This means that a modem is required in the television receiver to interface the FSK signals. This modem also allows transmission of the page-selection code in the reverse direction, again using FSK. The transmission speed in the reverse direction is 75 baud, a 390 Hz tone being used for binary 1 and 450 Hz for binary 0. The viewdata modem is also required to drive the line and field scan circuits of the CRT in the receiver. This means that the viewdata service, unlike the teletext service, does not require the provision of broadcast line and field synchronization. It is thus available outwith normal broadcast periods.

11.21 CONCLUSION

This chapter has outlined the fundamental engineering principles of broadcast television systems and has shown that, in general, the development of transmission standards has been largely influenced by requirements of compatibility. The advent of satellite television has broken this link and the prospect now exists for the development of the full potential of colour television using digital signal processing techniques, which has recently become possible at video frequencies. The next obvious development is the introduction of a high-definition television standard and associated solid-state, ultra thin, large screen display devices. The advantages of a single global standard are clear but whether such a standard will emerge it is not yet possible to predict.

REFERENCES

1. Patchet, G. N., *Television Servicing*, Vol. 2, Norman Price, 1971.
2. Sims, H. V., *Principles of PAL Colour Television*, Newnes-Butterworth, London, 1974.
3. Money, S., *Teletext and Viewdata*, Newnes-Butterworth, London, 1979.
4. Trundle, E., *Television and Video Engineer's Pocket Book*, Newnes, London, 1992.

PROBLEMS

11.1   The transmission standards for television in the USA are as follows:

number of lines/picture = 525
number of fields/picture = 2
number of picture/second = 30
field blanking = 14 lines
line blanking = 14 μs
displayed aspect ratio = 1.33:1

Differentiate between the transmitted and displayed aspect ratio and calculate a value for the former figure. What is the theoretical bandwidth of the transmitted video waveform?

Answer: 1.545:1, 6.37 MHz.

11.2 If the standards of the previous question were modified to accommodate 3 fields/picture what is the theoretical bandwidth of the transmitted waveform? Increasing the number of fields/picture produces a corresponding reduction in signal bandwidth. Comment on this statement and suggest why, in practice, there is a limit to the number of fields/picture.

Answer: 4.25 MHz.

11.3 The UHF television allocation allowance for channel 50 is 702 MHz to 710 MHz. A television receiver with an IF of 39.5 MHz is tuned to this channel. Calculate the frequency of the local oscillator in this receiver.

Using the channel allocation table (Table 11.2) determine which channel forms the image frequencies for channel 50.

Answer: 743.25 MHz, channel 60.

11.4 Three primary light sources *RGB* are designed to produce an output light power that is linearly proportional to a control voltage. When these three sources are used to match equal energy white light of intensity 1 lumen, the required voltages are $R = 6.9$ V, $G = 3.8$ V, $B = 8.6$ V. The same light sources are used to match an unknown colour, the corresponding voltage being $R = 5.6$ V, $G = 2.2$ V, $R = 1.5$ V. Find the trichometric coefficients of the unknown colour and its luminance. What are the chromacity coefficients of this colour on the colour triangle?

Answer: 0.81 *T*, 0.58 *T*, 0.17 *T*, 0.6 lumen, 0.52, 0.37.

Table 11.2 UHF channels and frequencies (British Isles)

Channel	Vision (MHz)	Sound (MHz)	Bandwidth (MHz)	Channel	Vision (MHz)	Sound (MHz)	Bandwidth (MHz)
Band IV							
21	471.25	477.25	470 – 478	29	535.25	541.25	534 – 542
22	479.25	485.25	478 – 486	30	543.25	549.25	542 – 550
23	487.25	493.25	486 – 494	31	551.25	557.25	550 – 558
24	495.25	501.25	494 – 502	32	559.25	565.25	558 – 566
25	503.25	509.25	502 – 510	33	567.25	573.25	566 – 574
26	511.25	517.25	510 – 518	34	575.25	581.25	574 – 582
28	527.25	533.25	526 – 534				
Band V							
39	615.25	621.25	614 – 622	54	735.25	741.25	734 – 742
40	623.25	629.25	622 – 630	55	743.25	749.25	742 – 750
41	631.25	637.25	630 – 638	56	751.25	757.25	750 – 758
42	639.25	645.25	638 – 646	57	759.25	765.25	758 – 766
43	647.25	653.25	646 – 654	58	767.25	773.25	766 – 774
44	655.25	661.25	654 – 662	59	775.25	781.25	774 – 782
45	663.25	669.25	662 – 670	60	783.25	789.25	782 – 790
46	671.25	677.25	670 – 678	61	791.25	797.25	790 – 798
47	679.25	685.25	678 – 686	62	799.25	805.25	798 – 806
48	687.25	693.25	686 – 694	63	807.25	813.25	806 – 814
49	695.25	701.25	694 – 702	64	815.25	821.25	814 – 822
50	703.25	709.25	702 – 710	65	823.25	829.25	822 – 830
51	711.25	717.25	710 – 718	66	831.25	837.25	830 – 838
52	719.25	725.25	718 – 726	67	839.25	845.25	838 – 846
53	727.25	733.25	726 – 734	68	847.25	853.25	846 – 854

11.5 The output voltages from a three-tube colour camera when viewing equal energy white light are adjusted such that $R = G = B = 1.0$ V. When the camera is directed towards an object of uniform colour the output voltages are $R = 0.56$ V, $G = 0.21$ V and $B = 0.75$ V. Calculate the amplitude of the resulting luminance signal when:
(a) gamma correction is applied after the formation of the luminance signal;
(b) gamma correction is applied to each colour separately.

What is the percentage difference in the luminance produced when these two signals are applied to a monochrome CRT? Assume the overall gamma is 2.2.

Answer: 0.639 V; 0.617 V; 7.42%.

11.6 When the output voltages of the camera of the previous question are $R = 0.5$ V, $G = 0.9$ V and $B = 0.8$ V, find the percentage saturation for this colour.

What would the values of the output voltages for a fully saturated colour of the same hue?

Answer: 44%, 0 V, 0.3 V.

11.7 The row addressing data in a teletext transmission is coded to have a Hamming distance of 4. If the probability of a single digit error is 1×10^{-4}, find the probability that an undetected error will occur in the coded address information. State all assumptions made.

Answer: 4.33×10^{-4}.

12 Optical fibre communications

Optical fibre communications present the most exciting, and probably the most challenging, aspects of modern systems. Fibres are exciting because they seem to offer so many benefits – low cost, enormous bandwidth, very small attenuation, low weight and size, and very good security against external interference. Physically, fibres occupy very little space, and they are so flexible that they can be used in places that would not be accessible to conventional cable.

Optical fibre is still expensive in the early 1990s, and there is a strong debate about whether it can justifiably be taken to every home. The cost of replacing copper would be very large, and it would probably not be recovered by a correspondingly large increase in domestic traffic. If video services appear, and are attractive to the user, then that balance of costs could change, and the unsatisfied demand could be tapped economically by using fibre.

There was a debate in the mid-1980s about whether to use monomode or multimode. That may have been resolved. The difficulties of managing monomode fibre have been overcome, and the very significant benefits that it has over multimode make it the unreserved first choice for communications applications. As noted later, the development of fibre amplifiers has given fibre communications a significant boost, making possible transmission over any terrestrial distance, and offering scope for extensive, low-loss optical fibre distribution systems.

Eventually, all-optical systems will appear, but at present the switching elements are at the research stage, and we will not consider them further here.

The optical components of a fibre communications system are, in simple terms, a light emitter, which initiates the optical signal, a fibre which transmits it, and a detector which receives it and converts it into an electrical equivalent. If several fibres need to be joined, end to end, the couplers must ensure that the fibres are correctly aligned and butted, to reduce any joining losses to a minimum.

Each of these components has essential ancillary parts; the detector and emitter are driven by stabilized voltages, and mounted in such a way that maximum transfer of light between them and the fibre is achieved. The fibre itself must be clad in some sort of protective coating and made up into a cable that will withstand the rigours of installation over long distance. However, although these factors are essential, they are, in a sense, of secondary importance, and we shall not be considering them further here. Our interest is confined to the operating principles of the basic devices, and, for further information on system details, reference should be made to the literature listed at the end of this chapter.

Before considering some aspects of a system, we will examine the way in which the fibres, detectors and emitters work.

12.1 OPTICAL FIBRE

An optical fibre is, in essence, a dielectric waveguide. It has been known for a long time that high-frequency electromagnetic energy can be transmitted along a glass or plastic rod and, indeed, observation shows that short rods are translucent to light. However, two factors prevented that knowledge from being used to produce useful light guides:

(i) energy leaked from the outside of the dielectric to the surrounding air, and
(ii) the attenuation was so large that worthwhile lengths could not be achieved.

The first difficulty, though virtually insurmountable at microwave frequencies, can be overcome in the optical and infrared parts of the spectrum by enclosing the guide in a cladding of similar material, but which has a slightly smaller refractive index. The boundary between the cladding and the core acts as a reflecting surface to the transmitted light. The second problem, that of high attenuation, could be reduced only by refining the methods of producing and drawing the glass so that the impurities and irregularities were reduced to a minimum. The attenuation now achievable in the laboratory is almost as low as possible, at about 0.2 dB/km.

Fibres of varying quality are used for communications, but when distances are significant, care is taken to ensure that the lowest attenuation possible is achieved. This involves choosing the best operating frequency for the particular fibre material, and ensuring that any contaminating elements are removed from the glass during manufacture. Before considering the loss mechanisms inherent in any fibre, we will look at the different fibres used, and examine, with the help of ray theory, the way in which light propagates along an optical waveguide.

12.2 STEPPED-INDEX FIBRE

As we noted earlier, light can be made to propagate down a fibre waveguide consisting of an inner dielectric, of refractive index n_1, and an outer cladding of refractive index n_2, if n_1 is slightly larger than n_2. In Fig. 12.1 we see the path taken by a beam incident on the end face of the fibre at an angle of incidence θ_0. Outside the fibre, the air is assumed to have a refractive index n_0.

By the laws of refraction,

$$n_0 \sin \theta_0 = n_1 \sin \theta_1 = n_1 \cos \theta_2 \qquad (12.1)$$

At the core–cladding boundary

$$n_1 \sin \theta_2 = n_2 \sin \theta_3$$

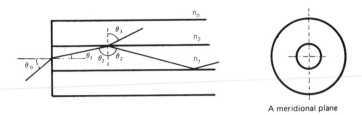

A meridional plane

Fig. 12.1 Path of meridional ray in fibre.

Since $n_2 < n_1$, total reflection will occur at the boundary for all values of θ_2 such that

$$\sin \theta_2 \leqslant n_2/n_1 \qquad (12.2)$$

From Eqn (12.1), the maximum value of θ_2 that will produce a reflected ray at the core–cladding interface is therefore given by

$$n_0 \sin \theta_{0\text{max}} = n_1 \cos \theta_2$$

where, from Eqn (12.2)

$$n_1 \cos \theta_2 = n_1 [1 - \sin^2 \theta_2]^{1/2} = n_1 \left[1 - \left(\frac{n_2}{n_1} \right)^2 \right]^{1/2}$$

Hence

$$\sin \theta_{0\text{max}} = \frac{1}{n_0} [n_1^2 - n_2^2]^{1/2} \qquad (12.3)$$

The value of $\theta_{0\text{max}}$ given by Eqn (12.3) is known as the maximum acceptance angle for the fibre. From that equation we can see that it is determined by the refractive indices of the core and the cladding. The relationship

$$\text{NA} = [n_1^2 - n_2^2]^{1/2}$$

is known as the numerical aperture of the fibre. It is a useful indicator of the launching efficiency of the guide, and we will consider it later in relation to light sources.

When the surrounding medium is air, $n_0 = 1$, and

$$\theta_{0\text{max}} = \sin^{-1}(\text{NA}) \qquad (12.4)$$

In θ_0 exceeds $\theta_{0\text{max}}$, θ_2 exceeds the value for which total reflection takes place at the core–cladding interface, and some of the energy will be refracted out into the cladding itself, where it is absorbed.

12.3 SKEW RAYS

The ray we have just discussed, with the help of Fig. 12.1, travels in a plane through a diameter of the fibre, and it is called a meridional ray. However, rays may also be launched into the guide in other directions, and these are

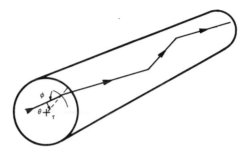

Fig. 12.2 Path of skew ray.

known as skew rays. Provided the angle of incidence of a skew ray falls within $\theta_{0_{max}}$ it will propagate along the fibre by total internal reflection, but, rather than following a path back and forth across a diameter, it will travel in a helix, as shown in Fig. 12.2.

12.4 MODES

The ray-theory approach to the mechanism of propagation along a fibre has the attraction of being simple, and so gives a good insight into fibre transmission. It does have some disadvantages, however. It assumes that the wavelength of the signal is extremely small, and therefore that the light can travel at any angle of incidence to the cladding, provided that the maximum acceptance angle is not exceeded. In fact that is not the case. As with all guides used to transmit electromagnetic waves, the boundary imposes conditions that restrict propagation to a series of modes. The details of these modes can be found by solving Maxwell's equations for the particular guide being studied. In the case of optical fibres, the solution is complicated by both the dielectric nature of the material, and its cylindrical geometry. We will not discuss these wave-theory solutions here, and for details of how they are obtained, reference should be made to one of the many specialist texts available.[1] We can note that as a result of solving Maxwell's equations, the various propagating modes in stepped-index fibre can be shown to have different wave velocities, a property that is called dispersion. As a result of this mode dispersion, a pulse of light launched on to the guide will, because it consists of energy in several modes, gradually broaden as it travels. This dispersion places a limit on the bandwidth of the fibre, a limit that is a function of the stepped-index profile, and not of the quality of the fibre material.

12.5 GRADED-INDEX FIBRE

The effect of mode dispersion can be minimized, and in theory eliminated,[2] by using a graded profile of the shape shown in Fig. 12.3. The difference between theory and practice is that it is very difficult to produce the exact profile

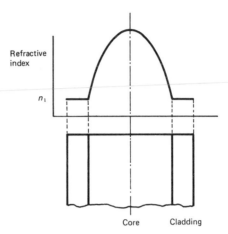

Fig. 12.3 Ideal graded profile to minimize mode dispersion.

required, and the effect of a slight deviation from it is marked. Not only is it difficult to produce the fibre to conform with theory, but there is bound to be some variation of profile with length, and this will allow dispersion to occur, with a consequent reduction in bandwidth. The ideal profile is almost a square law, i.e. the index of refraction of the core is given by

$$n^2(r) = n_1^2 \left[1 - \frac{2[n_1^2 - n^2(a)]}{2n_1^2} \left(\frac{r}{a}\right)^g \right] \tag{12.5}$$

where $n(r)$ is the refractive index at radius r, $n_1 = n(0)$ where the exponent $g \simeq 2$.

The graded-index changes the paths followed by the waves travelling down the fibre. In terms of the ray-theory of Fig. 12.1, the path is sinusoidal in shape, with the light travelling more quickly at the core–cladding boundary than in the centre of the guide, where the refractive index is at a maximum.

Skew waves will still follow a helical path, with their velocity increasing with distance from the centre of the fibre.

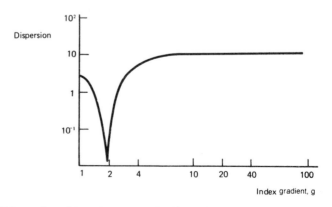

Fig. 12.4 Effect of profile index on mode dispersion.

The effectiveness of a particular profile on reducing the mode dispersion is also a function of wavelength, so it is possible to reduce the dispersion in a fibre by operating it at its most favourable wavelength.

We can see the effect of varying the profile index on the mode dispersion from Fig. 12.4. The sharp minimum shows why great importance is attached to producing fibre with the optimal value of g.

12.6 LOSS MECHANISMS

Attenuation of the signal as it travels along the guide is caused by several factors. The main ones are absorption, scatter and radiation; each is important, but as fibre production methods have improved, and the limits on the operating conditions of the fibre have been optimized, the attenuation is eventually determined by the scattering loss.

12.7 ABSORPTION

The contaminating elements in glass, particularly the transition metals (V, Cr, Mn, Fe, Co, Ni), have electron energy levels that will absorb energy from incident light. The amount of energy absorbed depends on the proportion of impurity atoms present. With care in the manufacture of the glass, and in the drawing of the fibre, these impurities can be reduced to about 1 part per billion, and at this level the loss is negligible.

The absorption produced by the transition metals is not very dependent on frequency, and therefore is not easily avoided by choosing a suitable wavelength for the light source.

The other main absorption mechanism is that due to the presence of water. The hydroxyl ion, CH^-, absorbs readily at 2.8 µm, and there are less strong,

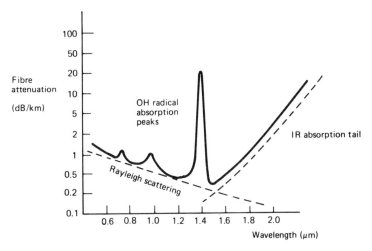

Fig. 12.5 Typical attenuation characteristic of fibres.

but still pronounced absorption peaks at wavelengths which are almost direct harmonics 1.4 µm, 0.97 µm and 0.75 µm. The position of the peaks can be seen clearly in an attenuation–frequency curve, such as that shown in Fig. 12.5. Some reduction in this absorption is obtained by pre-drying the powders used to form the glass. The strong frequency dependence of this mechanism allows an operating wavelength to be chosen that will avoid the severest levels of loss.

The two absorption mechanisms discussed are, in reality, complicated factors which depend on the composition of the glass and the method of fibre production. A deeper discussion is to be found in reference 1, and there are many specialist papers on the subject.

Optical fibres are sensitive to far-infrared radiation, and the loss peak in that region has a tail that reaches to wavelengths as low as 1.6 µm, as shown in Fig. 12.5.

12.8   SCATTER

The fine variations in atomic shape, occurring over distances very small compared with a wavelength, scatter the light by the classical Rayleigh mechanism, which in the atmosphere gives the blue colour to the sky. The loss coefficient is proportional to $1/\lambda^4$, and since it is unavoidable, provides a bound to the lowest attenuation which can be achieved. The Rayleigh loss curve is also shown in Fig. 12.5. The value of the loss is a function of the fibre and its constituent glass:

$$\alpha = \frac{8\pi^3}{3\lambda^4}[n^2 - 1]\, kT\beta \tag{12.6}$$

where T is the transition temperature at which any imperfections in the glass structure are 'frozen in', and β is the isothermal compressibility of the glass.

Other types of scatter can occur. In particular, Mie scatter, which is due to variations in the structure of the glass that occur at about one wavelength intervals, and waveguide scatter, which is caused by fluctuations in the geometry of the core along its length. These losses can be made insignificant if the fibres are produced with care.

12.9   RADIATION FROM BENDS

Energy is radiated at a bend in the fibre. The amount of radiation is usually small, but it is determined by the bend radius. As the radius is decreased there is a very rapid change, at a critical radius, R_c, from very little loss to very high loss. The value of R_c depends on the numerical aperture of the guide, and the radius of its core. Increasing NA and reducing the core size will help to decrease the minimum bending radius.

The mechanism that causes radiation results from the fact that the field pattern of the guided wave penetrates into the cladding. At a bend, the field at the outside of the bend has to travel faster than that at the inside, to maintain the phase relationship in the mode. At some distance from the fibre the velocity of the wave will reach the velocity of light; therefore the energy at a larger radius must be radiated. As the bend radius decreases, the radius at which radiation occurs also decreases, moving closer to the fibre. At the critical radius, half the light into the bend is lost. For many materials, the critical bend will be very small, and sometimes less than the limit imposed by the mechanical stress due to bending.

12.10 SINGLE-MODE FIBRE

As we have seen above, the principal reasons for attenuation and bandwidth limitations in multimode optical fibre are the various losses described, and mode dispersion. Multi-mode fibre is used because of its larger size compared with monomode fibre. The core diameter is usually 50–60 µm, with a cladding of 100–150 µm, and hence coupling into source or detector devices, and jointing fibres can be done efficiently, and easily.

Single-mode fibre has the attraction that there is no mode dispersion, and therefore the bandwidth available can be very high indeed. There is a limitation caused by dispersion, not of modes on the fibre, but of the various frequencies emanating from the source. The spread of frequencies generated by the emitter will depend on the type of device used, but there will be some, and each frequency will propagate at a slightly different velocity. This dispersion is called chromatic, and will decrease as the line-width of a source is reduced.

The size of a single-mode guide depends on the free space wavelength and the refractive indices of the core and cladding. For a particular waveguide, operating at a given wavelength λ_0, the cut-off radius for modes higher than the fundamental is a, where a is given by

$$a \leqslant \frac{2.405 \, \lambda_0}{2\pi(n_1^2 - n_2^2)^{1/2}} \tag{12.7}$$

and n_1 and n_2 are the refractive indices of the core and cladding, respectively.

12.11 DETECTORS

There are many devices available for detecting optical frequency energy. Some are vacuum tube, others solid state, either semiconductor or not. There are also several photosensitive effects employed to convert incident light into a proportional electrical signal.

In communications systems the power in the incident beam may be extremely small, of the order of 10^{-14} W, and this places additional requirements on suitable detectors. Although several devices have been used, and there is extensive research and development on new techniques, we will limit our attention to the PIN semiconductor, and its associated device, the avalanche photodiode (APD).

To be suitable for use in communications, the detector must be very sensitive, have high efficiency of conversion between light and electrical energy, respond very rapidly for high bandwidth, have low noise power, and good light-collecting properties. In addition, it should operate at low voltage, be easy to use, be robust and insensitive to changes in ambient conditions, have a long life, good reliability, and be inexpensive. Such an ideal specification is, of course, unattainable; in any real device compromises are necessary, and priorities must be settled between the various parameters for any application. Here we cannot look at the devices on the market; many companies produce a range of detectors, and the choice of which to use must be made by the system designer. However, we can examine some of the general features of photodiodes and the materials from which they are made.

In photodetection, light is incident on a semiconductor material that, because of the energy gap between the valence and conduction bands, will convert the light energy into electrical energy. On average, the photocurrent is related to the incident light energy by

$$I = \frac{\eta q}{h v} P \qquad (12.8)$$

where q is the electron charge, hv is the photon energy, P the optical power into the device and η is the quantum efficiency. η is the fraction of the incident power producing electron–hole pairs.

There is a statistical nature in the whole photodetection process that causes us to use mean values for the various parameters. These mean values, and the associated rms quantities, give us a long-term view of the device behaviour.

12.12   PIN DIODE

The diode consists of heavily doped p and n sections separated by an intrinsic layer, as shown in Fig. 12.6. The reverse bias across the device sweeps the

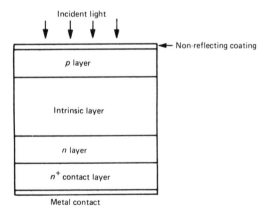

Fig. 12.6 Schematic of PIN diode.

carriers from the I region, leaving a depletion layer. Light falling on the depletion layer creates electron–hole pairs, and these carriers drift across the region to the terminals, where they leave as light-induced current. The device therefore appears to be a current source, across the depletion layer capacitance.

12.13   APD DEVICE

In an APD, the applied voltage is much higher, and there is a region where the electric field is in excess of 10^5 V/cm. At such a high level, photoinduced electrons are accelerated to a velocity at which they can ionize atoms in the intrinsic region, and produce an avalanche effect. In this way the device has an internal gain, M, which is the mean number of electrons at the output per photoinduced electron. The APD is therefore more sensitive than the PIN diode, but because of the stochastic nature of the avalanche gain, its noise level is higher.

12.14   SPEED OF RESPONSE AND QUANTUM EFFICIENCY

When discussing the PIN diode, we noted that the photoelectrons move across the depletion layer at the drift velocity produced by the applied voltage. This transit time limits the speed of response, and hence the bandwidth, of the device. If the depletion layer is narrow, a low response time can be achieved. However, the sensitivity of the device is related to the quantum efficiency, and this increases as the depletion width increases. Consequently, there is a conflict between these two parameters. In Si the depletion width is approximately 50 μm and the minimum response time is of the order of 50 ns, whereas in GaAs the depletion layer can be much narrower, because of the higher absorption coefficient of photons in GaAs compared with Si, and therefore it has a much faster response time of about 40 ps. If the depletion width is narrow, photoelectrons may be induced in the surrounding diffusion area and before being swept through the depletion layer, they will have to diffuse from the diffusion region, thus reducing the speed of response.

12.15   RELIABILITY

Silicon detectors have been developed over many years for use in the 0.8–0.9 μm range and improvements to the configuration and production techniques used have made available devices with projected lifetimes of many years, and high reliability is therefore readily available. For GaAs or other III–V devices, which are used at the longer wavelength of about 1.3 μm, experience is more limited, but long lifetimes are again anticipated. The reason for the limitation of Si to the shorter wavelength region is that its energy gap is 1.1 eV, and this cannot be achieved by photons of wavelength greater than about 1.1 μm. At the longer 1.3 μm wavelength, which is becoming increasingly attractive because of the development of very low-loss guides at that wavelength, Ge has high sensitivity but, because the ionization coefficients for electrons and holes are about the same, the noise level is high compared with that which can be achieved with GaAs, InP or some more complex compound such as InGaAs.

12.16 NOISE

The sources of noise in PIN and APD devices are well known. In PIN diodes noise is produced by:

  (i) random fluctuations in the photocurrent itself,
 (ii) background radiation from other than the intended light sources, and
(iii) current generated in the device when there is no light present, i.e. dark current.

In Si devices, noise produced by (ii) and (iii) can be made extremely small, leaving the shot noise of (i), and this provides a lower bound on the sensitivity of the device.

In APDs another source of noise exists – that due to the randomness of the gain process. It magnifies the shot noise by a factor $F \times M$ where M is the mean avalanche gain, as in Section 12.13, and F is a quantity called the excess noise factor. If the value of F is unity, the avalanche process is not itself noisy. In practice F is much greater than that. Smith[3] states that F is related to M, and a factor k, by the approximate relationship

$$F \simeq 2(1 - k) + kM \tag{12.9}$$

where k is the ratio of the smallest to the largest ionization coefficient in devices in which the avalanche is initiated by the carrier with the highest coefficient. In Si, for example, the avalanche is initiated by electrons that have a much higher ionization coefficient than holes, giving $k < 0.1$. By contrast, Ge is intrinsically more noisy because the electron and hole ionization coefficients are equal.

The choice between PIN and APD will depend on the priorities given to sensitivity and SNR, as well as on operating constraints. The APD can be much more sensitive than the PIN if the internal mean gain, M, is high. However, the shot noise is also a function of M, and, for some devices, this quantity will increase more rapidly than the gain as M is increased. An optimal value of M may exist at which, for a given SNR, the sensitivity is maximum. In practice, for applications in which good SNR is required, PIN devices are used, but if sensitivity is at a premium, an APD would be more appropriate. Compared with PIN diodes, avalanche devices have less attractive operating conditions. They require a much larger voltage, possibly of the order of several hundred volts, and they are sensitive to temperature variations. This means that more complicated driving circuitry is required to provide compensation if the temperature changes.

12.17 OPTICAL SOURCES

The special requirements of optical communications make many demands on optical sources, some of which may not be necessary for other applications. Apart from supplying sufficient power, at reasonable cost, over a long, reliable, lifetime, and with high efficiency, the source should have an emission wavelength that coincides with a loss minimum of the fibre. It should also show a linear output power versus drive current characteristic, emit over a narrow bandwidth, and be capable of transmitting high bit rates.

These constraints and requirements limit the types of suitable emitter to three:

(i) light emitting diode (LED),
(ii) semiconductor laser, and
(iii) solid-state-laser.

Here we will discuss only (i) and (ii), since they are the principal devices presently in use. The solid-state laser is likely to become an important source in the future.

Before considering the LED and the semiconductor laser separately, we will examine some general properties of these optical sources.

LED emitters are simple in their construction, and do not require complex drive circuitry. They are particularly suited to relatively short-distance links in which the bit-rate requirement is modest and the channel capacity low.

Alternatively, semiconductor lasers, which require higher drive currents and more complex circuitry, can produce higher power than LEDs, and are capable of high bit rates. They can therefore be used for high-speed, long-distance systems.

For both types of device, operating in the $0.8-0.9\,\mu m$ range, the most commonly used material is a doped GaAs, usually GaAlAs. Several III–V compounds have been used successfully. In general, to be suitable a material must have a direct energy gap in the region of $2\,eV$ so that, when electron–hole recombination takes place, a photon is released. The material must also be one in which it is possible to form a p–n junction very easily. This requirement rules out high melting point compounds. Ideally, the emission wavelength should be adjustable by changing the composition of the material. This requirement is usually satisfied by varying the mole fraction. For example, if x is varied in $Ga_{1-x}AsAl_x$, the emitted wavelength can be varied between that corresponding to GaAs and to AlAs.

The basic mechanism in both LEDs and lasers is that minority electrons in the conduction band recombine with valence band holes and release energy in the form of radiation, i.e. if the direct gap between the valence and conduction bands is Eg, the radiation produced by recombination is

$$vh = Eg \qquad (12.10)$$

where h is Planck's constant and v is the radiation frequency of the photon emitted.

Although p–n junctions in III–V materials are excellent sources of light, the basic compound will absorb light easily as it passes through, and therefore the separation between the active area and the outer surface of the diode must be very small, otherwise the semiconductor will absorb some of the energy produced in the active region, and thereby reduce overall efficiency.

Here we will discuss the performance features of sources, not the physical and material aspects of their operation. Information on the choice of compounds, and the physical mechanisms involved in the devices, is given in Section 16.2 of Miller and Chynoweth,[4] and Chapters 4 and 5 of Sandbank.[5]

12.18 LIGHT EMITTING DIODES (LEDs)

In its basic, homojunction form, an LED consists of a forward-biased p–n junction. Carriers are injected, under the influence of the electric field, from the majority to the minority side where they create an excess of minority carriers. As they diffuse away from the junction, recombination occurs, and if the energy gap Eg is approximately 2 eV, optical radiation will be emitted at a frequency given by Eqn (12.10). The optical energy radiates from the device, and is launched on to a fibre guide.

For efficient performance, the recombination process must have a very high proportion of radiating events, and the optical power must not be absorbed in its path from the junction to the surface of the device, or lost between the device and the fibre.

The quantum efficiency, defined as the proportion of radiating carrier recombinations, is a measure of the internal conversion between electrical and optical power. Values of about 80% can be achieved. However, all of the generated light cannot be used. Some will be absorbed by the semiconductor material itself, as we mentioned earlier. In addition, the semiconductor–air interface will not transmit all the energy incident on it, only that which falls within the critical angle, which is about 15° for III–V materials (see Fig. 12.7). In practice this means that only about 1% of the light generated will be useful. Some improvement can be achieved in this external quantum efficiency, η_c, by using a reflecting back plane.

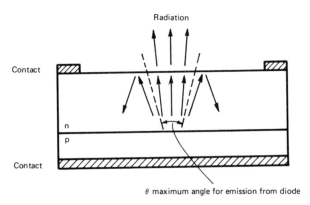

Fig. 12.7 Exit of light from an LED with planar geometry.

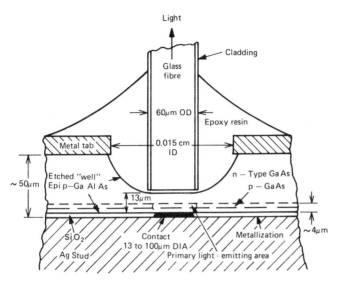

Fig. 12.8 Schematic of Burrus Light emitting diode. Reproduced from Dawson and Burrus, *Applied Optics* vol. 10, pp. 2367–2369, Oct. 1971.

The purpose of the diode is to provide a light source for fibre transmission, and one of the loss mechanisms occurs between the device and the fibre-end. The acceptance angle of the fibre is $\theta = \sin^{-1}(\text{NA})$, where NA is defined in Section 12.2. For sources of area less than that of the fibre-end, the efficiency is approximately equal to $\sin^2 \theta = (\text{NA})^2$. This results in a fibre-launching efficiency of about 5%. The total efficiency of coupling, given as the proportion of diode input power that is launched into the fibre as light, is therefore about 1% of 5%, or 0.05%.

Clearly, considerable effort is involved in improving this figure. Some mechanical changes to the diode can be made. The simplest diode structure has a planar geometry, as shown in Fig. 12.7. By forming a well in the upper material, the fibre-end can be placed near to the active junction, thus reducing significantly the absorption of light within the semiconductor material. If, in addition, the light-generating area is smaller than that of the diode, an epoxy resin layer between the device and the fibre will increase the launching efficiency. An example of this type of device, known as the Burrus diode, is shown in Fig. 12.8. Although it has a high launching efficiency, it has the disadvantage of a relatively large spectral width. In general, the range of spectral widths produced by LEDs is from about 100 to 400 nm.

12.19 LASERS

The other important light source is the semiconductor injection laser. It employs a similar mechanism to that of the LED, but operates with laser-type stimulated emission, and this produces a different output characteristic. The relationship between the output optical power and input drive current, is

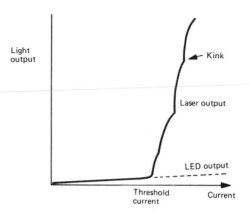

Fig. 12.9 Relationship between output optical power and input drive current for laser and LED.

shown in Fig. 12.9. We can note three features. First, the curve has two distinct sections, that below threshold and that above. The threshold level is closely related to the structure of the device. Below threshold, the action of the device is similar to that of a LED, and the output has a broad spectral width. Above threshold, the device operates under stimulated emission and the spectral width is very much reduced. The second notable feature is the steep gradient above threshold. This makes the device very fast, and this speed is exploited in high-speed digital systems. Finally, the output characteristic above threshold has non-linear features, called kinks. These kinks are thought to result from slight changes in current paths through the active region of the device, and they can be eliminated by using a stripe configuration.

Several structures have been used for injection lasers. In its simplest form, the device consists of an active p–n junction in GaAs, with planar terminal plates parallel to the junction, to which the supply voltage is applied. Two of the opposite faces, perpendicular to the junction plane, are cut to provide reflecting surfaces, between which the essential gain of the carriers across the junction can be achieved. Through one of these faces (or facets, as they are called) light is coupled to the fibre. The opposite, rear facet is sometimes used to monitor the performance of the device. Figure 12.10 shows such a simple laser. The active region has a slightly higher refractive index than the surrounding layers, and this causes a weak waveguiding mechanism, allowing the generated light to be constrained within a narrow region.

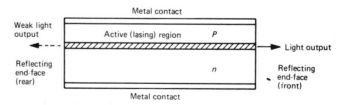

Fig. 12.10 Schematic of laser.

This waveguiding is a very useful feature because it allows the device to produce an intense light source that is more suitable for launching on to fibres than that from an LED, which produces a less directional output. To improve the containment of the light into the active region, a double-heterojunction (DH) structure is used. Figure 12.11 shows the main features. The heterojunction is an interface between GaAs and GaAlAs, and in the DH device GaAlAs is placed at each side of the GaAs active region. Apart from providing a stronger waveguiding action, the DH structure causes the threshold current to be considerably lower than that of the homojunction arrangement shown in Fig. 12.12. This reduction in threshold current is sufficient to allow the device to operate in a continuous mode, whereas in the homojunction structure the current density required to sustain the lasing action can only be achieved in a pulse mode.

In Fig. 12.12 we can see the stripe referred to earlier. Several experiments on the best width of stripe to use have been performed, and that most commonly available is about 20 μm across. The effect of the stripe is to confine the active region of the device to an area below the stripe, thus producing a two-dimensional waveguiding mechanism, and improving the stability and directivity of the generated light.

The DH GaAlAs semiconductor injection laser is only suitable for the lower, 0.8–0.9 μm, wavelength range. For the longer wavelength of about

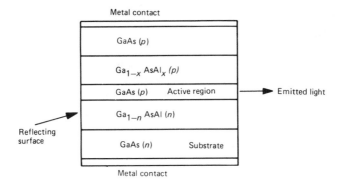

Fig. 12.11 Double heterodyne laser structure.

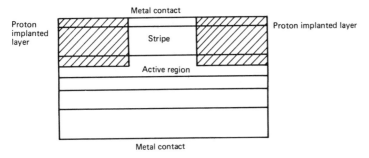

Fig. 12.12 Stripe laser.

1.3 μm, a different base material has to be developed. Present research suggests that the most promising is InP, with doped layers of GaInPAs forming the heterojunctions.

From a systems point of view, the injection laser offers the advantages of higher intensity, more directive output, with a faster response time and narrower spectral width than those of the LED. These advantages are gained at the expense of higher operating voltages and more complex drive circuitry, which is necessary to counteract the variation in threshold current produced by temperature changes and ageing in the injection laser.

12.20 COMMUNICATIONS SYSTEMS

Of the many applications of optical fibre systems, the one that will have, and indeed has had, the most significant effect on equipment development is long-distance telephony. Telephone operating companies around the world, stimulated by the paper of Kao and Hockham,[6] have investigated optical fibre systems in the hope that they could be used as a future trunk transmission system, with significant economic advantages over any alternatives. In the mid-1960s, it was clear that coaxial cable would not be the best medium to cater for the anticipated growth in telephone and television traffic, and considerable effort was directed to developing what seemed to be the most attractive alternative, overmoded circular waveguide. The special feature of an inherent reduction in attenuation with increase in frequency of the fundamental TE_{01} mode, made the circular waveguide approach very appealing, in spite of the essential expense of producing high-precision, helix-lined guides. At that time the major telephone administration research laboratories around the world were performing test trials on waveguide, before bringing it into service. Optical fibres were also interesting, but their high loss made them unable to challenge coaxial cable. Gradually, over the next five years, the considerable amount of research into the material and equipment aspects of fibre production began to be successful and the attenuation fell rapidly from 20 dB/km, or more, to 0.02 dB/km. A comprehensive understanding of the mechanism of transmission over fibre guide was being established, and it was becoming clear that there was no inherent loss mechanism to prevent transmission over very long distances. Gradually, the relative importance given to research into circular waveguide and optical fibre systems changed, and the big decision to commit the future to optical transmission was taken. The reasons why are, from our viewpoint, clear. The demand was for a large-capacity system that was cost effective, secure, comparatively easy to instal, and low-loss. Circular waveguide could be made low-loss, but otherwise it could not compete with fibre, and once the potential very low attenuation of fibre began to be realized, its other outstanding attributes made it the obvious choice. It is now highly possible that the transmission length between repeaters will be at least twenty times that which could be achieved with coaxial cable, and the small size and weight of fibre cable makes the installation a very straightforward operation. Rather than provoke additional difficulties, whose cost could be offset against its advantages, optical fibre is so light weight, flexible and slim that it can be threaded into ducts that had previously been thought to be full, and thus the lifetime of the trunk system can be extended. The cost of the fibre was, in the early days, very high. This was a reflection of the very high investment in development made by both the glass companies and the cable manufacturers, but cost is no longer a penalty. Once the large telephone administrations had decided to use fibre for all new trunk transmission, the per unit cost had to fall dramatically, to make it the cheapest medium on a capacity basis. The spin-off will be seen in other systems, as the technical properties of cheap fibres are exploited.

The system designer has several choices to make, once the decision to use fibre has been taken. The fibre type, refractive index profile, and operating frequency must be specified, and the types of emitter and detector must be chosen. The type of fibre to be chosen depends on the bandwidth required, the length of run, the future development of the system (called upgradability in reference 7), and the cost. There is a good case for installing as good a quality of fibre as possible, so that the system can be upgraded without having to replace the fibre. If high bandwidth is required, the choice lies between graded-index multimode or single-mode guides. As we saw earlier, the core diameter of single-mode fibre is about a tenth that of multimode, which is therefore easier to launch light on to, and easier to join. However, advanced designs of injection laser diodes, and microprocessor-controlled coupling jigs, have reduced the difficulties of single-mode working such that the comparison with multimode is much more balanced. Several administrations have plans for 500 Mb/s single-mode experimental trunk routes, and this development suggests that the attractive features of single-mode will outweigh its disadvantages. Another aspect of future systems that makes the choice of single-mode guide more appealing is the use of integrated processing devices, known as integrated optics. Several research laboratories have been working on integrated processing devices, such as switches, modulators and filters, for over a decade. Much of the effort has been in producing satisfactory materials, and making simple devices in a reproducible way. Gradually we are moving towards optical integrated circuits that will be used in monomode systems. When such devices are available, many additional communication applications will be possible.

As we have seen already, the choice of wavelength is governed by the behaviour of the fibre. A combination of low attenuation and low dispersion is desirable. Until 1982 the most commonly used wavelengths were $0.83 \, \mu m$ and $0.9 \, \mu m$, or thereabouts. After that time the lower dispersion available at longer wavelengths began to be exploited with silica-based fibres, operating at $1.3 \, \mu m$.

This change of frequency required new emitters and detectors. Silicon detectors had been clearly the best performers at the shorter wavelength, but at $1.3 \, \mu m$ it is necessary to use one of the III–V compounds, usually based on GaAs. The structure of the detector should offer a large area for light collection and the device should be sensitive, without being too noisy, as we discussed in Section 12.16.

The type of emitter used will depend on the application. The almost linear output versus drive current relationship of an LED makes it particularly

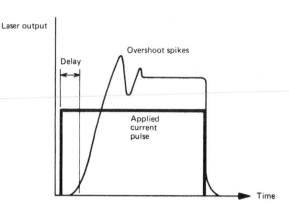

Fig. 12.13 Rise-time delay and output overshoot in laser response to pulse drive current.

attractive. It is also the obvious choice for low-speed, low-cost digital systems, particularly those using step-index fibre, which has a large value of NA. However, for launching on to graded-index fibre for high-speed systems, the injection laser might be preferred, and for single-mode systems it is essential. The small chromatic dispersion of the laser makes it necessary for long-distance communication systems. Semiconductor lasers are more temperamental than LEDs. Their performance characteristics change with temperature, and with age. To counteract these effects, some form of feedback is used, taking a monitoring signal from the rear facet of the laser, behind which is situated a simple photodiode. Two other problems with lasers are (i) rise-time delay and (ii) overshoot spikes, in response to a pulse input current (see Fig. 12.13). The dc bias used can affect the delay, and by biasing the device near its threshold level the delay is minimized. Here a compromise is required, for the spectral width increases as the bias increases, and the lifetime falls. In Section 12.19 we discussed the use of stripe conductors on the laser to stabilize its performance, and one of the benefits of this geometry is that the output oscillatory spikes are reduced.

The length of an optical fibre system will not be limited to that which can be achieved without intermediate amplification. In digital systems, regenerators will be required at intervals. The spacing will depend on the quality of the system, but it will be somewhere in the range 10–60 km. A block diagram of a regenerator is shown in Fig. 12.14. Essentially, it detects the incoming light signal, re-forms and re-times the pulses, and modulates an emitter for the next stage. Regenerators introduce additional problems for the designer. Supply voltages must be provided, and since regenerators are often placed in remote locations, the supply has to be fed along the optical cable. Monitoring, to allow remote fault finding from a control point at the end of a link, is usually included in the design. These features increase the cost of the system, so the gradual extension of the distance that can be covered without regenerators is a welcome result of fibre research.

Unlike systems that operate at lower radio frequencies, little attention has been paid to the use of free-space optical communications. This results from

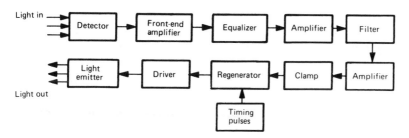

Fig. 12.14 Block diagram of optical regenerator.

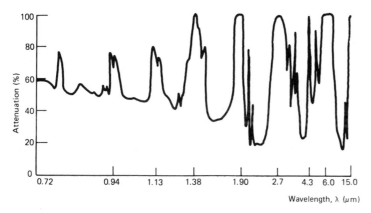

Fig. 12.15 Atmospheric attenuation plotted against frequency.

the heavy attenuation in the atmosphere due to moisture. Precipitation causes very large losses, but even under dry conditions there are attenuation peaks at certain frequencies (Fig. 12.15). In addition, the movement of air caused by localized heating produces refractive index variations that deflect the path of the light beam. Free-space links are limited to a maximum of about 2 km.

12.21 CONCLUSION

We have touched on some advantages that optical fibre systems possess. Many are attractive from the operations and management point of view, giving trouble-free performance. Some, such as no earth loops, and immunity to sparking or electrical interference, make communications links possible in environments that were previously very difficult, or even impossible. Apart from using fibres in such situations, we can expect them to form an integral part of the systems that will be required to satisfy the enormous increase in the quantity and type of traffic that will be generated as the revolution in office, computer and home communications takes place. Already there are designs for two-way, low-speed, data terminals in homes, and eventually all our domestic communications – telephony, television, automatic transactions, teletext, facsimile and systems not yet devised – will be transmitted over a single digital line – the optical fibre. Plans are being developed in the UK for national optical fibre cable television and fibre-based high capacity data systems. Similar systems are being developed in other countries. Coupled with digital representation of signals and integrated circuits of enormous density and complexity, optical fibre systems are transforming the aspirations of communications engineers, and encouraging a tremendous upsurge of innovation and enterprise. In fact, at present it seems that, for the first time in the history of telecommunications, we have more capacity and flexibility than we can use for any systems that presently come to mind.

REFERENCES

1. Midwinter, T. E., *Optical Fibres for Transmission*, Wiley, Chichester, 1979.
2. Kaminow, I. P., Marcuse, D. and Presby, H. M., "Multimode fiber bandwidth: theory and practice", *Proc. IEEE*, **68** (10), 1209–1213 (Oct 1980).
3. Smith, R. G., Photodetectors for fiber transmission system", *Proc. IEEE*, **68** (10), 1247–1253 (Oct 1980).
4. Miller, S. E. and Chynoweth, A. G., *Optical Fiber Communications*, Academic Press, New York, 1979.
5. Sandbank, C. P., *Optical Fiber Communication Systems*, Wiley, Chichester, 1980.
6. Wilson, T. and Hawkes, T. F. B., *Optoelectronics: An Introduction*, Prentice-Hall, London, 1983.
7. Kao, K. C. and Hockham, G. A., "Dielectric–fibre surface waveguides for optical frequencies", *Proc. IEE*, **113** (7), 1151–1158 (July, 1966).
8. Elion, G. R. and Elion, H. A., *Fiber Optics in Communications Systems*, Dekker, New York, 1978.
9. Wolf, H. F., *Handbook of Fibre Optics*, Granada, London, 1979.
10. Optical-fibre Communications, Special Issue, *Proc. IEEE*, **68** (10), (Oct 1980).

Packet switched networks $\boxed{13}$

13.1 INTRODUCTION

This chapter deals with the topic of packet switched technology which is now in common use on wide area networks (WANs), local area networks (LANs) and metropolitan area networks (MANs). The most intensive use of packet transmission at the present time (1994) is on local area networks and a substantial portion of the chapter is devoted to this topic. Packet switching was, however, initially developed for connecting computers separated by long distances and some consideration of WANs will also be given. Recent years have seen the development of WANs, LANs and MANs to carry integrated digital services including data, voice and video traffic.

The main difference between wide area and local area networks is that the former cover very large distances and have limited bandwidth and substantial propagation delay. Local area networks, on the other hand, cover distances up to a few kilometres and have relatively high bandwidths and low propagation delays. This means that the packet transmission techniques for the two types of network have some important differences. It is worth pointing out here that although WANs and LANs can be categorized separately, many LANs have gateways to WANs so that global communications often make use of combinations of both type of network.

A packet is essentially a block of data which is transmitted from a source to a destination. A typical packet format for Ethernet is shown in Fig. 13.1. The data is preceded by a header which contains the destination and source address and is followed by a cyclic redundancy check. The length of the data section (for Ethernet) can be anywhere between 46 and 1500 bytes (octets). The format of the packet will be discussed in detail later in this chapter; it suffices at this point to indicate that the channel between transmitter and receiver is occupied only during the actual transmission of the packet.

The actual transmission rate is much higher than the data rate of the source, which means that packet switching may be considered as a form of

Preamble	Destination address	Source address	Type field	Data field	CRC
64 bits	48 bits	48 bits	16 bits	8 N bits	32 bits

Fig. 13.1 Ethernet packet format.

time division multiplexing on demand, i.e. transmission is only requested when sufficient data to fill a packet is available at the transmitter. At other times the transmission medium·may be used to transmit packets between other sources and destinations. For this reason packetized systems are often referred to as ATDM or statistically multiplexed transmission systems; ATM (Section 10.44) is a form of packet transmission.

This contrasts with the public switched telephone network which allocates a circuit between caller and called party for the duration of the call. This is known as circuit switching and differs from packet switching in that the circuit remains allocated throughout any periods of inactivity between the communicating parties. Packet switched systems are thus generally more efficient than circuit switched systems. The analysis of packet switched networks is often very complex and, as a result, powerful software simulation tools have been developed to aid the design process. Discussion of these tools is outside the scope of a text of this nature; instead a simplified treatment will be given, where appropriate, and the results of simulations will be quoted.

13.2 WIDE AREA NETWORKS

The characteristic feature of wide area networks is that distances, and hence propagation delays, tend to be large. A typical topology is shown in Fig. 13.2 and it is usually the case that more than one path exists between a source and a destination node. Each node accepts data from computers or terminals connected to it, in a series of bursts, and assembles the data into packets with the appropriate address information.

The nodes then determine the routing, provide buffering and error control, and return acknowledgement information to the sender when the packet reaches the final destination; an early example of such a network is AR-PANET.[1] There will clearly be several hops between intermediate nodes and inter-node acknowledgements are given when a neighbouring node accepts a packet. Should an acknowledgement not be received within a specified interval (e.g. 125 ms) the packet is retransmitted. The end to end delay for short messages varies between 50 and 250 ms, depending on the packet length used. The delay can increase substantially under heavy network loads, when lengthy queues can form at intermediate nodes.

The topology of Fig. 13.2 is an example of a store and forward packet

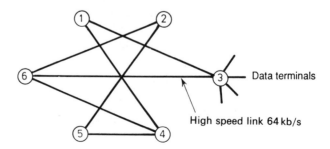

Fig. 13.2 Mesh connected wide area network topology.

switched network as each node stores data until a complete packet is received and then queues the packet for forwarding to a neighbouring node. Clearly physical links must be provided between the individual nodes and there are essentially two packet transmission techniques which can be identified. A **virtual circuit** transmission system is one in which a path is set up between soure and destination and all packets follow the same path and arrive in the sequence in which they were transmitted. A **datagram** transmission is one in which individual packets are transmitted by whatever route is available when they are presented for transmission. No fixed route is set up and each intermediate node decides on the appropriate path, according to some routing algorithm. In the case of a datagram the packets do not necessarily arrive in the order in which they are transmitted, and each packet must identify its position in the transmitted sequence.

The main feature of the datagram type of packet switched network is that the data links between nodes are used at near to their full capacity. Further, packets are scheduled for transmission so that there is no contention. Essentially each node queues packets for transmission and, since there is more than one outlet, each node can select an alternative route if the queue for the direct route becomes too long. There are many possible routing algorithms one of which is shown in Fig. 13.3.

In this example several priority ordered choices are stored in a routing table at each node. Considering Fig. 13.2, the first choice of a route between a source connected to node 1 and a destination connected to node 4 would be the direct route, i.e. 1 to 4. The second choice might then be via node 3. Once it has received an error-free packet from node 1, node 3 then selects the appropriate route for node 4. All of the nodes in a system of this type are referred to as packet switches.

In the topology of Fig. 13.2 the individual links are quite separate and routing algorithms are required. An alternative topology is one where each individual node is connected to a common broadcast channel and has to compete in some way for access, with all other nodes. This principle is used in most local area networks and in some wide area networks also. Wide area networks based on a single communications channel use either radio, to connect many geographically separate nodes, or satellite communications. This approach to wide area networks can be very cost-effective because it replaces the very expensive long distance dedicated links with a single wideband channel. A typical satellite system is shown in Fig. 13.4 in which each ground station transmits on a frequency f_{up} and receives on a frequency f_{down}. In such a system each ground station can monitor its own output for error-free transmission.

However, a single bus does introduce the problem of contention, i.e. several nodes requiring access to the channel at the same time. There is a number of techniques which have been developed to resolve such contention, and they can be categorized either as scheduled access or random access. The properties of these **access techniques** will be considered in some detail in this chapter. This will illustrate the suitability of specific access mechanisms for wide or local area application. Of particular interest will be the relationship between throughput, offered traffic and delay.

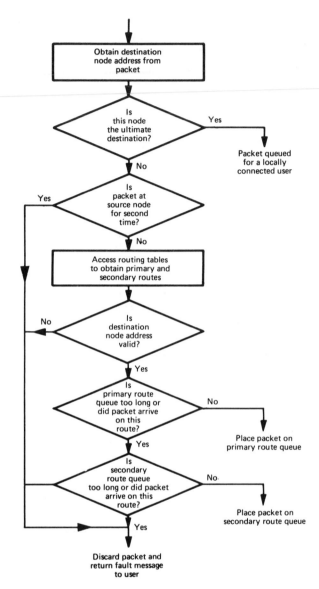

Fig. 13.3 Network routing algorithm.

13.3 THE POISSON DISTRIBUTION

Before any progress can be made in considering packet switched networks it is necessary to describe the statistics of packet arrivals on the network in any specified time interval T. To achieve this the time interval is assumed to be divided up into M intervals each of duration t seconds. As $t \rightarrow 0$ the probability of a packet actually arriving in the time interval t will be proportional to the actual value of t. Hence the probability of a packet arriving in t seconds is

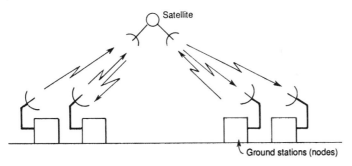

Fig. 13.4 Satellite packet broadcast system.

λt and the probability of no packet arriving in this interval is $(1 - \lambda t)$. A fundamental assumption is that the probability of a packet arriving in one interval t is completely independent of arrivals in all other similar intervals. The system is thus said to be memoryless and the probability that there will be no arrivals in the entire interval T is then

$$P(0) = (1 - \lambda t)^M = (1 - \lambda t)^{T/t} \qquad (13.1)$$

As $t \to 0$ then $M \to \infty$ and $P(0) = e^{-\lambda T}$ since $\lim_{x \to 0} (l - nx)^{1/x} = e^{-n}$

The probability of k arrivals in time T is the joint probability of one arrival occurring in k of the intervals t and no arrivals occurring in the remaining $M - k$ intervals, this is

$$P(k) = \frac{M!}{k!(M-k)!}(\lambda t)^k (1 - \lambda t)^{M-k} \qquad (13.2)$$

When $M \gg 1$ it may be shown that

$$M! \approx \sqrt{(2\pi)}e^{-M}M^{M+0.5} \qquad (13.3)$$

This may be verified by replacing M by any integer (greater than 2) in Eqn 13.3. Similarly if k is fixed

$$(M - k)! \approx \sqrt{(2\pi)}e^{-(M-k)}(M-k)^{M-k+0.5}$$

Substituting into Eqn (13.2)

$$k!P(k) = e^{-k}(1 - k/M)^{-(M+0.5)}(1 - k/M)^k (M\lambda t)^k(1 - \lambda t)^{M-k} \qquad (13.4)$$

But $\lim_{M \to \infty}(1 - k/M)^{-(M+0.5)} = e^k$ and if $M \gg k$, then $(1 - k/M)^k \approx 1$

i.e.

$$k!P(k) = (\lambda t)^k \frac{(1 - \lambda t)^{T/t}}{(1 - 1\lambda t)^k}$$

as

$$t \to 0, \quad (1 - \lambda t)^k \to 1 \text{ and } (1 - \lambda t)^{T/t} \to e^{-\lambda T}$$

Hence

$$P(k) = \frac{(\lambda t)^k e^{-\lambda T}}{k!} \tag{13.5}$$

This is known as the Poisson distribution and gives the probability of k arrivals in the interval T. In this expression λ may be interpreted as the average rate of arrivals per second.

13.4 SCHEDULED ACCESS TECHNIQUES

To illustrate the principles of scheduled access, consideration will be given to a single bus system of the type shown in Fig. 13.5 although it should be realized that the same arguments apply to ring topologies and the shared broadcast radio channel.

Scheduled access can be under centralized control or can be completely distributed. In the former case access is determined by a designated controller, and this will be considered first. The performance of such systems is usually measured in terms of access delay. This is the time which elapses between a packet being available for transmission at a node and commencement of transmission.

In order to avoid contention on the bus the controller poles each node in turn. If a node has a packet for transmission the transmission is commenced as soon as the polling signal is received. If a node has no data for transmission the controller is made aware and then poles the next node in the system. When the controller sends out a polling signal there will be a finite time which elapses before the polled node is able to reply (polling interval), followed by an interval required for data transmission (transmission interval).

The polling interval p_i will be a combination of the propagation delay on the network and the time required for the node to synchronize with the polling signal. The transmission interval t_i will depend on the amount of data a particular node has for transmission when polled. Both p_i and t_i are random variables, but it is clear that when the network extends over several hundred (or thousand) kilometres the polling delay p_i can greatly exceed the value of t_i. The efficiency of such a system is thus relatively poor when the propagation delay is very high.

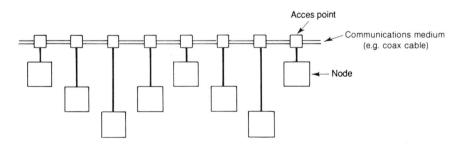

Fig. 13.5 Single bus communications channel.

A significant improvement can be achieved if central control is replaced with distributed control. When a node completes its transmission it essentially poles the next node on the network. The polling time is now, on average, significantly shorter than with centralized control, because the mean distance between a node and its immediate neighbour is much less than the mean distance between a designated controller and all other nodes. This type of distributed control is essentially the same as the control used on token passing LANs. The analysis for the token bus and token ring LAN is covered in Sections 13.8 and 13.9, respectively. Similar results are obtained for wide area token passing communications systems and the analysis will not be repeated here. It should be borne in mind, however, that propagation delay is usually insignificant in local area networks but this is not the case with long distance networks.

13.5   RANDOM ACCESS TECHNIQUES

It was stated in the previous section that when the utilization is low there can still be a considerable access delay as a result of the polling time. This problem can be overcome by allowing each node to transmit as soon as it has a complete packet ready. There are a number of variations on this theme but they all result in the possibility of contention. This occurs when two or more nodes each transmit a packet during a particular interval of time. One of the simplest random access methods is the ALOHA protocol developed by the University of Hawaii.[1]

13.5.1   The ALOHA and slotted ALOHA access protocol

The ALOHA protocol was developed for radio transmission and allows each user to transmit a fixed length packet as soon as it is formed. It relies on a positive acknowledgement to indicate that the packet was received without error. In the case of the single bus system of Fig. 13.5 the receiving node would be required to send an acknowledgement. In the satellite system of Fig. 13.4 the transmitting node can monitor the packet retransmitted by the satellite and this produces an automatic acknowledgement. Assuming that each packet requires a transmission time of P seconds then if such an acknowledgement is not received within a time $P + 2t_p$, where t_p is the maximum end to end propagation delay on a bus or the propagation delay between transmitting node and satellite, the transmitter retransmits its packet.

A packet will be successfully transmitted at a particular instant t only if no other packet is transmitted P seconds before or after t (i.e. the vulnerable period is $2P$); this is illustrated in Fig. 13.6.

The normalized channel throughput S (the average number of successful transmissions per interval P) is related to the normalized traffic offered G (the number of attempted transmissions per interval P, including new and retransmitted packets) by

$$S = Gp_0 \tag{13.6}$$

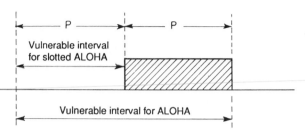

Fig. 13.6 Vulnerable period for ALOHA and slotted ALOHA.

Where p_0 is the probability that no additional packet transmissions are attempted in the vulnerable interval $2P$.

If it is further assumed that packet arrival times are independent and exponentially distributed with a mean arrival rate of λ per second, the probability of k arrivals in an interval of duration t is then a Poisson process given by

$$P(k) = (\lambda t)^k e^{-\lambda t}/k!$$

The probability of no arrivals in time t is $P(0) = e^{-\lambda t}$. In P seconds there will be $\lambda P = G$ arrivals, hence $\lambda = G/P$. Letting $t = 2P$ gives the channel throughput as

$$S = Ge^{-2G} \tag{13.7}$$

Equation (13.7) is valid for satellite systems where all nodes are on the earth's surface and are approximately the same distance from the satellite. For earth-based packet radio the satellite will be replaced by a base station and the expression must be modified to account for the fact that some nodes will be nearer the base station than others. In such cases there will be a maximum **difference** in propagation delay of t_{dm} between near and far nodes which has a normalized value of $a = t_{dm}/P$. In such circumstances Eqn (13.7) becomes

$$S = Ge^{-2(1+a)G}$$

note that for satellite systems the normalized difference in propagation delay $a = 0$.

The maximum throughput for a pure ALOHA channel is found by differentiating Eqn (13.7) with respect to G and occurs at $G = 0.5$, i.e.

$$S_{max} = 1/2e = 0.184 \tag{13.8}$$

The throughput characteristic for pure ALOHA is shown in Fig. 13.7.

Figure 13.7 shows that when the offered load G is low there are very few collisions and virtually all transmissions are successful and $S \approx G$. At higher values of offered traffic the number of collisions increases, which increases the number of retransmissions causing still more collisions, and so on. The characteristic is thus unstable and S actually begins to drop. The maximum throughput of the ALOHA protocol is thus limited to only 18% of the system capacity. However, this is still adequate for many purposes e.g. very bursty traffic, such as that produced by a human user at a remote terminal. If the

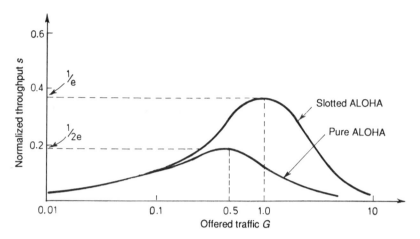

Fig. 13.7 Throughput characteristic for pure and slotted ALOHA.

system bit rate is, say, 1 Mb/s the useful capacity, including packet overheads, will be 180 kb/s.

In order to calculate the average time required to successfully transmit a packet, some knowledge of the retransmission procedure is required. A retransmission will be necessary if a node does not receive an acknowledgement within an interval $(P + 2t_p)$ after the first transmission. If a node retransmits immediately after this interval then the probability of a further collision will be unity since any other node involved in the collision will use the same retransmission procedure. To avoid a certain collision on retransmission some randomness should be introduced into the retransmission timing. One possibility is to define a retransmission interval $m \times P$ and to retransmit at time $n \times P$ where n is uniformly distributed between 1 and m. The average delay before a retransmission is attempted is then $P(m + 1)/2$. The time required to confirm a successful retransmission is thus $2t_p + P(m + 1)/2$.

If there is a total of E retransmission attempts the mean access delay is

$$t_a = P + 2t_p + E[2t_p + P(m + 1)/2] \tag{13.9}$$

The relationship between throughput and offered traffic is

$$\frac{S}{G} = \frac{1}{1 + E}$$

If all packet arrivals are independent (which will be true only for large values of m) then eliminating S from Eqn (13.7)

$$E = e^{2G} - 1 \tag{13.10}$$

The number of retransmissions can now be substituted into Eqn (13.9) to give the mean access delay. As an example consider the satellite system of Fig. 13.4 with a transmission rate of 1 Mb/s and a packet length of 8000 bits. The round trip delay for a synchronous satellite is approximately $2t_p = 270$ ms and the packet transmission time is 8 ms (note that the packet transmission time is much less than round trip delay, this is typical of WANs). If the normalized offered traffic is $G = 0.1$ and $m = 5$ then $E = 0.22$ and $t_a = 337.8$ ms. The major contribution to this access delay is the round trip delay of 270 ms. A typical round trip delay for a LAN would be of the order of 50 µs, and the corresponding access time would be $t_a = 8.413$ ms.

The throughput characteristic of pure ALOHA can be improved significantly if all transmissions are synchronized. The time axis is divided into intervals of time (or slots) of duration P and each user may transmit only at the start of a slot. Under these circumstances, if collisions do occur packets will overlap completely and the vulnerable period is reduced to P. The throughput of slotted ALOHA is then given by

$$S = Ge^{-G}$$

This has a maximum value when $G = 1$ given by

$$S_{max} = 1/e = 0.368.$$

The maximum throughput of slotted ALOHA is thus increased to 36.8% of the system capacity. The characteristic for slotted ALOHA is also shown in Fig. 13.7.

It is clear from Fig. 13.7 that slotted ALOHA has a similar instability when the offered load exceeds the value at which the throughput is a maximum. The access delay is developed in a similar way to that for pure ALOHA. There is, however, one important difference: a packet may be transmitted only at the beginning of a slot. This means that on average (assuming packet arrivals are independent) a packet will have to wait $P/2$ seconds until the beginning of the next slot. Hence, on average, the time required to ensure a successfull transmission is $1.5P + 2t_p$. The access delay for slotted ALOHA is thus

$$t_a = 1.5P + 2t_p + E\left[2t_p + \frac{P}{2}(m+2) \right] \qquad (13.11)$$

Hence the improved throughput of slotted ALOHA is achieved at the expense of a slight increase in access delay.

13.5.2 Carrier sense multiple access with collision detection

This access protocol is one of a group of protocols known as carrier sense multiple access, which differ from ALOHA in that a node listens to the channel and transmits a packet only if the channel is idle. Determination of whether a channel is idle or busy presents particular problems for radio systems because not all users within a radio network can hear all others, giving rise to 'hidden users'. There are methods of overcoming this problem[2], but as far as this text is concerned consideration of CSMA will be restricted to the bus system of Fig. 13.5. A coverage of several variants of CSMA is given by Kleinrock.[3] In this chapter we limit our attention to CSMA/CD which is the most widely used variant.

The CSMA/CD access protocol requires each node to continuously monitor the transmission medium, and as a result each node is able to detect a collision. Once a collision is detected all nodes stop transmitting immediately.

Retransmission is attempted at some later interval. CSMA/CD is more efficient than ALOHA since transmission is interrupted as soon as a collision is detected. The vulnerable period is also reduced to twice the end to end propagation delay on the network. This can be shown by considering the topology of Fig. 13.5. Assume a node at one end of the bus senses the bus idle and transmits a packet. A node at the other end of the bus will not be aware of this transmission until an interval t_p, equal to the end to end propagation delay, has elapsed. If it is supposed that the second node transmits a packet just prior to becoming aware of the first transmission then clearly a collision will occur. However the first node will not be aware of the collision until a further period t_p elapses; thus the minimum time required to detect a collision is $t_{sl} = 2t_p$. This argument assumes that each node will stop transmitting immediately upon detecting a collision. In practice there will be an additional period known as the jamming period. This will be ignored in the following analysis, but will be considered further in the description of Ethernet.

The characteristics of the CSMA/CD access technique are determined by assuming that a large number of nodes have a packet ready of transmission and that the length of each packet is the same (the restriction on packet length is imposed only to produce a tractable analysis and would actually be a disadvantage in practice). It is assumed that there are Q nodes with a packet ready for transmission and that the probability that any of the ready nodes will transmit within the interval t_{sl} is p. In order for this transmission to be successful one node only must transmit. The probability of a successful transmission is thus

$$P_r = Qp(1 - p)^{-1} \qquad (13.12)$$

The maximum value of P_r occurs when $dp_r/dQ = 0$, which gives $p = Q^{-1}$. In essence each node must determine the value of Q and it is assumed in the following analysis that this has been done (an approximate method of achieving this is considered in the Section 13.7 on Ethernet).

The probability of a successful transmission is thus

$$P_r = (1 - Q^{-1})^{Q-1} \qquad (13.13)$$

As $Q \to \infty$, $P_r = e^{-1}$. Thus when the network is sensed idle each node will transmit a packet with probability Q^{-1}. The probability that a node will experience a collision will be $(1 - P_r)$. The probability that the node will be delayed by J timeslots (t_{sl}) is thus the probability of $(J - 1)$ collisions followed by a successful transmission, i.e.

$$P_J = P_r(1 - P_r)^{J-1}$$

The average number of collisions is thus

$$J_{av} = \sum_{J=0}^{\infty} J \cdot P_J = 1/P_r$$

The mean access delay is thus $t_a = 2t_p \cdot J_{av}$ and the average time required to deliver a packet is

$$t_d = P + t_p + 2t_p \cdot J_{av}$$

i.e.

$$t_d = P\{1 + a(1 + 2/P_r)\} \qquad (13.14)$$

where $a = t_p/P$ is the ratio of propagation delay to packet transmission time.

If t_d is the average time to deliver a packet then the average number of packets delivered/second is $1/t_d$. The normalized throughput (the average number of packets delivered per interval P) is then

$$S = \frac{1}{\{1 + a(1 + 2/P_r)\}} \qquad (13.15)$$

The limiting case occurs when the number of nodes Q tends to ∞, in which case $1/P_r = e$. This characteristic does not collapse to zero, as is the case with ALOHA. Equation 13.15 is plotted as a function of a in Fig. 13.8.

Figure 13.8 shows that for a throughput of $S = 0.6$ the required value of a is 0.1. This means that the propagation delay t_p is restricted to 10% of the packet transmission time. It may be concluded from this that CSMA/CD is not suited to long-distance communication, with corresponding large propagation delays, unless the packets are excessively long or the data rate is low. An alternative way of presenting the characteristic is shown in Fig. 13.9 where S is plotted as a function of the number of users Q.

The significance of the parameter a is once again evident. If the packet size is equal to the smallest that will allow collision detection then $P = t_{sl}$ (i.e. $a = 0.5$) and the asymptotic value of S becomes 0.24. In typical applications with much longer packets and moderate loading the throughput is of the order of 80% and this increases to approximately 98% when relatively few users generate longer packets.[4]

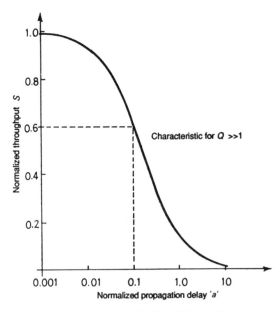

Fig. 13.8 Throughout characteristic for CMSA/CD as a formation of the normalized propagation time 'a'.

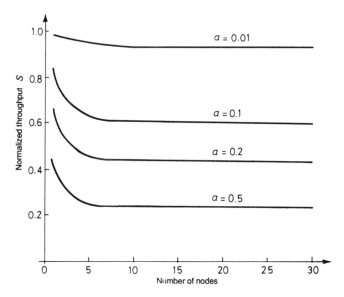

Fig. 13.9 Throughput/load for CMSA/CD.

The delay characteristic associated with CSMA/CD is very much influenced by the method of determining the transmission probability Q^{-1} and this is covered in Section 13.7 which is devoted to Ethernet.

13.6 LOCAL AREA NETWORKS

In this section we shall consider networks developed specifically for short-distance communication where the propagation delay is low and the bit rate is high. The three most common LAN topologies are shown in Fig. 13.10.

The star-connected network is a simple topology in which the central node performs the routing function for communication between nodes attached to it. For this reason the central node must have sufficient capacity to handle all simultaneous traffic and is a potential bottleneck. In particular a very high reliability of the central node is demanded, as a failure at this point completely destroys the network. In this chapter we shall be concentrating on the bus and ring structures and the access protocols which they employ.

It is useful at this point to consider some of the properties of bus and ring topologies as this will provide some insight to the reasons for their use in local area networks.

(i) A bus network is generally more reliable than a ring network as it tends to be purely passive. This means that failure of one or more nodes does not cause the failure of the network in total. In a ring network each node effectively repeats any transmission not destined for itself to the next node. This means that failure of any node in the ring can be catastrophic. One method of reducing the risk of network failure is to provide each

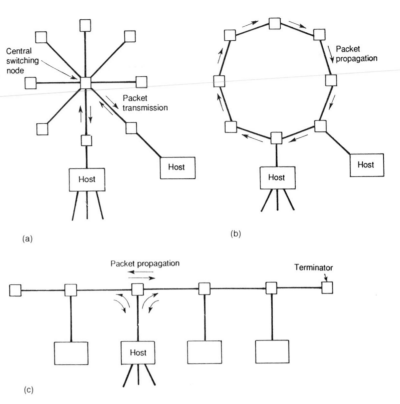

Fig. 13.10 LAN topologies, (a) star topology, (b) ring topology, (c) bus topology.

    node with bypass circuitry which is brought into action automatically on a node failure.

(ii) Additional nodes can be incorporated in bus networks without disrupting network operation. Ring networks require the insertion of new cable segments which requires the network to be taken out of service.

(iii) A bus network is limited (currently) to twisted pair or coaxial cable because of the need to transmit signals in two directions, bi-directional optical couplers would be required for optical systems. Ring networks tend to transmit signals in one direction only and lend themselves to optical fibre operation with all the inherent advantages of this medium.

(iv) Bus networks are prone to reflections and care is required to avoid impedance mismatch at tapping points.

(v) Fault isolation is very difficult in bus networks but is relatively straightforward in a segmented ring structure.

Ring networks tend to support some form of token passing for network access because of their sequential nature. Bus networks, on the other hand, are suitable for both random access and sequential (or polling) access. We shall consider two common examples of bus-based LANs and compare these with the alternative ring structure.

13.7 THE ETHERNET LOCAL AREA NETWORK

Ethernet is based on a bus network operating at 10 Mb/s baseband using the CSMA/CD access protocol. This local area network was developed for computer communications, but there has been significant interest in recent times for using Ethernet for packetized voice communications[5,6] and also video. The basic CSMA/CD protocol was covered in the previous section; however, there are some further points which should be considered with reference to Ethernet and its very close derivative IEEE 802.3. The hardware configuration of a typical Ethernet node is shown in Fig. 13.11. This consists of a microprocessor, buffer memory, an Ethernet data link controller (EDLC) and its associated Ethernet serial interface (ESI) and transceiver unit. This hardware performs both transmission and reception of Ethernet packets.

Any node with a packet ready for transmission monitors the medium for ongoing transmission. If there is no ongoing transmission the ready node waits for an interval of 9.6 μs after the bus becomes idle (the interpacket gap) and then transmits immediately. This differs from the CSMA/CD discussed previously where the node transmitted with a probability Q^{-1}. Collisions can occur during the interval $t_{sl} = 2t_p$ and this 'slot time' is set equal to 51.2 μs which limits the effective maximum length of the Ethernet bus to 2.5 km. (An Ethernet system is actually composed of a number of segments, see Fig. 13.13 for details.) If a collision is detected all nodes involved abort the current transmission attempt after transmitting a 32 bit jam pattern (010101···). This jam pattern ensures that all other nodes are aware of the collision. To prevent repeated collisions with contending nodes each node involved in a collision then generates a backoff interval given by

$$T_r = \text{RND}[0, 2^r - 1]t_{sl}$$

where RND[ ] is a uniformly distributed random number between 0 and $2^r - 1$ and r is the number of successive collisions that a particular packet has

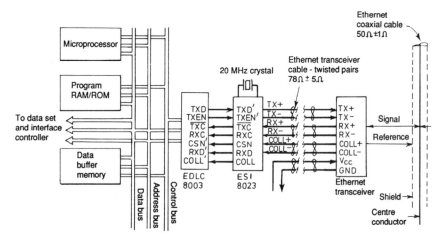

Fig. 13.11 CSMA/CD Ethernet node hardware.

experienced. The value of r is held at 10 should there be more than 10 but fewer than 16 collisions. On the 16th collision r is reset to 0 and a higher level collision warning is generated. The back-off algorithm is known as the (truncated) binary exponential back-off algorithm and is intended to facilitate the calculation of the transmission probability Q^{-1} which maximizes the chance of a collision-free despatch. In effect the binary back-off algorithm sets the probability of transmission as $1/2^r$, i.e. the number of ready users is estimated as $Q = 2^r$. For a highly loaded Ethernet with many collisions r will reach the value 10, which estimates the number of ready users as 1024. This is the maximum number of nodes specified for Ethernet. Clearly the optimum transmission probability is not implemented as when a new packet is available a transmission is attempted $9.6\,\mu s$ after the network becomes idle. Further, the Ethernet packet may have a data segment between 46 and 1500 bytes long, hence the constant packet assumption is also not valid.

In reality the performance of CSMA/CD with a binary exponential back-off is virtually impossible to analyse mathematically without making gross approximations. Hence performance measures are usually obtained by computer simulation. However it should be stated that Ethernet does exhibit a stable throughput for very high loads and this is greatly influenced by the parameter 'a' considered earlier. In the case of Ethernet the packet length can vary between 576 bits and 12 208 bits (including the overhead of 208 bits). With a transmission rate of 10 Mb/s and $t_p = 25.6\,\mu s$ the parameter 'a' can vary between 0.002 and 0.44. Of particular importance, especially for speech transmission, is the access delay which increases as the traffic on the network increases. A typical access delay/throughput characteristic for speech packets derived from computer simulation is shown in Fig. 13.12.

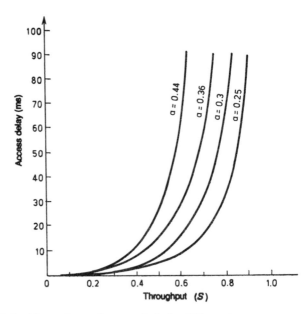

Fig. 13.12 Delay/throughput characteristic for Ethernet.

The reader's attention is drawn to the very rapid increase in delay which occurs at the higher network loads. For example when $a = 0.44$ the delay reaches 60 ms at a throughput of about 60%. The total delay in the case of speech is an important parameter. This will be composed of the access delay plus the packetization delay (the time required to collect sufficient speech samples to fill a packet).

A typical Ethernet network is shown in Fig. 13.13. This network is made up of several segments connected by repeaters or bridges. The maximum length of each segment is specified as 500 m and the maximum number of nodes per segment is 100. The maximum round trip delay for the whole system is specified as 51.2 µs.

13.8 THE TOKEN BUS

The random access nature of CSMA/CD is very attractive when the traffic levels are low because it is possible to transmit a packet almost as soon as it arrives. However, Fig. 13.12 illustrates that delays can become very large at high traffic levels owing to continued collisions between contending nodes. The contention problem can be avoided if transmission on the bus is

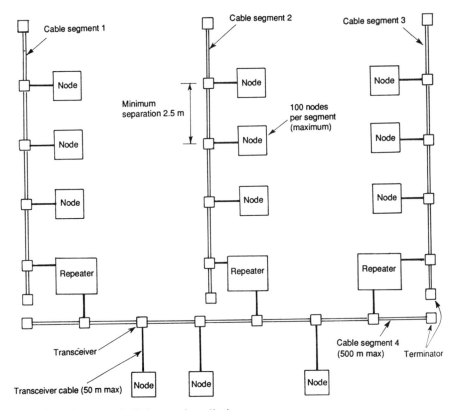

Fig. 13.13 Large-scale Ethernet installation.

scheduled and this is the basis of the token bus local area network, also known as IEEE 802.4. In effect each node waits for a token to be passed to it. When a node receives a token and it has data for transmission this data is transmitted in packetized form. The token is then passed to the next node and the process is repeated. If a node has no data for transmission the token is simply passed to the next node.

The token bus is less efficient than CSMA/CD at low traffic levels because a node must wait for the token even if the bus is free. At high traffic levels the token bus has a superior performance because contention (and the time required to resolve collisions) is avoided altogether. A simple token bus architecture is shown in Fig. 13.14. It should be noted that the token is passed from a node to the one with the next lower address, thus forming a logical ring on the bus. The token is passed only to active nodes thereby saving time. In order to receive the token an inactive node must request admission to the logical ring.

If it is assumed that the logical ring has been established the performance of the token bus LAN can be described as follows. (For information on the procedure of establishing the ring and admitting or deleting new nodes the reader is referred to Ramimi and Jelatis.[7])

A node collects packets for transmission and is able to transmit these packets when it receives the token from the previous node in the logical ring. In practice there is a maximum time allocated for transmission known as the token hold time. At the end of this period, or as soon as all waiting packets have been transmitted, the node passes the token to the next node on the logical ring. The token rotation time t_r is defined as the interval between successive arrivals of the token at a specific node. Evidently the token bus network will never be silent, as is possible with CSMA/CD, but will always be transmitting either message packets or a token.

In order to determine the delay characteristics of the token bus it will be assumed that token hold time is not limited, i.e. when a node receives the token it is able to transmit all its accumulated data. The token rotation time

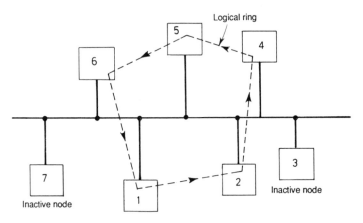

Fig. 13.14 Token passing local area bus network.

will thus be composed of the time to transmit N tokens (Nt_t) and the time required to transmit the total number of packets which accumulate during the token rotation time. If the mean packet arrival rate at each node is λ per second the total number of packet arrivals during the interval t_r is λNt_r. If the bus can transmit packets at a rate of μ/second the token rotation time is

$$t_r = Nt_t + \lambda N/\mu \qquad (13.16)$$

Defining the bus utilization as $U = \lambda N/\mu$, gives

$$t_r = Nt_t/(1 - U) \qquad (13.17)$$

The mean access delay will thus be $t_a = t_r/2$. Note: in order to ensure stable operation $U < 1$. The minimum token rotation time will occur when $U = 0$. In this case none of the N nodes has packets for transmission, when the token arrives, and the bus traffic consists merely of token transmissions. The bus utilization U is effectively the percentage of time that the bus is transmitting data packets. The factor $(1 - U)$ is then the percentage of the time that the bus is transmitting tokens. The normalized access delay (for $Nt_t = 2$) is plotted in Fig. 13.15.

As the ultilization approaches 1 the access delay approaches ∞. It may also be noted from Eqn (13.17) that the access delay is related directly to the time required to transmit the token t_t which in turn is proportional to the end to end propagation delay on the bus and the token length, and is inversely proportional to the data rate on the bus.

If the propagation time is large (as would be the case for a wide area network) the access delay can be reduced by increasing the data rate. Hence scheduled access is far more suited to long-distance communication than is CSMA/CD which (according to Eqn (13.15)) is limited to low data rates in order to minimize the value of the parameter 'a'. In Eqn (13.15) the characteristic is given in terms of the throughput S which is defined as the average number of successful packet transmission in a specified interval. The throughput S is related to the utilization U, i.e.

$$U = \frac{\text{mean packet arrival rate}}{\text{mean packet transmission rate}} = \frac{\text{mean arrival rate}}{S}$$

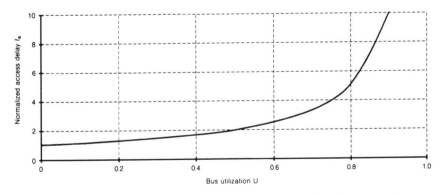

Fig. 13.15 Normalized access delay characteristic for the token bus network.

The IEEE 802.4 specification has been adopted for industrial use in the manufacturing environment. In this context a number of token bus networks operating at 1 Mb/s (known as carrier band) are connected via a bridge to a broadband cable operating at 10 Mb/s. These integrated systems operate under what is known as the Manufacturing Automation Protocol (MAP). A typical factory communications layout is shown in Fig. 13.16.

Each of the token bus subnets may be regarded as individual token bus networks in which the bridge to the broadband cable simply acts as an additional node. Tokens are not passed across the bridge so that the token passing on the subnets is independent from the token passing on the broadband cable. The subnets use phase coherent frequency shift keying with a logic low being represented by a carrier of 5 MHz and a logic high being represented as a carrier of frequency 10 MHz, the data transmission is omnidirectional. The broadband cable uses frequency division mutliplex transmission, one frequency being allocated for transmission and a second frequency allocated for reception. In this case all transmissions are unidirectional, a head end repeater being used for translating data between transmit and receive frequencies. The broadband cable is also able to handle other independent transmissions on a frequency multiplexed basis.

13.9 THE TOKEN RING

Ring networks are fundamentally different from bus networks in that the nodes on a ring network tend to act as repeaters and are therefore active

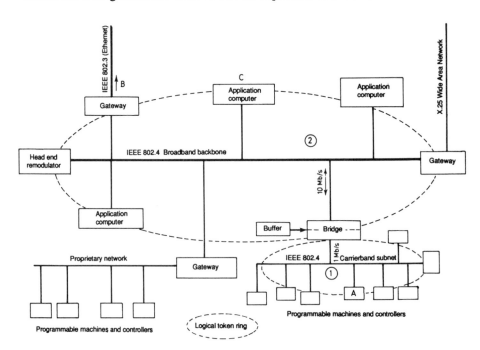

Fig. 13.16 Implementation of MAP in a factory environment.

devices. In the token ring access to the channel is controlled by passing a permission token around the ring. When the ring is initialized a designated node generates a free 8 bit token (11111111) which travels around the ring in one direction only until a node with packets ready to transmit changes the token to busy (11111110) and then transmits all its packets. The sending node is responsible for converting the token back to free, after it has travelled around the ring, and for removing its own packets. A further difference in ring structures is that each node adds delay to the circulating token and data. This is known as the node latency and is an essential requirement because each node must be able to store a bit in order to decide whether to change the last bit in a token (i.e. from free to busy or vice versa). Thus each node contributes a 1 bit delay and, if the number of nodes in a ring is large, this delay can become very considerable.

The token is usually 8 bits long which puts a lower limit on the number of nodes in a ring. If there are fewer than eight nodes on the ring the first bit of the token would return to the sending node before it had transmitted the last bit. Thus it is necessary to have sufficient delay on the ring to ensure that the complete token can be accommodated.

The mean access delay of the token ring may be obtained by assuming that there are N nodes on the ring and that each node has on average q packets ready for transmission. It is further assumed that packet arrivals have a Poisson distribution with a mean arrival rate, for all nodes, of λ packets/second. The mean arrival rate for the complete ring is thus $N\lambda$ packets/second. The time required for one bit to go all the way around the ring is termed the walk time t_w and is given by

$$t_w = N(\text{latency} + \text{propagation time between nodes}) \qquad (13.18)$$

The scan time t_s is the mean time between free token arrivals at a given node (this is equivalent to the token rotation time specified for the token bus LAN). This will consist of the walk time and the time required to service each of the packets ready for transmission. If the mean number of packets/second the ring can transmit is μ, the scan time is

$$t_s = t_w + N\lambda t_s/\mu \qquad (13.19)$$

Defining ring utilization $U = N\lambda/\mu$

$$t_s = \frac{t_w}{(1 - U)} \qquad (13.20)$$

The mean acquisition delay is then $t_s/2$ and this will approach ∞ as U approaches 1. This is similar to the characteristic for the token bus, which is not surprising as the token bus is, in effect, a logical ring. The token bus has the advantage that each individual connection to the bus is essentially passive.

13.10 METROPOLITAN AREA NETWORKS (MANs)

Local area networks are based on a shared bus which operates at a sufficiently high data rate to accommodate the needs of many users. However the LAN is restricted to a length of a few kilometres as the performance is limited by the normalized propagation delay. Metropolitan area networks are a relatively recent development which extend the operational distance of the shared bus up to distances in the range of 100 km. A number of standards have been developed for metropolitan area networks. The standard considered in this chapter was initially proposed by the University of Western Australia as the QPSX MAN.[8] This development was adopted as the basis of the IEEE 802.6 standard and is now known as the distributed queue dual bus metropolitan area network (DQDB MAN). FDDI is also considered in this category but is also often considered as a high-speed LAN.

The architecture of the DQDB MAN is based on two unidirectional buses as shown in Figure 13.17 and has been specified to operate at 155.52 Mb/s or 622.08 Mb/s. This allows full duplex communication between any pair of nodes connected to the network. The DQDB MAN is designed to support services requiring regular periodic access (isochronous services), such as digitized voice, and services requiring random access (non-isochronous services), such as packet switched data. For non-isochronous services the DQDB MAN provides a contention-free distributed queueing protocol which provides a mean packet access delay which is equal to that of a perfect scheduler. This section will concentrate on the mode of operation of the non-isochronous transport mechanism, with a brief description being given to the way in which the MAN handles isochronous services.

The principal components of the DQDB architecture are a slot (frame) generator, two unidirectional buses, a number of nodes (access units) and a slave slot (frame) generator. The slot generator creates a continuous stream of fixed size data slots on the A bus and the slave slot generator creates an identical pattern at the same rate on the Z bus. Access units are connected to both buses via read and write connections. The node reads information from the bus and writes information on to the bus by ORing its own data with that from upstream on the bus. This process is shown schematically in Fig. 13.18.

It will be noted from this figure that the data on the bus does not pass through the node, as is the case with the token ring. Hence failure of a node will not affect the operation of the remainder of the network.

The operation of the distributed queueing mechanism centres around two special bits in each slot known as the busy (B) bit and the request (R) bit. The B bit indicates that a slot is filled with user data and the R bit is used to indicate

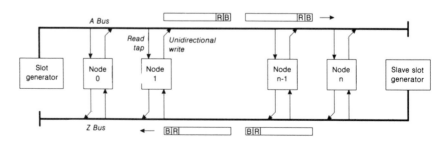

Fig. 13.17 Distributed queue dual bus MAN architecture.

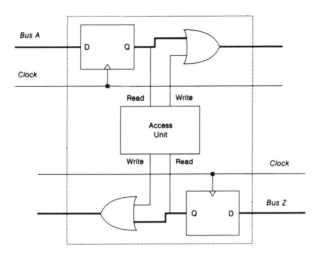

Fig. 13.18 Connection of nodes to the unidirectional buses.

that a node has a packet queued for transmission. By monitoring the B and R bits each node is able to keep a record of its position in the distributed queue. Thus when a node has a packet for transmission it will only access the bus when all other nodes that queued before it have transmitted. Access to the slots on the A bus are made using the R bit in slots on the Z bus and access to the slots on the Z bus are made using the R bit in slots on the A bus.

Consider a node n wishing to access the A bus. When the node has no packets for transmission it will be in the IDLE state and simply monitors the R bits which pass it on the Z bus. Each time an R bit which is set passes a given node on the Z bus, it indicates that a downstream node (a node with a number $> n$) on the A bus has a packet queued for transmission and is used to increment a request counter (RQ) within node n. The RQ counter is decremented each time an empty slot (i.e. one with the B bit clear) passes that node on the A bus, since that slot will be used to serve one of the downstream queued packets. Thus the value of the RQ count is at all times equal to the exact number of nodes downstream that have a packet queued for access.

When a sending node has a packet ready for transmission it will leave the IDLE state and if the value of the RQ counter is non-zero will enter a COUNTDOWN state. The contents of the RQ counter are transferred to a countdown counter CD and the RQ counter is cleared. The purpose of this state is to allow previously queued nodes to access the bus first. This is achieved by node n by decrementing the CD counter for each empty slot passing on the A bus. These empty slots will be used by nodes previously queued downstream of node n. While in the COUNTDOWN state the node continues to monitor any passing R bits which have been set on the Z bus and uses these bits to increment the RQ counter. These R bits do not affect the CD counter as they arrive after node n has queued a packet. In addition while in the COUNTDOWN state empty slots passing on the A bus decrement only the CD register and not the RQ counter as the empty slots are serving prior queued nodes.

When node n enters the COUNTDOWN state it will set the R bit in the first available slot on the Z bus. This R bit will be transmitted on the Z bus to all nodes which are upstream of node n on the A bus. This R bit indicates to the nodes upstream on the A bus that a request for an additional slot has been made on this bus. It is possible that a node could overwrite a R bit which has already been set by a downstream node. This would not affect the value of the R bit as the writing operation is achieved by a logical OR but the extra R bit generated by node n would not be detected by upstream nodes. To avoid this problem the read tap of each node occurs before the write tap and the node can detect whether the R bit is already set. In such a case the node would wait until a slot in which the R bit is clear passes it on the Z bus and it would then transmit a R bit.

The node remains in the COUNTDOWN state until the contents of the CD register reach zero and then enters the WAIT state. When in the WAIT state the node waits for the first empty slot on the A bus (i.e. the first slot with the B bit clear). It then sets the B bit and transmits its packet in the slot. If the node has no further packets for transmission it enters the IDLE state. If the node does have further packets for transmission it enters the COUNT-DOWN state. (Note that if a node wishes to transmit a packet on the Z bus it sends a R bit on the A bus and a similar action occurs with separate RQ and CD counters.)

The distributed queueing mechanism provides ideal access characteristics as a packet will gain immediate access if the queue size is zero or is required to wait only until previously queued packets are transmitted. Hence, unlike the other access mechanisms described in this chapter, capacity is never wasted while packets are queued and thus access delay is quaranteed to be a minimum. A typical performance characteristic of the DQDB compared with the token bus is shown in Fig. 13.19. The mean waiting time (access delay) for the DQDB is

$$t_a = \frac{P}{2(1-U)} \tag{13.21}$$

where P is the packet transmission time.

Comparing this equation, with Eqn (13.17) it should be noted that the limitation of the token bus is the transfer of the token as nodes with packets for transmission must wait on average for an interval of half the bus latency. This is not the case with DQDB as a packet may be transmitted as soon as other packets ahead in the distributed queue have gained access. Equation (13.21) reveals that the advantage of the DQDB increases as the size of the network and the operation speed increase, as P is not dependent on any bus latency, which is in sharp contrast to the token bus performance.

An extension of the DQDB architecture is the looped bus structure of Fig. 13.20 in which the end points of the linear bus structure are combined into a single head end node. It should be noted that data does not pass through the head end node so that the looped bus does not operate as a ring. There are two specific of advantages of this structure however:

(i) The clock generator for both the A and Z bus can be derived from a common point and synchronized with the PSTN clock.

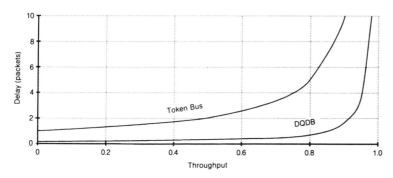

Fig. 13.19 DQDB throughput characteristic.

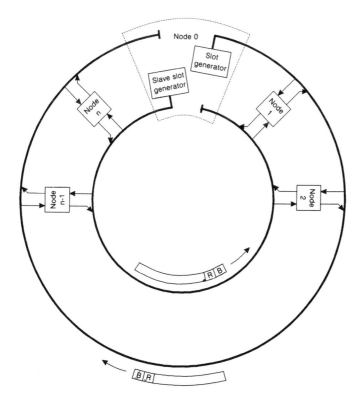

Fig. 13.20 Looped bus structure of the DQDB MAN.

(ii) In the case of a bus fault the network can isolate the fault and close the data buses through the head end of the loop. Hence the network can be reconfigured without the need for any redundant transmission hardware. This gives the DQDB looped structure a significant additional advantage in terms of reliability.

The discussions on the DQDB mechanism have concentrated on the non-isochronous transmission mode often termed the queue arbitrated (QA) mode. The DQDB also has another mode, known as the pre-arbitrated (PA) mode, which is used for isochronous services. The slots on the DQDB MAN can be made available either as QA slots or as PA slots. In the latter case the node at the head end of the bus manages the generation of PA slots which are divided into octets each of which may be allocated to particular access units for the transfer of isochronous service octets. Hence each slot can serve more than one access unit. Each octet within a PA slot has a unique offset position relative to the start of the PA slot payload. Each PA slot carries a special identifier and nodes which have requested and have been allocated an isochronous service will obtain from that identifer the offset of the slot which they been allocated. When slots are used in this way nodes are guaranteed periodic access and do not participate in the alternative distributed queueing mechanism.

13.11 OPEN SYSTEMS INTERCONNECTION REFERENCE MODEL

Packet switched networks were designed initially for computer communications. In such circumstances the computers themselves became an integral part of the communications system, performing tasks such as channel demand, addressing, routing, error checking, buffering, etc. Clearly the proliferation of various computer architectures could result in the creation of barriers to efficient communication. The early realization of this problem resulted in the adoption of an international standard for communication, in 1983, known as the Open Systems Interconnection reference model[9]. This model divides up the various aspects of communication into seven layers as shown in Fig. 13.21.

The OSI model was designed to facilitate global communications, primarily for computers, irrespective of the characteristics of the particular networks of which they are part. The model does not specify how systems are implemented but rather how they communicate. This means that many different networks, using products of different manufacturers, may be coupled together by mapping the OSI model onto the complete communications path. A brief description of each layer will now be given.

(i) The physical layer (layer 1)
   This layer deals with transmitting individual bits over a communications system. It deals with the way in which the bits are represented, e.g. voltage levels, bit duration, etc. on the transmission medium and whether the transmission may proceed simultaneously in two directions.

(ii) The data link layer (layer 2)
   The data link layer makes use of the network layer to ensure error-free communication between the network layer at each end of the communications link. In Ethernet, for example, the data link layer is responsible for transforming data into the packet format of Fig. 13.1. Layer 2 also

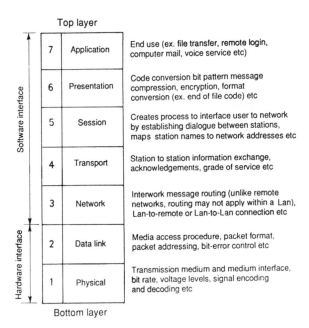

Top layer

7	Application	End use (ex. file transfer, remote login, computer mail, voice service etc)
6	Presentation	Code conversion bit pattern message compression, encryption, format conversion (ex. end of file code) etc
5	Session	Creates process to interface user to network by establishing dialogue between stations, maps station names to network addresses etc
4	Transport	Station to station information exchange, acknowledgements, grade of service etc
3	Network	Interwork message routing (unlike remote networks, routing may not apply within a Lan), Lan-to-remote or Lan-to-Lan connection etc
2	Data link	Media access procedure, packet format, packet addressing, bit-error control etc
1	Physical	Transmission medium and medium interface, bit rate, voltage levels, signal encoding and decoding etc

Software interface (layers 3–7), Hardware interface (layers 1–3)

Bottom layer

Fig. 13.21 Layered architecture of the OSI reference model.

handles acknowledgements from the receiver and responds to errors.

(iii) The network layer (layer 3)

This layer essentially deals with routing of packets over the communications link and ensures that the data is assembled in the correct order for presentation at the receiver. This layer would be concerned also with providing some form of error control.

(iv) The transport layer (layer 4)

This layer provides an end to end communication between host machines. It is not concerned with the hardware constraints of providing the actual connection, as are the lower layers, but is involved with such functions as how to create the most efficient use of the communications media for a particular throughput. The transport layer handles flow control, which is essential if a high speed transmitting host is communicating with a low speed receiving host.

(v) The session layer (layer 5)

This is the layer which deals with setting up the connection between users. This layer would handle, for example, the procedure for logging in on a remote machine and initiating a file transfer.

(vi) The presentation layer (layer 6)

This layer deals with the way in which the data is presented for transmission and is usually performed by subroutines called by the user. The presentation layer could, for example, handle data encription or the transmission of hexadecimal data in ASCII format.

(vii) The application layer (layer 7)

This layer deals with the way in which individual users' programs communicate with each other, i.e. the meaning of the transferred data.

The OSI model specifies seven layers, but not all communications connections will involve all seven layers. A special protocol known as X25 has been developed (actually before the OSI reference model) which implements layers 1, 2 and 3 of the OSI reference model. This protocol will be briefly described in the next section, as an example of the partial implementation of the OSI reference model.

13.12 THE X25 COMMUNICATIONS PROTOCOL

The X25 standard defines the interface between a host computer, known as the Data Terminal Equipment (DTE), and the carrier's equipment, known as the Data Circuit-terminating Equipment (DCE). When a host communicates, using X25, with a second host the DCEs at each end will be linked over an unspecified network by means of various switching centres. The connection between the two DCEs is given the general label of Data Switching Exchange (DSE) and is transparent to the X25 protocol. The relationship between X25 and the OSI reference model is shown in Fig. 13.22.

X25 defines three layers of communication, the physical layer (equivalent to the physical layer of OSI), the frame layer (equivalent to the data link layer of OSI) and the packet layer (equivalent to the network layer of OSI). The physical layer of X25 deals with the electrical representation of binary 1 and 0, timing, etc. and is described by a standard known as X21. This standard specifies how the DTE exchanges signals with the DCE to set up and clear calls, etc. and is illustrated in Fig. 13.23.

In many installations the DTE will be connected to the DCE over a standard analogue telephone line which may not support the full X21 specification.

The frame layer ensures reliable communication between the DTE and DCE. In effect the frame layer adds additional bits to the data stream produced by the packet layer to make the communication reliable. The additional bits and the data produced by the packet layer are combined into a frame to produce what is known as the link access procedure (LAP) protocol, a typical LAP format is shown in Fig. 13.24.

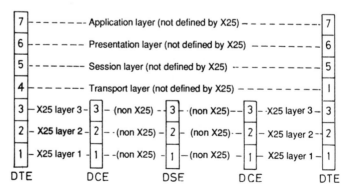

Fig. 13.22 Relationship between OSI reference model and X25.

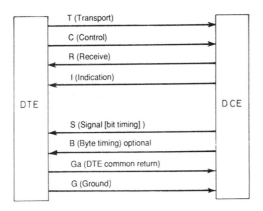

Fig. 13.23 Signal lines specified in X21.

Start byte	Address	Control	Data	Checksum	End byte
01111110	8 bits	8 bits	≥ 0 bits	16 bits	01111110

Fig. 13.24 LAP frame format.

The LAP frame structure will not be discussed in detail in this text but it should be noted that there are three kinds of frame known as information, supervisory and unnumbered. The reader is referred to Tanenbaum[10] for further information.

The third layer in X25 deals with the meaning of the various bits in the data field of the frame. The data within the frame is itself known as a packet, hence X25 packets are embedded in a LAP frame. There are actually 15 different types of X25 packet dealing with various aspects of call set-up and monitoring. As an example, when a DTE wishes to communicate with a second DTE a virtual circuit must be set up between them. The initiating DTE sends a 'call request packet' to its DCE which delivers this packet to the destination DCE which then passes the packet to the destination DTE. The calling DTE defines the logical channel on which communication is to take place by means of a channel field in the call request packet. If the distant DTE wishes to accept the call it returns a call accepted packet to the originating DTE and communication commences until a 'clear request packet' is sent to terminate the call.

There are many facilities available to X25, for example if the logical channel requested by the initiating DTE is in use at the far end the called DCE will substitute an unused channel. X25 is able to cope with collisions which could occur when a logical channel request from an initiating DTE coincides with an allocation of that channel (for a separate purpose) by the called DTE. The actual X25 protocol is quite complex and reflects the number of facilities available. A full description of X25 is given by Schwartz[11]. For the purpose of this text it is sufficient to illustrate the functional separation of the three defined layers of X25.

It is common practice to use X25 links for devices, such as simple terminals, which do not themselves have X25 capability. This has resulted in the widespread use of devices known as packet assemblers/disassemblers or PADs. A PAD is essentially a mulitplexer or concentrator which collects data from a number of asynchronous terminals and periodically outputs a correctly formated X25 packet made up of this data. The X25 packets may then be passed to a DCE in the normal fashion. The inverse operation is also required to convert X25 packets, from a host, into the asynchronous format for the individual terminals. The PAD contains significant intelligence and may itself be considered as an X25 DTE host.

13.13 CONCLUSIONS

This chapter has concentrated on the description of packet switched networks with a detailed consideration of local and metropolitan area networks. Some reference has been made to the integration of speech into local area networks. Packetization of speech and other real-time signals is a topic of intense interest both for wide area and local area communications. It is envisaged that all analogue telephone communications will eventually be replaced by integrated services digital networks. Major research projects currently being funded in Europe are directed towards replacing the fixed telephone by a completely mobile cellular digital radio telephone system. Initially this system is likely to be a PAN European system where any mobile user (which will eventually be all citizens) may communicate with any other mobile user without a prior knowledge of that user's physical location.

REFERENCES

1. Abrahamson, N., "The ALOHA System - Another Alternative for Computer Communications", *AFIPS Conference Proceedings*, 1970 Fall Joint Computer Conference, Vol. 37, pp. 281–285.
2. Dunlop, J., "Packet Access Mechanisms for Cellular Radio" *Electron. & Commun. Eng. J.*, 1993, Vol. 5, No. 3, pp. 173–179.
3. Kleinrock L *Queuing Systems Vol. II Computer Applications*, Ch. 5, John Wiley, 1976.
4. Shoch, J. and Hupp, J. A., "Measured Performance of an Ethernet Local Network" *Communications of the ACM*, Vol. 23, No. 12, 1980, pp. 711–721.
5. Dunlop, J. and Rashid, M. A., "Speech Transmission Capacity of Standard Ethernet Systems" *Journal of the IERE (UK)*, Vol. 55, No. 3, 1985, pp. 119–122.
6. Dunlop, J. and Rashid, M. A., "Improving the Delay Characteristics of the Standard Ethernet for Speech Transmission", *Journal of the IERE (UK)*, Vol. 56, No. 5, 1986, pp. 184–186.

7. Rahimi, S. K. and Jelatis, G. D., "LAN Protocol Validation and Evaluation", *IEEE Journal on Selected Areas in Communications*, Vol. SAC-1, No. 5, 1983, pp. 790–802.

8. Newman, R. M., Budrikis, Z. L. and Hullett, J. L., "The QPSX MAN", *IEEE Communications Magazine*, Vol. 26, No. 4, 1988, pp. 20–28.

9. Zimmerman, H., "OSI Reference Model-The ISO Model of Architecture for Open Systems Interconnection", *Trans. IEEE on Communications*, Vol. COM-28, 1980, pp. 425–432.

10. Tanenbaum, A. S., *Computer Networks* Prentice-Hall, 1981, Ch 4.

11. Schwartz, M., *Telecommunication Networks: Protocols, Modelling and Analysis*, Addison-Wesely, 1987, Ch 5.

PROBLEMS

13.1 A group of N stations share a 64 kb/s pure ALOHA channel. Each station, on average, outputs a 1024 bit packet every 50 seconds. What is the maximum value of N which may be accommodated?

Answer: 575.

13.2 A large number of nodes using an ALOHA system generate 50 requests/ second; this figure includes new and re-scheduled packets. The time axis is slotted in units of 40 ms.

  (i) What is the probability of a successful packet transmission at the first attempt?
  (ii) What is the probability of exactly four collisions followed by a successful transmission?
  (iii) What is the average number of transmission attempts per packet?

Answer: 0.135, 0.076, 6.39.

13.3 An Ethernet cable operating at 10 Mb/s has a length of 1.5 km and a propagation speed of 2.2×10^8 m/s. Packets are 848 bits long which includes an overhead of 208 bits. The first bit slot after a successful transmission is reserved for the destination node to access the channel to send an acknowledgement packet (data field empty). What is the effective (i.e. useful) data rate assuming no collisions? Assume an inter-packet gap of 9.6 μs.

Answer: 5.33 Mb/s.

13.4 One hundred nodes are distributed over a CSMA/CD bus of length 4 km. The transmission rate is 5 Mb/s and each packet is 1000 bits long including the standard overhead of 208 bits. Calculate the maximum number of packets/ second that each node can generate. Assuming an inter-packet gap of 9.6 μs and a propagation delay of 5 μs/km, calculate also the useful data rate of the system.

Answer: 45.45, 3.61 Mb/s.

13.5 The bit rate on the system of the previous question is increased to 10 Mb/s. Determine the effect of this increase on the number of packets/second that each node can generate and the useful data rate of the system. Repeat the calculation if the packet length is increased to 10 000 bits. Comment on the results.

Answer: 83.61, 6.62 Mb/s, 980.8, 9.6 Mb/s.

13.6 A coaxial cable of length 15 km supports the token bus protocol, the propagation speed being 2×10^8 m/s and the data rate being 10 Mb/s. There are 512 nodes connected to the bus each producing packets 12 kb long, with an overhead of 96 bits. The token, which is a standard packet with an empty data segment, is passed sequentially to each node. Assuming a latency of 8 bits at each node and that each node generates 6 packets per 10 seconds on average calculate the average token rotation time. What is the maximum rate of transmission of packets at each node?

Answer: 17.9 ms, 1.55 packets/s.

Satellite communications | 14

14.1 INTRODUCTION

For many years, the notion of satellite communications was a fantasy produced by the fertile mind of Arthur C. Clarke;[1] a brilliant idea but rather impractical. As with so many creative ideas, technology eventually caught up and now satellites are commonplace. For the communication engineer, however, they represent a challenging and stimulating field of work. Within satellite communication systems we find the whole gamut of technologies operating in a strange and demanding environment. Coupled with the essential requirement of demanding as little energy to run as possible, the system must be capable of withstanding the arduous journey from earth to orbit. Consequently a careful balance must be struck between the mechanical, structural, electronic, electrical and electromagnetic engineering requirements of the system[2]. However, in this chapter we will limit our attention to system considerations.

Satellite communications provide opportunities, and pose problems, in communication methods. Their large area of access (footprint) allows a single transmission to cover an enormous number of receivers, thus allowing broadcast signals to be transmitted simultaneously to large numbers of people. However, this feature itself creates difficulties; partly political and partly economic. National boundaries are no barrier whatsoever, and the charging mechanism required to allow the satellite operator to recover the cost of development and provide continuous support requires a novel solution.

Satellites can travel in a variety of orbits, basically eliptical in shape, but we will limit our attention to geostationary systems, in which the orbit is fixed, and essentially circular. A geostationary satellite maintains a fixed position relative to points on earth, and this allows a cheap receiving antenna to be set up once and then fixed in position.

The enormous attraction of satellites for a whole range of communication objectives has meant that there is considerable danger of overcrowding. More frequency bands will have to be allocated, and the satellites themselves will have to be pushed closer together so that more can be placed in the geostationary orbit. Recently there has been a compression from 3° to 2° separation.

In a sense, there is no new communications work in this chapter. What we have is an unusual application which produces unusual operating conditions and therefore requires novel developments of existing knowledge. In many of the following sections we will refer to earlier chapters in this book for explanations of principles, and we will limit our attention here to those features of communication systems which are peculiar to satellite applications.

14.2 GEOSTATIONARY ORBITS

The analysis of the satellite orbit is complex and rather beyond our scope. Excellent treatments are given elsewhere[2,3]. For our purpose, however, we only need to know that it is possible to place a satellite in a geostationary orbit at a height of approximately 36 800 km above the earth, and that the station can be maintained, provided it has the provision for slight adjustments of orientation and position by way of small rockets. Although we will concentrate on geostationary satellite communications, it should be realized that there are areas near the poles which can be reached only by satellites in other orbits. For example, satellites in tundra orbit are not geostationary, but they provide regular coverage for the extreme latitudes (Fig. 14.1).

For a geostationary orbit the angular velocity of the satellite is constant and therefore, since the angle swept per unit time (Kepler's third law) is also constant, the orbit must be circular. There is a simple relationship between the period and the radius of the orbit

$$a \propto (T^{2/3}) \tag{14.1}$$

and the equality is provided by Kepler's constant μ: i.e.

$$a = \left(\frac{T^2 \mu}{4\pi^2}\right)^{1/3} \tag{14.2}$$

Assuming that the period T is the same as a normal (sidereal) day $= 24.60.60$ s and taking $\mu = 3.986\,1352 \times 10^5\ \mathrm{km^3/s^2}$, the above equation gives the orbital radius to be

$$a = 42\,241.558\ \mathrm{km} \tag{14.3}$$

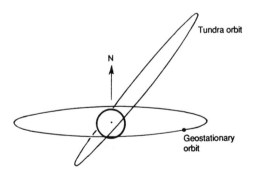

Fig. 14.1 Communication satellite orbits.

14.3   ANGLE FROM SATELLITE TO EARTH STATION

Again, the case of a geostationary satellite is a much simplified version of the general case. The point at which a line between the satellite and the centre of the earth intersects the earth's surface is called the sub-satellite point. There are two angles of intersect between the earth station and the satellite: the elevation and the azimuth. The elevation is the angle of the satellite above the horizon, from the earth station, and the azimuth is the angle between the line of longitude through the earth and the direction of the sub-satellite point. These are shown diagrammatically in Figs. 14.2 and 14.3.

From Fig. 14.2 the angle γ at the centre of the earth is given by

$$\cos \gamma = \cos L_e \cos (l_s - l_e) \tag{14.4}$$

where L_e = latitude of the earth station and $l_s - l_e$ = difference in latitude between the earth station and the sub-satellite point.

Applying the cosine rule

$$d_s^2 = r_e^2 + r_s^2 - 2r_e r_s \cos \gamma \tag{14.5}$$

From the sine rule

$$\frac{r_s}{\sin \psi} = \frac{d_s}{\sin \gamma} \tag{14.6}$$

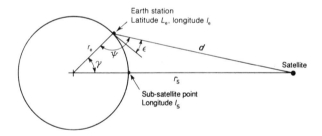

Fig. 14.2 Satellite position relative to Earth.

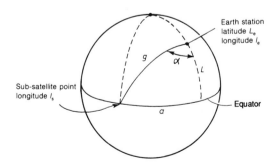

Fig. 14.3 Angle of azimuth.

giving

$$\sin\psi = \frac{r_s \sin\gamma}{d_s}$$

$$= \frac{r_s \sin\gamma}{[r_e^2 + r_s^2 - 2r_e r_s \cos\gamma]^{1/2}}$$

$$= \frac{\sin\gamma}{\left[\left(\dfrac{r_e}{r_s}\right)^2 + 1 - 2\left(\dfrac{r_e}{r_s}\right)\cos\gamma\right]^{1/2}} \qquad (14.7)$$

and since, from Fig. 14.2

$$\varepsilon = \psi - 90$$

$$\cos\varepsilon = \frac{\sin\gamma}{\left[1 + \left(\dfrac{r_e}{r_s}\right)^2 - 2\left(\dfrac{r_e}{r_s}\right)\cos\gamma\right]^{1/2}} \qquad (14.8)$$

Using $r_e = 6370$ km and $r_s = 42\,242$ km in Eqn (14.8) gives

$$\cos\varepsilon = \frac{\sin\gamma}{[1.02274 - 0.30159\cos\gamma]^{1/2}} \qquad (14.9)$$

Clearly the angle of elevation will depend on the latitude of the earth station via angle γ; ε will be large towards the equator and small towards the poles. A minimum value of about $5°$ limits the highest latitude. The azimuth angle $180 + \alpha$ in Fig. 14.3, is related to the latitude and longitude of the earth station, L_e and l_e and the longitude of the sub-satellite point l_s by

$$\tan^2\frac{\alpha}{2} = \frac{\sin(S - \gamma)\sin(S - L_e)}{\sin S \, \sin(S + L_e)} \qquad (14.10)$$

where

$$S = \tfrac{1}{2}(a + L_e + g) \qquad \text{and} \qquad a = (l_s - l_e)$$

The elevation and azimuth angles define the 'boresight' of the receiving antenna for a given receiver location. The azimuth angle uses the line from earth station to north pole as a reference.

It was noted earlier that towards the poles there is a restriction on the minimum angle of elevation considered suitable because the noise increases as $\varepsilon \to 0$. Another consideration is the additional attenuation produced by the increased distance to the satellite (slant height) with latitude.

From Eqn (14.5) the slant height is

$$d_s = [r_s^2 + r_e^2 - 2r_s r_e \cos\gamma]^{1/2}$$

14.4 SATELLITE LINKS

In satellite communications the direction of transmission is indicated by the term used: the downlink is transmission from satellite to earth station and the uplink is transmission in the opposite direction.

As we have mentioned before, communications by satellite impose enormous problems on the systems designer. In the uplink we have the availability of a high intensity transmission beam, because the transmitter is ground based and therefore power is not a problem, coupled with a sensitive receiver on the satellite, while in the downlink we have a low power source in the satellite and a highly sensitive receiver in the earth station. However, this presupposes that the earth station cost is not a constraint. For public utility applications such as telephony, cost will not be a limiting factor; the high cost of an expensive receiving station can be spread over a very large number of users. In direct broadcast television however, the story is different. A single user has to pay for the receiver and therefore its cost must be low. This means that the earth station will be 'basic'. The dish will be as small as possible and the amplifier as cheap as possible, consistent with obtaining an adequate signal for most of the year. Outage will be high, compared with that for a better quality receiver, but must still be at a reasonable level.

The preoccupation in satellite link design is the power budget; how much power can be obtained from the transmitter, how much of that power can be directed towards the receiver, how much power is lost over the link, and at the transmitter and receiver, and how much is left at the detector? What margin is necessary above the minimum detectable signal for the detector being used? Essentially the problem is to ensure that the signal level at the detector is large enough to produce a satisfactory output for a large part of the year.

Since the satellite distance is fixed, the attenuation (except for the effect of rain) is also fixed, and it is high. It cannot be avoided and so any improvement to be made must occur at the transmitter or receiver, and will be marginal, but very important. Fig. 14.4 shows the basic features of a satellite link.

We will discuss first the link transmission loss and the effect of antenna design.

Let P_t be the power from the transmitter

G_t the transmitter antenna gain

R the range from transmitter to receiver

P_r the received power

G_r the receiver antenna gain

We know from Chapter 7 that the power density, P, at a distance R from

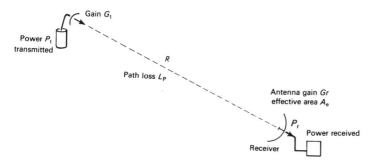

Fig. 14.4 Transmission link.

an isotropic source radiating total power P_t is

$$P = \frac{P_t}{4\pi R^2} \quad \text{W/m}^2 \tag{14.11}$$

The transmitting antenna of gain G_t will concentrate the power in the direction of its maximum lobe to produce a flux density

$$P = \frac{P_t G_t}{4\pi R^2} \quad \text{W/m}^2 \tag{14.12}$$

The receiver antenna, when properly lined up with the transmitter, intercepts power of density P. The effective area of the receiving antenna is $A_e = \eta A_r$ where A_r is the antenna aperture area and η is known as the antenna efficiency.

The power received = power density at the receiver × effective antenna area

$$= \frac{P_t G_t}{4\pi R^2} \cdot A_e$$

$$= \eta \frac{P_t G_t A_r}{4\pi R^2} \tag{14.13}$$

There is a simple relationship between the effective area and the antenna gain which is

$$G_r = \frac{4\pi A_e}{\lambda^2}$$

Hence

$$P_r = \frac{P_t G_t G_r \lambda^2}{(4\pi R)^2} \tag{14.14}$$

For a geostationary satellite, R is fixed, and for a particular application λ is also fixed. The expression naturally splits into three sections

$$P_r = P_t G_t G_r \left(\frac{\lambda}{4\pi R}\right)^2 \tag{14.15}$$

$P_t G_t$ is called the EIRP (effective isotropic radiated power), G_r is the receiver gain and $(\lambda/4\pi R)^2$ is a measure of the power dispersion between transmitter and receiver, known as the path loss, L_p. Thus

$$P_r = \frac{\text{EIRP} \times P_r}{L_p} \tag{14.16}$$

More simply, this can be expressed in dB as

$$P_r = (\text{EIRP} + P_r - L_p) \quad \text{dBW} \tag{14.17}$$

where

$$\text{EIRP} = 10 \log_{10} P_t G_t \quad \text{dBW}$$

$$G_r = 10 \log_{10} \frac{4\pi A_e}{\lambda^2} \quad \text{dB}$$

$$L_p = 20 \log_{10} \frac{4\pi R}{\lambda} \quad \text{dB}$$

The resultant power received must be sufficient to overcome losses not included in the above equation. As mentioned earlier, there will be precipitation losses, and there will be small losses at the transmitter and receiver which will need to be taken into account. For example, it has been assumed that all the power produced at the transmitter will be radiated from the transmitting antenna; that will not be so since there will be small losses in the transmitter circuits.

In Eqn(14.15) the received power was related to other system parameters by

$$P_r = P_t G_t G_r \left(\frac{\lambda}{4\pi R} \right)^2$$

This is known as the carrier power C. Of considerable interest is the ratio of this power to the noise power N at the detector. From the discussion in Chapter 4 the noise power at a receiver is

$$P_n = k T_s B G$$

where T_s is system noise temperature
$\qquad B$ is receiver bandwith
$\qquad G$ is receiver gain
$\qquad k$ is Boltzman's constant

and the power received due to the carrier is $P_r G$; thus the ratio C/N is given by

$$\frac{C}{N} = \frac{P_t G_t G_r}{k T_s B} \left(\frac{\lambda}{4\pi R} \right)^2 \tag{14.18}$$

assuming that the receiver has been replaced by an equivalent system characterized by T_s, and B

or

$$\frac{C}{N} = \frac{P_t G_t}{k B} \left(\frac{\lambda}{4\pi R} \right)^2 \frac{G_r}{T_s}$$

or, assuming the subscripts to be implicit, the carrier/noise ratio at a particular earth station is

$$\frac{C}{N} = \frac{P_t G_t}{k B} \left(\frac{\lambda}{4\pi R} \right)^2 \frac{G}{T} \tag{14.19}$$

where G/T is a measure of the goodness of the earth station. This quantity is widely used as a figure of merit to provide a means of comparison between different systems.

14.5 ANTENNAS

For satellite communications the ideal antenna performance is to have high gain, high efficiency (low loss), low side lobe levels, and narrow beamwidth. The beamwidth and side lobe levels are important in determining the interference from adjacent satellites and hence have a strong impact on the proximity which adjacent satellites can have in orbit.

The efficiency of the antennas used in satellite communications is extremely important and much care is taken to restrict any losses. Factors which reduce efficiency are wide ranging. The design of the antenna and its feed are obvious elements, but equally significant is the design of the waveguide and other components from, in the case of the receiver, the antenna feed to the first stage of the amplifier.

The degree to which clever antenna and feed design can be used to improve antenna efficiency will depend on the cost budget for a system. If the antenna forms part of an earth station carrying heavy traffic then a high cost can be tolerated, and sophisticated design techniques will be justified. However, for a simple broadcast television receiving station low cost and ease of manufacture are paramount, consequently antenna loss will be comparatively high.

The simplest antenna system, Fig. 14.5, is a straightforward dish paraboloid with a waveguide feed at its focus. The sources of loss are:

(i) poor illumination profile over the dish;
(ii) blocking by the feed;
(iii) long waveguide run in the feed;
(iv) high side lobe level.

As far as the antenna itself is concerned the paraboloid, in principle, allows a parallel beam of radiation from a distant source to be reflected to a focal point (section 7.18). However the ray theory approach on which that is based assumes that the wavelength is very small compared to the diameter of the antenna, but the wavelengths used in practice in satellite communications do not satisfy that condition. As a consequence there are diffraction effects which have a significant bearing on the performance of the system. Ideally, a uniform distribution of E-field amplitude is desirable at the antenna aperture, but this can only be achieved for an antenna of infinite diameter. Usually

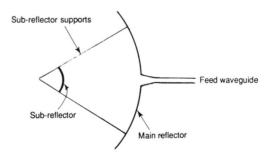

Fig. 14.5 Cassegrain dish antenna with direct feed.

some compromise is required between cost and performance. To reduce loss and limit cost, the direct feed arrangement is replaced by an indirect feed system, and in practice nearly all 'A' standard antennas for use in the INTELSAT system use this approach. Two methods are favoured, both based on the same type of principle and known after the sixteenth century inventors of the equivalent optical telescope. They are Cassegrain and Gregorian antennas. In both, the primary feed is at the centre of the paraboloid. This feed illuminates a secondary reflector; in the case of the Cassegrain of hyperbolic cross-section and in the case of the Gregorian of eliptical cross-section. These sub-reflectors in turn illuminate the main parabolic reflector. The sub-reflector and its mounting block some of the energy from the paraboloid. To reduce this effect the sub-reflector can be offset from the main axis as shown in Fig. 14.6.

The compromise between low side lobe level and high directivity (narrow beamwidth) is achieved by adjusting the illumination used across the aperture. To obtain the required distribution the primary feed, sub-reflector shape and dish shape can all be modified.[4]

Apart from the losses associated with diffraction effects and limited antenna aperture, there are other sources of antenna loss. In receiver mode the signal is transmitted from the primary feed into the receiver. For practical reasons the receiver is preferably placed away from the antenna, with the consequent waveguide run adding to the system loss. This loss can be reduced to a small value by using a beam waveguide, which is almost a free-space system enclosed in a large envelope, as shown in Fig. 14.7. This system is employed in a number of practical systems and in the example shown in Fig. 14.8 the beam waveguide can be seen clearly.

One type of primary feed which is of considerable interest is a corrugated pyramidal horn. The corrugations can be broadbanded and it has the advantage that it provides excellent cross-polarization rejection. In many systems the channel utilization can be increased significantly by using two directions of polarization to separate channels.

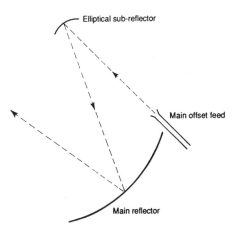

Fig. 14.6 Schematic of offset Gregorian reflector.

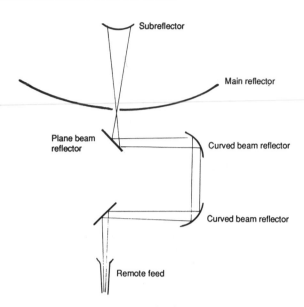

Fig. 14.7 Schematic of beam waveguide feed.

The function of spacecraft antennas is to direct the beam to those areas of the earth which have appropriate receivers. Although this may appear to be an obvious statement, it does emphasize the difficulties for the designer. Wasting power by sending a signal into areas with no receivers is undesirable, but to achieve coverage over specified areas only is also difficult.

With a standard paraboloidal dish the coverage on the earth's surface is circular. Since the angle subtended by the earth at the satellite is 17°, such a beamwidth will provide coverage over almost one hemisphere. To reduce the coverage area significantly requires a much larger antenna, and to provide a contoured (non-circular) coverage pattern involves the use of complex feed arrangements.

Gradually, linear array feeds are being introduced. They give the designer considerable flexibility in the range of patterns which can be obtained, at the cost of increased complexity. However, there are other benefits which may arise from these feeds. By making them electronically fed, the beam can be reconfigured to suit new users of the satellite who may require different coverage patterns.

Even further variability, and therefore options, can be achieved with antennas using linear arrays to illuminate a sub-reflector. The reflecting surfaces of both the sub-reflector and the main reflector can be distorted to modify the shape of the ground coverage profile.

There is a limit to the size of antenna array a satellite can accommodate. Large antennas can be deployed by using collapsible lightweight structures which can be opened out once the satellite is in orbit.

Offset feeds are often used, both to eliminate any blocking of the beam and to provide good mechanical support for the feeds. Several feeds, for different frequencies, may be mounted on the same support.

Fig. 14.8 Photograph of antenna and beam waveguide feed.

Earlier we discussed polarization of the beam to extend the utilization of the channel. This can be achieved by relying on the polarization from the feed horn being maintained, or by using a polarizer in the form of a grid of appropriately orientated wires in front of the reflector.

A future development in spacecraft antenna design, planned for business satellites in particular, is to provide space switching. Earlier we noted that additional capacity can be obtained by using polarization. Space switching will also provide more capacity over a given area by allowing frequency re-use on a spatial basis. Beams will be very narrow and once the area of

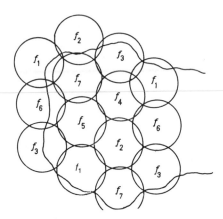

Fig. 14.9 Frequency re-use radiation 'foot-prints'.

potential interference has been exceeded the same frequencies can be used again (Fig. 14.9). This is the same basic idea as that used in cellular radio (Section 13.12).

14.6 ERROR CODING

It is in the nature of transmission channels that from time to time interference with the transmitted signal will produce errors. In digital systems these errors will result in bits being incorrectly received. Coding is a methodology that provides some protection against such errors.

Often the provision of error detection and correction is the responsibility of the satellite user; the operator merely supplies the link to meet some specification, but does not provide the means of ensuring accurate transmission. One reason for this division of functions between the operator and the user is that the operator is not concerned with the type of information being transmitted, and therefore cannot predict when designing the system the degree of correction the user will require.

Normally, speech channels are not error corrected; there is sufficient redundancy in speech to ensure that it is made intelligible in the presence of such errors (Section 1.13). In transmitting data, however, error detection is essential. There may be no redundancy in the information stream and consequently any error will result in wrong information being received and there will be no mechanism for making a correction.

Error detection is achieved by adding bits to the information packet in such a way that the receiver can, by using binary algebraic manipulation, determine whether errors are present (Section 5.6). The number of errors which can be detected in this way is related directly to the length of the information field, and the number of additional code bits provided by the encoder. The combination of information field and coding digits is called the code word. There is a trade-off, as mentioned in Chapter 5, between error detection power and transmitted capacity. Adding code bits effectively reduces the rate at which real information is transmitted, and therefore reduces the capacity of the link; the code bits represent redundancy.

Error correction is a different matter from error detection. It is possible to provide codes which will allow the receiver to detect and then correct a number of errors in a bit stream, but correction is more difficult to achieve than detection and therefore more additional bits are required. In practice this means that codes can correct a much smaller number of errors than they can detect. However, this type of correction, called forward error correction (FEC) has a significant role in satellite links. The very long transmission time between earth station and satellite discourages the use of retransmission for error correction, and therefore forward error correction in which the error is detected and corrected by the receiver is highly desirable for this application.

A simple method for correcting errors on short terrestrial links is to use a stop and wait protocol with automatic request for transmission (ARQ). The transmitter sends one packet at a time; when it is received correctly an acknowledgement signal (Ack) is returned. If the receiver detects an error no Ack is returned and the packet is retransmitted after a suitable time-out. The time-out is slightly longer than the expected time for the return of Ack, Fig. 14.10. Clearly the main drawback with this protocol is that the time between transmission of the packet and reception of the acknowledgement is wasted. For satellite applications, where that delay is about a quarter of a second, such wasted time cannot be tolerated. Instead a duplex link is used and packets are transmitted continuously. When an error occurs a retransmission mechanism causes those packets, and probably a few following, to be retransmitted.[5]

Clearly a number of mechanisms exist for handling the problem of error correction and detection. One is to provide such a strong link that the excess power over the minimum required for normal circumstances is sufficient to cope with nearly all the occurrences of poor transmission conditions. This is a very expensive and crude approach, and does not attack the problem of error correction directly. The second method is to arrange for all the incorrect packets to be retransmitted automatically (using ARQ). The long round-trip delay discourages complete reliance on this method so a third scheme is used in which a measure of forward error correction is provided at the receiver and if the number of errors exceeds the capacity of the FEC algorithm then ARQ

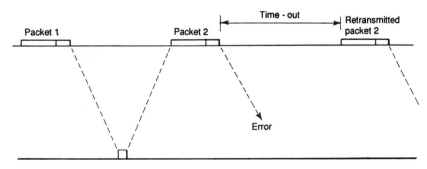

Fig. 14.10 Sample stop and wait protocol.

is used. A fourth approach would be to provide no error detection or correction facility; a satisfactory philosophy on channels of high reliability carrying information which has built-in redundancies such as voice and television.

Coding is designed to cater for channels with a low bit-error rate, consequently the improvement is best displayed in circumstances where errors are very occasional, and the error correction facility is rarely used. If the channel is very noisy the FEC system will break down, or the delay in providing the necessary retransmission will be intolerable. Indeed the receiver may not be able to decide whether or not it has received a packet. In such circumstances effective transmission breaks down. Practical systems are designed to ensure that breakdown occurs only within some specified limit, measured in terms of minutes per month.

14.6.1 Types of error

The incidence of errors is classified as random or bursty. Random errors occur as isolated erroneous bits in a stream of good bits; the probability of an error is the same for all bits, regardless of whether the previous bits were correct or not. Bursty errors are somewhat different, and in a sense more typical of some types of interference. As the name suggests, bursty errors turn up in blocks; a sequence of correct bits is followed by a sequence in which the incidence of errors is very high, followed by a stream of good bits. During an error burst the probability of bit error is high; during a sequence of good bits the probability of error is low.

Codes can also be classified into several types, but the principal division is between block codes and convolutional codes.

In a block coder the information is presented in a block of k digits, and this block is coded into a longer word of n digits; known as an (n, k) block code. The additional $k-n$ digits provide the redundancy required to enable error detection and correction to take place. Successive coded blocks are independent of each other.

Convolutional codes consist, similarly, of (n, k) blocks, but in addition there is a dependency between the codes used for successive incoming frames. The memory order of codes is denoted by m, the number of previous frames having an influence on the coder used for the present frame.

A feature of the decision making process in the detector is to determine if an error has occurred, and that could be based on a threshold-related decision (hard decision) or on a probability-related decision (soft decision). In the hard decision system there is no allowance for any doubt; either a bit has or has not been received. However, in soft decision decoders the doubt can be included as a probability and taken into account in the algorithm for decoding the received code word, although this approach can only be used in convolutional decoding.

14.7 INTERFERENCE AND NOISE

Background noise on the downlink has a significant effect on receiver performance since it reduces the C/N factor. This particular form of noise is represented as a noise temperature; and for an antenna that temperature is variable and depends strongly on the elevation of the boresight and the frequency of the channel. The quiescent noise temperature will be less than 70 K.

However, the direction of the antenna boresight is determined by the relative position of the earth station and the satellite. Consequently the antenna may pick up atmospheric noise from time to time well in excess of this value. Most noticeably, excess temperature occurs when the sun is behind the satellite, in which case the noise temperature will rise to over 10^4 K at frequencies up to about 10 GHz.

Interference increases the effective noise of the satellite channel as spurious signals from nearby links, or from terrestrial sources, are detected. However, there is also a reciprocal effect in which our channel interferes with other communication systems, both satellite and terrestrial.

14.7.1 Atmospheric attenuation

The link path between satellite and earth is not homogeneous to radio wave propagation, nor is it immune from interference from nearby radio systems. There is always some attenuation due to the excitation of electrons in the path of the radio wave, but at satellite frequencies it is a minor factor, accounting for about 0.5 dB loss. Much more important are those effects which are transitory in nature, and consequently unpredictable. On a 'clear day' there are none of these loss mechanisms, but at other times some or all of them may occur. We can consider such mechanisms in several categories: here we choose three; attenuation, depolarization and interference. The first two are properties of the radio link itself. The third results from other sources of radiation in the proximity of the link.

Attenuation

At satellite frequencies the most important variable attenuation mechanisms are those assciated with water droplets, hydrometeorites, in the form of rain, ice, snow, sleet and combinatioris of these. Rain is the most important type of precipitation from this point of view. Snow produces relatively little attenuation, but wet ice can impair the signal significantly.

There are several parameters associated with rainfall, but in this context the most important is rainfall rate; measured in millimetres per hour, it has a direct effect on attenuation.

The attenuation A can be related to the rainfall rate R and the effective path length, L by

$$A = kR^aL \quad \text{dB} \tag{14.20}$$

This simple relationship assumes that the rainfall rate R is constant over the whole of the path length L. Several very difficult problems arise in utilizing this relationship. The constants k and a are highly dependent on frequency and on rain drop size, as well as depending to a small degree on temperature. The relationships between k, a, frequency and drop size that are used in practice are empirical and have been tabulated by a number of organizations such as CCIR. Resort must therefore be made to such tables[6] to find values of A. However, knowing k and a is only part of the solution. R is impossible to determine precisely. The rainfall rate will vary along the length of the path in a statistical way, and even estimating the value of R at a particular point by using a ground-based measure of quantity of water versus time may be a poor approximation to the rate at which, over the same spot, raindrops are falling through the link path. The turbulence in the atmosphere precludes a straight fall of rain and hence complicates the assessment of fall rate.

To provide designers with a procedure to calculate A, CCIR have produced an algorithm based on tabulated values for k and a, and cumulative distributions for R. This gives, for example, a figure $A_{0.01}$ based on $R_{0.01}$, where the suffix indicates the percentage of the average year for which the attenuation will exceed the value $A_{0.01}$ given that the mean rainfall rate exceeded for 0.01% of the average year is $R_{0.01}$. In this way the designer can dimension the system to meet satisfactory performance specification for all but 0.01% of the year.

The ionosphere can produce absorption and other effects. At satellite frequencies the absorption is negligible unless there is electron activity in the ionosphere in the form of clouds of free electrons which interact with the earth's magnetic field, producing not only absorption but also depolarization due to the rotational effect on the wave, and a slight misalignment caused by variations in refractive index. If the variations are rapid they are called scintillations and these can produce variations of up to 10 dB in the received signal amplitude.

Depolarization

Previously, we noted that better use of the links can be achieved by sending out two orthogonally polarized signals. The success of this approach depends on the receiver being aligned to the correct polarization. Unfortunately there are several mechanisms which make this difficult to achieve consistently. The effect of changing the direction of polarization from that expected is called depolarization and that has two results. First, the intensity of the intended signal is reduced because the polarization of the incoming signal and that of the receiver are not matched. Second, the receiver can pick up some of the signal from the other polarization, hence effectively increasing the channel noise.

The main cause of depolarization is Faraday rotation, caused by action of the magnetic field and the ionospheric free electrons on the signal. The degree of rotation depends on both the electron density, and the length of path over which the electron cloud or clouds exist. The ionosphere does not exhibit the same electrical properties in all directions; it is said to be anisotropic. This lack of homogeneity itself results in depolarization.

As estimate of the rotation, φ, produced by the Faraday effect can be obtained from the formula

$$\varphi = \int k_1 \frac{NL}{f^2} B \cos \theta \, dl \tag{14.21}$$

where k_1 is a constant

   N is the ionospheric plasma electron density

   B is the magnetic flux density

   L is the path length

   f is the channel frequency

and θ is the angle between the magnetic field vector and the link

The integration is over the length of the path in the ionosphere. In practice N will not be constant and some suitable model will have to be chosen to determine its value as a function of L.

14.8 MULTI-ACCESS

As the number of earth stations increases, and satellite communications are more widely used, the multi-access techniques used become more significant. In principle the function of multi-access is to provide the many earth stations wishing to use a particular transponder with channel access, at the required bandwidth.

   In many respects the problems of multi-access in satellite systems are similar to those in local area networks, and the same mechanisms can be used in each. In particular, the traffic to be transmitted is a combination of voice and data and the same problems, and solutions, as discussed in Chapter 13 occur. There are, however, some additional issues in satellite systems, and they will be mentioned here.

14.8.1 FDMA

In FDMA the bandwidth can be divided on a multi-carrier or single carrier basis. In multi-carrier the transponder bandwidth is divided into sub-bands, each sub-band being assigned to an earth station. The sub-bands need not be all of the same width; the number of channels in each will depend on the earth station's requirements and the allocation they have been given. This arrangement gives the earth station some flexibility in the allocation of bandwidth within its own sub-band. The main disadvantage of this type of access is that the power output from the transponder must be reduced compared with the output that could be achieved with a single carrier arrangement. Except for a few satellites using solid-state amplifiers, the output device at the transponder is normally a travelling wave tube. It has a non-linear input–output characteristic, and this leads to serious inter-modulation products in the output. This effect becomes more serious as the number of carriers increases. To avoid the non-linearities from producing serious degradation on the output signal, it is necessary to impose an input back-off. This back-off has to be large enough to ensure that the output back-off brings the operating point to a position where the inter-modulating products are insignificant. The non-linear characteristic means, of course, that the output back-off will be smaller than the input back-off, Fig. 14.11. The amount of back-off required increases with the number of carriers in the system.

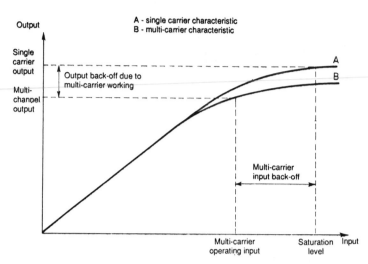

Fig. 14.11 Effect of multi-carrier working.

A single carrier FDMA arrangement, in which the available 36 MHz bandwidth is divided into uniform channels, allows more efficient use of the travelling wave tube output. If QPSK is used, and each channel is spaced by 45 kHz, there is room for 800 channels to avoid cross-channel interference, and some back-off is necessary because of the inter-modulation products between channels. Most satellites currently use FDMA access methods of one sort or another.

14.8.2 TDMA

Throughout communications systems design, the trend for many years has been to use digital representations of signals, and digital transmission. Its ease of transmission, and robustness in noise, makes it very appealing, and the same arguments in its favour in terrestrial systems also apply to satellite networks. As far as the TWT operating conditions are concerned, in digital transmission there is no reason why the full output should not be used: there is only one carrier, and since the transmission is digital, inter-modulation products are not a problem.

The discussion on TDMA in local area networks, in Chapter 13, is highly relevant here. The different characteristics of voice and data traffic, their different delay and error-correction requirements and the methods of accommodating them are all explained there.

Two issues are, however, of especial interest in satellite operations. The frame size, which has to be greater than 125 μs if we assume that normal PCM is used for speech, can have any value up to about 25 ms; the limit imposed by the maximum tolerable delay to speech traffic. In the *INTELSAT* system, for example, a frame size of 2 ms is used, with the frame structure shown in Fig. 14.12. The reference bursts are for synchronization and each traffic burst may be from a different earth station. The traffic bursts contain time multiplexed speech channels.

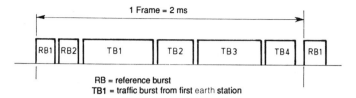

Fig. 14.12 Typical Intelsat frame.

The other issue is synchronization. The uplink packets arriving at the transponder from different earth stations must not overlap in time. Consequently the whole network related to one satellite has to be in synchronism. This can be achieved in different ways in detail, but it depends on the transponder sending out synchronizing bits to indicate the start of each frame, and to provide reference times to which earth stations relate their own transmission. The variable nature of the satellite position will produce some changes in this synchronization timing, and guard intervals are used to avoid overlap due to this cause. The guard interval adds to the transmission overhead, leaving less time available for revenue earning traffic. Other overheads which are included concern addresses, acknowledgement for data signals and network management.

14.8.3 Random access

The bandwidth allocated to each channel depends on the method of access employed. Fixed assignment means that at the outset the designer determines the bandwidth allocation, and there is no flexibility for the operator to adjust the system in use. Demand assignment does give the operator flexibility. Channel bandwidths can be adjusted to cater for the traffic mix, which may vary from time to time. For such demand assignment to be feasible, an overhead in terms of network management information is imposed.

An alternative approach is to use a random access mechanism. Instead of having a network management protocol within the system to ensure that both the channel allocation is satisfactory and there is no possibility of collision, random access is a free-for-all. Each station transmits at will, and runs the risk of collision. If collision occurs, the transmitting stations will back off, wait for some random time and then try again. The simplest random access system is the ALOHA, in which the utilization is low (about 18%); some considerable improvement is achieved with the slotted ALOHA (Section 13.5). It is worth remembering that the more sophisticated the access mechanism, the higher is the overhead and the more complex is the management protocol, so that there is a trade-off to be made.

14.8.4 Speech interpolation

One of the early developments in transatlantic submarine communications, when the demand immediately saturated the circuits provided, was to utilize the silent intervals in normal speech to allow channels to be shared. In normal speech about 45% of the time is not used, in the sense that it is occupied by silence intervals. Speech interpolation techniques allow these intervals to be utilized, and this same method can be applied to satellite links. In packet-based transmission systems, since there is no packet to be transmitted during silence intervals, speech interpolation is automatic.

REFERENCES

1. Clarke, A. C., 'Extra-terrestrial Relay', *Wireless World*, Oct. 1945.
2. Pratt, T. and Bostian, C. W., *Satellite Communications*, John Wiley, 1986.
3. Spilker, J. J., *Digital Communication by Satellite*, Prentice–Hall, 1977.
4. Evans, B. G. (ed.), *Satellite Communication Systems*, IEE London, 1987, Chapter 12.
5. Tanenbaum, A. S., *Computer Networks*, Prentice–Hall, 1981, Section 4.2.2.
6. Propagation in non-ionised Media. *CCIR, Recommendations and Reports*, Vol. V, 1982, ITU, Geneva.

Mobile communication systems 15

15.1 INTRODUCTION

In the fixed telephone network the final link with individual subscribers has traditionally been by means of a twisted pair cable, in other words the telephone handset is generally regarded as being immobile. Some minor changes to this situation became possible with the introduction of the cordless handset which allowed the subscriber limited mobility over a distance of a few metres. The concept of mobility has been extended from this limited movement to pan national and pan continental transit. Future forecasts suggest that the fixed telephone handset will eventually disappear completely and be replaced by mobile units allowing individual subscribers the facility of global travel with continuous personal communications.

This chapter considers the fundamentals of these new public mobile systems (there are numerous private mobile systems also) which are broadly divided into cellular mobile systems and cordless mobile systems. Although different in concept both cellular and cordless systems rely on radio transmission for the final link with the subscriber. The management of this radio resource is a fundamental requirement of mobile systems which therefore makes mobile systems substantially different from the traditional public switched telephone network. This chapter will consider elements of both cellular and cordless systems and will focus on their similarities and differences.

15.2 THE CELLULAR CONCEPT

Cellular systems are now well developed in many countries throughout the world. The systems in use in individual countries all have many similarities and also some significant differences. It is not possible to cover all the systems which are in use or are currently under development. To avoid this problem the coverage in this chapter will concentrate on European systems and will begin by considering what are known as 'first generation' analogue systems, and will then move on to 'second generation' digital systems.

The essential feature of all cellular networks is that the final link between the subscriber and fixed network is by radio. This has a number of consequences:

(i) radio spectrum is a finite resource and the amount of spectrum available for mobile communications is strictly limited;

(ii) the radio environment is subject to multipath propagation, fading and interference and is not therefore an ideal transmission medium;

(iii) the subscriber is able to move and this movement must be accommodated by the communications system.

The basic elements of a cellular system are shown in Fig. 15.1.

The mobile units may be in a vehicle or carried as a portable and are assigned a duplex channel and communicate with an assigned base station. The base stations communicate simultaneously with all mobiles within their area of coverage (or cell) and are connected to mobile switching centres (MSCs). A mobile switching centre controls a number of cells, arranges base stations and channels for the mobiles and handles connections with the fixed public switched telephone network (PSTN).

Figure 15.1 indicates that each base station (in a cluster) is allocated a different carrier frequency and each cell has a usable bandwidth associated with this carrier. Because only a finite part of the radio spectrum is allocated to cellular radio the number of carrier frequencies available is limited. This means that it is necessary to re-use the available frequencies many times in order to provide sufficient channels for the required demand. This introduces the concept of frequency re-use and with it the possibility of interference between cells using the same carrier frequencies.

Clearly with a fixed number of carrier frequencies available the capacity of the system can be increased only by re-using the carrier frequencies more often. This means making the cell sizes smaller. This has two basic consequences:

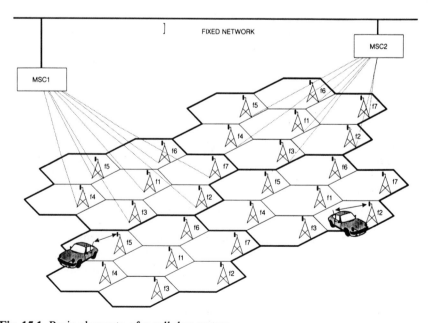

Fig. 15.1 Basic elements of a cellular system.

1. It increases the likelihood of interference (known as co-channel interference) between cells using the same frequency.
2. If a mobile is moving it will cross cell boundaries more frequently when the cells are small. Whenever a mobile crosses a cell boundary it must change from the carrier of the cell which it is leaving to the carrier of the cell which it is entering. This process is known as handover. It cannot be performed instantaneously and hence there will be a loss of communication while the handover is being processed. If the cell sizes become very small (microcells) handovers may occur at a very rapid rate.

It becomes clear therefore that frequency planning is a major issue in the design of a cellular system which must achieve an acceptable compromise between the efficient utilization of the available radio spectrum and the problems associated with frequency re-use.

15.3   TYPICAL CELL OPERATION

Each cell has allocated to it a number of channels which can be used for voice or signalling traffic. When a mobile is active it 'registers' with an appropriate base station. The information regarding the validity of the mobile (for charging etc.) and its cellular location is stored in the responsible MSC. When a call is set up either from or to the mobile the control and signalling system assigns a channel (from those available to the base station with which the mobile is registered) and instructs the mobile to use the corresponding channel. This channel may be provided on a frequency division basis (typical of analogue systems), on a time division basis (typical of digital systems) or on a code division basis (also typical of digital systems).

A connection is thus established via the base station to the fixed network. The quality of the channel (i.e. radio link) will be monitored by the base station for the duration of the call and reported to the MSC. The MSC will make decisions concerning the quality and will instruct the mobile and base station accordingly. As the mobile moves around the carrier-to-interference ratio on its allocated channel will vary. This is monitored and if it falls below some threshold (either as the mobile is about to leave the cell or as a result of co-channel interference), the mobile can be instructed to handover to the strongest base station. The handover algorithm is actually significantly more complicated than this simple treatment suggests. For example it must be able to cope with:

(i)   Whether the current loss in channel quality is due to short-term fading;
(ii) Whether a simple increase in power would be sufficient to restore the channel quality (this could however produce an unacceptable co-channel interference in other cells using the same frequency);
(iii) Whether the measurements from adjacent cells are valid (averaging is necessary to remove spurious fluctuations);
(iv) Whether the cell chosen for handover has spare channels available.

In analogue systems the fixed network (i.e. the base stations) performs all radio channel measurements and the mobile terminal is completely passive (this is known as network-controlled handover). In digital systems both mobiles and base stations make measurements and report these to the fixed network for handover decisions (this is known as mobile-assisted handover). When handover does occur communication is interrupted and the voice channels are muted. This interruption is usually short and largely unnoticed during voice communications. However, it does present serious problems if the mobile is transmitting or receiving data, especially when handover is frequent. When the call is completed the mobile releases the voice channel which can then be re-allocated to other users.

15.4 SYSTEM CAPACITY

The capacity of a system may be described in terms of the number of available channels, or alternatively in terms of the number of subscribers that the system will support. The latter measure takes account of the fact that each call has a mean duration and that not all of the subscribers will be trying to make a call at the same time.

The system capacity depends on:

 (i) the total number of radio channels;
 (ii) the size of each cell;
(iii) the frequency re-use factor (or frequency re-use distance).

The total number of voice channels that can be made available to any system depends on the radio spectrum allocated and the bandwidth of each channel. Once this number is defined a frequency re-use pattern must be developed which will allow optimum use of the channels. This, in turn, is closely linked with cell size.

The minimum distance which allows the same frequencies to be re-used will depend on many factors, for example:

 (i) the number of co-channel cells in the vicinity of the centre cell;
 (ii) the geography of the terrain;
(iii) the antenna height;
(iv) the transmitted power within each cell.

Assuming an omni-directional base station antenna it is appropriate to

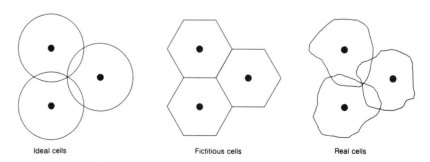

Ideal cells   Fictitious cells   Real cells

Fig. 15.2 Schematic representation of cells.

consider the cells as circles with the base station at the centre. (Such a model is actually only appropriate for a flat terrain with no obstacles). This means that the circular cells would overlap and this would make diagrams somewhat confusing. It is therefore common practice to represent the cells as non-overlapping hexagons which would fit into the corresponding circles. This fictitious model with the ideal and real models are shown in Fig. 15.2.

15.5 FREQUENCY RE-USE DISTANCE

When calculating the frequency re-use distance this is based on the cluster size K. The cluster size is specified in terms of the offset of the centre of a cluster from the centre of the adjacent cluster. This is made clearer by reference to Fig. 15.3. In this figure the cell cluster size is 7 and the centre cell is the cell marked 1. The next cell 1 is offset by $i = 2$ cell diameters to an intermediate cell and a further $j = 1$ cell diameter from that intermediate cell.

The cluster size is calculated from

$$K = i^2 + ij + j^2 \tag{15.1}$$

Common cluster sizes are 4 ($i = 2$, $j = 0$) and $7(i = 2$, $j = 1)$ for city centres, and $12(i = 2$, $j = 2)$ for rural areas.

The actual frequency re-use distance is D as shown in Fig. 15.3. Assuming circular cells of radius R (based on the hexagon shape), the frequency re-use distance may be determined from

$$D = R\sqrt{3K} \tag{15.2}$$

The corresponding re-use distances are given in Table 15.1.

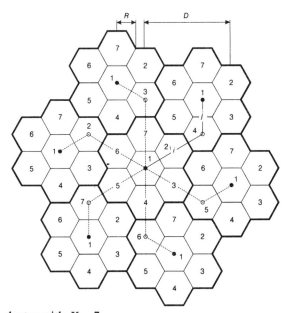

Fig. 15.3 Cell cluster with $K = 7$.

Table 15.1 Re-use distances for cellular systems

K	D
4	3.46 R
7	4.58 R
12	6.00 R

If all cells transmit the same power then as K increases the frequency re-use distance increases, thus increasing K reduces the probability of co-channel interference. However, in order to maximize frequency re-use it is necessary to minimize the frequency re-use distance. Hence the design goal is to choose the smallest value of K which will meet the performance requirements in terms of capacity and interference.

15.6 DETERMINATION OF CELL RADIUS

Figure 15.4 shows two cells using the same frequencies at a re-use distance D. Assuming that the power by each base station is fixed the received power at a distance r from the base station is proportional to $r^{-\gamma}$. For free space $\gamma = 2$, however, it is found that in the cellular environment a more appopriate value is $\gamma = 4$.

A mobile in one of the cells will receive a carrier-to-interference ratio (CIR) which, on average, is a function of $q = D/R$. (On average the CIR at a mobile receiver will be the same as at the base station receiver.) Note that the co-channel interference reduction factor (q) is independent of the actual power level P_0 which is assumed the same for all cells. The carrier to interference ratio within a cell depicted by Fig. 15.4 is thus

$$\frac{C}{I} = \left(\frac{R}{D}\right)^{-\gamma} = \left(\frac{R}{D}\right)^{-4} \tag{15.3}$$

For a fully developed cellular system based on the hexagonal model there will be six interfering cells in the first tier of surrounding clusters. (If $\gamma = 4$ it can be assumed that the interference due to cells in the second tier can be ignored.) The carrier-to-interference ratio in one of the cells will thus be

$$\frac{C}{I} = \frac{C}{\sum_{k=1}^{6} I_k} \tag{15.4}$$

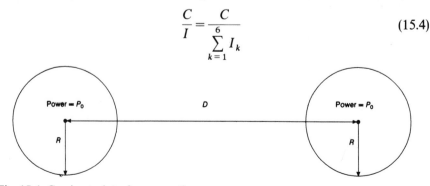

Fig. 15.4 Carrier-to-interference ratio.

Assuming local noise is much less than the interference level, this can be written

$$\frac{C}{I} = \frac{R^{-\gamma}}{\displaystyle\sum_{k=1}^{6} D_k^{-\gamma}} \qquad (15.5)$$

Assuming that all values of D_k are equal ($=D$), this becomes

$$\frac{C}{I} = \frac{R^{-4}}{6D^{-4}} \qquad (15.6)$$

In the cellular environment it is normal practice to specify that the CIR should be greater than 18 dB for acceptable performance ($= 63.1$); i.e.

$$q^4 = 6 \times 63.1 \quad \text{hence} \quad q = 4.41$$

For the hexagonal structure $q = \sqrt{3K}$, i.e.

$$K = 6.48$$

hence the cluster size for this CIR is 7. Note that this approximate analysis closely reflects the practical case based on statistical measurements.

Having established the cluster size it is then necessary to determine the cell radius. This is based on the number of available channels and the expected density of mobile subscribers (i.e. the average number of mobile subscribers/m^2).

To illustrate this point assume that the total number of channels available is 210. This means that the number of channels per cell is $210/7 = 30$. It is necessary to find the total offered traffic in the 'busy hour'. This is related to the average number of calls/hour and the mean duration of each call. Assume that there are W subscribers per cell and that during the busy hour a fraction η_c of these subscribers make or receive a call of duration T minutes; i.e. the total number of calls in the busy hour is $Q = \eta_c W$. The offered load is then

$$A = QT/60 \quad \text{erlangs}$$

To obtain the number of channels for this traffic it is necessary to attach a 'blocking probability' for each call. A typical value for this is 2%. The relationship between offered traffic, blocking probability and number of channels is given by the Erlang B formula which is usually represented in tabular form and is given in Appendix D. For example, 30 channels can support an offered traffic of 21.9 erlangs with a blocking probability of 2% (from the table). It is possible to relate this figure to the number of subscribers which the cell can support; i.e.

$$21.9 = Q\,T/60 = (\eta_c W T)/60$$

from which

$$W = 60 \times 21.9/(\eta_c T)$$

In this expression T is the mean call duration in minutes. Extensive measurements have indicated that for cellular systems $T = 1.76$ minutes.

If it assumed that 60% of the total subscribers make a call during the busy hour then

$$W = 60 \times 21.9/(0.6 \times 1.76) = 1244.3$$

Hence with 30 channels available the cell can support 1244 subscribers. If the cell radius is R metres then user density $= 1244/\pi R^2$.

It is now possible to calculate the cell radius required for an average user density. For example assume that the number of users/km^2 is 1600. This represents a user density of 1.6×10^{-3}/m^2; i.e.

$$1244.3/\pi R^2 = 1.6 \times 10^{-3}$$

which gives $R = 497.5$ m, i.e. the approximate cell diameter is 1 km.

15.7 SECTORING

In the previous section an approximate analysis indicated that in order to achieve a CIR of 18 dB it is necessary to plan frequency re-use on a cluster size of 7. This gives a value of $q = \sqrt{21} = 4.58$. However, it is found that in areas of high traffic density this value can be inadequate. The worst-case situation is illustrated in Fig. 15.5 (a) in which a mobile is on the boundary of its serving cell and interference is produced by all six interfering cells. In this case the distances from the mobile to interfering cells varies from $D - R$ to $D + R$. The carrier-to-interference ratio in this case is given by

$$\frac{C}{I} = \frac{R^{-4}}{2(D - R)^{-4} + 2D^{-4} + 2(D - R)^{-4}} = \frac{1}{2\{(q - 1)^{-4} + q^{-4} + (q + 1)^{-4}\}}$$

Substituting for q in this expression gives a CIR $= 17$ dB, which is less than the desired value. The situation can actually be worse than this and a more conservative CIR estimate would be about 14 dB which is 4 dB less than the specified value.

Clearly, increasing the value of K to improve the CIR would reduce the efficiency of frequency re-use and this is not an attractive option. An alternative is to reduce the co-channel interference in a cell by using directional antennas at the base station and dividing the cell into a number of sectors. A three-sector arrangement is shown in Fig. 15.5(b). The original frequencies allocated to the cell are then divided between the sectors. Reduction in interference is achieved by choosing a frequency re-use pattern such that the front lobe of any base station transmitter illuminates only the back lobe of its co-channel counterpart. What this means, in effect, is that the number of interfering base stations is reduced from six to two with a corresponding increase in CIR.

This is made clearer from Fig. 15.5(b) in which it may be seen that only base stations **A** and **B** actually cause interference to mobile **M**. This produces a reduction in interference of approximately 5 dB over the omni-directional case. Hence by employing sectored antennas in areas of high traffic density it is possible to achieve the required CIR values. It should be noted that the sectored approach effectively increases the cluster size to 21 without any increase in the frequency re-use distance, however, handovers may be required between sectors of the same cell.

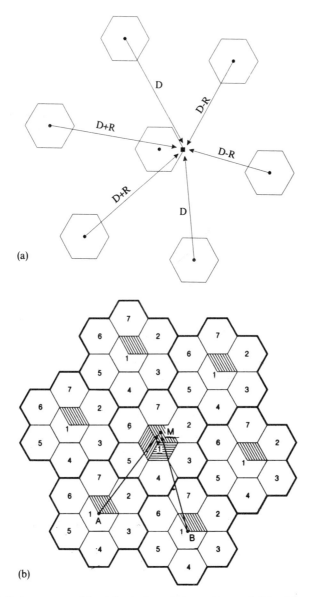

Fig. 15.5 Cellular system (a) with six interfering cells, and (b) with sectored base station antennas.

15.8 PROPERTIES OF THE RADIO CHANNEL

The radio channel in a cellular system has a major influence on the overall system design. This has already been evident in the way in which frequency re-use is implemented based on a radio attenuation proportional to D^4. Cellular radio systems are categorized by the fact that the height of antennas at both base station and mobile are usually low compared to the distance of separation. The model is shown in Fig. 15.6.

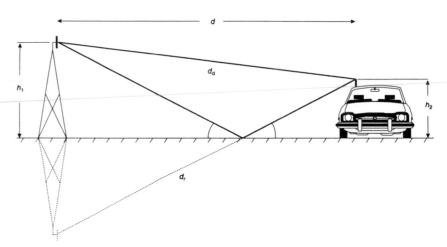

Fig. 15.6 Plane earth propagation model.

The analysis of Section 14.4 is the basis of the model in Fig. 15.6 and is repeated here for completeness. If it is assumed that an antenna radiates energy equally in all directions (isotropic antenna) it is possible to calculate the power density at a distance r from the antenna. If the antenna radiates a total power P_t the power at any distance r from the antenna is the power passing through the surface of a sphere of radius r. The surface area of the sphere is $4\pi r^2$ and the power received per unit area is thus

$$P_a = \frac{P_t}{4\pi r^2} \quad \text{watts/m}^2 \tag{15.7}$$

At sufficiently large value of r the wave becomes a plane wave. The power received by an antenna placed in this field is

$$P_r = P_a A_e \tag{15.8}$$

A_e is known as the 'effective aperture' of the antenna and is the equivalent power absorbing area of the antenna. The effective aperture of an isotropic antenna when used as a receiver can be shown to be $A_e = \lambda^2/4\pi$. Hence the power received by such an antenna is

$$P_r = P_a \times \frac{\lambda^2}{4\pi}$$

But $P_a = P_t/4\pi r^2$, i.e.

$$P_r = \frac{P_t \lambda^2}{(4\pi r)^2} \tag{15.9}$$

The isotropic antenna has unity gain in both the transmit and receive modes. A non-isotropic transmit antenna will have a gain of G_t and the product $P_t G_t$ is known as the effective radiated power (ERP). In mobile radio ERP is used as the standard method of quoting transmitted power. In effect, if the ERP is quoted as 100 W (50 dBm) and the antenna gain is 10 dB, the actual transmitted power would be 10 W (40 dBm). A non-isotropic receive antenna will have a gain of G_r and, in such cases, the received power would be given by

$$P_r = \frac{G_t G_r P_t \lambda^2}{(4\pi r)^2} \qquad (15.10)$$

This expression indicates that the attenuation is proportional to $(distance)^2$. In the case of mobile radio it is necessary to consider the height of both transmit and receive antennas above the earth's surface.

15.9 SPACE WAVE PROPAGATION

If the height of the base station antenna is h_1 and the height of the mobile antenna is h_2 the system may be represented as shown in Fig. 15.6, where the separation between transmitter and receiver is d. It is assumed that d is small enough to neglect the earth's curvature. Figure 15.6 shows that there will be both a direct and ground reflected wave. The direct path length is d_d and the reflected path length is d_r. It can be seen from the geometry of the system that

$$d_d = \sqrt{d^2 + (h_1 - h_2)^2}$$

Using the binomial expansion and noting that $d \gg h_1$ or h_2, the length of the direct path approximates to

$$d_d \cong d \left\{ 1 + 0.5 \left(\frac{h_1 - h_2}{d} \right)^2 \right\}$$

similarly

$$d_r \cong d \left\{ 1 + 0.5 \left(\frac{h_1 + h_2}{d} \right)^2 \right\}$$

The path difference is thus $\Delta d = d_r - d_d$, i.e.

$$\Delta d = \frac{2h_1 h_2}{d} \qquad (15.11)$$

The corresponding phase difference between direct and reflected path is

$$\Delta \phi = \frac{2\pi}{\lambda} \times \frac{2h_1 h_2}{d} = \frac{4\pi h_1 h_2}{\lambda d} \qquad (15.12)$$

The total received power is thus

$$P_r = P_t \left(\frac{\lambda}{4\pi d} \right)^2 \times |1 + \rho e^{j\Delta\phi}|^2 \qquad (15.13)$$

ρ is the reflection coefficient and for low angles of incidence the earth approximates to an ideal reflector with $\rho = -1$; i.e.

$$P_r = P_t \left(\frac{\lambda}{4\pi d} \right)^2 \times |1 - e^{j\Delta\phi}|^2 \qquad (15.14)$$

but $1 - e^{j\Delta\phi} = 1 - \cos\Delta\phi - j\sin\Delta\phi$, hence $|1 - e^{j\Delta\phi}|^2 = (1 - \cos\Delta\phi)^2 + \sin^2\Delta\phi$.

If $\Delta\phi \ll 1$ then $\cos\Delta\phi = 1$ and $\sin\Delta\phi = \Delta\phi$, i.e.

$$P_r = P_t \left(\frac{\lambda}{4\pi d}\right)^2 \left(\frac{4\pi h_1 h_2}{\lambda d}\right)^2$$

hence

$$P_r = P_t \left(\frac{h_1 h_2}{d^2}\right)^2 \qquad (15.15)$$

This is the 4th power law used in the frequency re-use calculation, and is known as the plane earth propagation equation. The loss is given by

$$\text{loss (dB)} = 40 \log_{10} d - 20 \log_{10} h_1 - 20 \log_{10} h_2$$

This means that the loss increases by 12 dB each time the distance is doubled. It should be noted that this equation is not dependent on frequency, which is a surprising result. This is a consequence of assuming that h_1 and h_2 are much smaller than d and that the earth is flat and perfectly reflecting. If the surface is undulating a correction factor, which is frequency dependent, must be included.

In the cellular environment it is quite likely that no direct path will exist between base station and mobile. The communication then depends on single or multiple reflections from buildings and surrounding objects. Under these circumstances field strength variations may only be derived from measurement and approximate computer modelling. In practice it is found that in the majority of cases the loss is close to that given by the plane earth propagation equation. However, it is important to realize that under these conditions the radio channel is subject to **fading**. The principle of fading can be demonstrated by reference to Fig. 15.7 in which it is assumed that there is no direct path between base station and mobile antennas.

If the phase difference between diffracted and reflected waves is a whole number of wavelengths the two waveforms will reinforce and the amplitude at the receiving antenna will (approximately) double. As the mobile moves the phase difference between the two paths changes. If the phase difference becomes a whole number of half wavelengths the two waves cancel producing a null. Thus as the mobile moves there are substantial amplitude fluctuations in the received signal known as **fast fading**. (There will also be a Doppler shift associated with the movement.) A typical variation of signal strength with distance is shown in Fig. 15.8.

Fast fading (due to local multipath) is also accompanied by a slower

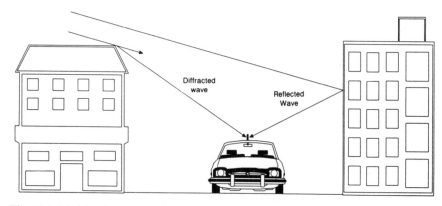

Fig. 15.7 Multipath propagation.

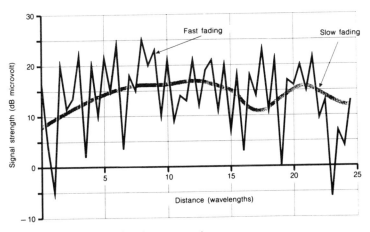

Fig. 15.8 Fading due to multipath propagation.

variation in mean signal strength known, as **slow fading**. Fast fading is observed over distances of about half a wavelength and can produce signal strength variations in excess of 30 dB. Slow fading is produced by movement over much longer distances, sufficient to produce gross variations in the overall path between base station and mobile. It should be noted that at the frequencies used in cellular radio a mobile moving at 50 km/h will experience several fast fades/second which will clearly effect the system performance. It should also be noted that fading is a spatially varying phenomenon which becomes a time-varying phenomenon only when the mobile moves.

It is clear that an exact representation of fading characteristics is not possible because of the effectively infinite number of situations which would have to be considered. Reliance therefore has to be placed on statistical methods which produce general guidelines for system design.

15.10 SHORT-TERM FADING (FAST FADING)

When a mobile unit is stationary the received signal strength will be formed by the vector sum of the various signals reaching the antenna and will have constant amplitude. When the mobile is moving it is assumed that the signal received will be the vector sum of N reflected signals of equal amplitude which arrive at the receiving antenna at a random phase angle ϕ_N. This is accepted as a reasonable model for the cellular environment where there is not usually a direct line of sight path between transmitter and receiver. (If there is a direct line of sight component this will alter the nature of the fading envelope and its statistics.) The addition of these components gives rise to a resultant with an amplitude (i.e. envelope) which varies in a random manner.

Applying the central limit theorem it can be shown that the received electric and magnetic field components have independent Gaussian distributions. This in turn leads to the conclusion that the envelope of the resultant received carrier has an amplitude which has a Rayleigh distribution given by

$$p_a(a) = \frac{a}{\sigma^2}\exp\left(\frac{-a^2}{2\sigma^2}\right) \tag{15.16}$$

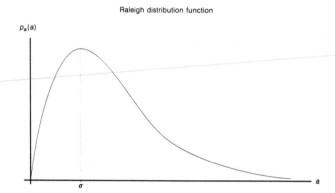

Fig. 15.9 Resultant carrier envelope distribution function.

In this expression σ^2 is the mean square value (i.e. mean power) of the carrier envelope and a is the instantaneous amplitude of the envelope. The distribution function is shown in Fig. 15.9. Note that the probability density function has a peak value of $0.6/\sigma$ at $a = \sigma$, where σ is the rms value of the received signal.

The corresponding cumulative distribution function (CDF) is

$$\text{prob}[a < A] = P_a(A) = 1 - \exp\left(\frac{A}{2\sigma^2}\right) \tag{15.17}$$

When the CDF is known it is possible to determine the average number of times per second that the signal envelope crosses a particular level in the positive direction. This is known as the level crossing rate. The level crossing rate is related to the velocity of the mobile v and the wavelength of the received carrier λ and, for a vertical monopole antenna, can be shown to be:

$$N(A_0) = \sqrt{2\pi} \frac{v}{\lambda} \rho \exp(-\rho^2) \tag{15.18}$$

where $\rho = A_0/\sigma$ and A_0 is the specified level.

The situation is shown in Fig. 15.10. It should be noted that $N(A_0)$ is a

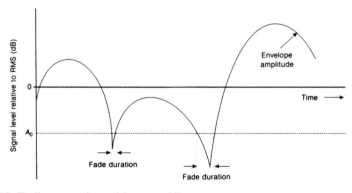

Fig. 15.10 Fading experienced by a mobile.

maximum when A_0 is 3 dB below the rms carrier level. This can be explained by the observation that if A_0 is low, the envelope is above this level for a large proportion of the time and hence the number of crossings per second decreases. A similar situation is observed when A_0 is set at a high level, the envelope being below this level for a large percentage of the time which again reduces the average number of crossings per second. At a carrier frequency of 900 MHz and a mobile speed of 48 km/h the level crossing rate at $A_0 = -3$ dB is $N(A_0) = 39$ per second. (In effect the number of fades per second is 39.) A further parameter of importance is the average fade duration. The duration of a fade is the interval of time that the envelope remains below the level A_0 and this is also shown in Fig. 15.10. The average duration of fades below the level A_0 is

$$\tau(A_0) = \frac{\text{prob}[a < A_0]}{N(A_0)}$$

However, $\text{prob}[a < A_0] = 1 - \exp(-\rho^2)$, hence the average fade duration for a vertical monopole is

$$\tau(A_0) = \frac{\lambda\{\exp(\rho^2) - 1\}}{\rho v \sqrt{2\pi}} \tag{15.19}$$

It is clear that fading is a frequency selective phenomenon. This effect is also apparent in the time domain and is measured in terms of delay spread. It has already been stated that the signal arriving at the antenna of a mobile (or base station) is the sum of a number of waves of different path lengths. This means that the time of arrival of each of the waves is different. If an impulse is transmitted from the base station, by the time it is received at the mobile it will no longer be an impulse but rather a pulse of width given by the **delay spread** Δ. The delay spread is, of course, different for different environments but typical values are given in Table 15.2.

The delay spread is an important parameter for digital systems as it limits the maximum data rate which can be sent. In general the time delay dispersion should be much less than the bit rate in a digital cellular system (without equalization). **Coherence bandwidth** is an additional parameter closely related to delay spread. In a wideband signal two closely spaced frequency components will suffer similar multipath effects. However, as the frequency separation increases the differential phase shifts over the various paths become decorrelated and the spectral components in the received signal will not have the same relative amplitudes and phases as in the transmitted signal. This is essentially frequency selective fading and the bandwidth over

Table 15.2 Delay spreads

Environment	Delay spread $\Delta(\mu s)$
Rural area	< 0.2
Suburban area	0.5
Urban area	3

which the spectral components are affected in a similar way (i.e. are correlated) is known as the coherence bandwidth. It is common practice to define the coherence bandwidth for a correlation coefficient of 0.5, in which case the approximate relationship between coherence bandwidth and delay spread is given by Eqn (15.20)

$$B_c = \frac{1}{2\pi\Delta} \qquad (15.20)$$

The radio channel provides a hostile environment for cellular radio and much of the system design deals with overcoming these difficulties. This will become clear when examples of both analogue and digital systems are considered.

15.11 FIRST GENERATION ANALOGUE SYSTEMS (TACS)

Analogue cellular systems are operated in many countries worldwide. It has already been pointed out that the systems in use all have some basic similarities, but also have some important differences. This section will concentrate on the total access communication system (TACS), employed in the UK and a number of other countries, which is a derivative of the advanced mobile phone system (AMPS) developed by AT&T for the USA.

TACS is an analogue FM system operating in the 900 MHz waveband providing 1000 duplex channels occupying the frequencies 890–915 MHz (25 MHz) and 935–960 MHz (25 MHz). The nominal channel bandwidth is thus 25 kHz. At the present time, in the UK, 600 channels have been divided between two operators (Cellnet and Racal) and the remaining 400 channels have been reserved for the second generation digital system (GSM). The lower frequency band is assigned for transmissions from mobile to base, the upper band is assigned for transmission from base to mobile. The channels are numbered consecutively in ascending order along the frequency spectrum of each band. The duplex spacing is the difference between the transmit channel frequency and receive channel frequency and maintained at 45 MHz in the TACS system. This is illustrated in Fig. 15.11 where it should be noted that mobile transmit channel 1 has a nominal carrier frequency of

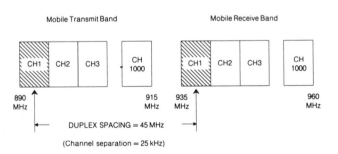

Fig. 15.11 Channel allocation in TACS.

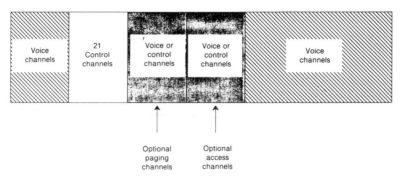

Fig. 15.12 Control and traffic channels.

890.0125 MHz, the corresponding base station transmit channel carrier being 45 MHz above this at 935.0125 MHz.

Each operator in the UK has been allocated 300 channels, 21 of which are a contiguous block of dedicated control channels. The first control channels of each block are channel 23 and 323 respectively. This structure is shown in Fig. 15.12. The channels are categorized as 'traffic channels' which carry voice or data, and 'control channels'.

The control channels carry the signalling information which is essential to the operation of the cellular system. The mobile unit has a single transmitter and a single receiver which means that once a voice channel has been allocated (at call set-up) any subsequent exchange of signalling information between mobile and base station must make use of the voice channel. When a mobile is not actively engaged in a call it is still necessary to transmit information periodically to the fixed network, for updating of the mobile's location, for example. Under these circumstances a control channel is used.

Although each mobile has only a single transmitter and receiver the channels which are used can be changed by instructions from the base station. There are actually four **signalling** paths between mobile and the base station. These are:

(i) the forward control channel (FOCC), from base to mobile;
(ii) the reverse control channel (RECC), from mobile to base;
(iii) the forward voice channel (FVC), from base to mobile:
(iv) the reverse voice channel (RVC), from mobile to base.

The FOCC and RECC are used to maintain contact between mobiles and base stations outwith a normal call and also for call set-up. The forward and reverse voice channels are used when calls are in progress. The FVC and RVC signalling messages are formatted in short bursts and are inserted from time to time into the voice path. Being of short duration the users are unaware of their existence. It will be apparent from Fig. 15.12 that, although there are 21 dedicated FOCCs and RECCs, provision is also made to use some of the voice channels for signalling information.

In a busy system using many of the available voice channels per cell the signalling data exceeds the capacity of the dedicated control channels. Hence it would not be possible for new users to access unused voice channels. Therefore provision is made to re-allocate some of the voice channels for signalling under high-load conditions. As far as system operation is concerned it is necessary for the FOCC channels to operate continuously in order that whenever a mobile enters a cell it can determine which channels are allocated for voice traffic and whether the RECC is in use. The RECC on the other hand is only activated occasionally when it is necessary to transmit information about mobiles. Such information is required when, for example, mobiles move from one cell to another or when a mobile wishes to initiate a call.

15.12 TACS RADIO PARAMETERS

15.12.1 Power levels

When considering frequency re-use it was assumed that all base stations operate at the same power levels. In the TACS system cell sizes range from a radius of 1 km (urban areas) to 15 km (rural areas) and the base station power level is chosen to ensure adequate coverage for the size of cell which it serves. The maximum ERP, for the larger cells, is restricted to 100 W.

The situation with the mobile is somewhat different as battery drain is an important consideration. In effect, the mobile operates at the lowest power level which will ensure an adequate link quality and can be instructed by the base station to alter power output as appropriate. It is also important to minimize the total ERP in the cellular environment to keep interference as low as possible, as a high power mobile can produce both adjacent channel and co-channel interference. Four classes of mobile are recognized, as in Table 15.3 (class 1 is a vehicle-mounted transceiver, class 4 is a hand-held portable). Note in this table that dBW is the power output relative to 1 watt, i.e.

$$ERP(dBW) = 10\log_{10}\left(\frac{\text{power}}{1\,\text{watt}}\right)$$

Base stations instruct mobiles to adjust power output levels to maintain an acceptable signal level at the base station by use of a 3 bit mobile attenuation code (MAC). The codes and the corresponding outputs are shown in Table 15.4.

Table 15.3 Mobile power levels

Class	Nominal ERP		Mobile output 1.5 dB antenna gain
1	10.0 dBW	(10 W)	8.5 dBW (7.0 W)
2	6.0 dbW	(4 W)	4.5 dBW (2.8 W)
3	1.6 dbW	(1.6 W)	0.5 dBW (1.1 W)
4	−2.0 dBW	(0.6 W)	−3.5 dBW (0.45 W)

Table 15.4 Mobile attenuation codes

Mobile power level	Mobile attenuation code MAC	Nominal ERP (dBW)			
		Class 1	Class 2	Class 3	Class 4
0	000	10	6	2	−2
1	001	2	2	2	−2
2	010	−2	−2	−2	−2
3	011	−6	−6	−6	−6
4	100	−10	−10	−10	−10
5	101	−14	−14	−14	−14
6	110	−18	−18	−18	−18
7	111	−22	−22	−22	−22

15.12.2 Modulation

The modulation used for voice signals is frequency modulation. This has superior SNR properties to AM, and it is also possible to take advantage of the FM capture effect to minimize the co-channel interference. The FM capture effect is fully described in Section 4.11 and Fig. 4.12.

In this figure, X may be interpreted as the carrier amplitude received from the serving base station and Y as the resultant co-channel interference from interfering cells using the same frequency. For values of $X/Y > 10(10\,\text{dB})$ X captures the system. Hence it can be seen that if the CIR is maintained at around 18 dB (which is the figure used to decide the cluster size) the capture effect of the FM transmission has significant advantage.

In the TACS system the bandwidth of the channel is restricted to 25 kHz. Assuming the audio signal has a bandwidth of 3 kHz this means that four significant sidebands are alowed giving a modulation index of approximately

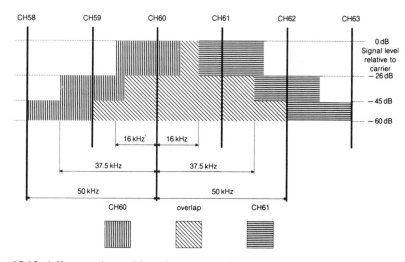

Fig. 15.13 Adjacent channel interference in TACS.

$\beta = 1.6$. The corresponding carrier deviation is

$$\Delta f_c = \beta f_m = 5\,\mathrm{kHz}$$

However this value is increased to 9.5 kHz ($\beta = 3.2$) in TACS to improve CIR performance, in the presence of co-channel interference. There is a penalty to be paid for this, however, as the signal bandwidth is no longer restricted to 25 kHz and **adjacent** channel interference will thus occur. The amount of interference allowed is shown in Fig. 15.13. The 26 dB bandwidth is specified as 32 kHz, the 45 dB bandwidth is specified as 75 kHz and the 60 dB bandwidth is specified as 100 kHz. Hence it is necessary to ensure, as far as possible, that allocation of adjacent channels in adjacent cells is minimized. This means that once the cluster size is fixed the channel allocation within a cluster should minimize the allocation of adjacent channels to adjacent cells. (A pre-emphasis characteristic of 6 dB/octave is used between 300 Hz and 3 kHz).

15.13 TACS MOBILE NETWORK CONTROL

It was shown in Fig. 15.11 that 21 channels from the available set are dedicated to the system management and call set-up functions. There are 21 forward control channels for base to mobile communication and spaced at 45 MHz above these there are a corresponding 21 reverse control channels for mobile to base communication. These channels are paired, i.e. if a mobile receives control information on a particular FOCC it responds on the corresponding REVC which is 45 MHz above.

Functionally there are three different types of control channel known as

1. Dedicated control channels (DCC)
2. Paging channels
3. Access channels

The DCC is the basic co-ordinating forward control channel for the network and is transmitted continuously. All mobiles are permanently programmed with the channel numbers of the DCCs and scan these channels at switch-on. This is necessary to obtain:

(i) basic information about the network;
(ii) the channel numbers of the paging channels.

The paging channels are used to:

(i) alert particular mobiles to an incoming call;
(ii) transmit channel numbers of the access channels in use in the mobile's locality;
(iii) transmit traffic area identities, etc.

Mobiles use the access channels to:

(i) obtain parameters about the required access procedure and the status of the access channels on the RECC (busy or idle);
(ii) acknowledge paging messages;
(iii) update the network with their locations by registering with the base stations offering the best radio path;
(iv) initiate outgoing calls.

All three types of functional control channels carry status information in sequences of data blocks known as **overhead messages** which contain a multiplicity of status information fields and instructions. When the control channels are multiplexed the overhead messages contain the relevant information in one continuous bit stream.

15.14. CALL MANAGEMENT IN TACS

The primary tasks of the mobile network are:

1. To have a record of the location of all active mobiles within the system at any given time. This is so that incoming calls may be directed to the appropriate cell.
2. To manage the handover process during calls as the mobile crosses cell boundaries.

In order to describe the call management procedure used in TACS it is first necessary to discuss the supervisory tones which are involved in the sequence for establishing a voice channel. Supervisory tones are necessary to monitor the progress of a call as the mobile moves in the Rayleigh fading environment. Fading will influence the level of the signal from a mobile's base station and also the level of co-channel interference from base stations operating on the same frequency in adjacent clusters. Note that if a mobile experiences a deep fade it is possible for a transmission from a different cluster to capture the receiver (FM capture effect) and an unwanted conversation would intrude. Such a possibility is prevented in TACS which uses two supervisory tones known as:

1. The supervisory audio tone (SAT)
2. The signalling tone (ST)

While a call is **in progress** a supervisory audio tone (SAT) is transmitted by the base station and re-transmitted by the mobile. Both the base station and the mobile require the presence of the SAT on the received signal to enable the audio path. The SAT is one of three frequencies (5.970 kHz, 6.000 kHz or 6.030 kHz).

In setting up the call the base station informs the mobile which SAT is appropriate to the *cluster* of cells which is handling the call. The mobile then re-transmits the defined SAT to confirm that the correct connection is made

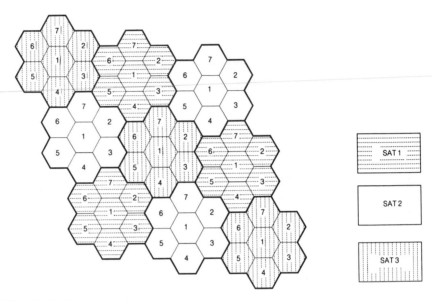

Fig. 15.14 Allocation of supervisory audio tones.

and continues to transmit this SAT throughout the call. The two other SAT frequencies are assigned to adjacent cell clusters as shown in Fig. 15.14.

If a co-channel interferer, originating in an adjacent cluster, captures the mobile receiver it will carry the wrong SAT and the audio output will be muted. The only time the SAT is itself muted is when data is being transmited from the mobile on the RECC. In such cases a digital colour code D'C'C' is transmitted instead. This is the digital equivalent of the SAT and is compatible with the signalling format used. The D'C'C' is also used on the FOCC during data transmissions.

15.14.1 Mobile scanning

When a mobile is powered up it scans the 21 control channels of its primary network. It then tunes to the strongest channel and attempts to read the overhead messages being transmitted. If successful the mobile receives information indicating the channel numbers of the paging channels in its location. If this information is not received correctly the mobile tunes to the second strongest channel and tries again.

If the second attempt is unsuccessful it is likely that the mobile is not within range of its primary network and so it repeats the scanning process on the secondary network (provided by a different operator). When the mobile receives the channel numbers of the paging channels it scans these looking for the two strongest signals. It tunes to the strongest signal and attempts to receive the overhead messages being transmitted on that channel. If successful it receives information about the traffic area in which it is operating and a number of parameters about the network configuration.

If unsuccessful the mobile tunes to the second strongest paging channel and again attempts to read the overhead messages. If this fails it restarts the complete scanning operation. From switch-on to locking on to a paging channel usually takes between 5 and 10 seconds. However this can be increased to 17 seconds if a second strongest control or paging channel is required. When the mobile finds a suitable paging channel it reads information concerning the access channels in use. The mobile then signals a registration on the RECC.

15.14.2 Registration

Registration is used by mobiles to inform the cellular network of their current location in order that incoming calls can be routed to the correct traffic area (i.e. group of cells). The routing within the mobile network is handled by special exchanges known as mobile switching centres (MSCs). Each MSC controls several cell clusters known as a traffic area and is connected to other MSCs and the fixed network, as shown in Fig. 15.1. The record-keeping activity of the MSCs requires that each mobile continually updates its location by a process known as **registration**.

The TACS system has two forms of registration known as **forced registration** and **periodic registration**. Either or both types of registration can be enabled by messages on the paging channels. With forced registration the mobiles are required to register every time they cross a new traffic area boundary. Whenever a mobile enters a scanning sequence to find a new paging channel it compares the received traffic area information with that of the previous paging channel. If there is a difference the mobile has crossed a traffic area boundary and so registers its location with the network. (This will also occur at power up since there is no previous traffic area.)

When it has registered the mobile updates its internal memory with the new traffic area. With periodic re-registration the mobile maintains a list of the last four traffic areas visited, together with a numeric indicator for each which tells the mobile when to re-register. The base station transmits a registration identity number on a regular basis on the paging channel. This number is incremented each time it is transmitted. The mobile compares the received number with the number in its memory appropriate to the current traffic area. If the two correspond the mobile intiates a registration with the network. When it has registered, the mobile increments the number stored in its memory and awaits the next registration period.

The network is able to control the rate at which mobiles re-register by varying the time between registration identity messages, or by changing the number by which mobiles increment their stored number. If a mobile finds itself in a traffic area for which it has no entry in its memory, it creates a new entry (deleting the oldest) and immediately registers with the network as for forced registration. Registration is achieved by the mobile performing a system access and sending a registration messsage. (The system access procedure is covered in the sub-section on call origination.)

When a mobile registers in a new traffic area its location is passed by the MSC, controlling the area, to the mobile's **home area location** MSC. This

MSC holds data on the location of all its active mobiles (all mobiles are allocated a home traffic area). Hence when a call is made to a mobile the system is able to decide, by interrogating the mobile's home traffic area MSC, which part of the network will page the mobile.

15.14.3 Call origination

When a mobile subscriber wishes to make an outgoing call the user enters the number manually from the keyboard (or automatically from an on-board memory) and initiates the call by pressing a SEND key. This causes the mobile to perform a system access in order to transmit its message to the system. The mobile first scans the access channels (whose numbers are indicated by overhead messages of the paging channels) in the same way as other scanning operations are carried out and chooses the two strongest. It attempts to read the overhead messages on the strongest access channel (of the FOCC) which contains parameters about the required access procedure. (If unsuccessful the procedure is repeated on the second channel.)

Once these parameters are read the mobile monitors the BUSY/IDLE bit stream being sent by the base station on the access channel. If this indicates an idle condition the mobile waits a random time and transmits its message on the mobile to base access channel which is 45 MHz above the base to mobile access channel to which it is tuned. The mobile continues to monitor the BUSY/IDLE bit stream transmitted by the base station. This is changed from IDLE to BUSY by the base station as soon as it receives the start of the message from the mobile.

The mobile checks the interval between its start of transmission and the transition from IDLE to BUSY. If this is too long or too short the mobile assumes that the IDLE/BUSY transition was caused by a message from another mobile and aborts its transmission. It then waits for a random time and attempts to transmit its message again. This is a form of collision resolution which is necessary on all random access systems.

When the mobile has completed its message it turns off its transmitter and continues to monitor the base-to-mobile access channel. For call originations the message from the base station is normally a speech channel allocation which contains a channel number and the SAT code (this is the digital colour code $D'C'C'$). On receipt of the message the mobile tunes to the required voice channel and transmits the SAT. If the SAT is correct the audio paths are enabled and the user can hear the call being set up. If the access was as a result of a registration the message received on the base-to-mobile access channel is a registration confirmation. On receipt of this message the mobile returns to the idle condition.

15.14.4 Call receipt

When an incoming call is received for a mobile its home area MSC is checked for the mobile's current registered location. A paging call is then transmitted on the paging channel of all base stations in the mobile's current traffic area. When a mobile receives a paging call it accesses the network in the same way as for a call origination, but the message sent to the base station informs it that this access is a result of receiving a page call. The mobile receives a voice channel allocation from the base station and checks the SAT received and loops it back to the base station. The base station then transmits an alert message to the mobile causing the mobile to alert the user to the incoming call and to transmit the 8 kHz signalling tone (ST). When the user answers, the 8 kHz tone is disabled and this enables the audio paths and the call proceeds.

15.14.5 Handover

Whenever the mobile is operating on a voice channel the base station monitors the received signal level. When this level falls below a threshold value the base station informs its MSC that handover to a nearby cell may be necessary. (Note that the nearby cell may be in the same traffic area or an adjacent traffic area.) The MSC then requests surrounding base stations to measure the signal strength of the mobile by using their special purpose measuring receivers. When the MSC receives the results of these measurements it decides whether any of the reporting base stations is receiving a stronger signal than the current base station.

If there is a better cell the relevant base station is requested to allocate a voice channel and the MSC requests the original base station to inform the mobile to tune to the new channel. (The new base station will be operating on a different set of carriers to the current one.) A short signalling message is sent to the mobile on the FVC giving the new channel number and the mobile tunes to the new channel. The duration of the signalling message is about 400 ms hence the user notices only a brief silence during the handover process.

15.14.6 Power control

If a mobile moves close to the base station a high signal level could result in inter-modulation in the base station receivers causing interference to other users. To avoid this happening the signal level is monitored and if found to be above a given threshold a message is sent to the mobile on the FVC to reduce its transmitter power level. The mobile acknowledges on the RVC and selects the appropriate power level. The opposite procedure is possible as the mobile moves away from the base station.

15.14.7 Additional services

The facility exists within the system to provide additional services such as three party calls, call diversion, etc. To request such facilities the user enters the appropriate code from the keyboard and presses the SEND key. The mobile sends a 0.4 s burst of 8 kHz signalling tone to the base station which responds with a digital signalling message on the FVC requesting the mobile to send its information. The mobile transmits the information to the base station in digital form on the RVC and then returns to the conversational mode while the network processes the request for the facility.

15.14.8 Call termination

When a mobile user finishes a call and replaces the handset the mobile transmits a 1.8 s burst of 8 kHz tone to the base station and then re-enters the control channel scanning procedure. If the other party on the PSTN clears down, a **release** message is sent to the mobile on the FVC. The mobile then responds by sending the 8 kHz tone after which it re-enters the control channel scanning procedure.

15.14.9 Protection of signalling messages

The radio channel is subject to interference (co-channel and adjacent channel) and fading and it is necessary to protect the signalling sequences against these imperfections. Two methods of data protection are employed in TACS and these are:

1. Repeat each transmitted data word several times and take a majority vote. This is designed to obtain a 95% accuracy in data word transmission.
2. Use an error-correcting code to improve the 95% accuracy to 99.9% accuracy.

Details of the coding scheme can be found in reference 1.

15.15 ALTERNATIVE ANALOGUE SYSTEMS

Although TACS has been considered in detail it has already been pointed out that it is not the only system currently in use. There are a number of others, e.g. AMPS used in the USA, NAMTS used in Japan and NMT used in Scandinavia.

These systems differ from TACS in relatively minor ways but they are not compatible. In NMT, for example, the frequency modulated carrier is restricted to a bandwidth of 25 kHz (compared to a value of 32 kHz in TACS). This means that in NMT adjacent channels can be used in the same cell, if required. However, because the resulting modulation index is smaller, the co-channel interference is significantly poorer. This means that NMTS requires a larger re-use distance than TACS.

The systems also differ in the number of channels dedicated to control functions. TACS has the most (21) and NAMTS has the least (1). The carrier spacing in AMPS is 30 kHz rather than 25 kHz, and so on. Hence there is no single standard in the analogue world. In the digital world, especially in Europe, considerable effort has been made to derive a common standard. The European standard is known as GSM and will be described in the next section.

15.16 DIGITAL CELLULAR RADIO

Since the introduction of first generation analogue systems an unprecedented demand for installations has occurred. This means that the systems are rapidly reaching capacity and ways of meeting the demand with second generation systems have been investigated. The capacity of the first generation systems is determined by the bandwidth occupied by the individual voice channels and the minimum CIR at which mobiles and base station receivers can operate.

Since the radio bandwidth allocated to mobile communications will always be finite, large-scale increases in capacity are only possible by utilizing the allocated bandwidth more efficiently. One example of such a technique is to use SSB instead of FM. However, such systems would not have the capture effect advantage of FM and the overall improvements likely to be achieved with SSB are not judged to be significant enough to warrant its introduction.

The alternative is to make the cells smaller which means that problems with frequent handovers would then have to be addressed. It is stressed that there is no solution which is optimum in all respects. However, there are a number of significant advantages to be achieved with digital systems which makes their adoption as second generation systems quite attractive. In assessing the available digital techniques the system(s) must satisfy the following criteria:

  (i) high subjective voice quality;
 (ii) low infrastructure cost;
(iii) low mobile equipment cost;
(iv) high radio spectrum efficiency;
 (v) capable of supporting hand held portables;
(vi) ability to support new services.

Clearly a significant advantage of digital systems is that digital signal processing can be used both to reduce voice bandwidth requirements and to extract signals in poor CIR conditions. There are other advantages, for example, digital transmissions can be encrypted to avoid unauthorized eavesdropping.

15.16.1 Advantages of broadband transmission

Digital transmission can be either narrowband or broadband and the actual choice is a compromise between many factors. It has been pointed out that the effect of multipath propagation is frequency dependent. If a null is produced at one frequency due to destructive addition a peak may be produced at a different frequency due to constructive addition. Hence if a signal has a bandwidth which is wider than the coherence bandwidth it follows that only some frequencies in the signal may experience a fade while others will not. Thus the use of wideband transmission can have some advantage in a Rayleigh fading environment (additional processing can sometimes recover the signal).

Four possible strategies which could be employed are:

1. **Narrowband**: in this case a single carrier is allocated to each channel (this is the simple frequency division multiple access (FDMA) used in the TACS system).
2. **Wideband**: in this case each channel uses all of the available bandwidth for

a fraction of the time (this is known as wideband time division multiple access (TDMA)).

3. **Intermediate**: in this case each carrier has a bandwidth which is made greater than the coherence bandwidth by time multiplexing a number of channels. The complete bandwidth is occupied by a number of such channels. This is usually referred to as narrowband TDMA or FDMA-TDMA.

4. **Spread spectrum**: in this case perfect theoretical signal extraction is possible in virtually zero CIR conditions.

The choice between these alternatives is not straightforward and each one has certain advantages and disadvantages. The advantages of FDMA (one channel per carrier) are:

1. The bandwidth of each carrier is considerably less than the coherence bandwidth. Hence equalization is not required as this will not produce any improvement in performance. Capacity increase is obtained by reducing the bit rate per channel and using efficient channel codes.

2. The technological advances required for such a system, over existing analogue systems, is modest. It is possible to configure a system so that subsequent advances in reduced rate speech coders could be easily accommodated.

3. The system is flexible and can be easily adapted to handle both large rural cells and small urban cells.

The disadvantages associated with FDMA (one channel per carrier) are:

1. The architecture is similiar to that used in analogue systems. Hence any capacity improvements would rely on operation at lower CIR values. Narrowband digital systems have only limited advantages in this respect which severely restricts any capacity improvements which would be achieved for a given radio spectrum allocation.

2. The maximum bit rate per channel is fixed at a low value. This is a considerable disadvantage for data communications, which is seen as an essential element in future systems.

The advantages of TDMA (narrowband and broadband) are:

1. It offers the capability of overcoming Rayleigh fading by appropriate channel equalization.

2. Flexible bit rates are possible, i.e. both multiples and sub-multiples of the standard bit rate per channel can be made available to users.

3. It offers the opportunity of frame-by-frame monitoring of signal strength and bit error rate to enable either base stations or mobiles to initiate handover.

The disadvantages of TDMA are:

1. On the UP link TDMA requires high peak power in the transmit mode. This is a particular problem for hand held portables with limited battery life.

2. To realize the full potential of digital transmission requires a significant

amount of signal processing. This increases power consumption and also introduces delay into the speech path.

Considering all these issues broadband digital systems appear to have the greatest potential but they also have the greatest demands on the associated signal processing requirements. Whether such systems will be developed for third generation systems is still the subject of intense research effort. Second generation systems have been developed, however, and the system adopted for Europe was chosen and specified by a specially formed group of the *Conference Europeene des Administrations des Postes et des Telecommunications* (CEPT) known as the Groupe Special Mobile (GSM). GSM is composed of about 40 members representing 17 European countries and its function is to co-ordinate and produce specifications for a Pan-European cellular mobile radio system.

The GSM carried out field trials of a number of competing systems in 1986 and chose as the standard a digital system based on a narrowband TDMA approach. The standard which has been developed is known simply as GSM, and a variation of this with low-power terminals is known as DCS 1800. The GSM standard has been adopted in countries worldwide and GSM has been renamed as Global System for Mobile communication. The choice of the GSM standard was based on the following ranked criteria:

1. Spectral efficiency (the number of simultaneous conversations/MHz/km^2).
2. Subjective voice quality.
3. Cost of mobile unit.
4. Feasibility of a hand portable mobile unit.
5. Cost of base station.
6. Ability to support new services.
7. Ability to coexist with existing systems.

In GSM the voice waveform is digitally encoded before transmission. As the system is based on TDMA individual users are given access to the radio channel for a limited period and transmit a burst of binary information. The GSM specification is very detailed and, in a text of this nature, it will be possible only to give an outline of the system. The natural place to begin is with a description of the radio interface.

15.17 THE GSM RADIO INTERFACE

The radio subsystem constitutes the physical layer of the link between mobile and base stations. The main attributes of the GSM interface are:

1. time division multiple access (TDMA) with 8 channels/carrier
2. 124 radio carriers in a paired band (890 to 915 MHz mobile to base station, 935 to 960 MHz base to mobile, inter-carrier spacing 200 kHz)
3. 270.833 kb/s per carrier
4. Gaussian minimum shift keying with a time bandwidth product $BT = 0.3$
5. slow frequency hopping (217 hops/second)
6. synchronization compensation for up to 233 μs absolute delay

7. equalization of up to 16 µs time dispersion
8. downlink power control
9. discontinuous transmission and reception
10. block and convolutional channel coding coupled with interleaving to combat channel perturbations
11. 13 kb/s speech coder rate using regular pulse excitation/linear predictive coding (RPE/LPC)
12. overall channel bit rate of 22.8 kb/s

The first pair of carriers in the GSM system are 890.2 MHz and 935.2 MHz, i.e. the spacing is 45 MHz which is the same as for TACS. GSM recommends that carriers 1 and 124 are not used due to energy of the modulated carrier lying outside the nominal 200 kHz bandwidth. The radio subsystem is the physical layer of the link between mobiles and base stations. Each cell can have from 1 to 15 pairs of carriers and each carrier is time multiplexed into 8 slots. The carriers and their associated time multiplexed slots form the **physical channels** of the GSM system. The operation of the radio subsystem is divided into a number of **logical channels** each of which has a specific function in terms of handling the transmission of information over the radio subsystem. Each of the logical channels must be mapped in some way onto the available physical channels. This is illustrated in Fig. 15.15.

The two categories are

1. Traffic Channels (TCH)
2. Signalling channels (Broadcast Control Channel (BCCH), Common Control Channel (CCCH) and Dedicated Control Channel (DCCH))

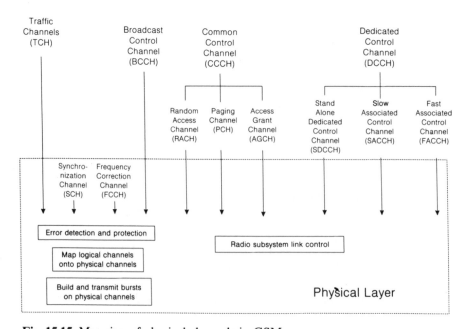

Fig. 15.15 Mapping of physical channels in GSM.

Some of these channels are divided into sub-channels. The sub-channels exist conceptually in parallel but the existence of one signalling channel may exclude the presence of another one. The Paging Channel (PCH) and Access Grant Channel (AGCH) are never used in parallel. The radio subsystem requires two channels for its own purposes. These are the **Synchronization Channel** (SCH) and the **Frequency Correction Channel** (FCCH).

The logical channels are mapped onto the basic TDMA frame structure which is shown in Fig. 15.16. The purpose of the radio subsystem is to provide to the **data link layer** a 'bit pipe' with a defined throughput, acceptable transmission delay, and a reasonable quality for each of the logical channels. To achieve this the physical layer performs a variety of tasks which can be grouped into four categories.

1. Create physical channels by building data bursts and transmitting them over the radio path.
2. Map the logical channels onto the created physical channels, taking into account the throughput needs of particular logical channels.
3. Apply error protection to each logical channel according to its particular needs.
4. Monitor and control the radio environment to assign dedicated resources and to combat changes in propagation characteristics by functions such as **handover** and **power control**.

Bursts of transmission from base station and mobile occur in the slots of the up and down carriers as shown in Fig. 15.16. The bit rate on the radio channel is 270.833 kb/s which gives a bit duration of 3.692 µs. A single timeslot consists of 156.25 bits and therefore has a duration of 0.577 ms. The recurrence of one particular timeslot on each frame makes up one physical channel. This structure is applied to both uplink and downlink. The numbering scheme is staggered by three timeslots to remove the necessity for the mobile station to transmit and receive at the same time. This is illustrated in Fig. 15.17. Data is transmitted in bursts which are placed in these timeslots. It is clear from Figure 15.16 that the length of the bursts are slightly shorter than the

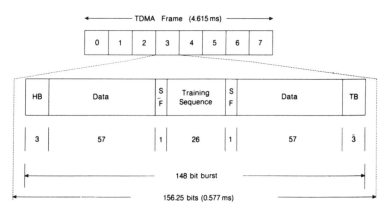

Fig. 15.16 Normal burst in GSM.

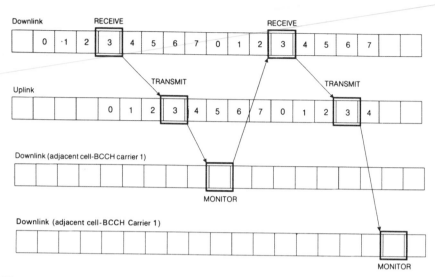

Fig. 15.17 GSM slots and scanning structure.

duration of the timeslots. This is to allow for burst alignment errors, time dispersion on the propagation path, and the time required for smooth switch on/off of the transmitter.

There are four types of burst which can occupy a timeslot, these are:

1. Normal burst (148 bits + 8.25 guard bits).
2. Frequency correction burst (148 bits + 8.25 guard bits).
3. Synchronizing burst (148 bits + 8.25 guard bits).
4. Access burst (88 bits + 68.25 bits), used to access a cell for the first time in case of call set-up or handover.

Figure 15.16 shows the data structure within a normal burst. It consists of 148 bits transmitted at a rate of 270.833 kb/s. Of these bits 114 bits are available for data transmission, the remaining bits are used to assist reception and detection. A training sequence (26 bits) in the middle of the burst is used by the receiver to synchronize and estimate the propagation characteristics. This allows the setting up of an equalizer to compensate for time dispersion produced by multipath propagation. Tail bits (3 bits) transmitted at either end of the burst enable the data bits near the edges of each burst to be equalized as well as those in the middle. Two stealing flags (one at each end of the training sequence) are used to indicate that a burst which had initially been assigned to a traffic channel has been 'stolen' for signalling purposes.

The burst modulates one carrier of those assigned to a particular cell using Gaussian minimum shift keying (GMSK). If frequency hopping is not employed, each burst belonging to one particular **physical channel** is transmitted using the same carrier frequency. A network operator can implement **slow frequency hopping**. It will be recalled that fading is frequency dependent, hence slow frequency hopping is one technique which can be used to overcome the problems of fading. If slow frequency hopping is implemented the frequency changes (among the set of carrier frequencies available within a cell) between bursts, this is discussed in more detail in Section 15.22.

15.18 MAPPING OF LOGICAL CHANNELS IN GSM

There are five different cases of mapping logical channels on to physical channels.

1. Mapping of a full-rate traffic channel (TCH) and its slow associated control channel (SACCH) onto one physical channel.
2. Mapping of two-half rate traffic channels and their two slow associated control channels on to one physical channel.
3. Mapping of the broadcast control channel (BCCH) and the common control channel (CCCH) onto one physical channel.
4. Mapping eight stand-alone dedicated control channels (SDCCH) onto one physical channel.
5. Mapping of four stand-alone dedicated control channels plus the broadcast control channel and the common control channel onto one physical channel. (This is for lower capacity than case 3).

This list indicates that there are at least two logical channels to be mapped onto each physical channel. The timeslots of the physical channel must therefore be assigned to the logical channels on a structured basis. For this purpose two **multiframe** structures have been defined:

1. A multiframe consisting of 26 TDM frames (resulting in a recurrence interval of 120 ms) for the TCH/SACCH cases 1 and 2.
2. A multiframe consisting of 51 TDM frames (resulting in a recurrence interval of 236 ms) for signalling channels, cases 3, 4 and 5.

15.18.1 Mapping of traffic and associated control channels

The mapping of a traffic channel and its SACCH is shown in Fig. 15.18 for both full-rate and half-rate channels. The full-rate TCH uses 24 frames out of the 26 available in the multiframe. One of the 26 frames is used for the SACCH and one remains idle. In this diagram TC0 represents a normal burst of duration 0.577 ms on a particular physical channel (i.e. carrier timeslot). TC1 represents the next burst of duration 0.577 ms on the same physical channel and will occur 4.615 ms after TC0. The duration of the multiframe is therefore $26 \times 4.615 \, \text{ms} = 120 \, \text{ms}$.

The gross bit rate per traffic channel is derived as follows:

Data bits per normal burst = 114 bits
Number of normal bursts per 120 ms multiframe = 24

$$\text{Gross bit rate} = \frac{24 \times 114}{0.12} = 22.8 \, \text{kb/s}$$

The SACCH uses 114 bits per 120 ms = 950 b/s

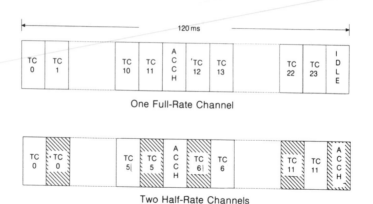

Fig. 15.18 Full- and half-rate channels.

The throughout of the physical channel is 114 bits per $4.615\,\text{ms} = 24.7\,\text{kb/s}$ (this includes the idle frame). In the case of the half-rate channel two half-rate channels share one physical channel, the idle frame is then used to accommodate the SACCH for the second half-rate channel. In this case each traffic channel occupies only 12 frames which results in a gross bit rate of $11.4\,\text{kb/s}$ for each (each SACCH uses $950\,\text{b/s}$).

GSM allows the possibility to provide additional signalling capacity if the one provided by the SACCH is not enough. The **Fast Associated Control Channel** (FACCH) steals capacity from the TCH by replacing bits from the TCH with bits from the FACCH. The stealing flags are used to indicate when TCH bits have been replaced by FACCH bits.

15.18.2 Mapping of the BCCH/CCCH

The TCH/SACCH structure is dedicated to one user (full-rate channel) or shared between two users (half-rate channel). The mapping of the BCCH and the CCCH uses a multiframe of 51 TDM frames and is shared by all mobiles currently in a cell. In addition all sub-channels transmitted on this structure are **simplex** channels i.e. they exist in one direction only. The sub-channels mapped onto this physical channel are:

1. The **Broadcast Control Channel** (BCCH, base to mobile), this provides general information about the network, the cell in which the mobile is currently located and the adjacent cells.
2. The **Synchronization Channel** (SCH, base to mobile) carriers information for frame synchronization and identification of the base station transceiver.
3. The **Frequency Correction Channel** (FCCH, base to mobile) provides information for carrier synchronization.
4. The **Random Access Channel** (RACH, mobile to base), this channel is used by the mobile to access the network during registration of cell set-up. The access is random and uses slotted ALOHA.
5. The **Access Grant Channel** (AGCH, base to mobile) is used to assign

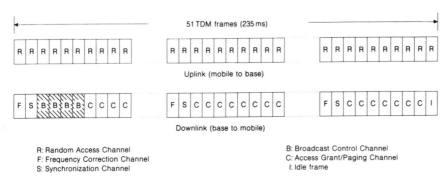

Fig. 15.19 Multiframe structure.

dedicated resources (a SDCCH or TCH) to a mobile which has previously requested them via the RACH.

6. The **Paging Channel** (PCH, base to mobile) is used to alert a mobile to a call originating from the network.

The mapping of these sub-channels onto a single physical channel using a 51 multiframe is shown in Fig. 15.19. This structure appears on timeslot 0 of one of the allocated carriers in the cell. This carrier is known as the BCCH carrier. The uplink of the BCCH/CCCH structure carries the Random Access Channel only as this is the only control channel which exists from mobile to base. The mobile can use any one of the 51 frames on timeslot 0 to access the network.

On the downlink the 51 frames are grouped into five sets of 10 frames (the 51st frame remains idle). The gross bit rate for the BCCH is 4 frames of 114 bits per 235 ms = 1.94 kb/s.

15.19 GSM MODULATION, CODING AND ERROR PROTECTION

There are several stages of coding and decoding in the GSM traffic channels and these are shown in Fig. 15.20. The speech coder used in GSM is known as a low bit rate coder and capitalizes on the inherent redundancy in the speech waveform. The speech coder is a regular pulse excited linear predictive coder with long-term pitch prediction. This coder produces a net bit rate of 13 kb/s and is a block coder which analyses speech samples in blocks of 20 ms duration. This results in an output block size of 260 bits.

Because the speech coder produces a reduced bit rate its performance is relatively sensitive to errors. Not all bits have the same significance, however. Of the 260 bits per block 182 (class 1 bits) are more sensitive to error than the remaining 78 bits (the class 2 bits). The channel coder introduces redundancy to protect against error and uses a combination of block coding and convolutional coding on the class 1 bits (the class 2 bits are not protected). This increases the number of bits per block from 260 to 456 and gives a gross bit rate of 22.8 kb/s. Note that $456 = 8 \times 57$, which is significant when considering interleaving.

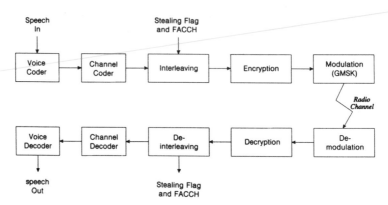

Fig. 15.20 Coding and decoding in GSM.

15.19.1 Interleaving

The channel coding is most efficient when bit errors are uniformly distributed within the transmitted bit stream. However errors due to fading cause errors to occur in bursts. The problem is reduced by a technique known as **bit interleaving**. The sequence of 456 bits is re-ordered and then divided into 8 blocks of 57 bits. Each block of 57 bits is transmitted in a normal burst on the TCH. This is known as an **interleaving depth** of 8. The individual bits in a block are allocated to the eight bursts as shown in Fig. 15.21. Since each normal burst has 114 data bits each burst contains bits from two separate speech blocks. Clearly it is necessary to have two speech blocks available before a normal burst can be formed. This requires an interval of 40 ms and hence the coding delay for GSM is of the order of 40 ms. The de-interleaving process is simply the reverse of the interleaving process and results in any burst errors being uniformly distributed in the reconstituted speech blocks.

The normal bursts are also used for transmitting messages on the FACCH.

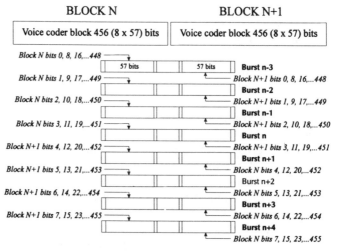

Fig. 15.21 Interleaving in GSM.

The FACCH signalling also occurs in blocks of 456 bits and undergoes the same interleaving process. To distinguish between a normal TCH and a FACCH transmission the stealing flags are set to 1 when the channel contains FACCH transmissions. Because of the interleaving process it is clear that a normal burst can contain both 57 bits of TCH and 57 bits of FACCH. Hence there are two stealing flags in each normal burst, one for each half burst. The control channels use a different error protection scheme with an interleaving depth of 4.

The radio path is subject to multipath propagation which can produce a delay spread of several microseconds. This becomes apparent at the receiver as inter-symbol interference. The training sequence of each burst is used by the receiver to estimate the multipath delay spread being experienced by that burst. This information is used to set up the appropriate equalizer coefficients. The delay spread is a dynamic parameter which changes from burst to burst. Placing the training sequence in the middle of a burst reduces the time over which the delay spread can change relative to that during the transmission of the training sequence.

15.19.2   Gaussian minimum shift keying (GMSK)

This is the form of modulation used in GSM and there are basically two problems which need to be addressed:

1. minimum bandwidth;
2. minimum error probability.

Standard frequency shift keying uses two separate carriers f_0 and f_1 to transmit binary 0 and binary 1. In order to produce the smallest error probability the carriers f_0 and f_1 must be orthogonal, i.e. they must have a correlation coefficient which is zero. In order to minimize the bandwidth of the transmitted signal it is necessary to determine the minimum difference between f_0 and f_1 which will produce orthogonal signals and this is called **minimum shift keying**. If the number of cycles of f_0 in the interval T (where T is the duration of 1 bit period) is n_0 then the number of cycles of f_1 in the same interval to achieve orthogonality must be $n_1 = n_0 + 0.5$, or $(f_1 - f_0) = 1/2T$. Hence MSK is effectively FSK with the minimum frequency difference between f_1 and f_0.

If MSK is considered in terms of the modulation of a single carrier frequency f_c the instantaneous frequency is given by

$$f_i = f_c + a f_d$$

where $a = \pm 1$ and f_d is the carrier deviation.

For MSK the carrier deviation $f_d = 1/4T$. When the keyed frequencies are separated by an amount less than the data bandwidth it is not appropriate to represent the spectrum as the sum of two independent sinc functions as these sinc functions will overlap to a considerable extent and will produce a 'composite' spectrum centred at the carrier frequency f_c. This spectrum has the form shown in Fig. 15.22 and it should be noted that this spectrum has a

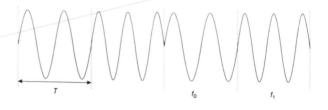

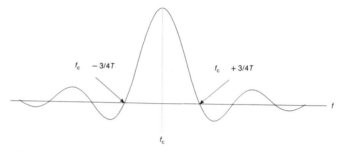

Fig. 15.22 MSK waveform and spectrum.

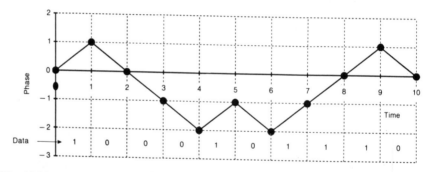

Fig. 15.23 Phase shift of a MSK carrier as a function of modulating waveform.

wider bandwidth than the corresponding ASK (or PSK) spectrum but that the sidelobes decrease at a faster rate.

Considering the MSK signal in terms of frequency modulation gives an expression for the modulated carrier of

$$v_c(t) = A \cos\left(2\pi f_c t + a \int_0^t f_d \, dt\right) \tag{15.22}$$

The phase of this carrier is thus a series of ramps as shown in Fig. 15.23.

Figure 15.22 demonstrates that there is a discontinuity in the MSK waveform at the edge of the binary interval caused by the rapid change from f_1 to f_0 and vice versa. The bandwidth of the signal can be reduced further by ensuring that the instantaneous change in digital signal levels from binary 0 to 1 (and vice versa) is smoothed out by passing the baseband signal through a filter with a Gaussian impulse response before modulation. This produces Gaussian minimum shift keying or GMSK. In the case of GSM the Gaussian filter was selected so that the product of filter bandwidth and modulating bit period $= 3 (BT = 3)$. The analysis of GMSK is actually quite involved but this simplified explanation illustrates the main properties of the signal.

15.20 HANDOVER IN GSM

The handover possibilities in GSM are more comprehensive than in TACS. The possible types of handover are:

1. Intra-cell handover – this occurs between traffic channels within the same cell.
2. Inter-cell handover – this occurs between traffic channels on different cells.
3. Inter-MSC handover – this occurs between cells belonging to different MSCs.

Handover may be used in a number of different situations known as **interference limited** and **traffic limited**, these are:

1. **To maintain link quality**. This is similar to the TACS situation, when the CIR falls below a given value the mobile will be required to handover to an adjacent cell which provides a stronger signal.
2. **To minimize interference**. Even though a mobile has an acceptable CIR ratio situations can exist where it may be causing unacceptable interference to a call in a co-channel cell. This interference may be avoided.
3. **Traffic management**. In an urban environment where cell sizes are small a mobile can possibly be served adequately from a number of cells. In such circumstances the network can request a handover in order to evenly distribute traffic throughout the cells (thereby avoiding congestion within particular cells). In order to implement such a traffic management policy it is necessary for the network to have a detailed description of the area in which the mobile is operating. Measurements on signal levels, interference levels, distances, traffic loading, etc. must therefore be collected and processed.

15.21 GSM HANDOVER MEASUREMENTS

The GSM system is able to assess the quality of both the uplink and downlink since these can be considerably different. The measurements performed in the GSM system are as follows:

1. The received signal level and received signal quality on the uplink, measured by the serving base station.
2. The received signal level and received signal quality on the downlink, measured by the mobile and reported to the network every 0.5 s by means of the slow associated control channel.
3. The signal level of the BCCH of adjacent cells. Adjacent cells are identified by the mobile by reading the base station identification code and frequency of the carrier. The results for the six strongest cells are reported every 0.5 s via the SACCH.

4. The distance of a particular mobile from its serving base station is determined from the adaptive frame alignment technique employed to cater for varying propagation delay within cells. This measurement is directly available to the base station.

5. The levels of interference on free traffic channels may be measured in the serving cell and possible target cells.

6. Traffic loading on serving and adjacent cells may be measured (by operations and maintenance functions).

The data generated by the handover measurements must be processed before any handover is initiated. The processing involves the following stages:

1. Averaging of measurements over several seconds to avoid the effects of fast fading.

2. Comparison of serving cells with predetermined thresholds which trigger the handover requirement. Handover is only initiated if link quality cannot be improved by increasing transmitted power.

3. If handover is required the best cell to handover to is determined from one of a number of algorithms, e.g. lowest path loss, strongest signal, acceptable signal level in a low traffic cell, etc.

4. The resources are then allocated and the actual handover signalling is initiated.

The detailed algorithms for handover implementation have not been defined by GSM but have been left open for manufacturers and operators. There is,

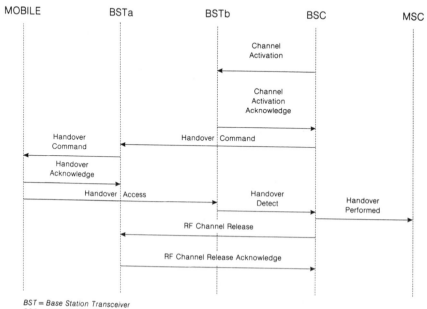

BST = Base Station Transceiver
BSC = Base Station Controller
(Note that it is usual for 1 BSC to control several BSTs)

Fig. 15.24 Handover sequence in GSM.

however, an optional recommendation which does contain the specification of a basic handover algorithm, the signalling sequence is shown in Fig. 15.24. The air–interface handover signalling has been designed such that the break in traffic which occurs during handover is minimized. Under most conditions the break is less than 100 ms which is only a barely perceptible break in speech.

To prevent the mobile station from exceeding the planned cell boundary while still using the same radio channel, a strategy can be applied which leads to a handover whenever an adjacent cell is entered which allows communication with less power. This is possible with GSM since the mobile listens to other base stations and takes measurements during the periods when it is not receiving or transmitting on an assigned traffic channel. The current base station receives measurements from the mobile, via the SACCH, and decides when handover should be initiated.

15.22 FEATURES OF THE GSM SYSTEM

There are a number of additional features of GSM some of which are closely linked with handover. The most important features are described next.

15.22.1 Adaptive frame alignment

It was illustrated in Fig. 15.17 that the mobile staggers its transmission by three timeslots after a burst from the base station. This means that there is a nominal delay of three TDM slots between transmit and receive frames at the base station. However, the propagation time between base and mobile depends on distance and it is possible for a burst from a mobile near the perimeter of a cell to overlap with a burst from a mobile close to the base station (on an adjacent timeslot). GSM calculates the timing advance required to ensure that bursts arrive at the base station at the beginning of their timeslots. This information is transmitted to the mobile on the SACCH. An alternative to adaptive frame alignment would be to use a long guard interval, which would be an ineffcient use of the radio resource.

The initial timing advance is obtained by monitoring the RACH from the mobile, which contains only access bursts with a long guard interval of 68.25 bit periods. This ensures that there will not be any overlap problems for a mobile separation from the base station of up to 37 km. The required timing advance is specified in terms of bit periods by a 6 bit number transmitted on the SACCH. This means that an advance between 0 and 63 bit periods can be requested. During normal operation when the TCH has been established the base station continually monitors the delay from the mobile. If this changes by more than 1 bit period the new advance will be signalled to the mobile on the SACCH. For cell radii greater than 35 km GSM specifies the use of every other timeslot. This allows for cell radii up to 120 km but does reduce capacity.

15.22.2 Adaptive power control

It has been stated in Section 15.5 that if the transmit power is constant then the mean CIR is a function only of frequency re-use distance. However, it is not necessarily desirable to work with constant power and the goal is rather to ensure that a minimum transmitted power is used on both the uplink and downlink in order to maintain adequate speech quality. This also has the advantage of conserving battery power for hand-held mobiles.

GSM specifies that mobiles must be able to control transmitted power on all bursts in response to commands from the base station. For a class 1 mobile (with a maximum power output of 20 W) there are 16 possible power levels separated by 2 dB (the minimum power level is 20 mW).

For initial access on the RACH the mobile is constrained to use the maximum power specified for the cell (as broadcast on the BCCH). After initial access the mobile power level is determined by the base and transmitted on each SACCH message block. The mobile will change by a 2 dB step every 60 ms until the desired value is reached. The mobile confirms its current power level by signalling this to the base station on the uplink SACCH.

As has been indicated previously adaptive power control is an alternative to handover.

15.22.3 Slow frequency hopping

It has been pointed out that the radio environment is subject to fast fading. If a mobile is moving with reasonably high velocity the duration of the fades will be short and the error correction procedures combined with interleaving will be suffcient to provide an acceptable service. However, if the mobile is moving slowly (or is stationary) the fade duration becomes longer and can exceed the interval over which bit interleaving is effective. This will result in errors in the class 1 bits of the transmitted encoded voice signal and will give rise to bad frames and degraded speech quality. If a mobile is stationary and in a deep fade, communication can be lost completely.

It has been noted previously that fading is frequency dependent. Hence a deep fade at one carrier frequency will be replaced by a strong signal at another carrier frequency. To overcome the problem of long duration fades the sequence of bursts making up a traffic channel are cyclically assigned to different carrier frequencies defined by the base station. Timing signals are available at the base and mobile to keep transmitters and receivers in synchronism on the defined hopping sequence. The result is that the positions of nulls change physically from one burst to the next. Hence the bit interleaving can correct errors even when a mobile is stationary. Another advantage of slow frequency hopping is that co-channel interference is more evenly spread between all the mobile stations.

15.22.4 Discontinuous transmission and reception (DT X)

During normal conversation a speaker is active for only about 44% of the time. The rest of the time the speaker is listening or pausing for breath. Measurements have shown that the percentage of time that both speakers talk simultaneously is very low (typically 6% of the active period). This means that a traffic channel will only be used in one direction for approximately 50% of any conversation. Advantage may be taken of this and voice activity detectors (VAD) are employed to suppress TCH transmissions during silent periods. This has two advantages.

1. the level of co-channel interference is reduced, on average, by 3 dB;
2. the battery life of the mobile can be significantly increased since it is not necessary to transmit a carrier during silent periods.

In practice it has been found that the silence periods are quite disturbing to the person at the other end of the link as the impression is given that the call has been disconnected. Hence a compromise is reached in which low level 'comfort noise' is synthesized during periods of silence. This requires periodic transmission of the background noise parameters during silence periods.

Discontinuous reception may also be employed to conserve battery power when a mobile is in the stand-by mode. The paging channel on the downlink CCCH is organized in such a way that the mobile needs to listen only to a subset of all paging frames. Hence a mobile can be designed to make the receiver active only when needed.

15.23 OPERATION OF THE GSM SYSTEM

15.23.1 Location registration and routing of calls

The GSM system is designed to accommodate international roaming. In order to provide for call routing the location of the mobile must be known. Registration is a fundamental requirement which assists in ensuring that calls are directed to the roaming subscriber. To achieve this geographical coverage, areas of national networks are divided into a number of **location areas**. The unique identity of each area is conveyed via a broadcast control channel. When a mobile is activated it selects the optimum BCCH and initiates a location updating procedure if the broadcast location area identification is different to the one stored in the mobile station before it was last de-activated. If the mobile has no BCCH information in its memory it searches all 124 carrier frequencies in the GSM system, making measurements of the average received signal strength on each. It then tunes to the carrier with the highest signal strength to determine whether this is a BCCH carrier. If it is, the mobile synchronizes with the carrier and reads the BCCH information. The mobile then initiates a location update procedure. (It should be noted that there is also a periodic registration procedure, similar to that in the TACS system, which is used to maintain accurate information concerning the status of mobile stations.) The location update is performed via the random access channel (RACH).

Each location area has a mobile switching centre with a **home location register** (HLR) and a **visitor location register** (VLR). The home location register is a database of all mobiles normally resident in that location area. The visitor location register is a database containing a record of all mobiles in the area which are not normally resident within that area. If a mobile enters a new location area, location updating is executed via the fixed network.

GSM then supports two alternatives:

1. The VLR immediately issues a mobile subscriber roaming number (MSRN) to be associated with the actual identity over the radio path (i.e. the international mobile subscriber identity [IMSI]). The international mobile subscriber identity and the mobile subscriber roaming number are then conveyed to the home location register of the mobile over the fixed network. At the end of this procedure the home location register contains the unique directory number of the mobile coupled with the international mobile subscriber identity and the current mobile subscriber roaming number.
2. In this case the network identity of the VLR or MSC, rather than the mobile subscriber roaming number, is reported to the HLR.

A call for a particular mobile is then routed to the appropriate home location register. In the first alternative the mobile subscriber roaming number is available at the HLR and the call is directed to the VLR, and the mobile is subsequently paged by transmitting the international mobile subscriber identity over the appropriate paging channel. In the second alternative the HLR signals the designated MSC and transacts for a mobile subscribers roaming number which is assigned by the VLR. Subsequently the mobile is paged with the assigned mobile subscriber roaming number.

15.23.2 Call establishment from a mobile

A channel is requested on the RACH and may be in contention with other mobiles. A slotted ALOHA protocol is used (see Section 13.5). If a request is received without a collision a dedicated control channel can be assigned by the network by a response on the access grant channel (AGCH). To minimize the probability of a collision during channel access a short access packet format is used which can be transmitted within 1 burst (see also the section on adaptive frame alignment).

The access burst contains a 7 bit random number which is used by the network in conjunction with the access slot number to address the originating mobile station for channel allocation. The full mobile identification is delivered once a dedicated control channel has been allocated. These channels are used for various functions such as authentication, etc. Detection of possible collision (or transmission errors) is performed within the network through a check of the received access burst. If a collision (or error) is detected the network aborts the procedure. If the mobile does not receive an access grant on the AGCH (which will be monitored 5 TDM slots later) a new access attempt will be made on the next slot with a given probability. (In effect this means that the mobile chooses a random number from 1 to n, which represents the next slot on which an access attempt will be made). It is possible that even when access bursts collide the FM capture effect will ensure that one packet is received without error.

When an access grant is received the mobile proceeds with the call set up on the allocated dedicated control channel by sending a SETUP message to the network. This contains addressing information and various network information. The network accepts the call establishment by returning a CALL PROCEEDING message on the SDCCH. In the normal call setup procedure the network will assign a dedicated traffic channel before it initiates the call establishment in the network. (The network may queue the traffic channel request up to a maximum queuing period.)

When called party alerting has been initiated an ALERTING message is sent to the mobile (over the SDCCH) and a ringing tone may be generated by the network and sent on the traffic channel to the mobile. When the call has been accepted at the remote end a CONNECT message is transferred to the mobile, indicating that the connection is established in the network. The mobile station responds by sending a CONNECT ACKNOWLEDGE message and then enters the active state. (Further signalling takes place over the SACCH or FACCH.)

15.23.3 Call etablishment to a mobile

In this particular case a paging message is routed to the traffic area in which the mobile is registered and transmitted on the paging channel. In responding to the page the mobile must first request a channel as in the previous case. When access grant is received from the base station the mobile responds with a CALL CONFIRMED message on the allocated dedicated control channel. A traffic channel is then allocated and the call proceeds (i.e. the mobile enters the active state).

15.23.4 Call release

Call release can be initiated either by the mobile or the fixed network, via the SACCH by sending a DISCONNECT message. If the release is initiated by the mobile the network responds with a RELEASE message. The mobile responds with a RELEASE COMPLETE message and releases the TCH. The mobile then enters the idle state and monitors the PCH.

There are many detailed feature of GSM which are not covered in this chapter. However the coverage of the radio interface has been in sufficient detail to emphasize the essential differences between analogue and digital cellular systems. Other technologies for digital mobile communications are developing in parallel with GSM based on cordless telecommunications technology. The two most prominent in Europe, at the time of writing are CT2 and DECT.

15.24 CORDLESS COMMUNICATION SYSTEMS

The mobile networks considered so far have been characterized by wide area coverage and the ability of the mobile unit to both initiate and receive calls. This means that it is necessary for the fixed network to keep a record of the location of individual mobile units and to handle the problems associated with mobility, such as handover. This clearly adds significant infrastructure and operating costs to both analogue and digital systems. A notable reduction in these costs can be achieved if the area of coverage is limited (i.e. low power terminals are possible) and the mobile is restricted to outgoing calls only. Further reductions in cost are possible if handover facilities are not provided. This is the principle of CT2 Telepoint, which allows the mobile to initiate calls by establishing access to the PSDN/ISDN via suitable base stations. The service available is reduced relative to that provided by full cellular networks but is still an attractive option because users of mobile telephones tend to initiate many more calls than they receive.

A fundamental feature of cordless communications is the Telepoint concept. This is essentially an extension of the fixed part (i.e. the base station) of the common domestic cordless telephone to handle a large number of mobile handsets using digital technology. The Telepoint service is a cordless payphone accessed by a portable terminal which is small enough to be carried at all times by the owner. The Telepoint base station is effectively an access point to the fixed network with the supporting administration and billing systems. The basic Telepoint concept is shown in Fig. 15.25 and it is interesting to note that both up and down channels of a particular mobile use the same carrier frequency. This means that data is compressed and transmitted in what is termed time division duplex (TDD) mode.

CT2 was the first cordless technology to reach the market and has been adopted as an interim European Telecommunications Standard prior to the availability of the Digital European Cordless Telecommunications Standard (DECT). It is necessary to understand that although CT2 was initially considered as a very basic service it is capable of development and may well, in future, be developed to support full mobility. The DECT standard

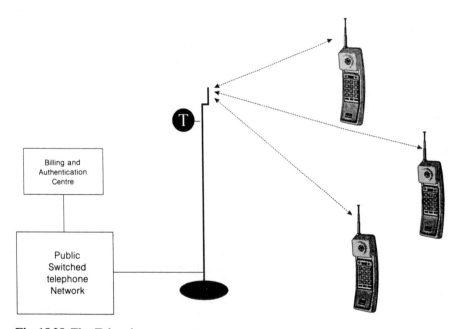

Fig. 15.25 The Telepoint concept.

supports many features which CT2 does not, including handover. DECT is primarily designed for the business environment, but it too is capable of considerable development although it is unlikely to provide the same degree of mobility as cellular systems.

15.25 CT2 SPECTRUM AND RADIO INTERFACE PARAMETERS

The band allocated to the service in the UK extends approximately from 864 MHz to 868 MHz, the channel 1 carrier being assigned to 864.15 MHz and the highest frequency carrier, channel 40, to 868.05 MHz. The nominal spacing between carriers is set at 100 kHz to support single channel per carrier, time division duplex operation.

The modulation format specified for CT2 is effectively GMSK with a bit rate of 72 kb/s. With this form of modulation the cost of the mobile units is kept low. The transmitters in the handsets are limited to a maximum output power of 10 mW although manufacturers may opt for a lower power down to a minimum of 1 mW. Provision is made for a further low-power setting at which the handset (or cordless portable part, CPP) can operate under instructions from the base (or cordless fixed part, CFP). The low-power setting is 16 dB $\pm$ 4 dB below the normal output.

Figure 15.26 illustrates the basic operation of the ping-pong time division duplex (TDD) process. Speech in both directions is sampled and coded at 32 kb/s. The coded voice is then transmitted at 72 kb/s which permits time compression into 1 ms bursts (or packets). There is a choice available to manufacturers to offer either 1 kb/s or 2 kb/s signalling and this is achieved by permitting two types of multiplex in the traffic channel.

15.26 CT2 CHANNELS AND MULTIPLEXING

CT2 relies on dynamic channel selection which means that the CT2 terminals must exchange signalling information to set up the voice channels before the principal traffic can be carried. However, in time division duplex operation, additional activity takes place which relates to bit and burst synchronization. This is essentially a layer 1 function of the OSI model (Figure 13.21) and, in CT2, the signalling responsible for channel selection and link initiation is

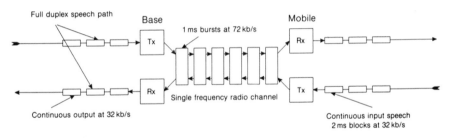

Fig. 15.26 Time division duplex transmission mode.

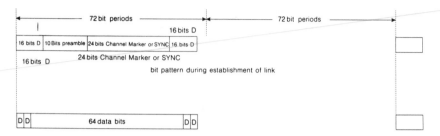

Fig. 15.27 Packet format during and after link set-up.

accommodated by dividing the digital traffic in each time division duplex frame into three logical channels: the D channel for signalling, the B channel for voice/data traffic and the SYN channel for burst synchronization.

When a link is being established and the CFP and the CPP are obtaining synchronization, there is no B channel and the pattern of bits in a packet is shown in Fig. 15.27. When a link has been established each packet contains 64 speech bits (representing 2 ms blocks of speech) and this pattern is also shown in Fig. 15.27. Clearly timing is all important in the TDD system and the CFP always ultimately takes responsibility for timing. This means that, when a portable (CPP) initiates a call, the base station (CFP) must reinitiate call establishment from the fixed end in order to impose its timing on the subsequent link traffic framing.

Since CT2 does not support a terminal registration all calls must begin by the portable acquiring access to a base station. This process requires a different multiplexing arrangement which takes account of two considerations:

1. the CPP on initiating the call has not acquired any bit or burst synchronization with base station activity;
2. because the CFPs and CPPs operate in time division duplex, the base station receivers cannot detect incoming signals from portables requesting access while the base transmitters are active.

To ensure that appropriate access requests are received by the base stations during the 1 ms periods when the base transmitters are silent, the CPP repeats a sequence of pre-amble bits followed by synchronization words for a series of five complete 144 bit frames followed immediately by two frame intervals during which the CPP listens for a response from the CFP. This procedure is necessary to guarantee that the base station receiver has an opportunity to detect the call request from the CPP and to respond.

15.27 CT2 VOICE CODING

The coder chosen for CT2 is the 32 kb/s adaptive differential pulse code modulation (ADPCM) standard (see Section 3.12) specified by CCITT (Recommendation G721, 1988). This clearly has a higher bandwidth than the

13 kb/s codec specified for GSM which uses a regular pulse excitation algorithm with a long-term predictor (RPE-LTP).

The two basic reasons for this choice are that the processing delay of the ADPCM coder is much less than the GSM equivalent and that such a coder is already widely available as a low-cost integrated circuit. Further, the speech quality is high and the coding is robust to radio-path variations. Processing delay is a major issue because CT2 remains essentially a cordless extension to the fixed network and therefore must conform to line system standards which permit a maximum round trip delay of 5 ms in the speech path. The TDD transmission scheme introduces a 1 ms delay and currently only the ADPCM codec can keep the speech processing element within the remaining 4 ms permitted. The issue of compatibility with line rather than cellular radio standards is important because of the fundamental difference between the Telepoint service and cellular radio.

15.28 THE DIGITAL EUROPEAN CORDLESS TELECOMMUNICATIONS (DECT) STANDARD

CT2 is a system designed primarily for voice communications. The main objective of the Digital European Cordless Telecommunications Standard, currently being produced by members of the European Telecommunications Standards Institute (ETSI), is to support a range of applications such as residential cordless telephone systems, business systems, public access networks (Telepoint) and radio local area networks. In addition DECT provides a system specification for both voice and non-voice applications and supports ISDN functions. DECT is a multi-frequency, TDMA-TDD cordless telecommunication system, however, the radio interface is significantly different to CT2 and DECT supports handover.

DECT has a layered structure similar to that of the OSI model and is divided into a control plane (for signalling data) and a user plane (for user

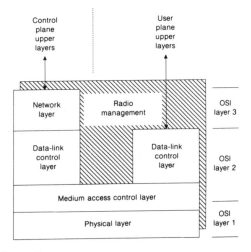

Fig. 15.28 DECT layered structure.

data). The layered description of the DECT standard is shown in Fig. 15.28. It should be noted from this figure that the DECT structure uses four layers for communication, between a DECT terminal and the DECT network, whereas the OSI model uses only three layers. The main reason for this discrepancy is that the OSI model does not adequately provide for multiple access to a particular transmission medium. (End-to-end communication is dealt with by layers above the network layer in both DECT and OSI.)

15.29   THE DECT PHYSICAL LAYER

The physical layer deals with dividing the radio transmission into physical channels. Its functions are as follows:

(i) to modulate and demodulate carriers with a defined bit rate;
(ii) to create physical channels with fixed throughput;
(iii) to activate physical channels on request of the MAC layer;
(iv) to recognize attempts to establish a physical channel;
(v) to acquire and maintain synchronization between transmitters and receivers;
(vi) to monitor the status of physical channels (field strength, quality etc.) for radio control.

In the DECT system ten carrier frequencies have been allocated in the band 1880 MHz to 1900 MHz the spacing between each carrier being 1.728 MHz. Each carrier is divided into 24 timeslots occupying a period of 10 ms, the duration of each slot being approximately 416.7 µs. As with CT2, DECT employs time division duplex transmission with slots 0 to 11 being used for base station to handset and paired slots 12 to 23 being used for the reverse direction. However, in DECT both slots in a pair can be used for transmission in one direction in response to unsymmetrical UP and DOWN traffic. Each slot transmits bursts of 420 bits in an interval of 364.6 µs which is 52.1 µs shorter than the slot duration. This guard space allows for timing errors and

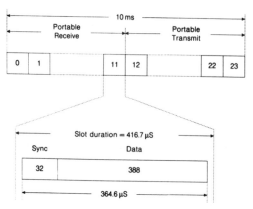

Fig. 15.29 DECT physical layer.

propagation dispersion. The sequence of one burst every 10 ms constitutes one physical channel which represents a mean bit rate of 42 kb/s. In fact voice is digitized at 32 kb/s ADPCM and compressed buffered and transmitted at 1152 kb/s, which is a considerably higher rate than the CT2 equivalent. The structure of the DECT physical layer is shown in Fig. 15.29.

15.30 DECT MEDIUM ACCESS CONTROL

The medium access control (MAC) allocates radio resource by dynamically activating and deactivating the physical channels which must accommodate the signalling channel (C-channel), the user information channel (I-channel), the paging channel (P-channel), the handshake channel (N-channel) and the broadcast channel (Q-channel). In addition the MAC layer invokes whatever error protection is appropriate for the service (in the case of speech there is no error protection). It should be noted that DECT differs from the standard OSI model in assigning error protection to the MAC layer which is the most efficient way of treating individual radio links.

The multiplex scheme used during a normal telephone conversation is shown in Fig. 15.30. In order to lock on to a particular base station the portable must verify the identity of the base station and receive call set-up parameters. This information is broadcast on the Q channel. The paging channel is used by the base station to initiate network-originated calls and is therefore broadcast to all mobiles in a cell. The signalling information on the C channel is for a specific mobile. The N channel is used to exchange identities of the portable and base stations at regular intervals. The multiplexed C, P, N and Q channels are transmitted in 48 bits of each burst, and capacity is allocated on demand while a minimum capacity for each channel is guaranteed. These 48 bits are protected by a 16 bit CRC and if transmission errors result an automatic repeat request (ARQ) procedure is used. The X bits in the packet are used to recognize partial interference in the I-channel independently of the user service.

15.31 DECT CALL MANAGEMENT AND DYNAMIC CHANNEL SELECTION

The DECT base station consists of one single radio transceiver that can change frequency from slot to slot. Hence each slot operates independently and can use any of the 10 DECT carriers. Thus unlike GSM, which uses a fixed frequency allocation, DECT uses a dynamic channel selection (DCS) procedure.

A physical channel is a combination of any of the DECT timeslots and any of the DECT carrier frequencies and every base station transmits on at least one channel, known as the beacon (when several channels are active there are an equal number of beacons in the cell). All active channels broadcast system information and base station identification. When a portable (known as the cordless portable part CPP) is in the idle mode it scans for the beacon of a

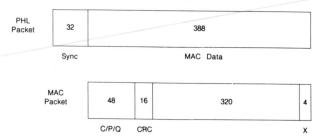

Fig. 15.30 DECT MAC layer.

nearby base station (known as the radio fixed part RFP) and locks onto the strongest channel. In the idle state the portable listens at 160 ms intervals for a possible paging call from the system. If the signal level drops below a fixed threshold the mobile will scan for another beacon and will lock onto one of appropriate strength.

In order for the portable to initiate a call set-up a number of exchanges are required between the CPP and the RFP and these are shown in Fig.15.31. The sequence of events is as follows:

1. An OUTGOING CALL REQUEST is transmitted by the CPP on a single channel which has been selected on the criteria of minimum interference. This is effectively the DCS procedure in which the portable selects a free channel with the minimum interference level. The transmission includes a field identifying the number of physical channels that the CPP envisages the call will require.
2. If the RFP receives this request and if the channel is free, half a TDM frame later (5 ms) the RFP transmits an OUTGOING CALL CONFIRMATION packet.

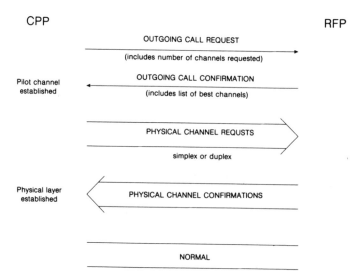

Fig. 15.31 Mobile-initiated call set-up sequence.

A 'pilot' link has now been established between the CPP and RFP which may occupy either a half or full rate physical channel. This link is sufficient for voice communications and is always duplex regardless of whether it is used for voice or data communications. Through this pilot link further physical channels can be activated so that the connection can support higher data rate services. If further physical channels are to be activated then the RFP will transmit a list of its available channels to the CPP.

3. The CPP may generate further physical channel requests with a high probability of successful confirmation by combining the RFP's channel information with its own signal strength measurements into a map.

   With reference to this map the CPP will transmit PHYSICAL CHANNEL REQUEST packets on sufficient channels to satisfy the link capacity that is required. The requests may be on half or full rate physical channels and include information as to whether they will be used as simplex or duplex channels.

   (i) SIMPLEX: When the channel is to be used as a simplex link then both uplink and downlink sections of the frame are used for transmission in one direction only, e.g. CPP to RFP. The ability to activate simplex physical channels permits efficient spectrum allocation when assigning several physical channels for a high data rate asymmetric connection. Data calls are often asymmetric in their bearer capacity requirements and simplex physical channels permit the DECT radio interface to reflect this.

   (ii) DUPLEX: Up and downlink are used for CPP to RFP and RFP to CPP transmissions, respectively, as with voice transmissions.

4. The RFP will transmit PHYSICAL CHANNEL CONFIRMATIONs on all the channels that it has received a request on and are acceptable to the RFP.

The procedure for network-originated calls is similar with the addition of a paging transmission to initially alert the CPP that a connection is required. This paging channel, which is a broadcast channel, will be multiplexed onto the beacon to which portable is locked. A suitable communications channel has now been established between the CPP and RFP for the connection.

15.32 HANDOVER IN DECT

DECT is designed for relatively small cells and supports handover between cells. The emphasis in DECT is for the handover procedure to be rapid without any interruption of service. While the portable is communicating on a particular channel it scans the other channels and records the free channels and identities of base stations that are stronger than the one it is currently using. Handover is initiated as soon as another base station is stronger than the current one in use.

   The current link is maintained in one timeslot whilst the new link is set up, in parallel, in another timeslot. When the new link is established the new base station requests the central control to make a switch from the old to new base station. As the old and new channels both exist in parallel on different time-slots, there is no break in service during the handover period. The DCS system is designed such that handover is completed before a significant loss of quality occurs.

15.33 CONCLUSIONS

This chapter has introduced the essentials of mobile communications and, in particular, has emphasized the design trade-offs in engineered systems. Unlike communications on the fixed network, mobile systems must exist within a hostile radio environment and this aspect determines the major design compromises. The chapter has also emphasized the importance of international standards, and European companies and administrations have played a leading role in this field.

Third generation universal mobile telecommunications are already being researched in detail, and many papers have already been published on this topic.[4] The accent of these programmes is the specification of a universal mobile telecommunications system (UMTS) in which a single portable can be used in any environment from rural macro-cells to indoor pico-cells. The mobile systems of the future will also offer many services such as voice, data and video transmission and will be an area of intense activity with a thrust towards global coverage.

There are many topics which have not been covered in this chapter, for example the fixed network signalling, the role of low earth orbit satellites, alternatives to TDMA (such as code division multiple access (CDMA)), etc. However it has been the objective of this chapter to introduce the fundamental elements of mobile communications which will make the study of these other topics more meaningful.

REFERENCES

1. Lee, W. C. Y., *Mobile Cellular Communication Systems*, McGraw-Hill, New York, 1989.
2. Parsons, J. D. and Gardiner, J. G., *Mobile Communication Systems*, Blackie, Glasgow, 1989.
3. Gardiner, J. G., "Second generation cordless (CT-2) telephony in the UK: telepoint services and the common air-interface", *Electron. & Commun. Eng. J.*, 1990, Vol. 2, No. 2, pp.71–78.
4. Dunlop, J., Cosimini, P., Maclean, D., Robertson, D. and Aitken, D., "A reservation based access mechanism for 3rd generation cellular systems", *Electron. & Commun. Eng. J.*, 1993, Vol. 5, No. 3, pp. 180–186.

PROBLEMS

15.1 A busy city centre traffic area occupies an area of 12 km^2. This area contains two six-lane motorways of mean length 3.5 km and 300 km of two-lane roads. Assume that the average spacing between vehicles in the rush hour is 10 m and that 45% of the vehicles are equipped with mobile telephones. Assume also that 80% of the cars make a call in the busy hour and that the average call duration is 1.9 minutes.

If the blocking probability is 2% find the offered load and the cell radius required for a cellular system with a cluster size of 7, if 300 channels are available.

What is the cell radius required if a cluster size of 4 is employed?

Answer 731.9 erlangs, 458 m, 618 m.

15.2 Assuming a hexagonal cell layout and a cluster size of 7, determine the reduction in carrier to interference ratio which occurs if second tier co-channel interference is taken into account. Repeat the calculation for a cluster size of 4.

Answer 0.5 dB, 0.5 dB.

15.3 Derive an expression for the 'antenna height gain' in a typical mobile system. In a cellular system the base station is operating at a frequency of 900 MHz. The base station antenna is at a height of 15 m and a mobile antenna is at a height of 3 m above the earth's surface. Assuming that $\sin q = q$ for $q < 0.6$ radian, find the minimum separation between mobile and base station which will ensure a 4th power attenuation characteristic.

If the base station antenna height is reduced to 10 m find the increase in ERP necessary to maintain the same value of received power at the mobile antenna.

Answer 2.86 km, 3.5 dB.

15.4 The mean C/I specified for acceptable operation in an analogue cellular system is normally assumed to be 18 dB. Derive an expression for the actual C/I experienced by a mobile at the limit of its serving cell, in terms of the first tier of interfering cells.

Assuming that all interfering cells produce the maximum interference calculate the worst case C/I if a cluster size of 7 is used. Determine the cluster size required to ensure that the worst case C/I would not be less than 18 dB (ignore multipath effects).

Answer 12.

15.5 A restricted GSM system is allocated a maximum of 40 paired carriers and operates at a fixed carrier to interference ratio of 12 dB. Assuming that one *physical* channel per cell is allocated to signalling calculate the cell radius required to support a user population of 26 561 in an area of 10 km² if the blocking probability is fixed at 2%. It is to be assumed that 78% of the users make a call of mean duration 1.7 minutes in the busy hour.

Answer 631 m.

15.6 A large number of nodes using an ALOHA access mechanism generates 50 requests per second; this figure includes new and re-scheduled packets. The time axis is slotted in units of 40 ms.

(a) Find the probability of a successful transmission at the first attempt.
(b) What is the probability of exactly four collisions followed by a successful transmission?
(c) Find the average number of transmission attempts per packet.

Answer 0.135, 0.0756, 7.39.

15.7 Minimum shift keying is a development of FSK designed to minimize bandwidth. If 10 MHz is used for binary 0 suggest a suitable carrier frequency for binary 1. Find the bandwidth of a MSK transmission if the carrier is switched periodically between 1 and 0 at a rate of 5000 times per second.

Answer 10.01 MHz, 30 kHz.

15.8 The duration of a random access channel slot in GSM is 0.577 ms and the periodicity is 4.615 ms. Calculate the number of accesses per second per cell which would be necessary for the average number of retransmissions attempts to be 2. Calculate also the mean delay in such a system between the initial transmission of an access burst and the reception of an acknowledgement on the access grant channel. Assume all mobiles are 1 km from the base station and that the capture effect may be ignored.

Answer 162, 12.7 ms.

APPENDIX A

FOUR FIGURE BESSEL FUNCTIONS

β	$J_0(\beta)$	$J_1(\beta)$	$J_2(\beta)$	$J_3(\beta)$	$J_4(\beta)$	$J_5(\beta)$	$J_6(\beta)$	$J_7(\beta)$	$J_8(\beta)$	$J_9(\beta)$	$J_{10}(\beta)$	$J_{11}(\beta)$	$J_{12}(\beta)$	$J_{13}(\beta)$	$J_{14}(\beta)$	$J_{15}(\beta)$	$J_{16}(\beta)$	$J_{17}(\beta)$	$J_{18}(\beta)$	$J_{19}(\beta)$	$J_{20}(\beta)$	$J_{21}(\beta)$	$J_{22}(\beta)$	$J_{23}(\beta)$	$J_{24}(\beta)$	$J_{25}(\beta)$
0.20	0.9900	0.0995	0.0050	0.0002																						
0.40	0.9604	0.1960	0.0197	0.0013																						
0.60	0.9120	0.2867	0.0437	0.0044	0.0003																					
0.80	0.8463	0.3688	0.0758	0.0102	0.0010																					
1.00	0.7652	0.4401	0.1149	0.0196	0.0025	0.0002																				
1.25	0.6459	0.5106	0.1711	0.0369	0.0059	0.0007																				
1.50	0.5118	0.5579	0.2321	0.0610	0.0118	0.0018	0.0002																			
1.75	0.3690	0.5802	0.2940	0.0919	0.0209	0.0038	0.0006																			
2.00	0.2239	0.5767	0.3528	0.1289	0.0340	0.0070	0.0012	0.0002																		
2.50	-.0484	0.4971	0.4461	0.2166	0.0738	0.0195	0.0042	0.0008	0.0001																	
3.00	-.2601	0.3391	0.4861	0.3091	0.1320	0.0430	0.0114	0.0025	0.0005																	
4.00	-.3971	-.0660	0.3641	0.4302	0.2811	0.1321	0.0491	0.0152	0.0040	0.0009	0.0002															
5.00	-.1776	-.3276	0.0466	0.3648	0.3912	0.2611	0.1310	0.0534	0.0184	0.0055	0.0015	0.0004														
6.00	0.1506	-.2767	-.2429	0.1148	0.3576	0.3621	0.2458	0.1296	0.0565	0.0212	0.0070	0.0020	0.0005													
7.00	0.3001	-.0047	-.3014	-.1676	0.1578	0.3479	0.3392	0.2336	0.1280	0.0589	0.0235	0.0083	0.0027	0.0008	0.0002											
8.00	0.1717	0.2346	-.1130	-.2911	-.1054	0.1858	0.3376	0.3206	0.2235	0.1263	0.0608	0.0256	0.0096	0.0033	0.0010	0.0003										
9.00	-.0903	0.2453	0.1448	-.1809	-.2655	-.0550	0.2043	0.3275	0.3051	0.2149	0.1247	0.0622	0.0274	0.0108	0.0039	0.0013	0.0004	0.0001								
10.00	-.2459	0.0435	0.2546	0.0584	-.2196	-.2341	-.0145	0.2167	0.3179	0.2919	0.2075	0.1231	0.0634	0.0290	0.0120	0.0045	0.0016	0.0005	0.0002							
11.00	-.1712	-.1768	0.1390	0.2273	-.0150	-.2383	-.2016	0.0184	0.2250	0.3089	0.2804	0.2010	0.1216	0.0643	0.0304	0.0130	0.0051	0.0019	0.0006	0.0002						
12.00	0.0477	-.2234	-.0849	0.1951	0.1825	-.0735	-.2437	-.1703	0.0451	0.2304	0.3005	0.2704	0.1953	0.1201	0.0650	0.0316	0.0140	0.0057	0.0022	0.0008	0.0003					
13.00	0.2069	-.0703	-.2177	0.0033	0.2193	0.1316	-.1180	-.2406	-.1410	0.0670	0.2338	0.2927	0.2615	0.1901	0.1188	0.0656	0.0327	0.0149	0.0063	0.0025	0.0009	0.0003				
14.00	0.1711	0.1334	-.1520	-.1768	0.0762	0.2204	0.0812	-.1508	-.2320	-.1143	0.0850	0.2357	0.2855	0.2536	0.1855	0.1174	0.0661	0.0337	0.0158	0.0068	0.0028	0.0010	0.0004	0.0001		
15.00	-.0142	0.2051	0.0416	-.1940	-.1192	0.1305	0.2061	0.0345	-.1740	-.2200	-.0901	0.1000	0.2367	0.2787	0.2464	0.1813	0.1162	0.0665	0.0346	0.0166	0.0074	0.0031	0.0012	0.0004	0.0002	
16.00	-.1749	0.0904	0.1862	-.0438	-.2026	-.0575	0.1667	0.1825	-.0070	-.1895	-.2062	-.0682	0.1124	0.2368	0.2724	0.2399	0.1775	0.1150	0.0668	0.0354	0.0173	0.0079	0.0034	0.0013	0.0005	0.0002
17.00	-.1699	-.0977	0.1584	0.1349	-.1107	-.1870	0.0007	0.1875	0.1537	-.0429	-.1991	-.1914	-.0486	0.1228	0.2364	0.2666	0.2340	0.1739	0.1138	0.0671	0.0362	0.0180	0.0084	0.0037	0.0015	0.0006
18.00	-.0134	-.1880	-.0075	0.1863	0.0696	-.1554	-.1560	0.0514	0.1959	0.1228	-.0732	-.2041	-.1762	-.0309	0.1316	0.2356	0.2611	0.2286	0.1706	0.1127	0.0673	0.0369	0.0187	0.0089	0.0039	0.0017
19.00	0.1466	-.1057	-.1578	0.0725	0.1806	0.0036	-.1788	-.1165	0.0929	0.1947	0.0916	-.0984	-.2055	-.1612	-.0151	0.1389	0.2345	0.2559	0.2235	0.1676	0.1116	0.0675	0.0375	0.0193	0.0093	0.0042
20.00	0.1670	0.0668	-.1603	-.0989	0.1307	0.1512	-.0551	-.1842	-.0739	0.1251	0.1865	0.0614	-.1190	-.2041	-.1464	-.0008	0.1452	0.2331	0.2511	0.2189	0.1647	0.1106	0.0676	0.0380	0.0199	0.0098

APPENDIX B

USEFUL TRIGONOMETRIC IDENTITIES

$$\sin(A + B) = \sin A \cos B + \cos A \sin B$$

$$\sin(A - B) = \sin A \cos B - \cos A \sin B$$

$$\cos(A + B) = \cos A \cos B - \sin A \sin B$$

$$\cos(A - B) = \cos A \cos B + \sin A \sin B$$

$$\sin 2A = 2\sin A \cos A$$

$$\cos 2A = \cos^2 A - \sin^2 A$$

$$\sin A + \sin B = 2\sin\left(\frac{A + B}{2}\right)\cos\left(\frac{A - B}{2}\right)$$

$$\sin A - \sin B = 2\cos\left(\frac{A + B}{2}\right)\sin\left(\frac{A - B}{2}\right)$$

$$\cos A + \cos B = 2\cos\left(\frac{A + B}{2}\right)\cos\left(\frac{A - B}{2}\right)$$

$$\cos A - \cos B = 2\sin\left(\frac{A + B}{2}\right)\cos\left(\frac{B - A}{2}\right)$$

$$\sin A \sin B = \tfrac{1}{2}[\cos(A - B) - \cos(A + B)]$$

$$\sin A \cos B = \tfrac{1}{2}[\sin(A + B) + \sin(A - B)]$$

$$\cos A \cos B = \tfrac{1}{2}[\cos(A + B) + \cos(A - B)]$$

$$\cos A \sin B = \tfrac{1}{2}[\sin(A + B) - \sin(A - B)]$$

$$\sin^2 A = \tfrac{1}{2}(1 - \cos 2A)$$

$$\cos^2 A = \tfrac{1}{2}(1 + \cos 2A)$$

$$\exp(j\omega t) = \cos \omega t + j\sin \omega t$$

$$\exp(-j\omega t) = \cos \omega t - j\sin \omega t$$

$$2\cos \omega t = \exp(j\omega t) + \exp(-j\omega t)$$

$$2\sin \omega t = \frac{1}{j}[\exp(j\omega t) - \exp(-j\omega t)]$$

APPENDIX C

NORMAL ERROR FUNCTION

$$\text{erf}(x) = \frac{2}{\sqrt{\pi}} \int_0^x \exp(-y^2)\,dy$$

x	erf x	x	erf x	x	erf x	x	erf x	x	erf x
0.00	0.000 000 000	0.80	0.742 100 965	1.60	0.976 348 383	2.40	0.999 311 486	3.20	0.999 993 974
0.01	0.011 283 416	0.81	0.748 003 281	1.61	0.977 206 837	2.41	0.999 346 202	3.21	0.999 994 365
0.02	0.022 564 575	0.82	0.753 810 751	1.62	0.978 038 088	2.42	0.999 379 283	3.22	0.999 994 731
0.03	0.033 841 222	0.83	0.759 523 757	1.63	0.978 842 840	2.43	0.999 410 802	3.23	0.999 995 074
0.04	0.045 111 106	0.84	0.765 142 711	1.64	0.979 621 780	2.44	0.999 440 826	3.24	0.999 995 396
0.05	0.056 371 978	0.85	0.770 668 058	1.65	0.980 375 585	2.45	0.999 469 420	3.25	0.999 995 697
0.06	0.067 621 594	0.86	0.776 100 268	1.66	0.981 104 921	2.46	0.999 496 646	3.26	0.999 995 980
0.07	0.078 857 720	0.87	0.781 439 845	1.67	0.981 810 442	2.47	0.999 522 566	3.27	0.999 996 245
0.08	0.090 078 126	0.88	0.786 687 319	1.68	0.982 492 787	2.48	0.999 547 236	3.28	0.999 996 493
0.09	0.101 280 594	0.89	0.791 843 247	1.69	0.983 152 587	2.49	0.999 570 712	3.29	0.999 996 725
0.10	0.112 462 916	0.90	0.796 908 212	1.70	0.983 790 459	2.50	0.999 593 048	3.30	0.999 996 942
0.11	0.123 622 896	0.91	0.801 882 826	1.71	0.984 407 008	2.51	0.999 614 295	3.31	0.999 997 146
0.12	0.134 758 352	0.92	0.806 767 722	1.72	0.985 002 827	2.52	0.999 634 501	3.32	0.999 997 336
0.13	0.145 867 115	0.93	0.811 563 559	1.73	0.985 578 500	2.53	0.999 653 714	3.33	0.999 997 515
0.14	0.156 947 033	0.94	0.816 271 019	1.74	0.986 134 595	2.54	0.999 671 979	3.34	0.999 997 681
0.15	0.167 995 971	0.95	0.820 890 807	1.75	0.986 671 671	2.55	0.999 689 340	3.35	0.999 997 838
0.16	0.179 011 813	0.96	0.825 423 650	1.76	0.987 190 275	2.56	0.999 705 837	3.36	0.999 997 983
0.17	0.189 992 461	0.97	0.829 870 293	1.77	0.987 690 942	2.57	0.999 721 511	3.37	0.999 998 120
0.18	0.200 935 839	0.98	0.834 231 504	1.78	0.988 174 196	2.58	0.999 736 400	3.38	0.999 998 247
0.19	0.211 839 892	0.99	0.838 508 070	1.79	0.988 640 549	2.59	0.999 750 539	3.39	0.999 998 367

APPENDIX C (Continued)

x	erf x	x	erf x	x	erf x	x	erf x	x	erf x
0.20	0.222 702 589	1.00	0.842 700 793	1.80	0.989 090 502	2.60	0.999 763 966	3.40	0.999 998 478
0.21	0.233 521 923	1.01	0.846 810 496	1.81	0.989 524 545	2.61	0.999 776 711	3.41	0.999 998 582
0.22	0.244 295 912	1.02	0.850 838 018	1.82	0.989 943 156	2.62	0.999 788 809	3.42	0.999 998 679
0.23	0.255 022 600	1.03	0.854 784 211	1.83	0.990 346 805	2.63	0.999 800 289	3.43	0.999 998 770
0.24	0.265 700 059	1.04	0.858 649 947	1.84	0.990 735 948	2.64	0.999 811 181	3.44	0.999 998 855
0.25	0.276 326 390	1.05	0.862 436 106	1.85	0.991 111 030	2.65	0.999 821 512	3.45	0.999 998 934
0.26	0.286 899 723	1.06	0.866 143 587	1.86	0.991 472 488	2.66	0.999 831 311	3.46	0.999 999 008
0.27	0.297 418 219	1.07	0.869 773 297	1.87	0.991 820 748	2.67	0.999 840 601	3.47	0.999 999 077
0.28	0.307 880 068	1.08	0.873 326 158	1.88	0.992 156 223	2.68	0.999 849 409	3.48	0.999 999 141
0.29	0.318 283 496	1.09	0.876 803 102	1.89	0.992 479 318	2.69	0.999 857 757	3.49	0.999 999 201
0.30	0.328 626 759	1.10	0.880 205 070	1.90	0.992 790 429	2.70	0.999 865 667	3.50	0.999 999 257
0.31	0.338 908 150	1.11	0.883 533 012	1.91	0.993 089 940	2.71	0.999 873 162	3.51	0.999 999 309
0.32	0.349 125 995	1.12	0.886 787 890	1.92	0.993 378 225	2.72	0.999 880 261	3.52	0.999 999 358
0.33	0.359 278 655	1.13	0.889 970 670	1.93	0.993 655 650	2.73	0.999 886 985	3.53	0.999 999 403
0.34	0.369 364 529	1.14	0.893 082 328	1.94	0.993 922 571	2.74	0.999 893 351	3.54	0.999 999 445
0.35	0.379 382 054	1.15	0.896 123 843	1.95	0.994 179 334	2.75	0.999 899 378	3.55	0.999 999 485
0.36	0.389 329 701	1.16	0.899 096 203	1.96	0.994 426 275	2.76	0.999 905 082	3.56	0.999 999 521
0.37	0.399 205 984	1.17	0.902 000 399	1.97	0.994 663 725	2.77	0.999 910 480	3.57	0.999 999 555
0.38	0.409 009 453	1.18	0.904 837 427	1.98	0.994 892 000	2.78	0.999 915 587	3.58	0.999 999 587
0.39	0.418 738 700	1.19	0.907 608 286	1.99	0.995 111 413	2.79	0.999 920 418	3.59	0.999 999 617
0.40	0.428 392 355	1.20	0.910 313 978	2.00	0.995 322 265	2.80	0.999 924 987	3.60	0.999 999 644
0.41	0.437 969 090	1.21	0.912 955 508	2.01	0.995 524 849	2.81	0.999 929 307	3.61	0.999 999 670
0.42	0.447 467 618	1.22	0.915 533 881	2.02	0.995 719 451	2.82	0.999 933 390	3.62	0.999 999 694
0.43	0.456 886 695	1.23	0.918 050 104	2.03	0.995 906 348	2.83	0.999 937 250	3.63	0.999 999 716
0.44	0.466 225 115	1.24	0.920 505 184	2.04	0.996 085 810	2.84	0.999 940 898	3.64	0.999 999 736
0.45	0.475 481 720	1.25	0.922 900 128	2.05	0.996 258 096	2.85	0.999 944 344	3.65	0.999 999 756
0.46	0.484 655 390	1.26	0.925 235 942	2.06	0.996 423 462	2.86	0.999 947 599	3.66	0.999 999 773
0.47	0.493 745 051	1.27	0.927 513 629	2.07	0.996 582 153	2.87	0.999 950 673	3.67	0.999 999 790
0.48	0.502 749 671	1.28	0.929 734 193	2.08	0.996 734 409	2.88	0.999 953 576	3.68	0.999 999 805
0.49	0.511 668 261	1.29	0.931 898 633	2.09	0.996 880 461	2.89	0.999 956 316	3.69	0.999 999 820

Leftmost block (x = 0.50–0.79):

x	value		
0.50	0.520	499	878
0.51	0.529	243	620
0.52	0.537	898	630
0.53	0.546	464	097
0.54	0.554	939	250
0.55	0.563	323	366
0.56	0.571	615	764
0.57	0.579	815	806
0.58	0.587	922	900
0.59	0.595	936	497
0.60	0.603	856	091
0.61	0.611	681	219
0.62	0.619	411	462
0.63	0.627	046	443
0.64	0.634	585	829
0.65	0.642	029	327
0.66	0.649	376	688
0.67	0.656	627	702
0.68	0.663	782	203
0.69	0.670	840	062
0.70	0.677	801	194
0.71	0.684	665	550
0.72	0.691	433	123
0.73	0.698	103	943
0.74	0.704	678	078
0.75	0.711	155	634
0.76	0.717	536	753
0.77	0.723	821	614
0.78	0.730	010	431
0.79	0.736	103	454

Second block (x = 1.30–1.59):

x	value		
1.30	0.934	007	945
1.31	0.936	063	123
1.32	0.938	065	155
1.33	0.940	015	026
1.34	0.941	913	715
1.35	0.943	762	196
1.36	0.945	561	437
1.37	0.947	312	398
1.38	0.949	016	035
1.39	0.950	673	296
1.40	0.952	285	120
1.41	0.953	852	439
1.42	0.955	376	179
1.43	0.956	857	253
1.44	0.958	296	570
1.45	0.959	695	026
1.46	0.961	053	510
1.47	0.962	372	900
1.48	0.963	654	065
1.49	0.964	897	865
1.50	0.966	105	146
1.51	0.967	276	748
1.52	0.968	413	497
1.53	0.969	516	209
1.54	0.970	585	690
1.55	0.971	622	733
1.56	0.972	628	122
1.57	0.973	602	627
1.58	0.974	547	009
1.59	0.975	462	016

Third block (x = 2.10–2.39):

x	value		
2.10	0.997	020	533
2.11	0.997	154	845
2.12	0.997	283	607
2.13	0.997	407	023
2.14	0.997	525	293
2.15	0.997	638	607
2.16	0.997	747	152
2.17	0.997	851	108
2.18	0.997	950	649
2.19	0.998	045	943
2.20	0.998	137	154
2.21	0.998	224	438
2.22	0.998	307	948
2.23	0.998	387	832
2.24	0.998	464	231
2.25	0.998	537	283
2.26	0.998	607	121
2.27	0.998	673	872
2.28	0.998	737	661
2.29	0.998	798	606
2.30	0.998	856	823
2.31	0.998	912	423
2.32	0.998	965	513
2.33	0.999	016	195
2.34	0.999	064	570
2.35	0.999	110	733
2.36	0.999	154	777
2.37	0.999	196	790
2.38	0.999	236	858
2.39	0.999	275	064

Fourth block (x = 2.90–3.19):

x	value		
2.90	0.999	958	902
2.91	0.999	961	343
2.92	0.999	963	645
2.93	0.999	965	817
2.94	0.999	967	866
2.95	0.999	969	797
2.96	0.999	971	618
2.97	0.999	973	334
2.98	0.999	974	951
2.99	0.999	976	474
3.00	0.999	977	910
3.01	0.999	979	261
3.02	0.999	980	534
3.03	0.999	981	732
3.04	0.999	982	859
3.05	0.999	983	920
3.06	0.999	984	918
3.07	0.999	985	857
3.08	0.999	986	740
3.09	0.999	987	571
3.10	0.999	988	351
3.11	0.999	989	085
3.12	0.999	989	774
3.13	0.999	990	422
3.14	0.999	991	030
3.15	0.999	991	602
3.16	0.999	992	138
3.17	0.999	992	642
3.18	0.999	993	115
3.19	0.999	993	558

Rightmost block (x = 3.70–3.99):

x	value		
3.70	0.999	999	833
3.71	0.999	999	845
3.72	0.999	999	857
3.73	0.999	999	867
3.74	0.999	999	877
3.75	0.999	999	886
3.76	0.999	999	895
3.77	0.999	999	903
3.78	0.999	999	910
3.79	0.999	999	917
3.80	0.999	999	923
3.81	0.999	999	929
3.82	0.999	999	934
3.83	0.999	999	939
3.84	0.999	999	944
3.85	0.999	999	948
3.86	0.999	999	952
3.87	0.999	999	956
3.88	0.999	999	959
3.89	0.999	999	962
3.90	0.999	999	965
3.91	0.999	999	968
3.92	0.999	999	970
3.93	0.999	999	973
3.94	0.999	999	975
3.95	0.999	999	977
3.96	0.999	999	979
3.97	0.999	999	980
3.98	0.999	999	982
3.99	0.999	999	983

APPENDIX D

BLOCKED-CALLS-CLEARED (ERLANG B)

A erlangs

Blocking probability

N	1.0%	1.2%	1.5%	2%	3%	5%	7%	10%	15%	20%	30%	40%	50%
1	0.101	0.0121	0.0152	0.0204	0.0309	0.0526	0.0753	0.111	0.176	0.250	0.429	0.667	1.00
2	0.153	0.168	0.190	0.223	0.282	0.381	0.470	0.595	0.796	1.00	1.45	2.00	2.73
3	0.455	0.489	0.535	0.602	0.715	0.899	1.06	1.27	1.60	1.93	2.63	3.48	4.59
4	0.869	0.922	0.992	1.09	1.26	1.52	1.75	2.05	2.50	2.95	3.89	5.02	6.50
5	1.36	1.43	1.52	1.66	1.88	2.22	2.50	2.88	3.45	4.01	5.19	6.60	8.44
6	1.91	2.00	2.11	2.28	2.54	2.96	3.30	3.76	4.44	5.11	6.51	8.19	10.4
7	2.50	2.60	2.74	2.94	3.25	3.74	4.14	4.67	5.46	6.23	7.86	9.80	12.4
8	3.13	3.25	3.40	3.63	3.99	4.54	5.00	5.60	6.50	7.37	9.21	11.4	14.3
9	3.78	3.92	4.09	4.34	4.75	5.37	5.88	6.55	7.55	8.52	10.6	13.0	16.3
10	4.46	4.61	4.81	5.08	5.53	6.22	6.78	7.51	8.62	9.68	12.0	14.7	18.3
11	5.16	5.32	5.54	5.84	6.33	7.08	7.69	8.49	9.69	10.9	13.3	16.3	20.3
12	5.88	6.05	6.29	6.61	7.14	7.95	8.61	9.47	10.8	12.0	14.7	18.0	22.2
13	6.61	6.80	7.05	7.40	7.97	8.83	9.54	10.5	11.9	13.2	16.1	19.6	24.2
14	7.35	7.56	7.82	8.20	8.80	9.73	10.5	11.5	13.0	14.4	17.5	21.2	26.2
15	8.11	8.33	8.61	9.01	9.65	10.6	11.4	12.5	14.1	15.6	18.9	22.9	28.2
16	8.88	9.11	9.41	9.83	10.5	11.5	12.4	13.5	15.2	16.8	20.3	24.5	30.2
17	9.65	9.89	10.2	10.7	11.4	12.5	13.4	14.5	16.3	18.0	21.7	26.2	32.2
18	10.4	10.7	11.0	11.5	12.2	13.4	14.3	15.5	17.4	19.2	23.1	27.8	34.2
19	11.2	11.5	11.8	12.3	13.1	14.3	15.3	16.6	18.5	20.4	24.5	29.5	36.2
20	12.0	12.3	12.7	13.2	14.0	15.2	16.3	17.6	19.6	21.6	25.6	31.2	38.2

21	12.8	13.1	13.5	14.0	14.9	16.2	17.3	18.7	20.8	22.8	27.3	32.8	40.2
22	13.7	14.0	14.3	14.9	15.8	17.1	18.2	19.7	21.9	24.1	28.7	34.5	42.1
23	14.5	14.8	15.2	15.8	16.7	18.1	19.2	20.7	23.0	25.3	30.1	36.1	44.1
24	15.3	15.6	16.0	16.6	17.6	19.0	20.2	21.8	24.2	26.5	31.6	37.8	46.1
25	16.1	16.5	16.9	17.5	18.5	20.0	21.2	22.8	25.3	27.7	33.0	39.4	48.1
26	17.0	17.3	17.8	18.4	19.4	20.9	22.2	23.9	26.4	28.9	34.4	41.1	50.1
27	17.8	18.2	18.6	19.3	20.3	21.9	23.2	24.9	27.6	30.2	35.8	42.8	52.1
28	18.6	19.0	19.5	20.2	21.2	22.9	24.2	26.0	28.7	31.4	37.2	44.4	54.1
29	19.5	19.9	20.4	21.0	22.1	23.8	25.2	27.1	29.9	32.6	38.6	46.1	56.1
30	20.3	20.7	21.2	21.9	23.1	24.8	26.2	28.1	31.0	33.8	40.0	47.7	58.1
31	21.2	21.6	22.1	22.8	24.0	25.8	27.2	29.2	32.1	35.1	41.5	49.4	60.1
32	22.0	22.5	23.0	23.7	24.9	26.7	28.2	30.2	33.3	36.3	42.9	51.1	62.1
33	22.9	23.3	23.9	24.6	25.8	27.7	29.3	31.3	34.4	37.5	44.3	52.7	64.1
34	23.8	24.2	24.8	25.5	26.8	28.7	30.3	32.4	35.6	38.8	45.7	54.4	66.1
35	24.6	25.1	25.6	26.4	27.7	29.7	31.3	33.4	36.7	40.0	47.1	56.0	68.1
36	25.5	26.0	26.5	27.3	28.6	30.7	32.3	34.5	37.9	41.2	48.6	57.7	70.1
37	26.4	26.8	27.4	28.3	29.6	31.6	33.3	35.6	39.0	42.4	50.0	59.4	72.1
38	27.3	27.7	28.3	29.2	30.5	32.6	34.4	36.6	40.2	43.7	51.4	61.0	74.1
39	28.1	28.6	29.2	30.1	31.5	33.6	35.4	37.7	41.3	44.9	52.8	62.7	76.1
40	29.0	29.5	30.1	31.0	32.4	34.6	36.4	38.8	42.5	46.1	54.2	64.4	78.1
41	29.9	30.4	31.0	31.9	33.4	35.6	37.4	39.9	43.6	47.4	55.7	66.0	80.1
42	30.8	31.3	31.9	32.8	34.3	36.6	38.4	40.9	44.8	48.6	57.1	67.7	82.1
43	31.7	32.2	32.8	33.8	35.3	37.6	39.5	42.0	45.9	49.9	58.5	69.3	84.1
44	32.5	33.1	33.7	34.7	36.2	38.6	40.5	43.1	47.1	51.1	59.9	71.0	86.1
45	33.4	34.0	34.6	35.6	37.2	39.6	41.5	44.2	48.2	52.3	61.3	72.7	88.1
46	34.3	34.9	35.6	36.5	38.1	40.5	42.6	45.2	49.4	53.6	62.8	74.3	90.1
47	35.2	35.8	36.5	38.1	40.5	42.6	45.2	49.4	53.6	54.8	64.2	76.0	92.1
48	36.1	36.7	37.4	38.4	40.0	42.5	44.6	47.4	51.7	56.0	65.6	77.7	94.1
49	37.0	37.6	38.3	39.3	41.0	43.5	45.7	48.5	52.9	57.3	67.0	79.3	96.1
50	37.9	38.5	39.2	40.3	41.9	44.5	46.7	49.6	54.0	58.5	68.5	81.0	98.1

APPENDIX D (Continued)

A erlangs

Blocking probability

51	38.8	39.4	40.1	41.2	42.9	45.5	47.7	50.6	55.2	59.7	69.9	82.7	100.1
52	39.7	40.3	41.0	42.1	43.9	46.5	48.8	51.7	56.3	61.0	71.3	84.3	102.1
53	40.6	41.2	42.0	43.1	44.8	47.5	49.8	52.8	57.5	62.2	72.7	86.0	104.1
54	41.5	42.1	42.9	44.0	45.8	48.5	50.8	53.9	58.7	63.5	74.2	87.6	106.1
55	42.4	43.0	43.8	44.9	46.7	49.5	51.9	55.0	59.8	64.7	75.6	89.3	108.1
56	43.3	43.9	44.7	45.9	47.7	50.5	52.9	56.1	61.0	65.9	77.0	91.0	110.1
57	44.2	44.8	45.7	46.8	48.7	51.5	53.9	57.1	62.1	67.2	78.4	92.6	112.1
58	45.1	45.8	46.6	47.8	49.6	52.6	55.0	58.2	63.3	68.4	79.8	94.3	114.1
59	46.0	46.7	47.5	48.7	50.6	53.6	56.0	59.3	64.5	69.7	81.3	96.0	116.1
60	46.9	47.6	48.4	49.6	51.6	54.6	57.1	60.4	65.6	70.9	82.7	97.6	118.1
61	47.9	48.5	49.4	50.6	52.5	55.6	58.1	61.5	66.8	72.1	84.1	99.3	120.1
62	48.8	49.4	50.3	51.5	53.5	56.6	59.1	62.6	68.0	73.4	85.5	101.0	122.1
63	49.7	50.4	51.2	52.5	54.5	57.6	60.2	63.7	69.1	74.6	87.0	102.6	124.1
64	50.6	51.3	52.2	53.4	55.4	58.6	61.2	64.8	70.3	75.9	88.4	104.3	126.1
65	51.5	52.2	53.1	54.4	56.4	59.6	62.3	65.8	71.4	77.1	89.8	106.0	128.1
66	52.4	53.1	54.0	55.3	57.4	60.6	63.3	66.9	72.6	78.3	91.2	107.6	130.1
67	53.4	54.1	55.0	56.3	58.4	61.6	64.4	68.0	73.8	79.6	92.7	109.3	132.1
68	54.3	55.0	55.9	57.2	59.3	62.6	65.4	69.1	74.9	80.8	94.1	110.0	134.1
69	55.2	55.9	56.9	58.2	60.3	63.7	66.4	70.2	76.1	82.1	95.5	112.6	136.1
70	56.1	56.8	57.8	59.1	61.3	64.7	67.5	71.3	77.3	83.3	96.9	114.3	138.1
71	57.0	57.8	58.7	60.1	62.3	65.7	68.5	72.4	78.4	84.6	98.4	115.9	140.1
72	58.0	58.7	59.7	61.0	63.2	66.7	69.6	73.5	79.6	85.8	99.8	117.6	142.1
73	58.9	59.6	60.6	62.0	64.2	67.7	70.6	74.6	80.8	87.0	101.2	119.3	114.1
74	59.8	60.6	61.6	62.9	65.2	68.7	71.7	75.6	81.9	88.3	102.7	120.9	146.1
75	60.7	61.5	62.5	63.5	66.2	69.7	72.7	76.7	83.1	89.5	104.1	122.6	148.0

76	61.7	62.4	63.4	64.9	67.2	70.8	73.8	77.8	84.2	90.8	105.5	124.3	150.0
77	62.6	63.4	64.4	65.8	68.1	71.8	74.8	78.9	85.4	92.0	106.9	125.9	152.0
78	63.5	64.3	65.3	66.8	69.1	72.8	75.9	80.0	86.6	93.3	108.4	127.6	154.0
79	64.4	65.2	66.3	67.7	70.1	73.8	76.9	81.1	87.7	94.5	109.8	129.3	156.0
80	65.4	66.2	67.2	68.7	71.1	74.8	78.0	82.2	88.9	95.7	111.2	130.9	158.0
81	66.3	67.1	68.2	69.6	72.1	75.8	79.0	83.3	90.1	97.0	112.6	132.6	160.0
82	67.2	68.0	69.1	70.6	73.0	76.9	80.1	84.4	91.2	98.2	114.1	134.3	162.0
83	68.2	69.0	70.1	71.6	74.0	77.9	81.1	85.5	92.4	99.5	115.5	135.9	164.0
84	69.1	69.9	71.0	72.5	75.0	78.9	82.2	86.6	93.6	100.7	116.9	137.6	166.0
85	70.0	70.9	71.9	73.5	76.0	79.9	83.2	87.7	94.7	102.0	118.3	139.3	168.0
86	70.9	71.8	72.9	74.5	77.0	80.9	84.3	88.8	95.9	103.2	119.8	140.9	170.0
87	71.9	72.7	73.8	75.4	78.0	82.0	85.3	89.9	97.1	104.5	121.2	142.6	172.0
88	72.8	73.7	74.8	76.4	78.9	83.0	86.4	91.0	98.2	105.7	122.6	144.3	174.0
89	73.7	74.6	75.7	77.3	79.9	84.0	87.4	92.1	99.4	106.9	124.0	145.9	176.0
90	74.7	75.6	76.7	78.3	80.9	85.0	88.5	93.1	100.6	108.2	125.5	147.6	178.0
91	75.6	76.5	77.6	79.3	81.9	86.0	89.5	94.2	101.7	109.4	126.9	149.3	180.0
92	76.6	77.4	78.6	80.2	82.9	87.1	90.6	95.3	102.9	110.7	128.3	150.9	182.0
93	77.5	78.4	79.6	81.2	83.9	88.1	91.6	96.4	104.1	111.9	129.7	152.6	184.0
94	78.4	79.3	80.5	82.2	84.9	89.1	92.7	97.5	105.3	113.2	131.2	154.3	186.0
95	79.4	80.3	81.5	83.1	85.8	90.1	93.7	98.6	106.4	114.4	132.6	155.9	188.0
96	80.3	81.2	82.4	84.1	86.8	91.1	94.8	99.7	107.6	115.7	134.0	157.6	190.0
97	81.2	82.2	83.4	85.1	87.8	92.2	95.8	100.8	108.8	116.9	135.5	159.3	192.0
98	82.2	83.1	84.3	86.0	88.8	93.2	96.9	101.9	109.9	118.2	136.9	160.9	194.0
99	83.1	84.1	85.3	87.0	89.8	94.2	97.9	103.0	111.1	119.4	138.3	162.6	196.0
100	84.1	85.0	86.2	88.0	90.8	95.2	99.0	104.1	112.3	120.6	139.7	164.3	198.0
102	85.9	86.9	88.1	89.9	92.8	97.3	101.1	106.3	114.6	123.1	142.6	167.6	202.0
104	87.8	88.8	90.1	91.9	94.8	99.3	103.2	108.5	116.6	125.6	145.4	170.9	206.0
106	89.7	90.7	92.0	93.8	96.7	101.4	105.3	110.7	119.3	128.1	148.3	174.2	210.0
108	91.6	92.6	93.9	95.7	98.7	103.4	107.4	112.9	121.6	130.6	151.1	177.6	214.0
110	93.5	94.5	95.8	97.7	100.7	105.5	109.5	115.1	124.0	133.1	154.0	180.9	218.0

APPENDIX D (Continued)

A erlangs

Blocking probability

112	95.4	96.4	97.7	99.6	102.7	107.5	111.7	117.3	126.3	135.6	156.9	184.2	222.0
114	97.3	98.3	99.7	101.6	104.7	109.6	113.8	119.5	128.6	138.1	159.7	187.6	226.0
116	99.2	100.2	101.6	103.5	106.7	111.7	115.9	121.7	131.0	140.6	162.6	190.9	230.0
118	101.1	102.1	103.5	105.5	108.7	113.7	118.0	123.9	133.3	143.1	165.4	194.2	234.0
120	103.0	104.0	105.4	107.4	110.7	115.8	120.1	126.1	135.7	145.6	168.3	197.6	238.0
122	104.9	105.9	107.4	109.4	112.6	117.8	122.2	128.3	138.0	148.1	171.1	200.9	242.0
124	106.8	107.9	109.3	111.3	114.6	119.9	124.4	130.5	140.3	150.6	174.0	204.2	246.0
126	108.7	109.8	111.2	113.3	116.6	121.9	126.5	132.7	142.7	153.0	176.8	207.6	250.0
128	110.6	111.7	113.2	115.2	118.6	124.0	128.6	134.9	145.0	155.5	179.7	210.9	254.0
130	112.5	113.6	115.1	117.2	120.6	126.1	130.7	137.1	147.4	158.0	182.5	214.2	258.0
132	114.4	115.5	117.0	119.1	122.6	128.1	132.8	139.3	149.7	160.5	185.4	217.6	262.0
134	116.3	117.4	119.0	121.1	124.6	130.2	134.9	141.5	152.0	163.0	188.3	220.9	266.0
136	118.2	119.4	120.9	123.1	126.6	132.3	137.1	143.7	154.4	165.5	191.1	224.2	270.0
138	120.1	121.3	122.8	125.0	128.6	134.3	139.2	145.9	156.7	168.0	194.0	227.6	274.0
140	122.0	123.2	124.8	127.0	130.6	136.4	141.3	148.1	159.1	170.5	196.8	230.9	278.0
142	123.9	125.1	126.7	128.9	132.6	138.4	143.4	150.3	161.4	173.0	199.7	234.2	282.0
144	125.8	127.0	128.6	130.9	134.6	140.5	145.6	152.5	163.8	175.5	202.5	237.6	286.0
146	127.7	129.0	130.6	132.9	136.6	142.6	147.7	154.7	166.1	178.0	205.4	240.9	290.0
148	129.7	130.9	132.5	134.8	138.6	144.6	149.8	156.9	168.5	180.5	208.2	244.2	294.0
150	131.6	132.8	134.5	136.8	140.6	146.7	151.9	159.1	170.8	183.0	211.1	247.6	298.0
152	133.5	134.8	136.4	138.8	142.6	148.8	154.0	161.3	173.1	185.5	214.0	250.9	302.0
154	135.4	136.7	138.8	140.7	144.6	150.8	156.2	163.5	175.5	188.0	216.8	254.2	306.0
156	137.3	138.6	140.3	142.7	146.6	152.9	158.3	165.7	177.8	190.5	219.7	257.6	310.0
158	139.2	140.5	142.3	144.7	148.6	155.0	160.4	167.9	180.2	193.0	222.5	260.9	314.0
160	141.2	142.5	144.2	146.6	150.6	157.0	162.5	170.2	182.5	195.5	225.4	264.2	318.0

162	143.1	144.4	146.1	148.6	152.7	159.1	164.7	172.4	184.9	198.0	228.2	267.6	322.0
164	145.0	146.3	148.1	150.6	154.7	161.2	166.8	174.6	187.2	200.4	231.1	270.9	326.0
166	146.9	148.3	150.0	152.6	156.7	163.3	168.9	176.8	189.6	202.9	233.9	274.2	330.0
168	148.9	150.2	152.0	154.5	158.7	165.3	171.0	179.0	191.9	205.4	236.8	277.6	334.0
170	150.8	152.1	153.9	156.5	160.7	167.4	173.2	181.2	194.2	207.9	239.7	280.9	338.0
172	152.7	154.1	155.9	158.5	162.7	169.5	175.3	183.4	196.6	210.4	242.5	284.2	342.0
174	154.6	156.0	157.8	160.4	164.7	171.5	177.4	185.6	198.9	212.9	245.4	287.6	346.0
176	156.6	158.0	159.8	162.4	166.7	173.6	179.6	187.8	201.3	215.4	248.2	290.9	350.0
178	158.5	159.9	161.8	164.4	168.7	175.7	181.7	190.0	203.6	217.9	251.1	294.2	354.0
180	160.4	161.8	163.7	166.4	170.7	177.8	183.8	192.2	206.0	220.4	253.9	297.5	358.0
182	162.3	163.8	165.7	168.3	172.8	179.8	185.9	194.4	208.3	222.9	256.8	300.9	362.0
184	194.3	165.7	167.6	170.3	174.8	181.9	188.1	196.6	210.7	225.4	259.6	304.2	366.0
186	166.2	167.7	169.6	172.3	176.8	184.0	190.2	198.9	213.0	227.9	262.5	307.5	370.0
188	168.1	169.6	171.5	174.3	178.8	186.1	192.3	201.1	215.4	230.4	265.4	310.9	374.0
190	170.1	171.5	173.5	176.3	180.8	188.1	194.5	203.3	217.7	232.9	268.2	314.2	378.0
192	172.0	173.5	175.4	178.2	182.8	190.2	196.6	205.5	220.1	235.4	271.1	317.5	382.0
194	173.9	175.4	177.4	180.2	184.8	192.3	198.7	207.7	222.4	237.9	273.9	320.9	386.0
196	175.9	177.4	179.4	182.2	186.9	194.4	200.8	209.9	224.8	240.4	276.8	324.2	390.0
198	177.8	179.3	181.3	184.2	188.9	196.4	203.0	212.1	227.1	242.9	279.6	327.5	394.0
200	179.7	181.3	183.3	186.2	190.9	198.5	205.1	214.3	229.4	245.4	282.5	330.9	398.0
202	181.7	183.2	185.2	188.1	192.9	200.6	207.2	216.5	231.8	247.9	285.4	334.2	402.0
204	183.6	185.2	187.2	190.1	194.9	202.7	209.4	218.7	234.1	250.4	288.2	337.5	406.0
206	185.5	187.1	189.2	192.1	196.9	204.7	211.5	221.0	236.5	252.9	291.1	340.9	410.0
208	187.5	189.1	191.1	194.1	199.0	206.8	213.6	232.2	238.8	255.4	293.9	344.2	414.0
210	189.4	191.0	193.1	196.1	201.0	208.9	215.8	225.4	241.2	257.9	296.8	347.5	418.0
212	191.4	193.0	195.1	198.1	203.0	211.0	217.9	227.6	243.5	260.4	299.6	350.9	422.0
214	193.3	194.9	197.0	200.0	205.0	213.0	220.0	229.8	245.9	262.9	302.5	354.2	426.0
216	195.2	196.9	199.0	202.0	207.0	215.1	222.2	232.0	248.2	265.4	305.3	357.5	430.0
218	197.2	198.8	201.0	204.0	209.1	217.2	224.3	234.2	250.6	267.9	308.2	360.9	434.0
220	199.1	200.8	202.9	206.0	211.1	219.3	226.4	236.4	252.9	270.4	311.1	364.2	438.0

APPENDIX D (*Continued*)

A erlangs

Blocking probability

N													
222	201.1	202.7	204.9	208.0	213.1	221.4	228.6	238.6	255.3	272.9	313.9	367.5	442.0
224	203.0	204.7	206.8	210.0	215.1	223.4	230.7	240.9	257.6	275.4	316.8	370.9	446.0
226	204.9	206.6	208.8	212.0	217.1	225.5	232.8	243.1	260.0	277.8	319.6	374.2	450.0
228	206.9	208.6	210.8	213.9	219.2	227.6	235.0	245.3	262.3	280.3	322.5	377.5	454.0
230	208.8	210.5	212.8	215.9	221.2	229.7	237.1	247.5	264.7	282.8	325.3	380.9	458.0
232	210.8	212.5	214.7	217.9	223.2	231.8	239.2	249.7	267.0	285.3	328.2	384.2	462.0
234	212.7	214.4	216.7	219.9	225.2	233.8	241.4	251.9	269.4	287.8	331.1	387.5	466.0
236	214.7	216.4	218.7	221.9	227.2	235.9	243.5	254.1	271.7	290.3	333.9	390.9	470.0
238	216.6	218.3	220.6	223.9	229.3	238.0	245.6	256.3	274.1	292.8	336.8	394.2	474.0
240	218.6	220.3	222.6	225.9	231.3	240.1	247.8	258.6	276.4	295.3	339.6	397.5	478.0
242	220.5	222.3	224.6	227.9	233.3	242.2	249.9	260.8	278.8	297.8	342.5	400.9	482.0
244	222.5	224.2	226.5	229.9	235.3	244.3	252.0	263.0	281.1	300.3	345.3	404.2	486.0
246	224.4	226.2	228.5	231.8	237.4	246.3	254.2	265.2	283.4	302.8	348.2	407.5	490.0
248	226.3	228.1	230.5	233.8	239.4	248.4	256.3	267.4	285.8	305.3	351.0	410.9	494.0
250	228.3	230.1	232.5	235.8	241.4	250.5	258.4	269.6	288.1	307.8	353.9	414.2	498.0
	0.976	*0.982*	*0.988*	*0.998*	*1.014*	*1.042*	*1.070*	*1.108*	*1.176*	*1.250*	*1.428*	*1.666*	*2.000*
300	277.1	279.2	281.9	285.7	292.1	302.6	311.9	325.0	346.9	370.3	425.3	497.5	598.0
	0.982	*0.988*	*0.994*	*1.004*	*1.020*	*1.046*	*1.070*	*1.108*	*1.176*	*1.250*	*1.430*	*1.666*	*2.000*
400	375.3	377.8	381.1	385.9	393.9	407.1	418.9	435.8	464.4	495.2	568.2	664.2	798.0
	0.986	*0.990*	*0.996*	*1.004*	*1.018*	*1.046*	*1.072*	*1.110*	*1.176*	*1.250*	*1.428*	*1.666*	*2.000*
450	424.6	427.3	430.9	436.1	444.8	459.4	472.5	491.3	523.2	557.7	639.6	747.5	898.0
	0.988	*0.994*	*0.998*	*1.006*	*1.022*	*1.048*	*1.070*	*1.108*	*1.176*	*1.250*	*1.428*	*1.668*	*2.000*
500	474.0	477.0	480.8	486.4	495.9	511.8	526.0	546.7	582.0	620.2	711.0	830.9	998.0
	0.991	*0.994*	*1.000*	*1.008*	*1.022*	*1.047*	*1.073*	*1.110*	*1.176*	*1.249*	*1.429*	*1.666*	*2.000*

600	573.1 *0.993*	576.4 *0.997*	580.8 *1.002*	587.2 *1.010*	598.1 *1.024*	616.5 *1.049*	633.3 *1.073*	657.7 *1.110*	699.6 *1.176*	745.1 *1.250*	853.9 *1.428*	997.5 *1.665*	1198. *2.00*
700	672.4 *0.994*	676.1 *0.998*	681.0 *1.004*	688.2 *1.011*	700.5 *1.025*	721.4 *1.050*	740.6 *1.073*	768.7 *1.110*	817.2 *1.176*	870.1 *1.250*	996.7 *1.433*	1164. *1.67*	1398. *2.00*
800	771.8 *0.997*	775.9 *1.000*	781.4 *1.004*	789.3 *1.013*	803.0 *1.025*	826.4 *1.050*	847.9 *1.074*	879.7 *1.111*	934.8 *1.172*	995.1 *1.249*	1140. *1.42*	1331. *1.67*	1598. *2.00*
900	871.5 *0.997*	875.9 *1.001*	881.8 *1.006*	890.6 *1.013*	905.5 *1.025*	931.4 *1.046*	955.3 *1.077*	990.8 *1.112*	1052. *1.18*	1120. *1.25*	1282. *1.43*	1498. *1.66*	1798. *2.00*
1000	971.2 *0.998*	976.0 *1.000*	982.4 *1.006*	991.9 *1.011*	1008. *1.03*	1036. *1.05*	1063. *1.07*	1102. *1.11*	1170. *1.18*	1245. *1.25*	1425. *1.43*	1664. *1.67*	1998. *2.00*
1100	1071.	1076.	1083.	1093.	1111.	1170.	1213.	1288.	1370.	1568.	1831.	2198.	

Index

How
to Peel
a Peach

How
to Peel
a Peach

AND 1,001 OTHER THINGS EVERY

GOOD COOK NEEDS TO KNOW

Perla Meyers

WILEY

JOHN WILEY & SONS, INC.

To Claude and Peggy who are blessed wth the gift of curiosity in and out of the kitchen

This book is printed on acid-free paper. ∞

Published by John Wiley & Sons, Inc., Hoboken, New Jersey
Published simultaneously in Canada

For general information on our other products and services or for technical support, please contact our Customer Care Department within the United States at 800-762-2974, outside the United States at (317) 572-3993 or fax (317) 572-4002.

Wiley also publishes its books in a variety of electronic formats. Some content that appears in print may not be available in electronic books.

LIBRARY OF CONGRESS CATALOGING-IN-PUBLICATION DATA:

Meyers, Perla.
  How to peel a peach : and 1001 things every good
cook needs to know / Perla Meyers.
    p. cm.
  ISBN-13: 978-0-7645-9738-1 (pbk.)
  ISBN-10: 0-7645-9738-8 (pbk.)
  ISBN-13: 978-0-471-22123-4 (cloth)
  ISBN-10: 0-471-22123-6 (cloth)
1. Cookery—Miscellanea. I. Title.
TX652.M473 2004
641.5—dc22   2003019728

Printed in the United States of America

10 9 8 7 6 5 4 3 2 1

Contents

Acknowledgments

This book has been years in the making, and I am grateful to the many people who have helped me acquire the ingredients, refine the formula, and finally bring the finished product to the table.

First of all, a special thanks to Jane Dystel, my intrepid literary agent, whose support has been unwavering throughout.

And to Carolyn Niedhammer, whose extraordinary editorial expertise has been invaluable to my shaping and defining of such an expansive subject.

To JoAnn Pappano, whose organizational skills and endless patience kept me on track.

To my editor, Susan Wyler, for her unflagging enthusiasm and commitment, and for her editorial guidance. To senior production editor, Andrea Johnson, for her attention to detail and for keeping track of all the pieces. And to art director, Jeff Faust, for coordinating all the creative impulses that became the jacket design.

I also want to thank Alice Medrich, Norman Weinstein, Jeremy Marshall, Susan Purdy, Arianne Daguin and Karen Lee for sharing their knowledge and patiently answering my questions.

My gratitude, too, to the thousands of students all over the country who have attended my classes and whose questions both inspired and formed the basis for this book. Their inquisitiveness continues to fuel my own endless curiosity about the world of food, its ingredients, and its preparation.

To my husband Robert whose cooking questions over the past 35 years contributed greatly to this project.

And, finally, loving thanks to my mother who has always shared and supported my love for food.

Whenever I put together the menus for a series of workshops, I always focus on dishes that raise questions, because, to me, that is the main reason why a cooking class makes sense and why many people feel that they need to take them. Most cooking classes tend to be recipe driven, with basic ingredients and simple techniques taken for granted. This is frustrating for me as a teacher because I am never sure of how much my students really know about food, how their interest in food has developed, and how much I can assume they know.

My best classes are those where we start at the beginning—by taking a field trip to a local grocery store, specialty store, or farmers' market. We break up into groups, and each group shops for one of the recipes that will be made in class that day, with the option of picking up additional ingredients for a vegetable dish of their choice. When we meet again, we go over how each group has gone about the task of marketing, with the group's notes and questions. To me, this is where cooking begins—with the intimate knowledge of one's ingredients, how to shop for them, store them, and cook with them.

I still remember my first day at the École Hôtelière, in Lausanne, Switzerland, where I first studied cooking. The chef started the class by saying, "The quality of your cooking more often than not depends on the choices you make before you begin." This has been my motto ever since I started to cook 35 years ago.

Today, few people grow up learning to cook at the side of their grandmother or mother. At a time when mothers hold full-time jobs and grandmothers live hundreds of miles away, we don't learn to cook instinctively, knowing how to peel a tomato, roast a chicken, or make a simple vinaigrette. At the same time, many of us are exposed to better and better restaurants, food magazines that tempt us to try something new for supper, and cookbooks with recipes for dishes that range from Arborio rice and truffle soup to zabaglione. We have access to specialty shops that introduce us to unusual ingredients from around the world. We are curious. We travel, we taste, and we read about food.

It is at this point that many people decide to make use of their own kitchens and see what it would be like to put together a meal. An aspiring cook might start with a cookbook or two. The photographs look great, and they make you feel inspired, but the fact is that none of the books tells you where to start. Yes, you know you need some utensils and some pots. Of course, you have some basic equipment, but then what?

The first trip to the market can be a dizzying experience. Why buy that big box of kosher salt when you have perfectly good salt around? And olive oil—virgin? Extra virgin? Light? Does that mean less fat? What is the difference between the bottle of balsamic vinegar that costs four dollars and the one for twenty-eight dollars? Your recipe calls for whole Italian plum tomatoes, but getting the crushed ones would eliminate a whole step, so wouldn't it make sense to get the crushed?

You plan to grill and your shopping list calls for skirt steak, but the meat department carries only flank. Are they interchangeable? Your grocery store does not carry either fresh thyme or fresh basil. Is dry okay? The pears feel like rocks, the lemons like golf balls, but the market is running a special on oranges. Should you change the menu? You soon realize that you have more questions than answers.

It all seemed so simple when you read the recipes, but now how important is it that you stick to them? Where can you improvise? These are the normal questions that all cooks have, whether beginners, chefs, or seasoned home cooks. The fact is that cooking is not hard if you are curious and flexible. But an intimate knowledge of ingredients is a must. No matter how creative a cook you are, you will always be handicapped by mediocre ingredients, whether the pantry basics, produce, meat, or seafood.

It is essential to understand that all meals start at the grocery store, a farm stand, or a green market, and that shopping for food is the single most important step in the preparation of any dish. Although the art of buying food carefully is familiar to most Europeans, particularly in countries where real markets still thrive, selective marketing is not part of the American shopping tradition today. This is partly due to lack of time, but it is also caused by inexperience.

I am always surprised to hear how many people are actually intimidated by the idea of shopping in a small specialty store. They prefer the anonymity of a supermarket. Perhaps they feel that they will be expected to buy something if they walk into a small shop, or they don't really know what they want and hesitate to talk to the shopkeeper. But small-business people, particularly those who deal in food, know their merchandise and are usually flattered to be asked questions, as they are eager to share their knowledge. Even if you do most of your shopping in supermarkets, occasionally make time to stop by a small meat shop, fish store, or produce market. Talking to the shopkeeper and forging an ongoing relationship can be a valuable learning experience. The same goes for farmers' markets, where you are likely to connect with people who care about their produce, meat, fish, poultry, and cheese and where you can get a real sense of seasonality and freshness.

Personally, I never think of marketing as a quick one-stop, because this is where the creative process begins. But it makes sense to start off with some knowledge, and that is what this book is about. The more you understand about an ingredient, the easier shopping will be.

You will find that the largest chapter of this book focuses on produce because buying good fruits and vegetables is such an important part of cooking. I still remember asking my mother what kind of potatoes she used for making her wonderful mashed potatoes.

"There are only two types of potatoes," was her answer. "Good ones and bad ones. I use the good ones."

These days it is not that simple. Most markets carry three or four types of potatoes—the same goes for onions and greens. What I try to do in this book is work through the hype and help you make the right choices when you go shopping for food. Once you get back in your kitchen with the best you can buy, it is important that you pay attention to proper technique, even if it is something as simple as roasting some carrots or sautéing a chicken breast.

To learn to cook well and gain confidence, you need to cook frequently. Cooking is about timing, about tasting, and most of all about feeling that it is worthwhile. It is not about a state-of-the-art kitchen or the best possible equipment. It is about loving to cook, not once a month or twice a year, not just to entertain or to impress a few friends, but because you like doing it. Creating a dish out of a few simple ingredients is ultimately extremely satisfying. That is why you should cook and that is why you should read this book.

I am often told that I am fortunate to be doing something that I really love to do, and it is true. Learning more and more about food, testing a new recipe, working with a new mushroom or fruit, cooking and savoring a new dish or one I have loved since childhood is a wonderful part of my life. I hope that by answering some of your questions, by being able to inspire you with my passion for good ingredients and the way to prepare them, I can make cooking a greater part of yours as well.

On a recent visit to a friend's house in New Mexico, I offered to make dinner. As we sat by the pool on a glorious spring day, we mapped out the menu. Asparagus risotto to start sounded like a good idea, followed by braised veal shanks with fresh peas and spring onions. Creamy mashed potatoes were also voted in. And to keep things rich, I convinced my friend that a marsala sabayon would make the ideal topping for local strawberries.

With an equipment list in hand, I decided to check out the kitchen before embarking on my shopping trip. What the kitchen cabinets revealed forced us to rethink everything. A couple of lightweight Revere Ware saucepans canceled the risotto, braising the veal shanks in a nonstick skillet seemed difficult at best, and attempting to make a sabayon in a makeshift glass double boiler would probably be a disaster.

Going back to square one, we decided on a nice salad and roast chicken accompanied by sautéed vegetables and roasted potatoes, which could be cooked in disposable foil pans. All that was kept of the original menu were the strawberries, but they would now be tossed in fresh orange juice, giving my hostess a chance to use her new toys: a juicer and a microplane zester.

The need to change the menu proved once again that without the right utensils, you are limited in what you can do. This does not mean that you need to invest in what the French call a *batterie de cuisine* to make a good meal. You can do wonders in a well-seasoned cast-iron skillet, and entire cuisines are based on grilling over small hibachis. Many a professional cook would laugh at the tables full of useless kitchen gadgets sold in cook shops and specialty stores these days.

When I left home for college in Switzerland in the early sixties, I raided my mother's kitchen for some cookware and tools. I did not plan to do very serious cooking, but by that time I was already hooked on the pleasures of preparing my own food and could not imagine relying on the school cafeteria. I remember settling on a couple of well-worn Le Creuset casseroles and a deep, cast-iron chicken fryer, some wooden spoons, and a few other odds and ends. The only thing I bought were two good knives, which I kept sharp by handing them over to my butcher every couple of weeks. All seemed well, and I coasted along, making some pretty tasty dishes that made me quite popular among my college friends.

Then I started cooking school and suddenly a world opened up to me. The right balloon whisk, rubber spatula, copper bowl, pepper mill, food mill, and best of all, mandoline— these were classic utensils that made all the difference. At the time we cooked in heavy tin-lined copper cookware, which had to be polished daily. Everything was done by hand—be it whisking a mayonnaise or beating egg whites or making a puree in a food mill.

All these tasks taught me the value of well-crafted kitchen utensils. The chef checked the edges of my knives daily, and my station had to be impeccably clean and organized, with chopped ingredients placed in bowls in order of

use. Gadgets were kept at a minimum and had to be placed in their exact place in the drawer, to be taken out only when needed.

At the beginning, it all seemed overwhelming and difficult, but I soon learned that the discipline would serve me well and that by learning to cook efficiently and by being organized, I was able to cooked more creatively. I learned that I did not need a drawer full of gadgets and bowls of all sizes, shapes, and materials; that flimsy saucepans were not worth whatever they cost; and that a good piece of cookware was worth the investment.

Soon I started to put together my own kitchen batterie, keeping my own style of cooking in mind. No need for various soufflé dishes or terrines, no need for individual quiche pans or copper molds, since I was not going to make soufflés or quiches or terrines. On my twenty-sixth birthday, however, my parents gave me a fabulous copper roasting pan and a 2-quart copper *faittout,* which means "does everything" and is, indeed, an all-purpose saucepan. (Thirty-five years later, these two pieces of cookware have traveled and moved with me and are still in constant use in my kitchen.) Slowly I put together the kitchen essentials I would need, adding and modernizing my equipment as time went on. My carbon steel knives gave way to stainless steel. Copper pots were replaced by enameled cast-iron and stainless steel saucepans, and old-fashioned iron skillets by some with a nonstick finish.

I can't stress enough that buying flimsy pots, pans, and other equipment is a waste of money. You really do not need that much, especially if you do not cook frequently, but buy the best there is, or at least the best you can afford. In some cases, if you feel that you can't handle the cost of a good piece of cookware at the time, wait until you can. When you settle for less quality, you will end up regretting it.

This does not mean that you need to waste money on "designer" tools. A simple well-crafted perforated spoon will do the same job as one imported from Italy or Switzerland. You do not need a state-of-the-art pasta pot or half a dozen food processor blades that you will never use. Do not buy knives or cookware in sets, but rather buy what you need and always keep in mind your own style of cooking. What is the point of investing in a beautiful copper bowl when all you will ever beat are 4 to 6 egg whites, which can be done perfectly well in a stainless steel bowl? And what is the point of buying an exquisitely crafted Japanese knife when you are never going to learn how to cut fish for sushi?

I'll confess that my kitchen has more whisks than I will ever need, more spatulas, and more gadgets—proof that I am a consummate consumer and that when you love to cook, it is almost impossible to pass a cookware shop or specialty store without being drawn in to purchase something that for the moment you cannot live without. But more and more I am actually streamlining my kitchen, getting rid of such excess items as a strawberry peeler or a clever but useless garlic press. At the same time, I am looking for new, useful tools. An Omega vegetable juicer suddenly makes sense, and I have replaced my old, rather heavy immersion blender with an updated lighter one. I recently discovered a wonderful All-Clad saucepan that has become my new risotto pot.

In the following Q & A's I try to answer some of the questions many of my students have asked me over the years. I am aware that cooks and bakers have different needs, and since baking is not on top of my list, I do not address in detail the needs of the "baking kitchen."

You'll find that your essential kitchen will change from year to year. As your cooking evolves, so will your needs and interest in cooking equipment. A cook who loves to experiment with dishes from the Far East will not need the same equipment as one who specializes in Italian dishes. You'll soon discover that acquiring new tools and learning to use them is part of the fun of cooking.

Equipment

■■■

Cutting Boards

I can never find a cutting board that feels good. What do you recommend?
I am very partial to wood, even though the days of wood in the kitchen are almost over. I hear all the arguments against wood, because if not cleaned properly, wood breeds bacteria. However, if you do take care of your chopping boards and clean them thoroughly, wood is still the very best surface to chop on and will not harm your knives. My choice is the J.K. Adams cutting board made of solid New England maple. It will not warp, but its drawback is

that at six pounds, it is heavy and hard to maneuver back and forth to the stove. If you want a lighter board, choose one made of wood composite. I find the best way to clean cutting boards is with a light detergent, scrubbing well with a scraper to remove all food residue. When a cutting board is in need of serious cleaning, sprinkle it with kosher salt and leave overnight. Then use half a lemon from which you have already extracted the juice to scrub and remove stains.

TIP Remember to oil wood cutting boards once every six months with mineral or almond oil to prevent them from splitting.

Knives

What should I look for in a good knife?

First, it should be made of top-quality, no-stain high-carbon steel. Second, it should have a full-tang handle. The tang is the part of the blade that extends into and forms part of the handle. When the tang is the same length and shape of the handle and you can actually see it, that is called a full tang. When you pick up the knife, it should have good balance, and the grip should feel comfortable.

I know that good knives are a must, but how many do I need?

You do not need a lot of knives. The ones you do buy should be of top quality. Although it may save you some money, do not buy knives in a set. Instead, buy the following knives:

Three 3- to 4-inch paring knives. It is very tempting to buy inexpensive paring knives, since they are available in every hardware store.

But a cheap knife with a poor blade and uncomfortable handle is of little use. Paring knives must not be too flimsy, and you must be able to sharpen them. A dull paring knife is a dangerous tool. Companies such as Henckels make rather inexpensive paring knives with plastic handles that have good blades. They sharpen easily, and I highly recommend them.

An 8-inch chef's knife and possibly a 10-inch one, if your hand is large. This is the most important knife in the kitchen. It should be made of high-carbon steel with a full tang and a plastic-impregnated handle. If all of this sounds complicated, just stick to the top brands such as Wusthof or Henckels or one of the other companies that make excellent knives. The key is to buy them from a reputable source who will guide you through the maze of knives available without trying to sell you the store.

A 10-inch slicer. This may not be a knife you will use daily, but even if you roast a turkey or a leg of lamb only a couple of times a year, this knife will come in very handy. Be sure not to use it for daily chores, since the blade does not lend itself well to chopping or mincing.

A serrated knife. This is a good knife to have on hand for slicing tomatoes, cucumbers, and even small loaves of bread. I find that I use it a lot. The best one on the market is made by Wenger, a Swiss company. It is carried by many good specialty shops.

A bread knife. Here you do not need to spend much money, so choose one that feels good in your hand and is not too heavy.

A cleaver. This can be made of either stainless steel or carbon steel. Cleavers must be heavy. Remember that their main function is to go through bones. Always select a heavier cleaver than you think you need. At the end of the day you will be glad you did.

Is there an easy way to keep my knives sharp?

The best way to sharpen a knife is with a sharpening steel. However, I find that while this is a key kitchen gadget, most people do not know how to use it properly. I therefore recommend two tools:

1. **A kitchen honing device.** The concept is similar to that of an electric knife sharpener, but it is inexpensive. If you get the Chantry-Victor, which I consider to be the best one on the market, you can run your knives through it every day and keep a nice sharp edge on them for a very long time. Be sure, however, to change the blade on the Chantry about twice a year. You can order replacements through the Internet. Companies such as Henckels and Wusthof also make this device, but the Chantry-Victor is by far the sturdiest and best one on the market.

2. **An electric knife sharpener.** This is an expensive tool, but one that I recommend very highly. Two small hardened steels are placed at the correct angle for treating both sides of the blade at the same time. As you draw the blade through it, the knife edge is realigned, ensuring a sharp cutting edge. The best one on the market is Chef's Choice.

TIP **Save an old paring knife for removing the plastic and metal bands on bottles and jars and cutting the plastic around wine bottles.**

Miscellaneous Kitchen Tools

With all the peelers on the market, is there one that does the job better than others?

The most inexpensive and easy-to-clean peelers are U-shaped and come in a variety of colors. These peelers tend to remove more than just a thin layer of the peel, so I usually do not use them for delicate vegetables such as asparagus. I very much like the old-fashioned Marvel Peeler, a combination peeler, grater, and corer, which you can still find in hardware stores. The ever popular Oxo brand makes a most attractive and sturdy peeler as well. Most people love that heavy black rubber handle, though I find it too space consuming.

TIP **Remember that peelers dull just like knives and since they cannot be sharpened, should be replaced once a year.**

I notice that you travel with your own pepper mill. Do cooks get attached to a mill as they do to their own knives? Do you have a favorite?

The key to a good mill is its mechanism, and its function is to grind pepper evenly. When you find a mill that will do that, it is worth traveling with. I like two mills; both happen to be French. First, the Perfex, which could also be called "the perfect." This is a stainless steel mill that holds almost two tablespoons of peppercorns. I like the design of this mill, with its pull-out chute, and its adjustable tension knob that allows you change the coarseness of the grind. The Perfex is sturdy and functional and will probably never need replacing. Another great classic is the Peugeot wooden pepper mill. The mechanism of this mill is probably the best one made and because of its durability, it is used by other pepper mill manufacturers as well. I like the beechwood finish but also recommend buying the dark wood one if you plan to have two mills, one for black and one for white peppercorns.

I have an old-fashioned ball-bearing rolling pin but now see many other types. Which do you recommend?

When it comes to rolling pins, I am pretty old-fashioned and like a wooden pin rather than those made out of marble or glass. Many good cookware stores now carry the French style, a straight pin with tapered ends made out of maple or beechwood. These are both ideal for rolling out a pie crust dough.

Which spatula is better for glazing a cake— one that is straight or one that has the blade set at an angle?

They are both the same. Just be sure that the spatula has enough flexibility to evenly coat the cake.

I am a serious gadget consumer and find myself with tons of stuff, but I never seem to find the right gadget when I need it. What are the must-haves?

Here is my "almost basic" list with a few tips:

1. **Garlic press.** Be sure to buy a good sturdy one. The most reliable is still the Susi, from Zyliss of Switzerland. What makes it so user friendly is that you do not need to peel the cloves and it is easy to clean. Another one that is high on my list is the Henckels stainless steel model, which also has a ginger press attachment. Stainless steel is considered the very best material, since it does not discolor or change the taste of garlic.

2. **Whisks.** Do not buy a whisk in a hardware store or the average supermarket. The whisk is one of the most important tools in the kitchen, almost right up there with your knives. So invest in quality. You will need two sauce whisks, such as the ones made by Best. These are elongated in shape with either wood or stainless steel handles. You will also need an egg white whisk. Personally, I much prefer the wood handle. The longer the handle, the better, since it allows you more freedom to move your hand up and down when you are whisking whites.

3. **Salad spinner.** In this age of constant spraying of lettuce in supermarkets, the spinner has become an essential kitchen tool to remove all that moisture. You can also use it for other greens, especially spinach and escarole. It is worth investing in a good one such as the Swiss Zyliss or the Oxo, which do an excellent job.

4. **Corer.** The only time you really need a corer is when you make a baked apple and want to fill the center with some butter, nuts, and spices. It is then you reach for this particular tool, and there is really nothing else that will do the job. Corers do not do a perfect job of coring an apple, and chances are that you will still have some seeds and bits of the core in the apple no matter which gadget you use.

5. **Melon ball cutters.** These gadgets come in many sizes and can produce both rounds and ovals. Once available only to professionals, they are now sold in good

cook shops. I highly recommend them because they are fun and can turn certain vegetables into decorative shapes. These may not be everyday gadgets and do not need to be top of your list, but if you do see them and wonder what to do with them, think twice and you will come up with many interesting ideas that can really change the appearance of a dish. Think of a round tiny zucchini or potato ball or an oval carrot ball.

So many recipes refer to a mandoline these days. Does it do anything that the food processor can't do?

The principle behind the mandoline is that you slide food over the blade, which allows you to slice it as finely and evenly as you like. This device is mostly used by restaurants for cutting vegetables and fruit into slices or strips. Although the manufacturers of food processors claim that you can do this well in a food processor, the results are not the same, and a mandoline is a great tool well worth investing in. You can now get an inexpensive mandoline made out of plastic that will do a very good job of slicing. But once you get beyond this and into serious cooking, you may want to invest in either the classic stainless steel French mandoline made by Bron or the new one by Matfer, also from France. This mandoline is easier to use and great fun. I highly recommend it.

TIP Mandolines come with guards, which I find hard and cumbersome to use. Instead, I use a kitchen towel to hold the food in place. Also it is key to remember to keep your hand flat to avoid cutting your knuckles on the blade!

Pots and Skillets

Which cookware set do you recommend?

It is tempting to buy a multiple-piece set of cookware. There are fewer decisions to make and you can save quite a bit of money. But there are large disadvantages. In many sets, one or several of the pieces duplicate the function of other pieces, and quite frankly, there is no culinary rationale for having matching pots and pans because no single kind of cookware does the best job for every kind of cooking. For a well-equipped kitchen, you should capitalize on the best features of each kind of cookware, even if the pots and pans are not matching. Of course, if you are starting from scratch, buy a small set of no more than five pieces plus lids, composed mostly of saucepans and possibly including an omelet pan. Supplement these pieces with cookware made of solid materials, such as cast iron or heavy-gauge stainless steel, suitable for specific purposes.

What size saucepans should I have? How heavy do they need to be?

I like to use easy-to-clean, medium-gauge stainless steel or anodized saucepans such as Cuisinart or All-Clad.

For starters, get three sizes: 1-, 2-, and 3-quart pans with lids. Now you are pretty much covered if you want to do anything from boiling water to cooking vegetables. When you get into the more serious cooking, the 3- and 4-quart heavy enameled saucepans from Le Creuset are a must. With these you can make more complicated sauces, risottos, and pilafs, and every kind of soup. If you plan to work with chocolate, sugar, or very delicate sauces, you should invest in one tin or stainless steel–lined copper saucepan with cast-iron handles. This is an expensive piece of cookware that is extremely high maintenance, but I wouldn't be without it because it heats up and cools down quickly and evenly, giving you complete control.

TIP Avoid flimsy light saucepans, since even water will burn in them. Also, as you are cooking, you will get a better idea of what size you use the most and you can build from there.

Is it necessary to have both a pasta pot and a stockpot?

A good-quality, heavy-gauge 8-quart stockpot can easily double as a pasta pot and I use mine all the time. The advantage of a pasta pot is the steamer insert, which is very helpful in draining pasta. However, if I had to choose one or the other, I would go with a quality stockpot such as the Calphalon commercial hard anodized nonstick stockpot or the Chantal enamel stockpot.

Are there some pots that are better suited for cooking on an electric stove?

Yes, the heavier stainless steel saucepans in which the steel or aluminum core is sealed between two layers of stainless steel are the best saucepans. They are also expensive. Another good choice is aluminum saucepans with a gray anodized coating that is guaranteed for life. I find them most useful because the handles stay cool.

What shall I look for in a skillet?

A heavy black cast-iron skillet is the best and most versatile piece of cookware I know. It can be used for anything, including oven-roasting. Cast iron needs getting used to, since it retains heat and does not cool quickly. Also, it is heavy, and the handle does get hot, but it is such a great piece of cookware that you will soon be willing to overlook these small inconveniences.

TIP For vegetable sautéing, pan-searing of fish, and egg cookery, I like heavy aluminum pans with a nonstick coating. All nonstick coating scratches and wears off, but the heavier skillets are better balanced with handles that do not heat up and overall are extremely forgiving. My favorite brand is All-Clad.

What size skillets do I need?

I recommend two 10-inch and one 12-inch skillet rather than 14-inch skillets. Larger skillets are hard to maneuver and they will never cook food evenly, especially if you are cooking on electric heat.

For the more serious cooks and those who cook with gas, the best sauté pan, or deep skillet, is made of heavy copper lined with either tin or stainless steel and with an iron handle. This fabulous piece of equipment is heavy but will do a great job on anything from slow-braising vegetables to cooking a cut-up chicken to perfection on top of the stove. The best size for the home cook is a 3-quart pan, about 10 inches in diameter.

For quick sautéing of small quantities, you should have at least two 8-inch skillets, preferably with nonstick coating. For specialty cooking such as stir-frying and crepe making, it is good to have a carbon steel flat-bottomed wok and a traditional black steel crepe pan.

My mother gave me her favorite cast-iron skillet. How can I use it so that food doesn't stick?

It needs to be seasoned properly, which you should do by covering the bottom with a thick layer of kosher salt, then adding ½ inch cooking oil. Place the pan on top of the stove. Heat the oil until it is almost smoking, then pour it out into a dry metal container. With a large ball of paper towel to protect your hand, wipe the surface of the skillet clean. Your pan is now perfectly seasoned. It will be practically maintenance free and can be used for anything, including oven-roasting. Its only drawback is that it should not come in contact with any acid foods such as tomatoes, lemons, or wine.

TIP Once a cast-iron skillet has been seasoned, never wash it with soap and never scrub it with anything but a soft sponge. If food gets stuck in it, treat the area again with a little coarse salt, rubbing it with a paper towel. Never soak in water and remember to reseason every 6 to 8 months or whenever food starts to stick.

I love to stir-fry but do not have a gas stove. Can you recommend a good pan to stir-fry in?
My favorite stir-fry pan is the original flat-bottomed wok by Joyce Chen. It is made out of carbon steel, which unfortunately cuts down on heat and makes the food steam slightly, but the flat bottom enables you to use it on any cooking surface. Also, the domed cover is a terrific asset, since it allows you to use the pan as a steamer. The pan comes in various sizes; I recommend the 12-inch size.

Do you recommend buying a double boiler?
I have a wonderful old aluminum double boiler that my mother-in-law gave me years ago and I use it often. But with the advantages of the microwave, you rarely need a double boiler. As for stovetop cooking, you can use a good saucepan with a heavy stainless steel bowl that fits snugly into it. If you are tempted to get a double boiler, do not buy a glass one, since they are poor conductors of heat.

Bowls

Mixing bowls come in glass, ceramic, and metal. Which are the most versatile?
Take your cue from restaurants that use only stainless steel with an occasional ceramic bowl for fruit salads, mixed salad, or dough. Heavy-duty stainless steel bowls are the best, but the only time weight really matters is when you are going to use them as the top part of a double boiler for making a custard or keeping something warm, such as a potato puree or a puree of vegetables. Otherwise, inexpensive stainless steel bowls are fine. The nesting glass bowl sets are very popular, but I am not much of a fan of glass.

TIP It is not a good idea to whip cream or egg whites in a glass bowl because it is hard to chill the bowl properly and evenly.

Do you recommend a copper bowl for beating egg whites?
I use my copper bowl all the time. A copper bowl is still the best tool for beating the perfect egg white because of the alchemy that takes place between copper and the protein in the egg whites and makes the foam more stable. Egg whites beaten in copper will also expand more when heated. Don't buy an expensive copper bowl, because they all do the same trick. Twelve inches in diameter is ideal.

Try this recipe for Sherry Citrus Sabayon In a bowl, combine ¾ cup cream sherry with 5 egg yolks and ½ cup sugar. Add 2 teaspoons orange zest and 1 teaspoon lemon rind and whisk the mixture until well blended. **Transfer to a copper bowl set over simmering water and whisk the mixture constantly until thick and smooth. Remove and transfer to a bowl. Chill until set. Whip 1 cup heavy cream until it forms soft peaks and fold into the mixture. Chill until ready to serve. Serve with poached pears, sliced strawberries, or raspberries.**

TIP **Before using your copper bowl, make sure that it is spotless by cleaning it with white vinegar and then rinsing it with soapy water. If there is even a speck of grease in the bowl, it will keep your whites from foaming.**

Roasting Pans, Baking Dishes, and Casseroles

I still have my old roasting pan that came with my oven, but I am ready for a new one. What do you recommend?

The most important feature of a roasting pan is weight, so look for a heavy roasting pan in which you can easily roast a large turkey, a leg of lamb, or a ham. My favorite pans are the enameled cast iron by Le Creuset and the All-Clad stainless steel roasting pan. Choose a pan 13 to 14 inches long, about 10 inches wide, and no more than 2 inches high. If the sides of a roasting pan are too high, the food will steam rather than roast. Make sure that the handles of the pan are comfortable, since you will have to handle the pan several times during cooking to baste your roast and check the pan juices.

When heavy roasting pans are full of food they become difficult to lift. Are there lighter versions that will do the job?

Roasting pans are meant to hold heavy ingredients and to be moved from the oven to the top of the stove for last-minute cooking such as deglazing or reducing a sauce. This cannot be done with a lightweight pan. Several companies such as Calphalon make lighter nonstick roasting pans but I do not recommend them, since I find them too flimsy. Nonstick is not a good choice when it comes to roasting pans, since the pan juices do not caramelize.

TIP **You must use the right size roasting pan for the amount of food you are cooking. If you use a large roasting pan for making a small roast or a single chicken, the pan juices will burn. On the other hand, if you use a pan that is too small for the size roast you are making, the food will steam rather than roast.**

Can I use a 12-inch black cast-iron skillet for roasting?

Absolutely. I do it all the time. A cast-iron skillet is perfect for roasting a chicken or half a leg of lamb. Because cast iron retains heat, you will have to watch your roast carefully and baste often so as not to burn the pan juices. Also remember that the handle will be very hot, so be sure to keep a couple of pot holders on hand.

What is the difference between a roasting pan and a baking dish?

The difference is only a matter of size and material. You need a roasting pan of a certain size for a turkey, leg of lamb, or brisket or if you are roasting more than one duck. But many foods such as potato gratin, roasted vegetables, pudding, or braised vegetables are cooked in a baking dish. You should invest in at least two or three baking dishes of various sizes.

There seems to be an endless choice of baking dishes. Which are the best sizes and which is the best material?

For best results, choose baking dishes that are shallow to allow the heat to circulate around the ingredients. Avoid glass, which is a poor conductor of heat. The best choice is porcelain, followed by glazed earthenware. I also like enameled cast iron, which is heavy and sturdy and perfect for roasting vegetables and a single chicken. Always check the handles; they should be practical and easy to handle. My favorite sizes are 9 by 11 inches (5 quarts) or 9 by 13 inches (6 quarts). Another good size is a rectangular 10-quart dish that measures 12 by 10 inches by 2¼ inches deep. These are standard sizes and easy to find in good cook shops in a variety of materials. They should be microwave proof and able to stand low direct heat.

There are so many casseroles to choose from these days. Which do you recommend and what are the best sizes?

My favorite casserole and one that I always recommend is the cast-iron one by Le Creuset. Yes, it is heavy, but it is a great piece of cookware that will last you a lifetime. The best sizes are the 4½- and 6-quart round casseroles. If you like to make short ribs, lamb shanks, or stew for a large number of people, then the 9-quart oval is also a good piece to have.

Electric Appliances

I am just starting to equip my kitchen and have limited space. Which appliances shall I start with?

I would start with a hand-held electric mixer that will mix, whip, and blend, and an immer-

sion blender. Combined, these will do the same jobs as a food processor and take up less space, provided that you are willing to use your knives for cutting and slicing and a four-sided grater for shredding. The immersion blender is the best tool for pureeing soups right in the pot and for making vinaigrettes and mayonnaise-based sauces.

My old electric hand mixer has finally given up. Which one do you recommend?

My two favorite hand mixers are the KitchenAid and the Cuisinart. The KitchenAid is strong and sturdy but not too heavy. It is also quieter than the old-fashioned mixers. The beaters are easy to clean, and I especially like the soft-start feature that eliminates splattering the dough or whipped cream all over the place. What is more, it comes in a variety of colors. Another equally good choice is the Cuisinart, which is designed for both left- and right-handed cooks. Its best feature is a balloon whisk that beats whites better than any hand mixer I have ever had.

Do you recommend the KitchenAid heavy-duty mixer? Many of my friends have it.

The KitchenAid is a favorite classic that many serious cooks rely on. It is great for baking and eases the task of making a double batch of cookies, making bread dough, or beating more than 6 egg whites. It also has the advantage of supporting attachments such as a pasta maker, citrus juicer, and meat grinder. However, it is not on my list for beginner cooks or those who are not interested in baking.

One of my wedding presents, a Crockpot, is still in the box. What is the best use for it?

Crockpots are great for cooking beans at a slow, even temperature with no worry about the

KITCHEN TOOLS

THE BASICS

2-cup glass measure

1-cup glass measure

1 set heavy, stainless steel dry measuring cups

2 sets measuring spoons, preferably stainless steel

Instant-read thermometer, preferably Taylor

Nutmeg grater

2 metal turner spatulas

Cake spatula with flexible stainless steel blade and wooden handle

2 flexible rubber scrapers

Soup ladle

Four-sided grater

Potato ricer or masher

Long-handled fork

Bulb baster, preferably glass rather than plastic

Bottle opener

Corkscrew

Food mill, either plastic or stainless steel

Basic can opener, not electric

2 stainless steel slotted spoons

2 stainless steel long-handled spoons

2 kitchen tongs, 1 short, 1 long

Vegetable peeler

Kitchen timer

Salad spinner, preferably Oxo or Zyliss

Wooden or metal pasta fork

3 to 4 wooden spoons

Balloon whisk

Sauce whisk

Reamer

Zester

Garlic press

Colander

2 sieves, 1 large, 1 small

Ice cream scoop

Kitchen shears

BAKING EQUIPMENT

2 9-inch round cake pans, heavy-duty aluminum or heavy-gauge steel with nonstick coating by Chicago Metallic

9-inch springform pan

2 standard 8½ by 4½-inch bread pans

4 nonstick 6 by 3-inch loaf pans

Bundt pan

8-inch pie pan

9-inch pie pan

2 baking sheets, preferably rimless, shiny heavy-duty aluminum

Jellyroll pan, preferably with a matte uncoated aluminum finish by Chicago Metallic

Flour sifter

2 10-inch quiche pans, preferably porcelain

Pastry brush with boar bristles

ELECTRIC APPLIANCES

Food processor

Blender

Electric hand mixer

Immersion blender

Toaster

Coffee grinders, one for coffee, one for spices

POTS AND SKILLETS

1½- to 2-quart saucepans

8-quart stockpot

Pasta pot with steamer insert

8-inch nonstick skillet

10-inch cast-iron skillet

10-inch nonstick skillet

12-inch heavy-gauge sauté pan

6-quart casserole

½- to 6-quart Dutch oven or braising pan

Roasting pan

2 baking dishes

Double boiler, optional

---

water boiling away and the beans burning. You can also use it for soups. Look for models in which the cooking pot is removable for easier cleaning.

I still have my very first food processor, but the blade is not sharp and the motor is finally giving out. What shall I buy to replace it?

Cuisinart, the pioneer of the food processor, still continues to make some of the best machines. The 14-cup-capacity basic machine would be my choice. The good news is that it now comes with an easy-to-handle cover that eliminates the need of using the rather complicated feed tube when you are not slicing or shredding. If you are looking for something smaller, Cuisinart also makes an 11-cup version ideal for smaller kitchen spaces or if you are not cooking large quantities. Another excellent choice is the KitchenAid Professional food processor, which features a top of the line blade made by Sabatier and comes with a small 3-cup bowl and blade that is perfect for chopping small amounts of ingredients such as parsley, a few cloves of garlic, or cheese. Although this machine comes with only two extra blades, they are really all you need. You can buy an additional five-disk set that includes a French fry cutter and a julienne disk.

I use my food processor all the time, but many recipes call for a blender. What is the difference? Do I need both?

The food processor is a terrific tool, but when it comes to a very fine puree or a homemade mayonnaise, the blender does a better and smoother job. The new blenders are much more powerful than they used to be, so if you want to make a summer smoothie, a gazpacho, or an herb mayonnaise, I would definitely invest in a blender such as the Kitchen Aid Ultra Power blender or the Waring professional blender.

To make blender mayonnaise Combine 1 whole egg, 1 egg yolk, 1 teaspoon Dijon mustard, 1 teaspoon red wine vinegar or 2 teaspoons lemon juice, and a pinch of salt in the container of the blender. With the machine running, add ⅔ cup oil (half grapeseed, half olive oil) in a very slow drizzle and blend until the mixture is thick. Correct the seasoning. Transfer to a container and chill. This makes 1 cup of mayonnaise that will keep for 4 to 6 days. For herb mayonnaise, add 2 tablespoons chives, parsley, dill, cilantro, tarragon, or basil or use a combination of herbs, such as chives and parsley, dill and chives, tarragon and parsley, or cilantro and parsley.

A friend gave me an immersion blender, and I have yet to use it. What exactly is it good for?

The immersion blender has been a household basic in kitchens throughout Europe for over twenty years. Its advantage is that it takes up little space and allows you to puree soups and vegetables right in the pot—no need to transfer foods to the container of a food processor, less cleaning up, and more counter space. I use mine all the time. The immersion blender makes great mayonnaise in seconds and is excellent for whipping cream as well.

No matter whether you are a beginner cook or a chef, a well-stocked pantry is the key to even the most basic cooking. Salt and pepper, oil and vinegar, herbs and spices, butter and eggs, flour and chocolate are just some of the ingredients that can make a significant difference in the results of any recipe. Knowing how to buy them and use them is important.

Smart shopping is fundamental to good cooking. Before you set out to make a dish, whether a simple salad or a complex ragout, you should learn to make smart choices about ingredients. Always ask yourself, why? What is the difference between kosher salt and fine salt? Do you need both? Should you use ground spices or whole? And which brand of, say, canned tomatoes should you buy?

Your pantry should contain all the right everyday ingredients needed to create good, flavorful food. But exactly what it should be stocked with has much to do with personal preference and the kind of cooking you plan to do. Because our food cupboards reflect our lifestyles and habits, no two are alike; no two contain exactly the same items. If you like to dabble in Chinese or Indian cooking, then an Asian ingredient shelf is a must. If, on the other hand, you prefer Mexican food, then several types of dried chiles are an important part of your pantry. And for those who like to bake, two or three types of flour, three kinds of sugar, nuts, vanilla beans, cocoa, dried fruit and a couple of types of chocolate are all key staples.

Because I love Italian and Spanish food and cook it often, I keep at least three kinds of pasta, excellent extra virgin olive oil, whole canned tomatoes, dried porcini mushrooms, and three types of rice in my pantry. Your pantry, on the other hand, may reflect your fascination with Thai or Indian cooking, your penchant for spur-of-the-moment salads, or your husband's love for peanut butter and preserves.

To keep a well-stocked pantry that feels "fresh," you need to shop carefully and often with a thorough understanding of ingredients. Personally, I follow no set schedule for restocking. Since I love to shop spontaneously and enjoy even a large supermarket, I am constantly replenishing my pantry shelves.

There are, I believe, certain guidelines for shopping wisely for the kinds of basics needed for everyday cooking. They can help us steer clear of trendy, overpriced, and out-of-season foods that don't deliver. While it is fun to browse through a good gourmet shop and check out the new items on the shelf, I try to focus on how I would use them. White truffle honey sounds great, but would it fit into my cooking repertoire? And how would I really use the beautifully bottled currant vinegar? If I feel that this will be an item that could just gather dust in my cabinet, I opt instead for a quality mustard or excellent peppercorns that I know I will use frequently.

I divide my pantry into a basic shelf, an ethnic shelf, and the refrigerator for perishable ingredients. It goes without saying that real basics like dried herbs, spices, dried beans, good oils, canned tomatoes, condiments, potatoes, garlic, and onions create the backbone of everyday cooking. But because our repertoire has expanded to include a wide variety of international and regional dishes, what used to be considered an exotic ingredient has likely become a pantry staple. Now fresh ginger, soy sauce, toasted sesame oil, coconut milk, and rice vinegar share the ethnic shelf with porcini mushrooms, truffle oil, and balsamic vinegar.

Your refrigerator should include fresh milk, yogurt, unsalted butter, and eggs. Add sour cream or crème fraiche if you use them frequently. Think of your freezer as part of the pantry, too, and use it for supplies like stock, frozen berries, nuts, ice cream, and pastry dough.

Since you will be spending time and effort on purchasing excellent ingredients, you will want them to be in top condition when you are ready to cook with them. The most important aspect of extending shelf life and maintaining quality is proper storage.

It is easy to recommend that you store spices, oils, potatoes, and other basics in a cool dark place. In a perfect world we would all have cool, dark basements, but the reality is that we keep things stored in the kitchen, which tends to be hot. That is why—short of ripe tomatoes and semi-ripe fruit—your best bet is to refrigerate everything you do not use on a constant basis. This includes rice, dried beans, nuts, fruity oils, and most spices. I keep my olive oil in the refrigerator.

For me, the best invention is the zippered plastic bag, which I use for everything, including the basic vegetables, which I try to shop for at least once a week. Do not let the vegetable bin become a catchall. Use produce while it is fresh and tasty, and keep it organized. Don't buy six lemons or limes if they are going to end up shriveled in the back of the bin. Buy what you need and use them, then buy more.

Be sure to buy a refrigerator thermometer and check the temperature often. If you are going to take up cooking seriously and want to be prepared to react spontaneously to a craving for your favorite dish, you'll need to keep your pantry up to date. Stale ingredients won't give you the results you want. Perla's rule: If it is out of date or out of taste, throw it out!

Admit to yourself that you'll never, ever use the rest of those pickled cactus pads you bought after that trip to Mexico five years ago.

Ask yourself why you seem to have an irrational, emotional attachment to that half bottle of rose water you used for a Moroccan recipe once. Admit that although you loved the food in Hungary, that can of paprika you bought there went flat two years ago. Throw it out!

And if you look in my pantry, please, don't ask me what I'm still doing with that enormous jar of miso mayonnaise, Aunt Sue's pear balsamic vinegar, the tiny bottles of fruit extract with pretty hand-painted labels, and the homemade jar of pickled asparagus that seems to have turned an unappetizing shade of chartreuse!

The Basic Pantry

■■

Anchovies

What should I look for when buying anchovies?

I always buy flat anchovies, since rolled ones are usually more expensive. The best anchovies are those packed in olive oil and imported from either Portugal or Spain. Avoid those imported from Taiwan, since they tend to be mushy and overly salty. Jarred Italian anchovies are more expensive than canned, but they are far more meaty and flavorful. Look for them in high-end grocery stores and specialty markets.

Is there an easy way to remove some of the saltiness of anchovies?

You can make anchovies less salty by soaking them in a little milk to cover for 10 to 15 minutes, then draining them well before adding to a dish.

Do anchovies give meat dishes a fishy flavor?

When mashed with a little garlic and fresh herbs, anchovies add a wonderful piquant flavor to a leg of lamb, a roasted chicken, or veal roast. In fact, used judiciously, anchovies act more like MSG—that is, enhancing flavor rather than adding a fishy taste.

Once I open a can of anchovies, I usually do not use the whole can. What is the best way to store them?

Anchovies will keep well in the refrigerator for several weeks. Just transfer them to a jar and cover with olive oil.

What is anchovy paste? How and when do you use it?

Anchovy paste is a mixture of pounded anchovies, vinegar, and spices. When all you need is a teaspoon or two, it offers a good alternative to canned anchovies. I often use anchovy paste to enhance the flavor of a mayonnaise, vinaigrette, or marinade. It is also mixed with soft butter for canapés and as a topping for grilled steaks.

Bouillon

When you can't make a homemade stock, which is better, bouillon cubes or canned stock?

I much prefer bouillon cubes or powder. But quite truthfully, I don't think either is good unless you boost its flavor. When using store-bought stock, simmer it for 20 to 30 minutes with some aromatic vegetables. I am a great believer in the extra flavor boost provided by a few extra chicken wings, which I keep in my freezer at all times. This will increase the cooking time to about 40 minutes, and, yes, it takes longer than opening a can but is still not as time consuming as making a big batch of stock from scratch.

Caperberries

Recently I sampled a wonderful dish of braised fish with olives and caperberries, which I had never tasted before. What exactly are caperberries and where can I get them?

A caperberry is the olive-sized fruit of the caper bush (rather than a bud, which is what a caper is). They are becoming very popular, but I have seen them for sale only in specialty stores, where they are stocked alongside capers and other condiments. You might try mail-ordering them from Zingerman's (see Sources, page 399).

Capers

What kind of capers are best? The salted ones or the ones in vinegar?

I much prefer the "meatier" salted capers. They are flavor packed and easy to handle. Simply rinse off the salt and use. For vinaigrettes and salads, I do use capers packed in vinegar. Your market may carry only the vinegar-packed variety, which in the end are really fine.

What's a nonpareil caper?

The small, young variety of caper called nonpareil is considered the most delicate. There are also larger capers available—some common ones are superfine, fine, and Capuchin. I find them less flavorful, but they are used extensively in Italian cooking, so I suppose it is really a matter of taste. Nonpareil capers are carried in grocery stores everywhere, so you should have no trouble finding them.

Cornichons

What are cornichons? Are they the same as gherkins?

Gherkins are small, young green cucumbers, used mainly for pickling, but they are also delicious raw. The smallest gherkins are called cornichons, which are harvested when very young and pickled. You may be most familiar with them as the tiny, tangy pickles traditionally served with a country pâté or sausages. I always add some diced cornichons to a tuna, egg, or chicken salad.

Cornmeal

I've seen blue cornmeal in specialty shops. Is it really derived from blue corn or is the coloring added later? Can I use it in place of yellow or white cornmeal?

Blue corn has been around for centuries and is grown and eaten by various Native American tribes for ceremonial use. Hybrids are now being cultivated to make the most of both yellow and blue corn. The blue color is actually purple pigment in the skin of the kernel, and the corn tastes much like yellow corn, depending on the variety. You can use blue cornmeal in recipes for cornbread, but add a pinch of baking soda to the batter to keep the blue looking blue, not murky. If you use blue cornmeal as breading, take care not to turn the oven heat too high. It tends to overbrown at temperatures higher than 350°F.

Mustards

There are so many different kinds of mustards in the grocery story. Which ones do I need?

If you are a mustard fan, one kind of mustard won't do. You should have on hand one good-quality Dijon mustard, a can of Coleman's dry mustard, and possibly a honey mustard if you like that taste. Other good choices are green peppercorn mustard and herb mustard. Mustard's affinity for meat and practically all salad dressings makes it an indispensable pantry basic.

What is the difference between Dijon mustard and other mustards?

The method of making Dijon mustard is quite different from methods used for other mustards. Traditionally, this mustard is made from seeds of a specific mustard plant called *Sinaplis alba* and is mixed with white wine. Dijon mustard has been made in the city of Dijon since the eighteenth century. The region is considered to grow the very best mustard seeds, and the variety *moutarde Blanc de Dijon* is stronger than other French mustards. I highly recommend Maille mustard, which is imported from France and when fresh has the right pungency and depth of flavor. If you can, avoid Grey Poupon Dijon mustard, which is made in California and has little flavor.

When do you use dry mustard?

A little dry mustard reconstituted with water goes a long way. I often add it to prepared Dijon mustard to kick up the flavor. Certain vegetables—especially carrots, greens beans, and celery root—need that extra punch, so whenever you make a mustard dressing to go with these vegetables, be sure to add some reconstituted dry mustard to the dressing.

If a recipe calls for Chinese mustard, can I substitute other mustard if I don't have any?

Chinese dry mustard is reconstituted for the mustard sauce that accompanies spareribs, dumplings, and other Chinese dishes. You can use Coleman's dry mustard.

Do I need to refrigerate mustard once the jar is opened? How long does it keep?

Once it's opened, definitely refrigerate mustard and even then try to use it within a few weeks. I consider myself a compulsive mustard buyer and never pass up an opportunity to buy it when I am in Canada or France. But I find that often the mustard does not hold up for more than a couple of months.

I recently opened a jar of mustard and it had a really off taste. How do you tell good mustard from bad?

Look at the jar and be sure to buy mustard only at a grocery store that sells a lot of it. If the mustard has separated and has even a tiny layer of oil at the very top, pass on it. Mustard has a limited shelf life and also loses taste and pungency if it has gone through extreme climate changes, as it does in the winter months. I recommend buying it at a market, not necessarily a specialty store, with a large turnover. Do not hesitate to return it if the taste is off.

Olive Oil

There are so many different brands of olive oil. Which do you recommend?

For everyday use other than in salads, Bertolli, Colavita, and DeCecco virgin or extra virgin oil are good all-around choices. These are mass-produced oils and none, in my opinion, is good enough to be used in salad dressings. But they are fine for sautéing vegetable and tomato-based sauces. For salad dressings, I particularly like the Spanish L'Estornell, the French Nicolas Alziari, and the Tuscan Antinori, which are full bodied and fruity.

Pure olive oil has been chemically refined. This strips it of most of its flavor and aroma. The result is a rather tasteless, colorless oil. Sometimes you may see pure olive oil with a slightly greenish tinge, which means that a little extra virgin olive oil has been added to give it some flavor. It is fine for cooking but does not hold up in salads where a good olive oil shines.

What is the difference between virgin and extra virgin olive oil?

Extra virgin olive oil is made from the first pressing of the olives. It is essentially the juice of the olive and to qualify as extra virgin it must be of low acidity, less than 1 percent. The percentage is always printed on the label.

Virgin oil is made with olives that have already gone through the first pressing. This means that the oil has a much higher acidity level and is less flavorful.

There is such a price difference in olive oils. Why is that?

When you buy an olive oil, check the label to see if it tells you where the olives have been harvested. Many less expensive oils are labeled "Product of Italy," but the olives actually come from various parts of the world and are machine processed in Lucca, Italy, resulting in an oil with no distinctive flavor or quality. First-press extra virgin olive oil requires a lot of hand labor; hence the high price.

When it comes to picking an expensive high-quality olive oil, much depends on how much you like an intensely flavored oil. I would start with a small bottle of very good oil from either France, Spain, or Italy and use it in a simple vinaigrette. Then decide if you think the price is worth it.

What exactly is light olive oil? Is it lower in calories?

Light olive oil is a marketing gimmick. It has nothing to do with calories and everything to do with lack of taste. There is really little point in using it except for certain baking.

Can you tell a good olive oil by its color?

The color of olive oil depends on the type of olives and where they come from. The color can range from a golden yellow to a deep green. Olives from each of the great growing regions—Tuscany, Puglia, Catalonia, Crete, and Provence—have their own flavor and color characteristics. The only way to learn and choose your favorite oil is by tasting oils from different regions.

I understand that olive oil is extremely high in fat. Can I use less or can I use another oil instead?

All oil is close to 100 percent fat and should be consumed in moderation. Some oils are better for you than others, however. Olive oil is the essential oil in the Mediterranean diet, and many studies have shown that this monounsaturated fat may be beneficial in reducing "bad" serum cholesterol.

What exactly does the term "fruity" mean when applied to olive oil?
This is a lovely way to describe an olive oil, because when you taste really good oil, it can remind you of freshly cut grass, apples, pears, and even melons.

I now see organic olive oils on the market. Are they better or safer? They seem expensive.
Because of the extensive use of pesticides on olive trees, organic olive oil is gaining popularity, and I buy it whenever I see it in my grocery store. Look for L'Estornell, a medium-priced, excellent extra virgin olive oil from Catalonia. Two excellent sources for organic olive oils are Zingerman's and Williams-Sonoma (see Sources, page 397).

I have been using the same extra virgin olive oil from Tuscany for years, and there seems to be a difference in taste from one bottle to another. Why is that?
This happens often. Olive oil is an agricultural product and much as with wine, its flavor is influenced by weather, climate, vintage, and so on. However, proper storage and handling also play a major role. If the oil tastes "off," return it to the store and exchange it for another brand.

Quite honestly I do not have much experience with olive oil. How can you tell a bad olive oil?
Bad oil will taste musty, rancid, and even bitter. It may leave you with a sharp unpleasant taste in the back of your throat.

What is the best way to keep olive oil?
All you need is a cool dark place in your kitchen or pantry. If this is not possible, then refrigerate the oil and allow 5 minutes for it to warm up and decongeal before you use it.

Vegetable Oil

Are there certain types of cooking oils that should I keep on hand?
In addition to olive oil, you will need corn, canola, or peanut oil and possibly grapeseed oil. Corn and peanut oil have more flavor, but canola oil, like olive oil, is monounsaturated. All three are also similar in the sense that they have a rather high smoking point and are, therefore, good for sautéing. Grapeseed is the least flavorful but has an even higher smoking point. It is also becoming increasingly popular for use in salads because it is high in linoleic acid, an essential fatty acid that is good at lowering cholesterol.

Specialty stores carry unusual oils that look so appealing. How do you use the citrus-, herb-, and nut-flavored oils?
Specialty oils are a wonderful addition to the pantry but should be bought, used, and stored carefully. Nut oils are best used in salad dressings with greens that complement them. Be sure to use mild vinegar so as not to overwhelm the nuttiness of the oil. Sesame oil is an excellent flavor addition to Asian foods. Citrus oils are excellent with cooked shellfish and in marinades. I am not a great fan of herb oils, since they are so specific in their taste and not

versatile enough for daily use. Rather, make your own herb oil as needed.

To make an herb oil Bring a large quantity of salted water to a boil. Fill a bowl with ice water and set aside. Add 1 heaping cup flat-leaf parsley, basil, or cilantro leaves to the boiling water and blanch for 20 seconds. Drain and immediately transfer to the ice water. When cool, drain and squeeze out as much water as possible. Transfer to a blender and add 2 cups canola oil in a thin steady stream. Blend for 3 to 4 minutes. Transfer to a jar and refrigerate.

TIP Nut oils have a very short shelf life, so be sure to refrigerate them as soon as they are opened. In fact, these oils will go rancid even when they have not been opened, so use them as soon as possible after purchasing them.

How long can I store cooking oil? Does it need to be refrigerated?

Cooking oils should be fine for up to 2 months, but they need to be stored in a cool, dark place. Refrigerating oils is definitely a good idea. Keep in mind that different oils have different shelf lives. In particular, olive and peanut oils, which are high in monosaturated fat, are best kept refrigerated.

Olives

Is there a good brand of canned or jarred olives?

I cannot recommend any brand of canned olives. They are simply tasteless. When it comes to jarred olives, choose green olives imported from Spain. Jarred black oil-cured olives are not a good choice, since they are quite bitter and very salty.

What kind of olives can I keep in my pantry as a snack?

The green pimiento- or anchovy-stuffed olives from Spain are excellent. Taste various brands and choose those with the most flavor and a crisp texture. But I prefer to buy loose olives and store them in the fridge. They keep for several weeks.

How long do olives keep?

It depends on whether they are brined or packed in oil. As long as the olives are covered in brine, they will keep for several weeks. You can add olive oil to oil-cured olives to extend their shelf life.

I love spicy olives and have tasted really good ones in various restaurants but cannot find them in my grocery store. Can I make my own?

You can flavor your own olives very easily. Just add a finely sliced jalapeño pepper, 1 or 2 sliced garlic cloves, and some fruity olive oil to your olives and let them marinate for 2 or 3 days. Get a crusty loaf of bread and a nice, slightly aged goat cheese and enjoy!

Peppercorns

What is the difference between black, white, green, and red peppercorns? How is each used?

Black and white peppers grow as berries on the pepper shrub. White pepper is the seed of the fully ripe berry. Black pepper results when the pepper berries are picked not quite fully ripe. Green peppercorns are the underripe berry of the *Piper nigrum* and are available dried and preserved in brine. Pink peppercorns, which are pungent and slightly sweet, are the dried berries from the Baies rose plant.

Black peppercorns are the most widely used, in salads, pasta dishes, marinades, rubs, and overall seasoning. White peppercorns are generally used with light-colored sauces where the dark-colored pepper would stand out. They are also used extensively Scandinavian cooking. Green peppercorns are used in a French butter sauce as well as the classic green peppercorn sauce that accompanies a steak. Pink peppercorns are more of a novelty item, but you may still see them used in vinaigrettes and light fish sauces.

TIP You can now find a peppercorn blend that includes Sichuan, pink, green, black, and white peppercorns. It is always best to toast this blend before transferring it to a mill; toast only as much as you think you will use at a time.

Specialty food catalogs offer all kinds of black peppercorns. Which do you recommend?

By far the best but also the most expensive peppercorns are Tellicherry, imported from the Malabar Coast of India. These are large and have a robust flavor, first hot, then with a sweet aftertaste. The Lampong from Indonesia are also excellent. They are smaller and have a sharper bite. Another good choice is Sarawak, from Malaysia. It is quite acceptable, but nothing comes close to the fruity and well-rounded taste of Tellicherry.

If you want to really understand the difference, get several types of whole black peppercorns and try them, freshly ground, with a salad or a pasta dish. You will soon be able to tell the difference.

TIP Peppercorns that are not labeled by variety or point of origin are most likely imported from Brazil and are of inferior quality. Avoid them if you can.

My grocery store carries peppercorns in various grinds—fine, medium, and coarse. Is it okay to use these instead of whole peppercorns?

There is a world of difference between freshly ground pepper and the ground spice you get at the grocery store. But if a recipe such as a marinade calls for as much as 1 tablespoon of ground pepper, or when you prepare a "rub" for a turkey, then using a medium or coarse grind of pepper makes sense.

Porcini

What exactly is a porcini mushroom?

Porcini is a fabulous-tasting fall mushroom that is rarely available fresh in markets in the United States. However, dry porcini, which have a rich woodsy aroma and a wonderful concentrated taste, are now available in gourmet shops everywhere. They are sometimes sold by their French name, *cèpes*. Make sure to choose those that are tan to pale brown and avoid those that look crumbly, a sure sign that the porcini are not "fresh."

I recently bought dry mushrooms and they were gritty even after several washings. Is there a secret to washing these mushrooms?

You probably bought the imported mushrooms from Chile that come in plastic containers simply labeled "imported dry mushrooms." These mushrooms, which are much less expensive than the Italian porcini, have an intense smoky flavor. I use them often for soups and risottos as well as in pasta dishes. They usually come with their stems attached, which contain a lot of grit.

To remove it, you will need to soak the mushrooms in warm water for 15 minutes. Once they are reconstituted, be sure to strain the mushroom broth through a double layer of cheesecloth.

I bought a large bag of porcini while I was in Italy, but by the time I wanted to use them, they were all broken. What is the best way to store dry mushrooms, and how long do they keep?

Dry mushrooms keep best in the fridge for as long as 6 months.

Salt

Is iodized salt better than plain?

I buy only salt that has not been iodized. Iodized salt is supposedly better for you because it has had various minerals added to it, but the result is a salt with a rather unpleasant taste that can easily change the flavor of a dish. People who eat fish regularly get enough iodine in their diet and do not need to use iodized salt.

What is the difference between coarse and kosher salt?

There is no difference. Mineral or rock salt is the common salt used in the United States and many other countries. It comes three ways: fine, iodized, and coarse, which is also called kosher salt. The two best-known kosher salts are Morton's and Diamond Crystal.

TIP When a recipe calls for 1 teaspoon salt, you need 1½ teaspoons coarse or kosher salt.

I find many recipes are confusing when it comes to salt. Many call for just salt, some call for coarse salt, and others call for sea salt. Is it necessary to have all three salts?

I use kosher salt in cooking and fine salt for last-minute flavoring, but I much prefer sea salt in salads. If you want to use just one kind of salt for everything, use sea salt, which has a less salty, sweeter taste than mineral salt.

What exactly is sea salt?

Sea salt is the mineral extracted by evaporating seawater. Mineral salt is extracted from mineral deposits in the earth.

The only sea salt I can find comes in large granules. How shall I use it?

The only way to use the granular sea salt is in a salt mill, which looks very much like a pepper mill. However, in cooking, it is far more practical to use fine sea salt, which is becoming increasingly available in gourmet shops everywhere.

Which is the best kind of sea salt?

This is really a matter of taste. I am a great fan of French sea salt from the Mediterranean and the coast of Brittany. Sea salt from Sicily has a wonderful taste as well and can be found in many good specialty stores. French Mediterranean sea salt in granular form is available in many upscale supermarkets around the country.

TIP When buying a salt mill, look for one with a good mechanism, such as the French Peugeot, and avoid the "see-through" plastic ones.

I recently saw several kinds of very expensive salt listed in a fancy food catalog. Which do you recommend?

The finest sea salt is the type raked by hand from the salt beds on the Atlantic coasts of Brittany and Dover, called *fleur de sel* in France and Malton salt in England. These salts are used on the table as condiments rather than in cooking. All you need are 3 or 4 grains of these salts to "up" the flavor of a grilled steak, a roasted piece of fish, or a salad. They are never used in cooking because they are so costly and also because of their intense salty flavor. However, it is fun to have some *fleur de sel* around. Nothing is more delicious than a super-fresh radish dipped in a little sweet butter with 1 or 2 grains of this salt. The same goes for a hard-boiled egg or leftover roasted turkey. *Fleur de sel* should be used coarse, and should be ground in a salt mill.

TIP Remember to use a pinch of salt in all cakes, muffins, and tarts. Salt makes sweet things taste sweet and gives them a complexity of flavor that makes all the difference.

Which is the best seasoned salt?

Seasoned salts are not a good idea. They are usually flavored with onion, garlic, and celery seeds and will change the flavor of any dish.

Spices

I have a cupboard full of spices but seem to have little use for many of them. What are the essential seasonings I need to keep on hand?

Here is my list of must-have spices:

■ **Caraway seeds** Besides being synonymous with rye bread, caraway seeds are a must in a goulash or a Viennese Goulash Soup (page 237), braised sauerkraut, and Ukrainian and Russian borscht. I also like the flavor of caraway seeds in a pork roast and roasted duck.

■ **Cardamom** Here is a spice that is a must in Indian cooking. It has a sweet floral aroma that enhances curry and is delicious in a rice pudding. It is also a key ingredient in garam masala, the Indian spice mix.

■ **Cayenne, red pepper flakes, whole chiles** A pinch of cayenne goes a long way to add a little bite to bland food. I usually add a pinch of cayenne to all egg-based dishes, especially scrambled eggs, omelets, and frittatas. A pinch of cayenne also perks up a potato salad, a slaw, or a tomato sauce.

When you are looking for real heat, you can use a teaspoon of red pepper flakes at the beginning of a preparation, but I prefer whole chiles, since you can regulate their heat. If you use them whole and remove once they darken in the oil, you get a light- to medium-spicy dish; if you break them up, you will get real heat. Try to find the Thai dried chile peppers, which come in small cellophane bags, rather than the Chinese ones that are larger and less seedy, and, therefore, less hot.

■ **Cinnamon, ground; cinnamon sticks** This is probably America's favorite spice. It is good when paired with apples, peaches, pears, and bananas in fruit desserts and also in quick breads and cakes. Ground cinnamon is also a good addition to Tex-Mex chili and to Greek lamb stews. Cinnamon sticks add great flavor to fruit compotes, homemade apple sauce, curries, and rice puddings, and are a must in the Indian spice mix garam masala.

TIP **Be sure to look for Ceylon cinnamon sticks rather than the cassia variety, which is what you get in the average grocery store. You will probably have to stop at a specialty shop or good cook shop for Ceylon cinnamon. It is also widely available by mail-order.**

■ **Coriander seeds** The whole seeds are best when lightly toasted. The somewhat soft seeds can be crushed in a mortar and pestle or placed between layers of paper towels and crushed with a heavy pan or rolling pin. Besides using them in Indian and Asian dishes, I like to sprinkle them over pan-seared salmon or tuna.

■ **Coriander, ground** I like to add ground coriander to curry in Indian preparations and to flavor a yogurt marinade.

■ **Cumin, ground** Ground cumin is a must in an all-American chili, in Mexican dishes, and in Indian marinades, especially those based on yogurt. I also like to use ground cumin in a honey-ginger vinaigrette and as part of a rub.

■ **Cumin seeds, whole** The seeds are best toasted before using, then must be ground in a spice grinder. I use cumin seeds extensively in marinades in Indian-type stews, and many Middle Eastern dishes. I also sprinkle some into a cucumber salad.

■ **Fennel seeds and Anise seeds** I put these together because they are so closely related. Both have a slight licorice taste and are good in many Middle Eastern and Indian dishes. I use fennel seeds more than anise—the seeds are softer and can be used whole in all dishes

made with fresh fennel, since they enhance the flavor of the vegetable. Both seeds can be used whole in a marinade for a whole fish, lamb chops, and pork tenderloins. Anise seed is better toasted; it adds an interesting flavor to a tomato sauce, to a Mediterranean seafood and tomato stew, and to biscotti and some cookies.

■ **Ginger, ground** I use ground ginger mainly in a cobbler topping, quick bread, and yogurt marinade. It is not interchangeable with fresh ginger.

■ **Juniper berries** Although I may only use juniper berries, which contribute an herbal, "gin" taste, a few times a year, I still find that I like to have them on hand for a flavorful sauerkraut and the occasional red wine marinade for a boned leg of lamb.

■ **Nutmeg, whole** Whole nutmegs are the olive-shaped hard seeds of the nutmeg tree. I much prefer to use freshly grated nutmeg rather than the ground spice, which has a short shelf life and is never as rich and aromatic as when you grind it fresh.

TIP **An inexpensive nutmeg grater is a rather clumsy gadget. I always feel that I am about to grate off the tip of my finger with it. Now you can get several nicely designed nutmeg graters that make grating nutmeg quick and easy.**

■ **Paprika** Paprika is one of my favorite and most frequently used spices. I use two kinds: the imported sweet Hungarian paprika (see page 32) that comes in a small can and Spanish smoked paprika (Pimentón de la Vera) that also comes in small cans but is not widely available. Do not purchase bottled paprika or even the one that comes in cans if it does not specifically say "Sweet Hungarian Paprika." All paprika should be refrigerated once it is opened.

■ **Peppercorns** Peppercorns are the most important kitchen spice. The best come from India, and you should look for ones labeled Tellicherry, Lampong, or Sarawak. Tellicherry is the very best. Chances are that you will not get any of these in the average grocery store and will have to stop at a specialty store or buy the peppercorns by mail. (See also pages 26–27.)

What is the difference between using whole and ground spices?

There is a world of difference in taste between a ground spice and one that you grind yourself. However, it is generally more practical to use a ground spice, and as long as it is fresh, it is fine. It is important to follow the recipe; if it calls for a whole spice, do as it suggests. For example, when making a sugar syrup, you should use a cinnamon stick rather than ground cinnamon.

How long do ground spices keep?

Much depends on the quality of the spices and how they are stored. Generally, I keep ground spices for only 4 to 6 months and then replace them. If the jars are tightly closed each time after you use the spice, that will help to keep them fresh. Curry and chili powder must be refrigerated as soon as they are opened.

TIP Sniff a ground spice and rub a little between two fingers. If it is highly aromatic, it is fresh.

Is it better to buy spices in jars or cans?

I buy them only in jars, since I usually opt for Spice Island, Wagner's, or Vanns, which come only in jars. I do not recommend buying spices in small cellophane bags or loose because they are less likely to be fresh.

How long can you keep whole spices?

If stored in a cool dry place, they should last for at least 2 or 3 years. When in doubt whether a spice has retained its flavor, toast a little in a skillet. If the spice is fragrant, it is fresh.

What is the best way to grind whole spices?

I suggest that you get an inexpensive coffee grinder and use it only for that purpose.

What is the best way to use garlic and onion powder in cooking?

I do not suggest you use either of them. Both impart a bitter taste to food and are hard to digest. When it comes to garlic, it is easy to mash a whole clove and add it to your seasoning. A couple of whole unpeeled onions cut in half will add the flavor you look for in a pan juice. You can also grate some raw onions into a spice mixture.

How long do bay leaves last? I rarely use them and am not sure if they are still good.

Whole bay leaves will last indefinitely if stored in a well-sealed container. Be sure to buy only Turkish bay leaves. They do not look as pretty as the California variety, which are slender and pointed, but they are much more flavorful. Make sure to remove the bay leaf before serving a dish because its sharp edges can get stuck in the throat, and people have been known to choke on them.

Why is saffron so expensive?

Saffron comes from the stigmas of a small purple crocus. An ounce of saffron contains 14,000 of these stigmas, each of which must be laboriously hand-picked and then dried.

***When a recipe calls for saffron, is there
another spice I can use instead?***

For a similar color, you can use turmeric, but
the unique taste of saffron cannot be replaced
by another spice.

***I bought some saffron a couple of years ago.
How can I tell if it is still good?***

If the saffron has become a very dark red,
chances are it is not fresh. Another good way to
tell is to toast a tiny bit in a skillet, shaking it
until the saffron becomes fragrant.

***What is the difference between ground chiles
and chili powder?***

Pure ground chiles are exactly that, while chili
powder usually has cumin added to it.

What exactly is Hungarian paprika?

There are various types of paprika available
from different parts of the world, with Hungarian considered the best. Hungarian paprika is
made out of special kinds of red peppers that
have a unique, slightly sweet flavor. It comes
both sweet and hot, but I prefer to buy it sweet
and add heat with a small chile pepper. Be sure
to buy Hungarian paprika only in cans labeled
"Sweet Hungarian Paprika." I also like the
smoked paprika now imported from Spain. It is
not widely available but is worth looking for.

***Curry powders differ widely in flavor—some
seem too hot, others too bland. Which do
you recommend?***

Curry powder is actually a very complex spice
consisting of up to twenty different spices,
herbs, and seeds. It varies from region to region
and according to the taste of the cook. A good
supermarket choice is the Madras curry im-

ported by Sun Company that comes in cans. It
should be refrigerated after opening.

***If a soup recipe calls for fresh herbs, is it
okay to use dried herbs instead?***

The taste of dried herbs generally is too aggres-
sive and can easily overwhelm the milder flavors in the soup. Make sure you add fresh herbs
only at the very end of cooking, because the
heat will kill their distinctive yet delicate flavor.

Tomatoes, Canned

***Many recipes call for whole canned
tomatoes, but since I only use them for
sauces and soups, doesn't it make sense to
buy them crushed instead?***

I really do not recommend buying crushed
or chopped tomatoes. The best tomatoes are
usually processed whole. Those that are packed
crushed or chopped usually contain a fair
amount of tomato paste, which gives the tomato
an unpleasant taste and texture.

***Is there a difference in taste between round
and plum tomatoes?***

The best canned tomatoes are plum tomatoes.
They are fleshier. I find that once you drain a
can of round tomatoes, you are left with little
tomato pulp and tons of juice. One of my
favorite brands of tomatoes, however, is Muir
Glen, which is a round organically grown
tomato that is packed with flavor and has very
good consistency. Unfortunately, they are not
widely available outside the West Coast, so you
may have to search for them.

***Many recipes call for a 2-pound can of plum
tomatoes, drained, or 2 cups, but often I do
not get 2 cups of tomato pulp out of a can.***

Is there a brand that really yields 2 cups of drained tomato pulp?

As long as you buy top brands of Italian plum tomatoes, preferably those imported from the San Marzano area, you will get plenty of tomato pulp. The worst offenders are brands such as Progresso and Vitello that contain tomatoes of inferior quality, with little pulp and lots of juice. Even the imported Italian tomatoes vary in quality. Some of my favorites are the organic Muir Glen from California and the Italian Lavalle, but these may not necessarily be available in your area. Try different brands, and you are bound to find a good one. You may have to seek them out at gourmet grocery stores rather than your local supermarket, which may carry only the large name brands.

What is the best way to store leftover canned tomatoes? Should I freeze them or just refrigerate them?

You can simply refrigerate them in a jar, and they will keep for as long as a week. If you have drained the tomatoes and are left with some pulp, you can add a sliced clove of garlic, some shredded basil leaves, and a few tablespoons good extra virgin olive oil. This adds wonderful taste and adds shelf life as well.

Can I make really good tomato soup with canned tomatoes?

You can make an acceptable soup with canned tomatoes, but unfortunately, it's almost impossible to duplicate the flavor of fresh, ripe tomatoes. I suggest making tomato soup when tomatoes are in season. If you don't have enough, supplement them with either organically grown canned tomatoes or those imported from the San Marzano region of Italy.

Tuna, Canned

What brands of canned tuna do you recommend?

Look for Progresso light tuna packed in olive oil. It is an excellent brand that is widely available. I think the flavor of the light tuna is more interesting than white albacore. Although it is called light, it is in fact rather dark in color and very juicy. If you are watching your fat intake, tuna packed in spring water is a good choice.

Is there a difference between tuna packed in vegetable oil and tuna packed in olive oil? It seems so much more expensive.

Olive oil adds more flavor to tuna, but if you can't get tuna packed in olive oil, you can buy it packed in vegetable oil. Once drained, cover it with virgin olive oil and let the tuna marinate in the oil for several days.

I have a terrific recipe for a niçoise salad that calls for freshly grilled tuna. Can I use canned tuna instead?

The traditional Provençal salade niçoise is always made with canned tuna, which gives the salad its special characteristics.

Vinegar

What kind of vinegar is best for salads?

Different salads require different vinegar, and much depends on the types of greens you use. With mild, buttery greens such as bibb and Boston lettuce, I usually use a milder vinegar like balsamic or Champagne vinegar, while romaine, frisée, and radicchio need a more assertive vinegar such as sherry or a wood barrel–aged red wine vinegar.

My supermarket carries a large selection of vinegar. How do I go about choosing one or two?

This is good news, because most supermarkets have very poor selections of vinegar. So check it carefully. If you like salads as much as I do, then I would start with three key vinegar varieties: a good wood barrel–aged red wine vinegar, preferably from France or the West Coast wine country, a Spanish sherry vinegar, and a balsamic vinegar. I also keep in my pantry some Japanese rice vinegar and a French Champagne vinegar, but these are not as important.

Besides the generic red wine vinegar by Heinz, I see red wine vinegar from Spain, Italy, and France. Which do you recommend?

The country of origin is not as important as the quality of the vinegar. Read the label carefully. For openers, it should state what kind of grapes were used and the percent of acidity. Pass on anything you generally see on the supermarket shelf, which is usually industrially made vinegar.

Should vinegar be very acidic or mild and sweet?

The usual acidity in vinegar is between 6 and 7 percent, and I always check the bottle before buying it. However, this is not a criterion you can apply to all vinegar, because so much depends on the way the vinegar is made, how it was bottled, and so forth. A poor-quality balsamic vinegar can be harsh in spite of being low in acidity, while a sherry vinegar may have a higher acidity yet taste smooth and mellow.

I cannot seem to get good vinegar and find the labels very confusing. Can you recommend a source for good vinegar?

Buying good vinegar can be more confusing than buying good olive oil. You'll find a thorough discussion in Ari Weinzweig's *Zingerman's Guide to Good Vinegar,* available from the Zingerman's catalog (see Sources, page 399). Other than aged balsamic and sherry vinegar, most vinegar is inexpensive, and it is well worth trying a few kinds to see what you like.

What exactly is sherry vinegar? Which one do you recommend?

I am extremely partial to sherry vinegar, since I grew up in Spain, but the truth is that good sherry vinegar is a fabulous product that will enhance anything from a sliced ripe tomato to the pan juices of a roasted chicken (see Provençal Chicken with a Sherry Vinegar–Garlic Essence and Concassé of Tomatoes, page 206). Quality sherry vinegar is made in the sherry-producing region of Jerez in southwest Spain. The most widely available brand is Sanchez Romate. Williams-Sonoma carries another excellent brand called De Soto. Some of the smaller lesser-known brands, such as Santa Maria, are available in different parts of the country.

I keep experimenting and believe that you should try various brands until you find one that really appeals to you.

I have seen balsamic vinegar at five dollars a bottle, at fifteen dollars, and even much more. What is the reason for the price difference?

In balsamic vinegar, there are no bargains. What you pay for is what you get. Making real balsamic is a very long process. Just think, a moderate-quality balsamic vinegar has to age for fourteen years. Add to this another forty to fifty years for great-quality vinegar, and you have the reason for the price difference. I would definitely stay away from the five-dollar bottles. The stuff is only balsamic by name and its age is about the time that it takes for it to get from Italy to your grocery store. A medium-priced balsamic should cost twelve to fourteen dollars, at which point you should be able to get some real flavor. The best way to determine if you are getting something worthwhile is to taste it.

Good specialty stores are now conducting tastings, and it is really important that you get to taste several vinegars before settling on one.

What should good balsamic vinegar taste like?
What you should look for is a complex, mellow, sweet and sour taste. A good balsamic has character. It should not be overly sweet or too acidic.

Ethnic Ingredients

Coconut Milk

Is there a difference between coconut milk and coconut cream?
Unsweetened coconut milk is used in many Southeast Asian dishes. It is made by soaking the grated coconut meat in water and straining the results. Coconut milk is available canned and frozen, or you can make it yourself from a fresh coconut. Coconut milk separates in the can with the thicker coconut cream rising to the top. If you need the cream for your recipe, skim this off. If your recipe calls for thick coconut milk, shake the can vigorously to mix the two together. Coco Lopez is sweetened coconut cream and is used for desserts. "Light" coconut milk is the same liquid with less of the fatty cream, but it also has less of the distinctive flavor and fragrance.

Does cooking with coconut milk require any special techniques?
After adding coconut milk or cream to your dish, heat the mixture gently, never letting it boil, or it will curdle. A bit of cornstarch will also help prevent curdling. If you open a can of coconut milk and find oil floating on top, discard it. This is an indication that the product was overheated during shipping. You can store opened coconut milk in the refrigerator for a few days, but do not let it sit at room temperature because it spoils easily.

Edamame

What are the salted pea pods served in Japanese restaurants? How do you prepare them?
They are fresh soybean pods, called *edamame*. Although you can find them fresh seasonally in Japanese grocery stores, your best bet is to buy them frozen, in which case they have already been steamed. All you have to do is to pop them into boiling water for 2 or 3 minutes, then refresh them in ice water and serve sprinkled with coarse salt.

Fish Sauce

Is fish sauce made from fish or for fish? How is it used?
Fish sauce is called *nuoc mam* in Vietnam, *nam pla* in Thailand, and sometimes fish gravy in

China. It is made from salted fish and is used in place of salt. Thai and Vietnamese brands are the best.

Garam Masala

What is garam masala? Where can I get it?
Good-quality garam masala is a sweet and often spicy blend of several spices that you can make yourself. The basic combination usually includes cumin seed, cardamom, black pepper, cloves, cinnamon, and occasionally hot pepper. You will find recipes for excellent garam masala In Julie Sahni's *Classic Indian Cooking* and in Madhur Jaffrey's *Indian Cooking*. You can buy prepared garam masala in good specialty stores or order it by mail from Kalustyan's (see Sources, page 402). Be sure to store the spice mix in a tightly covered jar and refrigerate it. As with all ground spices, the shelf life of garam masala is relatively short.

Ginger

What should I look for in fresh ginger, and is it a seasonal ingredient?
Always look for large ginger knobs that are unwrinkled, have a smooth skin, and are golden-beige in color. The best will have a yellowish-green interior and will smell spicy and gingery as you break off a piece. This is called winter ginger. Fresh stem ginger, also called spring or baby ginger, can be found during the spring in Asian grocery stores. It has a thinner skin and is milder than winter ginger.

Ginger will keep uncovered in the refrigerator bin for 4 to 6 weeks. I often buy a large piece, scrape off the peel with a spoon and then slice it with a paring knife or a mandoline and place it in a jar with rice vinegar to cover. Refrigerated, the ginger will keep indefinitely.

TIP For a delicious drink, combine 2 cups of water with 2 slices of ginger. Simmer for 5 minutes, strain, and serve. I find this to be great for colds and sore throats.

Can I substitute dry ginger for fresh ginger?
Powdered ginger does not add the snappy fresh flavor of knob ginger. Instead, it can easily overwhelm a dish with its spiciness, so be careful when using the powdered spice.

How do you juice ginger?
To make ginger juice, peel and slice the ginger root, then crush it with the side of a cleaver or a large chef's knife. Use a garlic press to squeeze the crushed slices. You can also use a ginger grater, which releases the ginger juice. These little graters are now available in cook shops everywhere.

Lemongrass

Some Asian recipes call for lemongrass. What does it look like and is there any substitute?
Fresh lemongrass looks like a long woody stalk with coarse leaves. It is pale yellow-green and cannot be eaten raw. When you add fresh lemongrass to a recipe, use only the bulb, 4 to 5 inches above the stem end; discard the tops. This softer part of the stem can be coarsely chopped or bruised with the side of a cleaver or a chef's knife and added to hot and sour Thai soups and to coconut milk–based and curry-flavored dishes.

Remember that lemongrass, just like bay leaf, is not eaten and is only used to flavor. If you cannot get fresh lemongrass, use the juice of a large lime and some lime zest rather than lemon.

TIP I have recently seen lemongrass sold in small packages containing 4 to 6 pieces. Stored in the crisper of the refrigerator, these will keep for several weeks.

Mirin

What is the difference between mirin and rice wine?

Mirin is a Japanese rice wine made from fermented rice and water. It is now widely available in Asian stores and many supermarkets. Unfortunately, most commercial mirin contains corn syrup, which masks its authentic flavor. For pure mirin, I usually go to a health food store or a good Asian market.

Chinese rice wine is also made from fermented rice but is aged for at least 10 years. It has a much drier flavor, with an alcohol content similar to a dry sherry. In fact, you can substitute Amontillado or Dry Sack sherry in recipes that call for Chinese rice wine. I often use Chinese rice wine in marinades and in deglazing a pan, while mirin is a lovely addition to sautéed vegetables that have a natural sweetness, such as snow peas, carrots, and beets.

What Asian sauces should I keep on hand for a quick stir-fry?

In addition to soy sauce I keep oyster and hoisin sauces, which are very flavorful additions to stir-fried vegetables, shellfish, and noodle dishes.

Rice Vinegar

Does rice vinegar differ greatly from other vinegars?

Both Chinese and Japanese rice vinegar are milder and less acidic than other standard vinegars you might have on your shelf. Japanese rice vinegar is pale to golden yellow and also comes seasoned with sugar. I use rice vinegar in a salad dressing to which I add sugar, sesame oil, and some grapeseed oil and toss it in a cabbage salad or with mixed greens. Pearl River Bridge is a good brand.

Sichuan Peppercorns

How important is it to use Sichuan peppercorns in recipes that call for them?

Sichuan peppercorns are not really pepper but rather small berries with a sharp spicy taste. They add an interesting flavor to highly seasoned marinades. I also like to mix them with coarse salt as a dipping salt for deep-fried quail or the Fried Cornish Hens "al Ajillo" on page 211. Just like cumin seeds, Sichuan peppercorns have to be toasted before being ground in a spice grinder or by mortar and pestle. You can also smash the peppercorns with a heavy iron skillet. Place the peppercorns in a small dry skillet and toast them over medium heat until very fragrant. Be sure to store the peppercorns in a vacuum-sealed bag to retain their fresh taste.

Soy Sauce

The Asian market where I buy soy sauce has many varieties. Are some better than others?

Selecting soy sauce has much to do with what you are going to use it for. For Chinese preparations, especially stir-fry, it is best to use a Chinese soy sauce that contains molasses and lends a touch of sweetness and gloss to the food.

In Japanese-inspired dishes, a good choice is Kikkoman, which is light in color and not too assertive. Tamari, the darkest of the soy sauces, has a strong flavor that can overwhelm delicate foods.

Are mushroom soy and superior soy better than ordinary soy sauce?

Superior, a soy sauce that contains molasses, is sweet and thicker than other brands. I sometimes use it in a marinade and for a last-minute coating for grilled fish steak or baby back ribs. But you must be careful not to use too much, particularly when it comes to grilling, since the sugar level in the sauce can easily make food burn.

Mushroom soy is infused with shiitake or straw mushroom extract; you can substitute it in any recipe that calls for plain soy sauce. If you are interested in the different flavors of soy sauce, buy one of each and experiment to see what you like best. Remember that soy sauce has an almost indefinite shelf life.

Is light soy sauce less salty?

Unfortunately not. Light means that the soy sauce is more delicate and lighter in color, but the salt content is pretty much the same. Kikkoman makes a "lite" version of its soy sauce with 40 percent less sodium.

Wasabi

I have not had much luck making wasabi from powder. It is always too runny and does not have the same texture as that served in restaurants. Is it possible to buy prepared wasabi?

You can now find prepared wasabi paste in every Japanese grocery store and many supermarkets, but I find the powder to be stronger, more flavorful, and less expensive. The key to a thick paste is to add water gradually drop by drop. Let it develop flavor for at least 10 minutes before serving.

The Baking Shelf

Chocolate

I asked Alice Medrich, author of *Extraordinary Chocolate Desserts*, to contribute some of her expertise on chocolate.

What is the best way to melt chocolate?

Several methods work well, so long as you pay enough attention to remove the chocolate from the heat while it is warm to the touch, rather than hot. Chop the chocolate into pieces about the size of almonds or smaller, then melt the chocolate in one of three ways: microwave at 50 percent power (use 30 percent for milk and white chocolates), using short increments of time. You can also use a double boiler with barely simmering water, or, and this is my favorite method of all, put the chocolate in a heatproof bowl set directly in a wide skillet of barely simmering water. In any and all cases, use perfectly dry containers and utensils, do not cover the chocolate, stir frequently, and try to remove the chocolate from the heat source before it is entirely melted, then stir to finish the melting.

How and for how long should chocolate be stored?

Store chocolate well wrapped in a cool, dry, odor-free environment. Refrigeration (or freezing) is not necessary unless you live in a very hot climate, in which case, let cold or frozen chocolate return to room temperature before you unwrap it; this prevents moisture from forming on the surface of the chocolate. Chocolate absorbs moisture and odor like a sponge. No matter how well you wrap it, do not store chocolate in the spice cupboard or in a closet with mothballs—it will absorb those and any other strong odors and flavors.

Dark chocolates (semisweet, bittersweet, and unsweetened), unless they contain some milk solids, have natural antioxidants that prevent spoilage. Properly wrapped and stored, these chocolates can keep for a year or more. Milk chocolate and white chocolate are stored similarly, but for a shorter period of time. The milk content of these chocolates turns sour, rancid, or just plain stale-tasting more quickly. Buy only what you will use within 2 to 3 months. To tell whether chocolate is still good, taste it.

How should chocolate be chosen for recipes?

The simpler the recipe, the less sugar it has in it and the more it will show off the flavors and quality of the chocolate that you use. Squares of ordinary cooking chocolate from the supermarket are fine when you are baking brownies for the soccer team, but when it comes to fine cooking and discerning company, your chocolate tortes, mousses, truffles, soufflés, and custards will taste even better with great chocolate—choose the same quality chocolate that you love to eat. For best results, however, stick to the type (unsweetened or semisweet or bittersweet) of chocolate called for in the recipe.

What are your favorite brands of chocolate?

I like Scharffen Berger, Valrhona, El Rey, and Callebaut chocolates.

Are there any other quick tips for buying, using, and storing chocolate?

Chop or cut chocolate with a heavy chef's knife or a knife with a serrated blade, or use the heavy-duty multipronged ice picks that are now being sold especially to break chocolate.

Make a point of tasting and enjoying some of the better-quality, stronger chocolates available. The labels on many premium imported and some domestic semisweet and bittersweet chocolates tell us what percentage of the bar, by weight, contains chocolate liquor (ground cocoa beans). The higher the percentage, the more intense the chocolate flavor and the less sweetness the chocolate can be expected to have. It is not uncommon to see imported chocolates labeled with percentages from 61 percent to over 70 percent, compared with standard American semisweet and bittersweet chocolates, which usually contain 50 to 55 percent chocolate liquor, although they are not so labeled. Stronger chocolates deliver strong chocolate impact and are delicious to eat. In recipes that do not call for strong chocolate, you may compensate by using less chocolate and adding a little extra sugar. If you want to substitute the stronger chocolate ounce for ounce in recipes, you will have to make some adjustments, such as adding extra liquid to sauces and glazes, baking cakes and tortes for a shorter period of time, or adding extra sugar to suit your taste.

Cocoa

What kind of cocoa powder do you recommend?

I use only Dutch process cocoa. Droste and Van Houten are two excellent brands. Droste is now available in most grocery stores so you should have no problem getting it.

What is "Dutch process" cocoa?

Cocoa that has been treated with alkali is called "Dutch" because the process was developed in Holland. The alkali removes some of the bitterness of the unrefined cocoa and neutralizes its natural acidity. Dutch process cocoa is darker and milder in flavor.

What is the best way to store cocoa?

I store cocoa in its box in a zippered plastic bag in the fridge. It keeps for more than a year.

I have cocoa around all the time. Can I substitute it for chocolate in cakes?

You can use cocoa as a substitute for chocolate in many recipes. Three tablespoons cocoa plus 1 tablespoon butter equals 1 ounce chocolate. Do not use instant cocoa, which contains sugar and is unsuitable for anything but a delicious hot drink.

Flour

I asked Susan Purdy, author of *Let Them Eat Cake* and *The Perfect Cake,* to contribute some of her expertise on flour.

Is it better to use bleached or unbleached all-purpose flour?

Whether labeled bleached or unbleached, most commercially milled American flour (with the exception of the King Arthur brand) is both bleached and conditioned. This process gives the flour a slightly different chemistry to produce a softer, more delicate crumb. Unbleached flour is often thought to be more nutritious, because some of the wheat bran is retained during milling and refining. Because of this, it is also slightly heavier than bleached flour. I use unbleached most often for pie crust, coffee cake, muffins, and cookies. To make the most delicate cakes, I always prefer to use bleached all-purpose flour unless the recipe calls for cake flour.

What is the difference between cake, pastry, and bread flour?

Each type of flour is milled for a specific use. All wheat flour contains gluten and gliadin as part of the wheat protein. When wheat flour is mixed with liquid and stirred or kneaded, the gluten and gliadin develop into stretchy stands that give dough both elasticity and strength. For tender, flaky pie crust or delicate pastries, soft-wheat, low-gluten flour, such as pastry flour, is desirable. For perfect pie crusts, I like to blend pastry flour half and half with bleached or unbleached all-purpose flour.

Cake flour is low-gluten flour that has been bleached to slightly increase its acidity, enhancing its ability to help cakes set faster, absorb liquid quickly, and retain moisture during baking. It makes cakes with a very delicate crumb; however, I find layer cakes with a more substantial, moist crumb do best with an all-purpose flour.

Bread flour is best for bread making because it is milled from high-gluten hard wheat. It has the greater elasticity needed to support the gases from the yeast.

How can I adjust a recipe to use self-rising cake flour?

Self-rising cake flour contains a mixture of baking powder, baking soda, and salt. I prefer not to adjust a recipe for self-rising cake flour. Use a recipe designed for this type of flour.

What is instant flour?

Instant flour is made to blend easily with water to make gravy without lumps. Sprinkle 2 teaspoons instant flour directly into the pan juices of a turkey or a roast chicken and whisk until the sauce thickens. Add 2 tablespoons butter for enrichment and you will have a beautifully textured sauce.

Is there a foolproof way of making a good pie crust?

Here are tips to keep pie crust light and flaky: keep all ingredients ice cold; add a minimum of moisture and a minimum of extra flour; handle the dough lightly; and add a little acidity, such as 1 tablespoon lemon juice or vinegar, to inhibit gluten development and keep the dough tender.

What is the reason for chilling pie dough?

Chilling dough causes the gluten to relax and ensures a tender crust. Gluten is the elastic portion of the protein in wheat flour; cold inhibits its development or relaxes it once it is developed, thus preventing dough from becoming tough.

Does flour always need to be sifted before use?

Follow the recipe. If the recipe calls for "1 cup sifted flour," you should sift before measuring. If the recipe says "1 cup flour, sifted," measure before sifting. Never count on flour to be pre-sifted, even though labeled as such. Before using flour that will not be sifted, stir it first to lighten and aerate it, spoon it into the cup, then sweep off the top with the back of a knife to level it. Don't tap the measuring cup; compacting the contents can cause a cup to hold up to 2 tablespoons more flour than needed, and this can toughen your baked goods.

Is it a good idea to store flour in the refrigerator?

I store bulk flour in the freezer and small amounts in the refrigerator. This prevents infestation, and flour is best used when cold because cold is an inhibitor of gluten development. The less gluten developed, the more tender the cake, pie crust, or pastry. For bread baking, bring flour to room temperature because bread dough needs warmth for good gluten development.

Gelatin

The last time I made my favorite gelatin dessert, which has always been reliable, it did not gel. What could have gone wrong?

Your gelatin may have been too old. Gelatin has a limited shelf life and should have a sell-by date, which unfortunately it doesn't. If you can't remember when you bought it, it is probably time for a new package.

I now see recipes calling for leaf gelatin. What is it and how do I use it?

Leaf gelatin is popular all over Europe and is now readily available here. Five sheets of leaf gelatin equal 1 package of powdered. The sheets have to be softened in a bowl of cold water, which may take up to a minute or two, then are added to whatever liquid you are using. The mixture needs to be heated to the point of melting the gelatin. The advantage of gelatin leaves is that they have practically an unlimited shelf life, and they will result in dishes with a much better texture than powdered gelatin would give.

TIP **Always measure the packet of powdered gelatin—it is supposed to contain 1 tablespoon but may contain more or less.**

I understand that gelatin contains animal products. I am a vegetarian. Is there any substitute?

Agar-agar is a flavorless dry seaweed that can be substituted for gelatin. It is available at Asian markets and natural food stores. Be careful of your recipe because you need to use much less of this product than of gelatin.

Leavening

What is the difference between baking powder and baking soda? Are they interchangeable?

Baking powder and baking soda are leavening agents that cause batter to rise through a chemical reaction that produces tiny bubbles of carbon dioxide. The two products are not interchangeable, because recipes for baked goods using them are more precise formulas than other recipes in which you can exercise your creativity by adding or subtracting a little of this or that. Baking soda produces the leavening gas when mixed with an acid, so it must be combined with acidic ingredients such as lemon juice, buttermilk, honey, molasses, or chocolate. It starts working as soon as it comes in contact with a liquid, so if your recipe calls for baking soda, have your oven preheated and mix the ingredients quickly. Baking powder is baking soda combined with one or more dry acids. The most widely available product in the United States is double acting baking powder that has one acid that begins working as soon as the liquid is added and another that doesn't begin to leaven until the batter is heated.

Some recipes rely on a more complex mixture of both baking powder and baking soda. Both of these products can lose their potency over time. If you don't bake regularly, check the expiration date on the container or test ½ teaspoon in ¼ cup warm water. If it fizzes, you can use it.

If you find yourself without baking powder and you're willing to take a little risk, substitute 2 teaspoons cream of tartar, 1 teaspoon baking soda, and 1 teaspoon salt per cup of flour in your recipe.

Seeds

I love the taste of sesame seeds and add them to many dishes, but they seem to lose their flavor. How can I get the nice toasty sesame flavor I am looking for?

Sesame seeds are best when toasted in a dry skillet until they turn a light brown and have a lovely, fragrant aroma. The best way to get that nice toasty taste is to use the seeds only as a garnish and not to cook with them.

What are black sesame seeds? Do they taste different from the white ones?

Sesame seeds come in both white and black. White sesame seeds have a lovely sweet nutty flavor; the black seeds are more bitter. Both should be toasted to bring out their flavor before they are added to a dish.

Are poppy seeds used in cooking? I like them on my bagels but never seem to see them anywhere else.

Poppyseed cake is an old-fashioned favorite. The seeds give a slightly crunchy texture to the cake.

Sugar

I store sugar in a plastic container but it always cakes. Is it better to keep the sugar in the original box?

Sugar cakes in humid surroundings whether it is stored in its box or a sealed container, but it can easily be mashed. For a small amount, use a fork; for larger amounts, I put the sugar in a zippered plastic bag and pound it with a rolling pin. You can also use a food processor or blender to break it up.

Try to find an airtight container. If you use the ones with a rubber gasket in the lid, chances are that your sugar will not cake.

What is the difference between powdered and confectioners' sugar?

Both names refer to the same product, which is refined granulated sugar that has been pulverized. Commercial bakers can choose from a range of textures, but the average consumer can usually get only the ones labeled XXXX or XXX, which is less fine. All powdered sugar contains 3 percent cornstarch to keep it from lumping, but to be on the safe side, sift powdered sugar before using.

What is icing sugar?

Confectioners' sugar is called icing sugar in Britain because it dissolves so easily and is often used to make icings and candy.

What is the difference between dark brown and light brown sugar?

Brown sugar is white sugar combined with molasses. The more molasses, the darker the sugar and the more intense the flavor. They are generally interchangeable.

TIP To prevent dark brown sugar from caking, place it in a container with a heel of rye bread; it will keep almost indefinitely.

I now see many types of gourmet sugars in my market, such as Demerara sugar and turbinado sugar. What kind of sugars are these and what are they used in?

Demarara sugar is a wonderfully delicate light brown raw sugar from the Demerara area in Guyana. It does not contain molasses and is delicious with cereals and as a topping for grapefruit. Turbinado sugar is raw sugar that has been steam-cleaned. It has a lovely blond color and a slight molasses flavor. It, too, can be used as a sweetener for bananas and grapefruit as well as a topping for cereals. I prefer it to brown sugar in crisps and cobblers.

Is it true that brown sugar is less fattening than white sugar?

Unfortunately, brown sugar has more calories than white sugar: white sugar has 770 calories per cup while brown sugar has about 820 per cup, but brown sugar is rather rich in calcium and potassium.

How do you make a sugar syrup?

A simple sugar syrup is made by combining 3 cups water with 1 cup sugar in a saucepan.

Bring to a simmer and cook for 2 or 3 minutes or until the sugar is dissolved. You can now store the syrup indefinitely.

If you plan to use the syrup for fruit, add a 2-inch piece of vanilla bean and leave it in the syrup. You can also add a cinnamon stick and a large piece of lemon rind. Be sure to remove the cinnamon stick and rind after 2 or 3 days.

What do you use a sugar syrup for?

Sugar syrup is used for freezing fruits, making sherbets, and stewing fruit (see Three-Berry Coulis with Nectarines, page 323). It is also used for sweetening lemonade and iced tea.

What is the difference between a sugar syrup and a caramel syrup?

A simple sugar syrup is nothing but a combination of sugar and water that is simmered until the sugar is dissolved. A caramel syrup is quite different: Here you combine 1 cup sugar with 3 tablespoons water and cook the mixture over high heat until it turns a hazelnut brown. Remove the pan from the heat and add ½ cup hot water, stirring the mixture until the caramel is dissolved. You now have a caramel syrup that keeps indefinitely. This syrup makes a great topping for ice cream and fresh fruit.

Many recipes call for superfine sugar, which I cannot get at my grocery store. Can I use regular sugar instead?

In many cases you can use regular sugar, or you can easily pulverize regular sugar in a blender.

Vanilla

What should I look for when buying vanilla beans?

Good vanilla beans should have a strong vanilla scent and be slightly oily. Depending on the variety, they may vary in length and thickness. Mexican beans and Bourbon beans from Madagascar are similar. Beans from Tahiti are shorter and plumper with a fruitier aroma. Always buy vanilla beans in glass vials so that you can judge the bean's quality.

Why are vanilla beans so expensive?

Vanilla beans are extremely costly to grow. First the beans must ripen on the vine for 9 months. Once they are harvested, they are put out to dry in ovens until they have shrunk by 400 percent, which can take months. It can take as long as four years for them to reach your grocery store. When you think that it takes about 5 pounds of uncured vanilla beans to yield 1 pound of cured beans, it is easy to understand their high cost.

How long can I store vanilla beans?

If kept sealed, vanilla beans will keep indefinitely. Even if they dry out, they can be used in a sugar syrup.

I am never quite sure how to use vanilla beans. Should they be kept whole or diced?

The best way to use a vanilla bean is to slit it open with a sharp paring knife and scrape out the seeds for use, as the seeds contain most of the flavor. Most recipes do not require more than a 2- to 3-inch piece of vanilla bean. Once the pod has been used for poaching fruit, it can be dried carefully with paper towels and reused. I usually bury a large piece of vanilla bean in my sugar container for several weeks. The sugar absorbs the taste of the vanilla bean, giving you an extra bonus when making any kind of dessert.

Is there a difference in taste between a vanilla bean and vanilla extract?

When a recipe calls for a vanilla bean, it is best to follow it, such as in sugar syrup or in flavoring cream or milk for a pudding. Vanilla extract is best used in cakes and pastry cream.

The Refrigerator

■■

Butter

Is there an important difference between salted and unsalted butter when it comes to cooking?

I use only unsalted butter in cooking and recommend that you do the same. Salt burns at low temperatures, which means that if you use it to sauté, it can easily burn your food. Salt acts as a preservative as well as a flavoring. The amount of salt added to butter differs from brand to brand, so you can never be sure how the salt in the butter will affect the taste of your food.

My family loves the taste of whipped butter. Can I use it in cooking as well?

Whipped butter is perfectly good for eating, but I do not recommend it for cooking, since it is full of bubbles and you would have to measure it by weight, which is a nuisance.

My market now carries several brands of European butter. Are they worth the extra expense?

I find that imported butter is well worth the extra you have to pay for it. European butter, with 90 percent butterfat versus 80 percent in American brands, is more suitable for sautéing, since there is less splattering and the butter can be heated to a higher temperature. It is also better for baking, especially in pastry used for tarts, where the higher fat content causes less shrinkage.

Does freezing butter change its texture or taste?

Once butter has been frozen, it will be slightly more watery and cause splattering when heated, but the taste will not be affected.

How should I store butter?

Butter is very delicate and will absorb refrigerator smells quickly. Put an opened bar of butter into a zippered plastic bag and then store it in the butter compartment of the refrigerator, never in the cold cut and cheese drawer.

What is ghee?

Ghee is clarified butter (page 384) that has been taken a step further by letting the butter turn a light hazelnut brown, which gives it a slightly nutty taste. Ghee has a higher smoking point than even clarified butter. It is used primarily in Indian cooking, in which spices are first fried over high heat.

GOAT CHEESE

I love goat cheese but am not sure which one to buy. Can you recommend any?

Buying good goat cheese can be a challenge because most supermarkets usually stock only the vacuum-packed logs. Most of these are made with powdered milk and frozen curd and usually taste chalky or just bland. If you have access to a good cheese store, look for imported farm goat cheeses or some of the domestic ones, such as Coach Farms or Laura Chenel, both of which are made with fresh goat's milk and are properly aged. Depending on where you live, you may find goat cheeses made by small local producers. These are well worth seeking out because, while they will be more expensive than the imported vacuum-packed ones, they are fresh and usually better tasting.

Is it true that goat cheese has much less fat than any other cheese?

Unfortunately, goat cheese is as high in fat as cow's milk cheese, but much depends on the aging of the cheese. When cheese is aged and the moisture evaporates, the fat gets concentrated, so fresh goat cheese is lower in fat than the aged varieties.

TIP If you can, try to avoid vacuum-packed goat cheeses, since these are never as fresh as the ones wrapped in paper.

MASCARPONE

So many recipes call for mascarpone these days. What exactly is that?

Mascarpone is an Italian cheese from the Lombardy region. Made from cow's milk, it is buttery rich and has the consistency of clotted cream or very soft butter. Good-quality mascarpone is now produced in Wisconsin and should become more readily available.

I now see mascarpone in the cheese department of my supermarket. What do you use mascarpone for?

Mascarpone is very similar in taste to very rich heavy cream. It is wonderful served with berries or fruit tarts. I like to enrich a risotto with 2 or 3 tablespoons of mascarpone. It is the key ingredient in a tiramisù, the very popular Italian dessert that is made with coffee-soaked ladyfingers and layers of sweetened mascarpone.

TIP Even though mascarpone usually has a sell-by date, be sure to taste it before using it. If the taste is not that of very sweet cream, it is not fresh. Once opened, mascarpone has to be used within a day or two, so turn it into a quick and delicious mousse by sweetening it and flavoring it with either instant coffee, vanilla, or an orange liqueur.

PARMESAN

My grocery store carries several kinds of Parmesan cheese. They vary greatly in price. What is the difference between them?

True Parmigiano-Reggiano is produced in the region of Emilia-Romagna, in Italy, and the words "Parmigiano-Reggiano" are always stenciled on the rind. There is also another import called Grana Padano, which is made in the style of Parmesan but is aged for only 4 months and is produced all over Italy. It makes a good grating cheese but overall lacks the flavor of real Parmesan. Still, if it is nice and moist, it makes for a good nibbling cheese with sliced apples and grapes.

Parmesan cheese is also produced in Argentina, Australia, and the United States. To me these products taste like sawdust and are too salty. You can tell domestic Parmesan by its rind, which is usually black. When you taste it, you will notice immediately that it lacks the characteristics of real Parmesan.

Sometimes the Parmesan I buy is mellow and lovely tasting and other times it is quite bitter. What exactly should I look for when buying Parmesan?

Always check the rind of the cheese. It should be cream-colored and there should only be a slight color difference between the rind and the cheese itself. Also, the cheese should have a crumbly texture. If the cheese is hard with a deep-colored rind, it has gone through major temperature changes and will have developed a bitter flavor.

Is it okay to buy grated Parmesan if I plan to use it as a topping for pasta?

I don't recommend it. Good Parmesan never comes grated, because the cheese loses its subtle wonderful taste within hours of being grated. Instead, buy a piece and grate it yourself. It is a cinch to do in the food processor. Don't even bother with any special blade—just use the steel blade and grate only enough for your dish.

What is the best way to store Parmesan?

The best way to keep Parmesan fresh is to wrap it loosely with a damp piece of cheese-cloth or 2-ply paper towel and then loosely in foil. Do not wrap the foil too tightly around the cheese or the Parmesan will become moldy within days.

I recently bought a rather large piece of Parmesan and a few weeks later it was covered with mold. Can the cheese still be used?

Surface mold does not affect the taste of Parmesan. Simply scrape it away, but next time buy less if you find that you do not use the cheese often.

TIP Parmesan cheese is relatively low in fat, since it is made from partially skimmed milk.

Is there a way to use Parmesan other than on pasta?

Parmigiano-Reggiano is a superb eating cheese that goes well with a number of fruits, especially pears, apples, figs, and melons. It is also lovely served in fine slivers as a topping for salads. Greens such as arugula, Belgian endive, and romaine have a particular affinity for Parmesan. Don't forget that grated Parmesan is perfect as a topping for a risotto and various soups, especially hearty bean soups and the classic minestrone.

TIP If you do not have a cheese slicer, you can use a vegetable peeler to make fine slivers of Parmesan.

RICOTTA

I have tasted wonderful fresh ricotta in Italy and the taste was totally different from what I get in my market. Is there anything I can do to improve its taste?

American ricotta is made from whole or partially skimmed cow's milk rather than the whey of sheep's milk. It is much blander, moister, and creamier and actually not very flavorful. Draining the cheese improves its consistency but not its flavor.

You can now get imported ricotta in some

specialty food markets but be sure to taste it before buying it. If it has a sharp prickly taste, pass on it because the cheese is no longer fresh.

What is the difference between ricotta and ricotta salata?

Ricotta salata is a dry pressed cheese. It is made out of lightly salted sheep's milk curd aged a minimum of 3 months. Good ricotta salata should have a sweet milky flavor. Be sure to taste it before buying it because it is often salty. I like to use ricotta salata grated on fresh pasta or as a topping for grilled or steamed vegetables.

Cream

HEAVY CREAM

Can I use light cream in recipes that call for heavy cream?

Light cream is not a substitute for heavy cream, since it contains only 18 to 30 percent milk fat, while heavy cream contains at least 36 percent. The high level of fat is important because it allows the cream to be boiled without separating and to be whipped. This is especially important when making custards and pastry creams. When air is beaten into heavy cream, the fat globules stick together and support the air bubbles, which in turn create whipped cream.

What is the difference between light whipping cream and heavy whipping cream?

Heavy whipping cream, also simply called heavy cream, has a fat content between 36 percent and 40 percent. It will double in volume when whipped. Light whipping cream has 30 percent to 35 percent fat. It sometimes contains stabilizers and emulsifiers to help it maintain volume.

I now see bottled heavy cream in the grocery store that is more expensive than the ultra-pasteurized cream in cartons. Is it worth the extra money?

The return of old-fashioned heavy cream that is not ultra-pasturized is great news and well worth the extra cost. Ultra-pasteurized cream is a product that has been heated at high temperatures to give it a longer shelf life, resulting in a major loss of flavor and a gummy texture. It also does not whip as well as regular heavy cream and you cannot use it to make crème fraîche.

Can I freeze leftover cream?

Freezing alters the taste and texture of heavy cream. However, ultra-pasteurized cream will keep for several weeks in the fridge.

TIP Madeleine Kamman in her book *The Making of a Cook* suggests stabilizing whipped cream by adding 2 level measuring spoons nonfat dry milk to each cup of plain cream before whipping.

When a soup recipe calls for cream, is yogurt an acceptable substitute?

Hot soups cannot be enriched with yogurt, which curdles when heated. Unfortunately, you really cannot beat the flavor and silkiness you get from cream. If you don't want to use it, simply omit it. It is delicious but not essential.

SOUR CREAM

Can I use light sour cream in recipes that call for full-fat sour cream?

You can use light sour cream if you add herbs and other flavorings to it. Otherwise, the taste is quite chalky and not too pleasant.

Can I substitute yogurt for sour cream in recipes?

I often use yogurt in Middle Eastern cooking and potato salad but always drain it first. A combination of yogurt and sour cream makes for a richer taste. Neither should be used in hot preparations, since both will curdle the sauce. That said, many Greek dishes call for yogurt in hot dishes with delicious though curdled results.

I recently added sour cream to enrich a sauce and it curdled. What did I do wrong?

If you have a choice, it is better to use crème fraîche rather than sour cream, since it does not curdle. If you can't get it, however, be sure to mix 2 teaspoons flour into the sour cream before adding it to a sauce.

CRÈME FRAÎCHE

So many recipes call for crème fraîche but my market does not carry it. Can I make my own?

Making your own is very easy. In a bowl combine 2 cups heavy cream with 4 tablespoons buttermilk. Whisk the mixture until well combined and transfer to a jar. Keep at room temperature for 12 to 24 hours or until very thick. Refrigerate and use the crème fraîche within 10 days.

I tried to make crème fraîche but the mixture remained liquid and never set. What did I do wrong?

To get crème fraîche to set you must use cream that has not been ultra-pasteurized. Unfortunately, this is not easy to get in certain parts of the country. In this case, you need to double the amount of buttermilk to 4 tablespoons per cup. Also, it is best to make it in a porcelain or ceramic jar, which holds the temperature evenly, rather than in glass or plastic, which are uneven heat conductors.

My recipe calls for crème fraîche to sit on the counter for 24 hours. Is there a quicker way to get it done?

Unfortunately, there is no quick method to make real crème fraîche. But for a quick way to get the same flavor, you can add a dollop of sour cream to whipped cream, which will give you the light sour taste of crème fraîche. This makes a delicious topping for fresh fruit but cannot be used in cooking the way crème fraîche can.

What brand of commercial crème fraîche do you recommend? How long does it keep?

Vermont Butter & Cheese Company crème fraîche is an excellent brand. All crème fraîche has a sell-by date and will be good for a week to 10 days after that.

Eggs

How do I know if the eggs I am buying at the supermarket are fresh?

The difficulty in shopping for really fresh eggs is that they are dated from the time they are packed rather than from the time they are laid. Your best bet is to buy AA or A graded eggs, which are the freshest. If the eggs in your market are not dated, then check the code that pertains to the date the eggs were packed. I try to buy my eggs either at a health food store that carries local farm eggs or at a supermarket that has a large turnover. You cannot always assume that farmers' markets carry fresher eggs. It is often not the case.

Is there a difference in taste between brown and white eggs? Which do you prefer?

There is no taste difference between white and brown eggs. The color of the shell has to do with the breed of the chicken. I prefer brown eggs because they have a more "country" look and because of their thicker shells.

Some of the eggs I buy have pale yolks while in others the yolk is golden. What does that mean?

The color of an egg yolk depends entirely on what the chicken is fed; the color does not affect its taste. Canadian eggs have pale yolks, which Canadians prefer. In the United States, we prefer deep yellow yolks, so most of our eggs come this way. Both kinds of yolks contain 34 percent fat and 16 percent protein, the rest being water.

Are organic eggs worth the extra money?

It really depends on taste and freshness. The very best-tasting eggs come from free-range chickens, those that stroll around freely pecking and scratching here and there for their food and living a normal chicken life. But these chickens lay very few eggs, so there are few free-range chickens out there. Organic eggs do not taste that much different from nonorganic ones, and therefore I do not feel that they are worth the extra cost. If you can find truly fresh eggs instead, they are definitely worth the high price.

I often lose count of when I bought the last batch of eggs. Is there a way to tell if they are still good?

The first thing to do is to check if the eggs are heavy for their size; the heavier the egg, the fresher it is. Then proceed to do the water test.

Place the eggs in a bowl of cold water. If the eggs remain horizontal, they are fresh; those that float almost vertically are not. The very best way to tell if an egg is fresh is to crack one open. The yolk should be centered in a dense, cohesive white. An older egg has a thinner white, is more watery, and the yolk may be off center. After 3 weeks the yolk will probably lie flat and the white will be watery. This egg is still considered fresh, but its taste and appearance do not compare to a truly fresh egg.

Food magazines suggest that eggs should be stored in their cartons. Why?

The shell of an egg is porous and can pick up refrigerator odors easily. Storing eggs in their carton protects them and keeps them fresh. In many refrigerators the egg rack in the refrigerator is in the door, an area that is too warm and unprotected.

TIP Store eggs with their broad rounded ends up. The rounded end of the egg is slightly less delicate and less likely to break when bumped accidentally.

Why is it sometimes so hard to peel a hard-boiled egg?

Believe it or not, a very fresh egg is harder to peel than one that is 1 to 2 weeks old. By that time the white has shrunk away from the shell slightly, making it easier to peel. On the other hand, only a very fresh egg will poach perfectly without the yolk breaking away from the white.

How do you keep hard-boiled or soft-boiled eggs from cracking during the cooking process?

Cold eggs crack more easily than eggs at room temperature. It is best to start to hard-boil eggs in cold water. When the water comes to a simmer, count 10 minutes, and the eggs will be done perfectly. Be sure to bring the eggs to room temperature before cooking soft-boiled eggs. Ease the eggs with a large spoon into lightly simmering water and then count 3 to 5 minutes, depending on how you like your eggs done.

The yolks of my hard-boiled eggs are often surrounded by a greenish tinge. What does this mean?

The greenish color is simply the sign of an overcooked egg. Be sure to time your eggs. A perfectly hard-boiled egg needs only 10 minutes from the time the water has come back to a simmer.

TIP Cold eggs separate better than warm ones because the egg white is more viscous. When using eggs for baking, always separate one egg at a time because even the tiniest speck of yolk will keep the white from whipping and absorbing air.

Is it necessary to stick to the size of eggs called for in a recipe?

Large eggs are the standard called for in most cookbooks. For many recipes, such as scrambled eggs, quiches, and custards, you can use any size egg. When it comes to baking, however, you need to measure the volume of the eggs according to this chart:

1 large egg, beaten = 3¼ tablespoons
2 large eggs, beaten = 6½ tablespoons
3 large eggs, beaten = 9⅔ tablespoons or ½ cup plus 1½ tablespoons
4 large eggs, beaten = 12¾ tablespoons or ¾ cup plus 1 teaspoon
5 large eggs, beaten = 1 cup

I often find myself with a lot of leftover eggs. Can I freeze them for later use?

Eggs freeze quite well. To do so, break them into a container. (I usually use 5 large eggs, which equals 1 cup.) With a fork, pierce the yolks and stir slowly to mix them. Be sure not to whip up any foam because the air bubbles will dry out the eggs as they freeze.

If you have any leftover yolks, you can freeze them, too. Be sure to add a large pinch of salt to the eggs you plan to use for cooking and a teaspoon of sugar to those you will use for baking.

Leftover whites can easily be frozen, and once thawed will whip beautifully.

Dried Fruit

How should I store dried fruit?

I store my dried fruit in zippered plastic bags in the fridge. When kept on a pantry shelf, dried fruit gets hard and leathery.

Any tips for buying dates?

Some dates are treated with corn syrup, which I find unnecessary, since dates are one of the sweetest foods I know. I buy only untreated dates, usually in a health food store or at a green market. If you have a choice of varieties, the Medjool is softer and creamier than the more common Deglet Noor.

PRUNES

Is it better to buy prunes pitted or unpitted?

Unpitted prunes are always my first choice, and the larger the better. They have much better flavor and they do not fall apart in cooking. Of course, if you serve a dish with unpitted prunes, be sure to warn your guests or family.

TIP Prune puree, which is available in most grocery stores, can be used instead of butter in baked goods. It adds the right moisture and an interesting flavor.

I know dried fruit is high in fiber, so I'd like to eat more of it, but except for adding raisins to my oatmeal cookies, I'm at a loss. Do you know some good ways to use dried fruit?

You're right, dried fruit is a terrific and delicious source of fiber. The highest-fiber dried fruits are dates and prunes, followed closely by apricots and raisins. One great way to use dried fruit is to make a compote.

In a saucepan, cover the fruit with water. Add a little sugar, a cinnamon stick, and a 3-inch piece of vanilla bean, and simmer for 25 to 35 minutes or until the fruit is very tender. Refrigerate in the poaching liquid.

You can also simmer dried apricots this way, puree them, then serve the resulting sauce over ice cream.

I see all kinds of dried fruits in the market these days: cranberries, sour cherries, even strawberries. What do you use them for other than a snack?

I usually use sour cherries and cranberries in muffins, quick breads, and especially in rice pudding, where they add a delicious chewy texture.

RAISINS

When a recipe calls for raisins, which ones should I buy?

It really depends on what you are going to make. I use golden raisins in some preparations and dark raisins in others. I also like currants. Now that you can get Muscat raisins, be sure to try them. It is a good idea to have a box of various kinds of raisins, since they also make wonderful healthy snacks. Make sure to refrigerate them if you do not use them regularly.

TIP If you find yourself with raisins that have dried out, just simmer them in water to cover until they plump up. If they don't, and stay crumbly, you will have to get a fresh batch.

What is the difference between a currant and a sultana raisin?

Sultanas are golden raisins, made from yellow sultana grapes (and sometimes, especially in the United States, from Thompson seedless yellow grapes). Currants are made from the tiny Zante grape, much smaller than raisins and sultanas.

Nuts

Nuts sold in cellophane packages are sometimes not as fresh as they should be. How can I ensure that I'm getting really fresh nuts?

Buying nuts can be most frustrating, because—contrary to European custom—we do not date our nuts. That means you have no idea how fresh they really are. If you buy nuts in bulk in a health food store you can, of course, taste them, which is an advantage. Canned nuts are usually more expensive, but vacuum packing keeps them fresher than the cellophane-packed varieties.

***Which nuts do you always keep on hand and
how do you store them?***

Except for pine nuts, walnuts, and pecans, I try
to buy nuts as I need them, and I highly recommend that you do the same. The best way to
store nuts is to keep them in the freezer and
toast them lightly before using. The exception
to this rule is pine nuts, which I keep in the refrigerator, because I use them on a regular basis.
Nuts contain natural oils that can eventually go
bad when the nuts are not properly stored.

***What's the best way to get the skins off nuts
like almonds and hazelnuts? Is this really
necessary?***

You really should remove the skins from nuts
such as almonds and hazelnuts because the
skins get tough during baking. This is quite
easy to do. Just place the nuts on a cookie sheet
and toast them in a 200°F oven for about 10
minutes, or until the nuts are nice and golden.
You should be able to smell a distinct nutty
aroma but not a burned one. Then transfer the
nuts to a clean kitchen towel and rub until
most of the skin comes off.

***When a recipe calls for blanched almonds
and I cannot find them in the store, can I
blanch them myself?***

Blanching almonds is very easy; in fact, it's a
good idea to buy almonds in their skins and to
blanch them yourself, since this way you will get
a fresher almond. To do this, put the almonds
in a saucepan with water to cover and cook for
2 to 3 minutes. Drain and rub off the skins
with some paper towels. Spread the blanched
almonds on a cookie sheet to dry at room temperature, or pop them into a 200°F oven for 5
minutes. Be sure to keep an eye on them—what
you want is blanched almonds, not toasted ones.

TIP When a recipe calls for ground almonds, use
slivered ones and be sure to add a little sugar
to them before grinding them up in the food
processor. The sugar will keep the nuts fluffy
and dry.

***I recently bought peanuts and they did not
taste "peanuty." Are there different kinds of
peanuts I should look for?***

Peanuts come two ways, either roasted or raw.
Raw peanuts do not have the familiar "pea-
nutty" taste, so look for roasted unsalted
peanuts for everyday cooking.

***I love to buy unshelled peanuts, but they
sometimes taste dry and even rancid. Is
there a way to tell really fresh peanuts?***

The first thing to do is taste as many peanuts in
the markets as you can. That is usually a pretty
good test, but most important is that the peanuts have clean unbroken shells and that the
nuts do not rattle when shaken; the rattle is a
sign that the nuts are not fresh. Consume peanuts quickly because they are high in fat and
become rancid quickly.

***When a recipe calls for ground peanuts as
in a Thai peanut sauce, can I use peanut
butter instead?***

In some cases you can, but in many Thai
recipes, it is the coarseness of the ground
peanut that gives the dish an extra dimension.
I usually prefer using ground whole unsalted
peanuts rather than peanut butter.

***What kind of nuts should I use to make
pesto?***

Classic pesto is always made with pine nuts, although there are versions of this wonderful
sauce that call for walnuts. You can call me a

traditionalist, but to me pine nuts, fresh garden-grown basil, excellent olive oil, and top-quality Parmesan are the four "must-have" ingredients for good pesto.

What are pignoli?

Pignoli is the Italian name for pine nuts, the most important ingredient in pesto. Unfortunately, most of the pine nuts you now see in grocery stores come from China and their quality is inferior to that of the Mediterranean variety. However, Chinese pine nuts are perfectly acceptable when toasted and added to rice dishes or used to garnish grilled vegetables. Mediterranean pine nuts are elongated in shape and have a nuttier, less oily flavor. It is easy to tell which is which, since real pignoli are twice as expensive as the Chinese nuts.

What is the best way to buy prepared chestnuts?

The best chestnuts now come vacuum-packed in either small or large jars and are available in markets everywhere. You can also order them by mail through specialty stores such as Zingerman's and Dean & DeLuca (see Sources, page 398). You will notice them more around Thanksgiving, when markets stock up on them. The jarred nuts keep for at least a year.

Olives

My market has just put in an olive bar, but I find choosing olives very confusing. Which olives do you recommend?

The nice thing about an olive bar is that it allows you to taste various olives. Generally, green olives are firmer with a sharper flavor than black olives. They are also less oily, since green olives are the unripe fruit. On the other hand, the flesh of black olives is softer and their flavor more delicate. Once you get hooked on olives, you are likely to choose more than one variety.

Most black olives seem to be very salty. Are some saltier than others?

The most common black olives carried by grocery stores are the black kalamata olives from Greece. These are quite fleshy, often large, and very salty. I prefer the smaller Italian black olives, especially the gaeta and the French niçoise olives. However, when a recipe calls for olives, I find that the pitted kalamata olives are quite acceptable. To remove their saltiness, just simmer in water for 2 or 3 minutes.

How long do olives keep?

It depends on whether they are brined or packed in oil. As long as the olives are covered in brine, they will keep for several weeks. You can add olive oil to oil-cured olives to extend their shelf life.

I love spicy olives and have tasted really good ones in various restaurants but cannot find them in my grocery store. Can I make my own?

You can flavor your own olives very easily. Just add a finely sliced jalapeño pepper, 1 or 2 sliced garlic cloves, and some fruity olive oil to your olives and let them marinate for 2 or 3 days. Then get a crusty loaf of bread and a nice slightly aged goat cheese and enjoy!

Is there a taste difference between green and black olives? Which are better suited for cooking?

There is a very definite taste difference between green and black olives, as well as texture differences. There is also a large flavor difference between various kinds of green olives and the many kinds of black ones. It is a matter of tasting as many kinds as you can and settling on the ones you enjoy. You can use both black and green olives in a variety of dishes. I personally prefer black ones because of their softer texture and milder flavor.

■ Cream of Asparagus and Spring Onion Soup ■ Avocado Salad in Ginger-Lime Vinaigrette ■ Roasted Beet, Watermelon, and Mango Salad ■ Pan-Seared Baby Bok Choy with Red Bell Peppers and Oyster Sauce ■ Skillet-Braised Broccoli with Garlic ■ Sauté of Brussels Sprouts with Bacon, Pine Nuts, and Sour Cream ■ Braised Red Cabbage with Orange Zest and Grenadine ■ Skillet Braised Carrots with Pine Nuts And Raisins ■ Puree of Celery Root, Potatoes, and Sweet Garlic ■ Creamy Celery and Stilton Cheese Soup ■ Corn and Chive Fritters with Smoked Salmon Sauce ■ Spicy Corn and Ginger Relish ■ Sauté of Cucumbers in Chive and Dill Cream ■ Fricassee of Roasted Eggplant and Two Mushrooms ■ Charcoal-Grilled Eggplant with Lemon-Scallion Mayonnaise ■ Skillet-Braised Belgian Endives ■ Ragout of Fennel, Tomatoes, and Potatoes ■ Sauté of Green Beans with Tuna, Tomatoes, and Red Onion ■ Kale, Potato, and Butternut Squash Chowder ■ Braised Leeks with Shiitake Mushrooms and Crème Fraîche ■ Portobello Mushrooms in Tarragon-Lime Butter ■ Wild Mushroom, Scallion, and Gruyère Bread Pudding ■ Bruschetta of Roasted Sweet Onions ■ Roasted Vidalia Onions with Garlic and Rosemary ■ Sugar Snap Pea Soup ■ Curried Bell Pepper and Zucchini Ragout ■ Mascarpone and Chive Mashed Potatoes ■ Gratin of Russet Potatoes and Sweet Onions ■ New Potato Salad in Mustard, Shallot, and Caper Vinaigrette ■ Puree of Sweet Potatoes and Carrots ■ Puree of Rutabaga and Potato with Crispy Shallots ■ Shallot and Cassis Marmalade ■ Sauté of Spinach with Scallions, Dill, and Yogurt ■ Acorn Squash, Apple, and Wild Mushroom Soup ■ Wild Mushroom and Parmesan Timbales

In 1973 my first book, *The Seasonal Kitchen*, was voted best cookbook of the year. The concept then seemed new, refreshing, and different from that of other cookbooks. Here was a book defined not just by recipes but by menus that focused on seasonality and freshness. It was about the way I grew up, the way I felt about raw materials, the way I thought about food, and the way I cook.

At the time I was often asked what type of cuisine I wrote about. I had to explain repeatedly that the seasonal kitchen was not a cuisine but a philosophy, and that every type of cooking fits into it. It was and continues to be what peasant and regional cooking is all about around the world. For millions of cooks, whether in Europe, the Middle East, or Southeast Asia, seasonal produce is the basis of most dishes because it is simply tastier and more economical. To this day, you can visit markets in France, Spain, Italy, or Istanbul, and even if you are not a gardener or a cook, you can tell what is in season simply by looking around you.

For Americans in the seventies, the concept seemed almost radical. After all, don't we live in a country where everything is available no matter the season? Is there a time when you cannot get apples or raspberries or tomatoes? They may have been grown in a hothouse, or lost taste and texture while being kept for months in cold storage, or picked green so they can be shipped from thousands of miles away, but they are on produce shelves everywhere. In the months after the book came out, I traveled across the country, conducting workshops and trying to make my point. At times I would make the same soup with frozen vegetables and with fresh ones to demonstrate the difference.

Americans do know what fresh produce tastes like. People all over the country have always loved farm stands. Everyone agrees that fresh vegetables bought at a local farm stand taste much better than grocery store produce.

Why, then, do so many consumers forget about seasonality the minute the weather turns cold? Why would anyone buy corn in a grocery store in the middle of the winter? Why bother with out-of-season tomatoes, melons, or asparagus? Why don't we think winter squash and greens, parsnips, mushrooms, and celery root deserve as much attention as the summer classics and are equally satisfying when fresh and in season? So, just as with your wardrobe, change your meals to include the many varieties of root vegetables, broccoli rabe, earthy spinach, and kale.

Seasons can change as often as month to month or even week to week, so be sure to plan your menus after you've done your produce shopping, rather than before. It is a waste of effort to build even a simple menu around a particular vegetable or fruit only to find that it is unripe, limp, or just not fresh. Instead, start with the best produce available. And you can be sure that you will find a recipe for it.

Often a perfect experience can come in simple ways. On a recent Sunday in the country, I made a Greek salad. The tomatoes were juicy and flavorful, the cucumbers crisp, the lettuce tasty, and the radishes crunchy. I had bought some delicious, not too salty feta cheese and excellent olives. These simple ingredients combined with a fruity olive oil made it all happen. Here was a salad for which there was no need for a recipe, and it was memorable.

It reminded me of my Greek friends who, whenever they make this classic salad, always enjoy it as if it were their first time. Why? Because the raw materials can change daily, and the way each cook uses them influences the result.

Just because a vegetable or a fruit is in season, it doesn't mean it is worth buying. It never ceases to amaze me when shoppers opt for less than perfect lettuce, bagged carrots, and fruit that will never ripen no matter how much time it spends in a bowl. You may have to look beyond your all-purpose market for the best produce. For example, if the carrots in your supermarket look limp and tired, if the radishes only come packaged, or if the cilantro or basil comes in small cellophane containers, try another market where the turnover is greater.

You'll find it worth your time to search out the best. What is the point of spending precious time and money working with produce in which the odds are against you from the start? Why put effort into a dish that cannot possibly deliver?

The reason for this chapter is to help you learn to buy the best produce by answering many of the fruit and vegetable questions I have heard over the years—questions such as "How do you tell a good celery root or parsnip?" And "Is it really worth buying carrots with their tops if you are going to discard them anyway?"

Another major concern today is the importance of shopping for food that has been grown organically. On a recent visit to the Berkeley

Bowl, one of the most exciting stores in northern California, I was thrilled to see the abundance of organically grown produce and to taste the difference between a commercially grown strawberry and an organically grown one.

There was a time when organic produce did not look as appealing as commercially grown varieties. Today the quality is far superior, and while you may not find absolutely perfect-looking fruits and vegetables, what you give up in looks, you will gain in taste.

Storage has become a big issue with the increasingly popular use of misting to keep vegetables looking fresh. Unfortunately for the consumer, supermarkets are ignoring the fact that all this moisture causes quick decay once these misted vegetables are bagged. I have now adopted the habit of shaking a bunch of parsley or lettuce several times before bagging it, often resulting in angry looks from a nearby shopper who was not anticipating a damp shower.

Because produce shopping is expensive and time consuming, allow yourself the time once you arrive home to store your vegetables properly. It is very frustrating to reach for what you thought was going to be fresh parsley or a bunch of arugula, only to find it rotten in its bag, or to find your leeks, celery, or carrots looking limp in the bottom of the crisper because you did not take the extra time to bag them.

A recipe for a fruit or vegetable dish is only a recipe; it is the quality of the produce that will greatly affect the outcome of your dish. Below you'll find many tips on how to keep your fruits and vegetables fresh and tasty. Then it is up to you to buy the freshest and to learn to distinguish the best from the merely acceptable.

Artichokes

Is there a secret to picking a good artichoke?
Choose an artichoke that feels heavy for its size. Lift a few off the pile and compare them to determine if they are firm and see if the leaves are tightly bound. Neither brown tips nor black streaks are signs of rotting but rather of frost damage that only affects the outer leaves. Artichoke growers call these "winter kisses."

What is the best season for artichokes?
The spring, starting in March and going through June, is the best time. September and October are a secondary season.

How do I clean an artichoke and prepare it for cooking?
Before you start, fill a large bowl with cold water and squeeze half a lemon into it to make acidulated water. Cut off the artichoke's stem and remove two or three layers of the bottom leaves. With a sharp knife, cut ½ inch off the top of the artichoke. Then, with a pair of sharp kitchen scissors, trim about ¼ inch off each leaf. Drop the artichokes into the acidulated water until ready to cook.

I love the taste of artichoke but have never cooked one. How do you cook and serve an artichoke?
The American globe artichoke, which is the most common variety that you will find in the

market, is best cooked whole in plenty of acidulated, slightly salted water.

Simmer it until the stem can be pierced easily with the tip of a sharp knife and an inner leaf comes out easily when it's pulled. This may take as long as 30 to 40 minutes. Be sure to drain the artichoke upside down on a double layer of paper towels or a large plate. Serve it warm with melted butter or a garlic-flavored butter.

What is the difference between the heart and the bottom of the artichoke? What is the fond?

The artichoke heart is the tender, meaty bottom of the artichoke, from which all the leaves and the fuzzy choke have been removed. The resulting flat disk is called the heart or bottom here, while in France and Spain it is called the fond.

How can I get to the heart, or bottom, of the artichoke in the easiest way?

First, choose the largest artichokes you can find. Break the stems off close to the base of the artichoke and then pull off three or four layers of leaves until you reach pale yellow-green leaves that fold inward. Slice off these leaves just above the heart and plunge the trimmed artichokes into acidulated water while you prepare the rest of the artichokes.

Cook in plenty of lightly salted water until the hearts, or bottoms, are easy to pierce with the tip of a sharp knife. Allow to cool in the water, drain, and gently scoop out the fuzz with a grapefruit spoon. Trim any remaining leaf ends.

Is there any use for the artichoke leaves when the recipes call only for the bottoms?

Unfortunately the leaves are of no particular value so I usually discard them.

Can artichokes be cooked successfully in the microwave?

A single artichoke microwaves beautifully. Trim the leaves and the stem end and then rinse the artichoke under cool tap water. Wrap the damp artichoke in plastic and microwave for about 10 minutes or until softened and an inside leaf pulls out easily when tugged. If you have a large microwave, you can cook two artichokes at a time by increasing the cooking time by a few minutes. If you want to cook more, it's easier to cook them in plenty of boiling water on top of the stove.

Is there a special method for cooking those baby artichokes sometimes available in the market?

Baby artichokes are wonderful braised. Trim the artichokes and cut ½ inch off the tops. Cut the chokes in half lengthwise and reserve. In a large cast-iron skillet, heat 3 tablespoons olive oil, place the sliced artichokes cut side down in the pan, season with a sprinkling of salt, and add 2 or 3 finely sliced garlic cloves and about 1½ cups water. Cover the skillet and simmer the artichokes for 15 minutes or until tender. Serve warm or at room temperature.

How long do cooked artichokes keep?

They'll store well for as long as a week if you put them in a covered container or sealed plastic bag and refrigerate. They also freeze well.

Asparagus

■■■

Is it better to buy thick or thin asparagus?

Thick stalks are more flavorful. Thin asparagus look pretty but not only are they harder to peel than thick stalks, they actually take longer to cook and can be quite stringy.

How can I tell if asparagus is fresh?

Check the bottoms of the stalks. They should be firm and moist, never dry or woody. The tips of the asparagus should have purplish-green, tightly closed clusters.

What is the best way to store fresh asparagus?

To store asparagus for more than 2 or 3 days, fill a quart-sized plastic container with one to two inches of water. Place the asparagus upright in the water and cover the tips with a double layer of slightly dampened paper towels, and keep them in the refrigerator. If you'll be cooking the asparagus within a day or two, simply store in the vegetable bin in a perforated plastic bag.

Is it really necessary to peel asparagus? The spears look so much nicer unpeeled.

Peeled asparagus spears cook more evenly. You can make them look attractive by using a swivel peeler that doesn't take off too much of the stalk. Make sure to rotate the asparagus as you peel to maintain their round shape. Otherwise, the stalks will look square.

TIP The best way to peel asparagus neatly is to place the stalks over a small, inverted stainless mixing bowl and rotate the stalks as you peel them.

When I prepare asparagus for cooking, how much of the stalk should I remove?

Cut 1 to 1½ inches off the bottom of the stalks. The best way to tell how much is to bend one of the stalks; it will snap at the point where the woody stem turns into a juicy stalk. This will give you an approximate idea of how much to cut off all the stalks.

My market bundles asparagus spears of different sizes, some thin, some thick. When cooking several bunches at one time, how can I make sure that some spears won't be overcooked while others are underdone?

Start with plenty of lightly salted, boiling water. Using ordinary kitchen string, tie same-sized spears together in bundles of 6 to 8 stalks—and leave one stalk of each size separate for testing. The bundles are easy to pull from the water when the test spear is done. Once the asparagus spears are cooked, spread them out on a double layer of paper towels. This keeps them from discoloring. If you plan to serve the asparagus at room temperature, it is best to shock them in a bowl of ice water to keep them bright and green. You can also steam asparagus, but it usually takes longer than boiling and the vegetable stays too crisp for my taste.

I like the idea of grilled asparagus but have had little luck with it. Can you help?

I find that uncooked asparagus does not lend itself well to grilling, and I suggest blanching the stalks in simmering water for 2 or 3 minutes until they are crisp-tender. Run them under cold water to stop further cooking. Dry on paper towels and then brush them with a little olive oil before putting them on a medium-hot grill. Grill the asparagus for 3 minutes, turning them once or twice until you can see grill marks. Use a large spatula to remove them from the grill and season with coarse salt and freshly ground pepper. The Emerald Mayonnaise that follows is an excellent accompaniment.

Emerald Mayonnaise

SERVES 6

3 tablespoons minced flat-leaf parsley
3 tablespoons minced scallions
2 large garlic cloves, crushed
4 flat anchovy fillets, drained
1 cup mayonnaise
½ cup crème fraîche or sour cream
Juice of 1 large lemon
Salt and freshly ground black pepper

Combine the parsley, scallions, garlic, and anchovies in a food processor or blender. Add the mayonnaise, crème fraîche, and lemon juice and process until smooth. Season with salt and a large grinding of black pepper, and chill for 4 to 6 hours or overnight.

Avocados

What are the marks of a good avocado?
Avocados should be evenly ripe or evenly unripe. The best way to tell is by cradling the fruit in the palm of your hand and squeezing it lightly. If ripe, it should be somewhat soft to the touch. Depending on where you live, it may not be easy to buy a ripe fruit or, worse, an evenly ripe one. If the fruit is soft in some places and hard in others, you had better pass on it, since it will never ripen evenly. Unripe avocados usually take 2 or 3 days to ripen. Simply place the fruit in a warm place in an earthenware or porcelain bowl. To hasten the ripening, put it in a brown paper bag with a banana.

Some avocados are large and smooth, others smaller with a pebbly skin. Is there a difference in taste?
There are, in fact, many varieties of avocados that come into the markets at various times of the year. They vary in taste from nutty and slightly bitter to sweet and buttery. The Hass avocado, which has a tough, hardy, and pebbly skin and turns almost black when ripe, is by far your best bet. It does not bruise easily, has lots of delicious pulp, and, when ripe, will keep well in the fridge for 2 to 3 days.

What is the best way to slice an avocado?
First, cut through the skin lengthwise, then give the fruit a light twist to separate the halves. Then (this I learned from the guacamole chef at Rosa Mexicana), hit the pit with the edge of a sharp knife, twist lightly, and the pit will come right out. Of course, the avocado must be perfectly ripe, which is not always the case. You can spoon the avocado out, but I find it equally easy to cut it in quarters and remove the peel with my fingers.

Is there any trick that will keep an avocado from turning dark?
Although many people, including chefs, believe that leaving the pit in the cut fruit or propping

it in the middle of a bowl of guacamole slows down the darkening process, I am not sure that it really helps. Sprinkling the fruit with fresh lemon or lime juice does delay discoloration. Plan to use the entire avocado at one time, and try to use it as soon as possible after slicing into it.

I love avocados, especially in guacamole. Is there any other simple preparation that is as good but has a different twist?

Good guacamole is hard to beat, but the Avocado Salad in Ginger-Lime Vinaigrette (page 109) is right up there with my favorite avocado recipes.

Beans

Is there a difference between snap beans, green beans, and string beans?

Snap, green, and string beans are really all the same bean. "Green beans" is the old-style name for them, which gave way to "string beans" (a misnomer, since new varieties of beans rarely need stringing). To avoid confusion, I refer to them as green beans, with the exception of wax beans. "Snap beans" to me are those grown in the garden or purchased at a farm stand that are picked even before reaching pencil-thin size. Test these by snapping one in half. If it breaks with a nice, clear sound, chances are the beans are fresh.

I see haricots verts in recipes and on restaurant menus. What are they?

Haricots verts simply means "green beans" in French but the term is used to indicate a slender type of bean that, while very popular in France, is just starting to be cultivated in the United States on a commercial basis. You can find them in specialty stores practically year-round, since they are imported and are quite expensive. Depending on the season and the turnover, haricots verts can be delicious and cook quickly and evenly, while at other times they take forever to cook and are quite tasteless. To make sure they stay green, "shock" them as soon as they are done (see below).

Recipes in food magazines suggest "shocking" vegetables. How do you do that and what does it do?

Shocking green vegetables is a restaurant technique that is well worth trying. Before blanching broccoli, asparagus, or green beans, place a large bowl of ice water next to the stove. As soon as the green vegetable is done, drain it in a colander and transfer to the ice water. This will help the vegetable retain its lovely green color. To heat, melt a little butter and olive oil in a skillet, add the green vegetable, season with salt and pepper, and toss gently in the butter just until heated through.

What is really the best way to prepare green beans for cooking? Should I snap off both tails and tips?

I usually snap off the tail only and leave the tip on, which I find looks pretty. Always make sure that the beans are not stringy. If they are, pull the string down along with the tail as you clean them. Haricots verts need to have only their tails removed.

What are Roma beans?

Roma or flat beans are wide, long, and meaty green beans that are unfortunately seldom available in the average supermarket. They are extremely popular in Mediterranean countries, particularly Spain and Italy. Roma beans do need stringing and are best when cut into 1-inch pieces, cooked until very tender, and then buttered.

What is the best method for cooking beans? Do you recommend blanching or steaming?

Personally I do not like to steam beans, since it takes longer and the result is usually a rather crunchy bean. To me, the best method is to drop the beans into plenty of lightly salted water and cook at a rapid boil, testing several times for doneness (tender with a slight bite). Run the beans under cold water to stop further cooking and they are ready for a number of preparations, even something as simple as a little fresh butter, a drizzle of good olive oil, or a nicely flavored vinaigrette.

Beets

I thought beets were always red, but sometimes I see yellow ones and even pink-and-white-striped ones. Are these true beets?

Several varieties of beets are available today—they are all true beets but have different colors and somewhat different flavors. You're probably familiar with the standard round, red beet (*Beta vulgaris*), which has been grown for over two thousand years. It's sweet and tender when cooked properly. The Chioggia, or candy-striped beet, has a bright red exterior and a wonderful ringed red-and-white interior. It's slightly mellower than the standard beet. The golden beet is deep yellow and slightly more oval in shape than the standard beet, with a denser texture and flavor. And, believe it or not, there is a white beet! It looks like a turnip and has a highly concentrated super-sweet flavor. While you can certainly use all these beets interchangeably, the Chioggia lends itself to raw dishes because of its amazing and surprising colors, the golden is best roasted and combined with other roasted root vegetables, and the white beet . . . best as a conversation piece, perhaps!

Is there any way to peel a beet without making a huge mess?

Garnet-colored beet juice does go everywhere. Always peel beets after they're cooked, as it's much easier to do at that stage. Here's my method for minimizing mess and cleanup: place raw scrubbed beets in the middle of a piece of heavy aluminum foil. Rub the beets with olive oil, seal the foil, and bake until a fork easily pulls out of the center of one of the beets. Let them cool slightly and then use a dry paper towel to rub off the skin. The oil makes the peel slip right off and the paper towel keeps your hands relatively beet-juice free. Another option is to wear rubber gloves.

Why do some beets cook in almost no time and others seem to take forever? Would peeling help?

Beets are a seasonal vegetable. They are at their best and freshest in spring and summer, and at that time of year beets cook in less than 30 minutes. Mature older beets can take up to an hour or more, so if you want young tender beets, stay seasonal. Peeling beets is not a good idea, since they will bleed their color into the cooking water and you end up with pink rather than deep ruby-red beets. Always cook them unpeeled, and be sure to leave about an inch of their tops so as not to cut into the beet, which also causes bleeding.

So many recipes call for roasted beets these days. What method do you prefer?

If you are cooking beets in cool weather, oven-roasting them at 350°F is a good idea. Be sure to wrap each beet in aluminum foil and place on a cookie sheet. Test the beets for doneness after about 45 minutes with the tip of a sharp knife; do not pierce all over, since the beets will start losing their juices. I prefer cooking young beets on top of the stove in plenty of water, since doing so does not require using an oven and goes rather quickly. Both baked and cooked beets will keep well refrigerated for several days.

Can beets be eaten raw?

Raw beets really show off their sweet, earthy flavor, and they are certainly striking. You can grate them right into a salad.

Whenever I try cooking the greens from my beets I get a stringy mess. What's the deal?

I cook only the leafy tops of very young beets, mostly those that I pull out of the garden when I'm thinning my crop. You can, of course, find young beets at farmers' markets during the spring and summer. Make sure the greens are tender and crisp. For best results, simply stir-fry the beet greens with a couple of sliced garlic cloves in a little olive oil and season with a generous grinding of black pepper. A cupful of diced or julienned cooked beets is a nice addition to the greens, as is a spoonful of sour cream or crème fraîche.

Broccoli

■■

I love broccoli and so does my family, but all I do is steam it. Do you have any other interesting suggestions for cooking broccoli?

One of my favorite methods is to skillet-braise it. See the recipe for Skillet-Braised Broccoli with Garlic, page 112.

Try adding other vegetables to the braised broccoli, such as a julienne of red peppers and shitake mushrooms.

When I serve raw broccoli with other raw vegetables as part of an appetizer spread, the broccoli often loses color, goes limp,

and remains uneaten. Do you have any
tricks to liven up broccoli when it is served
as a crudité?

I often pass on raw broccoli at parties for this very reason. To keep the broccoli bright green, drop the florets in a saucepan of hot but not boiling water for 2 minutes and then stop the cooking process by either running the drained broccoli under cold water or placing it in a bowl of water filled with ice cubes. This blanching and water-bath method really keeps the green color vibrant long after the party is over.

I see a variety of broccoli-like greens
in my produce department. What are
they?

There are several relatively new hybrids well worth including in your cooking repertoire. Broccolini is considered "baby broccoli." It is very flavorful and doesn't take as long to cook. Prepare it the same way you would broccoli with shorter cooking times. Broccoflower (a trademarked brand name) is another hybrid—a combination, as the name indicates, of cauliflower and broccoli. It looks and tastes much more like cauliflower than broccoli, and I find it to be just a novelty.

When I cook broccoli, the florets are too
soft by the time the stalks are done.
What am I doing wrong?

Try keeping the florets aside and cooking the stalks first, then adding the florets 2 or 3 minutes later. Also, be sure to scrape the stalks with a vegetable peeler or sharp paring knife and cut them into even sizes. The thinner and more uniform in size the stalks are, the more evenly they will cook, and the less likely you are to overcook them.

TIP When stir-frying broccoli, slice the florets in half to allow more of the floret to come into contact with the heated surface of the wok or skillet.

Can I microwave broccoli if I'm in a
hurry?

Yes, as a matter of fact, broccoli is one vegetable that does very well in the microwave. Place the broccoli in a microwaveable bowl with about 1 inch of water. Cover the bowl with either plastic wrap or a damp paper towel and microwave on full power for 3 minutes for florets, 6 minutes for stems, and 8 minutes for a full head. Be sure to allow the broccoli to rest for about 1 minute before carefully removing the covering. Pierce with a knife to see if it's done and return it to the microwave for another minute if it is not fully cooked.

I have plenty of recipes for florets, but I
often end up just throwing broccoli stems
away. Do you have any suggestions for
using them?

It's a shame to toss out perfectly good, tender broccoli stems. Raw, they make a nice crunchy addition to salads, but be sure to peel them with a vegetable peeler and then slice them finely. Toss with a well-seasoned vinaigrette, possibly made with a touch of sesame oil.

What is the difference between broccoli
and broccoli rabe?

The names confuse a lot of people, but these are two totally different vegetables—one is a cruciferous vegetable and the other is a type of bitter green. (The name broccoli rabe derives from the flowers at the tips of the greens, which resemble small broccoli florets.) Broccoli can be prepared in a variety of ways, but broccoli rabe, like all bitter greens, is best when sautéed simply in some olive oil with garlic. Broccoli rabe also sometimes goes by the names broccoli rapini or rape.

I am not sure how to prepare broccoli rabe. Do I remove all the leaves or cut off most of the stalk as with other greens?

I remove most of the stalk and some of the coarse leaves. Don't remove all the leaves or you will be left with hardly any greens.

I tried stir-frying broccoli rabe, but it was so bitter I could hardly swallow it. Is there any way to mellow the taste?

You can mellow the greens by blanching them in salted water for 2 minutes. Once they're drained, I usually like to sauté them in some virgin olive oil with a couple of finely sliced garlic cloves. Be sure to season the broccoli rabe with plenty of fresh cracked pepper.

Brussels Sprouts and Cabbage

■■

I have seen tiny little Brussels sprouts at Pike's Market in Seattle, but the ones I get at my grocery store are large and packed in cartons. Can I use frozen sprouts instead of the larger ones?

The fresh Brussels sprouts sold in most supermarkets are often a real disappointment, but the frozen sprouts are not a good substitute. You will do better quartering large sprouts or breaking them up into leaves and then sautéing them with some butter and broth. They are delicious this way. See Sauté of Brussels Sprouts with Bacon, Pine Nuts, and Sour Cream, page 113.

Whenever I cook Brussels sprouts, my whole kitchen smells like cabbage. Is there a way around it?

Only overcooked sprouts have that smell. Blanch the sprouts in water to cover for exactly

8 minutes. Drain and oven-braise, covered, seasoned with butter, salt, and pepper, for 25 minutes. You will get wonderful-tasting sprouts this way and your kitchen will not have that nose-wrinkling odor.

Do white and red cabbage taste different?

In their raw state, the cabbages taste similar. However, since they are usually prepared very differently, in a sense they have become two distinct cabbages. For example, red cabbage is traditionally teamed with beets in borscht or prepared in a sweet and sour sauce with red wine and vinegar (see Braised Red Cabbage with Orange Zest and Grenadine, page 114). Green cabbage can be braised in butter with leeks, shredded in a slaw, and added to a variety of hearty soups. It is more versatile, since it doesn't discolor other ingredients the way red cabbage does.

Should I buy a large or small head of cabbage?

Size is of little importance when it comes to cabbage; freshness is the key. Summer cabbages tend to be smaller with heads weighing 1 to 2 pounds, while winter cabbage is larger and can weigh as much as 3 pounds or more.

How can I tell a really fresh cabbage?

Very fresh cabbages still have their outer leaves attached and the leaves look green or dark red. As the cabbage ages, the outer leaves are trimmed off by the grocery store, or fall off, and the remaining leaves lose pigment, so that green cabbages look almost white and red cabbage is rosy white rather than that lovely deep red color of the fresh vegetable.

Does the age of the cabbage have anything to do with how much it smells when it's cooked? Are fresher cabbages milder?

The fresher the cabbage, the milder the taste, and I honestly think that cabbage has gotten a bad rap. Unless it's overcooked, cabbage is a mild vegetable that does not have a strong aroma.

I recently sampled the most wonderful, mild-tasting cabbage dish and was told it was made with Savoy cabbage. What is that?

This is indeed my favorite cabbage. Look for it in your market toward the end of summer. The heads have lovely ruffled green leaves, which are perfect for stuffing, and the inner leaves can be prepared in many interesting ways. Keep an eye out for the fresh heads of Savoy cabbage, which show up in markets during the winter months every once in a while. The only recipe for which Savoy is not suitable is American-style coleslaw. Otherwise, it is the king of cabbages.

What is mustard cabbage?

Mustard greens sometimes are called mustard cabbage. You will see them most frequently in the South, or in Chinese grocery stores, although I recently bought lovely fresh mustard greens in a Connecticut market in July. The leaves of this green are ruffled and bright green. They taste tangy, but the larger, mature leaves can be too strong for most tastes. Stick to very young mustard greens with leaves that are not full of holes. Use them in stir-fries the way you would other braising greens.

What is the difference between Chinese and Napa cabbage?

The terms are used interchangeably. There are actually two types of Napa, or Chinese, cabbage. One is football-shaped with a tightly packed head and green ruffled leaves. The other, less popular variety has long stems with compact leaves. Both are extremely mild and delicious in soups or stir-fries. I generally add some finely sliced garlic or a hot chile to this cabbage to enhance its taste. Napa cabbage keeps in the refrigerator for as long as 3 weeks and is good to have around for a quick vegetable.

I hear about bok choy all the time. Isn't it just another name for Chinese cabbage?

Bok choy is a leafy green Chinese vegetable that is a member of the cabbage family. It has dark green, spoon-shaped leaves and bright white stalks that come together at the root. It tastes something like very mild cabbage, but it is more about texture than taste.

Bok choy goes by many names: pak choi, "Chinese white" cabbage, white-mustard cabbage, celery mustard, and so on. It is not Chinese cabbage (which is also called Napa cabbage), although it is often mislabeled as such. Bok choy will keep for about 4 days in the refrigerator in a plastic bag.

Whenever I cook bok choy, it seems to shrink so much. How much do I need to serve four to six people?

Like all leafy greens, bok choy shrinks when cooked by releasing much of the moisture in its leaves. A 1½-pound head of bok choy will yield 4 scanty side-dish servings.

How should I cook bok choy? Are the yellow flowers edible?

To prepare bok choy for cooking, detach the stalks from the cluster at the base and then cut the leaves from the stalks. Cut the stems in 1½-inch pieces and stir-fry quickly in olive oil with some sliced garlic cloves until just tender. Add the cut-up leaves and continue sautéing until just wilted.

The lovely yellow flowers of bok choy are edible and look very pretty when scattered over the stir-fry.

Sometimes I see tiny bok choy in my market. How do you prepare it?

There are many different varieties of bok choy, although they all taste pretty much the same and respond well to the same cooking methods. Baby bok choy, sometimes called Shanghai cabbage, can be steamed whole either in a skillet on top of the stove or in the microwave. Plan on one to two per person. If you plan to chop it for a stir-fry, I would not spend the extra money for the baby variety, since it is more expensive. See Pan-Seared Baby Bok Choy with Red Bell Peppers and Oyster Sauce, page 111.

Carrots

■ ■

Is there any great difference between bunches of carrots sold with their tops and those bagged without them?

Other than carrots just dug from the garden, the best, sweetest carrots are those sold in bunches with their tops. If the tops look crisp and bright, you can be assured the carrots are fresh. Chop the greens off the carrots before storing them, leaving about 2 inches of stem—if you leave the greens attached, they will rob the carrots of some of their moisture and they will go limp in a day or two. Store carrots in a plastic bag and refrigerate them for up to 2 weeks.

Many recipes call for fresh carrots with their tops, but these are not always available in my market and are quite expensive. Can I use bagged carrots instead?

Bagged carrots are fine for stock and soups, and I always have a bag of these in my crisper, but fresh carrots with their tops attached are a much better and a fresher choice if you plan to eat them raw or as a cooked vegetable on their own.

Do you recommend the bags of baby carrots that are so popular these days?

Don't be fooled into thinking that these are really baby carrots—they are large carrots cut by machine to resemble baby carrots. To me, they are tasteless and uninteresting, so what you are paying for is appearance and not taste.

TIP To crisp flabby raw carrots, peel and put them in very cold water for about 20 minutes. They will perk right up and will be nice and crisp, ready for nibbling.

I live in New England and have been growing carrots for years, but truthfully the fresh carrots I buy in the market seem sweeter. Is this possible?

The further west you go, the sweeter the carrots. Carrots grown in the Northwest and California are the sweetest, followed by those coming from Florida and Arizona. Unfortunately, the soil in the Northeast does not produce a very sweet carrot.

For a good vegetable stock Start with 12 cups water, add 3 to 4 celery stalks with tops, 2 or 3 leeks, 3 to 4 carrots, a bunch of flat-leaf parsley, and a little salt. Cover the pot and bring to a boil. Reduce the heat and simmer for 45 minutes to an hour. Remove the stock from the heat and let the vegetables sit in the liquid for 2 or 3 hours. Strain the stock and discard the vegetables. The flavor of vegetable stocks is intensified when the vegetables sit in the broth for several hours after cooking. Use the vegetable stock right after it's strained, or refrigerate it for no longer than 3 days. If you plan to store it for longer than that, you can freeze it for several months. After a stock has been refrigerated for 2 or 3 days and you decide to freeze it, make sure you bring it to a boil first to kill any bacteria. Leave it uncovered to cool completely before freezing.

Some recipes call for carrots to be cored, others do not. When is it important to core a carrot?

Unless you use carrots for stocks or roasting, it is always a good idea to core them, especially if they are large and the core is very developed. Here is how you do it.

Peel the carrot. Cut it into 2-inch pieces crosswise and then quarter it lengthwise. Remove the core with a sharp knife, then cut the carrot into cubes. It is now ready for cooking or braising.

Is it always necessary to peel carrots before eating or cooking?

The only time I do not peel carrots is when I use them in stocks or when they are very young, right out of the garden. Be sure you don't remove too much of the peel, because most of the nutrients lie close to the skin. Small baby carrots, however, only need a thorough rinsing and possibly a light scrubbing.

Do you have any suggestions for carrot recipes that I can make in advance?

When I was growing up in Spain, one of my mother's best dishes was the Skillet-Braised Carrots with Pine Nuts and Raisins, page 116. This can be made up a day or two in advance. Another great make-ahead carrot recipe is the Puree of Sweet Potatoes and Carrots, page 143. This is my favorite side dish for a roast turkey, a chicken, or a pork roast.

Celery

■■

My market carries celery in loose bunches and in plastic packages. Which is the best way to buy it?

Buy celery only in bunches. It is less expensive and has the added bonus of the celery leaves—the most flavorful part of the celery stalk and a must for stocks and many soups.

Does the color of celery make a difference in taste? I see dark green bunches as well as many that are very pale.

There are two kinds of celery. Golden heart is bleached white and is therefore very light in color and has a very mild taste. Pascal celery has darker stalks and a more assertive flavor. Both are equally tasty. If you like milder celery, reserve the outer stalks for cooking and use the lighter-colored inner stalks for salads and snacks.

Celery seems to go limp in 2 or 3 days. How can I keep it crisp?

If properly stored, celery keeps well for 2 or 3 weeks. As soon as you buy it, cut about 3 inches off the leafy tops. Store these separately in plastic bags in the crisper and use the leaves to flavor stocks, soups, and stews. You may have to remove 2 or 3 of the outer stalks after a few days if they develop brown spots. Although I rarely freeze celery, you can do so. Cube it and store it in a zippered plastic freezer bag. The celery will lose its crunch but will still add good flavor to soups and stews.

My celery often freezes in the refrigerator and loses its crisp texture. Why does this happen?

Celery has a very high water content. If your refrigerator is cold, especially in the back, it will freeze rather quickly. Make sure you keep it away from the coldest section in the back of the refrigerator and store it in the crisper. If the bunch does not fit, cut it in half crosswise and store it in two separate bags.

I find I use celery only in chicken and tuna salad and for soup flavoring. This leaves me with lots of leftover celery. What can I do with it?

I like to use celery in creamy celery soup spiked with some blue cheese, for which celery has a great affinity. Try it in the recipe for Creamy Celery and Stilton Cheese Soup, page 118. It also makes a wonderful cooked vegetable that is worth trying.

Is it true that celery is high in sodium and not very healthful?

Celery is high in sodium, but it is also high in vitamins A and C and very low in calories. A large stalk has only 7 calories and ½ cup cooked celery can have as little as 11 calories. This makes celery a favorite with dieters.

Is there a difference between celery root and celeriac? What does celeriac look like?

Celery root, celeriac, and knob celery are names given to a wonderful, rather peculiar-looking vegetable. It is a cousin of the run-of-the-mill celery plant. The "knob" resembles an unwashed bumpy horseradish root about the size of a baseball and is covered with lumps and grooves. But don't be fooled. Under its unglamorous outer shell lies flavorful, crisp, cream-colored flesh that tastes somewhat like the finest celery but has a texture similar to a potato. If you're not familiar with this knobby root, it's definitely worth seeking out.

What should I look for when I buy celery root?

If you see celery roots with their tops, buy them. They usually come in bunches of 3 medium-sized roots. Unfortunately, particularly in winter and spring, the roots, which range from large to very large, are apt to be topless. But as long as they are really firm and crisp looking, they should be all right.

Also, look for roots with a distinct celery aroma. I cannot stress this enough. More than any other vegetable, celery root can be most disappointing if it is not truly fresh, because it will be hollow, stringy, and dry. As demand grows for this outstanding vegetable, you will have better luck buying it.

How long can I store fresh celery root?

Celery root keeps for at least 2 weeks. Leave the root uncovered or store it in a plastic bag and put it in the vegetable bin. As soon as you have the chance, make the Puree of Celery Root, Potatoes, and Sweet Garlic, page 117. You will be hooked on this vegetable for life.

Can celery root be eaten raw like celery?

Yes and no. You cannot nibble on celery root the way you can on other celery. On the other hand, the most popular way to serve celery root is grated raw and tossed in a very mustardy mayonnaise. Celeri-rave, or celeri rémoulade, is a classic French salad that is sold in many take-out shops in France. This wonderful salad is easy to make using the grating attachment for the food processor. By far my favorite version is in Julia Child's *Mastering the Art of French Cooking*, volume 1.

TIP Once peeled, drop celery root into acidulated water or toss with some fresh lemon juice to keep it from turning brown. To keep the root ivory-colored during cooking, add a couple of tablespoons of milk or 1 tablespoon of flour to the cooking water.

Chiles

■■■

Southwestern recipes call for types of hot peppers not available in my local market. Can I substitute one hot pepper for another?

There are numerous varieties of hot peppers, also called chiles. Believed to be the world's second most popular seasoning, topped only by salt, these are usually judged by their spiciness rather than flavor.

As a general rule, you can substitute one hot pepper for another, but some are hotter than others. Small, pointy chiles tend to be hottest, while larger, rounded ones are mild. Usually, you find that the smaller the chile, the hotter it is. This is because the seeds and ribs, which contain most of the heat-producing capsaicin, are most abundant and confined in the smaller varieties. A pepper's heat is also affected by growing conditions such as rainfall, temperature, soil conditions, and other variables, so one is not always identical to another.

Let taste be your guide. Buy a few different varieties and give them a try. But take care when you experiment—these fiery little pods are potent.

TIP **If you love chiles, acquaint yourself with the Scoville scale, which rates them according to their amount of capsaicin. Pure capsaicin would be 16,000,000 Scoville units. A cayenne chile— the common thin red pepper—is 30,000 to 50,000 units, and a jalapeño is 2,500 to 5,000 units.**

Is it just me, or do jalapeño peppers seem milder than they once were?

It's not your imagination. According to restaurateur and Mexican-food expert Rick Bayless, jalapeños are being grown for American palates, which are not as accustomed to heat as those of our southern neighbors. Some jalapeños are no hotter than green bell peppers, while others are richer tasting and offer only medium heat. In general, smaller jalapeños, called serranos, are hotter.

My market carries only serrano and jalapeño peppers. Will these substitute for poblano, Scotch bonnet, and others?

The only time you need to adhere to a certain pepper variety is when you are cooking a specific Mexican or New Mexican dish. Even then, you can make substitutions. But know that jalapeños and serranos, while hot, are not as hot as some other chiles. If you have tasted the dish made with a fiery Scotch bonnet (considered by some to be the hottest chile), you

may be disappointed. On the other hand, if the dish calls for mild poblanos and you substitute jalapeños, the dish will taste spicier than expected.

If fresh chiles aren't available, will the bottled ones do? What about dried chiles?

Keep in mind that the bottled and canned kind are usually pickled. Rinse them well before using them. If you are after heat alone, dried hot peppers are a good choice—especially the small Thai ones. But do use the whole ones rather than the flakes, because if you taste the dish and it is getting too spicy, it will be easy to remove the chile.

Is there a difference between ground chile and chili powder?

Ground chile is the pure ground dry chile pod minus the seeds. Chili powder is a combination of ground mild and hot chiles, oregano, and cumin. Sometimes garlic and salt are also added.

TIP **If you don't have fresh chiles on hand when a recipe calls for them, use a small dried chile. You won't get exactly the same flavor, but you will get plenty of heat.**

A jar of chili powder has been in my cupboard for a while and it's getting lumpy. Shall I toss it out?

That lumpiness is probably the result of natural oils in the chili powder, which is a sign of freshness. Chili powder should have a strong, spicy smell, and if you rub a pinch between two fingers it should leave a stain (it's those natural oils again). Chili powder keeps as long as two years when stored in an airtight container in the refrigerator, although it will lose some potency as it ages.

What's the best way to handle chiles so that my eyes don't sting for hours?

The volatile oils in chiles are capable of burning skin, especially the tender membranes around your nostrils, lips, and eyes. Many cookbooks suggest wearing rubber gloves when working with them. I don't, but I make sure to wash my hands with soap and warm water afterward. A vigorous scrub won't remove all the oils, but it helps. If you choose not to wear gloves, be careful not to touch your eyes, blow your nose (the oils soak right through a tissue), or touch any other sensitive skin tissue.

I am not a great fan of very spicy foods. Is there a way to tone down hot peppers?

The best way to reduce a fresh pepper's heat is always to start by removing the seeds and membranes. Once you have tasted the pepper itself, you can judge how spicy it is and add a few seeds at a time to adjust the spiciness.

Corn

What's the best way to select an ear of corn? Even local farm stands won't allow you to strip the ears back to look at the kernels.

I share your frustration, but I understand the farmers' need to protect their produce. Lots of shoppers strip back the husk of every ear of corn and discard those they don't want, which leaves the exposed kernels to dry out in the sun and air.

The key is to look for ears that are cool to the touch and have tight green husks and moist, green-tinged tassels. Avoid soft or shriveled ears with spots of decay or visible worm damage. Don't buy corn that has been sitting out in the sun all day. If you feel the need, peel back just enough of the husk to check if the kernels look plump and juicy and are packed tightly on the cob. Make sure to smooth the husk back into place if you decide to reject it. Also, pass on supermarket corn that's already been husked and wrapped in plastic. It is just not worth the trouble.

My market carries corn all year long, brought in from different parts of the country. Is it better from one state or another?

Some states, such as New Jersey, Pennsylvania, and places in the Midwest, are famous for their wonderful corn. But the secret to good-tasting corn is its freshness, not its origin. Corn that was grown just a mile from your house will be better than the very best corn shipped from another state. Unfortunately, supermarkets rarely buy from local growers, so the vegetable always has to go through some kind of storage, which at that point does not make it worth buying.

Is there a difference in flavor between yellow, white, and bicolor corn on the cob?

It's all a matter of personal preference. Some people like the super-sweet taste of crunchy white kernels, called Silver Queen in some regions. Others prefer the creamy texture and buttery flavor of yellow corn, which may be called Golden Bantam, or the in-between flavor of white-and-yellow corn, often called Sweet Sue. Names vary throughout the country, but every corn lover agrees that the fresher the corn, the better it tastes. When sweet corn is at its peak in midsummer, it's fabulous. Some people even eat it raw right off the cob or cut the tender kernels off the cob to add to a salad.

What is baby corn?

Those wee little ears you see either in the gourmet area of the produce section or packed in jars on the supermarket shelves are just standard ears in a dwarfed stage. They're grown close together so the ears have room to mature but never achieve their normal size. You can pickle them or use them in stir-fries, but be prepared to pay dearly for them. I think they're more about looks than taste.

If I can't get fresh corn for a recipe, do you suggest using canned or frozen corn?

Canned corn is almost a vegetable in its own right and is superior to frozen corn. Mind you, opinions differ here. Some cooks prefer using frozen corn when fresh is not available, while I prefer using canned. In the dead of winter when wonderful fresh corn is just a memory, use canned corn to make cornbread, a pudding, or fritters. See Corn and Chive Fritters with Smoked Salmon Sauce, page 120.

If I buy corn still in the husk, can I hold it in the refrigerator for a few days?

If you asked a farmer how to store corn, he'd tell you not to do it that way! As soon as corn is picked, the natural sugar in the kernels begins converting to starch, so the sooner you cook it, the better it tastes. However, on the more practical side, if you can't eat corn right away, wrap the unhusked ears in damp paper towels, put them in a plastic bag, and keep in the refrigerator for no longer than a day. Alternatively, you can husk the ears, boil them for 5 minutes, cool, and then slice the kernels from the cobs. Refrigerate the corn kernels for up to 2 days or freeze them.

What's the best way to cook fresh corn on the cob?

Drop the husked ears in enough rapidly boiling water to cover them, put a lid on the pot, then boil for 3 minutes. Remove the pot from the heat and let the corn stand in the hot water, still covered, for 5 minutes. Drain and serve the corn right away with butter and salt.

You can also microwave the unhusked ears for 5 minutes (for 2 ears) or 8 minutes (for 3 or 4). Let them stand for 5 minutes before husking. Use great care, as they will be very hot and emit quite a bit of steam.

I've heard about adding either salt or sugar to the corn water before cooking. What does this do?

Adding salt to the water will toughen the kernels, so I don't advise it. Some cooks suggest that a teaspoon or so of sugar will sweeten the corn, but if the corn is fresh, there is no need for the sugar; if the corn is starchy, no amount of sugar will help.

If a recipe calls for fresh cooked corn kernels, should I take the corn off the cob when it is raw or after it is cooked?

I usually take the corn kernels off the cobs after cooking. It is much less messy.

I've tried to grill corn and find I end up burning it much of the time. Are there any tricks to it?

I bow to the advice of John Willoughby, co-author of *The Thrill of the Grill*. He says to peel the husks back without detaching them from the ears. Remove the inner silk and then pull the husks up back around the corncob. Soak the corn in water for at least 15 minutes and then put it on the grill for about 15 minutes. You can serve the corn at this time, or, if you prefer, pull the husks off the corn, brush the ears with melted butter, and season with salt and pepper. Let the ears come into contact with the grill for a few minutes to give them a nice golden, toasted look and taste.

Cucumbers

Should I buy fat or thin cucumbers?

When you have a choice, definitely buy slimmer, medium-sized cukes. Large, fat cucumbers are full of seeds, and by the time you remove them and the watery flesh, you're left with very little.

What is the difference between a burpless, a gourmet, and an English cucumber?

There is no difference at all, but your confusion is understandable. These names and more are given to the slender, long green cucumber, often sold in plastic shrink-wrap. It is an English hybrid that has also been called European. Its advantage is that it never needs peeling and has so few seeds that it is often called "seedless." However, they usually still need seeding.

I noticed that the gourmets, or seedless cucumbers, are more expensive than the more common ones. Are they better?

During the winter months, you're better off paying the extra money for the seedless cukes.

For shape and color, I think they rate a "10," although they have somewhat less flavor than good summer cukes. They are my first choice for hot, cooked cucumber dishes, because they retain their shape and crunchy texture. See Sauté of Cucumbers in Chive and Dill Cream, page 123.

Many recipes call for seeded cucumbers. How do you seed a cucumber, and is it necessary?

Cucumbers are much more appetizing without their seeds, and removing them is easy. First, rinse the cuke well, peel it if you like, and cut it in half lengthwise. Next, scrape out the seeds and the watery flesh with a grapefruit spoon or the tip of a sharp knife. The seeds will pop right out and leave a hollowed-out, fleshy shell. Salt the cucumber lightly to draw out moisture and let it drain in a colander for about 30 minutes. Then rinse it under cool, running water and pat dry with paper or cloth towels.

Do all cucumbers have to be peeled, or only the waxed ones?

Unless you have a garden-fresh cucumber or a Kirby, peel the cucumbers that you intend to use in salads. The peel can be bitter, which can ruin a salad. For an attractive presentation, Kirbys and seedless cucumbers can be scored with the tines of a fork or peeled in alternating strips. In the wintertime, when waxed cucumbers are the only game in town, be sure to peel them.

What's a pickling cucumber?

Pickling cucumbers are also called Kirbys, my favorites! They are pale green and 4 to 5 inches long, which makes them a little larger than gherkins. Their skin is naturally bumpy and their shape irregular, but they have tons of flavor and plenty of delicious crunch. Because their seeds are so tiny, I rarely bother removing them, nor do I peel them when they are fresh and in season—especially if I buy them at a local farmers' market or roadside stand.

I also love using Kirbys in salads and for a quick, refreshing refrigerator sweet-and-sour pickle that takes absolutely no time to make.

To pickle Kirbys Layer 2 sliced medium-sized onions with 8 to 12 quartered Kirbys in a glass bowl. In a saucepan, combine 1 cup distilled white vinegar with 1 cup sugar, 4 tablespoons pickling spice, 1 teaspoon mustard seeds, and 1 tablespoon salt. Simmer until the sugar dissolves. Add a pinch of cream of tartar, stir, and pour over the vegetables. Cover and refrigerate for 24 hours, stirring from time to time. Serve chilled.

TIP Do not store cucumbers in plastic bags. Leave them loose in the crisper and be sure to use them within 2 or 3 days of purchase, especially if they are unwaxed.

Eggplant

Can you tell me how to choose a good eggplant?

A good seedless eggplant will be light in relation to its size, not heavy as most cookbooks suggest. The lighter the eggplant the fewer seeds it will have. Eggplants must have firm glossy skins without cuts or blemishes, and the stem should be green or at least greenish brown and not dark. The size of the eggplant does not matter. Some varieties are enormous while others are slim and tapered, but both can be delicious if really fresh.

The small and nearly seedless eggplants so common in Italy and Spain don't seem to be available in the States. Can I get the same delicious taste from large eggplants?

Unfortunately, the small varieties found in southern Europe are not grown in this country. During the early spring, you may be able to buy pricey baby eggplants imported from Holland. A little later in the spring, they will come from Mexico and will be far less expensive. These are tasty, but you can get the same flavorful results with larger eggplant, as long as it is quite fresh.

TIP **Always choose eggplants that weigh about 1 pound and pick those that are light for their size. The lighter the fruit, the less likely it is to be seedy.**

Are certain eggplants better for some dishes than others?

The common large purple eggplant is the one I use most often. It is ideal for grilling, sautéing, and roasting. It is the only eggplant you can use for classic dishes such as a ratatouille or Italian caponata. Asian eggplants, also called Japanese eggplants, are about the size and shape of a zucchini. They are less seedy with a creamier texture. They are good stir-fried, quickly sautéed, or grilled.

Is it necessary to peel eggplant?

Generally there is no need to peel eggplant because the skin is so flavorful, Plus, the peel keeps the vegetable from falling apart during cooking. Wash it well, dry it thoroughly, and then trim the stem and blossom ends before proceeding with the recipe.

If a recipe calls for peeled eggplant, a vegetable peeler does the trick. Make sure to peel it just before cooking or the flesh will darken.

Many recipes require salting and weighting eggplant slices before cooking. Is this step necessary?

This process does work, but I prefer submerging slices of eggplant in cold, salted water for an hour. This removes the bitterness and is far easier than using weights. Be sure to dry the slices well with an absorbent cloth towel—paper towels are not absorbent enough—before baking

or sautéing. Plant breeders are working on producing eggplant that is less bitter and does not need salting or soaking.

TIP **Test the freshness of an eggplant by pressing down on the flesh with your finger. If the indentation remains, the eggplant is not as fresh as it should be and may be pulpy and bitter.**

Eggplant seems to absorb an astonishing amount of oil during cooking, usually more than the recipe calls for. How can I avoid this?

When you fry eggplant, it does absorb great quantities of oil, especially if the eggplant slices have not been salted and weighted or submerged in salted water and thoroughly patted dry. The good news is that the eggplant releases the oil within 15 minutes of cooking, so let the cooked slices drain on a double layer of paper towels for 15 minutes before continuing with the recipe. You can also brush eggplant slices lightly with olive oil and bake them in a 350°F oven until done. They will not brown as nicely or be as tasty as when they are fried, but this is a good way to cut down on fat.

I love the taste of the roasted eggplant dish called baba ghanoush. Can I make this at home?

It is easy to make this classic, Middle Eastern dish at home either by charring the eggplant on a grill or roasting the whole vegetable over a medium-high gas flame while keeping a close watch on it. Turn the eggplant with tongs so it chars evenly on all sides. Do not prick the skin as you turn it, or all the delicious juices will escape. I've even roasted the eggplant directly over the coils of an electric stove with excellent results. In fact, the only technique I don't recommend is broiling—the eggplant will char, but you won't get the desired smoky flavor. See Charcoal-Grilled Eggplants with Lemon-Scallion Mayonnaise, page 126.

Do you have any new and unusual ways to serve eggplant?
Try filling hollowed-out fresh tomatoes with roasted eggplant pulp mixed with a touch of mayonnaise; serve it as a starter.

Endive

■ ■

Is there any way to tell which heads of Belgian endive are likely to be bitter?
Belgian endive needs to be kept in the dark. Good markets keep it in its original shipping box, wrapped in paper to protect it from light, which causes it to turn green and bitter. Avoid endive that is wrapped in plastic and make sure it is almost all white with just a tinge of yellow.

TIP To remove all traces of bitterness from endive, cut out its core with a paring knife. If you plan to cook the endive, add the heel of a loaf of stale bread to the cooking liquid, which will remove all bitterness.

Is there any way to serve endive other than raw in salads?
When cooked, endive is buttery soft and sweet. It can be braised in a skillet, oven-roasted, or blanched and served with brown butter. It is a perfect companion for sautéed sea scallops, salmon, chicken breasts, and all cuts of veal (see Skillet-Braised Belgian Endives, page 127).

Fennel

■ ■

What is Florence fennel? Is it different from the fennel used for fennel seeds?
Florence fennel—also called bulb fennel, or finocchio in Italy—is a crisp, anise-flavored vegetable that looks a little bit like celery only with a more swollen base and feathery fronds. It has become quite popular in recent years, with both home cooks and chefs, and has nothing to do with the feathery herb that produces the anise-flavored fennel seeds used in cooking and baking.

Is there a difference in taste between large and small fennel bulbs?
There is no difference in taste, but smaller fennel bulbs slice more evenly.

What is the best way to store fennel?
I chop off all the feathery leaves, right down to the bulb, before I store the fennel in the crisper in perforated plastic bags. You may want to keep some of the feathery greens in a separate bag to use for garnish.

How do you prepare fennel for cooking?

First, remove the stalks where they join the bulb and then remove the white, outer leaves encasing the bulb, which tend to be stringy and tough. Depending on its size, cut it into quarters or eighths. Use a vegetable peeler to remove strings and any brown spots. Instead of pushing the peeler away from your body, hold the fennel and work the peeler toward you. If you're not going to cook the fennel right away, put the bulb in a bowl of cold water. I save the stalks to add to soup later, and the wispy fronds can be minced and used as a garnish.

TIP When preparing fennel, don't cut too much off the flat end or the bulb will fall apart.

Can I serve fennel raw in a salad?

It is delicious in a salad, especially when mixed with arugula and Belgian endive, as the Italians do. It's also wonderful on its own, sliced and tossed with olive oil, pepper, and coarse salt. Make sure to slice it super thin—the best way to do this is with a mandoline, a kitchen tool described on page 11.

What are some simple ways to cook fennel?

My favorite way to prepare fennel is to roast it in the oven. It is also excellent braised. For a simple and tasty dish, blanch it for 3 to 4 minutes until the fennel is almost tender. Then sauté it on top of the stove in some butter and olive oil until nicely browned, or oven-roast it for 10 minutes.

What's the best way to prepare fennel to maximize its licorice flavor?

When I want that pronounced licorice taste, I add a teaspoon of fennel seeds to the dish before cooking, or I finish the dish with a few drops of Pernod, an anise-flavored liqueur imported from France.

Garlic

■■

I have seen both purplish-pink and pure white heads of garlic in the markets. Is there a difference between the two?

White garlic, which is rather mild, is the most common kind. It is grown primarily in California and, for my taste, is not nearly pungent enough. The pink variety is a Mexican garlic known as Spanish roja. It's smaller than the California garlic, but has plumper cloves and more flavor. Unfortunately, it's not available year-round, so your choice may be limited to white heads.

Either way, just make sure you select large, firm cloves that appear to be bursting out of their skins. These will be the freshest and easiest to peel. Whatever you do, do not buy the small, packaged garlic usually sold two to a box. These heads generally are not very fresh and have tiny cloves that are extremely hard to peel.

What is elephant garlic?

Heads of elephant garlic are two to three times larger than other varieties and look like a garlic lover's dream. But looks are deceiving. These gentle giants are so mild you might as well not use garlic at all—unless you want extremely mild garlic flavor for a salad dressing or marinade. Besides being nearly tasteless, elephant garlic is about three times more expensive than regular garlic. Recently I attended a press dinner at which the chef braised the oversized cloves in butter and cream and served them as a side dish to roast duck. No one at the table could identify the taste and we concluded we were eating a new, rare vegetable from another hemisphere.

TIP To keep garlic from sticking to the knife when mincing, sprinkle the garlic with coarse salt. This also heightens the flavor of the garlic.

I store garlic in one of those pots with little holes on the side, but it still dries out and shrivels. Is there a better way to store it?

Regardless of what every professional chef and cookbook author recommends, I find garlic does quite well stored in the refrigerator's crisper. Yes, it must be kept loose and not bagged, but the cloves keep for a long time this way. I also store whole heads in a shallow plastic bowl in the refrigerator with success. Just make sure it stays dry and your garlic will keep for at least 2 months.

Sometimes garlic cloves get little green sprouts in the middle. Are these edible?

That little green shoot is a sure sign that the garlic is mature and most likely will have a slight bitter flavor. If the sprout is tiny, as is often the case, either leave it or simply remove it with the tip of a sharp paring knife.

My market carries jarred peeled garlic cloves that look like a good idea. Is there much of a difference in flavor between these and fresh, unpeeled cloves?

I find there is an enormous difference. The peeled, prepared cloves taste strong and slightly bitter but lack the subtlety of fresh garlic. Since fresh garlic is available all year, I see no reason to use jarred garlic. However, if you are in a big hurry and need large quantities of peeled cloves, you can improve the flavor by blanching them for 2 or 3 minutes before adding them to the recipe.

What's the best way to peel garlic?

There are all kinds of clever devices on the market to help with this chore, but it's very easy to accomplish with nothing more than a chef's knife and cutting board. Place a few cloves on the cutting board, hold the flat side of a chef's knife over them, and give the knife surface a quick smack with the heel of your hand. Do this hard enough to crack the skin but not so hard you smash the clove. The skin slips right off. When a recipe calls for whole peeled cloves, blanch them for 2 or 3 minutes, drain, and then peel the loosened skin by hand.

How can I tone down the sharp taste of garlic?

One of the nicest ways to ensure that you get a delicate garlic taste is to blanch the whole cloves twice. Simmer the cloves in water to cover for

2 to 3 minutes, drain them, then simmer again in fresh water. Some cooks recommend blanching the cloves in a mixture of water and milk to keep the cloves white and tender. If you are looking for very mild garlic flavor in a salad dressing or marinade, try elephant garlic.

Do you recommend using a garlic press? Is there another way to get the same result?

Using a garlic press is just fine. I use mine constantly, especially when I want a strong garlic flavor. Make sure to use the garlic within 30 minutes of pressing, or it may develop a bitter, metallic taste. I recommend using a stainless steel garlic press rather than a plastic or aluminum one because it doesn't alter the taste of garlic as quickly.

Another way to mash garlic is to lay the cloves on a cutting board, sprinkle them with coarse salt, and then use a small sharp paring knife to shave each garlic clove until you have a smooth paste. This may sound time consuming, but it is really extremely easy and quick to do. The salt softens the clove and mashing it this way produces an intense garlic taste. Using a mortar and pestle is another good way to get a garlic paste. Put sliced cloves and some coarse salt in a mortar and work into a paste with the pestle. Or, use the side of a chef's knife to work the salt into the garlic until it becomes smooth and creamy.

I grow garlic chives in my garden. Can I use these in place of garlic cloves in recipes?

Garlic chives have a pleasing, mild flavor and are great as a tasty garnish. I love adding them to stir-fries or mincing them very fine and stir-ring them into sauces, but they are not a substitute for the real thing. They are more like pungent chives than garlic.

Is it possible to prepare whole roasted heads of garlic in the microwave?

You can microwave garlic, but for that sweet, creamy, nutty flavor you're after, I recommend using the oven.

Slice off the top of the head to expose the tips of the cloves. Arrange the tips facing up in an ovenproof dish. Pour a little olive oil over them, cover tightly with foil, and bake in a preheated 375°F oven for about 1 hour, or until the garlic is soft. When the heads are cool enough to handle, separate the cloves and squeeze the softened garlic out of the skin. This roasted garlic pulp is wonderful squeezed directly onto grilled or toasted peasant bread. Or put it in a small bowl and cover it with a thin layer of olive oil, and serve it with bread.

When you're short on time, microwaving is fine.

Trim the stem ends to expose the cloves. Put one head in a 2-cup measuring cup, add ¼ cup chicken broth and 1 tablespoon olive oil, cover tightly with plastic wrap, and microwave on high for 7 minutes. Remove from the microwave and let the garlic rest for about 5 minutes before using as described above.

TIP To remove the odor of garlic from your hands, hold them under running water and gently rub your hands along the flat side of a stainless steel knife or spoon. Rubbing the cut side of a lemon over your hands works well, too.

Greens, Leafy

In summer, the markets seem to be full of interesting greens that I am not familiar with. Which do you recommend?

My favorite summer green is Swiss chard. It's both a vegetable and a green because the ribs can and should be cooked separately and have a beetlike flavor. At many farm stands you can find both the white-ribbed Swiss chard and the red-ribbed variety. They are both delicious.

You may also find beet greens and sorrel in the summertime. These are more often found in farmers' markets than the supermarket. I simply steam the beet greens or stir-fry them with some sliced garlic and minced dill. Sorrel can be used in several wonderful soups and also in a creamy sauce to serve with pan-seared salmon fillets or grilled bass.

More and more you can now find a mix called braising greens, which, like the mesclun salad mix, is a combination of young greens.

These can be quickly stir-fried in a touch of extra virgin olive oil with 1 or 2 garlic cloves and minced fresh ginger.

Can you recommend interesting winter greens?

For my money, kale is the most interesting winter green. It's fabulous in soups, such as the Kale, Potato, and Butternut Squash Chowder, page 130, or sautéed with some pine nuts and finely diced smoked ham. Make sure the kale looks crisp and fresh and has dark glossy green leaves with a bluish tint.

If you live in the South, chances are you can find wonderful collard greens. These should be young, crisp, and velvety. Unfortunately, due to lack of turnover, impeccably fresh collard greens are hard to come by outside the Southeast or in other areas where they are popular.

Herbs

I often buy pretty herbs in the supermarket, but they seem to lack the flavor of the farm stand herbs I get in the summer. I wonder why?

Hothouse herbs are meeker than field-grown herbs, so if you're making a dish with supermarket herbs, you may have to double or even triple the amount called for in the recipe. This is especially true of cilantro, dill, and basil. A

way around this is not to chop these herbs finely but rather to snip them with scissors or, if the leaves are small, keep them whole to achieve the same flavor level. Since fresh herbs stand out most when added just before serving, you may want to try reserving about half of any fresh herb that the recipe asks you to incorporate during cooking and adding it at the end.

The flavors of my home-grown sage, thyme, and tarragon seem different from what is served in restaurants. Is it my imagination?

Most likely you are tasting varieties of these herbs different from the ones you grow at home. There are many types of thyme available to the gardener and at least three types of tarragon. The list of sage varieties is considerable as well. The "generic" types of these herbs are common sage, French thyme, and French tarragon, so you probably want to look for those names when buying seeds or plants. Sometimes recipes get more specific and call for lemon thyme, Mexican tarragon, or pepper sage, but you can always use the more common form of the herb.

TIP Be most careful when purchasing a tarragon plant for your garden if what you want is French tarragon and not the Mexican variety. Because the leaves are similar, you will have to taste them to be able to tell the difference.

What does fines herbes mean? Is it just a fancy French way of saying fine herbs?

Fines herbes is a family of herbs that includes chives, chervil, parsley, and tarragon. These complement each other and are key to many French preparations. A classic fines herbes mixture consists of 1 part tarragon; 2 parts chervil; 8 parts parsley; 1 part chives. Many recipes call for fines herbes to be chopped. In that case, you should chop the first three together, then snip the chives separately (to keep from mashing them) and combine all four. You can adjust these proportions to suit your own taste and the availability of certain herbs. For example, if you can't find chervil, just use more tarragon or fennel tops to make up for it.

Marjoram and oregano taste pretty much the same to me. Are they interchangeable?

It is fine to substitute these herbs for each other. Marjoram and oregano are very similar, although marjoram's flavor is much less assertive. Traditionally, oregano is used in Mediterranean dishes, while the cuisines of central Europe rely heavily on marjoram.

Some recipes call for flat-leaf parsley, some just for parsley. Is there a difference?

You will probably find two types of parsley in your grocery store, flat-leaf and curly. Curly has smaller leaves and tends to be rough in texture. I use flat-leaf parsley, sometimes called Italian parsley, almost exclusively.

Which herbs can be substituted for each other?

That's a complicated question. Some herbs, particularly parsley and chives, have an affinity for just about all foods, but others work only in specific dishes. When you're making a classic preparation, you should stick to the standard. For example, ravioli are traditionally served in a sage and butter sauce, and they taste heavenly that way, so there's no reason to replace the sage with cilantro, for example. A béarnaise sauce would not be the same if not made with fresh tarragon, but a cucumber or beet salad, on the other hand, is delicious when made with either dill or mint or a combination of herbs. So as long as you learn the tastes of herbs and which ingredients they complement, you can have fun and be creative cooking with them.

TIP Some herbs "marry" particularly well. You might want to try the following herb combinations:

dill, parsley, chives, and chervil

tarragon, parsley, chives, and chervil

basil, parsley, chives, and chervil

sage, parsley, and chives

rosemary, thyme, and parsley

oregano (or marjoram), thyme, and parsley

cilantro, mint, parsley, and chives

TIP There are two basic categories of herbs: character herbs and accent herbs. Character herbs are usually added to a dish during cooking or at the beginning and don't have any trouble asserting their flavors. Accent herbs are more delicate and are usually incorporated just before serving.

CHARACTER HERBS

bay leaves, marjoram, oregano, rosemary, sage, tarragon, thyme

ACCENT HERBS

basil, chervil, chives, cilantro, dill, mint, parsley

Within days after purchase, my fresh cilantro and basil turn into bags of rotten leaves. Can you suggest a way to keep these herbs fresh for a longer period of time?

Both basil and cilantro are delicate. The trick is to keep them dry and sealed in their bags. If you buy the herbs pre-washed in sealed bags, leave them in the bags until you're ready to use them. If you buy loose bunches, shake out any moisture, then remove the leaves and place them in a single layer on a paper towel. Top with another paper towel and continue to layer leaves and paper towels, ending with a sheet of paper towel on top. Roll up this stack and place in a plastic bag with a zipper closure. Seal the bag

carefully, expelling all air. They should keep for several days this way.

Should I keep fresh herbs in a glass of water or in a plastic bag in the vegetable bin?

The first thing to do is to shake the herbs vigorously to remove the excess water accumulated during all that misting in the supermarket. Then wrap the herbs loosely in paper towels and store them in a perforated baggy. I have not found the glass method to be too effective and only use it when I plan to use the herb as a garnish on the day I buy it fresh.

What's the best way to clean herbs to remove all the grit?

Some herbs are particularly sandy. With the leaves still on their stems, swish them in a bowl of water. Hold still for a moment to allow the dirt to fall to the bottom of the bowl, then pull out the herbs, rinse the bowl, and repeat until absolutely no sand appears in the bottom of the bowl. Then rinse one more time, just to be sure. Dry herbs thoroughly, either with paper towels or a dish towel. If you are using a large quantity of herbs, you may want to dry them in a salad spinner. Wet leaves are impossible to chop and quickly turn into mush.

My supermarket carries basil practically year-round. Is it worth the bother to grow my own?

Garden-grown basil has a stronger, cleaner flavor than hothouse basil. If you don't have a garden but live in the Northeast or Midwest, during the summer and well into the fall look for delicious field-grown basil at farm stands, which usually sell it in large bunches and at low prices. In Florida and California you may find it as early as February.

It would be easy to grow basil on my sunny patio, but there are so many different kinds. Which is best?

The best basil to grow on a patio is the all-purpose basil called fino verde. Its sweet, somewhat pungent flavor is ideal for all recipes calling for this popular herb. If you want to assemble a small basil garden, try a few of the tiny-leafed globe basil plants, plus a pot or two of purple basil, which is quite beautiful. You might want to try some ruffle basil too—its leaves are milder in flavor and can be added raw to salads.

Is there any way to freeze basil leaves without their turning black?

As a rule, herbs don't freeze well. If you want to save some basil for later use, grind the leaves to a paste and mix them with some olive oil. Cover the basil paste with a thin layer of olive oil before freezing.

What exactly is pesto?

Pesto is a basil paste from the Italian city of Genoa, used primarily as a sauce on pasta.

To make a traditional pesto

3 to 4 cups basil leaves, loosely packed

2 to 3 garlic cloves, crushed

2 to 3 tablespoons pine nuts

About ½ cup freshly grated Parmesan cheese

About ¾ cup extra virgin olive oil

Purée everything except the olive oil in a blender (or crush in a mortar and pestle, if you're a traditionalist). Drizzle in the olive oil with the blender running. You can store pesto, covered, in the refrigerator for 2 or 3 days. If you're planning to do that, omit the garlic and add it just before serving, and drizzle a thin layer of oil over the paste before you put it in the refrigerator to keep the basil from turning black.

TIP When is a pesto not a pesto? Classic pesto consists only of basil, garlic, pine nuts, cheese, and olive oil. Phrases like "cilantro-walnut pesto" or "tarragon pesto," often seen on restaurant menus, simply refer to herb pastes.

Do you have any suggestions for using pesto other than as a pasta sauce?

Pesto is wonderful stirred into soup (see White Bean and Chorizo Soup, page 292). You can also drizzle pesto over grilled vegetables or oven-roasted tomatoes, and it adds punch to rice pilaf when added just before serving. For an unusual potato salad, toss pesto with boiled, sliced potatoes while they're still warm, then serve at room temperature.

My large rosemary plant looks beautiful and smells great, but sometimes the flavor of the leaves is a little overwhelming. Any ideas for subduing it a bit?

Rosemary can be powerful, almost pungent. To avoid overpowering dishes with rosemary flavor, I use only the very young and tender rosemary leaves in most dishes. The older, needlelike leaves serve best to stuff a chicken or whole fish and are also good for marinades. Most herbs lose their flavor when finely minced, but rosemary becomes even more powerful, so if you're after delicate flavor, use minced rosemary sparingly.

Do those tiny thyme leaves really have to be stripped off their stems? The stems are so brittle that they break.

This is a necessary kitchen chore, as thyme stems are woody and unpleasant to eat. You can add whole sprigs to soups and sauces for flavor, as long as you remove them before serving. Try pinching the end of a thyme sprig with one hand and pulling straight down with the thumb and index finger of your other hand to scrape off the leaves. Also, if both the leaves and your hands are dry, you'll find the job easier.

My chive plants grow pretty purple flowers. Is there a use for them?

Chive blossoms are edible and make a lovely addition to a salad. Make sure that you cut the blossoms back two or three times a year to stimulate growth.

How are Chinese chives used?

These flat-leaf chives have a stronger flavor and less delicate texture. Unlike regular chives, they are best when lightly sautéed or incorporated into stir-fries. They are also nice added to an egg mixture for an omelet.

What are giant chives?

This recently introduced variety is sturdy and excellent for cooking, but it is rarely available in stores. Giant chives grow very successfully in window boxes.

Is there a difference between cilantro, Chinese parsley, and coriander?

Those are three names for the same thing—an herb that looks somewhat like flat-leaf parsley but has saw-toothed ridges around the edges of the leaves. This herb is used in cooking both Chinese dishes (hence Chinese parsley) and Mexican foods (hence cilantro, its Spanish name). Coriander usually refers to the dried version or the seeds of this plant.

Cilantro is one of my favorite herbs, but my family says it tastes like soap and refuses to touch it. Any suggestions for how to please myself and the rest of the family?

Isn't it amazing that an herb can provoke such intense emotions? Everyone's got an opinion about cilantro. The solution is simple: use whole leaves as a garnish rather than incorporating them into your dishes. That way you can sprinkle the cilantro on your own portion only. If you still want to try to sneak it into something, keep in mind that cilantro's particular flavor grows milder when the leaves are minced.

All of a sudden, so many recipes call for chervil. What exactly is it?

Chervil is a delicious herb that has long been used in French cooking and has recently experienced a boom in popularity here. It has feathery, delicate leaves and makes a nice addition to creamy sauces as well as to salads. Many commercial mesclun mixes include some chervil. Chervil's flavor—mild, with a hint of anise—is impossible to duplicate, so there is no real substitute for it. However, since chervil is often used as an element in fines herbes (see page 84), if I can't find any, I use the feathery tops of fennel to make up for its absence. Like dill, chervil thrives in cool places, so if you live somewhere that's not too warm, you may want to grow your own.

Are dill seeds a good substitute for fresh dill?

You would not be happy with the result. Dill seeds are best used for pickling and have none of the delicacy of dill fronds.

When a recipe calls for mint, does that mean peppermint or spearmint?

Peppermint is my first choice and spearmint my second, although you are more likely to find spearmint in the grocery store. I also like apple mint, especially for mint sauce. If you are interested in growing your own mint, remember that mint is a bully that will take over your entire garden if you let it, so it should be isolated in one flower bed.

When fresh herbs are not available in the supermarket, can the same herb be used in dried form?

Bay leaf, marjoram, oregano, and thyme can be used in cooking in dry form as long as you add them at the start of preparation. In other words, don't use dried herbs to flavor a finished dish, or in a vinaigrette or marinade. All other herbs should be used fresh, especially chervil, parsley, cilantro, basil, and tarragon, which lose their flavor when dried. If you can't find fresh rosemary, use fresh thyme.

Do dried herbs last longer if refrigerated?

They don't need to be refrigerated, but do date them and replace them twice a year. Dried herbs tend to have a short shelf life, so be sure to close their jars tightly and store them in a cool, dark place.

TIP You can use this quick test to check whether your dried herbs are still fresh. Place a small amount of the herb in the palm of one hand and rub it with your fingers. You should immediately detect the distinct aroma of that herb. If it doesn't have a smell, it won't have any taste either.

Jerusalem Artichokes

What are Jerusalem artichokes?

Jerusalem artichokes are root vegetables that look something like smooth pieces of ginger. While they are not related to globe artichokes, they taste enough like artichokes to make you notice. Jerusalem artichokes, also called sunchokes, are not as popular as they should be, and you may have trouble finding them. Their season runs from early fall to early winter. You'll have the best luck finding them in good specialty markets and farm stands, but more and more now you can find them in quality supermarkets.

What is the best way to store Jerusalem artichokes?

If you store them in a perforated plastic bag in the refrigerator, they will keep well for up to 10 days, after which they start to shrivel. It's best to cook them within 3 or 4 days of purchase.

My recipe for Jerusalem artichokes says to cook them for 20 minutes but they turn out mushy. What is the right way to cook this vegetable?

Jerusalem artichokes can be frustrating to cook because one minute they feel undercooked and the next they're overcooked. The best way to handle them is to pan-braise them in some butter with a touch of broth and seasoning. Cook them covered over low heat for 5 to 7 minutes or until just tender.

Jicama

My market carries jicama in all sizes. Does the size have anything to do with the taste?

The size of jicama has nothing to do with its taste. I prefer to buy a medium-sized jicama and use it up within 2 or 3 days. Large jicamas are harder to store and there is just so much jicama you can eat unless you really love it or are serving a crowd.

I have had some delicious, crisp jicama and some have a woody texture. Is there a way to tell a good jicama?

Look for jicama with relatively smooth, thin skin. Thick skin usually indicates a starchy jicama, instead of sweet, juicy flesh.

The only way I know to serve jicama is cut up as a snack or for a crudité platter. It's refreshing and crunchy this way, but do you have any other suggestions?

Jicama is delicious quickly stir-fried with snow peas or finely sliced red and yellow bell peppers.

Lettuce

The ready-made salad mixes are quite expensive. How can I make my own?

The key is to choose a selection of greens that go well together, keeping in mind that different lettuces have different characteristics. Some are buttery and mild while others are acid and bitter. As a rule, combine greens of different textures, flavor, and color. A nicely balanced mix would be bibb lettuce, arugula, Belgian endive, and raddichio.

What is the best way to store salad greens?

It is a real challenge to keep greens fresh these days because the constant misting in the market keeps them too wet and almost soggy. The key is to store them as dry as possible. The best way to do this is to wrap them unwashed in a double layer of paper towels and store them in a perforated plastic bag. The greens will keep this way for 3 to 4 days.

Which salad greens keep best?

The best-keeping green is, of course, iceberg lettuce, but it is not the most flavorful. Romaine has great flavor, plenty of crunch, and sturdy leaves. Before storing romaine, cut 1½ inches off the top and remove any wilted outer leaves. Store in a perforated baggie in the vegetable bin.

Lettuce from the farm stand is so gritty that it is practically impossible to clean. Is there a secret to getting rid of the dirt?

The best way to do this is to separate the leaves and soak them in plenty of warm rather than cold water, swish the lettuce gently, and then transfer to a bowl of very cold water. Lift it out gently and roll it in a kitchen towel or spin it dry. If there is any grit remaining in the lettuce when you serve it, you can always claim that the crunch is coming from freshly ground peppercorns!

Leeks

■■■■■■■■■■■■■■-■■■■■■■■■■■■■■■■■■■■■■■■■■■■■■■■■■■

Recipes frequently call for thin leeks, but all I ever see in the market are fat ones. Is there any difference in taste?

The smaller leeks are more tender, but there is no difference in taste between the two. It's a good idea to avoid leeks that are too large, though, especially in the spring, when they have a fibrous core that must be removed. Thin leeks are terrific for braising and serving whole, but they're only available in the fall, unless you live in the Northwest, where because of the climate they are available much of the year. Still, many specialty markets carry thin leeks, so when you see them, buy them. They're worth the expense if you plan to braise them. Either way, fresh leeks with dark green, crisp tops and white root ends are excellent for all recipes that call for diced or sliced leeks.

TIP Before you start to cook a green vegetable— snow peas, green beans, or asparagus—fill a large bowl with ice water. As soon as you drain the vegetables, pop them in the ice water. This will stop any further cooking and allow the vegetable to retain its beautiful green color.

Leeks tend to go limp after a couple of days in the refrigerator. Is there a way to keep them fresher longer?

Leeks will stay fresh much longer if you cut 3 inches off their tops. This also keeps precious nutrients from leaching back into the leaves during storage. Put the leeks in plastic bags and refrigerate them for up to 2 weeks. But don't discard the green tops. I keep the inner light green leaves in a separate bag and use them to flavor soups and stocks.

No matter how well I rinse leeks, some grit always remains. Is there a way to clean them thoroughly?

Leeks are really easy to clean once you know the secret. Begin by removing all but the first 2 inches of the green tops. Dice or slice the leeks lengthwise and rinse them in a colander under warm (not cold!) water for several minutes. Drain well. You'll have nice, clean leeks every time.

Can I treat leeks like other members of the onion family and serve them raw?

Even young leeks are too strongly flavored to eat raw. Even if you like strong onions, you'll find raw leeks too harsh tasting.

I use leeks most often in soups. Is there another way to cook them?

When leeks are poached, cooled, and dressed with a mustard vinaigrette, they make a great first course. You can always braise diced leeks in butter and broth and serve them alongside seafood, veal, or chicken (see Braised Leeks with Shiitake Mushrooms and Crème Fraîche, page 131).

Mushrooms

■ ■

Recipes frequently call for wild mushrooms, but I never see any in the market. Do I have to pick them myself?

Generally, recipes that call for wild mushrooms do not refer to true wild varieties, but rather wild mushrooms that are now being cultivated. These include enoki, oyster, portobello, shiitake, and cremini, which are also called brown mushrooms. You can easily find these in most supermarkets. Be sure to experiment with them to see which ones you like, since some are much more flavorful than others.

As for picking them yourself, that is definitely an activity best left to the experts, since amateur mushroom hunters can easily mistake toxic mushrooms for the edible varieties.

Are any true wild mushrooms available commercially?

The only true wild varieties available commercially are chanterelles, trumpet mushrooms, and the more costly morels, which are available in better markets at different times of the year. While most wild mushrooms appear in the fall, there are some spring varieties. Weather plays a major role in their availability. Also, keep in mind that since wild mushrooms are extremely perishable and generally do not travel well, you may not see a good assortment in your area. While you can get fabulous fresh chanterelles in the Northwest, the same mushrooms may never show up in a Northeastern market. Find out what kind of mushrooms grow in your area and seek out a market or farm stand that may carry them.

Are white button mushrooms a suitable substitute for wild ones?

Different mushrooms have different flavors and textures. Comparing them is like comparing zucchini to cucumbers. They belong to the same vegetable group and are somewhat similar in texture, but in fact are very different vegetables. So, while the white, all-purpose buttons are certainly pretty enough, they're very mild and unassertive compared to other mushrooms and won't give you the flavor that the wild varieties will.

I always see white mushrooms sold loose in a basket right next to the plastic-wrapped ones in a carton. Which is the better choice?

You'll notice that loose mushrooms are more expensive than packaged. That's because you can choose and pick each one individually according to its size and appearance, eliminating those that aren't in the best condition. But as long as the packaged containers are labeled "all natural," which means they haven't been treated with a chemical, they are basically the same.

Which of the grocery store mushrooms do you consider the most flavorful?

I'm particularly fond of the brown cremini mushrooms. They have a lovely, earthy taste and are not watery. The portobello, which is actually cremini's big brother, is also a flavor-packed mushroom that is good roasted at a low temperature (Portobello Mushrooms in Tarragon-Lime Butter, page 132) and even grilled. Shiitakes that are large and spongy are wonderful sautéed and served as a topping for a salad or julienned and added to stir-fries.

MUSHROOMS AT A GLANCE

Black trumpets Delicate texture, outstanding flavor, but better when dried, then reconstituted.

Button mushrooms Commercial mushrooms available fresh all year. Their taste is rather bland but you can mix with other "wild" mushrooms for a wonderful flavor.

Cèpes Also called porcini. Grow wild and are sometimes available fresh in the fall. Very perishable.

Chanterelles Resemble a curved trumpet or vase. Bright orange or apricot with an apricot-like aroma. These are very delicate, so clean gently with a damp towel. Fresh chanterelles require a longer cooking time than most other mushrooms.

Enoki Cultivated and available fresh all year. Resemble bean sprouts, slender and snow white. Very mild flavor.

Morels Grow wild and are available fresh in the early spring. Easily recognized by their conical shape and honeycomb markings. Should be heavy for their size and spongy to the touch. Have a meaty, intense flavor, wonderful sautéed in butter.

Oyster mushrooms Grow wild and also cultivated. Mild, buttery-tasting mushrooms with large, fanlike leaf and stem that somewhat resemble the shape of an oyster.

Shiitakes One of the best-flavored commercial mushrooms available. Stems are tough and inedible but caps have a great smoky flavor.

What is a porcini mushroom? I always see them dried but never sold fresh.

The porcini is a huge, fabulous-tasting, wild fall mushroom that looks like a giant brown common mushroom. But its meaty, silky texture is anything but common. Unfortunately, they're extremely perishable, so they're rarely seen fresh in markets and are very expensive. However, dry porcini, which have a rich, woodsy aroma and a marvelous concentrated flavor, are far from second best and are now available in gourmet shops everywhere. They're often sold by their French name, *cèpes*, or the Latin, *Boletus edulis*.

What exactly should I look for in shiitake and oyster mushrooms? Should they be large or small?

When buying shiitakes, try to pick the ones that are about 2 inches or less across the top. This is important because you eat only the caps, and if you're paying by the pound, you don't want any of the heavy, nonusable stems loading up the scale. The caps are slightly spongy in texture and have a pleasantly smoky flavor. Oyster mushrooms also have a wonderful silky texture, and you can use the entire mushroom.

Are black trumpet mushrooms safe to eat?

Yes, and they are wonderful. Despite its ominous name, this mushroom is excellent, although very delicate in texture. Black trumpets have an outstanding buttery flavor that is revealed only after they are dried and reconstituted.

What's the best way to clean mushrooms?

You can rinse mushrooms quickly, but never soak them, because they'll act like a sponge. The best way to clean the white ones is with a damp towel. Other mushrooms, such as shiitakes or portobellos, need only a quick rinsing; once you remove the stems, they are essentially ready to be cooked. In fact, mushrooms arrive in markets quite clean, since they're grown in peat instead of dirt. Don't bother with one of those fancy little mushroom brushes. You'll probably use it one time and then relegate it to the gadget drawer. Morels, on the other hand, need a thorough cleaning under running water, since there are so many nooks and crannies where dirt can hide.

TIP If you don't have access to good wild mushrooms, next time you make risotto, try this fool-the-eye trick. Remove and discard the gills from the bottom of a clean portobello and mince the mushroom. Sprinkle the raw portobello on top of the risotto, then drizzle with a bit of truffle oil. See if your guests can tell the difference.

What can I do with mushroom stems? I hate to throw them away.

The stems lend terrific flavor to soups and stocks. If you're not going to use them right away, trim them but do not rinse, and store in a paper bag in the refrigerator. Make sure to use them within a few days.

Onions

Is there a difference between yellow, white, and red onions?

Bulb onions, which come in yellow, white, and red, are the most readily available onions, and these are the ones we have to rely on through the winter months. All three types are actually similar in flavor, but depending on the time of year and where they're grown, there can be a distinct flavor difference. Some onions may be quite sweet, while others taste sharp. For most recipes, you can use yellow and white interchangeably. Both are pretty potent and have plenty of kick even after a few hours in a stew pot. You can also use red onions in many dishes that call for yellow ones, although the color will change the appearance of the dish.

What is the difference between thick-skinned onions and thin-skinned ones? I find the very thin-skinned onions hard to peel.

Like most other vegetables, onions are seasonal, and even though they're available in markets year-round, they vary quite dramatically. From April through August you'll generally find flat onions with paper-thin skins. These are called "new" onions, and even though they are quite sweet, they're hard to peel. New onions are very perishable, so I usually buy only a few days' supply and store them in a cool dark place or refrigerate. The onions that you'll find in the fall and throughout the cool-weather months are considered "old" or "storage" onions, since they have gone through a cool storage period.

These are round with thick skins. Stored properly in a cool well-ventilated place, these will last for 6 to 8 months, but their flavor will be sharp and more pronounced.

How do I tell which onions are sweet?
Sweet onions have a thinner, lighter-colored skin than the storage ones and tend to be more fragile. Signs in produce sections usually differentiate between the two and most producers also put stickers on each individual onion, such as "Texas 1015 Super Sweet," "Sweet Imperials," "Vidalia," etc. Another indication is price—sweet onions are premium products and prices are higher than for the regular bulb varieties with thicker skin.

What is the difference between a Bermuda onion and a Spanish onion?
There is no difference, and don't be fooled by the names. Neither the Bermuda nor the Spanish onion comes from its namesake country. Instead, both varieties are U.S. grown. The Spanish variety has a large, spherical shape and a mild, sweet flavor. I find it to be crisper than regular yellow storage onions. At one time Bermuda did actually grow a fabulous large, sweet yellow onion that you could occasionally get here, but that variety died out years ago. These days people regard the red onion as the "Bermuda," but they are mistaken.

What is the best onion to use in salads?
If you like a very mild fresh taste, use the green part of the scallion; otherwise, my first choice is always the red onion (also called Creole onion). The best red onions are imported from Italy and are available from the end of August through the fall and early winter. During the spring and summer, all red onions are grown in Texas. These are large and flat and have a very sharp taste. For a spring salad, try Florida or local spring onions that come with their greens attached, or the Georgia-grown Vidalia.

What makes Vidalia onions so popular? Is there a way to prepare them other than raw in salads and sandwiches?
Vidalia onions have enjoyed some great public relations. After peaches, Vidalia onions are one of Georgia's premier crops. For a while, they were hard to come by and rather expensive, making them the "in" thing. Now, because of proper storage, their spring season has been extended into summer and there are a lot more Vidalias available. They're still quite pricey and perishable and should be used within a few days of purchase. My favorite way to use them is to gently braise them in olive oil with a touch of rosemary and minced garlic and serve them as a side to roast chicken or steak. (See recipe, page 136.) You can also roast them in a slow oven and drizzle with a mustard vinaigrette before serving.

What is a Walla Walla onion?
The Walla Walla is to the Northwest what the Vidalia is to Georgia. It is basically the same onion, just grown in different locales. Both are large, sweet, seasonal onions, available from spring through summer, and are wonderful eaten raw with just a seasoning of salt and pepper and possibly a drizzle of olive oil. Another sweet onion that western chefs love to use if they can get it is the Maui onion from Hawaii. Eating this onion is truly a memorable experience, since it is so sweet you can almost eat it like an apple. If you see Mauis in the market, grab one (or two). Texas Super Sweets and Oso Sweets, the newest onion on the block from Chile, are also nice and live up to their names, but the Maui is magic.

ONIONS AT A GLANCE

SWEET ONIONS

Texas Sweets, Spring Sweets, and Texas 1015s Spring Sweets are the first spring sweet onions in the marketplace, arriving in March. The 1015s (named for the suggested planting date, October 15) arrive in mid-April. Both last through mid-June.

Vidalia From Vidalia, Georgia. Slightly squashed looking, thin-skinned with more water than storage onions. Their season is short, from April to June, but you may still see some as late as August, when the flavor becomes more sharp. Best in salads and sandwiches; make great onion rings.

Walla Walla From the Northwest. Available mid-June through late August.

Maui Available mid-February through late April. Hard to find in most markets east of the Rockies.

Sweet Imperial From California. Available late April through August.

Oso Sweet From Chile. Newest onion on the block. Available November through March.

BULB ONIONS

Yellow onion Known as the all-purpose onion, holds up well when cooked. Has a strong flavor, great in soups and stews.

Spanish Sweet A large, spherical yellow onion with a mild, sweet flavor. Crisper than regular yellow storage onions.

White onion Has a sharp taste but still milder and crisper than a yellow storage onion.

Red onion Similar in taste to the Spanish Sweet. Loses its bright color when cooked. Flavor ranges from mild to very strong. Usually at its best in salads or used raw in cold dishes.

Pearl onion Any bulb onion that measures no larger than 1½ inches in diameter is considered a pearl onion. Great in soups, stews. Sweeter than the larger bulbs.

Cippoline From Italy. A small, flat yellow onion found in many specialty markets. Mild.

When a recipe calls for "pearl onions," what does that mean? Can I use them in more than just soups and stews?

Most people think of pearl onions as only being white, but in fact, the tiny yellow ones and red ones now marketed in small mesh bags cook faster and are even sweeter than the white ones. All three colors are available in markets throughout the country. I love to braise them to use them as an accompaniment to roasted meats and chicken.

To braise pearl onions sauté ½ pound diced onions with ¼ cup diced bacon until lightly browned. Add 2 tablespoons brown sugar, ¼ cup chicken stock, ¼ cup red wine, and ½ cup raisins. Sprinkle with salt and pepper and cook over low heat until the onions caramelize and the liquid cooks down to a syrupy glaze, about 30 minutes. Cool and store tightly covered in the refrigerator. They will keep for weeks.

TIP When you can't find super-sweet onions, you can "tame" regular bulb onions, especially red ones, by soaking them in ice water for 30 minutes. Drain, then use raw or in your recipe. The onions will be crisp and less sharp tasting.

Pearl onions seem to take forever to peel. Is there a trick to peeling them faster?

You can speed up the process by dropping the unpeeled onions in boiling water for about 20 seconds, draining them, and then transferring them to a cold-water bath. When cool enough to handle, trim the root end with a sharp paring knife and gently squeeze the onion toward the stem, and the onion will pop out unblemished.

Peas

■■■

Are snow peas and sugar snap peas interchangeable?

Snow peas have flat, tender pods with tiny undeveloped peas and are completely edible. Sugar snaps are also completely edible, but they look like old-fashioned peas with fully developed peas popping through.

Snow peas and sugar snaps cannot be used interchangeably in every preparation. Both are good steamed, but snow peas are actually more interesting stir-fried. On the other hand, you can make a great Sugar Snap Pea Soup (page 137), which you cannot do with snow peas.

My market has just started to carry sugar snap peas. How do you prepare them and how long do they keep?

The simpler the preparation the better. I usually simply cook them in salted water for 3 to 4 minutes, then drain and butter them. I also like to add a sprinkling of dill, mint, parsley, or chives. Both sugar snaps and snow peas keep much better than old-fashioned peas—up to a week or more, depending on how fresh they are when you get them.

Fresh peas are almost impossible to find in my area. Can I use frozen peas instead?

There is no substitute for truly fresh peas, but here is the problem: peas, like corn, start to deteriorate and their natural sugars start converting to starch as soon as they are picked. By the time they get to the grocery store, they are several days or even weeks old. On the other hand, frozen peas are processed soon after harvest and flash frozen, which halts the deterioration. Another reason so many people prefer frozen peas is that no one has the time or patience to shell a bag full of peas—1 pound yields only 1 cup. Still, good fresh peas are in a league of their own.

Is it really necessary to string snow peas?

I usually string and trim the stems of both snow and sugar peas, since the fiber that runs along the spine of the pod is unpleasant to chew on. The good news is that more and more often snow peas come string-free, a nice bonus.

TIP **Buy snow peas at a market with a large turnover. Pass on those that come packaged or that are limp and soft.**

Food magazines occasionally feature a lovely garnish described as "pea shoots." Are they edible?

Pea shoots have a lovely delicate taste, and you will start seeing more and more of this "new" green in stores as their popularity increases.

Right now, you are most likely to find pea shoots at Asian grocery stores or at farmers' markets during the spring.

Peppers, Bell

■ ■

What is the difference between green bell peppers and the red, orange, and yellow ones?

All are varieties of bell peppers, but with some taste difference. Red peppers are fully ripened green peppers and taste much sweeter. Green peppers are fleshier and firmer and are my first choice for stuffed peppers. Yellow and orange peppers are thin-skinned and more perishable than red peppers, but their taste is similar and I use them more for color than flavor.

Why can't I get good green peppers in the supermarket? Is there a particular season for really flavorful green peppers?

Green peppers are available year-round in supermarkets everywhere, but in the fall and winter you'll find a thick-walled variety that is not particularly flavorful. The best green peppers are those that appear in markets during the spring and summer, when you get a smaller, thin-walled pepper that is juicy and delicately flavored. These peppers are ideal for stuffing and for eating raw in salads.

Why are red peppers so much more expensive than green peppers? Is there a time of year when they are less pricey?

For much of the year, red bell peppers are imported from Holland and are therefore pricey.

About mid-August, the domestic crop is harvested and prices come way down. Domestic red peppers are thin-walled and are a little harder to roast or grill, but I find that they are tastier than Holland peppers.

Can green peppers be used when a recipe calls for red or yellow peppers?

Green and red peppers have very different flavors and are not interchangeable. The taste of green peppers is very distinct when cooked and tends to dominate a dish. While green peppers are delicious stuffed and stir-fried, they are not good in pasta dishes or soups. Red peppers, on the other hand, can be used in innumerable dishes, both raw and cooked.

Is there an easy way to roast and peel bell peppers?

Roasting a pepper over a gas flame or a charcoal fire is still the very best way, and once you taste that delicious, smoky flesh, it's worth the trouble. You can pop peppers into the broiler, but this method cooks the peppers instead of roasting them and leaves them soft instead of crisp. When evenly charred over a flame or hot coals and wrapped in a damp paper towel for a few minutes, peppers are easy to peel. The skin will slip off once the stem separates from the flesh, or you can hold the peppers under cool running water and rub the skin away. Don't worry about removing every last bit of skin— little charred flecks are not the end of the world and give the pepper a rustic look.

Can peppers be roasted ahead of time and stored?

Cut the roasted peppers in wide strips, put them in a glass container, then add a thick layer of olive oil. Covered with a tight lid, they will keep for up to 2 weeks. Use them for quick appetizers, in salads and pasta dishes, or to top pizza.

What kinds of peppers are used for a stir-fry of peppers and onions like street vendors make?

There is no better aroma than that of peppers and onions cooking on a flattop. Urban street vendors serve this dish tucked into a hard roll with Italian sausage. The best peppers for this dish are the elongated, pale green frying peppers with thin skin, sometimes called sweet or Italian peppers. They cook quickly and have a mild sweetness that melds beautifully with onions. I often quickly grill them and serve them skin and all with a drizzle of olive oil as part of an antipasti table.

Potatoes

What are new potatoes? Should they be red or white?

A new potato is one that goes directly from the field to the market after harvesting in the spring, with no time in storage. There has been no time for its sugars to convert to starch, and consequently it is waxier and moister than other potatoes. Because new potatoes are not a variety, they can be white- or red-skinned, although most people prefer them red. They may arrive in the market as early as February, when Florida, Texas, and California ship their spring crops.

New potatoes do not mash well but are wonderful simply cooked in their jackets and served with a touch of butter. You can also use them in potato salad. Depending on the recipe, you don't have to bother removing their very thin skins.

Cookbooks say that the season for new potatoes is spring and summer. So what are the potatoes I see in my market during the winter months that are called "new potatoes"?

Calling small-sized potatoes that have been in storage for months "new" potatoes is a marketing gimmick. Some wholesalers call all red-skinned potatoes new potatoes, regardless of their size or age. Just to confuse the issue a little more, the Red Bliss potato, which is grown in the Midwest and harvested in the fall, also masquerades as a new potato. A true new potato will be moist and fresh tasting, but it's hard to know the difference until you taste one.

What is the difference between an Idaho and a russet potato?

There is no difference. Idaho is the largest producing region for russet baking potatoes. A large, Idaho-grown russet is considered the best of the lot. It has a fluffy, mealy texture that is perfect for baking. Other states also produce excellent russet potatoes. While they may be smaller, they are equally good and often much less expensive.

I am always confused as to which potato to use for which purpose.

Russet potatoes are oblong in shape with darkish brown skins and are best for baking. They can also be used in gratins and for French fries. Yukon Gold potatoes are my choice for mashing and potato salads, and red-skinned potatoes are good for just about everything else, such as in soups, sliced and fried, or cooked and turned into hash browns. All-purpose white potatoes bake just as well as russets. In fact, my favorite baking potato is the white-skinned variety grown in California, known as White Rose or Schafter White.

TIP Do not wrap potatoes in foil before baking. Because the foil keeps the moisture in the potatoes, they turn out soggy instead of dry and fluffy. Steak houses and other restaurants follow this practice simply because it's easy, not because it's good.

What potatoes should I use for mashing?

There is nothing better than the texture of mashed Yukon Gold potatoes to which you have added some butter and a touch of milk or sour cream. With their silky texture and fabulous flavor, they already taste buttery. Until recently these potatoes were considered gourmet items and were not widely available. Now you can get them anywhere, and you do not have to pay their weight in gold anymore, either.

Is there a specific kind of potato that is best for roasting?

New potatoes, red or white, and Red Bliss are my first choices for roasted potatoes. They retain a buttery fluffy texture inside, while the outside crisps up nicely.

What potato do you recommend for potato salads?

Red and white new potatoes are great for salads. Red Bliss potatoes are good as well, but my all-time favorite salad potatoes are fingerlings and very small Yukon Golds, which have now become widely available. Look for the ones bagged by Dole.

TIP Do not buy assorted fingerling potatoes. Although they look nice and colorful, they require different cooking times and do not mix successfully.

What is the best potato for making home fries?

Home fries are best when made with red-skinned potatoes; Red Bliss and Yukon Gold are excellent choices as well.

I now see blue potatoes in my grocery store. What is the best way to use them?

These vibrantly colored potatoes, which range from pale blue to deep purple, are just one example of the "heirloom" potatoes that have become popular in recent years. They are not particularly interesting, since they are rather dry and turn grayish when cooked. However, they will retain their blue color when deep-fried.

How are sweet potatoes related to other potatoes?

Botanically, they're not related at all. In the market, the tubers labeled sweet potatoes have yellowish or brown skin and yellowish or light orange flesh. Those labeled yams have coppery or purple skin and dark reddish-orange flesh. The two varieties are closely related and are best baked or pureed.

True yams are large starchy tubers with brown skin and white flesh which grow in the tropics. They can be boiled or baked and when mashed become gummy, a quality considered desirable by some African cultures in which they are a staple.

Shallots

■■

What are shallots?

Shallots are a member of the onion family and look rather like small, elongated onions with dark tan, papery skin. They often have double attached bulbs. If you can, buy them loose or in small mesh bags. Avoid those packed in cellophane-wrapped packages.

How are shallots different from onions?

Shallots have a much more distinctive and complex flavor, with a hint of garlic. Although they belong to the onion family and can be quite pungent, they actually sweeten fish stocks and tomato sauces and are essential in many preparations. Classic French preparations such as a beurre blanc and a sauce béarnaise simply would not be the same without shallots.

What can I use as a substitute for shallots if I don't have any on hand?

When they're not available, go ahead and substitute some finely minced scallions. Depending on the size, you can use the white parts of 6 to 8 scallions to equal one shallot.

TIP **Keep minced or sliced shallots covered with white wine in a jar in the refrigerator. They will remain fresh for weeks.**

When a recipe calls for a shallot, do I use the whole bulb or just a single clove?

The shallot grows as a small, single bulb or as a cluster of several cloves gathered into a bulb, a lot like garlic. Each single clove is considered to be one shallot.

How do you roast shallots?

Roasted shallots make a wonderful accompaniment to a pan-seared steak or added to the pan juices of a rib roast, pork roast, or roasted chicken. Another wonderful way to serve shallots is to sprinkle them with balsamic vinegar and a large grinding of black pepper and use as a topping for a well-seasoned salad.

To roast shallots Preheat the oven to 400°F. Toss 18 medium-sized unpeeled shallots with 3 tablespoons olive oil in a large bowl. Place snugly in a single layer in an ovenproof dish. Add ½ cup water, cover the dish tightly with foil, and place in the center of the oven. Roast the shallots for 40 minutes.

When done, the shallots will slip right out of their skin. Store the shallots covered with a layer of olive oil.

Spinach

■■

Sometimes fresh spinach has a wonderful, full flavor and other times it is bland and characterless. Are there different kinds of spinach?

There are basically two types of spinach widely available: the curly, crinkly variety, also called Savoy, and the flat-leafed variety, called New Zealand spinach (but grown in California).

Crinkly Savoy spinach has a far more assertive, true spinach flavor and is at its best and most available during cool-weather months. If you live in the Western states, and California in particular, chances are you will find only New Zealand (flat-leaf) spinach, which is very good in salads, but not particularly interesting when cooked.

Is it better to buy spinach in bunches or bagged?

You can buy very good bagged spinach as long as you know what to look for. The bag must feel bouncy, never soggy or wet, and the spinach should look dry and richly green. If it looks dark, wet, or yellow, pass on it. In theory, bagged spinach has fewer stems and is prewashed. However, I find that it still needs a thorough rinsing. Flat-leaf New Zealand spinach is sold only in bunches. It is usually much less gritty than the curly variety and easier to clean.

How do you get the dirt out of spinach? Even after rinsing in the colander, it remains gritty.

For starters, use warm water instead of cold. Fill a very large bowl with warm water, immerse the spinach, and swish it around briefly. Let it sit for a few minutes to give the sand and grit time to settle on the bottom of the bowl. Lift out the spinach and pour out the water.

For bagged spinach, you will need one or at the most two rinsings. Loose spinach will require as many as three to four. Flat-leaf spinach is easily cleaned in one rinsing.

What is the best way to cook spinach?

My favorite way is to blanch spinach in plenty of lightly salted boiling water for about 2 minutes, drain, and run it under cold water so it keeps its color. Sauté the spinach in a skillet over low heat in some butter and season it with plenty of freshly ground pepper and a pinch of freshly grated nutmeg. Pine nuts, raisins, finely sliced garlic, yogurt, and diced smoked ham all go well with sautéed spinach.

Another nice method that is particularly suitable for flat-leaf New Zealand spinach is to dry it thoroughly in a salad spinner once it's washed and then stir-fry it in some fruity olive oil with a couple of sliced garlic cloves. The spinach will wilt quickly and be ready in a couple of minutes. Season carefully and serve with seafood, poultry, or veal.

Blanched spinach keeps better than the raw, as much as 3 or 4 days in the refrigerator. You can also freeze it in a zippered plastic bag for several weeks.

Squash

■■

Summer Squash

Is there a difference between yellow squash and yellow zucchini? Are they interchangeable?

The pale yellow crookneck or straightneck squash are rather watery and mild flavored, while yellow zucchini have a lovely nutty taste and retain great texture when cooked. Unfortunately, yellow zucchini are not very plentiful, so if you can't find them, go right ahead and use yellow squash.

Which is the best summer squash for grilling?

They all work well, but yellow and green zucchini and pattypan squash do best, since they are less seedy and more flavorful than yellow crookneck squash. Cut the zucchini on the diagonal and make the slices a little thicker than you would for sautéing.

Baby squash are so attractive on a plate, but are they really worth the extra money?

I don't think so. They often taste bitter because they are picked prematurely to qualify as "baby." Also, unless your market has a large turnover, the tiny squashes are probably not as fresh as the average zucchini.

TIP **All the flavor of the green zucchini is in the peel. To get the best taste when sautéing or otherwise cooking zucchini, use the skin and a** small amount of flesh. To do this, cut the zucchini into logs 1½ to 2 inches long. Lay the logs vertically on a cutting board and slice the skin off with about a quarter inch of flesh. Put these zucchini rectangles flat on the board and cut them into strips or "batonnets," which will look like thin French fries.

What's the best way to prepare zucchini blossoms?

Delicate, deep-fried squash blossoms are one of the great treats of summer and are best eaten the same day they're picked. You can fry batter-dipped blooms plain or stuffed with smoked mozzarella and a bit of anchovy, as they do in Tuscany and Puglia, Italy. If you are fortunate enough to have a garden or access to a good farmers' market, zucchini blossoms will be easy to come by. Otherwise, seek them out where and when you can.

To make Fried Zucchini Blossoms **Combine 1 cup milk with 1 cup all-purpose flour and 1 large egg. Whisk to mix very lightly until just blended. Season with salt and a pinch of cayenne. Stir lightly; do not overmix. Pour vegetable oil into a pot to a depth of 2 to 3 inches and heat until very hot and simmering. Dip the blossoms in the batter and then deep-fry for 2 to 5 minutes until crisp and brown. Remove with a slotted spoon, drain on paper towels, and serve imme-**

diately with a little salt. This is enough for about 12 blossoms.

What's the best way to cook scalloped patty-pan squash? They have such an odd shape.

Choose small specimens, quarter them, and cut the quarters into slices about ⅛ inch thick. They are delicious simply sautéed or oven-roasted in a little butter and olive oil and then sprinkled with a little lemon juice and minced parsley or basil. If you are grilling, place the cut-up patty-pan squash on a large piece of foil and sprinkle with some fresh herbs, a drizzle of good olive oil, and a grinding of pepper. Wrap the foil into a package and cook the squash right along with your steak. The vegetable steams on the grill and tastes absolutely delicious.

Winter Squash

Recently more varieties of winter squash have been showing up at the farmers' market and grocery store, but it's hard to know which ones are worth experimenting with. Can you help?

I often experiment with different types of squash but always come back to my favorites: acorn squash, butternut, spaghetti squash, and buttercup. However, each region in this country has its own types of winter squashes and it is worthwhile experimenting.

Is there a difference between butternut and buttercup squash?

Butternut squash has a lovely buttery, nutty taste. It peels easily and has flesh tender enough to dice or slice. Its consistency is very different from that of buttercup, which is very sweet but stringy.

What is the mark of a good acorn squash? Is bigger better?

Both dark green and bright orange acorn squash are good choices. Look for those tinged with yellowish orange, an indication that they are ripe and sweet. Also, when it comes to acorn squash, large is better than small, and heavy is better than light. The lighter the squash, the more dehydrated and fibrous it is, so go for the heavyweight every time.

It is so hard to cut a raw acorn squash in half. Would it help to microwave it first?

I always bake the squash first for 35 to 40 minutes in a 350°F oven until it can easily be pierced with the tip of a sharp knife. You can also microwave it, but baking the squash does a more even job. At this point it is very easy to cut the squash in half and finish baking it with the seasoning of your choice. Remember that plenty of butter and brown sugar make for a delicious-tasting squash. This is one of my favorite ways to prepare acorn squash: Bake a medium-sized acorn squash until it is soft, let it cool a little, halve it, and scrape out the seeds with a grapefruit spoon. Put 1 tablespoon of unsalted butter and 1 tablespoon of brown sugar in each half, and season with salt and pepper. Return to the oven and bake for another 10 to 15 minutes, or until browned and fragrant. Serve half a squash per person.

Can all pumpkins be used for cooking?

Those grown for jack-o'-lanterns or other decorations are quite flavorless. Look for small, heavy pumpkins labeled Sugar Pies, Cheese, or Baby Bear pumpkins when you want to make a pie.

Is there a big taste difference between fresh and canned pumpkin?

Canned pumpkin is fine for pies, especially since you are going to add the usual spices. But be sure to begin with plain pumpkin puree, not those sold as pie fillings that are already spiced. You will want to add your own seasoning.

How do you prepare spaghetti squash?

You can handle spaghetti squash one of two ways. Either boil it in a large pot of water until tender, cut it open, and scoop out the pasta-like strands, or bake it. Baking is my preferred method, since it concentrates the squash's mild taste—and eliminates the boiling pot. Put the squash directly on the oven rack in a 350°F oven for about 1 hour, or until it can easily be pierced with a fork. Cool slightly and then cut in half. Scoop out the seeds, then pull out the fibrous flesh with a fork to separate the strands.

What kind of pasta sauce goes well with spaghetti squash?

Despite its name—which refers to the spaghetti-like strands of flesh inside the squash—spaghetti squash is not pasta, and I do not find that it marries well with tomato or any other full-flavored sauces. Instead, I treat it like winter squash rather than pasta and either braise it in butter and seasonings or toss the cooked strands in a butter and cream sauce flavored with mild goat cheese.

Here is a lovely side dish for roast turkey, duck, or pork tenderloin Combine 1 medium-sized cooked spaghetti squash with ½ cup heavy cream, 3 tablespoons unsalted butter cut into pieces, 2 tablespoons brown sugar, a pinch of nutmeg, salt, and pepper. Pour the mixture into a baking dish, cover tightly with foil, and bake at 350°F for 25 to 30 minutes. Serve hot.

Tomatoes

Is there a major difference in taste between round tomatoes and the oval ones called plum or Roma tomatoes? Are they interchangeable in recipes?

When perfectly ripe, both round and plum tomatoes deliver great taste, and for many recipes, either can be used. The plum tomato is meatier, less watery, and more suitable for tomato sauce. I also like a good, ripe plum tomato in salads, especially Greek salads. Some people object to the relatively thick skin of plum tomatoes. In general, a good tomato with distinct aroma and true freshness is the best bet, no matter what its shape.

TIP When you can't get good ripe tomatoes, switch to cherry tomatoes. They tend to be more flavorful and more easily available than vine-ripened tomatoes.

What are heirloom tomatoes?

The word "heirloom" does not refer to a variety but is a term to describe old-fashioned varieties of fruits and vegetables that have, for one reason or another, fallen out of favor but are now being "rediscovered." Tomatoes are the most popular heirloom, although you can find heirloom potatoes, apples, and other produce.

Heirloom tomatoes may be green, pink, yellow, or striped as well as red.

It's usually the most dedicated farmers who seek out heirloom seeds, so most of the heirloom tomatoes I've run across have been at farmers' markets. As a whole, their quality is very high. Brandywine tomatoes are my favorite heirlooms; they're extra juicy and spectacularly flavorful.

Do yellow and orange tomatoes have a distinct taste or are they mainly cultivated for visual appeal?

Yellow and orange tomatoes actually taste sweeter and less acidic than the regular red varieties, and they are very appealing when combined with red tomatoes in a refreshing sliced tomato platter drizzled with virgin olive oil and sprinkled with some herbs.

What are pear tomatoes? Are they the same as grape tomatoes?

Both these tomatoes are a little smaller than cherry tomatoes but can be used in precisely the same ways. Pear tomatoes bell out at the bottom to become pear shaped. They may seem new, but they have been popular for years in Italy. The pear-shaped Vesuvio tomatoes that grow near Naples, Italy, are particularly well known. Grape tomatoes are the shape of large grapes and taste incredibly sweet. These tomatoes have a slightly thicker skin than regular cherry tomatoes, which means they hold their shape better when sautéed. They also have a longer shelf life.

TIP For a quick hors d'oeuvre, serve a bowl of cherry or tiny pear or grape tomatoes alongside a small dish of vodka and a dish of coarse salt. Dip the tomato in the vodka first, then in the salt for an exciting nibble.

Can cherry tomatoes be used in recipes that call for all-purpose tomatoes?

It depends on the recipe. Cherry tomatoes can't be peeled or seeded and so are not recommended for any recipe that calls for peeled, seeded tomatoes such as a tomato fondue or sauce. They work perfectly in many other preparations, especially quick sautés, vegetable medleys, and many pasta dishes. The best news is that they provide more flavor than most other tomatoes.

What varieties of tomatoes do you recommend for growing in pots on my patio?

My favorite is the Carmello, which I consider to be the finest all-purpose variety. Besides being delicious, it's easy to grow and bears a lot of fruit. Early Girl is another good all-purpose tomato. The heirloom Brandywine, both red and yellow, is another good choice. The red ripens to a pretty pink and is big, meaty, and mild, while the yellow is very sweet with a creamy texture. For cherry tomatoes, try the Juliet Santa and the Sungold—a rich, juicy jewel that turns an intense golden orange and is as beautiful on the vine as in the mouth.

Will tomatoes ripen more quickly on a sunny windowsill?

Tomatoes ripen best in a dark, cool corner of the kitchen or when placed on a baking sheet, covered with newspaper, and stored under the bed. I do this often with the green tomatoes at the end of the summer growing season and it really works. In a sunny spot, they'll just get soft without ripening. Whatever you do, don't refrigerate tomatoes. It makes them woolly and tasteless and doesn't increase their shelf life.

Is there a trick to peeling and seeding tomatoes?

Peeling tomatoes can be quite simple if you cut a cross into the base of each tomato with a paring knife, then immerse them in boiling water for 30 seconds. The hot water causes the peel to un-furl from the body of the tomato, and you can lift it right off. Cut the peeled tomato in half cross-wise to expose the neat little seed pockets. Use your fingers to dig them out and the job is done.

How important is it to peel and seed the tomatoes when making homemade tomato sauce?

There are some preparations that absolutely re-quire peeled, seeded tomatoes, but these rarely call for more than 3 or 4 tomatoes. If you are cooking a large batch of sauce, just cut up the unpeeled tomatoes and once they are cooked run the mixture through a food mill. This will give you a nice smooth sauce which, if too thin, can be cooked down to a thicker consis-tency.

TIP Add a pinch of sugar to tomato-based sauces before cooking. The sugar counterbalances the acidity.

What's the best way to use sun-dried tomatoes? Are they a good substitute for fresh?

Sun-dried tomatoes have a wonderfully intense flavor but are not a substitute for fresh ones. They come packed in oil or dried. Drain the oil-packed tomatoes before chopping; recon-stitute the dried in warm water. Use them as an accent, like olives, or try them on a salad or pizza with crumbled goat cheese.

What is tomato confit? How is it made and used?

The term confit usually describes a particular method of using duck, goose, or pork fat to pre-serve certain foods, but in the past few years, the word has been co-opted for other foods. Today, it's common to see the term applied to dishes with tomatoes and other vegetables such as turnips and Belgian endive. Tomato confit is lovely served over crusty peasant bread, in pas-tas, as a filling for an omelet, or as a garnish for pan-seared seafood.

To make a Tomato Confit Cut ripe plum tomatoes in half lengthwise, gently squeeze to remove some of the seeds, and place cut side down on a baking sheet. Drizzle lightly with olive oil and season with coarse salt, freshly ground pepper, and a sprinkling of sugar. Bake in a preheated 275°F oven for 6 to 8 hours, or until the toma-toes have shrunk to half their size. Once cool, pack the confit in a large jar, cover with a thin layer of olive oil, and refrigerate.

Turnips

■■■

What is the difference between white and yellow turnips?

White turnips are delicate, mild-tasting spring root vegetables that keep well for as long as 10 days. Make sure to look for white turnips with a deep purple band about the root end. When they are fresh and young, white turnips can be eaten raw, much like radishes.

Yellow turnips, also called rutabagas, are large globes that mature in the early fall and are available through the winter. They are often sold waxed to increase their shelf life, and so will keep for up to 2 months in the refrigerator. Peel off the waxy skin before cooking.

What are Swedish turnips?

Recipes in old cookbooks, particularly those featuring Scandinavian recipes, refer to rutabagas as "Swedish turnips." This is because of the popularity of this hardy vegetable in Scandinavia, where turnips are a staple during the winter. Oddly, I saw big rutabagas called Swedish turnips in the southern state of Georgia.

What is the mark of good turnips? So many seem to be woody and have a sharp unpleasant flavor.

Turnips are a wonderful delicate root vegetable when they are really fresh. Unfortunately, because of lack of turnover in many parts of the country, the tops are removed to increase their shelf life. Even without their leafy tops, fresh turnips are easy to spot by their deep purple band and white skin. The lighter the band, the older the vegetable. Stay away from dull-looking turnips.

Are turnips and rutabagas interchangeable in recipes?

White turnips, with their delicate flavor, taste lovely combined with carrots, leeks, and potatoes, added to stew, or roasted alongside a plump chicken. On the other hand, rutabagas have a much more assertive taste and are best served on their own or in combination with potatoes, as in the Puree of Rutabaga and Potato with Crispy Shallots, page 144.

TIP If you have access to a good farmers' market where you can find unwaxed small rutabagas, you can substitute them for white turnips in soups and stews.

Cream of Asparagus and Spring Onion Soup

The marriage of asparagus and spring onions is heavenly, especially in this full-flavored soup. If you cannot get the Texas or Florida spring onions, you can use two bunches of scallions instead. A homemade chicken stock does make a difference in this delicate soup, but if you do not have the time to make it, a good-quality bouillon will do. Serve the soup hot or at room temperature accompanied by crusty bread and good butter.

SERVES 6

4 tablespoons unsalted butter

2 medium spring onions, all but 1 inch of
   green parts trimmed, diced

8 cups chicken stock or bouillon

3 tablespoons all-purpose flour

1½ pounds asparagus, trimmed, stalks peeled,
   asparagus cut into 1-inch pieces, stalks and
   tips separated

½ cup heavy cream

Salt and freshly ground white pepper

2 tablespoons minced fresh dill or chives

1. Melt the butter in a heavy 3-quart saucepan over low heat. Add the spring onions together with a couple of tablespoons of the stock and simmer, covered, until tender.

2. Add the flour and cook, stirring constantly, for 1 to 2 minutes without browning. Add the remaining chicken stock and whisk until well blended. Bring to a boil, reduce the heat, add the asparagus stalks, and simmer, partially covered, for 30 minutes.

3. Strain the soup and return the stock to the saucepan. Puree the vegetables in a food processor until smooth. Add the puree to the stock and whisk until well blended.

4. Add the asparagus tips and cream and simmer until just tender. Taste and correct the seasoning with salt and pepper and serve the soup hot or at room temperature, garnished with a sprinkling of dill or chives.

Avocado Salad in Ginger-Lime Vinaigrette

Here is an East meets West avocado salad that I simply adore. I like to serve fine slices of top-quality raw tuna interspersed among the avocado slices, or I garnish the salad with a large sprinkling of rock shrimp. A crumbling of aged goat cheese or finely sliced black gaeta olives or both is another nice variation.

SERVES 4 TO 5

Juice of 1 large lime

½ cup extra virgin olive oil

1 tablespoon soy sauce or more to taste

1 to 2 large garlic cloves, mashed

1 tablespoon grated fresh ginger

3 tablespoons minced scallions or chives

2 ripe Hass avocados, cut crosswise into fine slices

Tiny whole cilantro leaves

1. In a small bowl, combine the lime juice together with the olive oil and whisk the mixture until it emulsifies. Add the soy sauce, garlic, ginger, and scallions and whisk until well blended. Add a dash more soy if you like a saltier dressing. You may also add another clove of garlic.

2. Place the sliced avocados on a round serving platter and spoon the sauce over the slices. Sprinkle with the cilantro and serve at room temperature or lightly chilled.

Roasted Beet, Watermelon, and Mango Salad

When I first sampled a watermelon and roasted beet salad in a café on Venice Beach in California, I expected it to be gimmicky. It turned out to be amazingly tasty and refreshing, and I could see that the idea had some possibilities. I have since come up with this version, which includes mango, blue cheese, and pecans.

SERVES 4 TO 5

1 tablespoon sherry vinegar

1 tablespoon balsamic vinegar

¼ cup plus 2 tablespoons grapeseed oil

1 teaspoon Thai chili sauce, optional

Coarse salt and freshly ground black pepper

4 cups baby arugula, mâche, or mesclun greens

3 cups watermelon, cut into ¾-inch cubes

2 cups roasted beets, cut into ¾-inch cubes

1 mango, cut into fine julienne

4 tablespoons diced Maytag blue cheese

½ cup coarsely chopped pecans

Sprigs of mint

1. In a bowl, combine the two vinegars with the oil and the chili sauce. Season with salt and pepper and whisk the dressing until well blended. Taste and correct the seasoning.

2. Pour half the dressing into a large salad bowl. Add the greens and toss gently.

3. In another bowl, combine the watermelon, beets, and mango. Add the remaining salad dressing and toss gently. Season lightly with salt and pepper.

4. Divide the greens among individual dinner plates. Top each serving with some of the fruit and beet mixture. Sprinkle each serving with some of the cheese and pecans and add another grinding of black pepper. Garnish with the mint and serve immediately.

Pan-Seared Baby Bok Choy with Red Bell Peppers and Oyster Sauce

Until recently, the leafy mild-flavored bok choy was available only in Asian markets, greatly limiting its use. Now this lovely green, especially refreshing in its "baby" form, can be purchased in most grocery stores. I like to serve this quick stir-fry with anything grilled, such as a flank steak, shish kebabs, or fish steaks.

SERVES 4 TO 5

¼ cup olive oil

2 large garlic cloves, finely sliced

½ tablespoon finely sliced peeled fresh ginger

2 pounds baby bok choy, cut crosswise into 1-inch pieces

2 medium red bell peppers, stemmed, quartered, ribs removed, and finely sliced

Salt and freshly ground black pepper

3 tablespoons oyster sauce

1. In a large skillet, heat the olive oil together with the garlic and ginger. Immediately add the bok choy. (You may to have to do this in stages, waiting until one batch wilts to add more to the pan.) Add the bell peppers, season with salt and pepper, and cook over medium heat until the peppers are slightly soft and the bok choy wilted.

2. Add the oyster sauce and gently stir it into the vegetable mixture. Taste and correct the seasoning.

3. Transfer to a serving dish and serve hot.

Skillet-Braised Broccoli with Garlic

Broccoli is a wonderful year-round vegetable that we seem to take for granted, giving it little chance to shine. For a change of pace try it this way. Here the broccoli is skillet-braised with good olive oil and some garlic. For a variation, add 2 tablespoons sautéed pine nuts or a roasted pepper cut into fine julienne. Serve as an appetizer topped with some freshly grated Parmesan or as an accompaniment to a simple veal roast, pan-seared lamb chops, or the slow-braised chicken breasts on page 204.

SERVES 4 TO 6

1 large bunch of broccoli (about 1½ pounds)
3 tablespoons extra virgin olive oil
2 large garlic cloves, finely sliced
⅓ cup chicken stock or bouillon
Salt and freshly ground black pepper

1. Trim the broccoli stalks, removing all the leaves. Peel the stalks with a vegetable peeler and cut crosswise into ½-inch slices. Separate the tops into florets.

2. In a large heavy skillet, heat the olive oil over medium heat. Add the broccoli stems together with the garlic and cook for 1 to 2 minutes, stirring often. Add the florets and the stock; season with salt and pepper, and simmer, covered, for 10 minutes or until just tender. Taste and correct the seasoning. Serve hot.

VARIATION Core, seed, and finely slice a large red bell pepper and add it to the skillet in Step 2. Continue with the recipe.

Sauté of Brussels Sprouts with Bacon, Pine Nuts, and Sour Cream

Brussels sprouts can be prepared in the simplest of ways, but they must never be either undercooked or overcooked. The secret is to blanch them first before baking or sautéing to remove that cabbage taste. Here I team them with diced smoked bacon, but you can use smoked turkey or ham instead. Serve the sprouts as a side dish to a pork roast or pan-seared veal chops.

SERVES 4

3 tablespoons olive oil

½ cup slab bacon cut into ¼-inch dice and blanched

3 tablespoons pine nuts

2 pints Brussels sprouts (1½ pounds), trimmed and cut into ¼-inch slices

¾ cup chicken stock or bouillon

Salt and freshly ground black pepper

½ cup sour cream, optional

1 large garlic clove, mashed

I. Heat the olive oil in a cast-iron skillet over medium-low heat, add the bacon and pine nuts, and sauté until lightly browned. Add the sprouts together with ½ cup of the stock, season with salt and pepper and simmer, partially covered, for 6 to 8 minutes or until the sprouts are crisp-tender. Add the remaining ¼ cup stock if necessary to cook the sprouts.

2. Fold in the sour cream and the garlic, and cook until reduced to a glaze. Taste and correct the seasoning and serve hot.

Braised Red Cabbage with Orange Zest and Grenadine

Everyone has favorite holiday dishes and this is one of mine—although I am not quite sure which holiday I mean, since I try to make this dish as soon as the cool weather sets in and am still serving it by Memorial Day. Red cabbage braised with wine, grenadine liqueur, and plenty of orange zest develops a mellow sweet flavor that is a lovely accompaniment to roasted duck, pork, and venison.

SERVES 6

2 tablespoons butter

3 tablespoons peanut oil

2 medium onions, finely diced

3 tablespoons dark brown sugar

$\frac{1}{2}$ cup sherry vinegar

1 medium red cabbage, finely sliced
(5 to 7 cups)

Zest of 1 large orange, cut into fine strips

Coarse salt and freshly ground black pepper

2 cups red wine

$\frac{1}{2}$ cup grenadine

1 tablespoon flour

1. Preheat the oven to 350°F.

2. In a large heavy flameproof casserole, heat the butter and 1 tablespoon of the oil. Add the onions and cook over medium heat until soft but not browned. Add the brown sugar and sherry vinegar. Bring to a simmer, reduce the heat, and cook until most of the vinegar has evaporated. Add the cabbage and orange zest. Season with salt and pepper. Add the wine and grenadine. Bring to a boil, reduce the heat, cover, and transfer the casserole to the oven. Braise the cabbage for $1\frac{1}{2}$ to 2 hours, checking it every once in a while to be sure the juices do not run dry.

3. When the cabbage is very tender, remove from the oven and set aside.

4. In a small skillet, heat the remaining 2 tablespoons oil. Add the flour and cook over low heat for 1 or 2 minutes, stirring constantly, until the flour turns a hazelnut brown. Do not let the flour burn. Immediately add the flour mixture to the cabbage and stir well to blend thoroughly. Reheat the cabbage over low heat and correct the seasoning. Serve hot.

REMARKS This cabbage is best when made a day ahead of time. It will keep refrigerated for up to a week.

Skillet-Braised Carrots with Pine Nuts and Raisins

Carrots are one of the vegetables that we tend to take for granted and do not make enough use of. For me, they are a basic pantry vegetable that I reach for whenever I want to make a quick side to a roasted chicken, sautéed chicken breasts, or a stew. You can make the dish hours ahead of time and reheat it right in the skillet just before serving.

SERVES 4 TO 5

8 large carrots, peeled
1½ tablespoons olive oil
3 tablespoons pine nuts
3 tablespoons butter
1 teaspoon sugar
Salt and freshly ground black pepper
½ to ¾ cup chicken stock or bouillon
⅓ cup blanched raisins
Minced cilantro or flat-leaf parsley

1. Peel the carrots. Cut them in half lengthwise and then crosswise, to make 4 even pieces from each carrot.

2. In a small skillet, heat 1 tablespoon of the olive oil. Add the pine nuts and sauté for 1 minute, or until they are nicely browned. Do not burn. Set aside.

3. In a large heavy skillet, heat the butter and the remaining ½ tablespoon olive oil. Add the carrots, sprinkle with sugar, and season with salt and a large grinding of black pepper. Sauté over medium heat until the carrots are lightly browned. Add the stock, reduce the heat, and simmer, covered, for 8 to 10 minutes, or until just tender. Add the raisins and continue to cook for another 2 to 3 minutes, adding a little stock if necessary.

4. When the carrots are done, transfer to a serving dish and sprinkle with minced cilantro or parsley. Serve hot.

Puree of Celery Root, Potatoes, and Sweet Garlic

This creamy puree that combines the wonderful taste of celery root with potatoes and a touch of sour cream is addictive. Serve the puree as a side to a pork roast, braised lamb, or nice homey stew.

SERVES 6

1 cup low-fat milk

1 large celery root (about 1½ pounds), peeled and cut into 2-inch pieces

8 large garlic cloves, peeled

1 pound Yukon Gold or all-purpose potatoes, peeled and cubed

2 to 3 tablespoons butter

⅓ cup sour cream

Salt and freshly ground black pepper

1. In a large saucepan, bring 2½ quarts salted water to a boil. Add the milk, celery root, and garlic. Bring to a boil, reduce the heat, and cook for 10 minutes, or until the celery root is almost tender.

2. Add the potatoes and continue to cook until they are very tender.

3. Drain the vegetables. Puree with the butter and 2 tablespoons of the sour cream in the food processor until smooth. You may need a little more sour cream. Add salt and pepper to taste.

REMARKS The puree can be made well ahead of time and reheated in the microwave. Leftovers can be refrigerated and will keep well for several days.

Creamy Celery and Stilton Cheese Soup

I first had this delicious soup many years ago in a restaurant in Devon and was surprised to find out that it is an English classic. To many, its taste comes as a surprise because most of us do not associate celery with anything but salads or a raw snack. Here, with the addition of a creamy mild Stilton, celery has a chance to shine. It is best to make the soup a day ahead of time and reheat it just before serving.

SERVES 6

8 ounces Stilton cheese

4 tablespoons unsalted butter

6 large celery stalks, diced (about 3 cups)

1 large onion, minced

2 tablespoons all-purpose flour

6 to 7 cups chicken stock or bouillon

1 medium leek, white part only, cut into fine
   julienne

1 medium carrot, peeled and cut into fine
   julienne

1 large celery stalk, cut into fine julienne

1¼ cups heavy cream

Salt and freshly ground white pepper

1. Trim the Stilton cheese of its outer yellow-brown crust and use only the center. You should have about 6 ounces. Crumble the cheese.

2. In a large heavy saucepan, melt the butter over medium heat. Add the diced celery and onion, and cook, partially covered, until the onion is soft but not brown.

3. Add the flour, blend well into the vegetables, and cook for 1 minute. Add the stock all at once, bring to a boil, reduce the heat, and simmer, covered, for 25 minutes, or until the vegetables are very tender. Strain the soup through a colander set over a large bowl, reserving the vegetables and stock separately.

4. Transfer the strained vegetables to a food processor with ½ cup of the stock and process until very smooth. Whisk the pureed vegetables into the bowl containing the remaining stock and return the soup to the pan.

5. Add the julienned vegetables to the soup and simmer for 5 to 8 minutes, or until the vegetables are just tender.

6. In a food processor, combine the cream and Stilton cheese and process until smooth. Add the Stilton mixture to the soup, whisk until well blended, and just heat through. Do not bring the soup back to a boil. Add salt to taste and a large grinding of pepper, and serve hot.

Corn and Chive Fritters with Smoked Salmon Sauce

Corn is an ideal candidate for fritters. You can, of course, use fresh corn in season, but canned corn works just as well. Served with the delicious smoked salmon sauce, the fritters make an elegant appetizer, but I often serve them as an accompaniment to pan-seared scallops, shrimp, and salmon. You can vary the fritters by adding minced jalapeño and minced cilantro instead of chives.

SERVES 8 TO 10

One 11-ounce can corn kernels, drained
  (about 1½ cups)

½ cup whole milk

⅓ cup all-purpose flour

⅓ cup yellow cornmeal

½ teaspoon baking powder

2 extra large eggs

2 extra large egg yolks

4 tablespoons unsalted butter, melted

¼ cup minced fresh chives

¾ teaspoon salt or more to taste

Freshly ground black pepper

Corn or peanut oil or clarified butter
  (see page 384) for frying

Sprigs of fresh dill and chives

1. Combine the corn and milk in a food processor and process until smooth. Transfer to a large bowl, add the remaining ingredients except for the oil and the dill and chives, and whisk until well blended.

2. Cook the fritters in several batches of 4 to 6 fritters per batch. For each batch, heat 2 teaspoons oil or clarified butter in a large nonstick skillet over medium heat. For each fritter, add 1½ to 2 tablespoons of the batter to the hot oil, without crowding the skillet, and cook for 1 to 2 minutes per side or until lightly browned. Serve warm, garnished with a dollop of smoked salmon sauce and sprigs of dill and chives.

Smoked Salmon Sauce

MAKES ABOUT 1½ CUPS

2 teaspoons finely grated raw onion

1 cup crème fraîche

½ cup mayonnaise

Juice of ½ lemon

3 ounces smoked salmon

Salt and freshly ground white pepper

In a food processor, combine all the ingredients and process until very smooth. Season with salt and pepper.

Spicy Corn and Ginger Relish

Here is a wonderful piquant relish that I make with leftover fresh corn as soon as corn season begins. I usually double and triple the recipe because the relish keeps well refrigerated for 2 to 3 weeks. Serve the relish as a side with all grilled meats, especially a juicy burger or fish or chicken. It also complements a nice summer sandwich of finely sliced leftover flank steak or chicken breast.

SERVES 6

4 ears of fresh corn, cooked

¾ cup rice vinegar

¾ cup sugar

½ cup very finely diced red pepper

3 tablespoons very finely diced red onion

1 tablespoon minced fresh ginger

1 small fresh cayenne or jalapeño pepper, finely diced

⅓ cup grapeseed or peanut oil

Coarse salt and freshly ground black pepper

Small leaves of fresh cilantro or basil

1. Cut off the corn kernels from the cooked cobs and transfer to a bowl.

2. Combine the vinegar and sugar in a small saucepan and simmer until the sugar is dissolved and the mixture reduces to ½ cup and becomes syrupy.

3. Pour the vinegar mixture over the corn kernels. Add the remaining ingredients except the cilantro to the bowl, and let stand at room temperature for 4 to 6 hours before serving. Garnish with cilantro or basil and serve at room temperature as an accompaniment to grilled meats or fish.

Sauté of Cucumbers in Chive and Dill Cream

The large seedless cucumbers make a wonderful hot vegetable that takes minutes to prepare and is the perfect accompaniment to grilled or sautéed chicken breasts and every salmon preparation. I purposely do not peel the cucumbers because the peel contains so much taste and also allows the cucumbers to retain their shape.

SERVES 4

2 large seedless cucumbers
3 tablespoons unsalted butter
1½ teaspoons sugar
½ cup chicken stock or bouillon
Salt and freshly ground white pepper
½ cup crème fraîche
Minced fresh dill and chives

1. Cut the cucumbers lengthwise in half. Scoop out the seeds with a grapefruit spoon and discard. Cut the cucumbers crosswise into ¼-inch slices (half moons).

2. In a large skillet, melt the butter over medium heat. Add the cucumbers and sugar and sauté for 2 minutes, tossing constantly. Add the stock and season with salt and pepper. Bring to a simmer, reduce the heat, and cook, covered, for 5 minutes or until just tender. Remove the cover, raise the heat, and cook until all the moisture has evaporated and the cucumbers are nicely glazed. Stir in the dill and chives and correct the seasoning.

Fricassee of Roasted Eggplants and Two Mushrooms

The first time I had this dish was as part of an antipasti table in northern Italy. The fricassee had been made with local wild mushrooms and the taste was amazing. Even though I cannot duplicate that mushroom taste, I still love to make this dish toward the end of summer and serve it at room temperature, garnished with slivers of Parmesan and a drizzle of truffle oil.

SERVES 5 TO 6

2 medium eggplants, trimmed but not peeled and cut into ¾-inch cubes

4 to 5 tablespoons olive oil

2 shallots, minced

2 large garlic cloves, minced

½ pound cremini mushrooms, stems removed, caps cubed

Coarse salt and freshly ground black pepper

6 large fresh shiitake mushrooms, stemmed

3 to 4 tablespoons minced fresh parsley, preferably flat-leaf

1. Preheat the oven to 400°F.

2. Place the eggplant cubes in a single layer on a nonstick baking sheet. Drizzle with 1½ tablespoons of the olive oil and set the sheet in the center of the oven. Roast for 20 minutes or until tender and lightly browned.

3. Meanwhile, heat 2 tablespoons of the oil in a large heavy skillet. Add the shallots and half the garlic together with the cremini mushrooms. Season with salt and pepper and sauté over medium-high heat until the mushrooms are soft and lightly browned. Set aside.

4. When the eggplant is done, remove from the oven, transfer to a dish, and season with salt and pepper. Add the shiitakes to the baking sheet, drizzle with the remaining 1 tablespoon oil, season with salt and pepper, and roast for 10 to 15 minutes or until tender and lightly browned.

5. When the shiitakes are done, remove from the oven and quarter. Add the shiitakes together with the eggplant to the pan containing the cremini mushrooms. Add the remaining garlic and the parsley and cook the fricassee for 2 or 3 minutes, or until just reheated. Correct the seasoning and serve hot or at room temperature.

Charcoal-Grilled Eggplants with Lemon-Scallion Mayonnaise

Here is a variation of baba ghanoush, a popular Lebanese appetizer that is traditionally made by combining the flesh of smoked grilled eggplant with tahini, a Middle Eastern sesame paste, and garlic and plenty of lemon juice. This lighter and simpler version makes a delicious accompaniment to grilled shish kebabs, lamb chops, and marinated flank steak. It is best served slightly warm or at room temperature.

SERVES 4

2 medium eggplants

Juice of ½ lemon or more to taste

½ to ¾ cup mayonnaise

1 large garlic clove, mashed

¼ cup minced scallions

Salt and freshly ground black pepper

Small black oil-cured olives and ripe cherry or
grape tomatoes, halved

1. Prepare a charcoal grill.

2. When the coals are very hot, place the eggplants directly on the coals and grill until the skin is charred all over and the eggplants are quite tender.

3. Carefully transfer the eggplants to a cutting surface and, with a sharp knife, cut in half lengthwise. Scoop out the pulp, cut into cubes, and place the pulp in a shallow serving dish. Add the lemon juice, mayonnaise, garlic, and scallions, season with salt and pepper, and mix well. Taste and correct the seasoning, adding more lemon juice if necessary, and serve garnished with the olives and tomatoes.

Skillet-Braised Belgian Endives

Most people consider endives a green rather than a vegetable to be added to a salad. In fact, Belgian endives are delicious served hot, either skillet braised or roasted. Since they are available year-round and keep well in the fridge for a week or more, I consider them a basic that I reach for when I am looking for a flavorful and quick vegetable side dish. Serve the endives with pan-seared veal chops, roasted chicken, or sautéed salmon fillets.

SERVES 4 TO 6

3 tablespoons butter
2 tablespoons peanut oil
6 to 8 Belgian endives, cut in half
Salt and freshly ground black pepper
2 teaspoons sugar
½ to ¾ cup chicken or beef stock or bouillon
Minced flat-leaf parsley, optional

1. Heat the butter and oil in a large cast-iron skillet over medium heat. Add the endives, season with salt, pepper, and sugar, and cook until nicely browned. Add the stock and braise, covered, until tender and caramelized, 10 to 15 minutes.

2. Transfer the endives to a serving dish, garnish with the parsley, and serve at room temperature.

Ragout of Fennel, Tomatoes, and Potatoes

Fennel is fast becoming a staple on the fresh produce scene, and its affinity for fish preparations makes it one of my favorite vegetables. Gently braised in olive oil with potatoes and tomatoes, it is the perfect accompaniment to sautéed scallops, pan-seared bass or snapper, or oven-roasted mahi mahi. This vegetable ragout can be made up to a day ahead of time and reheated on top of the stove.

SERVES 4

1 large fennel bulb, trimmed

3 tablespoons olive oil

2 large garlic cloves, finely sliced

Coarse salt and freshly ground black pepper

½ teaspoon fennel seeds

2 ripe tomatoes, cubed

2 medium red potatoes, peeled and cubed

1 cup chicken stock or bouillon or water

2 tablespoons minced fennel tops

Freshly grated Parmesan, optional

1 tablespoon fruity extra virgin olive oil

1. Quarter the fennel bulb and cut crosswise into ¼-inch slices.

2. In a large heavy skillet, heat the olive oil. Add the fennel and garlic and sauté for 2 or 3 minutes or until lightly brown. Season with salt and pepper. Add the fennel seeds and tomatoes and bring to a simmer. Cook for 2 or 3 minutes, or until the tomatoes have released most of their juices.

3. Add the potatoes and stock and lower the heat. Cover the skillet and simmer the mixture until the potatoes are very tender, 20 to 30 minutes. Uncover, taste and correct the seasoning. If the ragout is still a little soupy, simmer, uncovered, for a few minutes.

4. Transfer the ragout to a serving dish, garnish with the fennel tops and Parmesan, and drizzle with the oil.

Sauté of Green Beans with Tuna, Tomatoes, and Red Onion

Here is a typical Niçoise appetizer that you can find in simple restaurants all over the South of France. Its flavorful simplicity is based on the freshness of the beans and the ripeness of the tomatoes.

SERVES 4

1 pound green beans, trimmed

3 tablespoons extra virgin olive oil

1 medium red onion, quartered and finely sliced

2 large garlic cloves, minced

6 ripe plum tomatoes, quartered

Salt and freshly ground black pepper

One 6-ounce can light oil-packed tuna, drained and flaked

2 to 3 tablespoons fresh basil, cut into fine julienne

Pitted small black oil-cured olives, cut in half, optional

1. In a large pot of boiling salted water, cook the green beans until tender, 5 to 7 minutes. Drain and rinse under cold running water; drain well. Cut the beans into 1½-inch lengths.

2. In a large heavy skillet, heat the olive oil over medium heat, add the onion and garlic, and cook until soft but not browned, about 10 minutes. Add the tomatoes, season with salt and pepper, and cook until most of the tomato juices have evaporated, stirring often. The tomatoes should still retain their shape.

3. Add the green beans and tuna and just heat through. Fold in the basil and correct the seasoning. Garnish with the olives and serve hot, directly from the skillet, or at room temperature.

REMARKS You may also omit the tuna and serve this as a vegetable side dish.

Kale, Potato, and Butternut Squash Chowder

Kale is a gutsy winter green that is wonderful in soups. Here it is teamed with potatoes, winter squash, and plenty of garlic. Diced chorizo or other smoked sausage adds character. I like to serve this soup for a Sunday lunch or simple supper, accompanied by crusty bread and good sweet butter.

SERVES 5 TO 6

3 tablespoons olive oil

5 large garlic cloves, minced

1 pound kale, stemmed, washed, dried, and torn into 1½-inch pieces

2 tablespoons unsalted butter

1 large onion, minced

Salt and freshly ground black pepper

2 cups butternut squash, cut into ¾-inch cubes

7 to 8 cups chicken stock or bouillon

3 medium all-purpose potatoes, peeled and cut into ¾-inch cubes

½ cup heavy cream, optional

2 cups diced smoked turkey sausage, or 1 cup diced chorizo

1. In a 4-quart saucepan, heat the olive oil over medium heat. Add the garlic and sauté until soft but not browned. Add the kale, tossing with the oil and garlic until just wilted. Remove the mixture from the pan and set aside.

2. Melt the butter in the pan, add the onion, and cook until soft but not brown. Return the kale-garlic mixture to the casserole and season with salt and pepper. Add the butternut squash, stock, and potatoes. Bring to a boil, reduce the heat, and simmer, covered, for 25 to 30 minutes, or until the vegetables are tender.

3. Add the cream and the sausage and bring to a simmer. Taste and correct the seasoning. Serve the soup hot.

Braised Leeks with Shiitake Mushrooms and Crème Fraîche

Tender buttery braised leeks are one of my favorite side dishes to all kinds of fish preparations, whether pan-seared or oven-roasted. The shiitakes add an interesting texture and flavor to this simple and flavorful vegetable dish. If you cannot get shiitakes, you can use cremini mushrooms instead, and for a variation add a sprinkling of dill or fresh tarragon. Serve as a side dish to a pan-seared fillet of salmon, grilled veal chops, or chicken breasts.

SERVES 4

4 medium leeks, all but 2 inches of green
   parts removed

4 tablespoons (½ stick) butter

2 teaspoons olive oil

½ pound shiitake mushrooms, stems
   removed, caps finely sliced

⅓ to ½ cup chicken stock or bouillon

Salt and freshly ground black pepper

2 tablespoons crème fraîche or heavy cream

2 tablespoons minced chives or flat-leaf
   parsley

1. Cut the leeks in half lengthwise and then crosswise into ¼-inch slices. Rinse thoroughly under warm water to remove all grit.

2. In a large heavy skillet, heat half the butter and the olive oil. Add the mushrooms and sauté until nicely browned. Remove with a slotted spoon to a dish.

3. Add the remaining butter to the pan. When it's hot, add the leeks and ⅓ cup stock. Season with salt and pepper. Cover the skillet and simmer for 4 to 5 minutes or until the leeks are just tender.

4. Add the mushrooms and crème fraîche or heavy cream. Simmer, uncovered, for 2 or three minutes. You may need to add a little more stock. Correct the seasoning. Add the chives or parsley and serve hot.

Portobello Mushrooms in Tarragon-Lime Butter

Until recently, I never thought of portobello mushrooms as being particularly interesting and found them rather leathery tasting, especially when grilled. Once I started slow-roasting them, I began to enjoy cooking them, especially in this simple preparation, which I serve as an appetizer or as a side dish to pan-seared salmon fillets or chicken breasts.

SERVES 6

8 tablespoons (1 stick) butter, softened
Juice of 1 lime
2 tablespoons minced fresh tarragon
Coarse salt and freshly ground black pepper
6 medium portobello mushrooms
2 tablespoons olive oil

1. Preheat the oven to 300°F. In a small bowl, combine 6 tablespoons of the butter with the lime juice and tarragon. Season the butter with salt and pepper and blend thoroughly.

2. Butter a large baking sheet with the remaining 2 tablespoons butter. Season the mushrooms with salt and pepper and drizzle with the olive oil. Place the mushrooms on the baking sheet and cover with foil. Roast the mushrooms for 25 minutes.

3. Remove the mushrooms from the oven. Top with bits of the tarragon butter and return the pan to the oven, uncovered. Roast for another 2 to 3 minutes or until the herb butter is just melted. Transfer the mushrooms to a serving platter and drizzle with the buttery pan juices. Serve hot.

Wild Mushroom, Scallion, and Gruyère Bread Pudding

I love all kinds of mushrooms, and now that you can get several types all year, it is fun to experiment with them. Here I use both porcini and cremini in a savory bread pudding that makes a delicious and elegant accompaniment to a roasted turkey, chicken, or pork. The pudding can be made a day ahead of time and reheated in a low oven.

SERVES 6 TO 8

3½ tablespoons butter

½ pound cremini mushrooms, finely diced

1½ cups finely sliced scallions, green tops included

Coarse salt and freshly ground black pepper

8 slices day-old Tuscan or other day-old peasant bread, sliced

4 cups milk

4 whole eggs

Pinch of cayenne

½ cups grated Gruyère cheese

1. Preheat the oven to 375°F.

2. In a large skillet, heat 3 tablespoons of the butter. Add the mushrooms and sauté until soft. Add the scallions, season with salt and pepper, and continue to cook until the scallions are just wilted.

3. Place the bread in a large bowl. Add the milk and soak the bread for a few minutes, pushing it down to submerge it. Remove the bread from the milk, squeezing it lightly to remove most of the milk, and tear the bread into bite-sized pieces and set aside.

4. Add the eggs to the milk. Season with salt, pepper, and cayenne and whisk the mixture until it is well blended.

5. Butter a 9 by 12-inch baking dish. Scatter half the bread on the bottom of the dish and top with the mushroom-scallion mixture, sprinkle with half the cheese, and top with the remaining bread. Add the egg mixture and the remaining cheese.

6. Place the dish in a larger baking dish. Pour enough hot water into the larger dish to come halfway up the sides of the smaller dish. Bake the pudding for 35 to 40 minutes or until a knife inserted in the center comes out clean. Serve hot or warm, directly out of the baking dish.

REMARKS If the bread is thoroughly dry, you may cube it prior to soaking it in the milk. A nice airy bread such as Tuscan bread is ideal for this recipe. Do not use a sourdough bread.

Bruschetta of Roasted Sweet Onions

When Vidalia onions are slow-roasted it brings out their wonderful sweet flavor. I usually do a batch whenever I plan to spend some time in the kitchen and the oven is available. You can store the onions for days and use them in many ways. I like to add a couple of spoonfuls to mashed potatoes or to the pan juices of a roasted chicken or turkey. They also make a wonderful topping for pan-seared salmon or swordfish steaks, or they can be used as part of an antipasti table. Place the onions in a jar, cover with a thin layer of olive oil, and re-frigerate until ready to use.

SERVES 4 TO 8

4 to 5 large Vidalia or other sweet onions
5 tablespoons fruity olive oil
Coarse salt and freshly ground black pepper
2 to 3 large garlic cloves, quartered
8 slices day-old peasant bread
8 pitted Kalamata olives, diced
Coarsely chopped basil or flat-leaf parsley
8 flat fillets of anchovies, optional

1. Preheat the oven to 325°F. Slice the onions crosswise into ¼-inch slices. Place in a single layer on a large cookie sheet. Drizzle with the olive oil. Season with salt and pepper and roast the onions for 2 hours or until they are lightly browned and very soft. Turn with a spatula two to three times during baking.

2. When the onions are done, transfer them to a colander set over a bowl and let the oil drain out of them for 30 minutes.

3. Turn the oven to broil.

4. Mash the garlic in a garlic press and rub a little into each bread slice. Drizzle each slice with a little of the onion oil.

5. Place the bread on a cookie sheet and broil until nicely browned. Watch the bread carefully so it does not burn.

6. Remove the bruschetta to a serving platter. Top each slice with some of the onions, sprinkle with olives, basil, or parsley, and top each slice with an anchovy fillet.

Roasted Vidalia Onions with Garlic and Rosemary

The season for Vidalia onions is becoming longer and longer, so I am now looking for interesting ways to use them. You can roast them with a variety of herbs, especially fresh oregano, marjoram, or, as here, rosemary. I usually double and triple the recipe, since the onions keep for several days and I like to serve them with practically everything, but especially with grilled meats and fish.

SERVES 4 TO 5

4 large Vidalia onions, cut into thick rounds
½ cup olive oil
2 teaspoons sugar
2 tablespoons minced rosemary
Coarse salt and freshly ground black pepper
1 large garlic clove, minced
¼ cup minced flat-leaf parsley
1 tablespoon balsamic vinegar, or more
   to taste

1. Preheat oven to 325°F.

2. In a large baking pan, combine the onions with the olive oil and toss gently. Spread the onions evenly in the pan and sprinkle with the sugar, rosemary, salt, and pepper. Add ½ cup water to the pan and set in the center of the oven. Bake for 3 hours, turning the onions once or twice to make sure they are roasting evenly.

3. When the onions are very tender, remove from the oven and transfer to a serving dish. Sprinkle with the garlic, parsley, and vinegar and toss gently. Taste and correct the seasoning, adding a little more vinegar to taste.

Sugar Snap Pea Soup

At the first sign of spring, heaps of fresh peas would appear at every market in Barcelona, and shelling peas at the kitchen table was part of my daily chores when I was growing up. It was well worth it because these jewels of the spring were used in many of my favorite dishes. I loved them simply cooked with bits of diced ham or in my mother's intensely flavored, superb pea soup.

Unfortunately, fresh peas are not easy to come by in this country, so I decided to try making the soup with sugar snaps, with the most delicious results. Be sure to make the soup a day ahead to allow it to develop full flavor, and whisk in some fresh butter just before serving.

SERVES 6

3 tablespoons butter

1 large bunch of scallions, diced (all but 2 inches of greens removed)

2 tablespoons flour

1½ pounds sugar snap peas

3 large sprigs of mint

8 cups chicken stock or bouillon

Salt and freshly ground black pepper

½ cup heavy cream or crème fraîche

2 tablespoons minced mint

1. In a large heavy saucepan, heat the butter. Add the scallions and 2 tablespoons water. Cook the scallions until very soft.

2. Add the flour and, with a wooden spoon, blend it thoroughly into the mixture. Add the sugar snaps, sprigs of mint, and chicken stock. Season with salt and a large grinding of black pepper. Bring to a boil, reduce the heat, and simmer until the sugar snaps are very tender.

3. Discard the mint and transfer the soup to a food processor. Puree until smooth and return to the saucepan. Add the cream or crème fraîche. Taste and correct the seasoning. Serve the soup either warm or at room temperature, garnished with a sprinkling of mint.

Curried Bell Pepper and Zucchini Ragout

I love curried vegetable stews, especially in the summer, when the green peppers are thin skinned and the zucchini is young and flavorful. Serve this delicious ragout either hot or at room temperature. You can vary it by adding ¼ cup blanched raisins to it and/or 2 to 3 tablespoons sautéed pine nuts.

SERVES 5 TO 6

¼ cup fruity olive oil

2 or 3 small zucchini, cut into ¾-inch cubes

Salt and freshly ground black pepper

2 large garlic cloves, mashed

1 tablespoon minced jalapeño pepper

2 teaspoons curry powder

2 medium onions, sliced

2 large green bell peppers, quartered, seeded, and finely sliced

2 red bell peppers, seeded, quartered, and finely sliced

4 plum tomatoes, diced, or 1 can Italian plum tomatoes, thoroughly drained and chopped

½ cup chicken stock or water, optional

Small cilantro leaves

1. In a large heavy skillet, heat 2 tablespoons of the olive oil. Add the zucchini and sauté until nicely browned. Season with salt and pepper. Remove with a slotted spoon to a side dish.

2. Add a little more oil to the skillet. Add the garlic, jalapeño, and curry. Cook for a minute, stirring constantly. Do not let the curry burn.

3. Add the onions and bell peppers and cook over low heat for 15 minutes, or until the mixture is soft and lightly browned. You may need a little more oil.

4. Add the tomatoes. Bring the mixture to a simmer and cook until the tomato liquid is mostly evaporated.

5. Return the zucchini to the pan and simmer for another 5 minutes. Correct the seasoning. Add the stock or water if more liquid is desired. Serve the ragout hot or at room temperature, garnished with the cilantro.

Mascarpone and Chive Mashed Potatoes

The good news is that mashed potatoes, that wonderful homey dish, is back in vogue. Not only can you make them at home but you can also now sample a variety of interpretations in many restaurants on both sides of the Atlantic. The bad news is that mashed potatoes taste good only when enriched with butter, sour cream, or mascarpone, and the more the better. Is it worth it? To me it is. Serve this version with just about any main course in this book. Also remember to double the recipe and have leftovers. The potatoes heat up beautifully in a low oven or the microwave.

SERVES 4

5 medium Yukon gold (about 1½ pounds),
  peeled and cut into eighths
Salt
4 tablespoons unsalted butter
2 tablespoons sour cream
¼ cup mascarpone
2 to 3 tablespoons minced chives
Freshly ground black pepper

1. Drop the potatoes into salted boiling water and cook until tender. Drain, pass through a food mill, and transfer to the top of a double boiler.

2. Add the remaining ingredients and mix well. Season with salt and pepper and serve hot.

Gratin of Russet Potatoes and Sweet Onions

You really can't beat the marriage of flavors of these two simple vegetables. They work together in many combinations, from an onion and potato soup to a zesty potato and onion salad. Here they take almost equal billing. Serve the gratin as a side dish to roast beef, grilled flank steak, or roasted leg of lamb.

SERVES 4 TO 6

4 tablespoons butter

2 to 3 tablespoons corn or peanut oil

2 large Spanish onions, quartered and finely sliced

Salt and freshly ground black pepper

4 large Russet potatoes, peeled and finely sliced

2 tablespoons finely minced fresh thyme, or 1 teaspoon dried thyme

1½ cups chicken stock or bouillon

2 tablespoons freshly grated Parmesan or Gruyère cheese

1. Preheat the oven to 350°F. In a large skillet, heat half the butter and the oil. Add the onions, season with salt and pepper, and cook over medium-high heat until lightly browned, 5 to 7 minutes. Lower the heat and continue cooking the onions for 40 minutes, stirring several times, until they are very soft and nicely browned. Do not let the onions burn; add more oil if needed.

2. Place a layer of potatoes, slightly overlapping, in a well-buttered baking dish and season with salt and pepper. Top with a layer of onions and some of the thyme. Make another layer of potatoes, onions, and thyme, finishing with a layer of potatoes. Season again with salt and pepper.

3. Add the stock. Cover the dish and place in the center of the oven. Bake for 50 to 60 minutes, or until the potatoes are tender and all the stock has been absorbed.

4. Sprinkle the cheese evenly over the potatoes and bake, uncovered, for another 10 minutes, or until the cheese is melted and browned. Serve directly out of the baking dish.

REMARKS The gratin can be made a day ahead of time and re-heated. Leftovers are delicious when reheated either in the oven or in the microwave. For a variation, use a mixture of fresh herbs such as thyme, rosemary, and parsley. For best results, cut the potatoes on a mandoline (see page 11). Not only will the potatoes cook more evenly, but the gratin will look much more attractive.

New Potato Salad in Mustard, Shallot, and Caper Vinaigrette

I love a good potato salad, but for years I rarely made it because I could not find the right potatoes. Now that fingerling potatoes are widely available, this zesty salad is tops on my list. Serve it often as an accompaniment to grilled or sautéed sausages or grilled chicken.

SERVES 5 TO 6

1½ pounds new potatoes, preferably Yukon Gold or fingerlings

1½ tablespoons sherry vinegar

6 tablespoons peanut or olive oil

1 tablespoon Dijon mustard

1 large garlic clove, mashed

Salt and freshly ground black pepper

1 large shallot, minced (about ¼ cup)

1 dill gherkin, finely diced

½ cup finely diced green or red bell pepper or a mixture of both

2 tablespoons tiny capers

1 to 2 tablespoons sour cream or mayonnaise, optional

1. In a large saucepan, combine the potatoes with water to cover. Cook for 20 minutes over medium heat until the potatoes are just tender. Do not overcook. Drain and cool completely. Peel, slice, and place in a large bowl.

2. In a small bowl, combine the vinegar, oil, mustard, and garlic; whisk the dressing until well blended. Season with salt and pepper and pour over the potatoes.

3. Add the shallot, gherkin, bell pepper, and capers to the bowl; toss the salad gently with a wooden spoon so as not to break the potatoes. Taste and correct the seasoning. Add the sour cream or mayonnaise if you like a creamier salad, and chill the salad for at least 2 to 4 hours or overnight. Bring the potato salad back to room temperature before serving.

Puree of Sweet Potatoes and Carrots

Sweet potatoes and carrots have a marvelous affinity for one another. Come fall, this puree becomes one my favorite sides. It makes the perfect accompaniment to a roasted turkey, pan-seared duck breasts, or a ragout of pork. For a variation, I often add grated ginger or season the puree with toasted cumin seeds and minced cilantro.

SERVES 6

3 large sweet potatoes (about 2 pounds),
   peeled and cut into 1-inch pieces

3 to 4 large carrots (about 1¼ pounds), peeled
   and cut into 1-inch pieces

3 tablespoons sour cream

2 to 4 tablespoons unsalted butter

2 tablespoons dark brown sugar

Salt and freshly ground black pepper

1. Combine the potatoes and carrots in a large saucepan with plenty of salted water to cover, and simmer until very tender.

2. Drain well and transfer to a food processor together with the sour cream, butter, and brown sugar; combine. Season with salt and pepper. Serve hot.

REMARKS You may keep the puree warm, covered, in a double boiler until serving.

Puree of Rutabaga and Potato with Crispy Shallots

Rutabaga is not a vegetable one likes at first taste, but when teamed with potatoes, sour cream, and butter in a delicious puree, it is a winner. The crispy fried shallots add texture to the puree, but they are not a must. You can also use small white onions cut into thick rounds instead of shallots. The puree can be made a day or two ahead of time and reheated either in the microwave or in a double boiler. Serve as an accompaniment to grilled, sautéed, or roasted meats and duck.

SERVES 6

> 2 medium rutabagas (about 3 pounds), peeled and cut into 2-inch pieces
>
> 1 medium all-purpose potato, peeled and cut into 1½-inch pieces
>
> 4 tablespoons unsalted butter
>
> 3 tablespoons sour cream
>
> Salt and freshly ground white pepper
>
> 1½ tablespoons peanut oil
>
> 3 large shallots, finely sliced

1. Preheat the oven to 200°F.

2. In a large saucepan, bring plenty of salted water to a boil, add the rutabagas and potato, and cook until very tender.

3. Drain the vegetables and transfer to a food processor. Add 3 tablespoons of the butter and the sour cream; season with salt and pepper and process until smooth. Transfer to an ovenproof dish, cover, and set in the center of the oven to keep warm.

4. In a heavy medium skillet, melt the remaining 1 tablespoon butter together with the oil over medium-high heat; add the shallots and sauté until they are nicely browned and crisp-tender, 4 to 5 minutes. Fold the shallots into the vegetable puree and correct the seasoning. Serve hot.

Shallot and Cassis Marmalade

I first had a taste of this delicious "marmalade" in the city of **Tours** in the Loire region, known for its superb goat cheeses. It was served as an accompaniment to warm goat cheese toasts and a simple salad. I like to serve the jam this way, but it also goes well with grilled salmon fillets, pan-seared tenderloin steaks, and sautéed calves liver. Caramelized shallots can be made in quantity and stored in a jar; they will keep for weeks. Sliced small pearl onions can be prepared in the same manner, but the subtle, more delicate taste of the shallot produces more interesting results.

SERVES 6

> 6 tablespoons unsalted butter
> 2 pounds shallots, finely sliced
> ¼ cup sugar
> 3 cups dry red wine
> ¼ cup cassis
> 1½ tablespoons sherry vinegar
> Salt and freshly ground black pepper

1. In a large cast-iron skillet, melt the butter over medium heat. Add the shallots and sugar and cook until the sugar has melted and the shallots begin to brown.

2. Reduce the heat, add the red wine, cassis, and vinegar, and cook for 25 to 30 minutes, or until all the liquid has evaporated and shallots are soft and nicely caramelized. Season with salt and pepper, and serve as an accompaniment to grilled or pan-seared fish.

REMARKS You may add ⅓ cup plumped currants and ¼ cup pine nuts, sautéed in a little olive oil, to the marmalade for a more unusual taste and texture.

Sauté of Spinach with Scallions, Dill, and Yogurt

Sautéed vegetables are often served as appetizers in Spain, and sautéed spinach is an especially popular starter in Catalonia. I like serving it this way, particularly in the spring, when spinach is at its best. If possible, use the curly variety, which is far more flavorful than the flat-leaf type of spinach. Serve as a side dish to pan-seared lamb chops, pork chops, or grilled or sautéed shrimp.

SERVES 6 TO 8

1 cup plain yogurt

2 bags (10 ounces each) spinach, stemmed and
   washed

2 tablespoons butter

1 tablespoons virgin olive oil

3 scallions (white and 2 inches of green), chopped

1 large garlic clove, crushed

Salt and freshly ground black pepper

4 tablespoons minced fresh dill

2 to 3 tablespoons finely diced feta cheese,
   optional

1. Place the yogurt in a fine sieve placed over a bowl and drain for 1 to 2 hours or overnight.

2. In a large saucepan, bring salted water to a boil. Add the spinach and cook for 2 minutes. Drain and run under cold water to stop the cooking.

3. In a large nonstick skillet, heat the butter and olive oil. Add the scallions and garlic and cook for 1 to 2 minutes, or until soft but not browned. Immediately add the spinach. Season with salt and pepper and cook for 2 to 3 minutes. Add the dill and yogurt and stir with a wooden spoon to thoroughly blend the mixture. Taste and correct the seasoning, adding a large grinding of black pepper. Add the feta and transfer the spinach to a serving bowl. Serve.

REMARKS If you can get imported Greek sheep's milk yogurt, be sure to use it, since it adds a special tart taste to the spinach and is especially good with feta. If you do use feta, season the spinach carefully, since feta is usually quite salty. Also, if you have the choice, use the French rather than the Greek or Bulgarian feta.

Acorn Squash, Apple, and Wild Mushroom Soup

Soups should be seasonal, and acorn squash, apple, and mushroom soup is just that. It spells fall, crisp cool weather, and an appetite for something hearty and full flavored. I like to serve this soup at Thanksgiving, but you don't need a holiday to enjoy it. Serve with crusty bread and some sweet butter.

SERVES 4 TO 5

2 medium acorn squash (about 1 pound each)

5 tablespoons unsalted butter

2 Golden Delicious apples, peeled, cored, and cubed

1 teaspoon granulated sugar

1 large onion, minced

2 tablespoons dark brown sugar

$\frac{1}{2}$ teaspoon ground ginger

$\frac{1}{4}$ teaspoon ground coriander

$\frac{1}{4}$ teaspoon ground cardamom

Large pinch of freshly grated nutmeg

8 to 9 cups chicken stock or bouillon

$\frac{1}{3}$ cup heavy cream, optional

$\frac{1}{4}$ pound fresh chanterelle or shiitake mushrooms, stemmed and cut into $\frac{1}{4}$-inch slices (if using shiitakes, discard the stems)

Salt and freshly ground white pepper

1. Preheat the oven to 400°F.

2. Place the acorn squash in the center of the oven and bake for 45 to 60 minutes, or until very tender when pierced with the tip of a sharp knife.

3. Remove the squash from the oven, peel, and remove the seeds. Dice the pulp coarsely and set aside.

4. In a large cast-iron skillet, melt 2 tablespoons of the butter over medium-high heat. Add the apples and sauté until lightly browned. Add the granulated sugar and continue to sauté until the apple slices are caramelized and very well browned. Remove from the heat and reserve.

5. In a 3½- or 4-quart saucepan, melt 2 tablespoons of the butter over low heat. Add the onion and cook until soft and lightly browned. Add the brown sugar and spices and continue to sauté for 2 minutes without letting the sugar burn.

6. Immediately add the acorn squash and stock. Bring to a boil, reduce the heat, and simmer the soup, partially covered, for 20 minutes. Add the apples and continue to simmer for another 15 minutes. Cool the soup slightly and then puree in batches in the food processor. Return the soup to the casserole and stir in the cream.

7. In a small heavy skillet, heat the remaining 1 tablespoon butter over medium-high heat. Add the mushrooms and sauté quickly for 1 to 2 minutes, or until the mushrooms are lightly browned. Season with salt and pepper. Add the mushrooms to the soup, and correct the seasoning. If the soup is too thick, thin it out with additional stock. Serve hot.

Wild Mushroom and Parmesan Timbales

Here is a lovely mushroom flan that can be served either as an appetizer or as a side to roasted chicken, veal chops, or a pan-seared steak. Instead of using ramekins, you can make the dish in a loaf pan and serve it sliced at room temperature over a bed of well-seasoned spinach greens.

SERVES 6 TO 8

4 tablespoons butter

1 large shallot, minced

4 garlic cloves, minced

1½ pounds cremini mushrooms, stems removed, caps minced

Salt and freshly ground black pepper

¼ cup minced flat-leaf parsley

2 tablespoons minced fresh thyme, optional

3 extra large egg yolks

1 cup heavy cream

½ cup finely grated Parmesan cheese

1. Preheat the oven to 375°F. Butter 6 to 8 ramekins.

2. In a large heavy skillet, heat the butter, add the shallot and garlic, and cook for 1 minute without browning. Add the mushrooms and continue to cook until the mushrooms are soft and all of their water has evaporated. Season with salt and pepper and add the parsley and thyme. Taste and correct the seasoning.

3. In a bowl, combine the egg yolks, cream, and Parmesan. Whisk thoroughly and season with a pinch of salt and pepper.

4. Add the mushroom mixture and blend well. Taste and correct the seasoning.

5. Spoon the mushroom custard into the prepared ramekins and place in a baking dish. Add enough hot water to reach halfway up the sides of the ramekins and place the dish in the oven. Bake for 25 to 35 minutes, or until the timbales are set and the tip of a knife inserted in the middle of the custard comes out clean.

6. Remove the ramekins from the baking dish and set aside for 5 minutes. Run a knife around each one and unmold onto a serving platter or individual serving plates. Serve hot.

■ Pan-Seared Arctic Char with Lemon-Chive Essence ■ Brook Trout en Papillote with Asparagus, Snow Peas, and Dill Butter ■ Stir-Fry of Calamari, Red Peppers, and Lemon ■ Fillets of Lemon Sole in a Lime, Jalapeno, and Caper Butter ■ Oven-Roasted Mackerel with Caper, Lemon, and Anchovy Butter ■ Steamed Mussels in Basil-Garlic Cream ■ Pan-Seared Red Snapper Fillets with Sweet and Spicy Asian Syrup ■ Pan-Seared Skate with Red Onion, Tomato Jam, and Brown Caper Butter ■ Roasted Sea Bass with a Fricassee of Fennel and Potatoes ■ Sauté of Swordfish with Shrimp, Snow Peas, and Three Bell Peppers ■ Sautéed Scallops with Red Bell Pepper and Chili Sauce ■ Spicy Shrimp and Ginger Fritters with a Ginger-Tamari Dipping Sauce ■ Fillets of Salmon in Teriyaki Marinade with Gingered Carrots ■ Tuna Tartare

No matter where I find myself shopping for seafood—in one of Barcelona's many fish markets, at the port in Nice, along the varied displays in New York's Chinatown, or even at my local seafood market—I find the amazing diversity of fish is one of the most exhilarating food experiences I know. And given the exceptional importance of freshness in relation to the seafood, it is also one of the most challenging.

I admit that buying top-quality fish is difficult for many shoppers. The distribution system is not ideal, and the flavor of farm-raised fish is a far cry from that of the wild species. But in spite of the fact that most Americans live far from an ocean and have to rely on local supermarkets for their fish, the consumption of seafood is growing.

The questions I am most often asked are "Where can I get good fish?" and "How can I tell when fish is fresh?" Buying fresh fish is not as simple as buying good beef or chicken, even for someone like me living in New York City, which has a fair share of good seafood markets and a bustling, lively Chinatown. The challenge becomes more difficult the farther you get from the ocean. Often in cities where I conduct workshops, the smell of ammonia and spoiled seafood hits me the moment I walk into the store. The fish look dead, with opaque flesh and dull skin. These are times when I want to change my menu to chicken or pasta. How frustrating not to have the choice of good fish and to have to make do with inferior quality.

The good news is that supermarkets around the country are rising to the challenge. I am often pleasantly surprised by fish counters that display lively-looking shrimp, fresh clams, and excellent salmon, laid out on beds of ice. I have found such a supermarket quite a long distance from my house, where the fish counter looks clean and inviting. Here the turnover is terrific, and I am always certain to find an excellent assortment of fish. As I approach the counter, I can almost smell the sea, with fish that look alive, sporting beautifully translucent flesh and shiny silvery skins.

It is at this kind of supermarket that you should try to build a relationship with the fishmonger, letting him or her know that you care about the quality of the fish you buy. Never ask the question "Is this fish fresh?" but rather, "What do you recommend today?" or "When did the scallops come in?"

I have learned to be spontaneous and try never to go with a specific recipe in mind. Instead, I choose the freshest-looking fish with the knowledge that cooking it is the easiest part of the meal. On days that the selection is particularly good, my creative juices start to flow, and I end up with way too much fish.

There is nothing better than finding yourself in a small coastal seafood market when the day-boats come in with fresh scallops or a beautiful line-caught cod. This is when you should be flexible. It is rare that you cannot substitute one fish for another in a given recipe, as long as you understand the texture differences. For instance, when a recipe calls for tuna, you can substitute swordfish, mako shark, or halibut steaks with equally good results. The same goes for recipes that call for striped bass, where sea bass, snapper, or grouper would work as well.

I have also learned that fish can't wait. As soon as I get home, I unwrap the fish and place it in a glass or porcelain baking dish surrounded with plenty of ice cubes, and store it in the bottom part of the fridge where it is coldest. If I can't cook it the day of purchase, then I know the following day is a must.

I am often asked whether freshly frozen fish is not a better choice than fish that is not truly fresh. If the fish has been frozen right on the boat as soon as it has been caught, chances are that you can get good results, but much depends on how it has been defrosted. Frozen fish tends to exude water, so the fish can't brown and is often dry and mealy. So unless I am dealing with shrimp or squid, I much prefer buying fresh fish or fish that is still frozen and looks as if it has been properly handled.

What makes fish an ideal food for today's busy cook is that it cooks quickly and marries beautifully with a wonderful variety of herbs and vegetables. When you need dinner in 30 minutes or less, think seafood. Since fish is naturally tender, all you need is a good nonstick skillet, preferably one that can be placed in the oven, and some practice to tell when fish is done.

When it comes to preparing fish, it is good to remember that just as with a good steak, less is more. Keep it simple, and keep it fresh. That is what all cooks who know their fish will tell you. I realize that this may sound simple, but, in fact, if you have never cooked a swordfish steak or a fillet of salmon, just remember that every type of fish can be pan-seared in just a little olive oil or butter. If the steak is thick, finish cooking it in a medium-hot oven; if it is a thin fillet, simply turn it over with a large spatula and finish cooking it on top of the stove, lightly covered.

For an even easier method, you can simply butter and season the fish of your choice, place it in a well-buttered dish covered with foil, and roast in a 350° to 400°F oven. If you want to be creative, you can always add a little white wine to the skillet, as well as a few pieces of chopped tomatoes and some herbs.

Most types of fish like the company of fresh vegetables, and some matches are especially delicious. Try salmon with beets, dill, or cucumbers; sea bass with tomatoes, oregano, and shallots; or pan-seared skate with a drizzle of brown butter, lemon, and capers. If you are planning a more elaborate sauce, make it first, since even a thick fish steak will not take longer than 12 to 14 minutes to cook.

I cannot think of any foods other than fruits and vegetables that offer as much variety, taste, and enjoyment as seafood with the added bonus of being low in fat and presenting a healthful alternative to meat. And, with more and more restaurants serving creative fish preparations, it is easy to be inspired and to duplicate them at home for a fraction of the price.

Fish

∎∎

How can you tell if a whole fish is really fresh?
The key is aroma. Very fresh fish smells clean, just the way the salty air at the beach does. There should be no smell of iodine. A whole fish should not look bruised in any way. Its skin should be firm, glossy, and shiny, not dull. Also, it must have its gills, which should be bright red and moist looking. Finally, the flesh of a fresh whole fish will adhere closely to the bones. If possible, press lightly on the flesh. It should be firm and return to its original shape. If a gentle nudge creates a permanent indentation, pass on the fish.

My market sells fish only in fillets and steaks. Is there any easy way to spot fresh fish when you cannot see the whole fish?
As long as the skin is still on, you should have no trouble telling if the fish is fresh. Inspect the fish carefully. Its skin should be smooth and unbroken, with an appealing silvery glint. Always ask to see it up close, since many fish counters have slightly rosy lights that make both the flesh and the skin look brighter. The flesh should not be flaky or separating and must have translucent, even coloring. Press gently on the flesh: If it leaves a dent, the fish is not truly fresh.

I often buy flounder that seems very fresh but has a strong iodine aroma. Is that okay?
Although in most fish that smell is a no-no, in flounder and skate you may detect a slight hint of iodine. That's due to their diet and will not affect their taste; however, it should never be an off-putting smell.

Cookbooks always say that you should look a fish in the eye to judge its freshness, but what exactly does that mean?
The eyes of a fresh fish are clear, bright, and flush with the surface of its head. Don't buy a fish with dull, cloudy, or sunken eyes; that's a sign of decay. When viewing a fresh fish and

a not-so-fresh fish side by side you will be able to tell the difference in an instant.

Are there any types of fish that are less fishy tasting?

Swordfish, mako, halibut, and very fresh tuna have a meatier mouthfeel and an unfishy taste. Tuna in particular can almost pass for a meaty steak.

Does fish benefit from marinating?

There are two very distinct cooking techniques that may both be referred to as "marinating." One is marinating the fish to give it more flavor before you grill or sauté it, in which case you want to let it marinate for only 30 minutes or less, because the acidity in a marinade "cooks" the fish. The other method is curing the fish in a salt, pepper, and herb mixture (see my book *Spur of the Moment Cook*, Morrow, 1996). In this case, you can cure it anywhere from 24 to 48 hours, depending on the recipe.

Cooking fish seems to smell up the entire house for days. Any suggestions for avoiding this?

Pan-searing with skin on is the worst offender in this department, as are oily fish such as salmon, mackerel, and bluefish. But there are wonderful odorless methods for cooking fish such as roasting, poaching, or baking in foil.

My husband does a lot of fishing and comes home with whole fish all the time. What's the best method for cooking a whole fish?

For small fish (up to 3 pounds), roasting or grilling will give you great results. When it comes to larger fish, and when we are talking about the freshest possible fish that has been caught that day, all pros agree that poaching is the number-one method. Old fishermen in Provence were famous for putting napkins over their heads to eat freshly poached fish because that way the aroma of the sea and that of the fish would mingle in those little tents, and they could enjoy the experience fully. All that a freshly poached fish needs is a sprinkling of salt, a grinding of pepper, a drizzle of olive oil, and a wedge of fresh lemon. Another great method for preparing a whole fish is roasting it in foil with some fresh herbs, seasoning, plenty of butter, and a sprinkling of lemon juice.

Is there really a simple way to tell when a whole fish is done?

Many cookbooks recommend measuring the fish at its thickest part and counting 8 minutes per inch. This is rather confusing and not accurate. To oven-roast a whole fish, heat the oven to 450°F and figure about 12 minutes per pound. Check the progress by making a small slit just behind the center bone. The flesh should be white and slightly flaky; if it is still translucent, it is not done. Alternatively, use an instant-read thermometer and cook the fish to between 140° and 145°F. This is still on the rare side, but it will continue to cook after you remove it from the oven. Many cookbooks will suggest cooking a fish up to 165°F, but at that point it is overcooked.

TIP When weighing a fish to decide how long to cook it, be sure to take the size of the head into consideration. A fish with a large head, such as a red snapper or sea bass, will weigh more than a branzino or mackerel.

When fish fillets are cooked with the skin on, they tend to curl up. Is there any way around that?

Try scoring the skin side of the fillets on the diagonal with a sharp paring knife. That way the skin doesn't tighten as the fish cooks, and the fillets remain flat.

What does cooking fish "en papillote" mean?
Fish cooked en papillote is enfolded in a parchment paper envelope together with herbs, butter, seasoning, and sometimes finely sliced or diced vegetables and a little wine, then baked in the oven, where it steams in its own moisture. Many chefs now use aluminum foil instead of the traditional parchment paper, with excellent results.

Is it a good idea to cook fish in the microwave?
The microwave might seem like a timesaver, but oven-braising produces a moister, more evenly cooked piece of fish, and the time difference is negligible.

Catfish

My market carries very fresh catfish, and I would like to know more about it.
Not so long ago, catfish was caught wild in rivers and lakes. It was an interesting fish that in spite of its sometimes muddy flavor had good taste. I often used it for fried fish. Now catfish is farmed on a major scale in the Mississippi River and has little character. Still, it is quite delicious fried in a light batter, and also takes well to a spicy rub or marinade, after which it can be simply broiled.

Grouper

I have had the most wonderful grouper in Florida but cannot find it in any market where I live. Can I use another fish instead?
Grouper is a distant relative of sea bass, which you can use instead. However, the texture of grouper is firmer and the flesh whiter. It also has a more lobster-like texture. Some fish markets sell grouper under names such as red grouper or black grouper, which may have a slightly different taste, but as long as the fillets are pearly white and fresh they are a good choice.

Halibut

I live in the Northwest and see wonderful halibut in the markets. How do you suggest using it? Can I substitute it in salmon recipes?
Halibut, which is at its best from March to September, is an extremely versatile fish, although personally I find it a little too mild flavored. Be sure to buy the fish very, very fresh. The flesh should not look opaque or cloudy, but rather translucent. When halibut is available only cut in steaks, I usually cube them for shish kebabs or pickle them to make a great dish.

And yes, you can use halibut for all preparations that call for salmon.

I recently pan-seared halibut, and it gave off a lot of liquid and would not brown. Is that the wrong preparation for this fish?
This does happen often with halibut, either because the fish has been frozen on the boat, or just because of this particular fish. Next time you cook halibut, I suggest oven-roasting it or broiling it very carefully so as not to overcook it. The grill is another excellent cooking medium for halibut.

Mahi Mahi

Is it true that mahi mahi is a name given to dolphin for marketing purposes? I loved the taste of the fish until I heard that. Is it true?

It is definitely not true. Mahi mahi, or dolphin fish, is not a mammal. Instead, it is a fish that is widely available in Hawaii and off the coast of Florida. It is usually available only in fillets. These should look bright and almost translucent. In Florida markets, I see wonderful-looking mahi mahi, while in New York it is often opaque—a definite indication that the fish is past its prime.

Mahi mahi is at its best when baked with spices, onions, tomatoes, and peppers.

The following is a nice rub for mahi mahi In a bowl, mash 2 large garlic cloves and combine with 1 teaspoon salt, 1 tablespoon coarsely ground black pepper, ½ tablespoon each thyme and oregano, 1 teaspoon sweet paprika or Spanish smoked paprika, and 2 tablespoons soy sauce. Rub the mixture on a 2-pound fillet of mahi mahi and marinate, covered, for 4 to 6 hours. Grill or broil the fish for 6 to 7 minutes on each side.

Salmon

What should I look for when shopping for salmon?

A fresh piece of salmon will boast a layer of glimmering silver skin so shiny you'd swear you could see your face reflected in it. If its skin is dull, don't even consider buying it. Also, with the exception of salmon steaks, try not to buy packaged salmon or skinless salmon.

What is the difference between Atlantic, Pacific, Chilean, and Norwegian salmon? Is one better than the others?

Not really. Some markets and restaurants label Atlantic salmon as Norwegian salmon and try to convince you that one is better than the others, but in fact, they are all farm-raised and pretty similar in flavor. At certain times of year, Atlantic salmon may be a little fattier.

Is there a difference between farm-raised and wild salmon? Is there any way to tell which is which if it isn't labeled?

As you might expect, wild salmon—whether Pacific or Alaskan—is less fatty and more intensely flavored than farm-raised fish. Although wild salmon is not abundant, it can certainly be found in markets in the Pacific Northwest and in specialty fish markets along the Oregon and Washington State coast. Very little wild salmon makes its way to other parts of the country, so it is really a matter of luck or knowing a chef or seafood market that can get it. Is it worth a try? Definitely.

Are there different types of wild salmon? Which is best?

Look for king, sockeye, and coho wild salmon, all of which are extremely flavorful.

What is a salmon trout? Is it some kind of hybrid?

A student once called me with a fish cooking question. She had bought a salmon trout and was following a recipe that called for poaching the fish 15 to 20 minutes. After 20 minutes, the fish still looked pink, so she poached it another 20 minutes. When, after 40 minutes of cooking, it still looked pink, she called me to ask how long it was going to take. My answer? You'd better feed that fish to your cat and start from scratch. What she did not realize is that a salmon trout is not a trout at all, but a small salmon, and the flesh will always be pink. Many markets now call salmon trout Arctic char to make it less confusing.

Do you recommend salmon fillets or salmon steaks?

Each has its place in a cook's repertoire. Thick steaks are moister and better suited for grilling, and they are also very good poached. However, salmon steaks have bones, which may be a problem for some people. The fillets are easier to handle and serve without breaking apart. Try both and see which you like better. It is easier to tell freshness in fillets, since they are generally sold with the skin on, and a salmon's skin is the key to gauging its freshness.

Many Japanese restaurants serve raw salmon. Is it safe to do that at home with store-bought salmon?

Yes, if it is super fresh, which means buying it at a reputable fish store. Remind the fishmonger that you plan to eat it raw and want sushi-quality salmon.

Is it better to pan-sear salmon with or without the skin?

No matter how you are going to cook it—even when preparing a recipe that calls for skinless fillets—always select salmon with the skin on. The skin's color and sheen are the key indicators of freshness, and markets that sell skinless fillets are usually trying to sell you a less than perfect product.

Once you have inspected the skin, you can ask the fish market to remove it. I prefer to cook salmon with the skin on and remove it once it is cooked, because there is a great deal of taste in the skin—actually considered a delicacy—and, just as with a steak, the fat keeps the flesh moist.

No matter what I do, I either undercook or overcook salmon. Is there a reliable way to tell when the fish is perfectly done?

Learning to cook the perfect salmon fillet can be frustrating, because once you've overcooked it, there's no going back. To test for doneness, use the knife method. Cook fillets until nicely browned on one side, then cover the skillet loosely with foil. Once the fillets lose their raw color and all you see is a deep pink dot the size of a dime, turn the fillets over and cook for 50 seconds. Insert the tip of a knife or a metal skewer into the center of the fillet, remove it, and place it on your wrist. If it's warm, the fish is done. If it's cold, the fish is rare, and if it's hot, the fish is well done and maybe overcooked. If all this sounds too complicated, use an instant-read thermometer and remove the fish from the heat when it reads 125° to 130°F. Do keep in mind, though, that the fish will continue to cook. For this reason, remove it from the stove and take it out of the pan when it is still somewhat rare.

TIP Grilled fish will be gently resistant when cooked, never rubbery or tough. The tricky part is testing grilled fish for doneness without cutting into it (which will cause it to fall apart). To do this, insert a metal skewer into the center of the steak or fillet and then carefully push it to one side while it is still inserted, and you should be able to glimpse the center of the fish without pulling it apart.

Some new recipes call for cooking salmon at 500°F. But the standard cookbooks call for 350°F. Which is right?

If you cook any fish steaks at 500°F, you will incinerate them. High-heat roasting can work for a large whole fish, but you have to time it carefully. Low-heat roasting is by far the best way to cook a whole salmon or any fish. See Roasted Sea Bass with a Fricassee of Fennel and Potatoes, page 178.

Skate

Skate is one of my favorite fish. I order it in restaurants all the time but have no idea how to prepare it. Can you help?

Skate has two flat, paired wings, one usually larger than the other. Have the butcher bone the skate, which means removing the central cartilage. You will now have two pieces, one thicker than the other.

To prepare skate in the most classic way, simply dredge the fish lightly in flour, season with coarse salt and freshly ground pepper, and sauté in a mixture of peanut oil and butter or clarified butter for 3 minutes on each side until nicely browned. Break off a little piece to see if the fish is done. To make sure you don't overcook the fillets, sauté the fish over medium rather than high heat. Serve the fish with a browned butter, lemon, and caper sauce.

To make Brown Butter, Lemon, and Caper Sauce
Melt 1 stick butter in a small heavy skillet or saucepan over medium-high heat until the butter turns a golden brown color and has the aroma of toasted hazelnuts. Remove from the heat. Add the juice of 1 lemon, 2 tablespoons well-drained capers, and 2 tablespoons toasted pine nuts. Swirl the butter and immediately pour it over the sautéed skate wings. Be sure not to let the butter get too dark and to remove the pan from the heat before adding the lemon juice.

Snapper

It seems that every time I go to the fish market, there is a different kind of red snapper available. What kind of snapper should I buy?

First of all, there is snapper and then there is Florida red snapper, which is in a class all its own. Florida red snapper commands a hefty price and is worth every bit of it, since it is a great-tasting fish. There is a quota when it comes to Florida red snapper, so when you see it toward the end of the month, it is not the real thing, since it can be fished for only the first three weeks of the month.

Other snappers, such as white Pacific snapper and New Zealand red snapper, which look very much like the Florida variety, are not in the same league. Never pay top dollar for fillets of snapper even if the skin has the right color, because chances are that you are not getting the Florida snapper variety, unless, of course, you are in Florida at a reputable market that fillets its own fish. In this case always try to buy some snapper bones and heads and make a fish stock. No fish makes a better stock than snapper. Fresh snapper fillets are good simply pan-seared in butter with some fresh herbs; they can also be oven-roasted with some white wine and shallots.

Sole

What is the difference between lemon sole and grey sole?

Grey sole is the mildest tasting of all flounders. It is very popular because it is so unfishy and has become the perfect choice for people who really don't like fish. Traditionally, lemon sole had to come from the Georgia Banks and weigh 3 pounds or more. It was a fish with excellent texture and taste. Now that the fish has become quite scarce, all flounders over 3 pounds are called lemon sole. There is not much difference in flavor between the various types of domestic sole.

Stock

I have trouble getting fish bones for fish stock. Do you have any ideas for substitutes?

In many instances, chicken stock works well in recipes calling for fish stock. For a more pronounced seafood flavor, add two to three handfuls of shrimp shells to the simmering stock, whether it's chicken stock or instant fish broth made from cubes, such as Knorr. Shrimp shells should be easy to get, even from the supermarket, but you may have to ask for them at the fish counter.

When is it important to use real fish stock, and can I use bottled clam juice instead?

I find that a full-bodied fish stock is a must whenever I make a delicate fish soup, seafood risotto, or a sauce for fish. On the other hand, intense flavorings such as saffron, lemongrass, or ginger work with instant fish broth with a fair amount of success. Bottled clam juice is extremely salty and does not provide the right seafood flavor. You can use diluted fish bouillon cubes, but be careful. These are also quite salty; I suggest that you add half again as much water to dilute them.

What are the best bones for making fish stock?

Snapper and bass bones make the best stock. If you live in an area where you have easy access to red or other snapper bones, you are in luck. Don't use sole, flounder, or the bones from other flatfish. They simply do not have enough flavor.

The most flavor is in the fish head. If you can get fish heads, you will be able to make great-tasting stock. I usually use an assortment of fish bones but try to include at least one or two snapper or bass fish heads.

TIP When buying fish heads, check to make sure they still have the gills—a sign of freshness. Then ask the fishmonger to remove the gills for you, since they are not easy to remove at home.

I recently made a fish stock with salmon bones, and the result tasted super fishy. Should I have done done anything differently?

Because salmon is extremely oily, salmon bones are not a good choice for making fish stock. As you discovered, they release a strong, almost unpleasant, fishy flavor.

Whenever I make fish stock, it looks cloudy and opaque. What am I doing wrong?

When fish stock is cloudy, chances are the fish bones and trimmings were not thoroughly rinsed. Place the bones and trimmings in a large bowl in the sink and trickle cold water over them until the water runs completely clear and the bones are snow white.

I have seen several recipes that call for court bouillon for cooking fish. What is this?

A court bouillon is a broth made with vegetables and herbs simmered in water with a touch of vinegar or wine, but without fish trimmings. I generally use court bouillon for poaching fish and braising large salmon fillets. What makes a court bouillon interesting to the seafood cook is that court bouillon can be used several times and frozen in between. After having poached fish in the bouillon a few times, you end up with a delicious full-bodied fish stock.

Swordfish

What are your thoughts on swordfish? I have seen it looking rosy pink and other times it has a grayish cast. Are both good? Is there a particular season for swordfish?

I discussed this with Jeremy Marshall, chef/owner of Aquagrill, one of New York City's most creative seafood restaurants. According to him, the key to the flavor of swordfish is the way it has been caught and at which stage of the game it reaches your fish market. The color varies from region to region, but it is not an indication of freshness. Swordfish can be both line caught and net caught. A line-caught fish is much more flavorful and juicy. Furthermore, the fishing boats go out for days at a time, which means that the last fish caught will come off the boat first and that fish will be fresher than others. If the fish market pays for top-quality swordfish, then you will be paying top dollar as well.

Since swordfish comes from different waters during different seasons, it can be good year-round. Personally, I find myself cooking swordfish more often in the summer, because I like the fish from the North Atlantic and at that time of year I may be lucky enough to find some line-caught fish.

I buy my swordfish at an excellent seafood market. Sometimes it is rich and juicy, while other times it is dry. Is there a way to tell quality by the color of the fish?

Unfortunately, you cannot determine the quality of swordfish by its color. Swordfish must be fatty to be good. The only way to tell is by cutting off a small piece. If the knife comes out clean, the fish is dry. If little bits of fish cling to the knife, it means the fish is fatty. I realize that at this point you have already bought the fish, but based on this little test, you can decide how to prepare it. When swordfish tests dry, I usually do not marinate or grill it. Instead, I pan-sear it in butter or olive oil.

Tilapia

Many seafood restaurants seem to have tilapia on the menu. What kind of fish is it and how should I use it?

Tilapia is a mild farm-raised fish that mostly comes from southern states; it is also imported from Israel or Hawaii. Most markets sell the fish in skinless fillets, but I always look for it with skin on, since the fish is more attractive this way and also cooks better.

Tilapia is excellent roasted whole, either in the oven or on the grill. Be careful not to overpay for tilapia. The fish is often called by other names such as St. Peter's fish, with inflated prices.

Tuna

What is the best tuna to use for tuna tartare?

The most widely available tuna is yellow fin, so chances are that you will be using it for both grilling and tartare. The key is not to use any of the sinew, so be sure to use only the part of the tuna steaks that do not contain it.

What should I look for in tuna? Is color an issue?

The lighter the tuna, the fattier. Most chefs prefer the fattier fish, but dark tuna is extremely flavorful. The most important issue is freshness rather than color. Look for bright translucent flesh and avoid tuna steaks that look opaque.

Whenever I make a tuna steak, it is overcooked by the time it is browned. How can I avoid overcooking tuna?

Always buy a large piece of tuna, enough to serve 4 to 6 people. This way you can buy a 4- to 5-inch piece and then cut it into rectangles or 2-inch cubes. The thicker the pieces, the better your chance of keeping the tuna from being overcooked. If you do not have a hot grill, then pan-sear the tuna quickly in a pan with plenty of very hot oil. The downside of this method is that the fish will release a strong fishy smell because its natural fat will start to burn.

Shellfish

Calamari

What is the difference between calamari and squid?

None. Italians call squid calamari. Spaniards call them calamares, and the British sometimes call them inkfish.

The only place near me that carries fresh squid sells them uncleaned. Can you tell me how to clean them?

Squid are easy to clean. First, pull off the head; most of the innards will follow. Then, work your forefinger deep into the body and pull out the quill (a hard piece of cartilage that feels like plastic) plus any remaining innards. Finally, peel off the skin (if this is difficult, hold the squid under cold running water and the skin will slip right off). Rinse the calamari well under additional cold running water and dry thoroughly with plenty of paper towels.

Is it better to buy small or large calamari?

Actually, large calamari tend to be slightly more tender and less rubbery, but size doesn't matter as much as how they are cooked.

Cook calamari either by sautéing them for no more than 30 seconds or by braising them covered for 1 hour or more to tenderize them. There is no middle way.

My market sells only frozen calamari. Is it okay to use them?

Of course fresh calamari are better, but you can get excellent results with the frozen ones. However, if you plan to freeze some for later use, make sure to ask at the fish market if they have been previously frozen (and hope the answer is truthful), since you should never refreeze seafood.

How much calamari is needed to serve four people?

It depends on whether you are buying them cleaned or uncleaned. Calamari lose about 25 percent of their weight during cleaning, so for a main course that will make use of the tentacles (the most flavorful part of the squid), I would start with 2 pounds to serve four.

Clams

Clams come in so many different varieties. Can you explain which is which?

Sizewise, the smallest clams are Pacific manila clams, followed by littlenecks and cherrystones, all of which are hard-shell varieties and plentiful. Another delicious hard-shell clam is the New Zealand cockle, which is generally only available in top-end fish markets and is pricier than either manila or littleneck. Soft-shell clams are native to the East Coast, from Long Island to Maine, where they are often called steamers. They are the most tender and, because of their scarcity, also the priciest.

Which are better—small or large clams?

Small clams are always more tender. If you like raw clams, the smaller types are best. Large clams have their place, too, though mostly in pasta dishes and soups.

What are the best clams for making clam sauce?

Cherrystone clams are a better buy than littlenecks, and since they are larger and meatier,

you can easily dice or chop them. Make sure not to overcook the clams, and remove them from the pot with a slotted spoon as soon as their shells open or they get tough and rubbery.

Is there a season when clams are fresher and less expensive?

Hard-shell clams are plentiful year-round, but I have found them to be less expensive in the summer, when they are easier to dig up.

I saw huge clams at the Pike Market in Seattle. What are they? Can you eat them?

This giant soft-shell clam is called a geoduck (pronounced "gooey-duck"). They generally weigh 3 pounds or more and have 18-inch necks protruding out of their 6-inch shells. The neck is edible when peeled and diced, but the meat is too tough to use in anything other than chowders or fritters. If you're eager to give geoduck a try, the best thing to do is to buy the flesh already cleaned and steamed.

Even after scrubbing my clams with a stiff brush, I still end up with tons of sand in my sauce. What's the best way to rid clams of sand?

A good scrubbing is the only way to remove sand on the outside of the shells, but for internal grit, the best thing to do is to soak the clams briefly in three changes of salted water, using sea salt or kosher salt.

Is there an easy way to open raw clams?

Shucking uncooked clams is tough, but if you put them in the freezer for 30 minutes, they will open slightly, allowing you to insert a clam knife or paring knife easily between the shells to pry them open.

Can I use canned or jarred clams in a pasta sauce?

You can, but the taste isn't the same. Since fresh clams are easy to come by, not that hard to clean, and quick to steam open, it's certainly not worth using canned clams. Many markets now carry chopped fresh clams, which are a good choice.

Is clam juice an acceptable substitute for fish stock?

Clam juice is not a good substitute for fish stock, no matter what some cookbooks say. It's too salty, and most bottled brands aren't very flavorful. Use a Knorr fish bouillon cube instead.

Lobsters

Which are sweeter, large or small lobsters?

Both large and small lobsters taste delicious, providing they have been cooked properly. Their size has no bearing on their taste.

When are lobsters in season? They seem to be especially expensive in the summer.

Fresh lobsters are available year-round, but spring lobsters are hard shelled, meaty, and reasonably priced, so late spring is the best time for buying and cooking lobsters. Although summer is considered lobster season, what you will find in summer are new-shell chicken lobsters that are very sweet and tasty and wonderful when steamed, but because of their high water content are unsuitable for grilling, roasting, or baking. So be extremely choosy when buying lobsters in the summer for anything other than classic steamed lobster preparations.

Which is better, a soft-shell or a hard-shell lobster?

The harder the shell, the better. In fact, give the shell a light squeeze and shake the lobster gently. If the shell rattles, it may be soft. Try to compare equal-size lobsters to determine if one feels heavier than the other. Go with the heavier lobster.

Is there a difference in taste between male and female lobsters? Is there any way to tell which is which?

The meat of male and female lobsters tastes pretty much the same. However, only the female has coral or roe. If you appreciate the coral, as I do, look for female lobsters. This is quite easy to do—just check the tiny claws under the tail. A female's claws will appear thin and feathery. In male lobsters these claws are thick and much less delicate looking. If you are still not sure, ask the fishmonger to pick female lobsters for you.

If a lobster is in a tank, does that automatically mean it's fresh, or is there something specific to look for?

In a fish market with large turnover—usually found in coastal areas—you should have no trouble. The further you get from the coast, though, the harder it is to get a truly fresh lobster, even if it is being held in a tank. Chef Jasper White, author of *Lobster at Home* (Scribner, 1998), suggests the following: First, check the antennae. If they are short or caked with an algae-like substance, the lobster has probably been in "the pound" for a long time, which affects the taste. Step back and examine the lobster in a general way. It should be lively. The claws should not droop. If you see a lobster that flaps its tail and swings its claws, buy it.

Why is it necessary to buy live lobsters?

Once a lobster is dead, its meat turns to mush, so it needs to be alive until it goes into the pot for cooking. This is because lobsters have very potent digestive enzymes that start decomposing the flesh the moment they die. The digestive organs are too complex to remove. To check whether a lobster is still alive, pick it up; if the tail curls under, the lobster is alive.

How should lobsters be stored between the market and the cooking pot?

Lobsters should be kept moist but not wet. Wrap them in damp sheets of newspaper. Be sure to keep them separate from each other. They should be stored in the coldest spot in your refrigerator, which is usually located at the rear of the bottom shelf. Store them this way for a few hours at the most. I wouldn't keep a lobster overnight. Absolutely do not store them on ice, as it kills them.

TIP A 1½-pound lobster provides 2½ to 3½ ounces of meat.

It is better to steam a lobster in 2 inches of salt water or to plunge it into lots of boiling water?

Steaming is preferable because it cooks the lobster more slowly, producing more tender meat. Also, steaming is more forgiving than boiling, meaning that it's harder to overcook your lobster that way. Most important, steam does not penetrate the shell the way water does, so lobsters that are steamed do not get watery the way

boiled ones do. To steam lobsters, fill a large pot or lobster pot with 2 inches of heavily salted water. Turn the heat to high, cover the pot, and when the water is boiling, place the lobsters in the pot and cover tightly. (The lobsters will be piled on top of each other—since the steam rises and cooks them—which is fine.) Steam for about 10 minutes per pound, or until the shells are bright red. A 1- to 1½-pound lobster will cook in 10 to 12 minutes.

What is the difference between the lobster roe and coral?

The lobster roe and coral are one and the same—the eggs of the female lobster. Green-black when raw, they turn a lovely coral shade when cooked and are absolutely delicious.

What is the tomalley? Is it edible?

Tomalley is the lobster's liver, which turns green when cooked. Lobster connoisseurs consider the tomalley a delicacy and cherish its rich flavor. It is often whipped into the melted butter that accompanies a steamed lobster and is also tasty spread on toast points. However, be extremely careful about tomalley. If a lobster has lived in contaminated water, the impurities will have been absorbed by its liver. If you are not sure about the water where a lobster was caught, you are better off discarding the tomalley.

Mussels

Which are best, large or small mussels?

Personally, I like the medium-sized mussels that are usually labeled "Mediterranean mussels," but small mussels can be extremely tasty as well. Some of the most wonderful small mussels come from Lopez Island in the state of Washington. The freshness of mussels will probably tell you more about their taste than either their size or the variety.

The mussels at my local market come packaged. Is it okay to buy them that way?

Avoid packaged mussels, unless they are in mesh bags. (Most farm-raised mussels are available in 2-pound bags.) It's hard to judge the freshness of packaged mussels, since the plastic holds them shut, and you can't really get close enough to judge them for yourself. If you have no other choice, be sure to rinse the mussels in three changes of salt water. This is called "purging," and it both rids shellfish of impurities and defines their flavor.

Is there a taste difference between farm-raised and wild mussels?

It is practically impossible to find wild mussels commercially. The only difference between one farm-raised mussel and another is the water in which they are raised. If they are in sandy beds, there is a lot of sand in the mussels. If they are raised in beds with current and water that is extremely cold, they will be cleaner and firmer. Looking at the mussel won't offer a clue, but you'll be very aware of it when you clean the mussels. And, yes, there is a great deal more taste in wild mussels than in the farm-raised varieties.

My fish man claims that open shells do not necessarily indicate bad mussels. Is he telling me the truth?

He's telling you a half-truth. If mussel shells are open but close up when you tap them, the mussels are still alive and edible. However, if they are open and don't shut when you tap them, they are no good. Unfortunately, since you will rarely find a fishmonger who will allow you to stick your hands into the fish case and tap away;

I buy only mussels with closed shells. The same goes for mussels you have stored in the refrigerator at home. If they are open, tap them to check for freshness.

What's the best way to store mussels at home? How long will they keep?

Get the mussels home as fast as you can and store them in a pan in the refrigerator covered with a damp (not dripping wet) towel to keep them cool and moist. Stored this way, mussels should last a couple of days. Remove the beards just before cooking. Simply yank them off. A tightly closed fresh mussel will not open when you tear off the beard. If a mussel does open when you tear off its beard, it's dead. Don't use it!

I recently bought mussels from a seafood truck, and they were covered with barnacles that were almost impossible to remove. How do you clean mussels?

The best way to remove barnacles is to rub one mussel against another. If this does not do the job, use a stiff brush. Definitely do not perform this task with a paring knife—you will blunt it permanently.

Is there a surefire way to get rid of the sand in mussels?

This is getting easier all the time, since most of the mussels you buy are farm-raised and therefore virtually sandless. But to be on the safe side, soak them in plenty of very cold water with 1 tablespoon flour added to it for 15 to 30 minutes. This helps disgorge the sand and also plumps them up.

Can you recommend a simple way to serve fresh mussels?

A wonderful and simple way to prepare mussels is to combine 1 cup dry white wine in a large heavy casserole with 2 tablespoons minced shallots, 2 minced garlic cloves, 2 tablespoons fresh thyme leaves or 1 teaspoon dried, 6 peppercorns, and 1 bayleaf. Add 4 to 5 pounds of well-scrubbed mussels. Cover the casserole and cook over medium heat until the mussels open. Serve in individual soup bowls with plenty of French bread.

Scallops

What's the difference between sea scallops and bay scallops? Are they interchangeable in recipes?

Sea scallops are easy to recognize, since they're more than twice the size of bays. They range from ¾ to 1¼ inches across and are about the same thickness. These mighty mollusks are slightly beige to pink in color and have a sweet aroma and a fuller flavor. Sea scallops are found on both the East and West coasts, but the Atlantic scallop is probably the best known and is available year-round. Their size makes them perfect for pan-searing, broiling, and grilling. Smaller bay scallops are also found on the East Coast, but they are seasonally available only from October through early winter. The best bay scallops—sweet and succulent—are from the New England and mid-Atlantic coasts, in the vicinity of Cape Cod and Long Island. Unfortunately, the season for these scallops is short and they are very expensive. Bay scallops are ivory to golden in color and measure about ½ inch across. They cook quickly, in 1 to 3 minutes, which makes them perfect for stir-frying, pan-searing, or marinating raw for seviche.

What are the really tiny scallops in the market year-round?

Those are calicos, which are harvested from the warmer waters of the Carolinas, Florida, and Central America. While they resemble bay scallops, they are vastly inferior. Calicos are shucked by a blast of steam before shipping, which not only destroys their flavor but also makes them prime candidates for overcooking—in which case they have the texture of pencil erasers. Sure, calicos are less expensive and plentiful, but the trade-off in taste isn't worth it.

Sometimes sea scallops are tinged orange. Is it okay to use them?

Don't be alarmed by that orange (sometimes pink) shading. It's from the roe and does not affect flavor or quality.

How can I tell if scallops are really fresh?

Trust your nose. You can identify a spoiled scallop with one whiff. Fresh scallops have a sweet, delicate odor. Any sulfur or iodine smell is a clear signal that these little guys are past their prime. If you're not able to smell them before buying, then by all means sniff before leaving the store and return them immediately if there's a problem. Also, if scallops are swimming in liquid, look opaque, or are even slightly discolored, definitely pass on them. To keep scallops looking fresh, many processors soak them in sodium tripolyphosphates (STP), which reduces water loss and prolongs shelf life. The problem, aside from the added chemicals, is that when they're cooked they release this liquid and end up steaming rather than sautéing. The key is to look for moist scallops that are firm, shiny, and translucent—not milky. If they appear too white, that's a sure sign they've been treated with STP. How to avoid this? Find a reputable fishmonger and insist on "dry," diver, or unsoaked scallops.

What is the best way to store scallops at home, and how long will they last in the refrigerator?

As with all seafood, it's best to cook scallops the day you buy them. But if that's not possible, open the plastic bag or container and rest it in a large bowl of ice. Don't place scallops directly on ice or they will become soggy and won't brown properly. Ideally, store them no more than one day after purchase.

Is it okay to use frozen scallops?

In a word, no. However, there are times when you'll be sold scallops that have been previously frozen without your knowing it. You'll discover this, though, as soon as the scallops hit the hot pan, as they'll exude a lot of juice but won't brown. When that happens, the best way to rescue dinner is to remove the scallops from the skillet as soon as they are cooked, and let the juices cook down to a glaze. Season with salt and pepper, add some minced fresh parsley and the juice of a large lemon, and spoon the sauce over the scallops. They will be delicious.

If bay scallops aren't available, can the larger sea scallops be cut into pieces?

Never halve or quarter sea scallops. Halved or quartered sea scallops will cook up tough, as the cutting process destroys their texture. Scallops can, however, be sliced crosswise.

Shrimp

Is there a difference between shrimp and prawns?

Usually restaurants and cookbooks refer to very large shrimp as prawns. To further confuse the issue, in England all shrimp, regardless of their size, are referred to as prawns. In other words, there's no real difference, other than size.

What should I look for when buying shrimp?

Good fresh shrimp should have no smell at all and not even a whiff of ammonia. The shells should be translucent and unbroken, and the shrimp should fill the shells tightly.

At different times of the year the shrimp in my market are pink, gray, orange, and even black. What's the difference in taste?

The color of shrimp has nothing to do with their flavor. Most of the shrimp we buy are farm raised and have been frozen and defrosted, so the key is not color but freshness.

How many medium shrimp are there in a pound?

There should be 35 to 40 shrimp in a pound of medium-sized shrimp. A smarter way to shop for shrimp is to know fishmongers' lingo. Fishmongers don't refer to medium and large shrimp. Instead they refer to shrimp as U-20 (which means there are about 20 in a pound), U-30 (about 30 in a pound), and U-40 (about 40 in a pound).

Are fresh shrimp worth the extra money?

Freezing has a major effect on both the taste and texture of shrimp, and fresh shrimp are undoubtedly worth seeking out. There are plenty of fresh shrimp in various parts of the country throughout the spring. You can get wonderful shrimp right out of the Gulf of Mexico on Florida's west coast, and the small Maine shrimp with their heads on that appear in markets in early spring are also succulent. In Louisiana, Texas, Georgia, North and South

Carolina, Oregon, and Washington State, fresh shrimp are more widely available. In most other parts of the country, however, the average grocery store rarely offers fresh shrimp. Consult a reputable fishmonger to find out more about what is available in your area.

What's the best method for freezing shrimp at home?

If you are lucky enough to live in a coastal area where you can purchase fresh shrimp (generally in Louisiana, Texas, Georgia, South Carolina, Oregon, Washington, Maine, and Florida), they can be frozen with excellent results. Place the fresh shrimp in water in zippered plastic bags and store them in the freezer. They will keep for about 3 months. Commercial flash-freezing is obviously better than what can be done at home, so don't expect quite the same results. Shrimp must never be frozen twice. The shrimp in almost all grocery stores has already been frozen and then defrosted before hitting the seafood case. If frozen again, they will turn tough and tasteless.

What is the best way to defrost large bags of shrimp?

If you are in a hurry, place the bag in a bowl and run cold water over it. Never use hot water, as it will begin cooking the shrimp. If you have the time, thaw them in the refrigerator.

Can I save time by buying precooked shrimp?

These tend to be dry and tasteless, and I've never understood what the advantage is supposed to be, since it takes no time at all to boil shrimp. In a large saucepan, bring salted water to a boil, add the shrimp, bring the water back to a boil, and simmer for 1 minute. Turn off the heat and let the shrimp cool in the water, about 5 minutes. That's all there is to it.

Is it really necessary to devein shrimp?

This is your lucky day—you now have permission to stop deveining shrimp forever. Yes, that little vein (actually the shrimp's intestinal tract) is not all that attractive, but it disappears when the shrimp are cooked. I never devein shrimp, except the very large blue-gray Thai prawns, which sport particularly unsightly veins.

What's a good marinade for shrimp, and how long should they marinate?

When marinating shrimp, the simpler the marinade the better. Try a mixture of lime juice, olive oil, fresh garlic, and some fresh thyme or oregano. Never marinate for more than 30 to 40 minutes, as the marinade contains an acidic component that "cooks" the shrimp and dries them out.

TIP Always use lime juice rather than lemon juice when marinating shrimp. Lime juice is sweeter and does not overpower the shrimp's delicate taste.

How do you keep rock shrimp from overcooking? They always seem to come out dry.

Cooking tiny, sweet rock shrimp is a challenge. One of the best ways to prepare them is to dip them in a tempura batter (made with 1 cup flour stirred gently into 1 cup of water or beer) and deep-fry them. This dish is very popular in the South, where it is known as popcorn shrimp. Another good way to use rock shrimp is to add them to a tomato-based pasta sauce and just heat them through (for a minute or two) so they don't get tough.

Pan-Seared Arctic Char with Lemon-Chive Essence

Arctic char used to be called salmon trout. It is a mild and delicious fish that takes well to creamy sauces and mild herbs such as dill and tarragon. I like to serve the char with simply boiled new potatoes tossed in a little butter, and some glazed cucumbers.

SERVES 4

LEMON-CHIVE ESSENCE

4 tablespoons unsalted butter

1 cup crème fraîche or heavy cream

½ Knorr fish bouillon cube

3 tablespoons fresh lemon juice

1 to 2 teaspoons finely grated lemon zest

1 tablespoon flour mixed into a paste with
   1 tablespoon softened butter

2 tablespoons minced fresh chives

2 tablespoons minced fresh dill or tarragon

Salt and freshly ground black pepper

FISH

4 fillets of Arctic char (6 to 7 ounces each),
   with skin on

Coarse salt and freshly ground black pepper

2 tablespoons clarified butter (see page 384)

2 large plum tomatoes, peeled, seeded, and
   finely diced

Glazed cucumbers

Minced chives

1. Make the lemon-chive essence: In a large heavy skillet, melt the butter over medium heat. Add the crème fraîche or heavy cream, bouillon cube half, and lemon juice and zest; bring to a simmer, and reduce slightly. Whisk in bits of the blended flour and butter until the sauce lightly coats the spoon. Add the chives, and dill or tarragon; season with salt and pepper, and keep warm.

2. Prepare the fish: Dry the fillets thoroughly with paper towels. Season with salt and pepper.

3. Heat the clarified butter in 2 large, nonstick skillets over high heat. Place 2 fillets in each skillet, skin side down, and cover loosely with foil. Cook for 2 minutes, slightly lower the heat, and cook for an additional 4 to 5 minutes without turning. You may turn the fillets once and cook for about 30 seconds if you want the tops to be browned.

4. Transfer one fillet to each of 4 individual serving plates, skin side up. Spoon some of the sauce around each portion and garnish the sauce with the diced tomatoes and glazed cucumbers. Sprinkle with the chives and serve at once.

Brook Trout en Papillote with Asparagus, Snow Peas, and Dill Butter

Here is a lovely way to prepare brook trout. The mild taste of the fish marries beautifully with asparagus and snow peas. Oven-baked with a fair amount of dill and chive butter, it becomes a delicately flavored dish that needs little in terms of accompaniment. A few boiled new potatoes tossed in parsley and a touch of butter make the perfect side dish.

I usually serve this dish right out of the foil; if you prefer a more elegant presentation, you can fillet the fish in the kitchen and transfer the fillets and vegetables to a warm platter or individual plates.

SERVES 4

2 cups finely sliced asparagus

2 cups fine julienne of snow peas

6 tablespoons butter, at room temperature

2 tablespoons minced fresh dill plus
  2 medium-sized dill sprigs

3 tablespoons minced fresh chives

Coarse salt and freshly ground black pepper

2 brook trout (about 1½ pounds each)

2 slices lemon, each cut in half

1. Preheat the oven to 350°F. In a saucepan, bring lightly salted water to a boil. Add the asparagus and snow peas. Return to a boil and cook for 50 seconds. Drain and immediately run under cold water; drain well.

2. Combine the butter, minced dill, and chives in a bowl and mash with a fork until well blended. Season with a sprinkling of salt and pepper and blend thoroughly.

3. Place each fish on a piece of heavy-duty foil. Place 1 sprig of dill and 2 pieces of lemon in each cavity. Season each fish well with salt and pepper and sprinkle with some of the asparagus and snow pea mixture, as well as placing some under each fish.

4. Divide the dill butter between the 2 fish. Cover with another piece of foil, crimping the foil to firmly enclose the fish. Place on a baking sheet and bake for 20 minutes. Remove from the oven and slide onto a large platter. With sharp scissors, cut open the foil and divide the fish and vegetables among 4 plates. Garnish with sprigs of dill.

Stir-Fry of Calamari, Red Peppers, and Lemon

You can get this zesty calamari dish in numerous tapa bars all over the Catalan coast and in many other parts of Spain. The tentacles are the most flavorful part of the calamari; if you can, be sure to include them. Remember that calamari must be cooked for only 30 to 50 seconds, or else they get tough. You can also make this dish with small whole peeled or unpeeled shrimp. Just cook them for an extra minute and serve with plenty of crusty bread.

SERVES 4

1 small lemon

3 to 4 tablespoons extra virgin olive oil

1 small dried red chile, broken into pieces

1 large red bell pepper, seeded and cut into
$\frac{3}{4}$-inch cubes

1 pound small calamari, cleaned and cut
crosswise into $\frac{1}{2}$-inch slices

Coarse salt and freshly ground black pepper

2 tablespoons tiny fresh cilantro leaves

1. Finely slice the lemon. Cut each slice into quarters.

2. In a large heavy skillet, heat 2 tablespoons of the olive oil over high heat until almost smoking. Add the chile and half the bell pepper and cook for 1 to 2 minutes, or until barely tender.

3. Add half the calamari and stir-fry for 30 seconds, or until just opaque. Immediately add half of the lemon slices and toss for 30 seconds longer, or until the lemon is just heated through. Transfer the calamari mixture to a serving dish.

4. Cook the remaining calamari and lemon in the remaining oil in the same manner. Transfer to the serving dish and season with salt and black pepper. Sprinkle with the cilantro leaves and serve immediately.

Fillets of Lemon Sole in a Lime, Jalapeño, and Caper Butter

SERVES 4

2 large fillets of lemon sole (about 2 pounds)

Coarse salt and freshly ground black pepper

Flour, for dredging

$\frac{1}{2}$ cup chicken stock or bouillon

10 tablespoons butter

2 tablespoons lime juice

2 teaspoons grated lime zest

1 tablespoon minced jalapeño pepper

1 tablespoon capers, well drained

2 teaspoons grapeseed or canola oil

Sprigs of flat-leaf parsley or dill

1. Cut the sole fillets into 4 even pieces. Season with salt and pepper. Dredge lightly in flour and set aside.

2. In a small heavy saucepan, heat the stock or bouillon. Cook over medium heat until slightly reduced. Reduce the heat and start adding 8 tablespoons of the butter a little at a time, whisking constantly until all the butter has been added and the mixture is well emulsified. Remove from the heat and add the lime juice, lime zest, jalapeño, and capers. Whisk until the sauce is well blended. Season lightly with salt and pepper and correct the seasoning.

3. In a large nonstick skillet, heat the remaining 2 tablespoons butter and the oil. Add the fillets and sauté over medium-high heat for 3 minutes. Carefully turn the fillets with a fish spatula and sauté for another 2 or 3 minutes. Do not overcook. Carefully transfer the fillets to individual plates or a serving plate and spoon the butter sauce over each fillet.

4. Garnish with the parsley or dill and serve.

Oven-Roasted Mackerel with Caper, Lemon, and Anchovy Butter

Boston mackerel is an East Coast spring delicacy. Because of its oiliness, it must be purchased absolutely fresh. Check the skin and gills. The skin should be silvery and shiny, the gills bright red. If you cannot get Boston mackerel, you can substitute smelts or fresh sardines in this simple but tasty preparation. These, however, will have to be pan-fried rather than baked.

SERVES 2 TO 4

8 tablespoons unsalted butter (1 stick)

2 tablespoons tiny nonpareil capers, drained

2 anchovy fillets, minced

2 tablespoons minced flat-leaf parsley

1 large garlic clove, crushed

Juice of 1 lemon

1 tablespoon grated lemon zest

4 Boston mackerel (1 to 1½ pounds each), cleaned, with heads left on

Coarse salt and freshly ground black pepper

2 tablespoons olive oil

2 lemons, quartered

1. Preheat the oven to 400°F.

2. In a small heavy saucepan, heat the butter over low heat until melted and just starting to brown. Immediately remove from the heat and add the capers, anchovies, parsley, garlic, and lemon juice and lemon zest.

3. Season the mackerel with salt and pepper and brush well with olive oil.

4. Place in a well-oiled roasting pan or baking dish, cover, and bake for 15 minutes. Remove from the oven.

5. Reheat the herb and anchovy butter until it just starts to bubble; do not let it come to a boil. Immediately pour the butter over the mackerel and transfer the fish carefully to individual serving plates. Garnish with the quartered lemons.

Steamed Mussels in Basil-Garlic Cream

Now that mussels are being farmed in several regions of the country, they are becoming more widely available. Unfortunately, farm-raised mussels lack the wonderful briny taste of their wild cousins and need plenty of flavoring. Make sure that the mussels are tightly closed when you buy them, and if possible ask to see the tag that indicates when the mussels were harvested and the sell-by date.

SERVES 4 TO 5

1 cup tightly packed fresh basil leaves

2 large garlic cloves, mashed

3 to 5 tablespoons olive oil

2 tablespoons unsalted butter plus 2 to 4 tablespoons unsalted butter, for enrichment, optional

$\frac{1}{4}$ cup minced shallots

$\frac{1}{2}$ cup dry white wine

1 large sprig of fresh thyme

1 sprig of fresh parsley

5 to 6 pounds fresh mussels, well scrubbed

1 cup crème fraîche

1 tablespoon flour mixed into a paste with 1 tablespoon softened butter

Freshly ground black pepper

1. In a blender or food processor, process the basil and garlic with enough olive oil to make a smooth paste.

2. Melt the 2 tablespoons butter in a large heavy saucepan over low heat, add the shallots, and cook for 1 minute without browning. Add the wine, herbs, and mussels and cook, covered, shaking the pan back and forth, until all the mussels open; discard any that do not. Transfer them with a slotted spoon to a bowl.

3. Strain the broth through a fine sieve. Return to the pan together with the crème fraîche and reduce by a third. Whisk in bits of the blended flour and butter until the sauce coats the spoon. Reduce the heat to low and whisk in the basil paste and butter for enrichment. Correct the seasoning by adding a large grinding of black pepper and keep warm.

4. Transfer the mussels to individual serving bowls, spoon some of the sauce over each portion, and serve at once.

Pan-Seared Red Snapper Fillets with Sweet and Spicy Asian Syrup

The first time I tasted this sauce in a San Francisco restaurant, I was intrigued by the interesting flavor combination and could not wait to make it. It has since become one of my favorite ways to serve a variety of fish—in particular, snapper and sea bass. The sauce can be made as much as a week ahead of time. I often double and triple it and keep it on hand for grilled tuna or salmon. Stir-fried snow peas or bok choy are both excellent accompaniments.

SERVES 4

ASIAN SYRUP

1 cup plus 2 tablespoons sake

1 cup plus 2 tablespoons mirin

2 tablespoons finely sliced peeled fresh ginger

2 large garlic cloves, very finely sliced

2 teaspoons finely sliced jalapeño pepper

2 tablespoons light soy sauce

¼ cup very finely diced red bell pepper

¼ cup very finely diced yellow bell pepper

2 tablespoons very finely diced zucchini, skin only

¼ cup corn kernels, optional

FISH

4 red snapper fillets (6 to 7 ounces each), skin on

Coarse salt and freshly ground black pepper

2 tablespoons peanut or canola oil

Sprigs of fresh cilantro

1. Make the Asian Syrup: In a heavy 2-quart saucepan, combine the sake, mirin, ginger, garlic, and jalapeño, bring to a simmer, and reduce until syrupy; this will take about 15 minutes. You should have about ⅔ cup syrup. Strain the syrup into a smaller saucepan. Add the soy sauce and vegetables and simmer for 2 or 3 minutes, or until tender. Keep warm.

2. Season the snapper fillets with salt and pepper. In a large nonstick pan, heat the oil. Add the snapper fillets skin side down and sauté for 3 minutes until nicely brown and crisp. Turn and cook for another minute. Do not overcook.

3. Transfer the fillets to a serving plate and spoon the sauce over them. Garnish with the cilantro and serve immediately.

Pan-Seared Skate with Red Onion, Tomato Jam, and Brown Caper Butter

SERVES 4

2 to 3 tablespoons olive oil

1 large red onion, cut in half and finely sliced

1 tablespoon minced fresh rosemary

Pinch of salt

1 teaspoon sugar

15 grape tomatoes, cut in half

1 large skate wing (about 2 pounds), or
    2 small skate wings, boned and cut into fillets

Freshly ground black pepper

Flour, for dredging

8 tablespoons butter (1 stick)

1 tablespoon grapeseed oil

Juice of ½ lemon or more to taste

2 tablespoons capers, well drained

2 tablespoons minced flat-leaf parsley

1. In a heavy skillet, heat the olive oil and add the onion and rosemary. Season with salt and sugar and sauté over high heat for 1 to 2 minutes. Lower the heat. Cover the skillet and continue to simmer the onion for 15 to 25 minutes, watching carefully so that it does not burn. You may have to add a little more oil.

2. Add the tomatoes and simmer, uncovered, for another 3 minutes. Set aside.

3. Season the fish with salt and pepper. Sprinkle the fillets lightly with flour.

4. Heat 1 tablespoon of the butter and half the grapeseed oil in each of two nonstick skillets. Add the fillets and sauté over fairly high heat for 3 minutes on each side. Transfer the fillets to a warm serving platter and set aside.

5. In a small saucepan, heat the remaining 6 tablespoons butter and cook until the butter turns a deep hazelnut brown. Add the lemon juice and capers and pour the butter over the fillets.

6. Reheat the onion jam and place large spoonfuls on individual plates. Top with a skate fillet, garnish with parsley, and serve immediately.

Roasted Sea Bass with a Fricassee of Fennel and Potatoes

Black sea bass is America's best fish. I am sure some would argue otherwise, but to me, nothing comes close to the sweet taste of this wonderful fish. Roasted whole on a bed of potatoes and fennel, it becomes a splendid one-dish meal. If you cannot get sea bass, try a whole red snapper, which will also be delicious.

SERVES 2 TO 3

6 tablespoons fruity olive oil

4 medium Yukon Gold potatoes, peeled and finely sliced

Coarse salt and freshly ground black pepper

2 medium fennel bulbs, finely sliced, tops removed and reserved

2 medium onions, finely sliced

1 teaspoon fennel seeds

1½ to 3 pounds sea bass or red snapper

1½ lemon, cut into wedges

1. Preheat the oven to 350°F.

2. In a large baking dish, heat 3 tablespoons of the olive oil. Add a layer of potatoes, season with salt and pepper, top with a layer of sliced fennel and onions, season with salt and pepper, and sprinkle with fennel seeds. Top with another layer of potatoes, fennel, and onions. Drizzle with a little oil, cover the dish with foil, and bake for 20 to 25 minutes, or until the vegetables are just tender.

3. While the vegetables are in the oven, season the fish with salt and pepper. Fill the cavity with some of the lemon wedges and some of the green fennel tops. Remove the vegetables from the oven, brush the fish with olive oil, and set the fish on top of the vegetables. Roast, uncovered, for 30 to 35 minutes, basting every 10 minutes with the oil that has accumulated in the dish.

4. Test the fish for doneness by inserting the tip of a sharp paring knife into the center. If the knife is warm, the fish is done.

5. Carefully transfer the fish onto a serving platter, garnish with more green fennel tops and lemon wedges, and serve the vegetables right out of the baking dish.

REMARKS A 2½- to 3-pound fish will serve only two to three people, so it is likely that you will have some vegetables left over. These are delicious reheated and served as a side to pan-seared salmon, shrimp, or scallops.

Sauté of Swordfish with Shrimp, Snow Peas, and Three Bell Peppers

Here is a quick stir-fry that only takes minutes to do once you have prepared the snow peas and peppers. I like to serve this dish with a side of steamed jasmine rice flavored with toasted pine nuts or a cumin- and scallion-scented couscous. Be sure not to overcook the fish. Swordfish is at its best when cooked somewhat rare.

SERVES 4

2 tablespoons soy sauce

2 tablespoons mirin

1 garlic clove, mashed

3 tablespoons grapeseed or canola oil

¾ pound large shrimp, peeled and cut in half

Salt and coarsely ground black pepper

1 pound swordfish, cut into 1-inch cubes, skin removed

¾ cup finely diced bell peppers (red, yellow, and green mixed)

2 cups snow peas cut into fine julienne

2 tablespoons dry sherry

2 tablespoons minced cilantro or mint, optional

1. In a small bowl, combine the soy sauce, mirin, and garlic and whisk well.

2. In a large heavy skillet or flat-bottomed wok, heat half the oil. Season the shrimp with salt and pepper and add to the hot oil. Cook for 1 minute or until the shrimp just turn a bright pink. Remove with a slotted spoon to a dish.

3. Add the remaining oil to the wok and when it's very hot, add the swordfish cubes. Season with salt and pepper and sauté over high heat for 2 minutes, turning the fish cubes to brown them evenly.

4. Add the bell peppers and snow peas and sauté for another minute, or until the snow peas are just tender. Return the shrimp to the pan together with the sherry and cook for 30 seconds. Add the soy mixture, bring to a simmer, and immediately transfer the swordfish and shrimp to a serving bowl.

5. Garnish with cilantro or mint and serve immediately.

Sautéed Scallops with Red Bell Pepper and Chili Sauce

Here is a quick flavor-packed sauce that goes well with grilled fish, steaks, scallops, and shrimp. I often make it with leftover roasted red, yellow, or orange bell peppers. The sauce will keep well for a week, so you can double and triple the recipe.

Since the best scallops appear in the market in the fall, I usually serve this dish accompanied by Fricassee of Roasted Eggplants and Two Mushrooms, page 124, or simply boiled and buttered new potatoes.

SERVES 4

3 tablespoons unsalted butter

1 small shallot, minced

2 garlic cloves, minced

1 large roasted red bell pepper, seeded and finely cubed

¾ cup chicken stock or fish bouillon

Salt and freshly ground black pepper

1 teaspoon Thai chili sauce or a few drops of Tabasco

2 tablespoons minced chives

1 pound large sea scallops

½ tablespoon canola oil

Whole chives and sprigs of flat-leaf parsley

1. In a small skillet, heat half the butter. Add the shallot and garlic and cook for 2 or 3 minutes or until soft but not browned. Add the bell pepper and stock. Season with salt and pepper, bring to a simmer, and cook for 5 minutes.

2. Transfer the mixture to a blender and puree until smooth. Add the chili sauce or Tabasco. The sauce should be rather spicy. Transfer the sauce to a bowl. Add the chives.

3. Season the scallops with salt and pepper. In a large non-stick skillet, heat the remaining butter and the oil. Add the scallops and sauté over high heat until they are nicely browned on both sides. Lower the heat and cook for another 2 minutes, or until the scallops are just done. Do not overcook. Transfer the scallops to a dish.

4. Place a dollop of sauce on individual serving plates. Top with 3 to 4 scallops and garnish with the chives and parsley.

Spicy Shrimp and Ginger Fritters with a Ginger-Tamari Dipping Sauce

I am always on the lookout for shellfish recipes, especially those for shrimp, which are so easily available these days. Here is a quick and delicious shrimp fritter that I like to serve as an hors d'oeuvre as well as a topping for the Avocado Salad, page 109.

SERVES 6

DIPPING SAUCE

3 tablespoons tamari or light soy sauce

Juice of 2 limes

1 large garlic clove, mashed

1 teaspoon freshly grated ginger

6 tablespoons extra virgin olive oil

3 tablespoons minced scallions

Freshly ground white pepper

FRITTERS

1 pound medium shrimp, peeled and cubed

¼ cup minced shallots

¼ cup minced scallions

Juice of 1 large lemon

1 tablespoon finely minced jalapeño pepper

2 tablespoons minced peeled fresh ginger

Coarse salt and freshly ground white pepper

Peanut oil, for frying

1. Make the dipping sauce: Combine the tamari or thin soy sauce, lime juice, garlic, and ginger in a small bowl. Whisk in the oil, slowly, until emulsified. Add the scallions, season with pepper, and whisk until well blended. Set aside.

2. Make the fritters: Combine the shrimp, shallots, scallions, lemon juice, jalapeño and ginger in a food processor and puree until smooth. Season highly with salt and pepper.

3. Heat ½ inch peanut oil in a nonstick skillet over medium-high heat. Shape the shrimp mixture with wet hands into 1½-inch disks and fry in batches for about 1 minute, or until nicely browned on both sides. Season with a little salt and serve as a topping for salads or as an hors d'oeuvre with the dipping sauce.

Fillets of Salmon in Teriyaki Marinade with Gingered Carrots

Teriyaki marinade seems to be the marinade of the moment, and it deserves to be popular because it adds such a nice flavor to chicken, beef, and especially salmon. I like to serve this simple dish with carrots or parsnips or butternut squash, which can be prepared the same way as the carrots. A side bowl of jasmine or basmati rice will complement the dish nicely.

SERVES 4

$\frac{1}{3}$ cup light soy sauce

2 tablespoons mirin

3 tablespoons rice wine vinegar

2 tablespoons brown sugar

1 large garlic clove, crushed

4 salmon fillets (about 7 ounces each)

3 tablespoons unsalted butter

4 large carrots, peeled and cut on the diagonal
   into thick slices

Salt and freshly ground black pepper

4 fine slices peeled fresh ginger plus
   2 teaspoons freshly grated

$\frac{1}{3}$ cup fish bouillon or water

Sprigs of fresh cilantro

1 avocado, finely sliced, optional

1. In a bowl, combine the soy sauce, mirin, rice vinegar, 1 tablespoon of the brown sugar, and the garlic. Whisk the mixture until well combined. Place the salmon fillets in a zippered plastic bag, add the marinade, and refrigerate for 2 to 4 hours.

2. Forty-five minutes before serving, preheat the oven to 400°F.

3. In a heavy cast-iron skillet, heat the butter. Add the carrots and season with salt and pepper. Sauté the carrots over medium heat until they are lightly browned.

4. Add the remaining 1 tablespoon sugar and the ginger and toss the carrots until they are nicely glazed. Add the bouillon or water. Cover the skillet and place in the oven. Roast the carrots for 10 to 12 minutes, or until tender and nicely glazed.

Remove from the oven and correct the seasoning with salt and pepper.

5. Turn on the broiler. Remove the salmon fillets from the marinade, place in an ovenproof dish, and broil 4 inches from the heat source for 4 to 6 minutes.

6. Place a mound of carrots on individual serving plates, top with a salmon fillet, and garnish with cilantro and avocado. Serve immediately.

Tuna Tartare

Tuna tartare is amazingly easy to make, provided, of course, that you get impeccably fresh tuna. Because I like it so much, I continue to order it in restaurants and have tried many versions. Right now I like this one the best. Don't be daunted by the long ingredient list. All the items except the tuna are pantry basics that you should have on hand at all times. Deep-fried wonton skins make a terrific accompaniment. Place a large spoonful of the tartare between two skins and serve, accompanied by marinated cucumbers or a simple sprig of cilantro.

SERVES 4 TO 5

1 pound very fresh tuna

2 to 3 cornichons, minced

1 tablespoon minced jalapeño pepper

2 tablespoons capers or more to taste

1 small shallot, minced

1 teaspoon finely grated lemon zest

Coarse salt and freshly ground black pepper

2 tablespoons sesame oil

1 teaspoon Thai chili sauce or Chinese Hot oil

2 tablespoons mayonnaise

Finely sliced seedless cucumber, left unpeeled, and small sprigs of dill or cilantro

1. Finely slice the tuna with a very sharp knife. Cut each slice into ¼-inch dice. Add the cornichons, jalapeño, capers, shallot, and lemon zest. Toss gently and season with salt and pepper. Add the sesame oil and chili sauce and toss again. Correct the seasoning. The tartare should be slightly spicy.

2. Fold in the mayonnaise and correct the seasoning. Serve lightly chilled but not cold. Garnish with the cucumber and herbs.

REMARKS Tuna tartare is at its best served within an hour. Although it will change color, it will keep well for several hours.

■ Broiled Chicken Legs with Herb Butter ■ Bouillabaisse of Chicken ■ Slow-Braised Chicken Breasts with Peas and Lemon-Dill Sauce ■ Provençal Chicken with a Sherry Vinegar–Garlic Essence and Concassé of Tomatoes ■ Ragout of Chicken Legs with Fennel Sausage ■ Roasted Chicken à la Flamande ■ Fried Cornish Hens "al Ajillo" ■ Roast Duck with Sautéed Pears ■ Fried Marinated Quail ■ Grilled Quail with Pineapple Caramel ■ Perfectly Roasted Turkey with a Mustard-Herb Rub

When I was growing up, chicken was a bird that was admired and treated with great respect. The purchasing of a chicken was taken very seriously; its cooking possibilities even more so. Both my mother and grandmother would only buy their chickens from one "chicken lady," Mercedes, at the Barcelona market. At home, we all gathered around to admire the bird's full breast and lovely shape. Settling on a cooking method took time. Would it be better poached or roasted? And how should we flavor it? With herbs, or perhaps just a touch of wine, or the simple way—with some good stock and a generous amount of butter? Some birds begged to be braised with root vegetables and whole cloves of garlic. Others were clearly born to be roasted.

And the discussion didn't end once the chicken was cooked. Was this chicken as good as the one from the week before? Mercedes had seemed slightly less friendly this time around, and maybe that was why her chicken wasn't as good. Should we—heaven forbid—take the unthinkable step of switching poultry sources? That would have meant buying from Mercedes' sister-in-law, who sold equally fine-looking chickens down at the other end of the market. This conversation would last for the length of the meal, yet when we were done, we'd agree that the chicken had been excellent.

My mother and grandmother would remain loyal to Mercedes, and we'd begin, almost immediately, planning our next meal.

If chicken was a religion in my family, then France's Bresse region—home of the famous Poulet de Bresse—was our cathedral. Five times each year, we'd make a pilgrimage from our home in Barcelona to Bresse, where we would always dine at L'Auberge Blanc. I still have vivid memories of the cozy dining room, decorated with large copper casseroles filled with fresh flowers and the flecks of sunlight that filtered through the lace curtains. Reservations for Sunday lunch were a must and had to be made months in advance, and the anticipation made the meal even more wonderful. Everyone there always seemed to be in a celebratory mood, all eager to sample the special bird after miles of travel, and we were never disappointed.

Here, the chicken could be ordered either poached *en garniture*, which included various seasonal vegetables simmered in flavorful broth, or perfectly roasted with homefried potatoes. Occasionally Mme. Blanc would also offer her trademark chicken *en cocotte*, a casserole-braised bird served with plenty of sweet garlic cloves, pearl onions, and cubed bacon.

When I moved to the United States in the seventies, the entire chicken experience changed. I still recall my first American supermarket outing. I stood in awe before a long poultry counter, studying whole chickens, cut-up chickens, family packs of wings, thighs, drumsticks, and even gizzards. Chicken livers by the pound, and skinless, boneless breasts were new to me. The choice was amazing; you could get as much chicken as you wanted, and in any cut or shape or size. I had visions of a succulent roasted bird and richly flavored stock.

I went into a chicken-buying frenzy and soon found myself up to my elbows in bones and wings, not to mention carrots, leeks, celery, and parsley. I cooked all day, with much anticipation, so what went wrong? Why was my stock greasy and flavorless? Why did the birds refuse to brown properly? And why were the livers watery and mushy? More was definitely not better, I learned. American chickens were not the same as their European counterparts. My quest for a flavorful bird was about to begin.

I sought out old-fashioned New York butchers, including several well-known kosher butchers. I trolled through Chinatown on a tip, seeking a fresh chicken from an unnamed storefront there. I tried to improve my luck through cooking methods as well, experimenting with marinades and spice rubs, but the results never rose to the heights of a simple Barcelona market chicken.

I soon learned why. A little research revealed that American chickens were almost all mass produced, raised on a diet that left them fatty and bland. And because of regulations, they were—and still are—soaked in water at length before being brought to market. Also, chickens that arrived "fresh" in grocery stores were actually partially frozen; hence their watery consistency and failure to brown. I almost despaired, but how could I give up on something that had been part of my repertoire for so long and that I knew was open to so many different preparations? And chicken represented something else for me—it represented family, and those leisurely meals and in-depth discussions around the table. So I kept on trying, never giving up on my favorite food.

These days, things are looking up—way up. Poultry farming has changed dramatically, primarily because of the newfound popularity of roasted chicken. As more and more chefs are searching for real flavor in poultry, the public benefits in turn. Organic chicken farming has had spectacular growth, and small producers around the country have found that there is a sizeable market for their free-range birds, despite their higher cost. While most supermarkets continue to sell mass-produced poultry, they are responding to consumer demand for other options. Recently, I walked into a Los Angeles supermarket and found four different organic brands available, two of them free-range as well. What this means is that we as consumers are now in a position to select the optimal bird for a juicy roast, a flavorful sauté, or a wonderful, rich stock. Never have our chicken options been better in this country than they are right now. In the more than twenty-five years since I stood in front of that supermarket counter, chicken has come a very long way.

If you can't locate top-notch chicken in your area, be sure to tell your supermarket manager or local butcher that you, like so many Americans, would be willing to pay a higher price for quality. While our birds do not quite compete with their European counterparts as yet, I firmly believe that before long they will, and eventually Americans will be able to enjoy the kind of bird that has made chicken one of the favorite foods of millions of people around the world.

Is there anything special I should look for when buying a chicken?

I think that you can learn to pick out a really good chicken. A fresh chicken should look plump and rosy. I would avoid buying chicken that is already wrapped, but if you have no other choice, check the package carefully and make sure there is no blood in it and the flesh of the chicken is not bruised or torn. After a while you will be able to tell a good fresh bird every time.

I used to go to a seamstress on the Lower East Side of New York who always seemed to be cooking the best chickens. As soon as I got off the elevator, I would get a whiff of her rich chicken soup or the aroma of a buttery roasting bird. She would treat me to a deliciously tender gizzard or the crunchy tip of a wing, and I begged her to introduce me to her butcher, whom she had been patronizing for fifty years.

At the butcher's, she carefully inspected several chickens until she found one that met with her approval. I roasted the chicken simply, and it was without a doubt the best one I've ever had. Of course, I kept returning to that butcher, and yes, the chickens I got from him were good, but somehow never as good as the one my seamstress had picked out for me that day.

What is a free-range chicken?

In Europe, a free-range chicken is literally a chicken that runs freely and finds its own food. While it's working for its food, the chicken develops some muscle rather than fat. But most free-range chickens here are raised differently. Although they're allowed to go out of their coops, their food is kept in the same place every day and is given to them. So American free-range chickens develop a little more muscle than those that are force-fed around the clock, but I have not found there to be a big difference in taste.

My market carries organic, free-range chicken as well as branded poultry from Perdue and Tyson. Is it worth the difference in price to buy organic or free-range?

It's all a matter of taste and testing. Once when I was in California teaching some classes, I did an experiment with some students. We went to the grocery store and scooped up one of every kind of chicken they carried. There were three brands of organic chicken (remember, this was California), two free-range chickens, and the usual branded chicken as well. We took them back to the school and roasted them all the same way, after which we had a tasting. Most students preferred the organic chickens, while others thought the free-range one was more intensely flavored.

It is important that you try several types of chickens in your area. Organic chickens must be raised in qualifying conditions. They can be vaccinated but not medicated; they must be fed organically and have access to the outdoors.

What is the difference between white- and yellow-skinned chickens?

Skin color is an indicator of feed and not of quality. I find yellow-skinned chickens to be very fatty and try to avoid them.

A lot of food magazines recommend using kosher chickens these days. What makes them better?

Kosher chickens are killed under rabbinical supervision with a special knife. They are then inspected for any defects or scars, any of which cause them to be rejected. After that, they are salted, which makes their texture somewhat firmer. Because of dietary laws, they have to be sold within 3 days of slaughtering, making them fresher than other chickens.

TIP When buying a kosher chicken, you can check the feet. They should be pearly white, an indication of freshness.

What is the difference between a broiler, fryer, and roaster? Is one better than the others?

These names have nothing to do with quality: they refer only to size. Broilers and fryers are pretty much the same size and weigh between 2½ and 3½ pounds, while roasters are larger and weigh between 4½ and 6½ pounds. To confuse matters further, these categories are different on the East and West coasts. On the West Coast, broilers and fryers weight 4½ to 5 pounds, or about a pound more than on the East Coast.

TIP If your chicken comes with a nice-looking liver, don't throw it away! It is a delicacy. Instead, combine it with milk to cover in a small bowl and let it sit for 30 minutes. Pat dry and sauté with a finely sliced onion in a little butter.

Is it better to buy chicken breasts on the bone or boneless?

Buying chicken breasts on the bone makes great sense because it gives you the extra bonus of having bones for stock. These I keep in a tightly sealed bag in my freezer. Once I have about 3 pounds of bones, I combine them with carrots, leeks, some celery, and parsley to make a quick stock.

To make quick chicken stock

MAKES 6 CUPS

2 to 3 pounds breast bones (bones from approximately 4 breasts)

1 chicken bouillon cube

1 carrot

2 celery stalks with tops

1 leek with 2 inches of greens attached, white part only cut in half lengthwise

1 medium onion, unpeeled

3 to 4 sprigs of flat-leaf parsley

Pinch of salt

Combine all ingredients with 8 cups water in a 6-quart pot. Bring to a boil, reduce the heat, and simmer for 30 to 45 minutes. Cool and strain. Store the chicken stock in either the freezer or the refrigerator. Bring the stock back to a boil every 2 days if you store it in the refrigerator.

TIP It is very important to skim a stock in the very early stages of cooking, because the impurities from the bones rise to the surface as soon as the stock comes to a boil. If the stock is not skimmed, it will remain cloudy and greasy. Also, a stock should not boil, but rather simmer. Bring it to a brisk simmer, reduce the heat, cover it partially, and then simmer gently. This prevents the stock from being too greasy and makes it easier to degrease it once it cools.

Can the bones from a roasted turkey or chicken be used to make stock?

I have never been happy using cooked bones to make stock. The initial roasting zaps most of the flavor from the bones. For a truly flavorful stock, I much prefer to begin with raw bones.

However, if you are roasting a turkey, it's a shame to throw away that entire carcass. Use it to make turkey stock, but boost the flavor with plenty of leek greens, carrots, and celery. Also, if possible, add some uncooked chicken wings, which are inexpensive and easy to find in every supermarket.

Can I use chicken stock for every kind of soup?

Chicken stock works for every kind of vegetable soup with two exceptions: You should not use it for traditional French onion soup, nor should you use it for classic borscht. Both of these soups are usually made with beef stock. The good news is that I have used chicken stock for seafood soups as well. So for me, it's the most important stock to have on hand.

How do you bone a chicken breast?

The best way to do this is with a boning knife, but you can also use a very sharp 4-inch paring knife. Place the chicken breast in front of you skin side up. Make a sharp slit in the center of the breast and keep your knife very close to the breast bone. Continue cutting the meat away from the bone and keep the knife as parallel to the bone as you possibly can. When most of the meat is detached, make a sharp cut lengthwise, completely detaching the breast meat from the bone. Remove the skin if you like and trim off any bits of fat or tendon.

TIP Once a breast is split, it's much harder to bone, so be sure to buy the whole breast.

If I can't cook a chicken right away, how long can I store it?

Fresh chicken will keep for up to 3 days. Be sure not to leave it in a package or a plastic bag, but rather rinse it lightly and wrap it in a kitchen towel. Do not use paper towels. I have noticed that many organic chickens now have a "cook-by" date, which is a good indication of freshness. Sometimes a "fresh" chicken will have a slight odor when removed from its package. If this does not go away after it is rinsed, be sure to return it to the store.

We hear so much about salmonella these days. What safety measures do I need to take when preparing chicken?

I find that following simple commonsense guidelines is all you need to do. I always rinse the chicken and dry it thoroughly. I also wash my cutting board, knife, and anything else that has come into contact with the raw chicken or its juices in very hot water. However, I do not believe in cooking chicken to death and find that all is well at an inner temperature of 165° to 170°F. Remember to refrigerate all chicken, even when marinating.

What exactly does brining do to a chicken?

Brining is a hot new trend, but it has pluses and minuses. The salt in the brine permeates the chicken so that the meat is evenly seasoned. Brining fans claim that this gives a chicken fuller flavor and firmer texture and keeps the breast meat moist and tender, but it can be difficult to control the level of saltiness. Personally, I like my chickens the old-fashioned way. If you're undecided, try preparing one brined chicken and one non-brined chicken—roasting both—and decide which you prefer. For the brine use ¼ cup kosher salt, ¼ cup sugar, and 2 quarts water. Whisk the mixture to blend thoroughly, then transfer to a zippered plastic bag together with the chicken. Refrigerate for 24 hours. Rinse the brine off and dry the chicken thoroughly with a cotton towel before roasting.

How can I make a roasted chicken that is both crisp and still juicy?

The problem starts with processing of chickens in the United States. American plants use the water-chilling process, in which chicken is dipped in an ice-cold chlorine and water solution to quickly decrease the body temperature and kill bacteria. This is why American chickens get to the market with very wet skins. It is therefore important to dry the chicken thoroughly with a kitchen towel before roasting so you will get a crisp skin.

Also try roasting smaller chickens. Instead of one 5- to 6-pound bird, choose two 2½- to 3½-pound chickens. Yes, it's more work, but the results are worthwhile.

Once the chicken is well seasoned, I roast it at 425°F in a heavy roasting pan without a rack and baste it every 10 to 15 minutes with the pan juices and some broth. You can also use a method that requires less basting: Roast the chicken for 25 minutes on each side, finishing with 10 to 15 minutes breast side up. Roasting a chicken breast side up the entire time will brown it more evenly, but you will lose some juiciness.

TIP **The European method of slowly decreasing the body temperature of chicken by using fans and cold air rather than water is slowly gaining popularity among processors of high-quality chickens. It is well worth checking out a good butcher in your area to see if you can find this type of bird.**

How can I keep a roasting chicken from sticking to the pan?

There are three steps I usually take. First, I dry the chicken thoroughly. That doesn't mean dabbing it with a paper towel. It means wrapping the chicken in an absorbent tea towel for an hour or even overnight. (Try to do this a day ahead.) Second, be sure to use a heavy-duty roasting pan. The heavier the roasting pan, the less chance that the pan juices will dry out and stick. Third, never let the chicken "fall asleep" in the oven. Instead, keep moving it around a bit and baste every 10 to 15 minutes to keep it from sticking.

TIP **When roasting a chicken, it is a good idea to remove the wing tips, especially if the pan is small. The tips will often stick to the side of the pan, so clipping them off gives the chicken less chance to stick.**

Do you recommend trussing a chicken for roasting?

I have gone through long periods in my cooking career when I swore by trussing and highly recommended doing it, but now I prefer roasting an untrussed bird. Still, I find that trussing can save the wings from sticking and is still the best way to cook a chicken on a rotisserie.

TIP **Using a bulb baster makes basting easy. Be sure to get a glass one that cleans easily, and always use the hot pan juices for basting.**

I love the taste and texture of a chicken that has been cooked rotisserie style. Can I duplicate that at home?

A brand-new gadget on the market, which looks like a giant skewer, will do this for you. The one I've seen is the Spanek Vertical Chicken Roaster. Although it makes a big mess in your oven, it produces a crispy, evenly browned bird. However, in addition to the messy cleanup, you also miss out on the delicious pan juices that you get from a traditionally roasted chicken.

Can all the various cuts of chicken parts be used interchangeably in a recipe?

Dark meat and white meat have different flavors. The dark meat (legs, thighs) is juicier and more flavorful, while the white meat (breast) has more subtle flavor, making it more of a blank canvas. In general, white meat goes with milder and lighter sauces such as an herb or cream-based sauce. Dark meat marries well with ethnic flavors, spices, curries, and Middle Eastern and Far Eastern spices and flavors. Then there are dishes that use the whole cut-up chicken and straddle both categories, in which the chicken is braised with tomatoes and herbs. There are also some new "cuts" available now, like drumettes (the fleshy part of the wings), which can be used like mini-drumsticks, and boned and skinned thighs, which I find to be the most flavorful part of the chicken. They can be braised in the oven, broiled, or grilled.

When a recipe calls for grilling chicken, can I use a broiler instead?

The broiler works very well for chicken that has some fat. The advantage of the grill is that you can move the chicken around so that it cooks evenly and thoroughly. The broiler will char the outside quickly while leaving it raw inside. Be sure to watch the chicken carefully and turn the broiler off. Also, grilling lends food a very special "grilled" flavor, which broiling does not.

What is the difference between chicken cutlets and chicken fillets?

In some areas, cutlets are flattened or butterflied chicken breasts. In other areas, the flattened tenderloins are called chicken cutlets.

Recently I have also seen tenderloins sold as both "tenders" and "chicken fillet tenders," which is confusing since the tenderloin and the fillet are one and the same.

Can I prepare chicken cutlets and breasts the same way?

The very best way to prepare a chicken cutlet is to bread it as you would the Viennese schnitzel. Just dredge it lightly in flour, dip it in beaten egg and unseasoned bread crumbs, and then fry it in a mixture of corn oil and butter just as you would a veal cutlet. However, I realize that a lot of people buy cutlets because they like their low-fat aspect, so if you are trying to avoid fat, season the cutlets with salt and pepper and pan-sear over high heat with just a touch of butter or oil. To give a cutlet more flavor, try flavoring it with herbs and some spices, such as a pinch of cumin, ground coriander, and freshly ground black pepper. Marinades such as teriyaki sauce work well, too, because they make the cutlet brown very quickly, so chances are you will not overcook it.

TIP Pan-seared chicken cutlets are an excellent topping for a mixed green salad.

Is there a trick to keeping chicken breasts moist?

The most important thing about buying a chicken breast is making sure the tenderloin is still attached. Unfortunately, it's hard to tell if that is the case by looking through the package, because many grocery stores roll the chicken breasts to give you the impression that the breast is plump. When the chicken breast is cooking, the tenderloin forms a layer between the pan and the rest of the meat and keeps it moist.

We eat chicken breasts frequently, so I feel I've made them every possible way. What do you think is the absolute best way to cook boneless, skinless chicken breasts?

Chicken breasts are best when sautéed in some butter and oil, preferably in a cast-iron skillet, then braised, covered, with some good chicken broth for 7 to 8 minutes. At this point you can add a number of interesting herbs and vegetables. I particularly like dill and tarragon with chicken, as well as sautéed wild mushrooms, cucumbers, and peas. Some lemon and heavy cream make interesting additions. Remember that chicken breasts are completely neutral and are open to almost endless possibilities.

I've tried grilling chicken breasts, but they always seem to dry out. What am I doing wrong?

You're probably not doing anything wrong; chicken breasts are low in fat, and they dry out easily when grilled. I'd recommend a gas grill rather than charcoal because a charcoal grill's heat is hard to control. Keep your gas grill on medium-high and cook a chicken breast—with the tenderloin attached—for just 4 minutes on each side, basting it with mixture of lemon juice, olive oil, and fresh herbs. You will be amazed at the delicious and juicy results.

How far in advance can I cook chicken breasts?

It's not really necessary to cook chicken breasts in advance: they cook in less than 10 minutes anyway. But if you brown chicken breasts on top of the stove, cooking them no more than 6 to 8 minutes, or until they are still slightly pink inside, you can prepare them 2 or 3 hours ahead of time. Later, you can reheat them either on top of the stove or by microwaving them for 1 or 2 minutes. Make sure the breasts have the tenderloin attached if you are going to do this; it will help to keep them moist.

I like to tenderize chicken by marinating it. Any suggestions?

Many recipes suggest that a marinade tenderizes chicken. That is not so; what it does is enhance the flavor. You can marinate chicken for as long as 24 hours or as little as 4 hours. For whole chickens and chicken legs, I usually reserve full-flavored marinades that contain strong spices such as hot chiles, ginger, and cumin. For chicken breasts, I use simple marinades that are based on olive oil, lemon juice, and herbs.

I see so many recipes for grilled chicken breasts, but I prefer the legs. Any suggestions for quick ways to prepare them?

Although I'm not a big fan of broiling, I think this method does an excellent job on chicken legs. Simply season them with salt, pepper, and a touch of cayenne, then run them under the broiler until well browned. Finish cooking the legs in a 400°F oven for 15 minutes. For chicken legs with a delicious sauce, see Broiled Chicken Legs with Herb Butter, page 201.

I see family packages of chicken wings in my market. Are there any interesting ways to use them?

There isn't a lot of meat on chicken wings, but they do make a very tasty appetizer when broiled, grilled, or deep-fried. I usually count about 3 wings per person, although many of my friends can consume a lot more than that.

TIP Think dipping sauces when grilling chicken wings; a zesty vinaigrette, a spicy jalapeño-flavored mayonnaise, and a mustard-garlic dressing are some flavorful ideas.

I like to stir-fry chicken. Which cut should I use?

One of the best quick-cooking cuts for chicken stir-fries is the chicken tenderloin, sometimes sold as "chicken tenders." The tenderloin is juicy, cooks quickly, and doesn't dry out as much as chicken breasts. When you want to cook chicken together with root vegetables, which takes a little more time, use boneless and skinless thighs, cut in half.

What's the best way to prepare chicken for chicken salad?

Contrary to what most recipes suggest, I don't believe that chicken breasts make great chicken salad. My first choice is a roasted chicken. If you prefer the breast, poach bone-in chicken breasts in a light bouillon with some aromatic vegetables such as leeks, carrots, celery, onions, and a couple of unpeeled garlic cloves. Do not overcook. Let the chicken cool completely in the broth before making the salad.

TIP For a terrific chicken salad, I often buy a ready-cooked rotisserie chicken and cut it up. Then, rather than binding it with mayonnaise, I like to toss the chicken in a mustard vinaigrette, adding some diced red bell pepper, a diced dill gherkin, and a little minced red onion, and let the salad marinate for an hour or two before serving.

Any tips for roasting a juicy turkey at Thanksgiving? I usually make a 24-pounder, and it is always somewhat dry.

I can easily say that I am a fanatic about not using large turkeys. I'd rather use two 10- to 12-pound birds than one that large. Remember that your turkey needs attention. You simply cannot leave it in the oven, go for a long walk, and then come home expecting a perfectly roasted bird. It can, however, be amazingly good if it is treated carefully. Choose smaller birds and don't stuff them. Season them well and baste every 15 minutes. Keep in mind, too, that you can't have it both ways. If you want juicy dark meat, you will have to sacrifice the breast somewhat. Smaller birds, 10 to 12 pounds, roast more quickly, which gives the white meat less of a chance to dry out.

Does stuffing keep a turkey moist?

That depends on the kind of stuffing you use. Since bread absorbs moisture, bread stuffing will draw out much of a bird's natural juices. Sausage stuffings are a good choice because they're somewhat fatty and will help baste the bird from the inside.

I find turkey very bland. Will a marinade make it tastier?

A turkey will absorb the flavors of a marinade and take on a somewhat different character. However, since turkey has such a mild flavor, I prefer a full-flavored dry rub (see Perfectly Roasted Turkey with a Mustard-Herb Rub, page 216). Keep in mind that at holiday times, even those who like to try new flavors usually prefer the simply roasted turkey that tradition dictates.

I've read that brining a turkey is a good idea. What is the advantage of brining and how do I do it?

Brining a turkey has become the favored method of many cooks, who feel that it renders a juicier bird and firms up the meat. Personally, I only brine smaller turkeys, because brining requires placing the bird in a large zippered plastic bag full of water, and that takes up a great deal of refrigerator space. To brine, coat the turkey with 1 cup coarse salt and transfer to a large zippered plastic bag. Fill the bag with water, place it in the fridge, and let sit for 24 hours. Rinse thoroughly, dry with an absorbent towel, and use it for any recipe of your choice.

I've heard you shouldn't leave a stuffed turkey at room temperature. Is that true?

Yes, I definitely suggest refrigerating a turkey once it has been stuffed, particularly if you use a sausage mixture. However, I do not believe you should concern yourself too much regarding the whole issue of not stuffing a bird until just before putting it in the oven. If it is refrigerated, it will be fine for several hours before roasting.

Is it better to roast a turkey breast side down or breast side up?

I have tried it both ways, and here's the difference: When you roast it breast side down, you may end up with a somewhat moister breast, but you're also running the risk of having the skin stick if you let the pan juices run dry. If you roast it breast side up, you absolutely must tend to it by basting every 15 minutes to keep it moist.

I often see a good buy at the supermarket on turkey breasts. Does it make sense to roast the breast alone?

When well seasoned and simply roasted, a turkey breast is delicious, and it also makes for a good-natured and inexpensive main course. It's more efficient, too, since it cooks more quickly than a whole bird. If you really like turkey, the breast is the perfect size for six or eight people. It benefits greatly from a rub or a marinade, after which I roast or grill it simply, basting often with chicken bouillon.

Is there an easy way to grill a turkey?

Absolutely. Follow the directions for indirect grilling on page 350. Place it off the heat and roast to 160°F, and then let it sit for 10 to 15 minutes before carving. I usually stick with small turkeys (12 to 13 pounds), which take about 1 hour 30 minutes to 1 hour 45 minutes and do not require basting.

TIP For a lovely smoky flavor, add some wood chips to the fire when roasting a turkey. Best woods are apple, lilac, and hickory. Be sure to soak the chips in water for 20 minutes before adding to the coals. Depending on the size of the turkey, you may have to add wood more than once.

I'm always looking for low-fat meats, so turkey burgers appeal to me. How should I prepare them?

Ground turkey is low in fat, but it is also low in flavor. It needs all the help it can get, so seasoning assertively is very important. Try this: Sauté a minced onion until just soft and incorporate it into the ground turkey together with some minced parsley, 1 mashed garlic clove, lots of freshly ground pepper, salt, a touch of cayenne, and a sprinkling of thyme. Pan-sear the burgers in some butter and oil or grill until just cooked, being careful not to overcook. Keep in mind, though, that unless the ground-turkey package promises that it is 5 percent fat or less, your fat savings are not that great.

TIP Ground turkey is especially good in meatloaf when it is cooked to well done with a lot of seasoning and possibly served with a tomato sauce. Substitute ground turkey for some or all of the ground beef in your favorite meatloaf recipe.

Is it better to buy an organic or a free-range turkey? How do you feel about frozen turkey?

I usually go with organic and have not found that the "free-range" tag warrants the difference in price. Also, there is no longer any reason to buy a frozen turkey. As long as you give your market a day's notice, it should be able to come up with a fresh one.

What are Cornish game hens? Can I prepare them the same way I would prepare a chicken?

In fact, you can do a lot with them, more than people think. Cornish game hens are not wild fowl; they are a small variety of chicken. On the practical side, they're a lovely dinner party entree, because they cook quickly. They have no distinctive flavor, and every preparation for whole chickens works with Cornish hens, as long as you adjust the cooking time. They do well on the grill, where they cook in 25 to 30 minutes, and when cut in half or quartered, they make very good finger foods. Cornish hens are also great roasted or braised.

I love the taste of squab and always look for it on restaurant menus. It does not seem to be as popular as it deserves to be and I would like to prepare it myself. How can I get it?

I agree with you. Squab is one of my favorite foods as well and I always order it when it is on the menu. Most good grocery stores will order squab for you if you give them one or two days' advance notice. Otherwise, you can order it from D'Artagnan (see Sources, page 398). The squabs they carry have plump breasts and are extremely flavorful.

Why is squab so much more expensive than chicken? Is it because it is wild?

Squabs are farm raised but take much longer to raise than chickens. A squab cannot be artificially inseminated, the birds are wed for life, and the courtship takes its time. It is followed by at the most nine baby chicks, which means that raising squabs is both time consuming and expensive. They come to market when they are twenty-eight days old and they must never have flown. If they have, they will be tough. This is a wonderful-tasting bird that is worth the splurge.

Can I substitute Cornish hens in recipes that call for quail or squab?

No. Quail and squab are more gamy, leaner birds that need a very short cooking period.

Is it okay to use frozen ducks? That's all my grocery store carries.

Yes. Frozen ducks are fine, so long as they don't show any signs of freezer burn. With most food, defrosting has to take place in the refrigerator.

Do remember to remove the bird from its plastic package as soon as it is defrosted and wrap it in an absorbent kitchen towel to get it nice and dry before popping it in the oven.

What exactly is a Long Island duck?

At one time, the eastern part of Long Island was home to many duck farms. Now that the building boom in that area has chased out farmers, ducks continue to be referred to as "Long Island ducks" when they are really Peking ducks raised in several other states, including Maryland, Delaware, and Pennsylvania.

Are duck breasts available individually like turkey breasts? Where can I get them?

The best duck breasts are those of Muscovy ducks, but sadly these are not yet available commercially. Moulard duck breasts are an equally fine substitute, which you can find in many supermarkets as well as by mail order. Many good grocery stores do carry vacuum-packed duck breasts, though. If you can't find a store near you, contact D'Artagnan, a company that provides many restaurants with specialty meats and poultry (see Sources, page 398).

I order duck in restaurants, but I've never tried to cook it at home. Is it complicated?

You're not alone. Many people seem intimidated by cooking duck, but in fact, it's one of the most good-natured foods you can prepare. It's literally foolproof. Whether you buy your duck frozen or fresh, medium or large, you almost can't go wrong if you follow this simple technique: roast it in a low oven for a long period of time (approximately 3 hours at 350°F). Besides all its other advantages, duck can be roasted well ahead of time and kept warm without a change in taste or texture. For me it's the number-one poultry in terms of flavor and ease of cooking. Try the Roasted Duck with Sautéed Pears, page 212, and you'll see what I mean.

***

TIP **If you prick the duck all over before roasting, roast it slowly at 350°F in a heavy-duty pan, baste it often with hot broth, and remove the fat from the pan with a bulb baster about every 15 minutes, you will end up with a duck that is very crisp with absolutely no fat underneath the skin.**

***

I'd like to buy a duck breast, but I don't know how to prepare it.

The secret to cooking a duck breast is to score the skin side first in a diamond pattern, then pan-sear it over low heat, skin side down, on top of the stove in a heavy cast-iron skillet until the skin is crisp and has released all or most of its fat. Then finish cooking in a 400°F oven for 10 to 12 minutes. Always serve duck breast rare or medium-rare. Since duck breast is a restaurant favorite, several cookbooks by top chefs contain wonderful recipes for it.

I love to grill chicken but have never attempted duck. Can it be done?

Because of its high fat content, duck does not lend itself to grilling; the fat melts and drips onto the coals, and then the flames flare up and burn the duck. I have, however, worked around this by poaching the duck first, then wrapping it in a towel and leaving it in the refrigerator overnight. The next day I slow-smoked it, and the results were delicious. Still, it's much easier and quicker to roast a duck in the oven, and the flavor is just as good. If you really crave that grilled taste, you can place a roasted duck on a gas grill for a few minutes after it is cooked through.

What is foie gras?

The ancient Egyptians noticed that ducks and geese stored extra calories in their livers just before they migrated, and that those livers were super tasty. Ever since, these birds have been cultivated for their livers, now known as foie gras, which can be incorporated into terrines or mousses, or sliced and cooked.

Can you use the liver that comes with a duck as a substitute for foie gras?

If only that were possible! Alas, the livers from store-bought ducks have very little flavor, and it's not really worth bothering with them. Do spring for some foie gras once or twice a year—it's a treat.

How do you cook foie gras?

Nothing can be simpler. The liver must be very cold. Dip your knife in hot water and make ½-inch slices. Heat a dry skillet to almost smoking, add the foie gras slices, and cook for 1 minute on each side. Drain on a double layer of paper towels before serving.

I finally made a Christmas goose and it was so tough. What could I have done wrong?

Usually it means that your goose was too old. For a goose to be delicious and tender, it needs to be young, so look for a smaller goose in the market. Be sure to roast the goose at 325°F and spoon off the fat every 15 to 20 minutes. Save the goose fat for sautéing anything from a steak to fish fillets. I usually freeze the fat in one-cup containers and use it all year as I need it.

Is it okay to buy a frozen goose? I cannot get a fresh one where I live.

Frozen geese are fine. Be sure to let them defrost in the refrigerator. This may take up to 2 days. Choose a goose that has a nice plump breast and plenty of fat that looks clean and unblemished.

Last Christmas a friend sent me two guinea hens and they are still in my freezer. How shall I prepare them?

Guinea hens are best simply roasted. Start by seasoning the hens with coarse salt, freshly ground pepper, and a sprinkling of thyme. Trussing the birds will keep the breast meat from drying out. Brown on top of the stove in a mixture of butter and oil and roast the birds on their sides at 400°F for 45 to 55 minutes, basting with good chicken stock every 10 minutes. The hens are done when the inner temperature is 145°F. Serve with a ragout of mushrooms, or a creamy risotto.

Broiled Chicken Legs with Herb Butter

Although I am not a great fan of the broiler, I find that I often turn to this recipe for a quick main course. Serve the chicken legs with a simple couscous and sautéed seasonal vegetables.

SERVES 4

8 chicken drumsticks

8 chicken thighs

Juice of 2 lemons

¼ cup olive oil

2 tablespoons fresh thyme or 2 teaspoons dried
  thyme

½ teaspoon cayenne

Coarse salt and freshly ground black pepper

5 tablespoons unsalted butter, at room temperature

2 tablespoons minced fresh parsley, plus small
  whole leaves

1 tablespoon minced shallot

1 tablespoon minced fresh tarragon

¾ cup chicken stock or bouillon

Lemon wedges

1. Place the chicken pieces in a large baking dish. Sprinkle with lemon juice, olive oil, thyme, and cayenne. Season with salt and pepper and chill for 4 to 6 hours.

2. In a bowl, combine the butter, minced parsley, shallot, and tarragon. Season with a pinch of salt and pepper and blend thoroughly.

3. Preheat the broiler. Transfer the chicken to a broiler or roasting pan and set 6 inches from the source of heat. Broil for 10 minutes on each side, making sure the chicken does not burn. Remove the pan from the oven, discard all the fat that has accumulated in the pan, and add the bouillon. Place a dollop of the herb butter on each piece of chicken.

4. Reduce the oven temperature to 400°F. Return the pan to the oven and continue to cook for another 5 minutes, or until the herb butter has just melted.

5. Transfer the chicken pieces to a serving plate. Spoon herb butter over them. Garnish with parsley leaves and lemon wedges.

Bouillabaisse of Chicken

What does a chicken bouillabaisse have in common with the great classic seafood stew? Not much. Still, the wonderful aroma and taste of saffron and a creamy garlicky aioli lend authenticity to this delicious chicken stew. Don't be put off by the length of the recipe; it is extremely easy to make and is even more flavorful reheated the next day.

SERVES 4

3 tablespoons olive oil

2 medium onions, quartered and finely sliced

1 small fennel bulb, quartered and finely sliced

3 large tomatoes, seeded and cubed

1 teaspoon dried thyme

1 teaspoon fennel seeds

Salt and freshly ground black pepper

8 chicken drumsticks

8 chicken thighs

10 cups chicken stock or bouillon

¼ teaspoon loosely packed saffron threads

3 small Yukon Gold potatoes, peeled and cubed

2 small zucchini, cubed

1 large red bell pepper, seeded and cubed

2 cups cubed butternut squash

Minced flat-leaf parsley

Bowl of Puree of Celery Root, Potatoes, and Sweet Garlic (page 117), as accompaniment

1. In a large heavy casserole, heat the olive oil. Add the onions and fennel. Cook over medium heat, stirring occasionally, until slightly browned.

2. Add the tomatoes, thyme, and fennel seeds. Season with salt and pepper and continue to cook for 5 to 8 minutes, or until most of the tomato water has evaporated.

3. Season the chicken with salt and pepper and add to the casserole together with the chicken stock. Add the saffron. Bring to a boil, reduce the heat, and simmer for 35 to 40 minutes, or until the chicken is tender but not falling apart.

4. Skim off the fat and add the potatoes. Simmer until the potatoes are almost tender, about 10 minutes. Add the zucchini, bell pepper, and squash. Continue to simmer until the vegetables are soft, about 10 minutes longer. Taste and correct the seasoning. Serve hot in deep soup bowls, garnished with parsley and topped with a dollop of the puree.

Slow-Braised Chicken Breasts with Peas and Lemon-Dill Sauce

Grilled, sautéed, baked, or roasted chicken breasts are popular fare in every one of these preparations. And chefs and home cooks are always looking for interesting ways to prepare them. The mild taste of the chicken breasts takes well to many herbs and vegetables, and even fruit and lemon are perfect partners. When fresh peas are not available, you can use asparagus tips or cubed mushrooms instead. You can also substitute a mixture of tarragon and chives for the dill depending on your mood. Serve the dish with a simple Parmesan-flavored risotto.

SERVES 6

$\frac{1}{2}$ cup heavy cream

Juice of 1 lemon

3 whole boneless, skinless chicken breasts, cut in half

Salt and freshly ground white pepper

All-purpose flour, for dredging

3 tablespoons unsalted butter

2 teaspoons peanut oil

$\frac{1}{2}$ to $\frac{3}{4}$ cup chicken stock or bouillon

1 tablespoon flour mixed into a paste with 1 tablespoon softened butter

$\frac{1}{2}$ cup cooked fresh peas, optional

2 to 3 tablespoons minced fresh dill

1. In a small bowl, combine the cream and lemon juice, mix well, and set aside to develop flavor.

2. Dry the chicken breasts thoroughly with paper towels. Season with salt and pepper and dredge lightly in flour, shaking off the excess.

3. Melt the butter together with the oil in a large heavy skillet over high heat. Add the chicken breasts without crowding the skillet and sauté until nicely browned on both sides. Add a little stock or bouillon, reduce the heat, and simmer, covered, for 7 minutes, or until the juices run pale yellow when the chicken is pricked near the bone. Add a little more stock or bouillon to the skillet if the pan juices run dry before the chicken is done. Transfer the chicken to a dish.

4. Add the remaining stock or bouillon to the skillet, bring to a boil, and reduce to ¼ cup. Add the lemon-cream mixture, bring to a boil, and reduce slightly. Whisk in bits of the flour-and-butter paste until the sauce lightly coats the spoon. Correct the seasoning.

5. Reduce the heat to low. Return the chicken to the skillet together with the peas and just heat through. Transfer the chicken to a serving platter, spoon the sauce over, and garnish with dill, chives, or tarragon. Serve hot with buttered orzo or couscous.

Provençal Chicken with a Sherry Vinegar–Garlic Essence and Concassé of Tomatoes

We tend to think of roasted chicken as "comfort food," something best suited for cool weather, to be enjoyed with a side dish of buttery mashed potatoes. This is a summer version, which feels cool and inviting because of the creamy tomato sauce added to the pan juices just before serving. Serve with butter-braised yellow zucchini and crusty French bread.

SERVES 6

2 large ripe tomatoes, seeded and cut into ½-inch dice

2 tablespoons extra virgin olive oil

⅓ cup plus 2 teaspoons sherry vinegar

3 tablespoons fresh thyme leaves

2 tablespoons minced fresh chives

Coarse salt and freshly ground black pepper

2 small whole chickens (about 2½ pounds each)

1 cup crème fraîche

2 teaspoons tomato paste

2 teaspoons Dijon mustard

2 garlic cloves, mashed

2 tablespoons unsalted butter

1 teaspoon canola or grapeseed oil

1 cup chicken stock or bouillon

2 tablespoons minced fresh flat-leaf parsley

1. Prepare the tomato concassé: In a small bowl, combine the tomatoes, olive oil, the 2 teaspoons vinegar, 1 tablespoon of the thyme, and the chives. Season with salt and pepper and mix well. Set aside to marinate.

2. Preheat the oven to 375°F.

3. Dry the chicken thoroughly with a kitchen towel. Sprinkle with the remaining 2 tablespoons thyme, the salt, and pepper; truss and set aside.

4. Combine the crème fraîche, tomato paste, mustard, and mashed garlic in a small bowl.

5. In a large heavy flameproof baking dish, melt the butter together with the canola oil over medium-high heat. Add the chicken and brown lightly on both sides. Add the remaining ⅓ cup sherry vinegar, bring to a boil, and cook until reduced to a glaze.

6. Add ½ cup of the stock to the baking dish, place in the center of the oven, and roast the chickens for 1 hour to 1 hour 15 minutes, or until the juices run pale yellow when the meat is pricked near the bone. Baste the chickens with hot stock every 10 to 15 minutes. Be careful not to let the pan juices run dry. When the chickens are done, remove them to a cutting board.

7. Thoroughly degrease the pan juices and return them to the baking dish together with any remaining stock. Place over medium heat. Whisk in the crème fraîche mixture and reduce slightly.

8. Add the tomato concassé and just heat through; do not let boil. Taste the sauce, correct the seasoning, and keep warm.

9. Carve the chickens and place on a serving platter. Spoon the sauce over the chickens and sprinkle with the parsley. Serve at once.

Ragout of Chicken Legs with Fennel Sausage

Here is a wonderfully gutsy, cool-weather chicken dish that is open to seasonal variations, such as roasted bell peppers cut into fine julienne or some shiitake mushrooms sautéed in a touch of olive oil. Serve the ragout with a side dish of soft polenta or buttery mashed potatoes.

SERVES 6

8 chicken drumsticks

4 tablespoons extra virgin olive oil

1 pound fresh sweet Italian (fennel) sausage

Coarse salt and freshly ground black pepper

$\frac{1}{4}$ cup minced shallots

3 garlic cloves, minced

$\frac{1}{2}$ cup dry white wine

3 large ripe tomatoes, peeled, seeded, and
    minced, or one 16-ounce can Italian plum
    tomatoes, drained and finely chopped

2 teaspoons tomato paste

2 tablespoons minced fresh thyme

1 teaspoon dried oregano

$\frac{1}{2}$ cup chicken stock or bouillon

2 teaspoons finely grated lemon zest

3 tablespoons minced fresh parsley

1 tablespoon capers, well drained

1. Ask the butcher to cut each drumstick in half crosswise and to cut off the tips. Dry the drumsticks well with paper towels and set aside.

2. In a large heavy skillet, heat 2 tablespoons of the olive oil over medium-high heat, add the sausage, and sauté until nicely browned all over. Remove from the pan and reserve.

3. Add a little more oil to the skillet, add the chicken pieces, partially cover the skillet, and sauté, turning, until nicely browned all over. Season with salt and pepper. Remove the chicken to a plate.

4. Add more oil to the pan if needed. Add the shallots and 2 of the minced garlic cloves and cook just until soft. Add the wine, bring to a boil, and reduce to a glaze. Add the tomatoes, tomato paste, thyme, and oregano. Bring to a simmer and cook, uncovered, until some of the tomato liquid has evaporated.

5. Return the chicken to the pan along with any juices that have accumulated on the plate and just 2 to 3 tablespoons stock. Cover the skillet tightly and simmer over low heat for 20 minutes.

6. Cut the sausage crosswise into ½-inch rounds and add to the skillet. Continue to simmer for another 10 minutes, or until the chicken is done and the juices run pale yellow when it is pricked near the home.

7. While the chicken is cooking, prepare the garnish. In a small bowl, combine the lemon zest, parsley, capers, and remaining garlic and mix well.

8. When the chicken is done, transfer it together with the sausage to a deep serving platter. Place the skillet over high heat, add the remaining stock, and reduce the pan juices until they heavily coat the spoon. Add the garnish and simmer the sauce for 2 or 3 minutes. Taste and correct the seasoning.

9. Spoon the sauce over the chicken and sausage, and blend carefully with 2 spoons.

REMARKS If you want more sauce, add a small amount of arrowroot or cornstarch mixed with a little stock to the pan to thicken the sauce. The sauce will be more flavorful, however, if allowed to reduce and thicken to the desired consistency without any thickening agent.

You can add some pitted black niçoise olives to the sauce or more capers to taste. If you are using fresh tomatoes, you may need a little more chicken stock, since spring tomatoes are not very ripe or juicy.

Roasted Chicken à la Flamande

The concept of cooking a vegetable alongside a roasting chicken has great appeal, because you get the benefits of the vegetable absorbing the flavors of the pan juices. Endives work perfectly like this, and the result is delicious. If you don't mind a touch of cream, add about $\frac{1}{3}$ cup to the pan juices just before serving. Sliced roast potatoes or hash browns and roasted parsnips make excellent accompaniments.

SERVES 3 TO 4

$3\frac{1}{2}$-pound whole chicken

Coarse salt and freshly ground white pepper

$1\frac{1}{2}$ teaspoons dried thyme

1 large garlic clove, mashed

3 tablespoons butter

1 tablespoon corn or peanut oil

4 large Belgian endives, trimmed and cut in half lengthwise

1 to $1\frac{1}{4}$ cups chicken stock or bouillon

1 teaspoon flour dissolved in 3 tablespoons cold chicken bouillon

1. Preheat the oven to 400°F.

2. Season the chicken with salt, white pepper, and thyme. Rub with the mashed garlic. In a large cast-iron skillet, heat the butter and oil. Add the chicken and cook over high heat, turning, until brown all over, about 7 minutes.

3. Add the endives cut side down and pour in $\frac{1}{3}$ cup of the stock. Place the skillet in the oven and roast the chicken for 45 minutes, basting every 10 minutes with a little more of the hot broth. Turn the endives over halfway through the cooking to be sure they do not burn. Season with salt and pepper.

4. When the chicken is done, quarter it and transfer to a serving platter. Garnish with the roasted endives.

5. Place the skillet over direct heat, add any remaining stock, and whisk in a little of the dissolved flour. Bring to a boil, whisking, to thicken the pan juices. Spoon over the chicken.

Fried Cornish Hens "al Ajillo"

Chicken fried *al ajillo* is one of southern Spain's most popular every-day dishes. Small chicken nuggets are rubbed with fresh garlic and quickly deep-fried in olive oil. Served hot with nothing more than quartered lemons and some fried garlic cloves, they make a delicious simple main course. Here is a variation that uses Cornish hens in-stead. The garlic-parsley oil makes a nice dipping sauce, which can also be used with grilled flank steak or pan-seared burgers.

SERVES 3 TO 4

    1 cup extra virgin olive oil

    2 to 3 garlic cloves, finely minced, plus 3 large
      garlic cloves crushed through a press

    3 tablespoons finely minced parsley

    2 teaspoons minced jalapeño or more to taste

    Coarse salt and freshly ground black pepper

    3 Cornish hens, cut into eighths, wing tips removed

    Pinch of cayenne

    Flour, for dredging

    2 quarts corn or peanut oil, for frying

    Sprigs of parsley

    2 lemons, quartered

1. Prepare the garlic-parsley oil: In a small bowl, combine the olive oil, minced garlic, parsley, and jalapeño. Season with salt and pepper. Stir until well blended and set aside until serving.

2. Thoroughly dry the Cornish hen pieces with paper towels. Rub each piece with some of the crushed garlic. Season gener-ously with salt and pepper and sprinkle very lightly with cay-enne. Dip the pieces very lightly in flour, shaking off any excess.

3. In a large saucepan or heavy casserole, heat the corn or peanut oil. Add the Cornish hen pieces a few at a time without crowding the pan. Deep-fry at 350°F for 6 to 8 minutes, or until nicely browned.

4. Remove to a double layer of paper towels and continue to deep-fry the rest of the Cornish hens, keeping a close eye on the oil, which must not burn. Transfer to a serving platter. Garnish with parsley and lemon quarters and serve accompanied by the garlic-parsley oil.

Roasted Duck with Sautéed Pears

Whenever I teach a roast duck recipe in one of my cooking classes, students are amazed at how easy and undemanding it is to prepare. All the ducks need is to be slow-roasted for as long as 2½ to 3 hours to release all the fat under the skin. The ducks can be roasted hours ahead of time and reheated under the broiler.

SERVES 6 TO 8

2 ducks (about 4½ pounds each)

Coarse salt and freshly ground black pepper

1½ teaspoons dried thyme

5 tablespoons unsalted butter

1 tablespoon peanut oil

1¼ cups chicken stock or bouillon

2 Bartlett pears, peeled and quartered

1 tablespoon sugar

2 tablespoons lingonberries in syrup

A few drops of lemon juice

2 teaspoons arrowroot mixed with a little
    broth, optional

Large pinch of fresh nutmeg

1. Rinse the ducks and dry thoroughly with paper towels. Season with salt, pepper, and thyme.

2. Melt 2 tablespoons of the butter and the oil in a large rectangular roasting pan, add the ducks, and brown on all sides on top of the stove. Discard all but 3 tablespoons fat and add ½ cup of the stock.

3. Place the ducks in the center part of the oven and roast for 2 hours 30 minutes, turning them once during the roasting time. Remove the fat every 15 to 20 minutes and baste with 3 to 4 tablespoons more stock.

4. While the ducks are roasting, prepare the pears. In a large heavy skillet, heat the remaining 3 tablespoons butter. Add the pears, season lightly with salt, pepper, and sugar, and sauté over fairly high heat until they are nicely glazed and brown. Set aside.

5. When the ducks are done, transfer them to a baking sheet. Set aside.

6. Thoroughly degrease the pan juices and add to the pears together with the lingonberries and drops of lemon juice. Taste and correct the seasoning.

7. If the pan juices seem too thin, add a little of the arrowroot mixture, just enough to thicken the sauce lightly.

8. Quarter the ducks and place on a serving platter. Spoon the pears and pan juices around the ducks, sprinkle the pears with nutmeg, and serve accompanied by soft polenta or the Puree of Rutabaga and Potato with Crispy Shallots, page 144.

Fried Marinated Quail

Quail makes wonderful eating. It is usually served grilled, but I like it just as much fried. Here is a dish to be enjoyed with friends when manners do not count, since it makes a great finger food. I usually serve quails as an appetizer, but they make a wonderful main course, as well, accompanied by mascarpone-flavored risotto, page 281.

SERVES 6 TO 8

⅓ cup tamari or soy sauce
1 tablespoon grated tangerine zest
⅓ cup fresh tangerine juice
1 tablespoon peeled and grated fresh ginger
1 tablespoon minced garlic
1 jalapeño, minced, or ⅛ teaspoon cayenne
1 teaspoon sugar
5 to 6 quails, quartered
3 to 4 cups corn oil, for frying

1. Combine all the ingredients except the quails and corn oil in a large bowl with 6 tablespoons water. Add the quails, turn, and marinate in the refrigerator overnight.

2. Drain the quails and place on a triple layer of paper towels and dry as well as possible.

3. In a deep cast-iron skillet, heat 2 inches of oil until very hot. Add half or one-third of the quail pieces and cook until nicely browned all over. Do not crowd the skillet and do not overcook. You will have to do this in 2 or 3 batches. Drain the fried quails on a double layer of paper towels and serve with a tiny bowl of coarse salt and the coarsely ground peppercorn mixture that follows.

Peppercorn mixture

MAKES ABOUT 2 TABLESPOONS

1 teaspoon Tellicherry peppercorns
1 teaspoon freeze-dried green peppercorns
1 teaspoon Sichuan peppercorns
1 teaspoon pink peppercorns or mixed colored peppercorns
1 teaspoon toasted cumin seeds, optional

Grind all the ingredients together in a spice grinder.

Grilled Quail with Pineapple Caramel

The slightly gamy but rich taste of quail benefits enormously from this sweet spicy marinade. Grill the quails and serve with a well-seasoned couscous or the Creamy Risotto with Vine-Ripe Tomatoes, Fresh Rosemary, and Mascarpone, page 281.

SERVES 4 TO 6

2 cups unsweetened pineapple juice

1 small serrano pepper, sliced

¼ cup brown sugar

6 black peppercorns, coarsely cracked

2-inch-long piece ginger, peeled and minced

2 garlic cloves, finely sliced

oil

Coarse salt and freshly ground black pepper

4 to 6 quails, cut in half lengthwise

Sprigs of fresh cilantro

1. In a bowl, combine the pineapple juice, serrano pepper, brown sugar, cracked peppercorns, ginger, and garlic. Whisk in the oil and season with salt and pepper.

2. Place the quails in a large zippered plastic bag and add the pineapple juice mixture. Transfer the bag to a baking dish and refrigerate for 12 to 24 hours.

3. Preheat the grill. While the fire is getting hot, remove the quails from the marinade and pat dry thoroughly with plenty of paper towels.

4. As soon as the fire is ready, place the quails skin side down and grill for 3 minutes. Turn and grill for another 3 minutes. Remove the quails to a dish and serve hot, garnished with the cilantro.

Perfectly Roasted Turkey with a Mustard-Herb Rub

Whenever I make a roasted turkey, friends ask me for the recipe. All it takes is getting a good fresh turkey, and it does not have to be either organic or free-range. Just season it assertively, baste it every 15 minutes with good stock, and do not forget it. I promise your friends will ask you for the recipe.

SERVES 8

10- to 12-pound fresh whole turkey

2 tablespoons Dijon mustard

2 teaspoons dried thyme

2 teaspoons dried marjoram

2 teaspoons imported sweet paprika

2 large garlic cloves, peeled and mashed

Coarse salt and freshly ground black pepper

2 tablespoons unsalted butter

1 teaspoon peanut oil

4 small onions, cut in half crosswise but
    not peeled

3½ cups chicken stock or bouillon

1 tablespoon flour mixed into a paste with
    1 tablespoon softened butter

Bouquet of fresh herbs

1. Preheat the oven to 425°F.

2. Dry the turkey thoroughly with kitchen towels. Combine the mustard, thyme, marjoram, paprika, and garlic in a small bowl and mix well. Rub the mixture all over the turkey (but not on the bottom), season with salt and pepper, and truss the turkey.

3. In a large heavy flameproof baking dish, melt the butter together with the oil over medium heat. Add the turkey, breast side up, and place the onions cut side down in the baking dish together with ¼ cup of the stock. Set in the center of the oven and roast for 1 hour 30 minutes to 1 hour 45 minutes, or until the internal temperature on a meat thermometer registers 165°F in the thigh and the juices run pale yellow when the meat is pricked near the bone. Add a little stock to the pan every 10 to 15 minutes and baste with the pan juices. If the turkey becomes too dark before it is done, tent loosely with foil and continue roasting.

4. When the turkey is done, remove it from the oven and transfer to a carving board. Remove and discard the trussing string and let the turkey sit for at least 5 to 10 minutes before carving. Remove the onions from the roasting pan, peel, and set aside.

5. Degrease the pan juices and return them to the baking dish together with the remaining stock. Bring to a simmer. Whisk in bits of the flour-and-butter paste until the sauce lightly coats the spoon. Taste and correct the seasoning. Keep warm.

6. Carve the turkey, place on a large serving platter, and surround with the roasted onions. Garnish with herbs and serve hot with the sauce on the side.

REMARKS You may puree the roasted onions in a food processor and add to the reduced sauce instead of thickening it with the flour-and-butter paste. The sauce will have a velvety texture and a delicious roasted onion flavor.

The onions can also be served as a side dish at room temperature, drizzled with a touch of sherry vinegar and extra virgin olive oil, and sprinkled with coarse salt and freshly ground black pepper.

■ Beef Tenderloin and Portobello Mushroom Kebabs ■ Roast Beef with Caramelized Onion Sauce ■ Marinated Grilled Flank Steak with Wilted Watercress ■ Pan-Seared Rib-Eye Steak in Herb and Shallot Butter ■ Viennese Goulash Soup ■ Boulettes Basquaise ■ Spring Ragout of Veal Shanks with Mushrooms and Peas ■ Braised Lamb Shanks with Vegetables Niçoise ■ Pork Tenderloin with Prunes and Port Sauce

Meat is back on American tables. If you have any doubts, just pick up a magazine or the food section of your local paper. Food magazines run recipes that call for short ribs, lamb shanks, and veal osso buco. Steak restaurants are opening everywhere, and even seafood menus offer diners several meat choices, such as rib-eye steaks and double pork chops. Unfortunately, because of years of consumer pressure to lower the fat in the American diet, producers have been breeding and raising animals to be leaner and, therefore, less flavorful. Pork was never meant to be lean, and you can't have chops that are both juicy and fat free.

I often opt to teach a beef stew at a cooking class, using the robustly flavored chuck. When my students see me degrease the sauce, the inevitable question comes up, "Can I use a less fatty cut? How about the tenderloin?"

I always find myself explaining that the tenderloin is not meant for stews. When, in another class, I brine and oven-roast a pork shoulder butt to be served with lentils and sautéed winter vegetables, occasionally a student will want to make the dish more elegant by using a boneless loin of pork. Yes, it will look good, but with today's lean pigs, chances are the meat will lack the rich, hearty flavor we are counting on to accentuate the other ingredients.

An important principle in cooking all meat is that tender cuts will remain tender only by being cooked quickly with dry heat. Overcook them, and they toughen. The tougher, more flavorful cuts require a marinade or rub to tenderize them and bring out the hearty flavors. These are best cooked slowly with moisture—stewed or braised.

So, how do you learn which cuts are tender and which are tougher? In earlier times, most grocery stores had butchers who could be relied on to give you this kind of information. You could count on their expertise and knowledge. Now most meat comes to the supermarket already cut, and few butchers have in-depth information about the product they sell. Today's shopper has to be much more knowledgeable and cautious to avoid disappointment.

My first recommendation to anyone interested in cooking meat of any kind is to learn the anatomy of the animals we eat. It is important to keep in mind that flavor and tenderness depend on where the meat comes from on an animal. The tenderness of a piece of meat depends on the amount of work a particular muscle does. Thus, it is not surprising that the same cuts that are tender in beef are also so in veal, pork, and lamb.

Unfortunately, the average grocery store is more than happy to provide the shopper with the most popular cuts, such as rib roast, tenderloin, strip steaks, chuck, and sirloin. But when it comes to wonderful flavorful and inexpensive cuts, such as oxtails, shoulder of veal, the tri-tip, or even short ribs or skirt steak, chances are that you will have to shop around or ask the butcher to get it for you.

Fortunately, distinguishing tender cuts from tough ones is not difficult, because the key barn animals we eat all use their muscles in pretty much the same way. The rib and loin sections are the most tender, because they run along the sides and back and do not get exercised as much. The rib-eye, the tenderloin, the rack of lamb, and the pork tenderloin come from this part of the animal as well and are perfect for roasting, pan frying, and grilling. The parts of the animal that get more exercise, the shoulder and legs, need to be cooked in moist heat for a longer period of time to become tender and juicy. The brisket, flank, and shoulder are tougher, but to me, they have more of that rich meaty taste. They should be reserved for braising, pot roasting, or stewing. As soon as you become familiar with these various cuts, you will be the best judge of what you should use for which type of recipe. When it comes to beef, choosing the right cut can be confusing because different cuts are sold by different names, depending on where you live.

When preparing pork, the challenge facing all cooks is to keep it moist, which can only be achieved by undercooking it slightly or by brining, or both. Here, too, it is the lesser cuts, such as the shoulder butt or the picnic roast, that will give you the most tender and flavorful results if you cook them slowly. If you are not squeamish about fat, also consider the robust flavor of spare ribs. Grill or slow-roast them, and they will be juicy and delicious.

On those days when you decide to go with lamb, you will be in luck, since good lamb is now widely available. American lamb from California, Colorado, Vermont, and Washington State is excellent. You may run into trouble when you decide to make a lamb stew, because butchers like to sell you the leg for everything. When used for kebabs or a stew, the leg meat becomes tough; it does not benefit from slow moist cooking.

Once you get your meat home from the market, don't just toss it into the fridge or the freezer. Take the extra few minutes to store it properly. Remove all meat from the package and, if you are not going to be using it the same day, wrap it loosely in freezer paper, waxed paper, or, best of all, brown butcher's paper, which you can easily get from the meat counter. Large cuts will keep well this way for 3 or 4 days, and small cuts such as veal and lamb chops or flank steaks will keep for 2 days. I have actually "aged" both rib-eye and strip steak in the cold part of the fridge for as long as a week with excellent results. Never wrap meat in foil or a zippered plastic bag, because the key to keeping it fresh is to allow air to circulate around the meat. Remember good meat is expensive, and you deserve to get your money's worth, but also remember the endless diversity of meat and the wonderful cooking opportunities it presents.

In the following Q & A's, I try to answer some of the many questions students have asked me over the years, and my advice continues to be the same: Try to break away from the common and the familiar and be willing to experiment. You will find it extremely rewarding.

Beef

■■

Many cookbooks and television cooks suggest asking the butcher for meaty beef bones for stock, but I have to rely on the supermarket for stock bones. What should I do?

It is important and actually rather easy to establish a relationship with the butchers at the supermarket. You will be surprised how helpful they usually are. Ring the service bell and ask for meaty bones for stock. Or look for inexpensive packages of extra bones called "soup bones." For these my first choices are beef shank bones and rib bones.

It's a good idea to build up a supply of veal and beef bones as well as chicken backs and wings, which you can keep in a sturdy plastic bag in the freezer. You can use these straight from the freezer whenever you want to make stock.

TIP **Invest in a heavy cleaver so that you can chop the beef, veal, or poultry bones into 3- or 4-inch pieces. These produce the most flavor in a stock.**

What exactly does the labeling of beef mean, and what should I look for?

The labeling of beef is becoming more and more confusing. Because there is very little real aged prime beef available anymore, you must make sure to choose Choice or Top Choice. Avoid any fancy names such as Butchers' Prime or Market Choice. This labeling is done by individual markets and has nothing to do with U.S.D.A. labeling. To be on the sure side when it comes to beef, try to buy certified Black Angus, which has a high degree of marbling and excellent beef flavor.

I always bought beef that was bright red but now hear that darker beef is a better choice because it means it has been aged.

In certain cuts such as a New York strip steak, a darker color can mean that the steak has been aged. Generally I look for a light cherry red color with marbled interior fat and white external fat, and a smooth tight grain in the meat.

Color is not a reliable indicator of quality, since many butcher cases use special lighting to give beef a misleading bright red hue.

Many of my friends buy their beef at wholesale club stores in vacuum-sealed bags, but the meat looks very dark to me. Is it really okay?

You can do very well buying beef this way, especially when it comes to the large cuts. The lack of oxygen in vacuum-packaged meat gives it a dark purplish color, but once the bag is opened, the meat will go back to its reddish color. Also, don't be alarmed if the meat gives off a strong odor. It will disappear after a few minutes of exposure to the air.

I find all the various beef cuts confusing. Can you tell me in simple terms what cut to use for what?

You do not need to be an expert to understand the important cuts of beef. The rib, short loin, sirloin, and tenderloin all come from the back area of the steer. They are all fine grained and tender and all can be used in oven-roasting, grilling, sautéing, and pan-searing. These cuts are best cooked rare to medium-rare.

For braising, the best cuts are the chuck roast and the brisket, which come from the shoulder and sides of the steer. These cuts are tougher and need long, slow cooking but deliver great taste in pot roasts and stews.

Other cuts, such as flank and skirt steak, take well to marinades and are best quickly grilled. Chances are your market will also carry the rump roast and the eye round. These less tender cuts need long, slow oven-braising, but I find them too chewy and too dry no matter what you do with them. Much of the round is

sold ground, and I even avoid that because I feel that the meat lacks the more fatty and interesting taste and texture of the chuck.

My market does not label its ground beef by fat content but rather by the cut. How can I tell the amount of fat the ground beef contains?

Ground chuck, which is my first choice, contains 20 to 25 percent fat while ground sirloin has 15 to 20 percent, and ground round 15 percent or less.

Unfortunately, labeling is inconsistent around the country and changes from area to area. In fact, the name of a cut is no guarantee that the meat is actually ground from a specific cut. Your best bet is to check the package and see if it is full of white specks, which is an indication of fat. Be sure to question your butcher about the fat content of the ground meat the market sells.

What is the secret of successful broiling? Why can't I get the same results as a restaurant does?

Meat broiled at home frequently ends up dry because a home range simply does not put out enough heat. Since the surface of the steak is not sufficiently seared, the flavorful juices escape. A home broiler reaches about 500°F while a restaurant range can get up to 700°F, and that makes a difference in taste and texture.

What should I look for in a great steak?

I think that you will find a lot of different opinions among steak lovers. Many beef eaters swear by the tenderloin and the almost fatless filet mignon. Others like myself much prefer the more assertive taste and texture of either the T-bone steak, which is the New York steak with a bit of the tenderloin attached, or the New York strip steak, which has the bone attached but not the tenderloin. This can be most confusing, since the New York strip steak is called by various names depending on what part of the country you are from. It can be called Kansas City steak, club steak, shell steak, or Delmonico. Another cut that makes a great steak is the rib-eye steak, either bone-in or boneless. This cut has more fat and is somewhat more chewy but very flavorful.

What is the difference between a tournedo, filet mignon, and chateaubriand?

A filet mignon comes from the small end of the tenderloin and is usually 1 to 2 inches thick. The tournedo is essentially the same piece of meat wrapped in bacon to add juiciness to this very lean cut. The chateaubriand is a French cut of the finest part of the filet mignon, usually reserved for restaurants.

What exactly is a porterhouse steak?

The porterhouse is the T-bone steak with a large portion of the tenderloin attached. (The T-bone is the strip steak with a small portion of the tenderloin attached.) The porterhouse is a favorite for the grill, which is the way I recommend using it, since it is too large for top-of-the-stove cooking.

What exactly is the London broil, and what is the best way to use it?

London broil is the name given to the top round. It is a rather tough cut that should be marinated and then quickly grilled. It looks somewhat similar to the sirloin steak, but the sirloin—although more pricey—is a much more interesting and flavorful cut.

Do you recommend tenderizing a steak?

I do not recommend using any meat tenderizer, because it changes the texture and flavor of beef.

What is the ideal thickness of a steak for pan-searing?

For pan-searing, a steak must be at least ¾ inch thick and up to 1¼ inches thick. If you cook on a gas range that puts out plenty of heat and have a well-seasoned black cast-iron skillet, you can go to as much as 1½ inches in thickness.

I always have trouble cooking a steak on top of the stove. Should I cook a thick steak on high or medium-high heat?

It all depends on the thickness and your pan. If it is a thick steak, it is important to pan-sear it in clarified butter on high heat in a heavy cast-iron skillet. However, if you are cooking on a gas range such as a Viking or a Garland, then medium-high heat is ideal.

TIP Always remove meat from the fridge an hour before cooking and never use a lightweight nonstick pan for pan-searing or cooking steaks or chops.

Many recipes suggest pan-broiling a steak or a burger; others call for sautéing. What is the difference?

Pan-broiling is done in a heavy skillet without oil or butter, while for sautéing you need to start with some fat. I start with a little butter and oil, and once the steak or burger has released some fat, I pour it off to make sure the meat continues to sauté and does not end up frying.

My market offers three or four choices of ground beef at a time. Which ground beef do you recommend?

Your first choice in ground beef should be the chuck. Because this cut has more fat in it than the sirloin or the top round, the burgers remain juicy and flavorful. Do not buy lean or extra-lean ground beef. You are not actually getting much less fat, but you are giving up a great deal of flavor.

TIP **If you have a food processor with a very sharp blade, I highly recommend grinding your own beef. Be sure to chill the blade and the bowl in the freezer for 30 minutes. Cut the meat and its fat into ¾-inch pieces and grind in small batches. The taste of freshly ground beef is far superior to anything you can buy at the market.**

Is it best to season steaks and burgers before cooking or after?

I believe in seasoning a steak with coarse salt and freshly ground black pepper prior to cooking. When it comes to burgers, mixing the salt and pepper into the raw meat makes a world of difference in taste.

What cut do you suggest using for a roast beef and how much should I get per person?

For great beefy taste, nothing beats a standing rib roast, also called prime rib. The entire roast is made up of ribs from the upper section of the back of the steer, which weighs over 16 pounds, but rib roasts are also sold in smaller pieces of 2 to 6 ribs. My favorite is a 3-rib roast from the back of the rib section, which is the small end. The meat from this section is lean, tender, and very flavorful. However, if you plan to serve

roast beef to a large group of people, you may prefer the large end of the rib. This cut has much more fat on it and is harder to carve.

Another terrific cut that I often use is the boneless rib-eye. This can run into quite a bit of money, but there is no waste, it is packed with flavor, and it is very easy to carve. Because it is boneless and very manageable, I like to serve it for a dinner party of six to eight people. If you prefer a roast with the bone in, you may want to go with a bone-in rib-eye.

You should always buy at least a 3-rib roast that serves four people with some leftovers for sandwiches. This cut will weigh 6 to 7 pounds. If you want to serve 12 people, you will have to go for a 5-rib roast, which weighs between 10 and 12 pounds.

TIP **Always buy a rib roast from the small end of the ribs rather than the large end. This cut is more tender, leaner, and more flavorful.**

Is there a surefire method for cooking a roast beef? I have tried three or four methods and have either undercooked or overcooked the roast each time.

Nothing can be simpler than making a roast beef, but I admit I overcooked several roasts before I got it right. Preheat the oven to 500°F. Season the roast with plenty of salt and pepper. Place it fat side down in a heavy roasting pan and sear it on top of the stove until nicely browned on all sides. Place the roast fat side up in the oven and roast for 15 minutes. Reduce the oven temperature to 350°F and continue to roast, counting eighteen minutes per pound or to an inner temperature of 115°F for rare and 120° to 125°F for medium-rare. Remove the roast from the oven and let it rest for

15 minutes. By this time the inner temperature should read 130° to 135°F, which will give you a perfectly cooked medium-rare roast.

Why do so many recipes these days suggest using an instant-read thermometer rather than tell you how many minutes per pound it takes to cook a roast?

The instant-read thermometer has been an absolute savior for many cooks, and I recommend using it with all roasts. Recipes should give you an approximate time, but so much depends on whether the meat has been cold or at room temperature, whether your oven is properly calibrated, and whether you are roasting in a gas or electric oven. However, I do not use the instant-read thermometer for steaks and chops of about 1½ inch thick. I prefer the knife method, in which you insert the tip of a very sharp paring knife into the center of the chop or steak while it is being pan-seared. If the tip of the knife comes out cold, the meat is underdone; if it is warm, it is medium-rare; and if the knife tip is hot, the meat is well done and over-cooked.

What cut of beef do you suggest using for kebabs?

There are several cuts I like for this preparation. The chuck steak from the rib section is delicious and inexpensive. It takes well to various marinades and to quick grilling. I also like to use the tenderloin, which may sound extravagant, but when using it for kebabs, a little goes a long way and you can either marinate it or not. A less tender choice but equally flavorful is the sirloin. However, this cut is only good when served rare.

Do not buy pre-cut shish kebabs. Chances are these nice-looking lean cubes have been cut from the top round, which cooks up dry no matter what you do to it.

What cut of beef do you use for making fajitas?

My choice and probably the only choice is the skirt steak. Be sure to marinate it for 4 to 6 hours or overnight (see the marinade, page 234), then grill directly over the coals on a hot grill for 2 to 3 minutes per side. Do not overcook.

My family loves a good beef stew. Sometimes the meat is tender after less than 2 hours, and other times it remains tough no matter how long it is in the oven. Why does this happen?

You are probably making the mistake of buying pre-cut stew meat. You have no idea what cut you are getting, since most meat markets try to get rid of lesser cuts. Instead, buy a piece of chuck and cut it up yourself into 1½- to 2-inch pieces, or have the butcher cut it up for you. It will be less expensive, and your stew will cook evenly every time.

What is the best way to cook short ribs?

Short ribs make the most delicious stew, and they take well to braising since they benefit from slow, long cooking. Have the butcher cut them into 3-inch chunks, brown them in a little oil, and then braise them, covered, with plenty of chopped onions, minced garlic, some carrots, a little celery, and some stock for 3 to 6 hours. Because short ribs tend to be fatty, they must be cooked a day ahead of time so that you can properly degrease the sauce. But nothing beats short ribs for fabulous beef flavor.

How do you cook oxtails and beef shank bones?

Both are delicious cuts that are used in soups and stews. You can cook oxtails or beef shank bones as you would chicken soup, with some carrots, leeks, and a couple of celery stalks for 2 to 3 hours. Chill overnight, degrease the broth, and return it a boil. Add ½ cup tiny pasta and cook until tender. Serve this soup as a simple one-dish meal.

Calf's Liver

Whenever I broil calf's liver, it becomes tough. What am I doing wrong?

Calf's liver is best seared in butter in a very hot cast-iron skillet no more than one to 2 minutes on each side, depending how thick the slice, and then seasoned with salt and pepper. Another good way to serve calf's liver is to cut it into 1-inch strips, quickly sauté them, and combine the liver with a soft onion jam.

TIP Avoid buying pre-cut liver and, if you can, check the color and make sure it is pale and rosy, and not blotchy. If the liver seems dark and has some bloody spots, place it in a rectangular dish and cover with milk. Soak for 1 to 2 hours, then dry well with paper towels.

Lamb

I like to make lamb kebabs. Sometimes I am very successful, other times the kebabs are tough. What is the reason for this?

Lamb kebabs are actually very forgiving, especially if they have been marinated, but you will get the most tender results from shoulder meat or the inexpensive sirloin. Most pre-cut kebabs are from the leg, which becomes tough if not cooked rare.

What is the best marinade to use for lamb kebabs?

Yogurt-based marinades work well for lamb. My favorite is a take-off on the Indian tandoori marinade: Combine 1½ cups plain yogurt with ½ cup olive oil, ¼ cup lemon juice, and 1 tablespoon each minced ginger, garlic, and serrano pepper. Add 1 tablespoon each ground cumin, coriander, and cardamom and 1 teaspoon each salt and freshly ground black pepper. Place all the ingredients in a food processor with 2 large whole bunches of cilantro, stems and all, and blend well. Place the lamb cubes in a zippered plastic bag, add the marinade, and refrigerate overnight. About 2½ pounds of lamb cubes will make 4 to 6 servings.

I followed a recipe for lamb stew, and although I cooked it for twice as long as the recipe called for, the meat was still tough. What did I do wrong?

It was probably the cut you used. Next time, insist on lamb shoulder and have the meat cut into 2-inch cubes. This is an unpopular cut with supermarket meat departments, who are not too happy to have to bone the whole shoulder for stew and prefer selling it as bone-in shoulder chops.

I see a lot of recipes that call for lamb shanks, but my market does not carry them. Can I use the leg of lamb instead?

The leg of lamb is not a substitute, since the two cuts require totally different cooking methods. The leg requires short roasting and is best served rare to medium-rare, while the shank needs to be braised in a low oven for as long as 3 to 4 hours until it is so tender that the meat almost falls off the bone.

Lamb shanks have recently become extremely popular. Your market should be able to get them for you if you order them in advance.

Pork

■■■

What is the difference between the pork loin and the pork tenderloin?

Although both cuts come from the part of the hog that runs on either side of the backbone, the two are very different. The pork loin is the most common and popular pork roast. It is sold either bone-in or boneless and usually weighs between 3 and 4 pounds. The tenderloin, or fillet of pork, is a thin cylindrical roast that weighs between ½ and ¾ pound. It is juicy and tender and is usually sold vacuum packed one or two fillets at a time. It is to the hog what the beef tenderloin or filet mignon is to the steer.

What is the best cut to choose when making a pork roast for a dinner party?

I am surprised by the consistent popularity of the pork loin roast, since this cut dries out no matter what you do to it. But you have two other excellent choices. First on my list is the pork tenderloin, which you can get from the oven to the table in 15 minutes. Another more humble cut of pork, but one that is loaded with taste, is the Boston butt or the shoulder butt. This cut is much fattier than the tenderloin and lacks the showy presentation of the pork loin roast, but it makes up for it in flavor.

My market carries various kinds of ribs. Which ones do you recommend?

For meatiness and delicious pork flavor, spareribs are the best. Make sure you buy heavier slabs, weighing more than 3 pounds. This will guarantee you the right meatiness. But make sure that they are not too fatty. For a leaner meal, try baby back ribs; these are relatively more expensive since they have almost no meat on them. Look for ribs that are the leanest and have the most meat on them. Plan on serving one slab for one to two people.

Country-style ribs are an inexpensive choice. They can be fatty but are delicious when marinated and grilled slowly over indirect heat (see page 350 for indirect heat information).

Do baby back ribs come from a baby pig?

Not at all. Baby back ribs are the bones of the loin ribs with the boneless pork chops removed. The name is one of the great pork marketing gimmicks, but since the ribs are tasty, we are willing to pay a high price for a bony cut with very little meat.

What is a good method for cooking pork chops so they don't dry out?

For years we shied away from pork because it was too fatty and "not good for you," with the result that farmers are growing larger, leaner pigs. The result is dry, tasteless pork. There is little you can do to pork chops these days to keep them juicy other than to flavor brine them and undercook them slightly so as to keep them somewhat more juicy.

Every cookbook I read seems to suggest a different inner temperature for cooking pork. What should it be?

The business of overcooking pork for health reasons goes back a long time and was drummed into a whole generation of cooks. In reality, trichinae, the parasites that cause trichinosis, are killed at 137°F or medium-rare. Lean cuts, such as loin chops, should be taken off the heat at that stage and left to rest for another 10 to 20 minutes, at which point the inner temperature will increase by 5 to 10 degrees. Don't be alarmed if the meat is still a faint pink in the center and the juices run pink as well. These colors are perfectly safe and it means the pork will be juicy.

Pork roasts, on the other hand, need to be cooked a little longer—to anywhere between 150°F inner temperature for the tenderloin to 160°F for a pork shoulder roast. Ignore all cookbooks that suggest that you cook pork to 180°F. At 160°F it is more than done, and at 180° it is inedible.

My butcher suggests very thin pork chops rather than thick ones, because they cook quickly and stay juicier. Is that true?

I prefer chops that are 1 to 1½ inches thick, but the key is that they must be uniformly cut.

Many butchers have the tendency to cut chops thick at the bone and thin at the periphery. Pass these by and request evenly cut ones; otherwise the outside part of the chop will be done while the meat near the bone will still be raw. This is also the problem with a thin bone-in chop. It is almost impossible to get the meat near the bone cooked properly without the rest of the chop tasting like cardboard.

Which is juicier—the rib or loin pork chop?

Rib chops have a little more fat and, therefore, stand a better chance of not drying out. Either way, buy the chops at least 1 to ½ inches thick.

TIP When choosing the pork loin roast, always buy it "bone in" rather than boneless, and have the butcher crack the bones for easier serving.

I recently got 1½-inch-thick pork chops from my meat market and tried to pan-sear them on top of the stove, but while they were practically charred on the outside, they were still raw inside. Is there a good method for getting them cooked through?

Whenever you have a thick chop or steak, you should cook it in clarified butter (see page 384) or rendered duck fat, which you can now buy at the meat counter of many supermarkets. (It is packed by D'Artagnan and comes in small plastic containers; see Sources, page 398.) Sear the chops in a heavy cast-iron skillet. Once they are nicely browned on both sides, and I mean deeply browned, turn down the heat only slightly, so that the chops continue to sizzle. Do not lower the heat to the point that the sizzling stops, because if the heat is too low, the chops will sweat, and the juices will exude, leaving the meat dry.

Cover the pan and continue to cook for 3 to 4 minutes. Use an instant-read thermometer after 2 minutes: If it registers 140°F, the chops are done. Let them rest for 5 minutes, which will bring the inner temperature up to 145 degrees. The chops may still be slightly rosy inside but, believe me, they are done.

TIP Whenever you sear any kind of chop or steak in a black iron skillet, be sure to remove the meat to a warm platter to let it rest. If left in the skillet, which retains heat, the meat will keep cooking.

How long should I marinate pork, and does a marinade tenderize it?

A marinade does not tenderize all types of meat, but it does add a great deal of flavor. For the marinade to be effective, however, you must marinate the meat overnight in a zip-pered plastic bag. For pork, I especially like the following:

Citrus Spice Marinade

MAKES ¾ CUP

¼ cup lemon juice
¼ cup orange juice
1 large onion, finely chopped
2 teaspoons ground cumin
2 teaspoons chili powder
1 tablespoon minced serrano pepper
1 tablespoon minced garlic
½ cup olive or grapeseed oil
1 teaspoon coarsely ground black pepper
1 teaspoon coarse salt.

Combine all the ingredients in a bowl and whisk until well mixed. Put 4 chops or a 2- to 3-pound shoulder butt roast in a zippered plastic bag, add the marinade, and refrigerate overnight.

Veal

What cut of veal is osso buco? What dishes is it used for?

Osso buco is not a cut but a classic northern Italian dish in which the cut-up veal shank is braised with white wine, onion, and tomatoes until very tender. Traditionally, it is served with a saffron-flavored risotto. While I like this dish, I find more interesting uses for the veal shank, such as a veal ragout—a traditional French dish in which the shank is cooked with a variety of mushrooms with spectacular flavor results (Spring Ragout of Veal Shanks with Mush-rooms and Peas, page 240).

The shank is also delicious braised whole in the oven and served with sautéed shitake mushrooms or roasted parsnips. Because the meat is cooked on the bone, it is done in a rel-atively short time and the collagen in the bone produces intensely flavored pan juices.

What is the right cut to use for a veal stew? When I buy pre-cut veal, the results are uneven. Some pieces are tender while others remain chewy and tough.

Stewing veal is expensive and you do not know what you are getting. Try to have your butcher cut up a veal shoulder roast, or buy boned veal shoulder and cut it up yourself into 2-inch cubes. It is not difficult at all. If you have a knowledgeable butcher, you could ask him for the meat from the neck area. This slightly gristly cut is full of collagen, so the meat is tender and flavorful, resulting in a delicious sauce.

TIP If you happen to find a good batch of veal scaloppine, remember that it freezes well for several weeks. Place a small sheet of waxed paper between each scaloppine, then wrap the whole package in foil and slip into a zippered plastic bag. Be sure to defrost in the refrigerator rather than at room temperature to avoid losing precious juices.

My market carries inexpensive ground veal. How is it used?

Veal meatballs are easy to prepare and, when combined with the Quick Tomato Sauce, page 274, are great tossed into freshly cooked tubular pasta, such as ziti, fusilli, or penne. They are also good served over creamy polenta. You can combine ground veal with an equal amount of ground pork and season with grated onion, salt, pepper, and a touch of oregano. Use it this mixture to stuff vegetables, such as green or red bell peppers or hollowed-out small zucchini. Drizzle with olive oil and roast the vegetables for 45 minutes at 350°F. Serve hot or at room temperature as a light main course.

Is it a good idea to marinate veal?
Try marinating veal chops for 2 to 3 hours in a mixture of olive oil and fresh lemon or lime juice with some minced fresh thyme, rosemary, or oregano. Veal has a delicate flavor, and a long, intensely seasoned marinade would compete with its lovely flavor.

Beef Tenderloin and Portobello Mushroom Kebabs

Few dishes look prettier than skewers of marinated beef or lamb tenderloin alternating with peppers, onions, and cherry tomatoes. Here is a slight variation on the classic, in which I use only peppers and portobello mushrooms. You may add onions if you like and a mix of red and yellow peppers. Be sure to use a long sturdy spatula to turn the skewers to prevent the food from rolling around. Another hint is to thread the kebabs on two skewers instead of one. Remember to soak the bamboo skewers in water for 20 minutes before threading them, so they won't burn. You will need 8 to 10 skewers.

SERVES 4 TO 5

2 pounds beef tenderloin

⅓ cup sake

½ cup mirin

½ cup soy sauce

⅓ cup canola oil

1 tablespoon minced peeled fresh ginger

3 large garlic cloves, crushed

1 dried hot chile pepper, crumbled

1½ teaspoons ground cumin

3 large portobello mushrooms, stems removed, caps cut into large cubes

2 tablespoons olive oil

Coarse salt and freshly ground black pepper

1 large red bell pepper, cut into 1-inch cubes

1 teaspoon coarsely cracked black pepper

1. Cut the beef into 1-inch cubes. In a bowl, combine the sake, mirin, soy sauce, and canola oil. Add the ginger, garlic, chile, and cumin. Whisk until well blended. Add the beef to the marinade and transfer to a zippered plastic bag. Marinate for 4 hours, or overnight.

2. The next day, light a hot fire in a barbecue grill. Toss the mushrooms with the olive oil and season with salt and pepper.

3. Thread the kebabs onto the bamboo skewers, alternating the beef cubes with mushrooms and bell pepper. Season with cracked pepper and grill for 4 minutes on each side. Serve with corn on the cob or Corn and Chive Fritters, page 120.

Roast Beef with Caramelized Onion Sauce

Most people believe that when it comes to a well-roasted piece of meat, less is more, and I tend to agree. Excellent meat needs little or no adornment. However, when you are looking for a simple change, caramelized onions slightly flavored with mustard and balsamic vinegar are an excellent choice. Serve the roast with a gratin of buttery potatoes or a puree of potatoes flavored with either celery root or rutabaga, and a well-dressed salad.

SERVES 4 TO 6

> 6-pound rib roast
>
> Coarse salt and freshly ground black pepper
>
> 4 tablespoons butter (½ stick)
>
> 1 tablespoon peanut oil
>
> 2 large Spanish onions, quartered and finely sliced
>
> 1 teaspoon sugar
>
> 1 tablespoon Dijon mustard
>
> 2 tablespoons balsamic vinegar
>
> ¾ cup beef stock or bouillon
>
> 2 teaspoons arrowroot mixed into a paste with a little stock

1. Preheat the oven to 500°F.

2. Rub the roast well with salt and pepper.

3. In a heavy roasting pan or large cast-iron skillet, heat 2 tablespoons of the butter. Add the roast and brown over medium-high heat on both sides. Place the roast top side down and continue cooking until the fat is well browned and crisp.

4. Spoon off all but 2 tablespoons fat from the pan. Place the roast in the oven and cook for 15 minutes.

5. Reduce the heat to 325°F and continue cooking for 45 minutes, or up to 18 minutes per pound.

6. While the roast is in the oven, heat the remaining 2 table-spoons butter and the oil in a large cast-iron skillet. Add the onions. Season with salt, pepper, and sugar and cook for 5 to 7 minutes, or until the onions start to brown. Lower the heat and continue to cook the onions for 45 minutes, stirring every few minutes and keeping an eye on them to be sure they do not burn, until they are very soft and lightly caramelized. Add the mustard and balsamic vinegar and continue to cook for another 3 to 4 minutes. Taste the onions and correct the seasoning.

7. When the meat is done, transfer it to a serving plate.

8. Discard all the fat from the pan. Place the pan over direct heat, add the stock, and whisk in the arrowroot mixture. Cook the pan juices until they heavily coat the spoon. Add the onions and stir into the sauce. Transfer the onion sauce to a serving bowl.

9. Carve the roast and serve on individual plates with a dollop of the onion sauce either on the side or over the roast.

REMARKS The onion sauce can be varied by adding minced fresh thyme or rosemary. If you really prefer your roast beef simple and unadorned, just follow the recipe, omitting the onion sauce altogether.

Marinated Grilled Flank Steak with Wilted Watercress

Here is a very interesting marinade that is actually a paste made out of shallots, garlic, and lemongrass. If you cannot find fresh lemongrass, you can use it dried and add the juice of a lime to the paste. You can also use this marinade with skirt steaks, beef kebabs, and a sirloin steak. Wilted watercress has a slightly bitter, assertive flavor similar to broccoli rabe, which you can use as well if you cannot get watercress in large bunches.

SERVES 4

MARINADE

3 large shallots

6-inch-long piece lemongrass, outer leaves removed

5 large garlic cloves

$\frac{1}{3}$ cup canola or grapeseed oil

3 tablespoons sesame oil

2 tablespoons nam pla (Thai fish sauce)

1 tablespoon soy sauce

2 tablespoons sugar

1 chile in adobo sauce, or 1 teaspoon pure chile flakes

STEAK

$1\frac{1}{2}$ to $1\frac{3}{4}$ pounds flank steak

3 tablespoons olive oil

2 garlic cloves, finely sliced

2 large bunches watercress, 2 inches of stems removed

Coarse salt and freshly ground black pepper

2 grilled red peppers, cut into julienne

1. Place all the marinade ingredients in a food processor or a small chopper and blend until finely chopped. Transfer together with the flank steak to a large zippered plastic bag and make sure the meat is well covered with the paste on all sides. Refrigerate overnight.

2. The next day, heat a gas grill and cook the steak over high heat for 3 minutes on each side. Remove to a platter and let the steak rest for 5 minutes before carving.

3. While the steak rests, heat 1½ tablespoons of the olive oil in a large nonstick skillet, add a few of the garlic slices, and, as soon as they start to brown, add half the watercress. Stir-fry for 2 minutes, or until just wilted. Transfer to a bowl and cook the second batch the same way. Divide among 6 serving dishes and set aside.

4. Slice the meat on the diagonal into thin slices and divide among the 6 plates. Garnish with grilled peppers and serve immediately.

Pan-Seared Rib-Eye Steak in Herb and Shallot Butter

Food trends come and go, much like fashion. Beef in green pepper-corn sauce was the "in" dish in the seventies; now it seems somewhat outdated. Personally, I think this combination of flavors deserves its place among the classics. Here, the addition of fresh tarragon and double-poached garlic cloves gives it a new dimension. When buying green peppercorns, avoid those that are pickled in vinegar and in-stead use the ones packed in brine.

SERVES 3 TO 4

3 tablespoons clarified butter (see page 384)

1½-pound rib-eye steak, 2 to 2¼ inches thick

Coarse salt and freshly ground black pepper

2 tablespoons minced shallots

½ cup Scotch whiskey

1 cup brown stock or beef bouillon

1 teaspoon brined green peppercorns, drained

2 teaspoons arrowroot dissolved in a little
    stock or bouillon

2 tablespoons minced fresh tarragon

12 cloves Double-Poached Garlic Cloves
    (page 388), optional

1. Heat the butter in a cast-iron skillet over high heat, add the steak, and sauté for 3 minutes. Turn and sauté for another 3 minutes or until nicely browned. Season with salt and pepper, reduce the heat, and continue to cook for 10 minutes. Turn the steak once more and cook for an additional 8 to 10 minutes for medium-rare. Transfer to a cutting board and let the meat rest for 5 minutes.

2. Discard all the fat from skillet, add the shallots, and cook for 1 minute. Add the Scotch and reduce to a glaze. Add the stock and green peppercorns and whisk in a little of the arrowroot mix-ture until the sauce lightly coats a spoon. Add the tarragon and double-poached garlic, and taste and correct the seasoning.

3. Slice the steak thinly on the bias, place on a serving plat-ter, and spoon the sauce over the steak. Serve at once with roasted portobello mushrooms, page 132.

Viennese Goulash Soup

Here is a hearty one-dish meal that is easy to make with ingredients that you can get in every grocery store. It is one of those dishes I fall back on when I crave something undemanding yet full of flavor. Be sure to make enough, because this soup gets better every day and also freezes well.

SERVES 4 TO 6

3 tablespoons vegetable oil
1 dried hot chile pepper
3 cups finely diced onions
3 garlic cloves, finely minced
1 tablespoon imported sweet paprika
2 tablespoons tomato paste
2 pounds beef chuck, cut into 1-inch pieces
Coarse salt and freshly ground black pepper
2 tablespoons caraway seeds
1 teaspoon dried marjoram
8 cups beef stock or bouillon
2 parsnips, peeled and cubed
2 medium red potatoes, peeled and cubed
Sour cream
Minced flat-leaf parsley

1. In a 6-quart heavy casserole, heat the oil. Add the chiles and cook until they turn dark. Add the onions and garlic and cook over medium heat until the onions are soft and lightly browned, about 7 minutes.

2. Stir in the paprika and tomato paste. Add the beef and season with salt and pepper. Mix well into the onion mixture and season with the caraway seeds and marjoram. Add the stock. Bring to a boil, reduce the heat, and simmer for 1 to 1½ hours.

3. Add the parsnips and continue to cook until they are tender, about 10 minutes. Add the potatoes and continue to simmer until the potatoes and beef are tender, about 10 minutes longer. Correct the seasoning. You may need quite a bit more salt, since the parsnips tend to sweeten the soup. Serve hot in deep soup bowls, garnished with a dollop of sour cream and some parsley.

Boulettes Basquaise

Boulettes **is the French a word for meatballs. I grew up with this recipe and still enjoy making it to this day. I sometimes make the meatballs with ground lamb instead of beef and vary the dish by adding a roasted pepper and a diced zucchini. The very best accompaniment to this homey preparation is mashed buttery potatoes or a well-flavored couscous. The meatballs can be made a day or two ahead of time and will reheat perfectly in the microwave.**

SERVES 4 TO 6

2 pounds ground beef, preferably chuck

1 small onion, grated, plus 1 large onion, quartered and sliced

2 garlic cloves, mashed

2 tablespoons minced flat-leaf parsley

1 teaspoon dried thyme

Salt and freshly ground black pepper

2 eggs, beaten

¼ cup unflavored bread crumbs

¼ cup club soda

2 tablespoons butter

1½ tablespoons olive oil

1 green bell pepper, diced

1 teaspoon hot paprika

One can (32 ounces) Italian plum tomatoes, well drained and chopped

1 teaspoon dried oregano

Minced flat-leaf parsley, for garnish

1. In a bowl, combine the ground meat, grated onion, 1 mashed garlic clove, 2 tablespoons parsley, and thyme. Season with salt and pepper.

2. Add the eggs and bread crumbs and work the mixture with your hands until smooth and well blended. Add the club soda and work the mixture again with your hands until well blended. If the mixture seems heavy, add a little more soda.

3. Form the mixture into round but somewhat flat meatballs about 1½ inches in diameter.

4. In a large heavy skillet, heat the butter and olive oil. Add the meatballs and sauté over medium-high heat until nicely browned all over. Carefully remove with a spatula to a dish.

5. Add a little more oil to the skillet and when hot, add the sliced onion, remaining garlic clove, and bell pepper. Sauté the mixture for 2 to 3 minutes, or until soft.

6. Add the paprika, tomatoes, and oregano. Season with salt and pepper and bring to a simmer. Return the meatballs to the skillet. Cover tightly and cook for 25 minutes. Correct the seasoning. The sauce should be highly seasoned. Transfer the meatballs to a deep serving bowl, garnish with parsley, and serve hot.

Spring Ragout of Veal Shanks with Mushrooms and Peas

The veal shank is suitable for many wonderful preparations. Most Americans are only familiar with the classic Italian osso buco, in which the veal is braised with tomatoes and wine and served with a saffron risotto. Here the shanks are braised with plenty of mushrooms, garlic, and herbs, and the dish is given a spring touch with the addition of fresh peas. The traditional accompaniment of saffron risotto works perfectly here, too. You can make this dish up to 2 days ahead of time and reheat it slowly in a low oven.

SERVES 6

3 tablespoons unsalted butter

1 tablespoon peanut oil

2 medium onions, finely chopped

3 large garlic cloves, crushed

Salt and freshly ground black pepper

3 small sprigs of fresh thyme

1 small sprig of fresh rosemary

½ cup dry white wine

4½ pounds veal shanks, cut crosswise into
    2-inch-thick pieces

1½ to 2 cups Brown Chicken Stock (page 389)
    or beef stock or bouillon

¾ pound cremini mushrooms, stemmed,
    caps cubed

1 cup cooked peas

⅓ cup heavy cream

1 tablespoon flour mixed into a paste with
    1 tablespoon softened butter

3 tablespoons minced flat-leaf parsley

1. Preheat the oven to 350°F.

2. In a large heavy casserole, heat the butter and oil. Add the onions and garlic and season with salt and pepper. Sauté the mixture over medium-high heat until nicely browned. Add the herbs and wine. Bring to a boil and cook until all the wine has evaporated.

3. Season the veal with salt and pepper. Add to the casserole together with 1½ cups of the stock and the mushrooms. Bring the mixture to a simmer. Cover the casserole tightly, first with foil and then the lid, and place in the oven. Braise the veal for 1 hour 45 minutes, or until fork tender. Check every 30 minutes to make sure that there is enough broth in the pot, adding the remaining ½ cup stock if necessary.

4. When the veal is done, place the casserole on top of the stove and add the peas. Heat through and transfer the veal to a deep serving platter together with the mushrooms and peas. Add the cream to the pan juices and whisk in bits of the blended flour and butter, just enough for the sauce to thicken slightly. Taste the sauce and correct the seasoning. Spoon the sauce over the veal and garnish with parsley. Serve hot.

REMARKS Brown cremini mushrooms are far tastier, with a better texture, than all-purpose white mushrooms. Cremini are now available in many supermarkets, where they are often labeled brown mushrooms. If you cannot find them, substitute all-purpose white mushrooms in this dish.

Braised Lamb Shanks with Vegetables Niçoise

Traditionally, lamb stews are synonymous with spring, but since the quality of American lamb is consistent year round, I find that a good lamb stew is equally delicious and welcome in the fall. This ragout, typical of the Nice region of France, makes use of the season's last ripe tomatoes and garden-fresh basil. The intense flavors blend and improve when allowed to develop for a day or two. Roasted bell peppers and sautéed cubed eggplant can be added to the ragout for additional texture and interest.

SERVES 6

6 whole lamb shanks

5 tablespoons unsalted butter

2 tablespoons olive oil

Salt and freshly ground black pepper

3 medium onions

3 large garlic cloves, minced

⅓ cup dry white wine

4 large ripe tomatoes, peeled, seeded, and chopped, or 1 can (28-ounce) Italian plum tomatoes, drained and chopped

2 tablespoons fresh thyme leaves

1 sprig of fresh rosemary

2 tablespoons fresh oregano leaves or 1½ teaspoons dried

2 cups Brown Chicken Stock (see page 389) or beef stock or bouillon

1 large leek, all but 1 inch of greens removed, leek cut into ½-inch cubes and washed

1 large zucchini, trimmed and diced, seedy center discarded

1 large red bell pepper, cut into ½-inch cubes

1 large yellow bell pepper, cut into ½-inch cubes

Finely grated lemon zest and juice of 1 lemon

4 tablespoons minced fresh basil

4 anchovy fillets, drained and minced, optional

2 teaspoons cornstarch, mixed with a little cold stock or water

3 tablespoons minced flat-leaf parsley mixed with 2 minced garlic cloves

1. Preheat the oven to 350°F. Dry the lamb shanks thoroughly with paper towels.

2. Melt 2 tablespoons of the butter with 1 tablespoon of the olive oil in a large cast-iron skillet over medium-high heat. Add the shanks and brown nicely all over. Transfer to an oval casserole or large, deep cast-iron skillet. Season with salt and pepper.

3. Reduce the heat to medium and add the remaining 1 tablespoon oil to the skillet. Add the onions and garlic and sauté for 5 to 7 minutes, or until nicely browned. Add the wine, bring to a boil, and reduce to a glaze. Add the tomatoes, thyme, rosemary, and oregano. Bring to a boil and transfer to the casserole containing the lamb. Add the stock, cover tightly, and place in the center of the oven. Braise the lamb for 1½ to 1¾ hours, or until tender when pierced with a fork.

4. While the lamb is braising, prepare the vegetables. In a large skillet, melt the remaining 3 tablespoons butter over medium-low heat. Add the leek, zucchini, and bell peppers, season with salt and pepper, and add 2 tablespoons water. Reduce the heat, cover, and simmer for 8 to 10 minutes, or until tender. Stir in the lemon zest, lemon juice, basil, and anchovies.

5. When the shanks are done, transfer them with a slotted spoon to a dish. Degrease the pan juices and strain them through a fine sieve back into the casserole. Place over high heat and cook until reduced by half. Whisk in enough of the cornstarch mixture to thicken the sauce so that it lightly coats a spoon. Correct the seasoning.

6. Return the lamb shanks and vegetables to the casserole and just heat through. Garnish with the parsley and garlic mixture and serve hot directly from the casserole with plenty of crusty French bread.

Pork Tenderloin with Prunes and Port Sauce

SERVES 6

2 dozen unpitted or pitted prunes

1 cup tawny port

¼ cup sherry vinegar

¼ cup sugar

1½ to 2 cups beef stock or bouillon

4 pork tenderloins (2¾ to 3 pounds)

Coarse salt and freshly ground black
   pepper

2 teaspoons dried thyme

2 tablespoons butter

1 tablespoon canola oil

About 1 tablespoon flour mixed into
   a paste with 1 tablespoon softened
   butter

1. In a nonreactive medium saucepan, combine the prunes with the port. Bring to a simmer and cook for 5 to 7 minutes.

2. In a small, heavy, nonreactive saucepan, heat the vinegar and sugar. Cook the syrup until it becomes a thick caramel and is reduced to 2 tablespoons. Immediately add ½ cup of the beef stock and boil for another 2 minutes.

3. Preheat the oven to 400°F.

4. Season the pork tenderloins with salt, pepper, and thyme. In a heavy roasting pan, heat the butter and oil. Add the tenderloins and sauté over medium-high heat, turning, until nicely browned all over, about 5 minutes. Add a little more stock to the pan and transfer to the oven.

5. Roast the tenderloins for 12 to 14 minutes, or until an instant-read thermometer registers 135°F. Remove from the oven and let rest.

6. Strain the prunes, reserving the port. Add the port to the roasting pan together with the sugar-vinegar broth and the remaining stock. Bring to a boil. Gradually whisk in the blended flour and butter. Cook the sauce until is slightly thickened, 2 to 3 minutes.

7. Add the prunes, taste, and correct the seasoning, adding a large grinding of black pepper. Slice the pork tenderloins into ½-inch slices. Place them overlapping on a serving platter and spoon the prune sauce over and around them. Serve hot.

■ Fettuccine with Jalapeño-Lime Cream and Shrimp ■ Fettuccine with Zucchini and Goat Cheese in a Ginger-Tomato Fondue ■ Bow Ties with Creamy Tomato Sauce, Bacon, and Radicchio ■ Linguine with Mussels in a Creamy Tomato-Saffron Fondue ■ Penne with Broccoli, Peppers, Tomatoes, and Smoked Mozzarella ■ Spaghettini with Tuna, Capers, and Black Olives ■ Quick Tomato Sauce ■ Catalan Chicken Paella ■ Oven-Baked Rice Pudding with Lemon and Cranberries ■ Risotto with Clams and Tomato-Saffron Fondue ■ Leek and Stilton Cheese Risotto with Mascarpone ■ Creamy Risotto with Vine Ripe Tomatoes, Fresh Rosemary, and Mascarpone ■ Braised Bulgur with Two Peppers and Lemon ■ Couscous with Spinach, Red Peppers, and Carrots ■ Corn-Studded Polenta with Parmesan ■ Cumin-Scented Quinoa with Yogurt and Mint ■ Chickpea, Tomato, and Cilantro Salad ■ Lentil Salad with Goat Cheese and Roasted Peppers ■ Stew of Pinto Beans and Shrimp with Fragrant Indian Spices ■ Ragout of White Beans with Bacon and Radicchio ■ White Bean and Chorizo Soup

One of my favorite memories is of the summer I spent with the Manfreddi family while I attended the University of Parma in Italy. La Nonna, the grandmother, made the best homemade pappardelle and gnocchi I have ever tasted, and I was lucky enough to learn firsthand about pasta from her. I watched how she made the dough by hand, with very basic ingredients—water, olive oil, flour, and salt—kneaded and rolled it to perfection, then cut and laid it out to dry on clean dish towels sprinkled with cornmeal. We ate pasta every day at the Manfreddis', and the varieties of shapes and sauces always amazed me.

When Sr. Manfreddi came home for lunch, he would shout from the bottom of the staircase, "*Mama, butta la pasta*" ("throw in the pasta"). His booming voice could easily be heard up to the fourth floor, and Nonna would immediately put a large pot of water on the burner. By the time Sr. Manfreddi walked through the door, the water would be at a rolling boil, and all of us would be seated at the big wooden table waiting for the pasta to be done. The rule is that pasta is never made to wait.

During my stay with the Manfreddis, I picked up some wonderful pasta-making techniques and recipes for homemade sauces. There's a whole art to preparing pasta, and Italians are very serious about cooking it properly and matching different shapes with the right sauce. The wonderful thing about pasta is its versatility. It is one of those foods that can fit anywhere on a menu. You can dress it up or dress it down. It can be an appetizer or an entree, and it can be served any time of year.

Pasta has become an American classic whose popularity continues to soar. There are lots of questions about pasta, especially since there are so many varieties and so many different ways of preparing it. Here are a few of the questions most frequently asked by my students all over the country.

What is the difference between dried and fresh pasta?

Dried pasta is simply fresh pasta that has been dried for a period of time in large commercial ovens. Usually dried pasta does not contain eggs and, in the case of imported Italian brands, is made with durum semolina flour. Fresh pasta generally does contain eggs; it is more pliable and requires a much shorter cooking time. I find that commercial fresh pasta, even from very good markets, is not very good. Which is why I suggest learning to make your own. Also, it is very important not to overcook fresh pasta, because it quickly turns mushy. Instead, look for dried egg pasta. There are several excellent domestic brands on the market these days as well as a number of artisinal imported ones. Test several brands, and when you find one you like, stock up on it.

I keep reading that dried pasta can be used for just about any pasta dish. Are there any preparations for which fresh pasta is much better?

The taste and texture of good fresh pasta is quite different from that of dried pasta. Fresh pasta is best for cream- or butter-based sauces, such as the classic Roman Alfredo sauce.

TIP To keep fresh pasta from cooking too quickly, dry it for a couple of hours by loosening the strands and spreading them on a large linen towel sprinkled with cornmeal. Drying the pasta overnight is even better. You can store it in your pantry in a large plastic bag.

What brand of dried pasta do you recommend?

DeCecco is a good all-around imported brand that is now available everywhere. Avoid domestic brands, because they tend to cook too quickly, and the result can be a rather mushy consistency.

What about the expensive, flavored varieties of pasta? Are they worth the extra money?

Some specialty pastas are simply trendy or gimmicky and aren't really worth the extra cost. The only purpose of vegetable-flavored pastas, especially those flavored with tomatoes, beets, or spinach, is to add color, not taste, to a dish. Some spices, in particular black pepper and saffron, will flavor it and, depending on the brand, can be quite good.

When you want color in a pasta dish, try a combination called *paglia y fieno*, "grass and hay." This is a classic mixture of white and green pasta usually served in a creamy Parmesan and butter sauce.

Many cookbooks and magazines call for pasta artisanale. What exactly is that?

Pasta artisanale, literally translated as "craft made," is available in many specialty stores and upscale grocery stores. Two excellent brands worth trying if you can find them are Rustichella d'Abruzzo and Benedetto Cavalieri. What is the difference? The large commercial pasta companies dry their pastas in enormous ovens for about an hour, producing noodles that are practically pre-cooked. *Pasta artisanale*, on the other hand, is generally produced by small companies who use better-quality flour and mineral water (versus tap water) and who dry the pasta for up to 12 hours at a much lower temperature. This produces pasta with a far superior texture. Good *pasta artisanale* can cost almost double the price of commercial brands, but it is definitely worth it.

TIP If you decide to serve a "designer pasta" colored with beets or squid ink, think about how it will actually look once it's cooked and sauced.

What is pasta asciutta? It appears in recipes and on restaurant menus, but I'm not sure what it means.

Pasta asciutta simply means dried pasta. It is a mixture of durum semolina flour and water that does not contain any eggs and, once formed into various shapes of pasta, is fully dried. This differs from fresh pasta, which may be dried later, but always contains eggs.

How much pasta should I cook per person? Is there an easy way to estimate how much is enough?

It depends on whether you are serving the pasta as an appetizer or a main course. If you plan on serving it as an appetizer, figure on 3 ounces of dried pasta per serving, or 5 ounces as a main course. Italians believe that when you eat pasta as a main course, you will be hungry within 2 hours, and they usually serve it as a starter.

Many recipes call for salting the pasta water; others do not. What's best?

Salt the water, adding about 1 tablespoon coarse salt to 14 cups water. If you forget to salt the water, don't worry. The seasoning in the sauce and the accompanying ingredients usually provide the necessary flavor. Some Italian cooks believe that seasoning the water gives you the option of using some of the pasta water for thinning out the sauce of a finished dish, but I find that unsalted water does just as well. Adding salt to make water boil faster is a myth.

What is the proper amount of water to use for cooking pasta?

Fourteen to 16 cups, or 4 quarts, water is about what you need to cook a pound of pasta. Why so much? You need to keep the noodles from sticking to each other and to the bottom of the pot. When pasta boils, it releases a fair amount of starch, and if there is not enough water, the pasta will cook unevenly. The key is to give the noodles as much room as possible to swirl around during cooking, and this means lots of water.

If you add a splash of olive oil to the water when cooking tubular pasta, such as ziti, fusilli, rigatoni, and penne, it will keep the pasta from sticking. However, many Italians and pasta purists frown on adding oil. They believe that oil makes the sauce slide off the pasta.

TIP If you do not add oil when cooking any of the tubular shapes, just make sure that you stir the pasta gently with a wooden spoon several times during cooking.

Recipes frequently call for cooking pasta "al dente." What exactly does that mean?

Literally it means "to the tooth," and when applied to pasta, it means the pasta should offer slight resistance when bitten. This does not mean it is half cooked, but that it keeps an interesting texture; overcooked pasta could be called "al mushy." Also remember that pasta continues to cook after draining and saucing. Is there a formula? Not really. The only way to master the timing is to practice, make a lot of pasta, and learn as you cook and eat.

I have heard that one of the best ways to tell whether pasta is done is to fling a couple of strands against the wall and see if they stick. Is there an easier way to tell when it is done? And is there a foolproof guideline?

Fortunately, you don't need to fling pasta around the kitchen to see if it's done. The best way to tell if pasta is ready is to taste the noodles several times during cooking. Keep in mind that pasta continues to cook even after it has been drained. A good rule of thumb is 7 to 9 minutes for flat, imported strands of pasta, and 10 to 12 minutes for tubular pasta.

TIP The best tool to use for testing pasta is a wooden or stainless steel pronged pasta fork. The noodles are easily caught in the fork and the wooden handle stays cool.

Can I substitute one shape of pasta for another? And do certain sauces call for particular shapes?

Within a general category, you can usually use any shape you want. Penne, ziti, and fusilli are shapes that can be used interchangeably in pasta recipes. So are strands of pasta like spa-ghetti, spaghettini, and linguine. Most sauces can be used on any shape of pasta except for the creamy Alfredo sauce, which is traditionally served over fettuccini.

I sometimes find myself with several boxes of pasta, each about a quarter full. Can I mix the various shapes?

If you have leftover penne, ziti, fusilli, or pennete, all of the same brand, it is possible to mix them. I wouldn't mix different brands, though, even if the shapes were the same, because cooking times vary from one brand to another. You can also use these small amounts in soups. Keep in mind that the older the pasta is, the longer it takes to cook.

TIP Italians use much less sauce on their pasta than we do in America. That's because they want to taste the pasta, and pasta tastes best if it's not swimming in sauce. Be frugal with the sauce and generous with the pasta. You can always add more sauce, but you can't remove it once it's been tossed with the noodles.

I would like to go beyond the standard tomato sauce over spaghetti. Do you have any suggestions?

First, stock your pantry with some basics: several shapes of dried pasta, canned plum tomatoes (preferably the brand imported from the San Marzano region of Italy), a chunk of good Parmesan cheese (Parmigiano-Reggiano), extra virgin olive oil, fresh garlic, onions, shallots, flat anchovy fillets, capers, oil-cured black olives, fresh lemons, crushed hot red pepper or dried chile peppers, canned tuna packed in olive oil, fresh flat-leaf parsley, and maybe some fresh or smoked mozzarella. Once you have these basic ingredients, you can make lots and lots of different sauces and match them with a variety of pasta shapes.

My homemade pasta cooks in less than 2 minutes, and it turns out mushy, even if I use special pasta flour. What am I doing wrong?

By its nature, homemade pasta is soft, particularly if it contains eggs. You can firm it up by drying the cut pasta on a rack for several hours. Be sure to use durum semolina flour to give the pasta the necessary firmness. You may want to try buying imported egg pasta. It cooks in 2 or 3 minutes but retains a more toothy texture.

When should grated cheese be served with pasta?

In this country, we assume that grated Parmesan should be served with nearly every pasta. In fact, many dishes do not call for cheese. Never add cheese to pasta with a seafood sauce. Many sauces that contain meat, butter, vegetables, and cream, however, are greatly enhanced by the addition of Parmesan. You can also use Asiago and pecorino, but to me nothing really compares with true Parmigiano-Reggiano imported from Italy. If it is not available, I prefer serving the dish without any cheese. Whatever you do, avoid domestic Parmesan or the one imported from Argentina, which comes with a black rind marked "imported."

TIP It's best to buy Parmesan in chunks and grate only enough for use in your meal. Once grated, Parmesan loses its unique aroma and texture within a day or two.

Is pasta fattening?

If you eat lots of pasta, it is right up there on the fat scale, especially if you lead a sedentary life. Pasta is a complete carbohydrate, which means it is great fuel if you exercise regularly and can burn it off. The other thing to remember is that when we are talking pasta, we are talking sauce, olive oil, butter, cream, and other flavorful toppings that add to the calorie count.

TIP Take a close look at the way Italians eat pasta. Their portions are small and so is the amount of sauce. If you can handle that, you can enjoy your pasta and eat it too.

Can I freeze leftover pasta?

There is not much point in freezing leftover pasta unless it is a baked dish like lasagna that can be reheated in the oven. You can freeze sauces, such as those made with tomatoes and meat, but it's best to cook the pasta at the last moment.

Grains

■ ■

Bulgur

I read a recipe that called for bulgur, and the instructions called for 5 minutes of cooking. Even after 45 minutes the grain was still tough. Why?

Sometimes bulgur and cracked wheat are mislabeled, and you probably bought cracked wheat.

True, they are the same grain, but bulgur has been processed so that it cooks much quicker. Cracked wheat can take as long as an hour or more to cook; bulgur, depending on the size of the grain, may only need to be soaked in water or cooked for 5 minutes. Bulgur comes in three sizes: fine, medium, and coarse. Only the fine bulgur can be left to soak up the water; medium and coarse bulgur should be steamed just like rice.

What grain is used for tabbouleh?

Bulgur is fine or medium-grain cracked wheat that has been par-boiled and dried, so it is essentially instant. For the salad, the soaked grain is mixed with parsley and flavored with lemon juice and olive oil. Bulgur is available in all health food stores and Middle Eastern markets.

Couscous

What is the difference between bulgur and couscous? Can they be used interchangeably?

If you are looking for a quick side dish flavored with spices or herbs, both couscous and bulgur can be prepared the same way. Just soak them in broth or water: they'll plump up to a fluffy consistency in very little time. Good quick couscous, especially the superior French varieties, is ready in about 10 minutes. Add a little butter and some melted scallions to your couscous, and you've got a lovely and effortless side dish. Add some diced tomatoes and sliced cucumbers to a bowl of bulgur, sprinkle it with fresh cilantro, lemon juice, and olive oil, and you're ready to serve a tangy salad that's refreshingly different.

I had a delicious fluffy couscous in a Moroccan restaurant, but the one I make is nothing like it. What am I doing wrong?

Making real couscous is very time consuming. If you want to know more about it, read Paula Wolfert's chapter on couscous in her cookbook *Mediterranean Grains and Greens.*

The instant couscous you are probably making can never have that wonderful fluffy texture, but it is quick and can be quite delicious when combined with herbs and spices.

What is the difference between couscous the dish and couscous the grain?

Couscous is not a grain but a tiny pasta made from semolina wheat. Couscous, the dish, is composed of fluffy-textured couscous served with a meat broth, cooked lamb, or poultry as well as vegetables. It is a popular dish in Morocco and Tunisia.

What is Israeli couscous?

Israeli couscous is not couscous at all. It is a novelty pasta in the shape of very tiny, pearly balls. It can be cooked in water just like pasta and flavored with butter, herbs, and spices. However, I prefer to sauté it first in a little oil until nicely browned and then cook it in some broth, like a risotto. It makes for a lovely side dish, which can be made seasonal with bits of asparagus, tomatoes, zucchini, red peppers, and fresh herbs.

Polenta

What exactly is polenta?

The word *polenta* refers to a cornmeal product imported from Italy, and to the northern Italian dish of the same name. When cooked, it resembles cornmeal mush or cooked grits. Somewhat like cooked pasta or rice, it is rather bland but takes well to many toppings and sauces.

The three classic preparations usually associated with polenta are sautéed chicken livers, grilled quail, and sautéed peppers and sausages. From here, you can be as creative as you want and use polenta as a bed for tomato- and meat-based sauces in place of pasta.

I understand that polenta is time consuming and difficult to cook. Is there an easy way to make it?

Actually polenta is quite easy to prepare, and although the traditional method does take 25 minutes, you can now get quick-cooking polenta that is just about foolproof and can be cooked in as little as 10 to 12 minutes. Here's how to do it.

To make quick polenta Bring 3¼ cup skim milk or water to a simmer. Season with a pinch of salt. Add ¾ cup quick-cooking polenta in a slow steady stream and cook over moderate heat, stirring constantly, for 10 to 12 minutes, or until thick and smooth. Correct the seasoning, then stir in 2 to 3 tablespoons of unsalted butter for enrichment.

I see various kinds of polenta in the markets. What should I buy? Does it matter if it's fine or coarse?

The texture of polenta varies, depending on how and where the corn has been milled; some varieties will absorb more liquid than others. I personally like a medium-coarse polenta that retains a nice texture when cooked, but I suggest you try various brands and types of polenta to discover what you like best. The finer the polenta, the more baby-food texture it has.

Coarser-textured polenta is more interesting to me.

I recently bought a bag of polenta that smells stale. Does this mean it's gone bad?

Fresh polenta should smell sweet. If you have had the box for a while or it's languished on the store shelf, there is a good chance that the polenta has become stale and may even have an unpleasant cardboard smell. Do not use it because the staleness is difficult to disguise.

If I cannot get imported Italian polenta, can I use cornmeal instead?

Yes, but I would use only stone-ground cornmeal. It has wonderful taste, and when I am willing to give it the time it needs, I actually prefer it to imported Italian polenta. To maximize the taste of cornmeal, you must stir it constantly for anywhere from 25 to 35 minutes and then enrich it with butter and freshly grated Parmesan.

I love pan-fried polenta, but whenever I attempt to make it, the slices fall apart. What am I doing wrong?

If you plan to pan-fry polenta, you have to start with a firm mixture, one that has not been enriched with either butter or cheese. In this case, I use a ratio of 1 cup imported polenta or cornmeal to 3½ cups skim milk. If you use water, use only 3 cups. Spread the cooked polenta about ¾ inch thick in a buttered pan and chill for several hours before slicing and frying.

I don't recommend grilling polenta. It is not easy, and I usually opt for frying. Even at my favorite restaurant in Florence, famous for their grilled squab with polenta, they use the grill only to give the polenta squares decorative grill marks before transferring it to a flat pan to finish cooking.

I have tried to make polenta several times, and it was always lumpy. Any suggestions?

There are two simple, very important steps to take when making polenta. First, the liquid cannot be too hot. Second, add the polenta in a fine stream, never all at once.

How long can I store cooked polenta?

Polenta cooked in water or broth will keep for as long as 4 to 5 days, but when made with milk, it will keep only for 2 or 3 days. Be sure to wrap it well to keep it from picking up refrigerator smells.

The specialty market in my neighborhood carries prepared polenta logs. What do you think of them?

Now that you can get quick-cooking polenta that cooks in about 10 minutes, there is no reason to bother with these. First, they are expensive for what they are, and second, they have a rubbery texture that cannot be changed.

Quinoa

In recent years, I've read a lot about quinoa but am still not sure what it is or what it goes with. Can you give me some ideas how to use it?

Quinoa is a protein-packed grain that is now becoming increasingly available. The cooked grain is similar in texture to tapioca; its taste is unique. You can find it either imported from South America or in a domestic variety. On the East Coast, it is usually available only in health food stores and specialty food markets, but west of the Rockies it is quite popular and widely available in supermarkets.

What is the best way to cook quinoa?

First, you must rinse the grain well under cold running water, just as you would rice. Bring 2 cups vegetable broth or chicken bouillon to a boil. Add 1 cup of quinoa, cover, and simmer for 12 to 15 minutes, or until all the liquid is absorbed and the quinoa is tender. 1 cup dry quinoa will serve four to five people.

I recently cooked some quinoa, and it had a bitter aftertaste. What did I do wrong?

You didn't do anything wrong. That off-taste probably came from saponin, a natural chemical that often occurs on the outside of the quinoa. Usually this harmless coating is removed during processing, but sometimes you will still find quinoa that has some saponin clinging to it. Simply rinse quinoa in a fine-mesh strainer until the water runs clear.

Rice

I still remember one of my mother's favorite weekday dishes from when I was growing up in Barcelona, Spain. It was a simple rice preparation, what she called a "poor man's paella," made with bits of vegetables and some sausage or chicken. We would sit down to lunch, and invariably my mother would exclaim, "I like potatoes but I love rice," and so did I. This recipe was forever changing depending on the season or what was fresh at the market or in our garden. It was good at room temperature and just as flavorful as a cold snack. It made a good side dish as well as a fine picnic dish. It taught me to appreciate rice years before I ever knew what a risotto or a biryani was.

Many years later, after having tasted many varieties of rice and learning a variety of rice cooking techniques, I would always encourage my students to give rice a chance. Now that we can easily find good-quality Arborio rice in most grocery stores, making a risotto at home should not be daunting. And a bowl of simply steamed jasmine or basmati rice is an easy and delicious side dish to just about any entree.

Having grown up in Barcelona, I know nothing can top the taste of saffron-tinged paella rice that's slow-cooked on the grill. But there is much more to rice than even the best of paellas, from a spicy shrimp pilaf to a Thai braised rice in lime-coconut milk. Rice is quick to prepare, goes with everything, and is so very satisfying.

Even risottos, which have always demanded a great deal of time and attention from the cook, are possible to create with a short-cut method that doesn't short cut the rich, intense taste. I think my simplified version on page 281 is perfect for the everyday cook. I'm no longer rigid about how I make risotto or how I serve it. I often make it the main course, preceded by a soup or a salad or a quick skillet sauté of shellfish. When I choose a grilled or pan-seared fish as my main course, I balance it with a hearty risotto as an appetizer.

Whenever I teach a class I try to include a quick rice dish, and I am amazed by the response. Many of my students have only cooked converted or flavored rice, or they feel that rice is hard to make. For some reason, they are intimidated by it and do not realize how good-natured the grain really is. Rice is more forgiving than pasta. As long as it is not mushy, it is fine. In many cases it can be prepared in advance, as it reheats well.

In other cultures, cooks are familiar with their typical rice dishes because they make them almost daily. Venetian or Milanese cooks can eye their risottos and get the perfect result without measuring the broth. An Indian can make a perfect biryani, and a Japanese cook knows exactly how to make sushi rice. Of course, they've been practicing for years. Each of these cooks is looking for a familiar texture and taste. One of the difficulties facing American cooks is that they embrace so many cuisines and need to change gears every time they are cooking a different type of rice with a different technique. The cook who makes a fluffy pilaf of rice one day, a risotto the next week, and a rice pudding as well needs to understand the difference that each type of rice brings to a dish.

Sometimes the selection is not really critical. Basmati and Texmati are very similar, although Texmati does not have quite the aroma of the real thing. But if you are in Texas, then Texmati will do just fine. If you cannot get La Bomba paella rice, you can make good paella with Arborio, and if your store does not carry Superfino Italian rice, then go with a lesser brand.

Experimenting with many international cuisines makes rice-cooking a challenge; it also makes it fun. With three types of rice—a long-grain such as basmati or jasmine, a short-grain such as Arborio, and a medium-grain such as La Bomba—you can make endless wonderful rice dishes.

Remember that rice needs to be stored properly. Although cookbooks tell you rice keeps indefinitely, some types draw bugs and develop an "off taste" after a while. To be sure, store rice in an airtight container, and in the case of Arborio, keep it refrigerated.

The important thing is to give rice a chance and enjoy the fabulous flavors it can deliver.

Buying rice can be very confusing. Can you recommend one or two types of rice that can be served with everything?

When you consider that there are over 7,000 varieties of rice produced around the world, you would think that choosing two would be practically impossible to do. However, since most grocery stores offer only a limited choice, the task is not difficult.

Start with one type of flavorful long-grain rice, such as long-grain Carolina rice or jasmine rice, both of which can be simply steamed. I also recommend a short-grain rice such as the Italian imported Arborio. Arborio superfino is excellent; Carnaroli and Vialone Nano are even better.

There are three preparations that require short-grain rice: classic risotto, Spanish paella, and rice pudding.

I have recently seen medium-grain rice in my grocery store. It comes in 10-pound bags and is very inexpensive. What do you use it for?

Years ago, you could get medium-grain rice only in western states, especially California, where it is grown. But now it is becoming widely available, especially in areas with large Hispanic communities. Medium-grain rice is a compromise kernel, since it is shorter than long-grain, but not as dense or glutinous as traditional short-grain. It is amazingly flavorful and works well in all sorts of dishes, especially in paella and rice pudding. In fact, I have met chefs in California who use medium-grain rice for risotto and swear by it.

How important is it to stick to the exact variety of rice called for in a recipe?

It is always a good idea to stick to the type of rice called for in a recipe. If the recipe calls for long-grain rice, you have many choices, such a long-grain basmati or jasmine. When a recipe calls for short-grain rice, you can use either a medium-grain California rice, Spanish-style Valencia, or Arborio, which is creamier and has a more interesting taste.

What is instant rice? Is there a particular brand that you recommend?

I am not a fan of instant rice; it lacks taste and texture. And there is really no reason to use it, since rice cooks in 15 minutes on top of the stove, or in less than 10 minutes in the microwave.

What does enriched rice mean? Does it mean it has more calories?

Enriched rice means that the rice has been spiked with iron, calcium and assorted vitamins. It should not be rinsed, since you would be rinsing some of the vitamins and minerals down the drain. It has the same amount of starch and calories as other types of rice.

What exactly is converted rice?

Converted rice is parboiled rice that has been steamed and pressure-cooked before milling. This process supposedly conserves vitamins, making it higher in vitamin content than regular rice. In spite of its obvious popularity, I am not fond of converted rice, because it lacks character and is not as aromatic as other types of rice. Also, it takes longer to cook and has a tendency to lose its nice texture when left on the stove for a period of time.

What is the difference between basmati and Texmati rice? Are they from different regions?

Basmati is a fine-textured long grain rice. It has been grown for thousands of years in the Punjab region of northern India, in Pakistan, and in the foothills of the Himalayas. It has a distinct aroma and a nutlike flavor. When cooked, it is soft, with the grains remaining separate. Texmati is a hybrid grown in Texas, and Calmati comes from California. They have neither the assertive flavor of true basmati nor the fluffiness of other domestic long-grain rices.

What is paella rice?

Paella rice is a short-grain rice similar to Arborio, but much less expensive. You can substitute Arborio in any recipe that calls for paella rice or use a medium-grain California rice.

What kinds of flavored rice do you recommend?

I really do not recommend using flavored rice. It is so easy to season rice that there is no reason to pay the extra money for something you can do better yourself. Using flavored rice sounds easy, but it is often harder to adjust the seasoning than to start from scratch.

What is risotto rice and how do you make risotto?

Risotto is not a type of rice but a preparation that is usually made with Arborio rice, a grain unique to the Po Valley in northern Italy. It is shorter and rounder than other short-grain rice and contains the right degree of starch needed to create the creamy texture that binds a risotto. In this classic northern Italian dish, the kernels should be tender on the outside but still retain a touch of chewiness on the inside.

My market does not carry Italian Arborio rice but does carry Italian-style rice. Is it okay to use for risotto?

In a pinch it is fine, but it is worthwhile to stock up on Arborio rice, which you can easily get through a good mail-order resource. Be sure to buy Arborio superfino, which is more expensive than the regular rice but has larger, plumper kernels.

Recipes for risotto sound long and quite difficult. Can you recommend any shortcuts?

A properly cooked risotto takes 25 minutes of almost nonstop stirring to make and should be served as soon as it is done. However, I have developed a method that requires only 10 minutes of stirring and works beautifully (see Creamy Risotto with Vine-Ripe Tomatoes, Fresh Rosemary, and Mascarpone, page 281.) Do not be tempted by recipes that suggest making a perfect risotto in the microwave. If you have ever tasted a really good risotto, you will agree that it is well worth spending 10 to 12 minutes to make this exquisite dish.

What is a rice pilaf and what kind of rice do you use to make it?

Pilaf is a classic Turkish peasant dish. The rice is first sautéed in some butter until the grains turn translucent, then baked in either vegetable or chicken broth. Toward the end, currants or raisins are gently folded into the rice. It is served as an accompaniment to grilled meats and seafood.

In classic French cooking, pilaf is simply braised rice that is either baked in the oven or cooked on top of the stove over very low heat.

Herbs such as parsley, dill, and basil can be added toward the end of cooking, as can cooked peas, bits of asparagus, and other parboiled vegetables.

Many pilafs are started with a small amount of minced shallots or finely minced onions that are first sautéed in a little butter, which adds more taste to the rice. I usually use long-grain rice to make a pilaf but have also used converted rice, which takes a little longer to cook.

I like to cook simple Chinese food and never know what kind of rice to make with it.

Look for short-grain Asian rice, such as White Rose, which is starchy and has a sticky consistency when cooked. In Japan and China, this rice is known as glutinous rice and is considered best when it is young, because that is when it cooks into a soft mass ideal for picking up with chopsticks. Look for California-grown short-grain rice rather the Japanese brands, which are much more expensive.

I love Thai food and find the rice in Thai restaurants especially tasty. Is there a special rice or does the cooking method make a difference?

The rice of choice in Thai cooking is jasmine rice. It is exceptionally aromatic, and now that it is grown in the United States, it is widely available.

What is the best rice to use for rice pudding?

Many recipes suggest using long-grain rice, which results in a lighter-textured pudding. Spanish and French rice puddings are usually made with a short- or medium-grain rice, such as Arborio, then oven baked. These puddings are denser and creamier.

Many recipes suggest rinsing rice well, others say not to. What do you suggest?

When it comes to rinsing rice, every cook has an opinion. As a rule, I never rinse rice for a risotto or paella because it is the starch in the rice that binds the ingredients and gives these preparations their special character. I do rinse all long-grain rice because it makes for a fluffier consistency.

Some recipes recommend soaking rice for up to 8 hours. Is this a good idea?

In many Asian countries, rice is soaked anywhere from 30 minutes to 8 hours. Soaking does produce fluffier rice, and I often do it when I remember, but it is not a must.

Is there a foolproof recipe for cooking rice?

Here is the basic recipe using long-grain rice: In a heavy 2-quart saucepan, bring 2 cups lightly salted water to a boil. Add 1 cup rice. Cover the pan tightly, lower the heat, and simmer for 15 minutes, or until the rice is fork tender. Remove from the heat and season to taste with salt and pepper. I often use chicken bouillon instead of water, for additional flavor. I also like to flavor the cooked rice with a tablespoon or two of butter, but this is not a must.

What should the consistency of rice be like when it is done?

If you go around the world asking what sort of texture perfectly cooked rice should have, you will get different answers everywhere you go. Basmati or jasmine rice should be light and quite dry, with separate fluffy grains. Sushi rice must be soft and sticky. Generally I like the mouthfeel of separate grains—tender but not mushy, with just a little bite left in them. Uncover the rice two or three times during its cooking and taste to see if it is done. Slightly overcooked rice is not a disaster, but undercooked rice is unpleasant to eat.

What is the best way to reheat rice?

It depends on the rice. Both long-grain and converted rice can easily be reheated in a low oven or in the microwave. You can also reheat Arborio rice, but it is not advised; there is a definite change in texture and it gets somewhat gummy.

What exactly is brown rice?

Brown rice is the whole, unpolished grain with the outer husk intact. It is much chewier than white rice and takes twice as long to cook, because each kernel is enclosed in a delicate layer of high-fiber bran. The bran adds a nutty flavor to the rice, plus, of course, a slew of vitamins and minerals. But it is the fiber that is the main reason to include brown rice in your diet.

I know that brown rice is healthier than white rice, but I am always put off by the long cooking time. Is there any way around it?

Not really. I soak brown rice overnight. This makes for more tender rice, but does not shorten the cooking time, which, I admit, is long. Try cooking the rice in the oven. Although the cooking time is the same, somehow it does not feel as long as when the rice is cooked on top of the stove.

TIP If you do not have a saucepan with a tight-fitting lid, wrap the lid with a tea towel, which keeps steam condensation inside the lid while the rice cooks. This is also a good way to put a risotto on hold and keep it from drying out.

I have been saving a bag of Arborio rice for over a year, and when I opened it, it was full of bugs. How should I store rice?

All varieties of rice can be stored for a long time. But it is always best to transfer the rice to airtight jars. Keep the rice in a cool place, or refrigerate it, and it will last indefinitely.

Wild Rice

How is wild rice related to other types of rice?

Wild rice is not actually rice at all but the seeds of an annual aquatic marsh grass that grows naturally in the northern Great Lakes area in both the United States and Canada. There is also commercial production in California and the Midwest. It is the only cereal native to North America, and the only wild grass plant that produces a grain large enough for use as a food.

Because of the way wild rice is harvested, you'll need to clean it before cooking. Put it in a bowl of water, swish it around, and strain off any plant material that floats to the top.

What is the best way to cook wild rice? I find the kernels stay too hard or become mushy.

There is no real formula for cooking wild rice because the amount of liquid needed and the cooking time can vary enormously from one variety of rice to another. I usually use 3 cups water to 1 cup wild rice and allow at least 45 to 50 minutes of cooking. If the rice is tender but has not absorbed all the liquid, you can strain it off. Don't worry about letting the rice sit. It is better to give yourself plenty of time, and if the rice is done before you are ready to serve it, you can keep it hot in a warm oven or reheat it in the microwave.

Unfortunately, there are no bargains when it comes to wild rice. The real stuff is very expensive and very delicious. Organic hand-picked wild rice from the Great Lakes region is a superb-tasting grain that is difficult and time consuming to harvest and, therefore, commands the mighty price ticket.

To be on the sure side, buy only organic wild rice and look for kernels that are long with black-brown hues. I find it best to buy wild rice from a reputable mail-order source, since supermarket brands are usually inferior and never seem to cook evenly.

What wild rice mixes do you recommend?

I do not recommend any wild rice mixes, although I realize that they are quite popular. Good wild rice has a unique flavor and texture that is best served on its own.

Beans

With the continuing interest in peasant cooking of every type, beans are gaining in popularity, and I am delighted. I grew up with wonderful bean soups, lentil ragouts, and chickpea salads. The bean stand in the Boqueria Market in Barcelona was one of my favorite stops for many years, and strolling through the aisles with my small cone-shaped bag of warm white beans drizzled with olive oil continues to be one of my favorite things to do when I am there.

That beans are extremely versatile and packed with nutrition is widely known by now, so why is it that beans are have been so slow to gain the popularity they deserve? For starters, legumes in general, and especially chickpeas and beans, require advance preparation and do not fit easily into the lifestyle of the spur-of-the-moment cook. They need to be soaked in water overnight, and while some cookbooks recommend a cooking method in which the beans are brought to a boil for a minute and then left

to soak briefly before cooking, this method does not work as well as overnight soaking.

Also, even though they are dried, legumes need to be fresh. In spite of the fact that we are led to believe that beans have an indefinite shelf life, this is far from the truth. The older the bean, the tougher the shell, and the longer it takes to cook, resulting in mushy, unevenly cooked beans, which can be less than digestible. In Europe, packages of beans are dated and are meant to be consumed within a year of purchase.

Some types of beans are easier to buy fresh than others. Pinto or cranberry beans as well as black beans are usually delicious if you can buy them in a Hispanic neighborhood, because the turnover is such that the beans are fresh. If you have access to a Middle Eastern store, the same holds true for chickpeas. Other beans, especially great northern and cannellini beans, can be more of a problem. Try a good mail-order resource, but always check first to find out if the resource dates their beans. If not, you are much better off buying your beans at a health food store that has a large turnover.

Avoid buying beans in fancy cook shops or specialty stores, because, again, chances are that they are not fresh. There is a good chance that a lot of effort will have gone into the packaging and not into the product.

Since beans are often called the caviar of Tuscany, I am always on the lookout for them when I go to Italy. To me beans are worth that kind of search, because when they are good they are as good as caviar and equally memorable.

The choice of beans can be very confusing. Which of them do you recommend having in the pantry?

I usually stock up only on the beans I use regularly. There is no point in hoarding beans, which is something I used to do, because they do not keep long. Here is a list of my basic pantry beans:

- Black beans
- Pinto beans
- Great Northern beans or cannellini beans
- Red kidney beans
- French Puy lentils

When a recipe calls for a specific type of bean, can you substitute one type for another?

It depends entirely on your recipe. If you are making something as simple and gutsy as chili, you can use red, pinto, or pink beans, but if you are making a French cassoulet or the French garbure (a traditional soup made with cabbage, root vegetables, and beans), you will need a good-quality white kidney bean, such as Sois-son or Great Northern. The famous Italian soup pasta e fagioli is traditionally made with cannellini beans, but it works with Great Northern beans. Black beans, also called turtle beans, are more problematic because of their distinct color and taste, and cannot be substituted in recipes.

Are organically grown beans a better choice?

Organically grown beans are usually fresher but also pricier. If a bag of beans is clearly marked with an expiration date and appears fresh, it doesn't matter if they are organically grown. For mail-order sources of organically grown beans, see Sources, page 397.

Can dry fava beans be substituted in recipes that call for fresh?

Fresh fava beans, one of the oldest members of the bean family, are just starting to show up in supermarkets around the country, but the going is slow. Your best bet for fresh is in Italian neighborhoods, specialty markets, and greengrocers during the spring and early summer. The dried bean, which is very popular in the Middle and Far East, is practically another vegetable.

Cook fava beans only when you see them fresh at your market in the spring. Choose those that have bright green, velvety-textured skin with no black spots.

TIP Unlike peas, which are ready to eat out of the pod, fava beans have both an outside pod and a tough skin that needs to be removed to expose the delicate bean inside. Blanch the shelled beans in a pot of boiling water for 30 seconds, and this protective skin will slip right off.

Can I use canned beans, such as kidney beans and chickpeas, in bean soup?

I use canned beans as an accent when they are needed mainly for texture, not flavor, as in minestrone or winter cabbage soup. But for real bean soups, such as the White Bean and Chorizo Soup, page 292, you must begin with dried beans that have been soaked overnight.

What is the difference between a garbanzo bean and a chickpea?

Just the name. Garbanzo bean is the Spanish name for chickpea; the Italians call it *ceci*, and the French *pois chiche*.

Why do I have more trouble cooking chickpeas than other legumes?

As with most legumes, chickpeas should be less than a year old if they are to cook quickly. It is best to buy them from a Middle Eastern or Latin market or health food store. Chickpeas older than a year will practically never soften, no matter how long you cook them. Be sure to soak them in water to cover for 24 hours with about ¼ teaspoon baking soda added to the water. Drain and then cook them in plenty of unsalted water.

Can canned chickpeas be used in recipes that call for the dried legume?

Sometimes that works, especially when the chickpeas are only used for additional texture. Use the Mexican or Goya brands, which are firmer than the American varieties.

TIP Spring water is a good choice for cooking beans because it is softer. Many Spanish cooks believe that it makes a more tender legume.

What is the difference between the basic brown lentil and green, yellow, pink, and red lentils? Are they interchangeable in recipes?

The brown lentil you see in your grocery store is the common lentil most often consumed in the West. It is a good-natured legume, extremely high in protein; it needs no presoaking, and it cooks quickly. It deserves to be high up there in everyone's cooking repertoire because it is healthful, inexpensive, and extremely versatile. Pink, yellow, red, and green lentils, except for the French Puy lentils, can be classified as Indian lentils, since it is in India that they are consumed the most. All Indian dals are made from split lentils and many of India's vegetarian dishes are based on this legume.

French green lentils, so often mentioned in cookbooks and food magazines, are not available in my grocery store. Can I substitute brown lentils for the French ones?

The taste and texture of brown lentils is quite different from the French ones. You can easily get Puy lentils through a good mail-order source (see page 397). Since they have a long shelf life, they are good to have on hand.

TIP When shopping for any type of bean, look them over carefully. If you find many that are broken and split, it means that they are old and will stay tough no matter how long you cook them.

Black beans are hard to find dried in my area. Are canned acceptable?

Personally, I do not recommend canned black beans because they are usually mushy and often too salty. Good-quality dried black beans will look dark and glossy. They are available through mail-order sources (see page 397).

Do all beans need soaking? Is there a shortcut that you can recommend?

All presoaked beans have a better texture and better overall flavor than those that have not been soaked. If you find yourself pressed for time, here is a shortcut: Cover the beans with water, bring them to a simmer, then let them stand off the heat for 1 to 2 hours. Drain, cover with fresh water by an inch, and cook over low heat either on top of the stove or in the oven.

At what point in cooking should beans be seasoned?

Never salt beans until they are almost tender, or they will be tough. You can flavor the bean cooking water with a whole unpeeled onion and a whole unpeeled head of garlic. You can also use fresh herbs, such thyme, rosemary, and bay leaf.

Beans also must be tender before being combined with a tomato mixture, because the acidity in the tomato will keep the cell walls of the bean from softening.

I find beans very hard to digest. Is there any way to get around it?

There is no question that beans are hard to digest. Be sure to soak your beans 12 to 14 hours or overnight, changing the water at least once and using fresh, cold water to promote the softening of the starches. Do not use hot or boiling water to start cooking soaked beans because they will not cook evenly. Always cover the beans with at least 2 inches of water. This helps extract the oligosaccharides, a string of sugar molecules that our normal digestive enzymes can't deal with. Beans have high levels of these compounds because they are seeds, and plants store high levels of certain sugars in their seeds to ensure their survival.

I always add 3 to 5 drops of Beano, a natural enzyme available in all health food stores, for each portion of beans.

Fettuccine with Jalapeño-Lime Cream and Shrimp

Pasta has moved beyond the classic preparations and can now be found prepared with a variety of interesting ingredients. I first sampled this dish in a Santa Fe restaurant, and it has been one of my favorites ever since. For a variation, use grilled diced chicken instead of the shrimp.

SERVES 2 TO 3

3 tablespoons olive oil

½ pound peeled medium shrimp

Coarse salt and freshly ground black
  pepper

1 cup green bell pepper, cored and
  cubed

1 cup yellow bell pepper, cored and cubed

1 cup red bell pepper, cored and cubed

1 medium red onion, cubed

1 tablespoon minced jalapeño pepper,
  seeds removed

Juice of ½ lime or more to taste

½ cup heavy cream

½ pound egg fettuccine

Minced fresh cilantro or basil

8 grape tomatoes, cut crosswise in
  half, optional

1. In a large heavy cast-iron skillet, heat 2 tablespoons of the olive oil. Add the shrimp and sauté over high heat until nicely browned and slightly charred. Season with salt and pepper. Remove to a cutting board and, when they are cool enough to handle, cube the shrimp.

2. Add the remaining 1 tablespoon oil to the pan. Add the bell peppers and red onion. Season with salt and pepper and cook over medium-high heat until soft and slightly charred, 3 to 5 minutes. Be sure not to burn the peppers.

3. Lower the heat and add the jalapeño, lime juice, and cream. Bring to a simmer and correct the seasoning. Set the sauce aside.

4. In a large pot, bring plenty of salted water to a boil. Add the fettuccine and cook for 3 to 4 minutes, or until just tender. Drain and return to the pot.

5. Pour the pepper-and-shrimp sauce over the pasta; toss with two spoons. Add a large grinding of black pepper and the minced cilantro, and serve in individual soup bowls sprinkled with the sliced tomatoes, if using.

Fettuccine with Zucchini and Goat Cheese in a Ginger-Tomato Fondue

Here is a lovely pasta dish packed with flavor and texture. Be sure to use good-quality dried egg fettuccine here. DeCecco is acceptable, but Rustichella D'Abruzzo is a sturdier and much more flavorful pasta that is ideal for this recipe. You can vary the dish by using red, yellow, and green peppers and yellow zucchini instead of green. I also like to add some gaeta olives to the dish for additional piquant flavor.

SERVES 4 TO 5

4 tablespoons extra virgin olive oil

3 small zucchini, quartered lengthwise and cubed

1 red bell pepper, cored and cubed

1 small dried hot red pepper, broken into pieces

2 large garlic cloves, minced

1 tablespoon minced fresh ginger

1 large shallot, minced

1½ cups Creamy Tomato Sauce (page 268)

1 tablespoon minced fresh oregano, optional

Salt and freshly ground black pepper

½ pound egg fettuccine

4 tablespoons minced fresh basil

1 cup crumbled mild goat cheese

Freshly grated Parmesan cheese

I. Heat 2 tablespoons of the olive oil in a large heavy skillet over medium-high heat. Add the zucchini and bell pepper, and sauté quickly until lightly browned, about 3 minutes, Transfer to a dish with a slotted spoon.

2. Add the remaining 2 tablespoons oil to the skillet and when hot, add the hot pepper and cook until dark. Remove and discard the pepper. Add the garlic, ginger, and shallot and cook for 1 minute without browning. Add the tomato sauce and oregano, if using, and season with salt and pepper; simmer, covered, for 15 minutes.

3. Bring plenty of salted water to a boil in a large saucepan, add the fettuccine, and cook for 3 to 5 minutes, or until just tender. Drain well and return the pasta to the pan.

4. Mix the tomato sauce and zucchini-and-pepper mixture together with the basil and goat cheese and toss lightly; the goat cheese should just be warm, not melted. Taste and correct the seasoning. Transfer to a serving bowl. Serve at once with a bowl of the Parmesan.

Bow Ties with Creamy Tomato Sauce, Bacon, and Radicchio

A quick basic tomato sauce is probably the most important sauce to have on hand when making pasta. I usually make a large batch of it and freeze it in 1- and 2-cup containers. Here the sauce is enriched with a touch of cream, and when tossed into pasta with some sautéed bacon and radicchio, it makes for a delicious main course.

SERVES 4

2 tablespoons butter
½ cup diced pancetta or blanched bacon
1 tablespoon olive oil
1 large garlic clove, finely sliced
1 large head of radicchio, cored and cut into eighths
Salt and freshly ground black pepper
1 cup Quick Tomato Sauce, page 274
¼ cup heavy cream, optional
½ pound bow ties (fusilli)
2 to 3 tablespoons freshly grated Parmesan cheese

1. In a large heavy skillet, melt the butter over medium heat. Add the pancetta or bacon and sauté for 2 minutes, or until crisp. Remove with a slotted spoon to a dish and reserve.

2. Add the olive oil to the skillet and when hot, add the garlic and radicchio. Season with salt and pepper and sauté for 2 minutes, or until just wilted. Transfer to a side dish.

3. Heat the tomato sauce, and add the cream; taste and correct the seasoning. Keep warm.

4. In a large pot, bring salted water to a boil. Add the pasta and cook over high heat for 8 minutes, or until just tender. Drain, reserving ½ cup of the pasta water.

5. Return the pasta to the pot together with the tomato sauce, pancetta, and radicchio. Toss with two spoons. Add a large grinding of pepper and 2 tablespoons of the Parmesan. If the sauce seems too thick, add a little of the reserved pasta water. Taste and correct the seasoning. Serve hot in deep soup bowls.

REMARKS When choosing radicchio, be sure to buy light, large, fluffy heads. Separate the leaves, folding them in half lengthwise and cutting out the white triangle with a sharp knife.

Linguine with Mussels in a Creamy Tomato-Saffron Fondue

Here is a terrific gutsy pasta preparation. It can be made more elegant by shelling the mussels, leaving 6 to 8 in their shells as a garnish. Other pasta shapes, such as spaghetti or dried egg fettuccine, can be used as well. Serve this as a main course preceded by a well-seasoned salad or a soup.

SERVES 4 TO 5

1 large shallot, thinly sliced, plus 2 large shallots, minced

¾ cup dry white wine

2 sprigs each of thyme and flat-leaf parsley

6 whole black peppercorns

2 to 3 pounds fresh small mussels, well scrubbed

2 tablespoons unsalted butter

1 teaspoon extra-virgin olive oil

1 small dried hot red peppers, crumbled

2 large garlic cloves, minced

4 ripe medium tomatoes, peeled, seeded, and chopped

1 tablespoon minced fresh thyme

Salt and freshly ground black pepper

¼ teaspoon saffron threads

½ cup crème fraîche

3 to 4 tablespoons fine julienne of fresh basil

1 pound fresh linguine

Tiny leaves of fresh basil, for garnish

1. In a large flameproof casserole, combine the sliced shallot, wine, thyme and parsley sprigs, and peppercorns. Bring to a simmer over medium heat. Add the mussels and simmer, covered, shaking the pan, until the mussels open, 5 to 7 minutes; discard any that do not. With a slotted spoon, transfer the mussels to a large bowl and reserve. Strain the mussel broth through a double layer of cheesecloth and set aside.

2. Melt the butter together with the olive oil in a heavy skillet over medium heat. Add the hot peppers, minced shallots, and garlic and cook for 1 minute, stirring constantly. Add the tomatoes and minced thyme and season with salt and pepper. Reduce the heat and simmer, partially covered, until all the tomato liquid has evaporated.

3. Add the reserved mussel broth and saffron; simmer until slightly reduced. Transfer the mixture to a food processor and puree until smooth. Return the sauce to the skillet. Add the crème fraîche and julienne of basil. Heat just through. Keep warm.

4. Bring plenty of salted water to a boil in a large pot. Add the fresh pasta and cook for 3 to 4 minutes, or until just tender. Immediately add 2 cups cold water to the pot to stop further cooking. Drain well and return the pasta to the pot. Add the warm sauce and reserved mussels and toss gently. Taste and correct the seasoning.

5. Transfer the pasta to individual serving bowls and garnish each portion with a few of the reserved mussels in their shells and tiny basil leaves. Serve at once.

Penne with Broccoli, Peppers, Tomatoes, and Smoked Mozzarella

I always have some homemade tomato sauce on hand either in the refrigerator or the freezer, which makes it much easier to come up with a pasta dish like this almost on the spur of the moment. Smoked mozzarella is also best kept in the freezer; let it thaw just enough so you can grate it into the finished pasta dish. Leftover pasta is delicious at room temperature the next day. Just sprinkle with additional minced fresh basil, a few drops of good olive oil, and freshly ground pepper.

SERVES 4 TO 5

5 tablespoons extra virgin olive oil

1 large shallot, minced

4 large garlic cloves, thinly sliced

10 ripe Italian plum tomatoes or one 32-ounce can Italian plum tomatoes, drained

1 tablespoon fresh oregano or 2 teaspoons dried

4 tablespoons fine julienne of fresh basil

Coarse salt and freshly ground black pepper

1 bunch of fresh broccoli

½ to ¾ cup chicken stock or bouillon

1 red bell pepper, cored, quartered, and thinly sliced

1 yellow bell pepper, cored, quartered, and thinly sliced

½ pound imported penne

1 cup finely diced smoked mozzarella

Freshly grated Parmesan cheese

1. In a heavy 2-quart saucepan, heat 3 tablespoons of the olive oil over medium heat. Add the shallot and 2 of the garlic cloves and cook until just soft. Add the tomatoes, oregano, and basil, season with salt and pepper, and simmer, covered, for 25 minutes, or until all the tomato water has evaporated and the mixture is quite thick. Transfer the mixture to a food processor and process until smooth.

2. Trim the broccoli into florets. Peel the stalks with a vegetable peeler. Remove any leaves and slice the stalks crosswise into ½-inch slices.

3. In a large cast-iron skillet, heat the remaining 2 table-spoons olive oil over medium heat. Add the remaining 2 garlic cloves and the broccoli florets and stems and season with salt and pepper. Add ½ cup of the stock, cover, and simmer for 5 to 7 minutes, or until the broccoli is tender. Be careful not to over-cook. Set aside off the heat.

4. Bring water to a boil in a steamer. Add the bell peppers and steam, covered, for 2 to 3 minutes, or until just tender.

5. In a large pot, bring salted water to boil. Add the penne and cook until just tender. Immediately add 2 cups cold water to the pot to stop the penne from further cooking and drain well.

6. Return the pasta to the pot. Add the broccoli, peppers, and tomato mixture and season with salt and pepper. Add the smoked mozzarella, toss gently, and serve immediately in shallow soup bowls accompanied by a bowl of the Parmesan.

Spaghettini with Tuna, Capers, and Black Olives

Here is a wonderfully simple pasta dish I make almost weekly and vary according to what I see fresh in the market and seasonally. Some interesting additions could include cubed and sautéed eggplant, a couple of fire-roasted red peppers, or a medium zucchini cubed and sautéed in a little oil. Now that pitted black olives are widely available, this dish can be put together in a matter of minutes. A well-seasoned salad is the perfect accompaniment.

SERVES 2 TO 3

- $\frac{1}{2}$ cup extra virgin olive oil
- 4 tablespoons minced flat-leaf parsley
- 3 large garlic cloves, minced
- $\frac{1}{2}$ cup minced fresh basil, optional
- 1 teaspoon dried oregano
- 4 medium tomatoes, diced, or 14 grape tomatoes, cut in half
- One 7$\frac{1}{2}$-ounce can tuna packed in olive oil
- 4 flat anchovy fillets, drained and minced
- $\frac{1}{2}$ cup oil-cured black olives, preferably kalamata, diced
- Salt and freshly ground black pepper
- $\frac{1}{2}$ pound spaghettini
- 1 cup cubed smoked mozzarella, optional

1. Heat the olive oil in a large heavy skillet. Add 2 tablespoons of the parsley, and the garlic, basil, and oregano. Cook just until the garlic is soft, about 2 minutes.

2. Add the tomatoes and cook for another 3 to 4 minutes, or until the tomatoes are soft. Add the tuna, anchovies, and olives and heat through. Season with salt and pepper and set aside.

3. In a large pot, bring salted water to a boil. Add the pasta and cook for 7 to 8 minutes, or until it is just tender. Drain and return to the pot. Add the sauce and toss with two forks. Add the remaining 2 tablespoons parsley and the mozzarella and serve immediately.

Quick Tomato Sauce

A quick basic tomato sauce is a wonderful sauce to have on hand if you are a pasta fan. I usually make a large batch of it and freeze it in 1- and 2-cup containers. The sauce can be enriched with a touch of cream, which gives it a more mellow, delicate flavor that works especially well with homemade or egg-based pasta.

MAKES 2 CUPS

2 tablespoons olive oil

1 tablespoon butter

⅓ cup minced shallots

2 large garlic cloves, minced

1 dried hot red pepper, optional

One 32-ounce can Italian plum tomatoes,
    drained and chopped

1 teaspoon dried oregano

1 teaspoon sugar

Salt and freshly ground black pepper

¼ cup heavy cream and 2 tablespoons butter,
    optional

1. In a 2-quart heavy saucepan, heat the olive oil and butter. Add the shallots, garlic, and hot pepper. Sauté the mixture until soft but not browned. Add the tomatoes, oregano, and sugar. Season with salt and pepper. Bring to a boil, reduce the heat, and simmer for 30 minutes, stirring from time to time.

2. Transfer the tomato sauce to a food processor or a blender, add the cream and butter, and puree until smooth. Taste and correct the seasoning. Transfer the sauce to a covered jar and refrigerate for up to 5 days, or freeze for longer storage.

Catalan Chicken Paella

Although a traditional paella calls for a combination of seafood and poultry, this classic Spanish dish changes according to the region. In the Pyrennees it is often made with rabbit, chicken, and chorizo sausage. Here I use only chicken, but if you can get fresh rabbit do try it, because it gives the dish another flavor dimension that is quite interesting. Serve with a crusty loof of bread.

SERVES 6

3½ cups chicken stock or bouillon

¼ teaspoon loosely packed saffron threads

4 small whole chicken legs, cut in half at joint to make 4 thighs and 4 drumsticks

4 tablespoons extra virgin olive oil

Salt and freshly ground black pepper

1 or 2 small dried hot red peppers, broken into pieces

1 large onion, quartered and thinly sliced

3 large garlic cloves, minced

2 cups peeled, seeded, and chopped Italian plum tomatoes or one 16-ounce can Italian plum tomatoes, drained and chopped

1 red bell pepper, cubed

1 green bell pepper, cubed

1¼ cups Arborio rice or medium-grain Spanish

½ pound smoked chicken, cut into ½-inch cubes

½ cup cooked peas

Lemon wedges

Tiny leaves of flat-leaf parsley

Thinly sliced pimientos

1. In a small saucepan, combine the stock and saffron threads. Bring to a boil, reduce the heat, cover, and simmer for 20 minutes. You should have 3 cups saffron stock; if not, boil to reduce the stock to 3 cups.

2. Preheat the oven to 350°F. Dry the chicken pieces thoroughly with paper towels. In a large, deep cast-iron skillet, heat 2 tablespoons of the olive oil over medium-high heat. Add the chicken pieces and brown nicely on all sides. Remove the chicken from the skillet and drain on paper towels. Season with salt and pepper.

3. Discard all but 1 tablespoon of fat from the skillet. Add the hot peppers; cook until dark and discard. Heat the remaining 2 tablespoons oil in the skillet. Add the onion and garlic and cook, stirring often, for 10 minutes, or until the onion is soft and nicely browned.

4. Add the tomatoes and bell peppers. Season with salt and pepper and continue to cook until the tomato water has evaporated.

5. Add the rice, stirring it thoroughly into the onion and tomato mixture. Add the reserved saffron broth and browned chicken pieces. Bring to a boil, cover tightly, and place in the center of the oven. Braise for 20 minutes. Add the smoked chicken and peas and continue to cook for 5 to 10 minutes longer, or until the rice is tender.

6. Remove from the oven, correct the seasoning, and garnish with the lemons, parsley, and pimientos. Serve at once.

Oven-Baked Rice Pudding with Lemon and Cranberries

For an easy and satisfying weekend dessert, few recipes are as delicious and homey as a rice pudding. Once placed in the oven, it needs no attention, and the result is a no-nonsense dessert that will be devoured in a matter of a day. I often vary the pudding by adding dried cherries or raisins. The strawberry sauce, page 395, adds a refreshing finishing touch.

SERVES 6

$3\frac{1}{2}$ cups whole milk

1 cup sugar

$\frac{1}{2}$ cup long-grain rice

Zest of 1 lemon, cut into fine julienne

$\frac{1}{4}$ cup dried cranberries

2 extra large eggs, separated

1 teaspoon vanilla extract

Ground cinnamon or freshly grated nutmeg

1. Preheat the oven to 325°F.

2. Heat the milk in a large ovenproof saucepan together with all but 2 tablespoons of the sugar, and stir until dissolved. Add the rice, lemon zest, and cranberries; cover loosely with foil and bake for 2 to $2\frac{1}{2}$ hours, stirring once or twice. The rice should be quite tender and all the milk absorbed. Remove from the oven and whisk in the egg yolks and the vanilla.

3. Beat the egg whites with the remaining 2 tablespoons sugar until stiff but not dry; fold into the rice. Transfer to a serving dish, sprinkle with cinnamon or nutmeg, and serve at room temperature or slightly chilled.

Risotto with Clams and Tomato-Saffron Fondue

If you like spaghettini with clams you will love this risotto. The rice can be infused with the delicious briny flavor of clams and ripe tomatoes. Serve the risotto as a light main course preceded by a seasonal salad or a quick sauté of seasonal vegetables, or as an appetizer. Leftovers reheat beautifully in the microwave, but I doubt that you will have any.

SERVES 4

2 dozen littleneck clams
4 tablespoons extra virgin olive oil
1 dried hot red pepper, crumbled
½ cup white wine
2 cups fish stock or bouillon
2 tablespoons minced shallots
2 garlic cloves, minced
3 ripe tomatoes, peeled, seeded, and chopped
¼ teaspoon powdered saffron
Coarse salt and freshly ground black pepper
1 cup Carnaroli or Arborio rice
2 tablespoons minced flat-leaf parsley

1. Wash the clams thoroughly, scrubbing well with a hard brush.

2. In a large saucepan, combine 2 tablespoons of the olive oil, the hot pepper, and wine. Add the clams. Cover the saucepan and simmer the clams until they open. Discard any unopened clams. Remove the clams from the broth, discard the shells, and dice the clams. Reserve the broth and the clams separately.

3. Measure the clam broth, adding enough fish bouillon to make 5 cups.

4. In a 3½-quart heavy saucepan, heat the remaining 2 tablespoons olive oil. Add the shallots, 1 clove of the garlic, the tomatoes, and saffron. Season with salt and pepper and cook the mixture over medium heat until all the tomato juice has evaporated.

5. Add the rice and cook for 1 minute or until it turns opaque. Add 2 cups of the clam broth. Cover the pan and simmer over low heat for 12 minutes, or until the rice is barely tender.

6. Start adding more broth, ¼ cup at a time, stirring constantly and switching to fish bouillon as needed. When the rice is done, it should be soft but still slightly chewy. Add the clams and taste and correct the seasoning.

7. Add the parsley and the remaining garlic clove and simmer for another minute or two. Serve immediately in deep bowls accompanied by a bottle of extra virgin olive oil.

Leek and Stilton Cheese Risotto with Mascarpone

Leeks and rice make a wonderful combination, especially when teamed in this buttery risotto, which gets a kick from Stilton cheese. If you can't find good Stilton, try another blue, such as Danish or Maytag. Serve the rice as a starter or a side dish to a veal roast or pan-seared veal chops. Pass a bowl of grated Parmesan cheese on the side.

SERVES 6

½ cup mascarpone

2 ounces Stilton cheese, diced, or more to taste

2 tablespoons unsalted butter

2 cups minced leeks, rinsed and drained

5 cups chicken stock or bouillon

1½ cups Arborio rice

Salt and freshly ground black pepper

3 tablespoons freshly grated Parmesan cheese

Minced flat-leaf parsley

1. Combine the mascarpone and Stilton in a food processor and puree until smooth.

2. In a heavy 3-quart saucepan, melt the butter over low heat. Add the leeks and 3 tablespoons of the stock and braise, covered, until tender, about 5 minutes. Add the rice, season with salt and pepper, and stir well with the leek mixture. Add 2 cups of the stock, cover tightly, and simmer over very low heat for 10 minutes.

3. Raise the heat to medium and uncover the saucepan. Gradually add the remaining stock, ¼ cup at a time, stirring constantly, until each addition has been absorbed, for the next 10 minutes; you may not need all of the remaining stock. The rice should be tender on the outside but somewhat chewy on the inside.

4. Add the Stilton mixture and the Parmesan and fold gently. Taste and correct the seasoning. Garnish with parsley and serve at once.

Creamy Risotto with Vine-Ripe Tomatoes, Fresh Rosemary, and Mascarpone

Here is a wonderful "tomato season" risotto, which I make as soon as I get really good tomatoes at my farm stand or when my own are finally ready for picking. Other herbs, especially basil and chives, are good additions. Be sure not to cook the risotto any longer once the tomatoes are added. Serve the rice either as an appetizer or as a side dish to grilled pork tenderloins, quail, or flank steak.

SERVES 4

2 tablespoons extra virgin olive oil

1 medium onion, finely diced

$\frac{1}{4}$ cup dry white wine

$1\frac{1}{2}$ cups Arborio rice

4 cups hot chicken broth or bouillon

Salt and freshly ground black pepper

2 large ripe tomatoes, seeded and diced

$\frac{1}{3}$ cup mascarpone

2 tablespoons julienne of fresh basil, plus sprigs for garnish

1 tablespoon minced fresh rosemary

2 to 3 tablespoons coarsely grated Parmesan cheese, plus some for sprinkling

1. Heat the oil in a heavy 3-quart saucepan over low heat, add the onion, and cook until soft. Add the wine and reduce to a glaze. Add the rice and cook for 1 minute, stirring constantly.

2. Add 2 cups of the broth or bouillon, set over the lowest possible heat, and simmer, covered, for 10 minutes. Raise the heat to medium, uncover the saucepan, and add the remaining broth or bouillon $\frac{1}{4}$ cup at a time, stirring constantly, for the next 10 minutes, until each addition has been absorbed; you may not need all of the remaining broth. The rice should be tender on the outside but still slightly chewy on the inside. Season with salt and pepper.

3. Fold in the tomatoes, mascarpone, herbs, and Parmesan, correct the seasoning, and serve immediately in shallow soup bowls, garnished with sprigs of basil and accompanied by grated Parmesan.

REMARKS If you cannot get mascarpone, you can still make this risotto successfully by adding 2 tablespoons of heavy cream or crème fraîche to the rice. If getting good Parmesan is a problem, it is better not to use it at all, since this is a rather delicate risotto that could easily be overpowered by a stronger or salty cheese.

Braised Bulgur with Two Peppers and Lemon

Bulgur is a favorite grain in the Middle East, where it is used much like rice but also as stuffing for various vegetables. It is also the main ingredient for the famous tabbouleh. Bulgur comes two ways: fine or medium grained. I use medium-grained bulgur for everything, since I find its texture to be more interesting. Serve this flavorful dish as a side to shish kebabs or grilled fish steaks, veal chops, or chicken. Make enough to have leftovers, since this dish can easily be reheated.

SERVES 4 TO 6

2 tablespoons virgin olive oil

$\frac{1}{2}$ cup minced red onion

1 teaspoon minced jalapeño pepper, seeds included

1 medium red bell pepper, finely diced

1 cup medium-grained bulgur

$2\frac{1}{2}$ cups chicken stock or bouillon

Juice of $\frac{1}{2}$ lemon or more to taste

2 tablespoons minced flat-leaf parsley or cilantro

Salt and coarsely ground black pepper

1. In a saucepan, combine the olive oil, onion, jalapeño and bell pepper. Stir in the bulgur. Add the stock. Cover tightly and simmer over low heat for 20 minutes, or until tender.

2. Add the lemon juice and parsley; season with salt and pepper and more lemon juice if it needs it. Serve hot or at room temperature.

Couscous with Spinach, Red Peppers, and Carrots

Couscous makes a wonderful side dish to many preparations. I like to serve it with all types of grilled or pan-seared fish steaks, lamb chops, or a butterflied leg of lamb. You may add several spices to the couscous, depending on your mood. Cumin, saffron, and curry powder all work well. Leftover couscous can be turned into a delicious salad by adding the juice of a large lemon, some fruity olive oil, and a mincing of flat-leaf parsley.

SERVES 5 TO 6

3½ cups chicken stock or bouillon

¼ teaspoon saffron

2 tablespoons unsalted butter

½ cup finely diced red bell pepper

½ cup finely diced carrots

2 large scallions, minced

1½ cups couscous

Salt and freshly ground black pepper

2 cups fresh spinach leaves, washed, dried and
    cut into fine julienne

1. In a small saucepan, combine the stock and saffron. Bring to a boil, reduce the heat, and simmer, covered, for 20 minutes, or until the stock is infused with the saffron and is reduced to 2 cups.

2. Melt the butter in a 2-quart saucepan over low heat. Add the bell pepper and carrots. Cook, covered, until just tender, 5 to 6 minutes. Remove the cover and add the scallions and a little broth. Simmer for 2 to 3 minutes until the scallions are just wilted.

3. Add the couscous to the pan together with the broth and season with salt and pepper. Bring to a boil, cover tightly, and let stand off the heat for 10 minutes, or until all the stock has been absorbed. Return to the stove over very low heat, add the spinach, and fold gently until the spinach has just wilted. Correct the seasoning. Serve at once as an accompaniment to grilled lamb or swordfish steaks.

Corn-Studded Polenta with Parmesan

Fresh corn adds a crunchy texture to this lovely creamy polenta. You can be quite creative with this recipe by substituting grated Cheddar or smoked mozzarella for the Parmesan; or add some heat to it with 1 tablespoon minced red or green chile pepper. When fresh corn is not in season, I use canned corn with excellent results. If you make the polenta ahead of time, keep it warm in the top part of a double boiler and whisk in 2 tablespoons of butter just before serving.

SERVES 4 TO 6

3¼ cups skim milk, or use half whole milk
    and half water

Salt

¾ cup semolina or yellow cornmeal

5 tablespoons unsalted butter

2 cups cooked corn kernels

⅓ cup freshly grated Parmesan cheese

Freshly ground black pepper

1. In a heavy 3½-quart saucepan, combine the skim milk or whole milk and water and 1 teaspoon salt and bring to a slow boil. Sprinkle in the semolina or cornmeal very slowly to avoid lumping, whisking constantly, until all has been added. Reduce the heat to very low and simmer, covered, for 20 minutes, stirring often. A skin will form on the bottom of the pot; do not be alarmed.

2. Remove from the heat, add the butter, corn, and Parmesan, and stir until well blended. Taste and correct the seasoning, adding a large grinding of black pepper, and serve at once.

REMARKS A trick for making polenta without lumps is to place the cornmeal in a grated cheese shaker and shake the cornmeal slowly into the hot liquid.

VARIATION Add 4 ounces crumbled goat cheese in addition to the Parmesan and 2 tablespoons fresh thyme leaves to the finished polenta and just heat through.

Cumin-Scented Quinoa with Yogurt and Mint

When I first tasted quinoa I was not particularly keen on it, but after a trip to Ecuador, where quinoa is used extensively in soups, fritters, and desserts, I started to appreciate this interesting and healthy grain. Now I often add it together with root vegetables to a home-made chicken stock or serve it as a side dish with grilled chicken or pan-seared lamb chops.

SERVES 4 TO 5

1 cup quinoa

2½ to 3 cups chicken stock or bouillon

Salt

2 tablespoons olive oil

½ cup finely diced red onion

2 teaspoons minced serrano pepper

1 teaspoon ground cumin

⅛ teaspoon ground coriander

Freshly ground black pepper

½ cup plain yogurt

2 to 3 tablespoons minced mint

1. In a saucepan, combine the quinoa with 2½ cups of the stock. Season lightly with salt and cook, covered, for 10 to 12 minutes, or until the quinoa is tender and all the broth has been absorbed.

2. In another heavy saucepan, heat the olive oil. Add the onion and serrano pepper and sauté for 2 or 3 minutes, or until the onion is soft but not browned. Add the cumin and coriander and cook for another minute. Add the quinoa, season with salt and pepper, and simmer for 2 to 3 minutes.

3. Add the yogurt and mint and blend well into the quinoa. Taste and correct the seasoning. Serve hot.

REMARKS When made ahead of time, the quinoa may get quite thick. Thin it out with a little broth or add more yogurt. It should have the texture of thick oatmeal. If possible, use the imported Greek yogurt, which has a wonderful thick texture and delicious flavor.

Chickpea, Tomato, and Cilantro Salad

Chickpeas are amazingly versatile legumes that are delicious both hot and cold. I like to serve this salad as a starter, often topped with cubed feta or goat cheese, but it also makes an excellent accompaniment to grilled flank steak, salmon steaks, or grilled shrimp. If you plan to use canned chickpeas, be sure to use a Spanish brand such as Goya.

SERVES 4 TO 5

4 to 5 cups cooked chickpeas

2 ripe tomatoes, seeded and finely cubed

1 cup diced red bell pepper

½ cup diced red onion

1 to 2 tablespoons diced jalapeño pepper

1½ tablespoons sherry vinegar

Juice of 1 lemon

8 tablespoons extra virgin olive oil

Coarse salt and freshly ground black pepper

3 tablespoons tiny cilantro leaves

1. In a bowl, combine the chickpeas, tomatoes, bell pepper, onion, and jalapeño pepper.

2. In another bowl, combine the sherry vinegar, lemon juice, and olive oil. Season with salt and pepper and whisk the mixture until smooth. Combine with the chickpea and tomato mixture and toss gently. Add the cilantro, taste, and correct the seasoning. Cover and chill for 2 to 4 hours before serving.

Lentil Salad with Goat Cheese and Roasted Peppers

Early on I knew that there was a trick to making a good lentil salad. My grandmother taught me this little trick and I never forgot it. The key to the success of all legume salads is that the dressing is poured over the warm legumes and that the beans or lentils should then remain over low heat for as much as 30 minutes before serving. Any dried bean or lentil dressed the day before will taste even better, so always plan on making a little more for tasty leftovers. Since excellent goat cheese is hard to get outside large metropolitan areas, I suggest you use something as easily available as Montrachet, but freeze it lightly so slices it neatly. Drizzle this cheese with olive oil and some fresh thyme to give it the flavor it needs. If you are pressed for time, you can use jarred roasted red bell peppers.

SERVES 4

1 cup green Puy lentils

Salt

1 medium shallot, minced

1 large garlic clove, mashed

2 tablespoons balsamic vinegar

6 tablespoons extra virgin olive oil

Freshly ground black pepper

4 slices of goat cheese, cut about ½ inch thick

Minced fresh thyme, for garnish

2 roasted red bell peppers, peeled, cored, and finely sliced

1. In a saucepan, combine the lentils with water to cover, season lightly with salt, and cook over medium heat for 15 minutes, or until the lentils are tender but not falling apart.

2. While the lentils are cooking, prepare the vinaigrette. In a bowl, combine the shallot, garlic, balsamic vinegar, and olive oil. Whisk the dressing until thoroughly emulsified. Add the lentils and toss. Season with salt and pepper. Let stand for 30 minutes Taste and correct the seasoning.

3. Divide the lentil salad among 4 salad plates, top each portion with a slice of goat cheese, sprinkle with thyme, and drizzle with olive oil. Add a mound of roasted peppers and serve accompanied by crusty bread.

Stew of Pinto Beans and Shrimp with Fragrant Indian Spices

All legumes take well to spices, but pinto beans are especially delicious when teamed with curry and other Indian spices. Here is a gutsy one-dish meal that is best made a day or two ahead of time. Serve it accompanied by chilled yogurt, a bowl of fragrant jasmine rice, or a refreshing cucumber salad. Try to find Greek yogurt, which has a more interesting texture and taste than domestic varieties and is now quite widely available.

SERVES 6

4 tablespoons olive oil

1 cup finely diced slab bacon, blanched

2 small dried hot red peppers, broken into pieces

$\frac{1}{2}$ pound medium shrimp, peeled

Salt and freshly ground black pepper

2 large onions, quartered and thinly sliced

3 large cloves garlic, minced

2 teaspoons minced fresh ginger

1 teaspoon tomato paste

3 to 4 large ripe tomatoes, peeled, seeded, and chopped

1 tablespoon curry powder, preferably Madras

$\frac{1}{2}$ teaspoon ground cumin

4 to 5 cups Cooked Pinto Beans (page 392)

2 medium red bell peppers, roasted, peeled, and thinly sliced

2 tablespoons minced flat-leaf parsley

1. In a large heavy skillet, heat 2 tablespoons of the olive oil over medium heat, add the bacon, and cook until almost crisp. Remove to a dish with a slotted spoon.

2. Add the hot peppers; cook until dark and discard. Add the shrimp and cook until just lightly browned. Season with salt and pepper; set aside.

3. Add the remaining 2 tablespoons olive oil to the skillet and, when hot, add the onions and cook, stirring constantly, until they begin to brown, about 5 minutes. Reduce the heat and continue to cook, partially covered, stirring occasionally, for 30 to 40 minutes, or until soft and nicely browned.

4. Add the garlic, ginger, tomato paste, tomatoes, curry powder, and cumin; season with salt and pepper and simmer, uncovered, for 20 minutes, or until all the tomato juices have evaporated.

5. Add the pinto beans and roasted bell peppers and continue to simmer for another 10 minutes. Add the shrimp and bacon and just heat through. Taste and correct the seasoning, garnish with parsley, and serve hot or at room temperature with crusty French bread.

Ragout of White Beans with Bacon and Radicchio

Here is a wonderfully tasty bean dish that is a delicious accompaniment to lamb, pork, and grilled fish steaks. You can cook the beans 2 or 3 days ahead of time and store them in their cooking broth. You can also serve this as a one-dish meal by adding some smoked diced kielbasa or chorizo sausage to it.

SERVES 6

2 tablespoons virgin olive oil

1 head of radicchio, cored and cut into eighths

Salt and freshly ground black pepper

2 tablespoons butter

1 cup finely cubed blanched bacon, or 1 cup
   cubed pancetta

3 tablespoons minced shallots

2 large garlic cloves, minced

2 tablespoons tomato paste

4 cups cooked or canned cannellini or other white
   beans, cooking broth or liquid reserved

2 cups bean cooking broth or chicken bouillon

2 tablespoons minced flat-leaf parsley

1. In a large nonstick skillet, heat the olive oil. Add the radicchio, season with salt and pepper, and sauté until limp and lightly browned.

2. In a large cast-iron skillet, heat the butter. Add the bacon and sauté for 2 or 3 minutes, or until almost crisp. Remove with a slotted spoon to a dish.

3. Add the shallots and garlic to the skillet and sauté for 1 minute or until soft and lightly browned. Add the tomato paste and beans together with the reserved bacon. Add 1 cup of bean broth and season with salt and pepper. Cover the pan and simmer over low heat for 15 to 20 minutes, adding a little more broth if the beans seem dry.

4. Add the radicchio and toss gently into the beans. Taste and correct the seasoning.

5. Transfer the beans to an oval serving dish, garnish with parsley, and serve as an accompaniment to grilled lamb chops, a roast leg of lamb, or pan-seared or grilled sausages.

White Bean and Chorizo Soup

This soup is really a one-dish meal. You can start or follow it with a well-seasoned salad and finish with a bowl of fruit and a nice piece of cheese. Perfect for Sunday nights and simple suppers.

SERVES 4 TO 6

½ cup extra virgin olive oil

1 large onion, minced

6 garlic cloves, minced

½ cup minced fresh parsley

3 ripe tomatoes, peeled, seeded, and chopped

1 tablespoon tomato paste

1 teaspoon dried oregano

4 to 5 cups Cooked White Beans (page 394)

6 to 7 cups chicken stock or bouillon

Salt and freshly ground black pepper

⅓ cup uncooked thin spaghetti, broken up

2 cups tightly packed fresh basil leaves

½ cup freshly grated Parmesan cheese, plus
   some for sprinkling

1 cup thinly sliced chorizo sausage

1. Heat 3 tablespoons of the olive oil in a large, heavy flame-proof casserole. Add the onion, 2 of the garlic cloves, and the parsley and cook for 2 to 3 minutes, or until the onion is soft but not browned. Add the tomatoes, tomato paste, and oregano, and continue cooking until all the tomato juices have evaporated. Add 2 cups of the beans and 1 cup of the stock, season with salt and pepper, cover, and simmer for 10 minutes.

2. Remove the casserole from the heat and let the bean and tomato mixture cool. Place it in a blender and puree until smooth. Pour the puree into the casserole, add the remaining stock and beans, and season with salt and pepper. Add the spaghetti and simmer for 10 to 12 minutes, or until the spaghetti is done. If the soup seems too thick, add more stock.

3. In a blender, combine the remaining 4 garlic cloves, the basil, the remaining olive oil, and the Parmesan. Blend the mixture until smooth.

4. When the soup is done, whisk in the basil mixture. Taste and correct the seasoning. Add the sliced sausage and heat through, but do not let the soup come to a boil.

5. Serve the soup hot, with a bowl of freshly grated Parmesan cheese and crusty bread.

■ Spiced Applesauce ■ Three-Berry Coulis with Nectarines ■ Caramelized Banana Tatin ■ Blueberries in Lemon Sabayon ■ Classic Cherry Clafoutis ■ Chocolate and Pear Tart ■ Terrine of Citrus Fruit with Strawberry Sauce ■ Cranberry, Ginger, and Orange Chutney ■ Crab and Mango Salad ■ Poached Pears in Sherry Sabayon ■ Sliced Oranges in Red Wine–Cinnamon Coulis ■ Stewed Peaches in Muscat, Cinnamon, and Vanilla Syrup ■ Pear Crisp with Brown Sugar and Ginger ■ Persimmon Mousse ■ Pineapple Caramel Compote with Oranges ■ Italian Plum and Almond Tart ■ Vanilla-Scented Pots de Crème with Raspberries ■ Rhubarb and Strawberry Pecan Crisp

On a trip to the South of France in early June two years ago, I soon realized that it was the beginning of the melon season. Along the roads, stand after stand displayed rows of the beautiful small Cavallon melons—always showing one cut in half to assure you that the melons were ripe.

The Cavallon melon is named after a small town in Provence where it is grown and prized for its superb taste and texture. Since I was working in the kitchen of a local restaurant that week, I was looking forward to sampling the region's prized melons.

On my second day, I joined the chef on his weekly trip to the Antibes market. The intoxicating aroma of ripe melons was in the air. As we walked through the market, I was hoping that Lucien would choose some for the restaurant, but he didn't. We stopped at several stands, but without even a sniff, he decided the melons were not yet ripe enough. However, as we were about to leave, he spotted a small heap of rather unattractive melons that looked as if they were about to burst open. "These," he said, "are perfect!" We bought all the melons they had, and I will never forget the taste of the melon sorbet we had that evening or of the fresh slices that appeared on the breakfast table the next morning.

It was a lesson in restraint. After decades of purchasing fruit, always following the calendar, trusting and believing that when fruit is in season, it must be good, I realized how much more there is to fruit than eye appeal. Americans are especially vulnerable. We live in a country that produces masses of fruit of every kind. Not only do we expect to find almost every fruit year-round, we want our fruit to look perfect with taste to match. Unfortunately, we rarely get what we are hoping for, at least not when it comes to summer fruit—or more precisely to store-bought summer fruit.

The most frustrating aspect of fruit shopping is having to wait for ripe fruit to arrive in the market. I have spent many a summer unable to make a good peach cobbler and many spring seasons without sampling a single juicy strawberry. Because today's fruit is less about cooking than about eye appeal and instant pleasure, many people are only vaguely aware of the seasonal importance of fruit. And yet, is there anything more rewarding than a lovely ripe peach for dessert or a baked apple bursting with buttery juices, or a ripe piece of cool melon?

Today's fruit has been bred primarily to withstand shipping and handling. Flavor is secondary, and people who do not live in areas where a specific fruit is grown may never get to taste a truly sweet, tree-ripened example. Many of our favorite fruits are packed green, are gassed along the way to ripen them visually, and arrive at market looking quite lovely but tasteless. We are led to believe that 2 or 3 days on the countertop will ripen them. Of course, all stone fruit—peaches, nectarines, and plums—will ripen eventually, but that does not mean that they will be flavorful or juicy. Unfortunately, even the most intense sugar syrup cannot add true flavor to any fruit.

I still remember one spring when I was conducting a workshop in Fresno, California, and had my fill of the ripest, juiciest apricots I had had in years. Come spring, I still look longingly at apricots, which are my very favorite fruit, but after a sniff, I move on. Even though that experience was more than ten years ago, I have rarely had a ripe apricot since.

Unless you live in Florida, Arizona, or parts of California, you may not even be aware that oranges or lemons have a season; or that limes are better in the summer, while lemons are juicier in the winter months. I have had students searching for tangerines in July, and others looking for an interesting recipe for raspberries in January.

The good news is that with the growing proliferation of farmers' greenmarkets, we can now get wonderful local berries in the spring and juicy, flavor-packed apples and pears well into late fall. Also, because of consumer interest, more large supermarket chains are beginning to offer local varieties of fruit, making it easier for the cook to make a fruit tart or cobbler on the spur of the moment.

With education, product availability, and greater sophistication about food, we are going back to buying fruit on the basis of taste rather than eye appeal. We are learning to trust our instincts and to ask questions. And the more we ask, the more we learn about fruit, saving ourselves both money and disappointment. I may not be able to answer all your questions, but the dialogue that follows should lead you to a much better understanding of how to buy, cook, and enjoy fruit.

Apples

■■

Which is your favorite cooking apple?

Different preparations require different cooking apples. It all depends upon whether you want the apples to retain some shape or to become a smooth puree, and whether you are looking for a flavor that is sweet or tart. For sautéed apples and tarts, my favorite is the Golden Delicious, but when I'm cooking a chunky applesauce or baking a pie, I look for Jonathans, Cortlands, or Empires. For making applesauce without sugar, try Braeburns and Golden Delicious. I try not to get stuck on one type or another but look for those that are tastiest and in the best condition.

Apples are available all year long and are grown nearly everywhere, but this does not mean they are always top quality. For example, if the Golden Delicious look bruised and mealy, I'll go with Granny Smiths or switch to McIntoshes for applesauce. For a more complex flavor, try a combination of varieties. A mixture of all those flavor and aroma molecules can be something quite splendid.

I love the soft, tender texture of a good baked apple, but mine explode in the oven. Am I using the wrong apple?

To prevent the apples from bursting open, peel about a third of the skin from the top of the apple and fill the cavity loosely. When it comes to baked apples, you can't beat Rome Beauties for texture and flavor, although they require about 10 minutes longer in the oven than other apples, so adjust recipes accordingly. These are perfect candidates because they can take a lot of heat without collapsing or cracking.

What apples make the best applesauce?

Macs are considered a good sauce apple because they're widely available, but they tend to dissolve during the long, slow stewing required for applesauce, rather than retaining a little texture. I get much better results with Cortlands, which stay chunkier, but your best bet is to do a little research and try some of the regional varieties in your area.

What is the best way to keep apples from discoloring?

Discoloration is caused by exposure to oxygen and a naturally occurring enzyme, polyphenol oxidase, also known as tyrosinase, that is released when the apple is cut. To keep browning to a minimum, rub the exposed area with the cut side of a lemon. This works for several hours. If you are going to use sliced or chunked apples for fruit salad, be sure to add some kind of acidity to the apples, such as orange juice or a sprinkling of lemon juice. If you plan to sauté the apples, don't bother. Sautéing itself browns the apples and camouflages any discoloration, and cooking, of course, immediately stops any chemical reaction.

What is the best way to store apples, and how long do they keep?

Although it is lovely to have a big bowl of apples on a table or counter, if your goal is eating rather than decorating, keep your apples in the refrigerator. Apples purchased in season—that is, late summer through the first hard frost—will keep for quite a while. Years ago, people used to store them for months in an

APPLES AT A GLANCE

VARIETY	DESCRIPTION	USES	SEASON
Braeburn	Crisp, moderately tart, spicy, juicy, tender skin	Fresh, sauce, pie	November–January
Cortland	Fine-textured, mild, tart-sweet, juicy, thin skin	Fresh, sauce, pie, baking	October–December
Empire	Mildly tart, crisp, juicy, thick skin, mealy if overripe	Fresh, good in sauce, pie, cooks quickly	September–November
Fuji	Tangy-sweet complex flavor, snappy crisp texture	Best fresh, takes longer than average to cook	January–April
Gala	Sweet with tart accent, crisp, juicy, tender skin	Best fresh, creamy in sauce, needs little sugar	January–April
Golden Delicious	Rich, sweet aromatic flavor	Fresh, sauce, pie, needs little sugar	September–October
Granny Smith	Balanced tart and sweet, firm, crisp, juicy	Fresh, sauce, pie, falls apart when baked	Sweetest after mid-October
Gravenstein	Aromatic, tart-sweet, crisp, juicy	Good fresh, juicy, sauce, bakes quickly	July–September
Jonathan	Rich, tart, distinctive flavor, thin skin	Good fresh, smooth juicy sauce, retains shape when baked	August–November
McIntosh	Mildly tart, aromatic, juicy, tough skin separates from flesh	Best fresh, dissolves in sauce, falls apart in pie	September–October
Red Delicious	Sweet and aromatic with hint of tartness	Best fresh, flavor weakens when cooked	September–March
Rome Beauty	Mild flavor with little acid, somewhat mealy, sugar enhances flavor	Only fair fresh, best for baking, pies	September–November

unheated garage. Today, wholesalers keep apples to be used in the fall and early winter in cold storage. Those scheduled to go to market after the first of the year are held in a controlled atmosphere of reduced oxygen. This keeps them in good shape. However, once taken out of storage, the apples deteriorate quickly, losing crispness and flavor ten times faster at room temperature than they do at 32°F. When purchasing apples during the winter, refrigerate them right away and consume them as soon as possible.

Do you have a recipe for a quick apple dessert?

Apples sautéed in butter and sugar make a great quick dessert served with either vanilla ice cream or sugared crème fraîche. Use Golden Delicious or Granny Smith. Here is how to to sauté them: Peel 4 apples; core them and cut into eighths. In a large skillet, melt 3 tablespoons butter over medium-high heat. Add the apples to the pan and sprinkle with 2 tablespoons sugar. Sauté until nicely browned on all sides, 5 to 7 minutes. Add a sprinkling of grated fresh nutmeg and serve warm or at room temperature.

Bananas and Plantains

Often the bananas in my market are hard and green. Why is it so difficult to buy a ripe banana?

Bananas often arrive in markets hard and green because ripe bananas are very fragile and can easily be damaged during packing and transport. On arrival, they are gassed with ethylene, which begins the ripening process. This triggers the synthesis of naturally occurring enzymes. Fortunately, bananas ripen perfectly off the tree. Be sure that they are at least greenish-yellow when you buy them. At that point they will take as long as 3 days to ripen at room temperature. During the ripening process, bananas go from 25 percent starch and 1 percent sugar, to 20 percent sugar and 1 percent starch, making them sweet and succulent. Don't be concerned about a few brown spots: this just means that the banana is perfectly ripe.

What is the best season for bananas?

Bananas are in the markets all year, but their selection can be spotty in the winter. This is when you find more jade green bunches than nice yellow ones. A deep green color is an indication that the bananas have been chilled and will never ripen properly.

Is it necessary to wait until bananas turn black to make banana bread?

Overripe bananas, with their soft, super-sweet pulp, make terrific banana bread. But there is no need to wait until they turn black on their own. You can help them along by simmering the whole, unpeeled fruit for 3 to 4 minutes in water to cover until the skin turns black before continuing with your recipe.

What are plantains? They look like large bananas, but I am not sure how to use them.

Plantains look like overgrown, green bananas, but unlike their sweet cousins, plantains are eaten as a vegetable, rather like a tropical potato. The two fruits have very different textures, too. Bananas are soft and buttery and are good both raw and cooked; the starchier plantains must be served cooked. Plantains hold their shape well and are best sautéed or fried. Here's how: Peel and slice the plantains into 1-inch-thick slices. Lightly dust them with flour, shaking off the excess, and sauté for 2 or 3 minutes in equal parts butter and peanut oil. Sprinkle with salt and pepper and serve as a side dish for grilled chicken, pork, or lamb.

When a recipe calls for plantains, can I use bananas instead?

No. You won't get acceptable results because the two, while related, have very different textures and react differently when cooked. Bananas are delicious raw, lightly sautéed, or baked. Plantains, on the other hand, have a firmer texture and can never be eaten raw. They are best fried or baked.

Will plantains ripen at room temperature, like bananas?

Yes. To ripen plantains, leave them in a bowl at room temperature for several days. Their degree of ripeness determines how best to cook them. Fully ripe, black-skinned plantains are best for mashing or frying. Starchier, less ripe fruit, with yellow-green skin—with or without brown spots—are best served fried.

If I refrigerate bananas and plantains, will they last longer?

Refrigerating ripe bananas affects the color of their skin but not their flavor, and they will keep for a couple of days longer. If you want to store bananas for any length of time, freeze them, peeled and wrapped in plastic. Once defrosted, they are great for making banana breads and muffins.

Refrigeration stops the ripening of plantains and will allow you to keep them for about a week. Since they will be cooked anyway, their texture in the raw state is not that important. Wrapped tightly in plastic, peeled plantains freeze well.

TIP Bananas ripen faster when stored in a brown paper bag. Adding an apple to the bag ripens the bananas even faster, because the ethylene gas given off by the apple signals the bananas to ripen.

How do I use those red bananas and baby yellow bananas available in some markets?

They are wonderful for eating raw and also for cooking. Red bananas, which signal their ripeness by taking on a purplish tinge, are deliciously sweet and creamy. The little apple, or finger, bananas turn dark yellow when ripe and remain quite firm. They are excellent sautéed and caramelized in butter and sugar and served with ice cream. Unlike other bananas, these have slightly crunchy seeds, which give them an unusual, interesting texture. Here's a tasty way to prepare both red and finger bananas: Preheat the broiler. Peel the banana and slice it in half lengthwise. Put it on a well-buttered baking dish, dot with bits of butter, and top with a sprinkling of sugar. Broil for a few minutes, until lightly browned. Sprinkle with grated orange zest and serve warm.

Blueberries, Raspberries, and Strawberries

■■

Is there a difference between wild blueberries and the ones in the supermarket? Do the wild ones taste better?

Visually, wild blueberries, often called huckleberries, are quite different. Dark blue and very small, they have an intense flavor that the cultivated berries lack, and some purists prefer them. Unfortunately, they are rarely available outside the Northwest, except in Maine during the summer season. If you do come across wild blueberries at a farmers' market, be sure only to rinse them lightly and never to soak them.

TIP All berries should be rinsed quickly; they become waterlogged if soaked.

Is it all right to wash blueberries, or does that bruise them?

Blueberries are not particularly delicate; the new cultivated varieties are actually quite thick skinned. A quick rinse will not bruise them.

I usually find a fair number of pale, unripe blueberries in a box. Will these late bloomers ripen if I leave them at room temperature?

Once picked, unripe blueberries, or any other berries for that matter, will not ripen or get any sweeter. Unripe berries have their uses. They are higher in pectin than ripe berries, so if you plan to make jam, preserves, compotes, pies, or cobblers, they will help thicken them. Pectin is a naturally occurring carbohydrate that "sets" or jells cooked fruit.

How can you tell a really fresh blueberry?

Color is the number-one clue. The best blueberries have a velvety, true blue color and look as if they were dusted with very fine powder. This lighter surface color, known as a berry's "bloom," is natural and not due to spraying, as many people think. Above all, it is an indicator of freshness. It will not rinse off, but it does fade as berries get older. Dark, shiny blueberries are not as fresh, firm, or flavorful.

Raspberries are my favorite fruit, but they're so expensive! Is there any way I can use frozen ones instead?

Frozen raspberries make a terrific dessert sauce. Once the berries are pureed, pass the sauce through a fine-mesh sieve to remove the seeds.

At what time of year are raspberries least expensive?

Depending on where you live, June, July, and August are peak months in the United States. If you happen to live in or travel through the Northwest or Michigan during the summer, you can pick up whole flats of sweet, juicy raspberries for a fraction of the price they cost elsewhere.

I like the idea of serving raspberries for dessert. Do you have any suggestions for serving them other than with a sprinkling of sugar?

While fresh unadorned berries are good, raspberries become even more appealing when teamed up with something creamy. I like mine served as a topping for Vanilla-Scented Pots de Crème (page 342).

Why is it getting harder than ever to get really juicy strawberries?

Growing strawberries has become big business, with California producing more berries than anywhere else in the world. Growers focus more on size than taste; large strawberries are easier to pick, pack, and ship. Unfortunately, bigger doesn't mean better. These giant "gourmet" berries are frequently hollow and tasteless. You can still find good strawberries, but it takes effort. Be sure to shop seasonally and locally and look for berries in the spring and early summer, when they are at their best.

Do you have any tips for buying strawberries?
First of all, try to buy them loose. If that is not possible, always check the bottom of the carton and pass on those in which there are only a few ripe strawberries. Once you get them home, be sure to transfer strawberries to a bowl as soon as you can, since one moldy berry can easily ruin the rest in a very short time. Also, check the fragrance of strawberries. If you smell more container than berry, chances are they will taste like the container as well.

Do you know a quick method for making chocolate-covered strawberries?
Chocolate-covered strawberries are easy and fun to make.

To make chocolate-covered strawberries Line a baking sheet or sheet pan with waxed paper. Select 16 well-formed strawberries with stems attached. Rinse and dry them, but leave the leafy stem in place. Melt 6 ounces bittersweet or semisweet chocolate and 2 tablespoons butter in the top of a double boiler set over medium-low heat. Holding a berry by its stem, dip it in the chocolate so that the bottom half is coated. Transfer the berry to the prepared baking sheet. Repeat with the remaining berries. When all are dipped, refrigerate for at least 10 minutes to set the chocolate.

Are there any desserts in which frozen strawberries can be used instead of fresh ones?

Frozen berries are fine for a strawberry sauce. Use the unsweetened fruit and add just enough sugar to taste with a few drops of fresh lemon juice. Serve this delicious sauce with sliced oranges or cantaloupe or drizzled over ice cream (see Strawberry Sauce, page 395).

What is strawberry coulis?
Coulis is a French term for sauce, and it most often refers to a fruit puree. But the term has become so popular that it is sometimes used for tomato sauce as well.

Recipes sometimes call for "macerating" berries. What does this mean?
To macerate means to soften by marinating and refers exclusively to fruit preparations. Strawberries are macerated by sprinkling them with sugar and letting them sit for 30 minutes or so to release their juices. Other fruits may be tossed with liqueur or juice and left to macerate. Macerated strawberries and raspberries are often used as an accompaniment to cakes and ice creams, but they are also delicious on their own topped with a little sugared crème fraîche. To macerate strawberries, rinse them

quickly; hull and slice them. Sliced berries absorb sugar more readily. If they are very small, you can leave them whole. Place in a bowl and sprinkle heavily with sugar. Let them sit at room temperature for 30 to 60 minutes before serving.

TIP Refrigerate berries for storage, but never serve them chilled. They taste best at room temperature.

I have bought strawberries in June at a pick-it-yourself stand, hoping for juicier berries, but with no luck. The berries can't get any fresher than that, so why don't they taste better?

Just because berries are in season does not mean they are good. So much depends on soil and weather. After a wet spring, they will be watery and less tasty. If you live in Florida, chances are that the berries will not be as sweet as those grown in California. Keep in mind that less-than-perfect strawberries still make excellent jams and sauces and if you have had fun at the berry patch, you've had a good day.

Are the large, fancy strawberries sold with their stems more flavorful than those sold in the cartons?

Absolutely not. These fancy berries are usually hollow and less juicy than their small, unassuming cousins. But they are lovely as a garnish, dipped in chocolate or powdered sugar.

Should strawberries be hulled before or after being rinsed?

Always rinse strawberries before removing the stems or else they become waterlogged. With a small paring knife, cut down and around the stem end. This way, you can leave them whole and hollowed out, or you can slice and sugar them.

Cherries

When is the best time to buy cherries?

Cherry season is short. It begins in June and ends in late July. Even when cherries are plentiful in the markets, they are pricey—unless, that is, you live in Michigan or the Northwest, the two premier cherry growing areas in the country. I have bought flats of cherries for next to nothing at a pick-your-own farm in Oregon. As the cherry season tapers off, cherries of various shades of red show up in the bins. This is because produce markets tend to mix up varieties to extend the season. You might find some late-harvest Bings mixed with Lamberts and Lari-ans. The latter, however, lack the flavor and firmness of Bings.

I know there are different varieties of cherries, but I have never seen them named. How can I tell the difference?

The season opens with Tartarian and Burlatt cherries, which are not especially sweet. But as soon as the Bings, Lamberts, and Raniers arrive in the markets, you are in for a treat. Bings are dark colored, crisp textured, and juicy—characteristics that put them in a class by themselves. Lamberts, which ripen a little later, are similar to Bings, but they are smaller. Raniers, which are cream colored with rosy cheeks, are juicy and meaty. They bruise easily and are not as readily available. No matter; even Raniers do not compare in taste to a good Bing cherry.

I used to love Queen Anne cherries, but I rarely see them in the market anymore.

Queen Annes are now used mainly for maraschino cherries and have been replaced in the market by the Raniers, which are bigger, firmer, and sweeter.

I have a recipe that calls for tart cherries and another one that calls for pie cherries. What is the difference and where can I get them?

The major variety of tart cherry grown in the United States is the Montmorency. Tart cherries, which are sometimes called pie cherries or sour cherries, are seldom sold fresh. They are usually canned or frozen, so your best chance to get them fresh is at local farmers' markets in Michigan, Utah, New York, Wisconsin, Washington, Oregon, and Pennsylvania. They are mostly available in July.

If I buy cherries that are not sweet, will they develop more flavor if they are kept for several days at room temperature?

Cherries do not ripen once they are picked and will not get darker or sweeter. So taste a couple of cherries at the store and buy only those that are crisp and sweet. Also be sure to pick those with green rather than black stems

Does the size of cherries have anything to do with their taste?

Bigger is better. Large, firm cherries may be more expensive, but they taste great. For cooking, however, smaller cherries are fine.

Do I need to pit cherries for a pie or cobbler? How do you go about doing that?

A cherry pitter is a handy gadget sold in kitchenware shops and through kitchenware catalogs. It is inexpensive and small enough to be stored easily. It is really indispensable when pitting cherries for pies or cobblers, but I often use unpitted cherries in a compote and for the Classic Cherry Clafoutis (page 327). Just be sure to warn family and guests.

Cranberries

I love making cranberry bread, muffins, and coffeecake, but I can find fresh cranberries in the markets only around Thanksgiving. Why is this?

That's because fresh cranberries are a seasonal crop, harvested between October and December. They start appearing in markets in late September and last only until the end of the year. But the good news is they're great keepers. Fresh cranberries last up to a month in the refrigerator, and frozen berries will keep for as long as a year.

Is there a difference in flavor between fresh and frozen cranberries?

Once cranberries are cooked, it's almost impossible to tell the difference between fresh and frozen, whether cooked into a sauce or baked into a muffin. Around Thanksgiving, I always make it a point to buy an extra bag or two to pop into the freezer. This way, I have some on hand when I get a craving for a cranberry tart in the middle of July. And they are so easy to use. You don't even have to thaw them first; just give them a quick rinse under cold running water, and they're ready to go.

Since all cranberries come bagged, how do I pick fresh berries?

Just follow the bouncing berry. Cranberries are called "bouncing berries" because in the old days, barrels of cranberries were overturned down a flight of stairs. Fresh berries bounced like mad down the steps. Overripe or underripe berries never made it past the top steps and were discarded.

Of course, today the berries are locked in plastic bags, but you can inspect them before you buy. Look for berries that are dry, plump, firm, and a bright, dark red. Apply gentle pressure on a few through the bag—if they squish easily, their bouncing days are over. Pass on any bag that contains mostly pink berries; these are underripe. The closer to Thanksgiving you can buy them, the better. This is peak cranberry marketing time, and so you are apt to get the cream of the crop.

Can I use cranberries raw, or do they have to be cooked?

Cranberries are exceptionally tart, so plan to pucker if you pop a raw one in your mouth. Raw berries are harmless, but taste much better cooked. Their delightful tartness is what I love in breads, sauces, pies, and even in turkey stuffing. A lot of people like a quick relish made with chopped raw cranberries mixed with sugar and orange peel, but I prefer cooked cranberry sauce, which softens the skins. These little bright berries are showcased in the Cranberry, Ginger, and Orange Chutney (page 332). It is the best possible accompaniment to a roasted turkey or duck, and it sure beats the canned stuff that accompanies many holiday dinners.

Is there a taste and quality difference between large and small cranberries?

The New England cranberries called early blacks are rather small but more flavorful than the larger ones grown in the Midwest and the Northwest. They are also more perishable. If you open a bag and find that the berries are small, use them right away. Larger berries, on the other hand, keep well in the refrigerator for 4 to 6 weeks.

Why are dried cranberries so much sweeter than fresh cranberries?

Practically all dried cranberries have some sweetener added. In most cases this is sucrose, common table sugar, but if you search you might find some that are sweetened with apple juice.

What is the best way to use dried cranberries?

In many recipes, you can use dried cranberries just like raisins. For instance, sprinkle some into rice or couscous or stir them into your favorite muffin or scone batter. Dried cranberries look like bright red raisins. Once they are plumped, they make a wonderful garnish for roasted duck, turkey, or pork. Another idea is to add dried cranberries to chutney and relish recipes along with the fresh ones. The tartness of dried cranberries can also enhance salads, especially when the salad includes some blue cheese and toasted nuts such as walnuts or pecans.

Dried cranberries are great for snacking because they're high in vitamin A and fiber and low in sodium and calories.

Grapefruit

■■

What is the difference between pink, red, and white grapefruit?

Quality, not color, is what counts. Chose fresh, firm fruit that feel heavy for their size. Pink and red grapefruit are generally sweeter than white varieties, but they are also about 20 percent more expensive.

Which are better grapefruit, the ones from Texas or the Florida varieties?

Both are excellent. It comes down to the quality of the particular variety. Florida's Indian River grapefruit are superb, and so are Texas Ruby Reds and Star Rubys. The price and availability of these depends on where you live and the time of year. The season starts with Florida grapefruit; the ones from Texas appear a little later. Also, you will find more Florida grapefruit in the Northeast, and Texas, Arizona, and California fruit in the Midwest and along the West Coast.

TIP To enjoy their full flavor, let refrigerated grapefruit return to room temperature before serving.

Should grapefruit be thin- or thick-skinned? Does shape matter?

Both the thickness of the skin and the shape of the fruit indicate the quality of the grape-fruit. Thin skin is better, and a flatter, disk-shaped grapefruit is better than a round one with a pointy stem end. However, if you live in a grapefruit-producing state, you may find wonderful-tasting fruit that is thick skinned and very juicy.

I love the refreshing taste of grapefruit, but I know only one way to serve them: halved and eaten for breakfast. Do you have any other suggestions?

There's no reason not to eat a good, juicy grapefruit as you would an orange: peel it and enjoy the segments. I also like grapefruit salad, where the sections are served with soft, buttery greens and a slightly sweet dressing. For a simple dessert, try broiled grapefruit: Preheat the broiler. Halve a grapefruit and set the halves in a shallow broiling pan. Sprinkle with brown sugar and broil just until the sugar starts to melt and caramelize and the fruit juices bubble.

TIP Use grapefruit zest as you would orange zest—to flavor muffins, pound cakes, a fruit salad, or a sabayon. Be sure not to include any of the white membrane, or pith, which is very bitter. It is always best to blanch grapefruit zest for 2 minutes before using it in any recipe.

Lemons and Limes

■ ■

I have trouble finding really juicy lemons in the summer. Why?

Lemons and almost all citrus fruit except limes are at their best in the winter, so I always use limes instead of lemons at that time of year. Also, you will find that during the summer, limes have twice as much juice as lemons. If you like the combination of lemon and berries as much as I do (see Blueberries in Lemon Sabayon, page 326), use a little more lemon juice than the recipe calls for and add some lemon zest for extra kick.

TIP For maximum juice, let lemons sit at room temperature for at least 30 minutes. Before cutting and squeezing, firmly roll the lemon under the palm of your hand on the countertop to break up some of the juice-filled membranes. Or pop the lemon in the microwave for a few seconds to warm it up and release its juices.

Is it better to buy large lemons or small ones?

When it comes to lemons—or more accurately, when it comes to juicy lemons—bigger is not better. Smallish thin-skinned lemons are much juicier, but you get more zest out of the large thick lemons; so choose lemons with an eye to the dish you are preparing.

There often seems to be a bitter flavor when I add lemon peel to a dish. What am I doing wrong?

Chances are that you are including some of the white pith, which is extremely bitter. Be sure to use only the outer yellow zest. Also, be sure to

get a good zester, and if in doubt, before using the zest, parboil it in a little water for 2 to 3 minutes to remove any traces of bitterness.

I hear a lot about Meyer lemons. What makes them so special?

Meyer lemons are very large and very juicy; they are much less tart than other lemons, but still not exactly a fruit you'd eat out of hand like an apple. They are too smooth skinned to provide good zest. More and more Meyer lemons now appear in specialty fruit stores, but most of the time the only way to get your hands on Meyer lemons is to befriend someone with a tree in his or her yard.

TIP Only need half a lemon? Immerse the other half in a glass of water and store it in the refrigerator. Change the water every other day, and it will keep for a week to 10 days.

Can I substitute lemons for limes in marinades and vice versa?

As a general rule, yes, but I prefer limes to lemons in seafood marinades because limes are less tart and more fragrant. I also prefer the milder taste of limes in vinaigrettes to be drizzled over shellfish and avocados.

In Mexico I have tasted wonderful, pebbly little limes that are quite different from the limes I buy at the grocery store. How can I duplicate their taste?

Unfortunately, their taste cannot be duplicated, but Mexican limes (called *limones*) are becoming more available, particularly on the West Coast and in the Midwest. You can usually find these little green gems in Latin grocery stores everywhere. Their unique flavor enhances seafood and poultry, and they can be used in any recipe that calls for Key limes.

How much juice can you expect to get out of the average lime or lemon?

In summer, a single lime will yield at least ¼ cup juice while a lemon will yield at the most 2 tablespoons; but in the winter it is the reverse, and a juicy lemon will yield at least ¼ cup juice.

Mango

■■■

Do you have any tips for picking a good mango?

Be sure to buy mangoes only during their peak season from early spring through late summer. Especially avoid buying mangoes in the winter. At the market, use your nose. Ripe mangoes have a tropical, fruity aroma, and they give slightly to the touch. Although mangoes come in all sizes, you'll find that the larger the mango, the more fruit you'll get in relation to the big pit.

Books and magazine articles on mangoes recommend specific varieties, but my supermarket doesn't label them. How can I tell the good ones by sight?

Picking a mango by variety is really a problem, since there are more mango varieties in the world than there are apples.

Avoid the beautiful but stringy Tommy Atkins variety. You can easily spot this mango by its attractive red skin and lovely oval shape. Instead go for the Haden. This is a fairly small mango with yellow skin and red cheeks. While it is less showy than the Tom Atkins, it rates a "10" as far as flavor and texture are concerned. Other good mango varieties that you can occasionally find in June and July are the Keitt and

the Kent. Both are fairly large with green skin and reddish cheeks. If you want to learn more about mangoes, take a look at *The Great Mango Book*, by Allen Susser (Ten Speed Press, 2001).

I understand that Mexican and other Central American mangoes are heavily sprayed with chemicals and that they are unsafe to eat. Is that true?

There is no question that you must rinse a mango thoroughly before serving, but once you do so and peel it, the mango is perfectly safe to eat. However, there are more and more organic mangoes in the markets these days, so if you have the choice, do buy those.

Can a mango be ripened at room temperature? Does it get sweet?

As with most fruit, your best bet is to buy a mango that is at least semiripe. Mangoes do ripen within 2 or 3 days at room temperature, but I never feel that they become as sweet as the ones that are picked ripe off the tree.

TIP Make sure that when you buy a ripe mango its skin is taut and never flabby, and pass on any mangoes with soft spots.

Once a mango is sliced, how long will it keep?

A ripe mango will keep for as long as a week. Once it is cut, it will keep for 2 or 3 days in the refrigerator.

When I cut into a mango, I always seem to hit the pit. Is there a right way to cut into a mango?

There is no neat way to slice a mango. One way is to first peel it, either with a vegetable peeler or, if the mango is very ripe, by scoring the skin into quarters and pulling it off. Then cut the mango on either side of the long flat seed that runs lengthwise. The other way is to cut the unpeeled mango in half, doing your best to avoid the pit. Score the flesh into small squares, turn the half inside out, and the cubes will pop up and can be cut easily from the skin.

I live in Florida, and we get heaps of ripe mangoes at the farmers' market. Do you have any interesting mango recipes?

Mangoes can be used in various interesting ways. You can use them in a mango "soup," in which the fruit is pureed and combined with diced strawberries and finely sliced mint. I often serve sliced mango with finely sliced prosciutto, much the same way you would serve melon.

TIP One of the fun ways to enjoy a really ripe mango is to roll it back and forth on the counter as you would a lemon, then make a small incision at the stem end and suck out the nectar-like pulp.

Melons

■■

My market carries cantaloupes practically year-round, but I still have a hard time picking a good one. Is there a trick to picking a good cantaloupe?

This reminds me of one of my favorite melon stories. Several years ago I was standing at a farmers' market in Connecticut. It was early spring, and there was still a light frost on the ground. A woman picked up a melon and asked whether it was locally grown. Without missing a beat, the salesperson said, "Yes, it is grown locally in California."

We tend to forget that all melons, including cantaloupes, are seasonal and are one of the most disappointing fruits during the winter months. A good cantaloupe is fabulous at its peak from June through September. If you live on the West Coast, you can probably get a jump on the season and find a good ripe melon as early as April, but not much before that.

What is the difference between a cantaloupe and a muskmelon?

The terms cantaloupe and muskmelon are used interchangeably in the United States. True cantaloupes *(cantalupensis)* are mainly found in southern Europe, especially Provence. They are smallish and have a hard scaly rind and orange or green flesh. Muskmelons have a netted rind.

MELONS AT A GLANCE

VARIETY	CHARACTERISTICS	SEASON
Cantaloupe	Ripe when fragrant and skin is yellow under netting. Flesh ranges from golden to orange. Best comes from California and Arizona.	June–November
Casaba	Ripe when skin is pale yellow. Freckles indicate high sugar content. Flesh is white, juicy, but not very flavorful.	July–December
Charentais	Rind gray green and slightly ribbed. Ripe when it has a sweet aroma. Flowery taste, bright orange flesh. Imported from Europe.	June–September
Crenshaw	Very large melon. Ripe when skin is golden and velvety and flesh is golden yellow. Sweet and juicy.	August–December
Galia	Looks like a pale, smooth cantaloupe. Light green flesh. Sweet and juicy with smooth texture. Best imported from Israel.	November–February
Honeydew	Ripe when very fragrant and skin is the color of butter, velvety and slightly tacky.	August–October
Juan Canary	Ripe when skin is bright yellow with no green and feels slightly waxy. White flesh. Sweet, but perfectly ripe ones are rare.	June–October
Persian	Large round melon. When vine ripened, skin shows orange through heavy netting. Flesh is deep orange, sweet and musky.	August–September

I live in the Northeast, and my farmers' market carries local muskmelons in late August. They smell heavenly but have no taste. Why is this?

It is simply a matter of soil and sun. No matter how we look at it, California probably grows the world's best cantaloupe melons. I am often tempted to buy local muskmelons at my farmers' market in Connecticut, but unfortunately they do not deliver tastewise. If you find yourself tempted by the wonderful aroma but the melon seems to lack sweetness, a sprinkling of fresh lemon juice and sugar will do the trick.

What kind of melons are good in winter?

The only good winter melon is the Israeli Galia. It looks somewhat like a cantaloupe but has green flesh. You can now find Galia melons imported from Latin America. I don't think they are as sweet as those grown in Israel.

Is there a trick to picking a good honeydew? Frequently they are hard and tasteless.

A good honeydew is a great fruit, but it is not easy to come by. I have seen heaps of honeydews at supermarkets without a single ripe one. The reason is that honeydews, much like ripe tomatoes, are too delicate to withstand rough handling.

So what to do? If possible, buy your melon

at a good greengrocer, one that handles ripe fruit and knows how to pick a good melon. But if that is not possible, look for three things: skin color, skin texture, and a slight softness at the blossom end. A truly ripe honeydew is not green or white outside; it is the color of creamy butter. The skin must be velvety. If it is too smooth and shiny, it is not ripe. And finally, the blossom end must give to very gentle pressure, the kind of pressure you would apply to a ripe avocado.

Last, if you do not want to be disappointed, buy honeydews only during their peak season, which is August and September, when there is a good chance you will find a perfectly ripe one.

TIP When you shake a honeydew and the seeds go "slurp," don't buy it. It is overripe, and overripe melons can actually cause an upset stomach.

I am totally confused by all the different varieties of melons I see in the market in the summer. Which do you recommend?

Picking a good melon even in peak season can be a challenge, especially in California and Arizona, where the choices can be overwhelming. Here are some of my favorites:

■ **Crenshaw** This is a huge gold-colored melon that is sweet and juicy with an almost buttery texture. Many markets offer a half Crenshaw, which is a good way to buy it, since you can easily see if it is ripe. If the skin is not golden yellow, pass on it. Best in August.

■ **Persian melon** This, too, is a fabulous melon if you can get a ripe one. Be sure to check the skin carefully. The color peeking through the webbing must be orange, not green.

■ **Orange-fleshed honeydews** This is a rather new hybrid that combines cantaloupe flavor with the more flowery taste of the honeydew. Here again, it is not easy to tell ripeness. The skin must have a velvety texture. If it is slick and hard, pass on it.

■ **Charentais** This is a French import that you can now find in high-end grocery stores. Not as delicious as the famous French Cavallon melon, it is still well worth looking for. The skin of these firm-fleshed, flowery-tasting melons is slightly ribbed, with a gray green rind. Check the blossom end for ripeness and make sure it has a sweet floral aroma.

Oranges

■■

I love freshly squeezed orange juice, but some of the oranges I buy are full of seeds while others aren't. Is there a way to tell which ones are seedless?

Many of the better markets will display a cut orange so you can see what you're getting.

Other than that, there's no way of telling. But the fact is, if you want delicious juice you will have to accept seeds. One of the best juicing oranges is the Florida Pineapple variety. It's packed with seeds, but it's also the sweetest and juiciest.

CITURS AT A GLANCE

FRUIT	CHARACTERISTICS	SEASON
GRAPEFRUIT		
Marsh	White flesh, good flavor, very juicy. Few seeds.	April–November
Red Blush	Red flesh, good flavor, very juicy. Few seeds.	mid-January–November
Star or Texas Ruby	Deep red flesh, with some red on peel. Very sweet.	mid-January–May
Pomelo	Very large fruit, good sweet-sour flavor. To serve, peel and segment.	December–May
KUMQUATS		
Nagami	Small, oblong, orange fruit. Entire fruit is edible with mildly sweet rind. Good for preserving, marmalades.	December–mid-March
LEMONS		
Lisbon and Eirela	Highly acid and juicy, thick peel. Good for zesting or grating.	Year-round, best in winter and spring
Meyer	Slightly sweeter than other lemons. Very juicy, thin skinned.	Year-round
LIMES		
Key or Mexican	Small green to yellow fruit, Highly acid and aromatic, very juicy. Few to many seeds.	July–December
Bears or Persian	True lime flavor, very acid, few seeds. Picked and used green. Ripe fruits are yellow.	Year-round, peak June–August
MANDARINS		
Clementines	Sweet and juicy, easy to peel. Many seeds.	November–mid-April
Satsuma	Mild, sweet, and juicy, easy to peel. No seeds.	November–mid-April

FRUIT	CHARACTERISTICS	SEASON
MANDARINS (*continued*)		
Dancy tangerines	Sprightly rich flavor, moderately juicy and easy to peel. Many seeds.	November–mid-April
Honey tangerines	Small but very rich and juicy.	January–March
ORANGES		
Blood	Red flesh, rich, sweet, distinctive flavor. Few seeds.	December–May
Skaggs Bonanza navel	Medium-large fruit, rich and sweet. Easy to peel, no seeds.	November–May, peak January–March
Washington navel	Large fruit, rich flavor, moderately juicy. Easy to peel, no seeds.	November–May, peak January–March
Pineapple or Arizona Sweet	Small fruit, rich, aromatic, juicy, Used in commercial juices.	November–March, peak January–February
Seville	Juicy and sour, easy to peel. Many seeds. Best for marmalade.	November–March
Shamouti or Jaffa	Fragrant, sweet, juicy. Few or no seeds.	mid-December–May
Valencia	Sweet and juicy. Hard to peel, few seeds. Best for juice.	February–October
HYBRID CITRUS FRUIT		
Calamondin	Cross between mandarin and kumquat. Small round fruit, juicy with acidic flavor. Good for preserves.	December–September
Tangelo–Minneola	Cross between mandarin and grapefruit. Easy to peel, few to many seeds.	January–mid-May
Temple (also called Tangor)	Cross between mandarin and orange. Rich, spicy flavor, juicy. Easy to peel. Many seeds.	Late December–March

Which are better oranges, those from Florida or California?

It depends on where you live, the time of year, and whether you want an orange for eating or juicing. Oranges are at their sweetest and juiciest from November until April, when both the Florida and California seasons are in full swing. The California oranges look better because their skins are cleaner and brighter. They're also easier to peel and are great for eating. But don't be fooled. If it's juice you're after, the Floridas make the best juice, even though their skin is thinner and they're harder to peel. A good rule of thumb is to choose Californias for eating and stick to Floridas for juice.

What is the easiest way to make orange zest?

The perfumed, pigmented outermost layer of all citrus fruits is the zest, and that's where the fruits' essential oils are. Cooks value the zest for its intense flavor; it adds a lot of depth to many dishes. The best way I know to remove the zest is to use a zesting tool, but a swivel-bladed vegetable peeler works well, too. First wash and dry the fruit well. Then remove the colored peel in long strips, but don't dig in too deep or you'll get some of the fruit's bitter, white membrane, or pith. Lay the strips on a cutting board and use a very sharp paring knife to cut the zest into fine slivers. Be sure to put the slivered zest immediately into a bowl of water to prevent it from drying out.

What is the best orange for zesting?

I prefer using the Florida Temples or tangelos, which appear in markets around February or March. All three are great for zesting, but if they're not available, I find that a thick-skinned California navel works just fine.

Can orange zest be stored?

Orange zest keeps well for a day or two immersed in cold water and refrigerated, but if poached and simmered in sugar syrup, it will keep almost indefinitely.

To preserve orange zest Combine the zest of 2 large oranges with I cup water and simmer for 2 minutes. Drain and reserve. In a small saucepan, combine I cup water and ¾ cup sugar. Bring to a simmer and stir until the sugar dissolves. Add the blanched zest and simmer for 2 to 3 minutes. Transfer the zest and its syrup to a jar, let cool, then cover and refrigerate.

TIP Add a tablespoon or two of grenadine to the sugar syrup. It will turn the zest a pretty crimson color. Use this lovely bright red zest as a garnish for pork roast, poached pears, sliced oranges, or sugared strawberries.

Many Latin recipes call for sour orange juice. Does that require a certain kind of orange?

What you want are Seville oranges, a nubby, rough-skinned variety that has bitter juice. These are the oranges used for marmalade. Seville oranges are available for only a short time in just a few specialty markets.

To make I cup of sour juice Combine 6 tablespoons each of orange and grapefruit juice with 4 tablespoons lemon juice and 2 teaspoons freshly grated grapefruit zest. This will give you about the same amount of bitterness as a Seville.

What's so special about blood oranges?

From the outside, blood oranges look like regular oranges with a slight reddish blush. But slice one open, and you'll see they range from a beautiful deep pink to speckled red to burgundy. Sections of tart-sweet oranges are a lovely addition to an endive and arugula salad, simply drizzled with a light sherry vinegar dressing.

Should I avoid oranges that have a green tinge?
Bright color doesn't mean better flavor when it comes to oranges. The green coloration has nothing to do with ripeness; it only shows that the fruit has been through some cold nights before picking.

Papayas

How can I tell when a papaya is ripe?
A ripe papaya will have bright gold skin and give slightly when you press it with your thumb. You can quicken the ripening process by placing it in a brown paper bag at room temperature for a few days.

Papaya seeds are so beautiful that I hate to throw them away. Can I use them for something?
No matter how many times I cut open a papaya, I am surprised and delighted to see the beautiful, black, pearl-like seeds nestled together at its core. Papaya seeds are, indeed, edible and contribute a peppery flavor, but their greatest talent is visual. Try adding papaya seeds to a vinaigrette or tossing them into a fruit salad to punch up the flavor.

What is a green papaya, and how do you use it?
A green papaya, common in both Vietnamese and Thai cuisine, is simply the unripe fruit. It has a pleasant bite and a cucumber-like texture and is usually cubed or grated into salads.

Peaches and Nectarines

What exactly are nectarines? Can I use them in recipes that call for peaches?
There seems to be a difference of opinion over whether a nectarine is a fuzzless peach or a cross between a peach and a plum, but, yes, you can substitute nectarines in many recipes that call for peaches, especially in a cobbler and a compote.

I now see nectarines in my market practically year-round. Is there a real nectarine season?
Ripe nectarines are difficult to come by even at the height of their season, which is from June through August. The nectarines you see during the winter months are Chilean imports; because of long shipping and handling, they are never as sweet as the California- or Georgia-grown summer varieties. Also, depending on where you live, you may find fabulous tree-ripened Washington State–grown nectarines in September. These have superb flavor and juice.

Can unripe nectarines and peaches ever become really sweet?

It is getting harder and harder to get ripe stone fruit. Nectarines, like peaches, are extremely fragile, and with the large demand for this summer fruit and the mishandling by self-serve consumers, few markets can afford to bring in tree-ripened produce.

Unripe nectarines will ripen at room temperature in 2 or 3 days, but they are rarely as sweet as naturally ripe fruit. One clue to quality, however, is size. Don't buy small fruit, which has probably been picked before it has had a chance to fully mature; instead, go with medium to large nectarines. Also, avoid oversized fruit, which has probably been bred for appearance.

If you have access to a gourmet produce market, you actually may have a chance of getting tree-ripened nectarines. Just make sure they are not bruised, and do not refrigerate them until they are fragrant and slightly softened.

What is the easiest way to peel a peach without sacrificing some of the fruit?

The best way to peel peaches is to blanch them briefly in simmering water. First, cut a small cross at the blossom end with the tip of a sharp paring knife. Then immerse them in a large pot of simmering water for about 30 seconds. Remove them promptly and drop the fruit immediately into a bowl of ice water to stop further cooking. The peel should slip off easily.

Which peaches are more flavorful—the white or the yellow variety? Is there a flavor difference between cling and freestone peaches?

In the past few years, white peaches have been making a comeback, and they are my first choice for pickling and compotes. Generally, yellow peaches have a more intense peach flavor, but on the whole, the flavor depends more on ripeness than color. Both cling and freestone peaches can be absolutely delicious or flavorless, depending on their stage of ripeness.

Pears

■■

I understand that some pears are better for eating and others for cooking, but I am confused as to which is which. Can you help?

The Bartlett pear is good for both eating and cooking. For strictly eating, a ripe Comice is excellent. Bosc pears, with their rather crunchy texture, are good for eating and poaching, but not for baking; Bosc pears marry particularly well with cheese. The most common winter pear, the Anjou, is unfortunately often shipped so unripe that it is green and hard. If you do come upon some good Anjou pears—still firm, but yellowish green in color and quite fragrant—use them for cooking rather than for eating.

Is there one really terrific pear I can always count on?

The Bartlett is by far the best all-round pear. However, Bartletts are pretty much out of season by Christmas, so look for them as early as late summer and throughout the fall when they're at their best.

PEARS AT A GLANCE

VARIETY	DESCRIPTION	USES	SEASON
Bartlett (red or yellow)	Sweet and juicy	Eating, poaching, baking, roasting	August–December
Anjou	Juicy and slightly peppery	Sautéing and poaching	October–April
Comice	Very juicy; crunchy, buttery texture	Dessert	October–March
Bosc	Crunchy and buttery texture	Eating, poaching, roasting	October–March, best in October
Forelle	Crunchy and sweet; somewhat one-dimensional	Eating, roasting	October–April

Are yellow Bartlett pears better than red ones?

I much prefer yellow Bartletts, but when I say yellow I don't mean greenish-yellow, which indicates that they are underripe, or deep golden yellow, which means they are overripe and therefore mealy. It is a challenge to find a Bartlett that is just the right color and consistency, but when you bite into it and the juice runs into your mouth, you know it's worth it.

What is the best way to poach pears?

There are three ways to poach pears: in a water-based syrup, in a red wine syrup, or in a white wine syrup. The sugar to liquid ratio is the same for all three types.

To poach pears As a guideline, for 4 to 6 pears, you should use 1½ to 2 cups sugar for 4 cups water, white wine, or red wine; use less if the pears are quite ripe and sweet. I add a cinnamon stick, a 2-inch piece of vanilla bean, and a long strip of lemon zest to the poaching liquid.

Poach for 8 to 10 minutes, testing pears with a fork every few minutes. Some may be done before others. Chill pears in the poaching liquid.

TIP To keep pears from discoloring, place a circle of parchment inside the casserole, using it to cover the pears in the bowl as well.

I live in an area with lots of pear orchards and fruit stands. What are some interesting ways of using pears?

When it comes to pears, don't underestimate the power of sugar to enhance flavor. Combine them with cranberries in a luscious tart. Use them instead of apples in your favorite crisp recipe. Make a spiced pear butter out of very ripe ones to serve as a topping for pancakes, crepes, or waffles. Simply slice and serve your favorite ripe pear with some blue cheese or a ripe Brie for dessert.

For a refreshing salad, roast peeled pear halves in red wine, sugar, butter, salt, and pepper with a pinch of nutmeg until tender, then fan out the cooled slices on a bed of fresh greens and top with a dollop of sugared crème fraîche. Or use the unpeeled hollowed-out roasted halves as edible containers for all kinds of sweet and savory fillings, such as a sweetened ricotta cheese or a sauté of leeks, walnut, and goat cheese mixture.

What are Asian pears and how do I serve them?

Asian, or Japanese, pears are actually a variety all to themselves. While they have the same granular texture and taste of a soft, ripe pear, they look and even taste more like a crisp, juicy golden apple. They're best served raw on salads and are used frequently in Asian cooking, especially in spicy noodle dishes where their sweet, cool taste helps temper the fire. They're also delicious sautéed with sliced toasted almonds or preserved ginger and served as a side dish.

I now see charming-looking small greenish pears in the market. What exactly are they?

They are the tiny, spicy Seckels, which are pricey but worth it. They have a dull olive-green color with a bit of a blush and are perfect for eating when you just want that spicy pear taste. But don't confuse the succulent Seckel with the Forelle, another miniature variety that is usually half yellowish green fading into a deep red. While cute as can be, the Forelle is virtually tasteless unless it's roasted unpeeled in butter and plenty of sugar, but then just about everything is good prepared this way.

I like Bosc pears, but because of their brown skin I can never tell when they are ripe.

The color of Bosc pears does change slightly. The unripe pears will be quite green under the brown skin. When they are ripe, they become a yellowish brown with a distinct pear aroma.

TIP A pear is ripe when a light squeeze at the base of the neck yields lightly to pressure. To ripen hard pears, put the unripe fruit in a paper bag for 2 to 3 days.

Persimmons

■■

I have heard that freezing an unripe persimmon ripens it overnight. Is that true?

Unfortunately, that is not the case. When the unripe persimmon is defrosted, it will have a mushy consistency but it will still be unripe.

I can never find a ripe persimmon in my market and have tried to ripen them at room temperature for as long as 3 or 4 days. Is there a way to ripen a persimmon quickly?

Not really. In fact, 3 days is a short time for an unripe persimmon. It may take as long as 6 days, and you can never rush it, either, because in order to be really sweet and flavorful, the persimmon has to be almost mushy. When choosing persimmons, pick those that are already slightly soft; otherwise they simply take forever to ripen and sometimes never do.

There seems to be more than one kind of persimmon around. Is one variety better than the other? Which one is sweeter?

Hachiya is the most common variety, but the round Fuyu is becoming increasingly available. They have very distinct flavors, and the Fuyu does not lend itself to puddings. The major advantage of the Fuyu is that you can eat it even when it is hard, and I always choose it when I see it in the market. Besides that, it is also almost tannin free and does not have the astringency that is common in a less than perfectly ripe Hachiya persimmon.

Should a persimmon be served peeled or unpeeled?

The ripe persimmon is very hard to peel. To serve, slice it lengthwise, then free the flesh from the peel with a sharp paring knife, much as you would prepare a grapefruit. Score the flesh into small cubes and drizzle with a squeeze of lime juice. You can eat the skin of a very ripe Hachiya persimmon, but the skin of the Fuya is rather hard and somewhat bitter.

Pineapple

I always try to select a good pineapple, but the ones I end up with are still hit or miss. Any tips?

For a pineapple to be sweet, it must be picked when ripe, since, unlike bananas, the fruit has no starch reserve that can be converted into sugar. This means that a pineapple will not ripen once it is off the plant. The fruit should be slightly soft to the touch, and the skin color should be full and rich looking (color will vary somewhat depending on the variety and time of year). The base of the pineapple should give off a distinctively fruity aroma. Finally, disregard the old wives' tale about pulling a leaf from a pineapple's crown—it doesn't work. You should, however, look for fresh, crisp leaves.

Del Monte has a brand called Hawaiian Gold that is consistently sweet, with a deeper tropical flavor. You may pay a bit more, but you will be able to count on great flavor.

Is a Hawaiian pineapple a specific variety, or does that just refer to the place where it's grown?

The latter. Almost all of the pineapples sold in the United States are grown in Hawaii, although Costa Rica and Honduras also have large pineapple crops. By far the most common variety of pineapple sold here is the Smooth Cayenne variety, although occasionally you can find the Red Spanish pineapple, which is delicious.

Pineapples seem to be available all year. Is there a season when they are better?

Because sun and heat are the prime promoters of sweetness in pineapples, it follows that the sweetest fruit will come to market in the warmest months. Hawaii has somewhat cooler weather between December and April, so pineapples harvested then generally have a lower sugar to acid ratio, which leads to fruit that is less sweet.

I recently bought a small "baby" pineapple that was the sweetest one I have ever had. Does size have anything to do with flavor?

You most probably tasted a pineapple from Africa's Ivory Coast. These are usually much smaller than Hawaiian varieties and are deliciously sweet.

Sometimes I get a pineapple with little brown seeds in it. What are these?

They're just what you think they are, pineapple seeds. While most varieties of pineapple grown for sale are seedless, there are a few seeded varieties; sometimes different varieties are grown close together, and they cross pollinate, unbeknownst to the growers. Pineapple seeds are small and brown and resemble apple seeds. They are perfectly harmless, but you wouldn't want to eat them.

TIP Whenever you buy a pineapple, check the green tops and do not buy one with a double crown, which usually is an indication of a double core as well.

What's the best way to store a pineapple for a few days? Should I refrigerate it or leave it out on the counter?

If you are not going to use a pineapple right away, the best thing to do is to cut out the flesh, place it in an airtight container, and refrigerate it. (Although you may find, as I do, that you can't stop yourself from snacking on pineapple chunks.) Since a pineapple won't ripen further once it's picked, leaving it at room temperature may allow the skin to become more golden, but it will not become any sweeter.

What is the best way to cut up a pineapple?

The traditional way is to quarter the whole unpeeled fruit right through the green crown. Lay the quartered pieces on their sides and slice off the hard core. Next, place the pieces flesh side up, cut the flesh away from the skin with a sharp paring knife, and then cut crosswise into $1/2$-inch pieces.

I have a recipe for a gelatin dessert that uses canned pineapple. Since I don't like to use canned fruits when I can find fresh, I tried it with fresh pineapple instead. It was a disastrous blob. What went wrong?

Pineapple contains an enzyme called bromelain, which breaks down the protein in gelatin so it never sets. (Some people also think bromelain helps digestion, but this has never been proven.) Canned pineapple is processed at a high temperature, which destroys bromelain. Boiling breaks down bromelain, so either cook your fresh pineapple before adding it to gelatin or reach for a can.

Plums

■■■

What is the best eating plum?

My favorites are the bright red or dark red-skinned, large Japanese varieties. The dark reds are quite round, have very small pits, and, when perfectly ripe, are loaded with juice and a refreshing tart-sweet flavor. Look for Laroda,

considered the quintessential plum and a real winner because this one does it all—it's great for eating, cooking, and jam. Also El Dorado, Queen Anne, and Friar are outstanding. Excellent bright red varieties are usually more oval in shape and can be identified by the "Rosa" in their names. They are the Santa Rosa, which is packed with flavor, Queen Rosa, and Simka Rosa. Unfortunately, many markets don't label their plums, which makes it difficult to identify them.

What is an Italian Plum?

For marketing purposes, European plums are often called Italian plums, or prune plums. This does cause some confusion, possibly because the Latin word for plums is *prunus* and in the English language the words *prunes* and *plums* are used interchangeably. They're oval-shaped, purple in color, and can be as small as 1 inch in diameter or up to 3 inches in diameter. Sometimes they can be very sweet but not juicy and are best reserved for cooking and baking.

There are so many varieties of plums throughout the summer. Which are good for eating and which are better for baking?

Plums seem to come in every size, shape, and color imaginable, and that does make the choice confusing. There are two clear distinctions of plums: European and Japanese. The more common eating plums are the Japanese varieties that come in red, yellow, and green. These are essentially eating plums, since they practically dissolve when cooked. The small, ovoid purple European varieties that have a green flesh are best for baking and poaching; they make excellent tarts. These are the plums that are dried to make prunes.

TIP To minimize the confusion when you're shopping for plums, remember that the Japanese varieties are simply called plums, and the European, English, or Italian varieties are called prune plums.

Are green plums as juicy as the red ones?

If you buy a green plum, make sure it's a Kelsey or Greengage, or you'll be sorely disappointed. Other green varieties are bitter and tough. Otherwise, chances are that red plums will be juicier and more flavorful. Greengage plums, are originally from England and are best eaten when fully ripe. They have a juicy, yellow flesh. Slice and toss with fresh greens for a wonderful salad.

What's the best season for plums?

The Japanese varieties come from California and start arriving in markets in May, but the peak season is July through September. Don't buy the first plums that come on the scene. By July, they'll be so much more juicy and flavorful. Prune plums don't show up until mid-August and are best in early to late fall.

Spiced Applesauce

I love a simple buttery applesauce and usually serve it warm with a side dish of sugared crème fraîche. I also flavor it with prepared hot horseradish and a touch of sour cream and serve it as a side dish to boiled beef. If you feel ambitious, you can use some applesauce as a filling for a butter crust, top it with additional finely sliced Golden Delicious apples, and bake it until the apples are nicely browned.

SERVES 6

4 pounds McIntosh apples or a mixture of
   Macs and Jonathans

½ cup granulated sugar

2-inch piece lemon zest

1-inch-long cinnamon stick

2 to 3 tablespoons brown sugar

2 tablespoons unsalted butter, optional

Freshly grated nutmeg

Crème fraîche, as accompaniment

1. Core and quarter the apples, but do not peel.

2. Combine the apples, ½ cup water, and the sugar in a large saucepan. Add the lemon zest and cinnamon stick and cook for 15 to 20 minutes, or until the mixture is very soft.

3. Discard the cinnamon stick. Pass the apple puree through a food mill. Add the brown sugar, the butter, and freshly grated nutmeg. Taste for sweetness and serve warm accompanied by a bowl of sweetened crème fraîche.

Three-Berry Coulis with Nectarines

I start making this compote in June when good blueberries, black-berries, raspberries, and strawberries become more readily available. Although I call it a coulis, the French word for a fruit sauce, this is more of a compote, especially with the addition of nectarines or peaches. Serve the coulis as a sauce with vanilla ice cream or by itself as a fruit soup topped with sugared crème fraîche. You do not have to wait for good nectarines; just the berries are delicious by themselves.

SERVES 6 TO 8

2 ripe nectarines or freestone peaches
2 pints blueberries
1 pint blackberries
¾ to 1 cup sugar
1 pint strawberries
Small scoops of peach sorbet
Small sprigs of mint

1. In a pot of boiling water, blanch the nectarines or peaches for 30 to 50 seconds. Drain and remove the skins. Cut each nectarine or peach into eighths and set aside.

2. Thoroughly rinse the blueberries and remove all stems. In a large heavy saucepan, combine the blueberries, blackberries, sugar, and ½ cup water. Bring to a simmer and cook for 15 minutes.

3. Add the strawberries and simmer for 5 minutes. Add the nectarines or peaches and cook for another 5 minutes, or until the nectarines are just soft.

4. Transfer the mixture to a bowl and chill for several hours before serving. Serve in individual bowls, topped with scoops of peach sorbet and sprigs of mint.

Caramelized Banana Tatin

A tatin is essentially an upside-down fruit tart. Apple Tatin is one of France's most famous desserts, but the same technique can be used for pears or, as here, bananas. Choose bananas that are just ripe or else they will become mushy as they bake. To save time, you can use store-bought pastry. It will likely be less flaky but still delicious.

SERVES 6 TO 8

½ cup sugar

4 large semiripe bananas

3 tablespoons unsalted butter

½ recipe Basic Tart Dough (page 385)

1. Preheat the oven to 375°F.

2. In a small saucepan, combine the sugar and 3 tablespoons water. Bring to a boil and stir once to dissolve the sugar. Continue to boil without stirring until the syrup turns hazelnut brown.

3. Immediately remove the saucepan from the heat and pour just enough of the caramel into the bottom of a 9-inch round, nonstick cake pan to cover it completely. As you are pouring, turn the cake pan at the same time to create an even layer of caramel. Set aside to cool completely.

4. Peel the bananas and slice them on the diagonal into ¼-inch slices. Dot the caramel with the butter and arrange the banana slices in a single layer, tightly overlapping them to create a circular pattern.

5. Roll out the dough on a lightly floured surface to about 9½ inches in diameter and about ⅛ inch thick. Place the dough over the bananas to cover completely, tucking the edges into the pan. Prick with a fork and place in the center of the oven.

6. Bake for 30 to 35 minutes or until the dough is crisp and nicely browned. Removed the pan from the oven. If there are any excess pan juices, remove them from the cake pan with a bubble baster and reserve.

7. Invert the tatin carefully onto a platter and let cool.

8. Place the banana juices in a small saucepan, bring to a boil, and reduce until syrupy. Serve the tart cut into wedges, with a drizzle of the reduced juices and a small scoop of vanilla ice cream on top.

Blueberries in Lemon Sabayon

This recipe was first published in 1973 in my cookbook *The Seasonal Kitchen*, **and made the cover of** *House & Garden* **magazine that year. It is still one of my favorite desserts. I have since adjusted the lemon sabayon, which in the original recipe was quite fragile and could not be made ahead of time. The revised version will keep well for 3 to 4 days. It can also be frozen and served as an iced lemon mousse.**

SERVES 6

1 quart blueberries, washed and stemmed
1 cup sugar
5 extra large eggs, separated
Juice of 2 large lemons
2 teaspoons finely grated lemon zest
1 cup heavy cream, whipped
Tiny leaves of fresh mint

1. Place the blueberries in a glass serving bowl. Sprinkle with ¼ cup of the sugar and set aside.

2. In a double boiler, combine the egg yolks and remaining ¾ cup sugar. Whisk until fluffy and pale yellow. Add the lemon juice and zest and whisk until well blended. Set over simmering water and whisk constantly until the mixture is thick and heavily coats a spoon, about 5 minutes. Be careful not to overcook, or the eggs will curdle. Immediately transfer the sabayon to a stainless steel bowl and let cool completely.

3. Beat the egg whites in a large bowl until they form firm peaks. Add them to the cooled sabayon and fold in gently but thoroughly. Add the whipped cream and again, fold in gently but thoroughly. Cover and chill until very cold.

4. Spoon the sabayon over the berries, garnish with mint leaves, and serve at once.

Classic Cherry Clafoutis

Cherry clafoutis is a classic French dessert that appears in pastry shops all over France when fresh cherries are in season. Unfortunately, it is rarely very good, but I don't give up and continue to sample it whenever I am in France during cherry season. Last year, I discovered a delicious cherry clafoutis at a market café in St. Remy and begged the owner for the recipe. Here it is.

SERVES 6 TO 8

$\frac{1}{2}$ pound Bing or sour cherries

$1\frac{3}{4}$ cups sugar plus 2 tablespoons

1 cinnamon stick

4 eggs

3 tablespoons flour

$\frac{1}{2}$ cup heavy cream

$\frac{1}{2}$ cup milk

2 teaspoons vanilla extract

One 9-inch partially baked pastry crust

1. Preheat the oven to 350°F.

2. Remove the stems from the cherries. Combine $2\frac{1}{2}$ cups water, $1\frac{1}{4}$ cups of the sugar, and the cinnamon stick in a large saucepan. Bring to a boil, reduce the heat, and add the cherries. Simmer for 25 to 30 minutes, or until the cherries are very tender. Let the cherries cool completely in the syrup; then drain thoroughly, reserving the syrup. Pit the cherries and set aside.

3. In a bowl, combine the eggs, $\frac{1}{2}$ cup of the sugar, and the flour. Whisk until thoroughly blended. Add the cream, milk, and vanilla extract. Taste. If you like it sweeter, add the remaining 2 tablespoons sugar. Strain the custard into a bowl to make sure it is completely smooth.

4. Add the cherries to the custard and pour into the prebaked crust. Be sure to distribute the cherries evenly.

5. Bake the clafoutis for 30 minutes, or until the custard is set and a knife inserted in the middle comes out clean.

6. Serve slightly warm or at room temperature.

REMARKS The clafoutis can be reheated in a low oven the next day. For a nice addition to this homey dessert, I usually cook the cherry syrup until it is well reduced, fold in a little crème fraîche, and serve it as a sauce. Other fruits, especially sautéed summer peaches and roasted Bartlett pears, can be used instead of the cherries.

TIP Since I pit the cherries after they are poached, no pitter is needed; gently squeeze the fruit and the pit will pop right out.

Chocolate and Pear Tart

Many people like the taste of raspberries or strawberries with choco-
late, but to me the marriage of pears and chocolate in unbeatable,
whether the pear is simply poached and drizzled with hot chocolate
sauce or baked in this delicate yet easy-to-make tart. Although all
tarts are best baked and served the same day, this one can be made
a day ahead of time and still be quite delicious. Make sure the pears
are at least semiripe before making the tart.

SERVES 6

2 tablespoons unsalted butter

3 large, slightly underripe pears, preferably Bartlett,
   peeled, cored, and cut into ½-inch wedges

5 tablespoons sugar

½ cup heavy cream

1 large egg

1 partially baked 10-inch Basic Tart Shell (page 386)
   in a pan with a removable bottom

5 ounces bittersweet chocolate, preferably Lindt or
   Tobler, coarsely chopped

½ cup sliced blanched almonds

Lightly sweetened crème fraîche, as accompaniment

1. Preheat the oven to 375°F.

2. In a large skillet, melt the butter over medium-high heat.
Add the pears, sprinkle with 2 tablespoons of the sugar, and sauté
quickly until nicely browned. Reduce the heat and cook until tender, about 15 minutes. Drain the pears in a colander and set aside.

3. Combine the heavy cream, egg, and 1 tablespoon of the
sugar in a bowl and whisk until smooth.

4. Sprinkle the bottom of the partially baked tart shell with the
chopped chocolate. Arrange the pears in a decorative pattern
over the chocolate and drizzle with the cream-egg mixture.
Sprinkle the top with the sliced almonds and remaining 2 tablespoons sugar. Place in the center of the oven and bake for 30 to
35 minutes, or until the custard is set and the almonds are nicely
browned. Remove from the oven and let cool.

5. Cut the tart into wedges. Serve with a bowl of lightly sweetened crème fraîche.

Terrine of Citrus Fruit with Strawberry Sauce

Here is an impressive yet simple dessert that is really nothing more than "upscale" Jell-O. Although I like the terrine look of this refreshing dessert, you can also spoon the mixture into martini glasses and serve topped with a dollop of sugared crème fraîche and a light grating of lime or orange zest.

SERVES 6 TO 8

> Flavorless oil (preferably almond oil) for
>   the terrine
> 1½ tablespoons unflavored gelatin
> 2 cups fresh orange juice, strained
> 2 pink grapefruit
> 4 navel oranges
> Strawberry Sauce (page 395)
> Sprigs of fresh mint

1. Lightly oil the inside of a 1½-quart terrine, preferably porcelain.

2. Combine the gelatin and orange juice in a small saucepan and let stand 5 minutes, or until the gelatin has softened.

3. Peel the grapefruit and oranges with a sharp knife, removing all traces of the rind and white pith. Cut each segment away from the membrane on either side, then cut each segment in half crosswise.

4. Heat the orange juice and gelatin mixture over medium-low heat, stirring constantly until the gelatin has completely dissolved. Remove from the heat and transfer the mixture to a large stainless steel or glass bowl. Chill until the mixture begins to thicken but is not set. (The thickened juice will keep the fruit suspended throughout the gelatin mixture instead of letting it sink to the bottom of the terrine.)

5. Fold the fruit into the gelatin mixture and pour into the oiled terrine. Cover and chill overnight.

6. The next day, remove the terrine from the refrigerator and run a hot knife around the inside of the terrine to loosen the molded fruit. Invert onto a sheet of plastic wrap and shake to free the molded fruit from the terrine. Wrap completely in the plastic wrap and place in the freezer for 3 to 4 hours or until just frozen.

7. To serve, cut the molded fruit into ½-inch slices and place 1 slice on each individual dessert plate. Let sit for a few minutes, then wipe excess liquid from the plate with a paper towel. Surround with strawberry sauce and garnish with sprigs of fresh mint.

Cranberry, Ginger, and Orange Chutney

I love the puckery sour taste of cranberries. Cooked with fresh orange juice, fresh ginger, and cinnamon, cranberries take on a chutney-like texture, which I find works well as a condiment with roasted turkey, duck, and pork. It also goes well with French toast, pancakes, and as a filling for crepes. It is a good idea to double the recipe, since the chutney will keep for at least 6 months.

MAKES 8 CUPS

- 3 packages (12 ounces each) fresh cranberries
- 3 semiripe Golden Delicious apples, peeled, cored, and cubed
- 1½ cups golden raisins
- 1½ cups sugar
- 1¼ cups freshly squeezed orange juice
- 2 tablespoons finely grated orange zest
- 3-inch-long cinnamon stick, or 2 teaspoon ground cinnamon
- 2 tablespoons minced fresh ginger
- ¼ teaspoon freshly grated nutmeg
- ⅓ cup Cointreau, optional

1. Place all ingredients except the liqueur in a large nonreactive saucepan. Bring to a boil; reduce the heat to low and simmer, uncovered, stirring often, until the mixture thickens, about 25 minutes.

2. Remove the saucepan from the heat, stir in the liqueur, and let the chutney cool to room temperature. Cover and refrigerate overnight. Serve lightly chilled.

REMARKS The chutney will keep in the refrigerator, in tightly covered jars, for months. You may also "pack" the chutney in pint jars, leaving ½ head space, and process them for 20 minutes.

Crab and Mango Salad

Aqua Grill is one of New York's best restaurants, and this salad is just one of their delicious seafood salads. It is best to dress the crab right before serving, but you can prepare the fruit and make the vinaigrette hours ahead of time. A few baby greens tossed in the dressing would make a lovely accompaniment.

SERVES 4

THE VINAIGRETTE

2 tablespoons finely minced shallots

Juice of ½ orange plus 1 tablespoon orange zest

1 tablespoon lemon juice plus 1 tablespoon
    lemon zest

1 tablespoon sherry vinegar

1 tablespoon soy sauce

1 tablespoon honey

½ cup grapeseed oil

½ to 1 teaspoon Thai chili sauce

Salt and freshly ground black pepper

THE SALAD

1 pound lump crabmeat

1½ cups finely diced jicama

1 large ripe mango, peeled and diced

2 tablespoons minced fresh cilantro

Tiny leaves of cilantro and mint, for garnish

1. In a bowl, combine the shallots, orange juice and zest, lemon juice and zest, vinegar, soy sauce, and honey. Whisk until well combined. Slowly add the oil and whisk until the vinaigrette is lightly emulsified. Stir in Thai chili sauce to taste. Season with salt and pepper.

2. In another bowl, combine the crab with just enough vinaigrette to bind it. Season with salt and pepper to taste.

3. In a third bowl, combine the jicama with the mango and cilantro. Season with salt and pepper.

4. To serve, place 2 tablespoons of the mango-jicama mixture in the center of individual serving plates. Top with 2 spoonfuls of the crab and garnish with small cilantro and mint leaves. Drizzle a little of the leftover vinaigrette around each serving and serve the salad slightly chilled.

Poached Pears in Sherry Sabayon

1½ cups sugar

3-inch-long cinnamon stick

1 vanilla bean, split

6 large pears, preferably Anjou or Bosc, peeled

3-inch piece lemon zest

Sherry Sabayon (recipe follows)

Tiny leaves of fresh mint, for garnish

1. Start by poaching the pears: In a large flameproof casserole, combine 5 cups water, the sugar, cinnamon stick, and vanilla bean. Bring to a boil over high heat and cook for 5 minutes.

2. Add the pears and lemon zest, reduce the heat, and simmer, partially covered, until the pears are tender when pierced with the tip of a sharp knife, 15 to 20 minutes. Remove the cinnamon stick and vanilla bean and set the pears aside to cool completely in their poaching liquid.

3. Drain the pears well and pat dry on paper towels. Cut in half lengthwise, core, and place cut side down on round shallow plates in a decorative pattern. Top each portion with a large dollop of sabayon. Garnish with tiny leaves of mint and serve at once.

Sherry Sabayon

MAKES ABOUT 2½ CUPS

4 extra large egg yolks

¼ cup sugar

⅔ cup cream sherry

1 cup heavy cream, whipped

1. In the top part of a double boiler, combine the egg yolks, sugar, and sherry; place over simmering water and whisk constantly until the mixture is thick and heavily coats a spoon.

2. Immediately place the top of the double boiler in a large bowl of ice and stir constantly until the sabayon is thick and cool. Chill for 2 hours.

3. Fold in the whipped cream gently but thoroughly, cover, and chill the sabayon until serving time.

Sliced Oranges in Red Wine–Cinnamon Coulis

Whenever I have some leftover red wine, I turn it into syrup, which keeps indefinitely. Besides pairing with oranges, it is equally delicious with grapefruit segments, baked apples, or stewed prunes. Make sure to remove the cinnamon stick after a few days, or it will give the wine a bitter flavor.

SERVES 4 TO 5

3 cups red wine, preferably Bordeaux

1½ cups sugar

1 cinnamon stick

3-inch piece vanilla bean

6 black peppercorns

two 1-inch pieces of lemon zest

4 large navel oranges, peeled and sliced
   crosswise into ¼-inch slices

2 tablespoons Grand Marnier

Zest of 1 large navel orange, cut into fine
   julienne strips, and small whole
   strawberries, for garnish

1. In a heavy nonreactive saucepan, combine the wine, sugar, cinnamon, vanilla bean, and peppercorns. Add the lemon zest and bring to a boil, reduce the heat, and simmer until the wine is reduced by half and the mixture is syrupy. Discard the lemon zest, transfer the wine syrup to a bowl, and chill for 4 hours or overnight.

2. Divide the orange slices among 4 or 5 glass serving dishes. Sprinkle with Grand Marnier. Spoon the wine coulis over the oranges and garnish with the orange zest and whole strawberries. Serve chilled.

Stewed Peaches in Muscat, Cinnamon, and Vanilla Syrup

I love peaches and cannot wait until July, when I can pick them at our local orchard. Unfortunately, many people are less fortunate and often spend entire summers without tasting a single juicy ripe peach. Here is a good way to use fruit that is less than perfectly ripe. However, don't fool yourself into thinking that you can use hard unripe fruit even when stewed with sugar, because while the compote will be sweet, it will lack that real peach flavor. Be sure to cut a thin cross into the top of each peach before dropping it into the water (or before blanching).

SERVES 6

5 large semiripe or ripe peaches or nectarines
2½ cups Muscat wine
½ cup sugar
3-inch-long cinnamon stick
3-inch piece vanilla bean
two 1-inch pieces lemon zest
2 tablespoons peach liqueur, optional
Honey-sweetened ricotta, optional
Tiny mint leaves

1. In a large saucepan, bring plenty of water to a boil. Add the peaches or nectarines and cook for 1 minute. Drain and peel; the skin should slip off easily. If it does not, cook for another minute. Slice the peaches into ½-inch pieces and reserve.

2. In a nonreactive medium saucepan, combine the Muscat wine with the sugar, cinnamon, vanilla, and lemon zest. Add the peaches and poach for 5 to 8 minutes, or until the fruit is just tender and the syrup is lightly reduced. Chill the peaches in the syrup until serving.

3. Serve in individual glass bowls, topped with a drizzle of peach liqueur and a dollop of sweetened ricotta and a few tiny mint leaves.

REMARKS The compote will keep for at least a week. Remove the cinnamon stick and vanilla bean after a day.

Pear Crisp with Brown Sugar and Ginger

Everyone loves a good apple crisp, but I find the taste of pears to be more interesting and am always on the lookout for good pears to make this crisp. Bartletts are my first choice, but ripe Boscs or Anjous work well too. Serve with sugared crème fraîche or vanilla ice cream.

SERVES 6

6 ripe medium pears, preferably Bartlett
¼ to ⅓ cup granulated sugar
1 teaspoon ground ginger
1 tablespoon candied ginger, minced, optional
⅔ cup tightly packed dark brown sugar
¾ cup all-purpose flour
8 tablespoon (1 stick) unsalted butter

1. Preheat the oven to 375°F.

2. Peel the pears with a vegetable peeler, core, and quarter. If the pears are large, cut them into eighths. Place the pears in a bowl and sprinkle with one-fourth cup of the granulated sugar (or more, depending upon the sweetness of the pears), the ground ginger, and optional candied ginger.

3. In a small bowl, combine the brown sugar, flour, and 6 tablespoons of the butter. Work the mixture with your hands until it resembles cornmeal.

4. In a large oval flameproof gratin dish, heat the remaining 2 tablespoons butter over low heat, add the pears in one layer, and sprinkle with the brown sugar topping. Place the dish in the center of the oven and bake for 30 minutes, or until the topping is golden and crisp.

5. When the crisp is done, remove from the oven and let cool for 30 minutes before serving. Serve with vanilla ice cream.

REMARKS You may substitute Golden Delicious apples, peeled and cored, for the pears.

Persimmon Mousse

A ripe persimmon needs little enhancement, but it is not an easy fruit to serve with style. Whenever I can, I simply dice the fruit and combine it with some sugared crème fraîche and freshly grated nutmeg. Remember that persimmons are sold quite firm, and it can take as long as 10 days to ripen them.

SERVES 4

4 ripe persimmons, peeled

1 cup crème fraîche

3 to 4 tablespoons sugar

1 tablespoon minced candied ginger,
  optional

Freshly grated nutmeg and mint leaves,
  for garnish

1. In a food processor or a blender, combine the persimmons, crème fraîche, and sugar. Blend gently; do not overprocess. Add the minced ginger. Transfer to individual glass bowls and top with a generous grating of fresh nutmeg and mint leaves.

2. Cover and chill for 2 to 4 hours. Serve in martini glasses.

Pineapple Caramel Compote with Oranges

A sweet ripe pineapple needs little enhancement, but occasionally it is fun to use it creatively. Here the cubed fruit is tossed with an orange-flavored caramel and combined with fresh orange segments for a light dessert.

SERVES 4 TO 5

½ cup sugar

1 pound finely cubed pineapple

2 tablespoons Grand Marnier

2 navel oranges, peeled and cut into segments
  or cubed

Fresh sprigs of mint, for garnish

1. In a small heavy saucepan, combine the sugar with ¼ cup cold water. Bring to a boil and cook without stirring until the syrup turns hazelnut brown. Do not let it get too dark. Immediately remove the caramel from the heat and pour in ¼ cup hot water. Whisk the caramel until it becomes a smooth syrup. Pour over the pineapple, toss gently, and chill for several hours.

2. An hour or 2 before serving, add the Grand Marnier and the oranges and toss gently.

3. To serve, transfer the compote to individual glass serving bowls. Garnish with sprigs of fresh mint and serve chilled.

Italian Plum and Almond Tart

This was one of my mother's favorite recipes and I still make it as soon as fresh prune plums appear in the market in early fall. Unfortunately, the season for these small fleshy sweet plums is rather short, so take advantage of it either by making this luscious tart or by poaching them in a wine, sugar, and cinnamon syrup and serving them with sweetened crème fraîche.

SERVES 8

1 cup finely ground slivered almonds, lightly toasted

½ cup plus 2 to 3 tablespoons sugar

1 teaspoon lemon zest

½ teaspoon ground cinnamon

¼ teaspoon almond extract

1 extra-large egg

One 9-inch unbaked dessert tart shell (see Remarks, page 387), in a removable-bottom tart pan, frozen

2 dozen small Italian prune plums, halved and pitted (quartered if large)

2 tablespoons unsalted butter

Lightly sweetened crème fraîche or mascarpone mixed with a little whipped cream and honey

1. Preheat the oven to 400°F.

2. To make the almond paste: Combine the almonds, the ½ cup sugar, and the lemon zest in a food processor and process until the almonds are finely ground. Add the cinnamon, almond extract, and egg and process until a smooth paste is formed. Spread the almond paste in an even layer in the bottom of the prepared tart shell.

3. Arrange the plum halves very close together in a decorative pattern over the almond paste. Sprinkle with 2 to 3 tablespoons sugar, according to the sweetness of the fruit, and dot with the butter. Set the tart shell on a baking sheet and place in the center of the oven. Immediately reduce the heat to 375°F and bake for 1 hour to 1 hour 30 minutes, or until the crust is evenly browned and the juice from the plums has evaporated.

4. Remove the tart from the oven and let cool completely on a wire cake rack. Remove the tart shell carefully from the tart pan, transfer to a serving platter, and let cool to room temperature. Serve cut into wedges, with a dollop of sweetened crème fraîche or mascarpone.

Vanilla-Scented Pots de Crème with Raspberries

When I was growing up, pots de crème was a familiar weekday dessert at our house. The little custards, flavored with coffee, chocolate, or vanilla, were served in decorative covered porcelain ramekins especially designed just for this dessert. Today, I still love making these custards, but I simply bake them in 6-ounce porcelain ramekins. The toppings vary according to the season, but my favorite is always some kind of seasonal berry. Strawberries, raspberries, and blackberries work best, but if you flavor the custard with some lemon zest, then blueberries are a good choice as well.

SERVES 6

¾ cup whole milk

½ cup heavy cream

2 extra large whole eggs

2 extra large egg yolks

⅓ cup sugar

1 teaspoon vanilla extract

Toppings: Fresh raspberries or other spring berries, such as blueberries, raspberries, or sliced strawberries

1. Preheat the oven to 350°F. Heat the milk and cream together in a saucepan and keep warm.

2. Combine the whole eggs, egg yolks, sugar, and vanilla in a bowl and whisk until fluffy and pale yellow. Add the warm milk-cream mixture and whisk until well blended. Pour the custard into six 4-ounce porcelain ramekins.

3. Place the ramekins in a heavy baking dish, fill the dish with enough boiling water to reach halfway up the sides of the ramekins, and bake for 15 minutes or until the tip of a knife comes out clean. Do not overcook; the centers should remain slightly soft. Remove from the oven and from the water bath and let cool. Refrigerate for 2 to 4 hours before serving.

4. Top with fresh raspberries or other spring berries, and serve lightly chilled.

Rhubarb and Strawberry Pecan Crisp

The marriage of rhubarb and strawberries seems to be unbeatable, whether in a compote or in this delicious crisp. Serve the dessert with lightly sugared crème fraîche or good vanilla ice cream.

SERVES 8 TO 10

2¼ pounds rhubarb, trimmed and washed

1 cup plus 2 tablespoons granulated sugar

2 pints strawberries, rinsed and hulled

1 cup plus 3 tablespoons all-purpose flour

1 cup packed light brown sugar

2 teaspoons grated fresh ginger

½ teaspoon ground ginger

½ cup coarsely chopped pecans

9 tablespoons unsalted butter, cut into small
   pieces, at room temperature

1. Preheat the oven to 375°F. If the rhubarb stalks are more than 1 inch thick, split them in half lengthwise. Using a sharp knife, cut the rhubarb stalks into 1-inch pieces.

2. Place the cut rhubarb in a bowl, add the 1 cup granulated sugar, and toss well. Let sit for 20 minutes, or until the rhubarb starts to release some liquid and the sugar is moist.

3. Depending on the size of the strawberries, cut them in half or quarters. Toss with the remaining 2 tablespoons granulated sugar and add to the rhubarb. Add the 3 tablespoons flour and toss gently. Spoon the fruit mixture into a 1½-quart casserole.

4. Combine the remaining 1 cup flour, the brown sugar, fresh and ground ginger, pecans, and butter in a medium bowl. Mix well with your fingertips, crumbling any lumps. Work the mixture gently until it resembles coarse crumbs. Sprinkle the crumb topping evenly over the fruit mixture.

5. Set the dish in the center of the oven. Bake for 50 minutes to 1 hour, or until the topping is golden and the filling is bubbling. Serve warm or at room temperature.

REMARKS The crisp can be prepared up to 8 hours in advance and refrigerated. Bring to room temperature before serving.

■ Lemon-Herb Grilled Chicken "Under a Brick" ■ Lime-Marinated Chicken Kebabs ■ Smoked Chicken Wings with Spicy Peanut Sauce ■ Lemon- and Thyme-Scented, Apple-Smoked Cornish Hens ■ Fresh Fennel with Grilled Fennel Sausage ■ Butterflied Leg of Lamb with Shallot, Garlic, and Lemongrass Rub ■ Grilled Lamb Chops in Shallot and Herb Butter ■ Citrus-Marinated Grilled Pork with Sweet-and-Sour Red Onion Confit ■ Grilled Quail with Pineapple Caramel ■ Brochettes of Swordfish Grilled in Sesame Marinade ■ Grilled Salmon Kebabs in Dill and Mustard Marinade ■ Grilled Scallops with Mango and Jicama Salad ■ Grilled Shrimp with Mushrooms and Bell Peppers ■ Middle-Eastern Char-Grilled Skirt Steaks in a Cumin and Cilantro Marinade ■ Grilled Marinated Cremini Mushrooms ■ Grilled Eggplant ■ Grilled Green Tomatoes in Basil Vinaigrette ■ Roasted Red Bell Peppers ■ Grilled Tomato Salsa

I am not quite sure what it is about grilling that makes people embrace the fire with a confidence they would never have with other cooking methods. More than 80 percent of American households own at least one type of grill, but unfortunately too many of them sit unused after a few unsuccessful experiences. No matter how reassuring grill cookbooks try to be, the fact remains that grilling is an inexact science, prompting more questions from my students than any other cooking technique. To begin with, you are dealing with a live fire and its constantly changing dynamics, which means that while it is exciting and fun to grill, it is also very challenging. Temperature varies with the type of grill you own, fuel source, even the weather.

For many cooks outside the developed world, grilling is the only way to produce hot food. Often it is a way to make a living. In China, Morocco, Thailand, Greece, and New York City, street vendors cook on a grill with the most rudimentary resources, producing delicious, succulent morsels of meat, sausage, and poultry. In many countries, including some regions of the United States, grilling lies at the very core of a culture's culinary identity. What would a Malaysian street festival be without skewers of chicken satay or a Mexican fiesta without roasted corn on the cob, or a Texas barbecue without sizzling ribs?

Having grown up in Europe, I came to grilling by participating in the all-American experience, the backyard cookout. When I first met my husband, we would spend many summer weekends with his family in Pennsylvania. On Saturdays we would make our way through the Amish farmers' market, ending up with wonderful fresh corn, juicy vine-ripened tomatoes, plump sausages, and jars of homemade pepper slaw. Our last stop was always at Michael's Meat Locker, where the sides of aged beef hanging made us all think of wonderfully succulent, perfectly grilled steak.

By late afternoon, our friends would arrive, bringing side dishes and dessert, and over a glass of cider or wine, my husband would start the grill. Things rarely went smoothly. Often the fire was too hot, which resulted in charred steaks and overly browned sausages. Other times the fire was not hot enough, and we would stand around wondering what to do. "Let's add more starter fluid," or "It needs more charcoal," or "Take the steak off for a while" were some of the suggestions. But somehow it all worked out, and everyone had a good time. I can still remember the lively conversations, satisfying aroma of searing meat, and sense of well-being that only an outdoor cookout can provide.

That said, most people would love to have their grilled steak or pork chops be juicy and tender rather than something akin to shoe leather; so if you want to grill, it's a good idea to learn some techniques. Successful grilling is determined by some very basic rules that when properly applied will lead to a delicious outcome almost every time.

When buying a grill your first decision is whether to go with charcoal or gas. Nothing beats a charcoal grill for infusing food with the characteristic flavor that comes from the interplay of wood and smoke. Be sure to invest in a big one, since the larger the surface, the more room you have to move foods around and keep them from burning. It also allows you to cook all kinds of additional vegetables to serve alongside the meat or have on hand for another meal.

When using charcoal, be patient and wait until the coals are covered with a fine gray ash before adding the food. Over the years I have been to many cookouts where "let's throw something on the grill" was a recipe for disaster, because the fire was not ready. I have since concluded that waiting for the right moment to put the steak, chicken, or chops on the grill is the key to success.

The next challenge is to choose the right cuts and the grilling technique you plan to use. Uncovered direct grilling is similar to broiling, except that the source of heat is on the bottom. This method is perfect for small cuts of meat, fish steaks, or chops. If you love thick pork chops or double veal chops, be ready to use the grill cover or to create a small oven by tenting the chops with a pie plate or by finishing them in the oven. The indirect method is more like roasting in that the heat from the fire—gas or charcoal—radiates from the sides of the grill but not directly beneath the food. This is the best method for small roasts, whole chickens, or turkeys. Also, once you have achieved a smoky taste, it is safer to finish cooking food indoors or over indirect heat rather than taking the risk of burning your steak or chops by keeping them directly over the fire. Whenever I make a small fire for a quick grill I choose foods that do not require much time on the grill. Best choices are single rib lamb chops, medium-sized shrimp, small fish steaks, or chicken breasts.

Always season assertively when grilling, using plenty of coarse salt and freshly ground pepper. When you've chosen an exceptional piece of meat, these seasonings will allow the true taste to come through. If you're looking for more exotic flavors, spices, rubs, or marinades can turn even chicken parts into a delicious feast.

Now with the great choice of fruit woods, you can vary the flavors by adding aromatic chips or twigs to the fire. To me there is nothing more wonderful than going out into the garden and snipping off some lilac or fresh apple wood twigs to sprinkle over the coals to add dimension to a roast pork or Cornish hens. For those who do not live near an orchard, there is now a wealth of excellent mail order resources available.

Learning when food is done is key to grilling, since the time between when a piece of meat or fish is perfect and when it is overcooked is very short. In the following pages you'll learn several ways to do that. It's very important to have your side dishes ready and made ahead of time, so that you can give all your attention to the grill. Even the pros concede that grilling is an unpredictable art. But when all the theories have been debated and the techniques experimented with, the real purpose of grilling is to have a good time in your backyard, porch, or at the beach. There is something about being outside and smelling the rich aromas of food as it cooks that makes this form of cooking so much fun and enjoyable.

What is the difference between grilling and barbecuing?

We frequently use these terms interchangeably, but in fact they refer to two different techniques.

Grilling is done rather quickly at high heat, either directly over coals or indirectly off to the side. If the grill has a lid, it is kept off. Choose only tender cuts to grill because the cooking is done quickly and does not have time to break down the fibers. When grilling meat, watch it closely so it does not burn.

Barbecuing involves cooking meat very slowly in a covered grill or smoker, surrounding the food in hot smoke until it becomes tender.

I am about to buy my first outdoor grill. Do you recommend charcoal or gas grills?

Grilling on gas and charcoal are very different experiences. Gas is cleaner, simpler, and more predictable, while charcoal grilling is more challenging as well as more work. When it comes to the flavor of the food, I don't think there is any comparison: food cooked properly over hardwood charcoal is far more flavorful. It has that real grill taste that you just cannot get from a gas grill.

What type of charcoal grill do you recommend?

To me, there is only one great charcoal grill, and that is the Weber kettle. It has everything a good charcoal grill needs. You can control the cooking temperature by opening the vents on the bottom of the grill and in the lid. As long as you buy the medium- or large-sized grill, you can easily use either direct or indirect grilling methods.

When you shop for a grill, keep this checklist with you:
- Are the legs sturdy?
- Is the grill made of heavy-gauge metal?

- Does it have a tight-fitting lid with vents in it?
- Does the grill have vents in its base?
- Is the grate hinged or easy to lift so that you can add more coals to the firebox easily?

Some grills have wire side baskets that hold coals for indirect grilling and a shallow ash-catching plate for easy cleaning.

The Weber kettle grill pretty much fits the bill, although older versions have no side baskets. But I don't find that much of a deterrent.

There is room for only a very small charcoal grill on my balcony. Can I grill successfully on one of these small models?

Street food around the world is cooked on small grills with great success. If you watch food vendors on the sidewalk in Vietnam or Thailand, you will see them fan the fire to increase the heat. These small grills, which often are hibachi-style, are very good for grilling small cuts of meat, poultry pieces, and seafood. Because they do not come equipped with lids and because they are just too small, they cannot be used for indirect grilling or for grilling large cuts or whole birds.

What should I be looking for when shopping for a good gas grill?

Heat and more heat. The more BTUs the better—35,000 is good, 40,000 is better, and 60,000 is the best. High BTU means a hotter grill. These grills get sufficiently hot that they can be used uncovered. This means you can sear thick cuts of beef on the grill as well as foods that don't have any fat at all, such as vegetables or fish fillets.

Remember that you are looking for a grill, not an oven. Make sure that it comes with adjustable controls so that you can set it on high, medium, or low. Three adjustable heating zones are important, too. Many grills come with only two, which I do not consider to be enough. You should be able to move the food around and keep some things hot over a lower heat while others are cooking over high.

A warming rack is another nice feature. This, too, allows you to keep some foods warm while others are grilling. If you want to keep the main course warm while you are serving an appetizer or drinks, this is extremely useful.

Will I get the same results using my built-in stovetop grill as I do grilling outdoors?

Indoor grills are wonderful for preparing low-fat foods, but you will never get that assertive grilled taste that you get from a very hot gas grill or a charcoal grill. When grilling indoors, stick to thin chops, fish fillets, sliced vegetables, and boned chicken breasts.

Recently I've heard that charcoal briquettes are not safe and produce an off taste. Do you agree?

Briquettes are a reliable product, which is good to know since they are so much easier to find than lump charcoal. Briquettes are made of charcoal, coal, and starch and so do not burn as cleanly as lump charcoal, which is nothing more than carbon.

The food will taste fine if cooked over charcoal briquettes, but it will never have the distinctive, pure grilled flavor of food cooked over natural lump charcoal. The irregular pieces are sold under various names, including charwood, lump charcoal, and natural charcoal.

This burns hotter than briquettes and while you may have to wait a little longer before the fire is ready, the natural charcoal gives food fabulous flavor. It does not impart any unpleasant odor, so when the fire dies down, you can add more charcoal without worrying about fumes. This is one of the best things about lump charcoal as far as I am concerned.

If you use briquettes, shop carefully. Some of them are impregnated with chemicals to make them self-lighting. Avoid these at all costs. Stick to known brand names. Inexpensive briquettes can burn "dirty" and smell unpleasant.

Mesquite seems to be the trendy wood for grilling. What is it and where do I find it?

More natural lump charcoal is made from mesquite than from any other wood. This might be because the mesquite tree, considered a "weed tree," grows with little help from human cultivation in arid regions of both North and South America and is easy to gather. Mesquite, which has a light distinctive aroma, is also popular for charcoal because it burns hot and stays hotter longer than other charcoals.

How much charcoal do I need to grill a butterflied leg of lamb or steak to feed four to six people?

For most grilling needs, start with about 5 pounds of charcoal. If you also plan to grill some shrimp and vegetables for appetizers, and will be grilling for longer than 45 minutes, you will need more.

To determine how many coals to light for direct-heat grilling, lay the charcoal in a single layer under the area where the food will be. Toss in a few extras chunks for good measure, and you have your fuel. If you will be grilling the food over indirect heat (more on this later in the chapter), you will probably need more charcoal. Mentally divide your grill into three sections. In one outside section, pile the coals 4 to 5 inches deep. Create a middle section with one layer of coals. Leave a third of the grill without coals. This allows you to cook food at different temperatures as required.

What is the best way to light a charcoal fire?

If your grill is near an electrical outlet, electric starters work well. They are inexpensive and easy to find at hardware stores. Unfortunately, these devices have a short life span—they break easily, particularly if you leave them in the grill for more than 20 minutes. It's good to have more than one on hand, since they tend to break at the most inconvenient times. (For this same reason, I recommend you have an extra tank of propane gas on hand if you use a gas grill!)

You might also try the chimney, which is reliable and very popular. If you like to grill as much as I do, you should have one. These are cylindrical metal containers with room on the bottom for crumpled newspapers and on the top for charcoal. The newspaper easily ignites, which in turn ignites the charcoal. When the coals at the top of the chimney are covered with light gray ash, tip them out of the chimney into the firebox of the grill.

What about using starter fluid to light charcoal?

I stay away from it when I'm at home, but when I am camping—one of my favorite things to do—or when we picnic down by the river near my house, I use it. If you let it burn off, it's okay. Follow the directions on the can carefully. Give the fluid time to seep into the coals before you light them and never squeeze any fluid onto hot coals. Once the coals are lit, they will have to burn for at least 30 minutes to reach the correct temperature. By then, any noxious fumes in the fluid will be long gone.

I hear about "direct" and "indirect" grilling. This sounds confusing. What do they mean?

Direct grilling is exactly what it implies. You put the food directly over the hot coals, which, for this kind of cooking, can range from hot to medium-hot. Direct grilling is best for steaks, chops, chicken breasts, fish steaks, and vegetables—in other words, foods that cook quickly and need little time on the grill.

Indirect grilling is really a form of roasting. It does, however, add a nice grilled flavor. I usually start by searing the food directly over hot coals (direct grilling) and then move it to a cooler part of the grill to finish cooking. Once the food is moved off the coals, I cover the grill and keep the vents open. Indirect grilling works best for foods such as loin of pork, bone-in leg of lamb, or a whole turkey or chicken.

How can I tell when the coals are ready?

Unfortunately, there's no foolproof method—like any other cooking technique, the more you do it, the better you become. Practice is important, but so is patience. Most fires made from charcoal can take as long as 40 minutes until they are ready. A good rule of thumb is to count on 35 to 40 minutes for a hot fire, 45 minutes for a medium fire, and about 50 minutes if you want a low fire.

The recipe should tell you how intense the fire should be. Watch the coals and when they are covered with a fine gray ash, check their temperature by using the "hand test": Hold your hand 5 inches above the coals and if you can hold it for 6 seconds before yanking it away, the fire is low; if you can hold it over the coals for only 5 seconds, the fire is medium-hot; and 1 or 2 seconds means a very hot fire.

TIP Patience is a virtue, but by being too careful, you may find that the fire has literally burned away before you have even begun cooking! If this happens, add a few more coals and give the fire another 10 to 15 minutes to regain its proper temperature.

How do I prepare the grill for indirect grilling?

Most cookbooks suggest that you make two piles of coals on opposite sides of the grill and leave a space in between them for a drip pan. Each pile of coals is ignited, and the food is positioned over the drip pan.

I usually don't do this. Instead, I arrange the coals on one side of the grill, put the food on the other side, and rarely bother with a drip pan. Although you may lose pan juices—which are minimal at best—my technique is simpler and much more straightforward. Remember that you can always divide the smoldering coals and set a drip pan between them, which is a useful technique when grilling a turkey or large roast.

Once you try indirect grilling, it will begin to make sense and you won't find it difficult. In fact, it will expand your grilling repertoire.

Can I use a gas grill for indirect grilling?

Absolutely, as long as you can turn off part of the grill. The best gas grills come with three zones, or burners, which allows plenty of choices. Even if your gas grill has only two burners, you can grill indirectly by turning off one element and putting the food on that side.

The food on my gas grill seems to cook slowly and does not develop a true "grill flavor." How do I get the food going? Can I sprinkle it with oil or is that dangerous?

For years I had the same problem. I realized it was because my gas grill just did not get hot enough. Even when I used marinades with large quantities of oil, the food didn't taste grilled. After a bit of research I discovered that in order to be effective, the gas grill must heat up to 600° or 700°F, which means the grill should have at least 35,000 BTU. If your grill does not generate enough heat to begin with, it may not maintain the heat when the lid is opened, resulting in food that is baked, rather than grilled.

As far as using oil while grilling to "get the food going," all you will accomplish is flare-ups—which are not only dangerous, but result in burned, bitter-tasting food.

I like the taste of smoked food and have experimented with some wood chips, but I am still not exactly sure how to use them.

Start by soaking the wood chips for 20 to 30 minutes in a bucket or similar container. This will allow them to smoke before they burn. Then make packages from heavy-duty foil with about two handfuls of wet chips. Perforate the packages in two or three places and set them on top of the coals. The size of the food dictates how many foil packages you will need. For example, if you are grilling a small turkey, you will need at least two packages of chips, while if you are grilling a butterflied leg of lamb, one may be enough. When you grill something like chicken breasts, quail, or fish steaks, which cook quickly, simply scatter the wet wood chips directly over the hot coals—this will impart the smoky flavor you're after.

When should I add wood chips to the grill?

This depends on two factors. The first is the choice of chips; some impart a more intense fla-

vor than others—particularly hickory. The second factor is the size of the piece of meat, poultry, or fish you are grilling. For large cuts of meat, I add the wood chips (wrapped in foil packages) 20 to 30 minutes into the grilling process so the smoke does not overwhelm the food. I usually add more packages once or twice before the food is done.

For small cuts, such as chicken breasts, quail, or fish steaks, I sprinkle wet chips directly on the coals right at the beginning, because the grilling time is so short.

The taste of hickory can be overwhelming. What other wood chips do you recommend?

If you don't care for hickory, you have other wonderful choices and can change them seasonally. For instance, in the cold winters of New England, I use heavier woods such as oak, apple, and maple for more robust foods. Toward spring, I switch to lighter woods such as lilac, alder, and cherry. Alder and mesquite are great for summer grilling because of their pleasingly mild flavor. Dried herb twigs, such as rosemary and fennel stalks, are good choices as well.

Do wood chips impart enough smoke to actually smoke food?

Using wood chips and hot coals to grill food is not the same as smoking food, which is a cool cooking method. The food will taste smoky, but it will be grilled, not smoked.

To smoke food, you need to set aside a lot of time—usually 5 or 6 hours—and cook the food over very low coals. Be prepared to tend the fire closely. It's important to keep the coals at 140° to 150°F and add soaked chunks of hardwood and wood chips every 20 to 25 minutes. Add only enough to produce smoke, but not so many that you risk extinguishing the fire.

Additionally, you will have to add live (already lit) coals to the fire every 45 minutes or so. That means you need a separate grill or chimney filled with smoldering coals because the key to proper cold smoking is to maintain an even temperature at all times.

If you like to smoke foods, consider investing in a commercially made smoker, which I find to be a great tool.

When a recipe calls for grilling, can I use a broiler instead?

The broiler works best for foods with some fat, such as pieces of chicken on the bone or chops. It does not work as well for delicate foods such as vegetables, scallops, and skinless fish fillets. Basically grilling and broiling are two different cooking techniques. One cooks the food from below, the other from above, so there will be a difference in taste and texture. Also grilling lends food a very special "grilled" flavor, which broiling does not.

What is the purpose of marinating meats and seafood before grilling?

Marinades are popular because they are fun to concoct and add a certain flavor dimension to the food. But, unless they are packed with intensely flavored herbs and spices (hot peppers, cumin, cardamom, cinnamon, cloves, coriander, and so on), their flavor tends to dissipate during cooking.

Which foods should be marinated? Does marinating improve the flavor?

Traditionally, marinades have been used in countries where the quality of the raw ingredients (fish, poultry, meat) was not always of utmost freshness. The marinades were used to mask or perk up dull flavors. This is not the case today, although we still tend to marinate lesser cuts of meat, especially flank and skirt steak and a butterflied leg of lamb.

A good rule of thumb is that the higher the quality of the raw ingredient, the less marinating it needs. There is nothing more wonderful than tender baby lamb chops simply enhanced with a touch of rosemary and some garlic, or a great steak properly seasoned with coarse salt and freshly ground pepper before it's grilled over hot coals. Of course, with truly fresh fish, less is more—a grinding of sea salt and black pepper and a squeeze of fresh lemon is all it needs. This is not a hard-and-fast-rule. If you prepare chicken or lamb often, marinades are good ways to add a spectrum of tastes.

Ironically, marinades can have an adverse effect on seafood and fish. If these are marinated for too long, they become dry and rubbery. While meat and poultry can withstand hours in a marinade, seafood and fish should be removed after 30 minutes.

Fish steaks, such as swordfish and tuna, are good when marinated for short periods of time, but I prefer salmon simply grilled and lightly smoked with wood chips, served with an interesting sauce on the side.

TIP Don't be tempted to use the marinade as a sauce unless you cook the marinade, too. This means bringing it to a boil and then letting it simmer briskly for at least 5 minutes to destroy any harmful bacteria that the raw food might have imparted. For this same reason, don't baste food with the marinade in which it has soaked except during the very early stages of cooking.

If I am in a hurry and have no time to plan ahead, can I marinate something for 1 or 2 hours even if a recipe calls for 12 to 14 hours?

Depending on the food, even short periods of marinating will give you excellent results; in fact fish and shellfish only need 30 minutes. In recipes that call for long marinating times in the refrigerator, you can get pretty good results by marinating at room temperature for 1 to 3 hours, as long the temperature of the kitchen does not exceed 65°F.

TIP If you want to marinate for a short period of time, it is best to use yogurt-based marinades in terms of taste and texture. Be sure to use yogurt with a live, active bacterial culture.

Recipes usually say to marinate in non-reactive pans. What is does nonreactive mean?

A nonreactive pan is one that will not react with acidic ingredients, which include vinegar, tomatoes, citrus juice, and wine, all commonly used in marinades. The most common reactive containers are non-anodized aluminum, unlined copper, and cast iron, all of which should be avoided. They can impart a metallic taste to the food and cause it to darken. A pan with an aluminum core is fine for marinating if it is coated with stainless steel. Glass, ceramic, and plastic are good materials for containers used to hold marinades.

Over the years, I have marinated in many different types of containers and have become a fan of zippered plastic bags. Once the food and the marinade are enclosed in the bag, they mingle in close quarters. It's easy to turn the food, allowing you to distribute the marinade evenly. Also the pliable bag fits more easily in a crowded refrigerator.

If I put a steak or chicken on the grill and there is a flare-up, should I move it to a cooler part of the grill or remove it entirely?

I suggest taking it off the fire right away and giving the fire time to cool down a little. Even if you move it to a cooler part of the grill, it will still cook and possibly burn. It will also throw off your timing. This is why I feel so strongly about exercising patience and waiting for the coals to reach the correct temperature. They should be covered with gray-white ash before starting to grill. If you are using a gas grill and there is a flare-up, it is best to remove the food and turn down the heat.

Can every vegetable be grilled?

Most vegetables are delicious grilled while a few fare less well. My favorites are peppers, zucchini, onions, eggplant, and mushrooms. I am not a fan of potatoes, leeks, and winter squash, such as butternut, but they too will soon join the ranks of the all-popular classics. As with everything, I suppose, it's a matter of personal taste.

I have tried to cook spareribs on the grill, but always end up with a charred mess. What is the secret?

There are two methods that will allow you to grill spareribs so that the meat is tender but the ribs are not charred. The first is to roast the ribs in a 350°F oven for about 45 minutes to cook them, then place the racks directly over medium-high coals for about 15 minutes, until brown and crisp. The other method is to grill them over low indirect heat in a covered grill for about 45 minutes before finishing them over direct heat for the same 15 minutes.

TIP Never brush barbecue sauce on the ribs until the last 2 minutes of cooking. The sugar and tomatoes in the sauce will quickly caramelize and burn the surface of the meat. Even when you apply the sauce during the final minutes of grilling, watch the ribs very carefully to avoid charring.

What is the best way to cook chicken on the grill—whole, halved, or in pieces?

My two favorite ways are whole or cut in half and grilled "under a brick" (see page 356). When I grill a whole chicken, I cook it first over indirect heat for about 35 minutes and then set it directly over medium-high coals to continue cooking until the internal temperature of the dark meat is 165°F. Then I remove it from the grill, cover it loosely with foil to keep warm, and let it rest for 10 to 15 minutes before serving. During this time, the internal temperature of the dark meat will rise to the correct 180°F.

If you have your heart set on grilling cut-up, bone-in chicken pieces, start them over medium-high direct heat, turning them once or twice until the skin is browned. Move them to the cooler side of the grill for 15 to 20 minutes, or until cooked through. Quickly sear them again over direct heat, if necessary, for a crispy finish.

I keep reading that cutting into meat causes it to lose its juices. Is this true?

Certain cuts of meat such as skirt and flank steaks as well as a butterflied leg of lamb can be tested by cutting into them. You may lose some juice, but there should be no loss of flavor. Cutting into poultry is also not a problem, but I never cut into smaller cuts of meat, such as steaks and chops. Instead I use a meat thermometer, which has become the "tool of the grill" or the tip of the knife method, which will give you an accurate reading every time.

TIP Insert the tip of a sharp paring knife into the thickest part of a piece of meat, chicken, or fish. Test it against the inside of your wrist (the same place you test the heat of the milk in a baby's bottle). If it feels cold, the food is still raw; if it's just warm, it is medium-rare; and if the tip of the knife feels hot, the food is cooked through and may actually be overdone.

I am always nervous when I grill fish because I can never tell when it's done.

It takes experience and a lot of overcooked and undercooked fish before you learn how to tell when fish is perfectly cooked. Never trust the exact times given in a recipe, because so much depends on the heat of the grill. When grilling fish steaks, remove them from the grill at about the time you think they may be done. Once off the fire, use the "tip of the knife" method employed by all professional chefs to determine doneness. For salmon, tuna, and swordfish you want the fish to be medium-rare.

This is not the optimum method for a whole fish, which should be opaque all the way through. Always test a whole fish in a couple of places and especially near the bone where a little translucency is okay.

What is the best way to grill a whole fish?

Unless you have incredible expertise, a whole fish is challenging to cook on the grill, especially if it weighs more than 2 to 2½ pounds. You can roast a fish that is wrapped in foil fairly easily, because you can easily remove it from the grill and test the flesh for doneness. Also, the fish will not stick.

My salmon steaks or fillets tend to fall apart on the grill. What's the secret to keeping them whole?

If you want your grilled salmon to look pretty as well as taste delicious, make sure that your fillets or steaks (either cut is okay in this case) still have their skin attached. Also, cook them through almost all the way before turning them. If you turn them more than once, they are sure to fall apart.

I have heard you can cook food buried in the ashes. Is this safe when you use charcoal briquettes?

When the charcoal has burned down to the point of ashes, it is quite safe to cook foods nestled in them. Whatever chemicals may have been in the charcoal have burned off long ago. Unless the food has a protective skin (like a potato or corn on the cob), wrap it in heavy-duty aluminum foil before burying it in the ashes.

Is it necessary to clean the grill after each use?

It depends on what grill you use. A charcoal grill should not be emptied of ashes since they allow for better conduction of heat. However, if the ashes are wet, you are better off starting from scratch. As for the grill itself, a quick scrubbing with a grill brush will do. When using a gas grill, be sure to scrape food particles off the rack every time you use the grill.

Lemon-Herb Grilled Chicken "Under a Brick"

The method of cooking chicken under a brick is popular in many countries around the Mediterranean. The Italians call it Chicken Fra Diabolo, in Spain it is Pollo Planchado, and in Russia it is called Truck Chicken, "because the chicken looks like it has been run over by a truck." No matter, the result is delicious and a great addition to one's summer cooking repertory. You can use Cornish hens with equally good results.

SERVES 2 TO 4

2 to 4 small poussins (about 1¼ pounds each)

MARINADE

1 tablespoon soy sauce

Juice of 1 large lemon or lime

1 tablespoon minced fresh thyme

1 tablespoon minced fresh rosemary

3 large garlic cloves, smashed

⅓ cup fruity olive oil

1 teaspoon coarsely ground black pepper

2 teaspoons diced serrano chile, optional

2 to 4 sprigs fresh thyme

2 to 4 tiny sprigs fresh rosemary

Salt and freshly ground black pepper

Large pinch of cayenne pepper

Lemon wedges, a sprinkling of *fleur de sel*,
    rosemary sprigs, for garnish

1. Split the chickens in half and cut off the backbone. Place the chicken between 2 sheets of waxed paper and pound until flattened evenly. Set aside.

2. Combine all the marinade ingredients. Transfer to one or two large zippered plastic bags. Add the chickens, turn to coat, and marinate for at least 4 hours or overnight.

3. The next day, remove the chickens from the marinade and dry thoroughly with paper towels. Carefully lift the skin, separating it from the breast meat without tearing, and insert some fresh minced thyme and rosemary under the skin. Set aside.

4. Heat a gas grill to medium-high. Place the chickens skin side down on the grill rack and top each half with a brick wrapped in foil. Grill for 10 to 12 minutes, or until the skin is nicely browned but not burned.

5. Remove the bricks, turn the chickens over and again top with the bricks. Grill for another 12 minutes, or until cooked through but still juicy. Transfer the chickens to a serving platter, season with salt and pepper, and a sprinkling of cayenne. Garnish with a sprinkle of *fleur de sel*, quartered lemons, and rosemary sprigs, and serve hot.

REMARKS Be sure to check the chicken after the first 5 minutes. If the skin seems to be cooking too quickly, turn off one side of the grill and move the chicken to the cooler side; return the chicken to the hot part of the grill once it has been turned over.

Lime-Marinated Chicken Kebabs

I don't know of any grilled food that is more popular than chicken, and deservedly so. Chicken cooked on the grill is both versatile and satisfying. Here I use just the thighs, but you can also add some drumsticks and drumettes, the fleshy section of the wing. A teaspoon of turmeric or a pinch of real saffron will add a brilliant yellow color to the chicken, which makes for an interesting change.

SERVES 6 TO 8

3 medium onions, coarsely chopped
1 cup fresh lime juice
1 cup extra virgin olive oil
2 cups cilantro leaves and stems
Coarse salt and freshly ground black pepper
½ teaspoon turmeric
Large pinch of powdered saffron, optional
16 to 18 boneless chicken thighs

1. In a food processor combine the onions, lime juice, olive oil, and cilantro. Puree until blended but still coarse. Season with coarse salt, turmeric, saffron, and lots of black pepper. Transfer the mixture to a large zippered plastic bag. Add the chicken and refrigerate for 4 to 6 hours or overnight. Let the chicken return to room temperature 30 minutes before grilling. Drain and scrape off the marinade.

2. Make a charcoal fire on one side of the grill and when the coals are white, place the well-drained chicken pieces directly over the coals. Grill for 5 minutes on each side, or until the thighs are nicely browned but not charred. Transfer to the cooler side of the grill, cover, and continue to cook for another 10 to 12 minutes. Serve hot, accompanied by a platter of grilled peppers of assorted colors and Charcoal-Grilled Eggplants with Lemon-Scallion Mayonnaise (page 126).

Smoked Chicken Wings with Spicy Peanut Sauce

Whenever I grill, I think of various ways to use the hot coals. Vegetables or chicken wings are delicious served as part of a cookout or as finger food with drinks for dinner the next day. If you cannot get hickory chips, use any other fruitwood or omit it altogether.

SERVES 4

1 medium onion, quartered

4 large garlic cloves

2 tablespoons peeled fresh ginger

1 small jalapeño pepper

2 cups plain yogurt

$\frac{1}{2}$ to 1 tablespoon tandoori seasoning,
   (see Sources, page 397)

2 teaspoons ground cumin

Coarse salt and freshly ground black pepper

2 cups hickory chips, soaked in water for
   1 hour

20 chicken wings, cut in half at joint and
   tips removed

Spicy Peanut Sauce (recipe follows)

1. Combine the onion, garlic, ginger, and jalapeño in a food processor and grind to a fine paste. Add the yogurt, tandoori seasoning, and cumin and process until well blended. Transfer to a large zippered plastic bag, add the chicken wings, and seal the bag. Refrigerate for 6 hours or overnight, turning the bag often.

2. Prepare the grill with the charcoal placed to one side only. Open all the vents.

3. When the coals are white, remove the wings from the marinade and scrape off any bits. Sprinkle the wings with salt and pepper and place on the grill directly over the cool side of the grill. Drain the hickory chips and add to the coals. Cover and continue to grill for 15 to 20 minutes, or until the juices run pale yellow.

4. Transfer the wings to a serving platter and serve hot with individual bowls of warm peanut sauce on side.

Spicy Peanut Sauce

MAKES ABOUT 1½ CUPS

1 tablespoon peanut oil

2 teaspoons minced garlic

2 teaspoons minced peeled fresh ginger

2 tablespoons tomato paste

⅔ cup chicken broth

1½ teaspoons Chinese chili paste

2 tablespoons hoisin sauce

2 tablespoons creamy peanut butter

2 tablespoons toasted sesame oil

Heat the peanut oil over low heat in a skillet. Add the garlic, ginger, and tomato paste and cook for 1 minute. Add the broth, 1 teaspoon of the chili paste, the hoisin sauce, and peanut butter, whisk until well blended, and simmer for 3 minutes. Add the sesame oil and the remaining chili paste and keep warm.

Lemon- and Thyme-Scented, Apple-Smoked Cornish Hens

Cornish hens each serve one person nicely and are perfect dinner party food. They make a beautiful presentation and are easier to grill than regular-sized chickens because they take less time to cook. Fresh lilac or apple wood added to the coals infuses the birds with a lovely flavor. Serve with a side dish of grilled vegetables. For this recipe, you will need both charcoal briquettes and some fresh apple wood.

SERVES 4

4 small whole Cornish hens (about 1 to
   1¼ pounds each)

Juice of 1 large lemon

2 large garlic cloves, mashed

2 tablespoons minced fresh thyme leaves, plus
   sprigs for garnish

2 tablespoons minced fresh flat-leaf parsley

1 teaspoon coarsely ground black pepper

6 tablespoons extra virgin olive oil

Coarse salt

1. Marinate the hens: Place the hens in a large zippered plastic bag. In a small bowl, combine the lemon juice, garlic, thyme, parsley, and pepper. Whisk in the oil and pour the marinade over the hens. Seal the bag, place in a shallow dish, and refrigerate overnight, turning the hens in the marinade several times.

2. Prepare the fire: Use a medium-sized charcoal grill. Open all vents. Place 30 briquettes to one side of the lower grill and, using an electric or cylindrical starter, ignite the charcoal. Set a rectangular, disposable drip pan beside the coals and position the cooking grill in the kettle with one handle directly over the pile of coals. This will allow you to add briquettes through the opening by the grill handle during smoking.

3. Remove the hens from the marinade, truss them, and season with coarse salt. Place the hens on the grill, breast side up, directly over the drip pan. Add a few twigs of apple wood to the pile of coals, cover the kettle, and "roast" the hens for 1 hour, or until the juices run clear and a meat thermometer registers an internal temperature of 165°F. The hens will need no basting or turning, but every 15 minutes add 6 to 8 briquettes and a few twigs of apple wood to the pile of burning coals, through the opening on the cooking grill.

4. When the hens are done, transfer them to a cutting board and let rest for 5 minutes. Cut the hens in half and place 2 halves on each of 4 serving plates. Garnish with fresh thyme and serve either warm or at room temperature.

Fresh Fennel with Grilled Fennel Sausage

Fennel-flavored sausage has a natural affinity to the sweet and slightly licorice taste of fresh fennel bulbs. Together they make a lovely and easy supper course when served with crusty peasant bread or buttery mashed potatoes. I like to poach the sausage prior to grilling; this removes much of the fat and also eliminates flare-up. However, if the sausage is very lean, you can put it directly on the grill.

SERVES 4

1½ pounds sweet Italian fennel sausage

2 small fennel bulbs, tops removed, bulbs quartered

2 to 3 tablespoons extra virgin olive oil

Coarse salt and freshly ground black pepper

2 tablespoons coarsely chopped flat-leaf parsley

1. In a deep skillet, combine the fennel sausage with water to cover. Bring to a boil, reduce the heat, and poach the sausage for 3 minutes. Drain and reserve.

2. In a large saucepan, bring salted water to a boil. Add the fennel pieces, return the water to a boil, reduce the heat, and simmer for 3 minutes. Drain and set aside.

3. Preheat a gas grill. Place the fennel on a rack and sprinkle very lightly with olive oil, salt, and pepper. Transfer to the grill together with the sausage. Cover and grill over medium-hot heat for 5 minutes on each side, or until nicely browned.

4. Transfer the sausage and fennel to a serving dish. Drizzle with the remaining olive oil and parsley and serve accompanied by crusty bread or garlic mashed potatoes.

REMARKS Two grilled red bell peppers cut into ½-inch slices are a delicious and colorful addition to this simple dish, and excellent olive oil is a must!

Butterflied Leg of Lamb with Shallot, Garlic, and Lemongrass Rub

Lamb is a wonderful meat for grilling, and it takes well to many marinades. Make sure that the meat is actually butterflied and not just boned, and that the pieces are as evenly thick as possible. You are looking for a thickness of 1 to 2 inches. You can then cut the lamb into three to four sections, which allows you to remove the thinner pieces off the grill as they are done.

SERVES 5 TO 6

1 lemongrass stalk (bottom 6 to 7 inches)

3 large shallots, coarsely chopped

5 large garlic cloves, coarsely chopped

1 jalapeño pepper, coarsely chopped

2 tablespoons chopped fresh oregano, or 1 teaspoon dried

½ cup peanut or grapeseed oil

3 tablespoons Asian sesame oil

3 tablespoons Thai fish sauce

1 tablespoon sugar

1 tablespoon coarsely ground black pepper

4- to 5-pound butterflied leg of lamb

1. Combine the lemongrass, shallots, garlic, jalapeño, oregano, peanut oil, sesame oil, and fish sauce in a food processor and blend until finely minced. Add the sugar and pepper and blend well.

2. Rub the lamb thoroughly with the paste and place in a zippered plastic bag. Marinate for 6 hours, or overnight.

3. Build a fire on one side of a grill that can be covered. Remove the lamb from the marinade and pat it dry with paper towels. Place the lamb directly over the hot coals and sear until well browned, 3 to 4 minutes.

4. Move the lamb to the cool part of the grill. Cover, leaving the vents open, and cook for 15 to 20 minutes. After 10 minutes check the thinner pieces by cutting into one and if done, remove from the grill.

5. Remove the rest of the lamb from the grill when it is done to your liking. Let the meat rest for 5 minutes before serving. Serve accompanied by the Chick Pea, Tomato, and Cilantro Salad (page 287).

REMARK If you are using an instant-read thermometer, it should read 135°F for medium-rare.

Grilled Lamb Chops in Shallot and Herb Butter

A juicy grilled lamb chop needs little in the way of a sauce. However, the flavor is highly enhanced by this compound butter that melts nicely into the chop. You may vary the herb, keeping in mind that lamb has a particular affinity to character herbs such as rosemary, sage, oregano, and of course, garlic and shallots. Serve the chops with Mascarpone and Chive Mashed Potatoes (page 139) or Braised Leeks with Shiitake Mushrooms (page 131).

SERVES 2

6 tablespoons unsalted butter, softened

2 tablespoons minced shallots

2 tablespoon minced flat-leaf parsley

2 teaspoons minced rosemary

1 large garlic clove, minced

Coarse salt and freshly ground black pepper

4 double loin lamb chops, cut 1¼ to 1½ inches thick

sprigs of fresh rosemary

1. Combine the softened butter, shallots, parsley, rosemary, and garlic in a small bowl. Mash with a fork to blend thoroughly. Season with salt and pepper. Refrigerate for 30 minutes.

2. Bring water to a simmer in a 2-quart saucepan, set a heavy 10-inch dinner plate on top of the pan, and put half of the herb butter in it. Keep warm over low heat.

3. Preheat a gas barbeque grill to 500° or 600°F. Place the chops on the grill and cook for 2 to 3 minutes on each side. Be careful not to char them. Check for doneness with an instant-read thermometer: the chops should have an internal temperature of 120°F. If they are well browned and not cooked to medium-rare, turn off one side of the grill and place the chops on the cool side for another 2 to 3 minutes, or until done. Remove from the heat and season with salt and pepper.

4. Transfer the chops to the plate with the melted herb butter, cover tightly with foil, and let the chops absorb the butter, about 2 minutes. Serve at once, topped with the remaining herb butter and garnished with sprigs of rosemary.

Citrus-Marinated Grilled Pork with Sweet-and-Sour Red Onion Confit

Pork takes beautifully to citrus and spices. I particularly like the shoulder butt, which is an inexpensive cut usually available in Asian markets and German grocery stores. Grilled over fruitwood and served with melted onions, it is a wonderful early fall dish. The onions can be prepared several days ahead and reheated. The pork roast can be grilled 2 or 3 hours ahead of time and served at room temperature. Depending on the season, a side dish of **Puree of Sweet Potato and Carrots** (page 143), **Roasted Vidalia Onions** (page 136), or a plateful of grilled shiitake mushrooms would complete this dish beautifully.

SERVES 4 TO 5

⅓ cup olive oil

Juice of 1 orange

Juice of ½ lemon

1½ teaspoons dried oregano

1 teaspoon ground cinnamon

½ teaspoon ground coriander

½ teaspoon ground cardamom

Coarse salt and freshly ground black pepper

2 medium onions, thinly sliced

3 large garlic cloves, minced

1 tablespoon minced peeled fresh ginger

2½- to 2¾-pound pork shoulder butt, rolled and tied or 2¾-pound boned loin of pork, rolled and tied

Sprigs of fresh thyme

Red Onion Confit (recipe follows)

1. In a small bowl, combine the olive oil, orange juice, lemon juice, oregano, cinnamon, coriander, cardamom, and salt and pepper. Whisk until well blended. Add the onions, garlic, and ginger and stir to mix. Place the pork butt in a large zippered plastic bag, pour the marinade over the pork, and seal the bag. Set the bag in a shallow dish and refrigerate for 24 hours, turning the pork several times in the marinade.

2. The next day, remove the pork from the marinade and dry thoroughly on paper towels. Strain the marinade and reserve.

3. Prepare the charcoal grill: Place charcoal briquettes or charwood on one side of the grill. When the coals are very hot and white, divide in half and carefully push an equal amount to each side of the grill. Set a disposable rectangular pan, slightly larger than the pork butt, in the center of the lower grill between the piles of coals. Position the cooking grill in the kettle with the handles directly over each pile of coals. This will allow you to add briquettes through the openings by the grill handles during "roasting."

4. Sprinkle the pork butt with salt and place it on the center of the cooking grill directly over the roasting pan. Add a few green twigs of lilac or apple wood to each pile of coals, cover the kettle, and cook the pork for about 1 hour 10 minutes, or to an internal temperature of 155°F on a meat thermometer. Every 15 to 20 minutes, add a little more charcoal and a few twigs of wood to each pile of burning coals. Baste often with the reserved marinade; stop basting at least 5 minutes before the meat is done, so no "raw" marinade is left on the meat.

5. When the pork is done, transfer to a carving board and let rest for 5 minutes before carving. Reheat the red onion confit.

6. Cut the pork butt crosswise into ½-inch slices and place in an overlapping pattern on a serving platter. Garnish with the thyme and serve warm or at room temperature with the red onion confit on the side.

REMARKS You can use other fruit woods, such as alder or cherry. These come in chips and are easy to obtain through mail order.

Red Onion Confit

SERVES 4 TO 5

3 tablespoons rendered duck fat or
  2 tablespoons butter and 2 tablespoons
  corn oil

6 large red onions, quartered and thinly sliced

2 teaspoons sugar

2 tablespoons fresh thyme leaves

Salt and freshly ground black pepper

1 to 2 tablespoons sherry vinegar

1. In a nonreactive, large heavy skillet, heat the duck fat or a mixture of butter and oil over high heat. Add the red onions, sprinkle with the sugar, and cook for 3 to 4 minutes, stirring constantly, until the onions begin to brown.

2. Reduce the heat to low, add the thyme, and season with salt and pepper. Continue to cook, partially covered, until the onions are soft and caramelized, 40 to 45 minutes.

3. Add the sherry vinegar to the onions and cook, stirring often, until it has evaporated, 1 to 2 minutes. Set aside at room temperature for up to 3 hours or refrigerate in a covered container for up to 1 week.

REMARK Duck fat packed by D'Artagnan is now widely available in many supermarkets and most specialty shops.

Brochettes of Swordfish Grilled in Sesame Marinade

To me, swordfish tastes best grilled. Here it is marinated in a teriyaki-like marinade and quickly grilled. A quick sauté of baby bok choy with red peppers or the Avocado Salad on page 109 and couscous tossed with cilantro make perfect accompaniments.

SERVES 6

¼ cup soy sauce

¼ cup rice vinegar

1 tablespoon minced peeled fresh ginger

2 large garlic cloves, minced

1 tablespoon sugar

1 tablespoon minced jalapeño pepper

2 tablespoons minced lemongrass, optional

⅓ cup fruity olive oil

2 tablespoons toasted sesame oil

3 to 6 drops Tabasco sauce

2½ to 3 pounds swordfish steaks cut into 1-inch cubes

2 bell peppers (a mix of red, yellow, orange, or green)

1 large white onion, cut into 1¼-inch cubes

Coarse salt and freshly ground black pepper

Small leaves of fresh cilantro

1. Combine the soy sauce, vinegar, ginger, garlic, sugar, jalapeño, and lemongrass in a medium mixing bowl. Whisk in the olive and sesame oils in a slow stream. Add the Tabasco and swordfish cubes and marinate for 2 to 4 hours. While the swordfish is marinating, soak 6 wooden skewers in water.

2. Prepare a hot fire in a charcoal grill. Remove the swordfish from the marinade and thread 4 cubes, alternating with the bell peppers and onion, onto each wooden skewer. Sprinkle with a little salt and pepper. Grill the swordfish brochettes over very hot coals 3 to 4 minutes per side, or until browned outside and almost opaque in the center. (Test by removing one piece of swordfish and checking it for doneness.)

3. Transfer the brochettes to serving plates and garnish with cilantro.

Grilled Salmon Kebabs in Dill and Mustard Marinade

Now that good fresh salmon is available everywhere, I find that I am always looking for new and interesting ways to prepare this fish. Here the salmon is marinated and grilled on skewers. Depending on the season, I like to serve the kebabs with sautéed cucumbers, a few simply boiled potatoes, and some roasted beets. In the summer, a cool cucumber and radish salad would make a lovely side dish as well.

SERVES 4 TO 5

1 large bunch of dill, large stems removed, dill
  coarsely chopped

¾ cup peanut oil

Juice of 1 large lemon

1 cup sliced scallions

1½ tablespoons sugar

1 heaping tablespoon Dijon mustard

1 teaspoon coarsely cracked black pepper

1 teaspoon ground coriander

2 pounds boneless, skinless center-cut salmon
  fillets, cut into 1-inch cubes

1 large red onion, cut into 1¼-inch cubes

Coarse salt and freshly ground black pepper

Spigs of dill

1. In a food processor combine all the marinade ingredients and puree the mixture until smooth. Transfer to a zippered plastic bag. Add the salmon cubes and coat well in the marinade. Chill for 3 to 4 hours.

2. When ready to grill, remove the salmon cubes from the marinade and wipe dry with paper towels. Skewer the fish cubes onto metal or bamboo skewers, alternating with pieces of red onion.

3. Grill the fish for a total of 6 to 8 minutes, turning the skewers after 3 minutes. The kebabs should be nicely browned but not charred. Test for doneness with the tip of a sharp paring knife: it should come out warm, not hot. Season with salt and pepper.

4. Remove the kebabs to plates and garnish with dill sprigs.

REMARKS If using bamboo skewers, remember to soak them in water for at least 30 minutes before using.

Grilled Scallops with Mango and Jicama Salad

Mangoes and scallops are a wonderful combination and they are in perfect harmony in this rather sweet and spicy salad, which I like to serve with a side dish of grilled corn on the cob. Start by making the mango salad, then let it chill while you prepare the scallops.

SERVES 4 TO 5

1½ pounds large sea scallops
1 tablespoon grated lime zest
Juice of 1 lime
Juice of 1 orange
¼ cup grapeseed oil
2 tablespoons Thai fish sauce
2 teaspoons Thai chili sauce
2 large garlic cloves, crushed
Mango and Jicama Salad (recipe follows)
Whole mint leaves, cilantro leaves, or whole
  chives

1. Combine the scallops with the lime zest, lime juice, orange juice, grapeseed oil, fish sauce, chili sauce, and garlic. Toss to mix. Cover and refrigerate for 45 to 60 minutes.

2. Preheat a gas grill. Remove the scallops from the marinade and grill for 3 minutes on one side. Turn over and grill for 2 minutes on the other side. Do not overcook. Remove to a platter and season lightly with salt.

3. Place a mound of mango salad on individual serving plates and top with 3 to 4 scallops. Garnish with mint or cilantro leaves or whole chives.

Mango and Jicama Salad

SERVES 4 TO 5

2 ripe mangoes, peeled and cubed
2 cups finely diced jicama
Juice of 1 orange
1 fresh hot red chili pepper, finely diced
  (about 1 tablespoon)
3 tablespoons minced fresh mint

Salt and freshly ground black pepper

2 tablespoons minced cilantro, optional

3 tablespoons grapeseed oil

In a bowl, combine the mangoes, jicama, orange juice, hot pepper, and mint. Season very lightly with salt and a grinding of pepper. Add the cilantro and toss lightly. Taste and add more hot pepper to taste. Cover and chill until 20 minutes before serving.

Grilled Shrimp with Mushrooms and Bell Peppers

Come summer I start to seriously experiment with all kinds of kebabs. I find them appealing and very practical, since once the vegetables, meat, poultry, or shellfish is threaded onto the skewers, the main course is basically ready. All you need as an accompaniment is an interesting grain—be it rice, couscous, polenta, or bulgur.

SERVES 6

1 teaspoon whole coriander seeds

1 teaspoon whole cumin seeds

$\frac{1}{2}$ cup olive oil

$\frac{1}{4}$ cup freshly squeezed lime juice

1 small yellow onion, minced

1 bay leaf

3 large garlic cloves, finely sliced

$\frac{1}{2}$ teaspoon coarse salt

Freshly ground black pepper

30 large shrimp, peeled and tails left attached

1 large red onion

1 large red bell pepper, seeded and cut into
    1-inch squares

1 green bell pepper, seeded and cut into
    1-inch squares

$\frac{3}{4}$ pound cremini (brown mushrooms), cut
    into large cubes

3 tablespoons minced fresh cilantro or
    flat-leaf parsley

1. Make the marinade: In a dry heavy skillet, combine the coriander and cumin seeds and place over medium heat. Toast gently for 1 minute, or until the seeds are lightly brown and fragrant. Transfer the seeds to a mortar and pestle or a spice grinder and finely crush.

2. In a medium bowl, combine the crushed toasted spices, the olive oil, lime juice, minced onion, bay leaf, sliced garlic, salt, and pepper to taste. Whisk until well blended. Add the shrimp, stir to coat, cover with plastic wrap, and refrigerate for 30 to 45 minutes.

3. Prepare a hot fire in a charcoal or gas grill. Soak 6 long wooden skewers in water for at least 30 minutes.

4. Cut the red onion into eighths. Remove the shrimp from the marinade; reserve the marinade.

5. To assemble the kebabs, alternate the shrimp, onions, bell peppers, and mushrooms on the skewers, starting with a shrimp, then adding a piece of onion, pepper, and mushroom, ending with a shrimp. Use 5 shrimp per skewer. Sprinkle the kebabs with a little coarse salt and a grinding of pepper.

6. When the coals are very hot, place the shrimp kebabs on the grill directly over the hot coals and grill for about 2 minutes on each side, until the shrimp turn bright pink tinged with brown.

7. Remove the kebabs from the grill and transfer each to an individual serving plate. Sprinkle with minced cilantro or parsley and serve immediately.

Middle-Eastern Char-Grilled Skirt Steaks in a Cumin and Cilantro Marinade

Skirt steak is an inexpensive cut of meat that is outstanding prepared on the grill. It is best when marinated in a zesty marinade such as this one and served with the Chickpea, Tomato, and Cilantro Salad (page 287) or the warm Lentil Salad (page 288). Double the recipe if you can because leftover skirt steak makes wonderful sandwiches.

SERVES 6

> 1 heaping tablespoon ground cumin
>   seeds
>
> 3 bunches fresh cilantro, washed
>   thoroughly and dried
>
> 4 large garlic cloves
>
> 2 tablespoons cracked black pepper
>
> 2 chipotle peppers in adobo sauce,
>   or more to taste
>
> ¾ cup olive oil
>
> Juice of 2 limes
>
> Coarse salt
>
> 3 pounds skirt steak
>
> Sprigs of fresh cilantro

1. In a blender or food processor, combine the cumin seeds, cilantro, garlic, black pepper, hot peppers, olive oil, and lime juice. Season lightly with salt and process until smooth.

2. Cut the skirt steak into ½-pound pieces and place in a large zippered plastic bag. Pour the marinade over the pieces and seal the bag. Set the bag in a shallow dish and refrigerate for 24 hours, turning the steaks in the marinade once or twice.

3. Prepare a hot fire in a charcoal grill. Remove the steaks from the bag and wipe off the excess marinade.

4. When the coals are red hot, sprinkle each side of the skirt steaks with a little coarse salt and set the steaks over the hot coals. Grill 2 to 3 minutes per side, until nicely browned for medium-rare, 130° to 135°F internal temperature on a meat thermometer.

5. Remove the steaks from the grill, transfer to a cutting board, and let sit for a few minutes before slicing. Cut the meat across the grain on the bias into thin slices. Garnish with cilantro.

REMARKS For a more intense cumin flavor, toast 2 tablespoons whole cumin seeds in a small skillet until fragrant and crush in either a mortar and pestle or a spice grinder.

Grilled Marinated Cremini Mushrooms

I am always looking for ways to use the fire after the main course is done and these mushrooms are ideal for this use, since they only take 3 to 4 minutes to cook, and the dressing can be made well ahead of time. I like to serve these mushrooms as an appetizer over young greens or as a side dish for grilled veal chops, chicken, or a sirloin steak. It is a good idea to double or even triple the recipe and let some of the mushrooms marinate for a day or two.

SERVES 4

¼ cup plus 2 tablespoons extra virgin olive oil

2 tablespoons lemon juice

1 tablespoon balsamic vinegar

1½ teaspoons Dijon mustard

1 large garlic clove, mashed

Coarse salt and freshly ground black pepper

1 pound cremini mushrooms, stems removed

2 tablespoons minced fresh thyme, oregano,
   or marjoram

1. In a bowl combine the ¼ cup olive oil, the lemon juice, balsamic vinegar, Dijon mustard, and garlic. Whisk the vinaigrette until well blended. Season with salt and pepper.

2. Brush the mushrooms with a little of the remaining olive oil. Season with salt and pepper.

3. Place the mushroom on a rack over medium-hot coals and grill for 2 minutes on each side. Remove to a serving dish. Drizzle the vinaigrette over the hot mushrooms and let them stand at room temperature until cooled before serving.

Grilled Eggplant

Grilling eggplants is a wonderful way to use a hot fire. The intensely flavored, smoky pulp of grilled eggplant can be used in many delicious ways. It is the base for the popular baba ghanoush in which the eggplant flesh is combined with tahini (sesame paste), lemon juice, and garlic. I often substitute mayonnaise or yogurt for the tahini which makes for a lighter dip.

I also use the grilled eggplant pulp as a base for soup and in spicy eggplant fritters. Be sure to choose eggplants that are light for their size. Heavy eggplants are usually seedier and may have a bitter flavor.

MAKES ABOUT 2 CUPS

2 medium eggplants, unpeeled

1. Prepare a charcoal grill.

2. When the coals are very hot, place the eggplants directly on the coals. Grill, turning often, until the skin is charred on all sides and the eggplants are quite tender. Be careful not to char beneath the skin.

3. Remove the eggplants carefully from the grill and transfer to a cutting surface. Scoop out the pulp and use as directed in your recipe (see headnote for suggestions).

Grilled Green Tomatoes in Basil Vinaigrette

I love the crunchy refreshing texture of green tomatoes and often use the hot fire to grill a few slices, which I marinate in a variety of vinaigrettes. You can vary this simple preparation by using cilantro or parsley instead of basil or use a combination of two of your favorite herbs. You can serve the tomatoes as an appetizer sprinkled with some aged goat cheese and diced oil-cured black olives or diced smoked mozzarella and tiny capers.

SERVES 6

4 to 6 medium-sized green tomatoes
$\frac{1}{2}$ cup plus 2 tablespoons extra virgin olive oil
Coarse salt and freshly ground black pepper
2 tablespoons sherry vinegar
1 to 2 large garlic cloves, mashed
$\frac{1}{2}$ cup basil leaves
2 tablespoons minced shallots or red onion

1. Prepare a charcoal or gas grill.

2. Cut the tomatoes crosswise into $\frac{3}{8}$-inch slices. Brush with 2 tablespoons olive oil. Season with salt and pepper. Grill the slices over a hot fire for 2 minutes on each side, or until nicely browned. Transfer to a shallow serving dish and reserve.

3. In a blender, combine the vinegar, 1 clove of the garlic, the basil leaves, and remaining $\frac{1}{2}$ cup olive oil. Puree until smooth. Season with salt to taste. Add a little more garlic, if desired.

4. Spoon the vinaigrette over the tomatoes and sprinkle with shallots or red onions and a generous grinding of black pepper. Marinate for 3 to 6 hours before serving.

Roasted Red Bell Peppers

I still remember a time when red bell peppers were only available in the fall. Now that we can get Holland peppers year-round, it is fun to make a batch of roasted peppers whenever you plan to cook on the grill. They will keep for several days in the refrigerator, covered with olive oil, in a tightly covered jar.

At different times of the year, many grocery stores carry both the Holland peppers and domestic, or Mexican, peppers, which have thinner skins and are more unevenly shaped. These peppers are usually less expensive and very flavorful but are harder to peel evenly.

1. *Charring the peppers outdoors:* Prepare a charcoal grill.

2. When the coals are red hot, place the peppers directly on top of the coals and grill until the skins are blackened and somewhat charred on all sides.

3. As soon as the peppers are done, remove them from the grill and wrap each pepper in a damp paper towel. Set aside to cool completely.

4. When the peppers are cool enough to handle, peel off the charred skin, core, and remove all seeds. The peppers are now ready to be used in a recipe.

5. *Charring the peppers indoors:* You may also roast the peppers indoors on either an electric or gas stove. For an electric stove, place the peppers directly on the coils of the burner over medium-high heat to char on all sides. For a gas stove, pierce the peppers with the tines of a long fork through the stem end. Set over a medium-high flame and char on all sides. In both cases remove from the burner, wrap in a damp paper towel, cool, and peel.

REMARKS You may also roast green bell peppers, the variety peppers (jalapeños, serranos, yellow, and orange), and fresh tomatoes in the same manner. Tomatoes do not need to be wrapped in a damp towel; just set aside to peel.

Grilled Tomato Salsa

The slightly smoky taste of charcoal-grilled tomatoes adds a wonderful dimension to this classic salsa. I often combine the salsa with some yogurt and a little mayonnaise and serve it as a sauce with grilled fish steaks. For a variation substitute cilantro for the rosemary and add 8 to 10 diced pitted black olives to the salsa.

SERVES 4 TO 5

4 to 6 medium-ripe large tomatoes
2 jalapeño peppers
3 tablespoons red onion, minced
1 large garlic clove, mashed
$\frac{1}{3}$ cup extra virgin olive oil
Juice of 1 lime
2 to 3 tablespoons minced cilantro
Coarse salt and freshly ground black pepper
2 tablespoons minced fresh basil, optional

1. Heat a small charcoal grill. Add a handful of wet wood chips and wait until the fire exudes a smoky aroma. Place the tomatoes and jalapeño peppers on a rack directly over the coals. Cover the grill and roast the tomatoes and peppers for 3 to 4 minutes, or until lightly charred. Remove from the grill and let stand until cool enough to handle. Peel the tomatoes and peppers and remove the seeds. Finely dice both vegetables. Transfer to a bowl.

2. Add the red onion, garlic, olive oil, lime juice, cilantro, salt and pepper to taste, and basil. Toss the salsa gently to mix. Taste and correct the seasoning. Serve at room temperature.

Beurre Manié

A *beurre manié* is used in both classic and peasant cooking to thicken sauces. It is a flour-and-butter paste that can be formed into a ball and held successfully for several weeks. You will rarely need an entire beurre manié in any recipe in this book, since it is preferable to reduce the sauce naturally before thickening it so as to intensify the flavor.

MAKES 8 PORTIONS

8 tablespoons unsalted butter (1 stick),
  slightly softened
½ cup all-purpose flour

1. Combine the butter and flour in a food processor and process until smooth.

2. Refrigerate the mixture just long enough so that you can shape into balls with your hands. Divide the mixture into 8 equal parts and shape each into a ball.

3. Place the balls in a tightly covered jar and refrigerate until needed. The beurre manié can also be frozen for 2 to 3 months.

Clarified Butter

Clarifying butter is a simple technique that removes the milky residue and impurities that make butter burn quickly. It works best when done with at least 8 ounces of butter. Once clarified, the butter can be refrigerated in a tightly sealed jar for at least 2 weeks or frozen for several months. Clarified butter is used for sautéing delicate foods such as fish fillets, fritters, or anything that is breaded. You can also use it for sautéing in any recipe that calls for regular butter.

1. Melt any desired amount of the butter in a heavy saucepan over low heat. As soon as the butter is melted and very foamy, remove from the heat and carefully skim off all the foam.

2. Strain the clear yellow liquid through a fine sieve, discarding the milky residue on the bottom of the pan. The strained, clear yellow liquid is clarified butter. It can be stored in a covered jar in the refrigerator for 2 to 3 weeks.

Basic Tart Dough

$1\frac{1}{2}$ cups all-purpose unbleached flour

$\frac{1}{4}$ teaspoon salt

9 tablespoons unsalted butter, cut into
   12 pieces and chilled

3 to 5 tablespoons ice water

1. In a food processor, combine the flour, salt, and butter. Pulse quickly until the mixture resembles oatmeal. Add 3 tablespoons of the ice water and pulse quickly until the mixture begins to come together. Do not let it form a ball.

2. Transfer the mixture to a large bowl and gather it into a ball. If the dough is too crumbly, add the remaining ice water, 1 tablespoon at a time, and work it quickly into the ball.

3. Flatten the ball into a disk, wrap it in foil, and refrigerate for at least 30 minutes, or until firm enough to roll. The dough is now ready to be used.

Basic Tart Shell

MAKES TWO 9-INCH OR ONE 11-INCH SHELL

2 cups all-purpose unbleached flour

¼ to ½ teaspoon salt

12 tablespoons unsalted butter, cut into 12
   pieces and chilled

4 to 6 tablespoons ice water

1. In a food processor, combine the flour, salt, and butter. Pulse quickly until the mixture resembles oatmeal. Add 3 tablespoons of the ice water and pulse quickly until the mixture begins to come together. Do not let it form a ball.

2. Transfer the mixture to a large bowl and gather it into a ball. If the dough is too crumbly, add the remaining ice water, 1 tablespoon at a time, and work it quickly into the ball.

3. Flatten the ball into a disk, wrap it in foil, and refrigerate for at least 30 minutes.

4. Roll out the dough on a lightly floured surface, into a round about ⅛ inch thick. Roll the dough around the rolling pin, then unroll it into a 9- or 10-inch tart pan with or without a removable bottom. Press the dough into the bottom and sides of the pan, being careful not to stretch the dough. Fold the excess overhang into the pan and press it firmly against the sides. This will create a double layer of dough, which will reinforce the sides of the shell. You now have an unbaked tart shell.

5. Prick the bottom of the shell with a fork, cover with aluminum foil, and place the shell in the freezer for 4 to 6 hours or overnight. At this point the dough can be wrapped tightly and frozen for up to 2 weeks.

6. Preheat the oven to 425°F. Remove the foil and line the frozen shell with parchment paper or buttered foil. Fill with pie weights or large dried beans and place it on a cookie sheet. Bake the shell in the center of the preheated oven for 12 minutes, or until the sides are set.

7. Remove the paper and beans and continue to bake for 4 to 6 minutes longer, until the shell is set but not browned. Remove

the shell from the oven and place the pan on a wire rack to cool until needed. The tart shell is now partially baked. For a pre-baked shell, after removing the paper and beans, bake for 10 to 12 minutes, or until golden brown.

REMARKS For a dessert tart shell, use only a pinch of salt instead of ½ teaspoon. Add 2 tablespoons of confectioners' sugar along with the flour, salt, and butter to the food processor and proceed with Step 1.

Double-Poached Garlic Cloves

MAKES 2 TO 3 HEADS OF GARLIC CLOVES

The garlic may be cooked days in advance. Simply place the poached garlic in a jar, cover with olive oil, seal tightly, and store in the refrigerator until needed. The garlic will keep for up to 3 months.

2 to 3 heads of garlic, separated into cloves
and peeled

1. Bring plenty of water to a boil in a medium saucepan. Add the garlic and return the water to a boil. Cook for 1 minute.

2. Drain, discarding the poaching liquid. Return the garlic cloves to the saucepan, add fresh water to cover, and again bring to a boil. Simmer for 15 to 20 minutes or until the cloves are tender when pierced with the tip of a knife. Drain again. The garlic is now ready to be used in any recipe that calls for it.

Brown Chicken Stock

5 tablespoons corn or peanut oil

2 large onions, quartered and sliced

20 to 24 chicken wings

3 to 4 pounds beef shanks, meaty beef neck bones,
   or short ribs of beef, or a combination

3 small onions, skins on, cut in half crosswise

3 large carrots, peeled and diced

3 large celery stalks, diced

1 parsley root with greens attached, peeled,
   optional

1 teaspoon dried thyme

1 bay leaf, crumbled

6 whole black peppercorns

3 large sprigs of flat-leaf parsley

1. Preheat the broiler.

2. In a large heavy skillet, preferably cast iron, heat 3 tablespoons of the oil over medium-high heat. Add the sliced onions, and cook, stirring constantly, for 2 to 3 minutes, or until they begin to brown. Reduce the heat to low, partially cover, and cook for 25 to 30 minutes, or until the onions are soft and nicely caramelized; do not let them burn.

3. While the onions are browning, prepare the chicken wings and beef: Cut each chicken wing at its joints into 3 pieces. Place the chicken pieces in a single layer in a roasting pan, set under the broiler 4 to 6 inches from the source of heat, and broil until nicely browned on all sides; be careful not to burn them. Transfer the wings to an 8- to 10-quart stockpot and reserve.

4. Place the beef shanks, bones, and/or ribs in the roasting pan and broil until nicely browned on all sides, again being careful not to burn them. Transfer to the stockpot.

5. When the onions are done, transfer them also to the stockpot and add the remaining 2 tablespoons oil to the skillet. Add the unpeeled onions, cut sides down, and cook for 5 to 8 minutes, or until very brown. Add them to the stockpot.

6. Add the carrots and celery to the skillet and cook, stirring often, for 10 to 15 minutes, until nicely browned; they will not brown evenly. Transfer to the stockpot together with the parsley root, if using, thyme, bay leaf, peppercorns, and parsley sprigs. Add enough water to cover by 2 inches, 14 to 16 cups. Bring to a boil, reduce the heat, and simmer, partially covered, for 3 to 4 hours, skimming often, until the liquid becomes a deep hazelnut brown.

7. When the stock is done, strain it through cheesecloth into a large bowl and let it cool completely. Refrigerate, uncovered, overnight.

8. The next day, thoroughly degrease the stock, place it in a large casserole, and bring slowly to a boil. Remove from the heat and again let cool completely. Degrease again if necessary. Transfer to containers or jars, cover tightly, and refrigerate or freeze.

REMARKS The stock will keep for several months in the freezer. If refrigerated, it will keep for up to 10 days and must be brought back to a boil every 2 to 3 days.

White Stock

4 to 5 pounds meaty beef bones, preferably
    beef shanks or beef short ribs

2 large carrots, peeled and cut in half

2 to 3 celery stalks with leaves, cut in half

2 leeks, trimmed and washed

2 medium onions, peeled

6 whole black peppercorns

1 parsley root, optional

1 large sprig of fresh thyme

2 large sprigs of fresh flat-leaf parsley

1½ teaspoon salt

1. Combine all the ingredients in a tall and large but narrow stockpot with 3 quarts water. Bring to a boil, reduce the heat, and simmer, partially covered, for 3 to 4 hours, skimming the surface several times.

2. Strain the stock through a double layer of cheesecloth into a large bowl and set aside at room temperature, uncovered, until completely cool. Place in the refrigerator overnight.

3. The next day, remove and discard all the fat from the surface. Pour the stock into a flameproof casserole or large saucepan. Bring to a boil, remove from the heat, and let cool completely. Refrigerate for up to 1 week, bringing the stock back to a boil every few days, or freeze in covered containers for up to 1 month.

Cooked Pinto or Red Kidney Beans

1 pound dried pinto or red kidney beans

1 small onion, skin on

1 large sprig of fresh thyme

1 large sprig of fresh parsley

6 whole black peppercorns

½ pound smoked pork shoulder butt, quartered

Salt

You can cook beans by either of the following two methods.

METHOD #1

1. Place the dried beans in a large bowl with plenty of water to cover. Let soak overnight at room temperature.

2. The next day, preheat the oven to 325°F.

3. Drain the beans and place in a large, heavy flameproof casserole with a tight-fitting lid. Cover with water by 2 inches and add the onion, thyme, parsley, peppercorns, and pork butt.

4. Bring to a boil on top of the stove and cook for exactly 1 minute. Immediately cover tightly and place the casserole in the center of the oven. Cook the beans for 1 hour to 1 hour 15 minutes, or until just tender. Do not overcook and do not add salt until the beans are almost done. Remove the casserole from the oven and let the beans cool in their cooking liquid.

5. Discard the vegetables and herbs. Store the beans in their liquid. They will keep in the refrigerator in their cooking liquid for up to 5 days. You will need both the cooked beans and the bean cooking liquid in most recipes.

METHOD #2

1. Do not soak the beans at all. Instead, place them in a large, heavy flameproof casserole with a tight-fitting lid. Cover with water by 2 inches and add the onion, thyme, parsley, peppercorns, and pork butt.

2. Bring to a boil on top of the stove and cook for exactly 1 minute. Immediately remove the casserole from the heat, cover, and set aside for 2 hours at room temperature. Drain the beans, add fresh water to cover by 2 inches, and continue to braise the beans in the oven as in Method #1, Step 4.

REMARKS Smoked pork shoulder butt is available in all supermarkets. It is often called a daisy ham or porkette.

Also use either of these two methods to cook white beans, kidney beans, black beans, and flageolets. Cooking times for each will vary.

Cooked White Beans

MAKES ABOUT 6 CUPS

1 pound dried white beans

1 small onion, skin on

1 carrot, cut in half

1 celery stalk with tops

Pinch of salt

4 to 6 whole black peppercorns

You can cook white beans by either of the following two methods.

METHOD #1

1. Place the beans in a bowl, cover with 2 inches of cold water, and let the beans soak overnight.

2. The next day, drain the beans. Place the beans in a large saucepan and cover with water by 2 inches. Add the onion, carrot, celery, salt, and peppercorns.

3. Bring the beans to a boil on top of the stove. Reduce the heat to medium-low and cook for 1½ hour or until tender. Do not overcook. The beans may be left in their cooking water for 3 to 4 days.

METHOD #2

Place the dried beans in a saucepan with 2 inches of water to cover. Bring to a boil on top of the stove for exactly 1 minute. Remove from the heat. Cover the saucepan and let the beans sit for 2 hours. Drain, add fresh water, and proceed with the recipe as above.

Strawberry Sauce

2 packages (10 ounces each) frozen
  strawberries in syrup

¾ cup red currant jelly

1½ to 2 tablespoons lemon juice

2 tablespoons kirsch or Grand Marnier

Sugar, optional

1. Defrost the strawberries and drain well, reserving the juices. Place the berries with the juice of 1 package in a food processor and puree until smooth. (Reserve the remaining juice for another use or discard.)

2. Place the jelly in a small saucepan and set over low heat. Stir until the jelly is completely dissolved and add to the strawberry puree. Stir in 1½ tablespoons of the lemon juice, and the kirsch or Grand Marnier. Correct the sweetness of the sauce by adding either a little sugar or a little more lemon juice. Strain the sauce through a fine sieve into a bowl. Cover and chill until serving time.

KITCHENWARE

Broadway Panhandler
Excellent selection of knives, cookware, cookbooks, and bakeware
477 Broome Street
New York, NY 10013
1-866-COOKWARE or 212-966-3434
www.broadwaypanhandler.com

Sur La Table
Excellent assortment of cookware, gadgets, and small electrical appliances
800-243-0852
www.surlatable.com

Williams-Sonoma
Excellent selection of cookware and private-label spices
P. O. Box 7456
San Francisco, CA 94120-7456
800-541-2233
www.williams-sonoma.com

SPECIALTY FOODS

Aidell's Sausage Company
Huge variety of all-natural sausages
1625 Alvarado Street
San Leandro, CA 94577
Tel: 800-AIDELLS
Fax: 510-614-2846
www.aidells.com
E-mail: info@aidells.com

Asian Pantry (part of chefshop.com)
Assortment of Asian ingredients, including a variety of soy sauces, rice and curry pastes
Tel: 877-337-2491
Fax: 206-282-5607
www.chefshop.com
E-mail: shopkeeper@chefshop.com

Atlantic Game Meats

Superb quality of venison, including marinated venison steaks and roasts

Tel: 207-862-4217

www.atlanticgamemeats.com

Bob's Red Mill

Organic flour, whole grains, excellent polenta

5209 SE International Way

Milwaukie, OR 97222

Tel: 800-349-2173

Fax: 503-653-1339

www.bobsredmill.com

chefshop.com

Excellent choice of dried beans, including Arrocina rice beans from Spain, organic cannellini beans, porcini mushrooms, Italian canned albacore tuna in olive oil, and various types of capers

Tel: 877-337-2491

Fax: 206-282-5607

www.chefshop.com

D'Artagnan

The best source for game, rabbit, squab, poussins (baby chickens), quail, duck fat, duck breasts, duck foie gras, and chorizo sausage

280 Wilson Avenue

Newark, NJ 07105

Tel: 800-327-8246

Fax: 973-465-1870

www.dartagnan.com

Dean & DeLuca

Produce, grains, oils, vinegars, mustards, teas, and spices

560 Broadway

New York, NY 11101

Tel: 877-826-9246

Fax: 800-781-4050

www.deandeluca.com

Earthy Delights

Excellent source for dried beans and lentils

1161 E. Clark Road, Suite 260

DeWitt, MI 48820

Tel: 800-367-4709

Fax: 517-668-1213

www.earthy.com

E-mail: info@earthy.com

El Paso Chile Company

Dried chiles, southwestern spices, etc.

Tel: 888-472-5727

Fax: 915-544-7552

www.elpasochile.com

Kalustyan's

Large international selection of grains, rice, unusual spices, and nuts

123 Lexington Avenue

New York, NY 10016

Tel: 212-685-3451

Fax: 212-683-8458

www.kalustyans.com

Leech Lake Wild Rice Co.

Organic wild rice that is hand harvested and packaged by Leech Lake Indian Reservation

Tel: 877-246-0620

Mozzarella Company

Wonderful ricotta, cacciota, and fresh and smoked mozzarella

2944 Elm Street

Dallas, TX 75226

800-798-2954

www.mozzco.com

Penzeys Spices
Great selection of spices and spice blends, including Mexican spices such as achiote and Mexican oregano
Tel: 800-741-7787
Fax: 262-785-7678
www.penzeys.com

The Spanish Table
Spanish imports: rice, spices, beans, oils and vinegars
1427 Western Avenue
Seattle, WA 98101
Tel: 206-682-2827
Fax: 206-682-2814
E-mail: tablespan@aol.com

Spice Island
Top-quality dried herbs and spices
Tel: 800-635-6278

Todaro Bros.
Italian imports, pasta, rice, oils, and cheeses
555 Second Avenue
New York, NY 10016
Tel: 877-472-2767
Fax: 212-698-1679
www.todarobros-specialty-foods.com

Zingerman's
Quality foods including cheeses, aged balsamic vinegar, large selection of olive oils
422 Detroit Street
Ann Arbor, MI 48104-1118
Tel: 888-636-8162
Fax: 734-477-6988
www.zingermans.com
E-mail: toni@zingermans.com